Beam, Loading, and Bending Moment Diagram	Equations for Slope and Deflection
	$0 \leq x \leq \frac{L}{2}$: $$\theta = \frac{P}{16EI}(4x^2 - L^2)$$ $$y = \frac{P}{48EI}(4x^3 - 3L^2 x)$$ $$\theta_A = -\frac{PL^2}{16EI}; \quad \theta_B = \frac{PL^2}{16EI}$$ $$y_{\max} = -\frac{PL^3}{48EI}$$
	$0 \leq x \leq a$: $$\theta = \frac{Pb}{6EIL}(3x^2 + b^2 - L^2)$$ $$y = \frac{Pb}{6EIL}(x^3 + b^2 x - L^2 x)$$ $a \leq x \leq L$: $$\theta = \frac{Pa}{6EIL}[L^2 - a^2 - 3(L - x)^2]$$ $$y = \frac{Pa(L-x)}{6EIL}(x^2 + a^2 - 2Lx)$$ $$\theta_A = -\frac{Pb}{6EIL}(L^2 - b^2)$$ $$\theta_B = \frac{Pa}{6EIL}(L^2 - a^2)$$ For $a \geq b$: $$y_{\max} = -\frac{Pb}{9\sqrt{3}EIL}(L^2 - b^2)^{3/2}$$ $$\text{at } x = \left(\frac{L^2 - b^2}{3}\right)^{1/2}$$
	$$\theta = -\frac{M}{6EIL}(3x^2 - 6Lx + 2L^2)$$ $$y = -\frac{M}{6EIL}(x^3 - 3Lx^2 + 2L^2 x)$$ $$\theta_A = -\frac{ML}{3EI}; \quad \theta_B = \frac{ML}{6EI}$$ $$y_{\max} = -\frac{ML^2}{9\sqrt{3}EI}$$ $$\text{at } x = L\left(1 - \frac{1}{\sqrt{3}}\right)$$

Beam, Loading, and Bending Moment Diagram	Equations for Slope and Deflection
	$0 \le x \le a:$ $$\theta = \frac{M}{6EIL}(-3x^2 + 6aL - 3a^2 - 2L^2)$$ $$y = \frac{M}{6EIL}(-x^3 + 6aLx - 3a^2x - 2L^2x)$$ $$\theta_A = \frac{M}{6EIL}(6aL - 3a^2 - 2L^2)$$ $$\theta_B = \frac{M}{6EIL}(L^2 - 3a^2)$$
	$$\theta = -\frac{w}{24EI}(4x^3 - 6Lx^2 + L^3)$$ $$y = -\frac{w}{24EI}(x^4 - 2Lx^3 + L^3x)$$ $$\theta_A = -\frac{wL^3}{24EI}$$ $$\theta_B = \frac{wL^3}{24EI}$$ $$y_{\max} = -\frac{5wL^4}{384EI} \quad \text{at } x = \frac{L}{2}$$
	$0 \le x \le a:$ $$\theta = -\frac{w}{24EIL}[4Lx^3 - 6a(2L - a)x^2 + a^2(2L - a)^2]$$ $$y = -\frac{w}{24EIL}[Lx^4 - 2a(2L - a)x^3 + a^2(2L - a)^2x]$$ $a \le x \le L:$ $$\theta = -\frac{wa^2}{24EIL}(6x^2 - 12Lx + a^2 + 4L^2)$$ $$y = -\frac{wa^2}{24EIL}(L - x)(-2x^2 + 4Lx - a^2)$$ $$\theta_A = -\frac{wa^2}{24EIL}(2L - a)^2$$ $$\theta_B = \frac{wa^2}{24EIL}(2L^2 - a^2)$$
	$$\theta = -\frac{w}{360EIL}(15x^4 - 30L^2x^2 + 7L^4)$$ $$y = -\frac{w}{360EIL}(3x^5 - 10L^2x^3 + 7L^4x)$$ $$\theta_A = -\frac{7wL^3}{360EI}$$ $$\theta_B = \frac{wL^3}{45EI}$$ $$y_{\max} = -0.00652\frac{wL^4}{EI} \quad \text{at } x = 0.5193L$$

Structural Analysis

THIRD EDITION

Aslam Kassimali
Southern Illinois University—Carbondale

THOMSON

Australia · Canada · Mexico · Singapore · Spain · United Kingdom · United States

THOMSON ™

Structural Analysis, Third Edition
by Aslam Kassimali

Associate Vice-President and Editorial Director:
Evelyn Veitch

Publisher:
Bill Stenquist

Sales and Marketing Manager:
John More

Developmental Editor:
Kamilah Reid Burrell

Production Service:
RPK Editorial Services

Copy Editor:
Shelly Gerger-Knechtl

Indexer:
Aslam Kassimali

Proofing:
Jackie Twoney

Production Manager:
Renate McCloy

Creative Director:
Angela Cluer

Cover Design:
Vernon Boes

Compositor:
Asco Typesetters

Printer:
Quebecor World

North America
Nelson
1120 Birchmount Road
Toronto, Ontario M1K 5G4
Canada

Asia
Thomson Learning
5 Shenton Way #01-01
UIC Building
Singapore 068808

Australia/New Zealand
Thomson Learning
102 Dodds Street
Southbank, Victoria
Australia 3006

Europe/Middle East/Africa
Thomson Learning
High Holborn House
50/51 Bedford Row
London WC1R 4LR
United Kingdom

Latin America
Thomson Learning
Seneca, 53
Colonia Polanco
11560 Mexico D.F.
Mexico

Spain
Paraninfo
Calle/Magallanes, 25
28015 Madrid, Spain

10006807

IN MEMORY OF *AMI*

Contents

6   Deflections of Beams: Geometric Methods   228

7   Deflections of Trusses, Beams, and Frames: Work–Energy Methods   277

8   Influence Lines   339

9   Application of Influence Lines   403

10   Analysis of Symmetric Structures   427

PART THREE   ANALYSIS OF STATICALLY INDETERMINATE STRUCTURES   461

11   Introduction to Statically Indeterminate Structures   463

Appendix C   Computer Software   837

Preface

The objective of this book is to develop an understanding of the basic principles of structural analysis. Emphasizing the intuitive classical approach, *Structural Analysis* covers the analysis of statically determinate and indeterminate beams, trusses, and rigid frames. It also presents an introduction to the matrix analysis of structures.

The book is divided into three parts. Part One presents a general introduction to the subject of structural engineering. It includes a chapter devoted entirely to the topic of loads because attention to this important topic is generally lacking in many civil engineering curricula. Part Two, consisting of Chapters 3 through 10, covers the analysis of statically determinate beams, trusses, and rigid frames. The chapters on deflections (Chapters 6 and 7) are placed before those on influence lines (Chapters 8 and 9), so that influence lines for deflections can be included in the latter chapters. This part also contains a chapter on the analysis of symmetric structures (Chapter 10). Part Three of the book, Chapters 11 through 18, covers the analysis of statically indeterminate structures. The format of the book is flexible to enable instructors to emphasize topics that are consistent with the goals of the course.

Each chapter of the book begins with an introductory section defining its objective and ends with a summary section outlining its salient features. An important general feature of the book is the inclusion of step-by-step procedures for analysis to enable students to make an easier transition from theory to problem solving. Numerous solved examples are provided to illustrate the application of the fundamental concepts.

A CD-ROM containing computer software for the analysis of plane frames, continuous beams, and trusses is attached to the back cover. This interactive software can be used to simulate a variety of structural and loading configurations and to determine cause versus effect relationships between loading and various structural parameters, thereby enhancing the students' understanding of the behavior of structures. The software shows deflected shapes of structures to enhance students' understanding of structural response due to various types of loadings. It can also include the effects of support settlements, temperature changes, and fabrication errors in the analysis. A solutions manual, containing complete solutions to text exercises, is also available for the instructor.

A NOTE ON THE REVISED EDITION

In this third edition, 37 new solved examples have been added to increase the total number by about 30%. The number of problems has also been increased to bring the total to over 600, of which about 40% are new problems. The chapter on loads has been revised to meet the provisions of the ASCE 7-02 Standard, and the treatment of the force method has been expanded by including the topic of the three-moment equation. The force method is now covered in two chapters (Chapters 13 and 14), with the new Chapter 14 containing the three-moment equation and the method of least work. There are many other minor revisions, including some in the computer software, which has been recompiled to make it compatible with the latest versions of Microsoft Windows. Finally, some of the photographs have been replaced with new ones, and some figures have been redrawn and rearranged to enhance clarity.

ACKNOWLEDGMENTS

I wish to express my thanks to Bill Stenquist of Thomson Engineering for his constant support and encouragement throughout this project, and to Rose Kernan for all her help during the production phase. Thanks are also due to Jonathan Plant and Suzanne Jeans, my editors for the first and second editions, respectively, of this book. The comments and suggestions for improvement from colleagues and students who have used previous editions are gratefully acknowledged. All of their suggestions were carefully considered, and implemented whenever possible. Thanks are due to the following reviewers for their careful reviews of the manuscripts of the various editions, and for their constructive suggestions:

Ayo Abatan
Virginia Polytechnic Institute and State University

Riyad S. Aboutaha
Georgia Institute of Technology

Osama Abudayyeh
Western Michigan University

Thomas T. Baber
University of Virginia

Gordon B. Batson
Clarkson University

George E. Blandford
University of Kentucky

Ramon F. Borges
Penn State/Altoona College

Kenneth E. Buttry
University of Wisconsin

William F. Carroll
University of Central Florida

Malcolm A. Cutchins
Auburn University

Jack H. Emanuel
University of Missouri—Rolla

Fouad Fanous
Iowa State University

Leon Feign
Fairfield University

Robert Fleischman
University of Notre Dame

George Kostyrko
California State University

E. W. Larson
California State University/Northridge

L. D. Lutes
Texas A&M University

Eugene B. Loverich
Northern Arizona University

David Mazurek
US Coast Guard Academy

Ahmad Namini
University of Miami

Arturo E. Schultz
North Carolina State University

Kassim Tarhini
Valparaiso University

Robert Taylor
Northeastern University

C. C. Tung
North Carolina State University

Nicholas Willems
University of Kansas

John Zachar
Milwaukee School of Engineering

Mannocherh Zoghi
University of Dayton

Finally, I would like to express my loving appreciation to my wife. Maureen, for her constant encouragement and help in preparing this manuscript, and to my sons, Jamil and Nadim, for their enormous understanding and patience.

Aslam Kassimali

Part One

Introduction to
Structural Analysis and Loads

1

Introduction to Structural Analysis

1.1 Historical Background
1.2 Role of Structural Analysis in Structural Engineering Projects
1.3 Classification of Structures
1.4 Analytical Models
   Summary

Marina City District, Chicago
Photo courtesy of Hisham F. Ibraham/Photodisc Green

Structural analysis is the prediction of the performance of a given structure under prescribed loads and/or other external effects, such as support movements and temperature changes. The performance characteristics commonly of interest in the design of structures are (1) stresses or stress resultants, such as axial forces, shear forces, and bending moments; (2) deflections; and (3) support reactions. Thus, the analysis of a structure usually involves determination of these quantities as caused by a given loading condition. The objective of this text is to present the methods for the analysis of structures in static equilibrium.

This chapter provides a general introduction to the subject of structural analysis. We first give a brief historical background, including names of people whose work is important in the field. Then we discuss the role of structural analysis in structural engineering projects. We describe the five common types of structures: tension and compression structures, trusses, and shear and bending structures. Finally, we consider the development of the simplified models of real structures for the purpose of analysis.

1.1 HISTORICAL BACKGROUND

Since the dawn of history, structural engineering has been an essential part of human endeavor. However, it was not until about the middle of the seventeenth century that engineers began applying the knowledge of mechanics (mathematics and science) in designing structures. Earlier engineering structures were designed by trial and error and by using rules of thumb based on past experience. The fact that some of the magnificent structures from earlier eras, such as Egyptian pyramids (about 3000 B.C.), Greek temples (500–200 B.C.), Roman coliseums and aqueducts (200 B.C.–A.D. 200), and Gothic cathedrals (A.D. 1000–1500), still stand today is a testimonial to the ingenuity of their builders (Fig. 1.1).

Galileo Galilei (1564–1642) is generally considered to be the originator of the theory of structures. In his book entitled *Two New Sciences*, which was published in 1638, Galileo analyzed the failure of some simple structures, including cantilever beams. Although Galileo's predictions of strengths of beams were only approximate, his work laid the foundation for future developments in the theory of structures and

FIG. 1.1 The Cathedral of Notre Dame in Paris Was Completed in the Thirteenth Century
Photo courtesy of the French Government Tourist Office

ushered in a new era of structural engineering, in which the analytical principles of mechanics and strength of materials would have a major influence on the design of structures.

Following Galileo's pioneering work, the knowledge of structural mechanics advanced at a rapid pace in the second half of the seventeenth century and into the eighteenth century. Among the notable investigators of that period were Robert Hooke (1635–1703), who developed the law of linear relationships between the force and deformation of materials (Hooke's law); Sir Isaac Newton (1642–1727), who formulated the laws of motion and developed calculus; John Bernoulli (1667–1748), who formulated the principle of virtual work; Leonhard Euler (1707–1783), who developed the theory of buckling of columns; and C. A. de Coulomb (1736–1806), who presented the analysis of bending of elastic beams.

In 1826 L. M. Navier (1785–1836) published a treatise on elastic behavior of structures, which is considered to be the first textbook on the modern theory of strength of materials. The development of structural mechanics continued at a tremendous pace throughout the rest of the nineteenth century and into the first half of the twentieth, when most of the classical methods for the analysis of structures described in this text were developed. The important contributors of this period included B. P. Clapeyron (1799–1864), who formulated the three-moment equation for the analysis of continuous beams; J. C. Maxwell (1831–1879), who presented the method of consistent deformations and the law of reciprocal deflections; Otto Mohr (1835–1918), who developed the conjugate-beam method for calculation of deflections and Mohr's circles of stress and strain; Alberto Castigliano (1847–1884), who formulated the theorem of least work; C. E. Greene (1842–1903), who developed the moment-area method; H. Müller-Breslau (1851–1925), who presented a principle for constructing influence lines; G. A. Maney (1888–1947), who developed the slope-deflection method, which is considered to be the precursor of the matrix stiffness method; and Hardy Cross (1885–1959), who developed the moment-distribution method in 1924. The moment-distribution method provided engineers with a simple iterative procedure for analyzing highly statically indeterminate structures. This method, which was the most widely used by structural engineers during the period from about 1930 to 1970, contributed significantly to their understanding of the behavior of statically indeterminate frames. Many structures designed during that period, such as high-rise buildings, would not have been possible without the availability of the moment-distribution method.

The availability of computers in the 1950s revolutionized structural analysis. Because the computer could solve large systems of simultaneous equations, analyses that took days and sometimes weeks in the pre-computer era could now be performed in seconds. The development of the current computer-oriented methods of structural analysis can be attributed to, among others, J. H. Argyris, R. W. Clough, S. Kelsey,

R. K. Livesley, H. C. Martin, M. T. Turner, E. L. Wilson, and O. C. Zienkiewicz.

1.2 ROLE OF STRUCTURAL ANALYSIS IN STRUCTURAL ENGINEERING PROJECTS

Structural engineering is the science and art of planning, designing, and constructing safe and economical structures that will serve their intended purposes. Structural analysis is an integral part of any structural engineering project, its function being the prediction of the performance of the proposed structure. A flowchart showing the various phases of a typical structural engineering project is presented in Fig. 1.2. As this diagram indicates, the process is an iterative one, and it generally consists of the following steps:

1. ***Planning Phase*** The planning phase usually involves the establishment of the functional requirements of the proposed structure, the

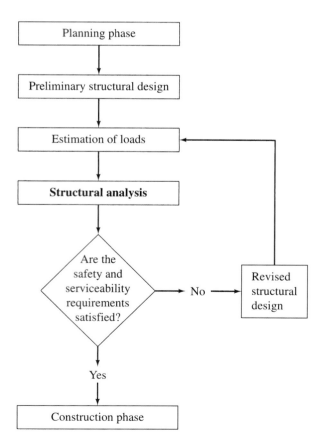

FIG. 1.2 Phases of a Typical Structural Engineering Project

general layout and dimensions of the structure, consideration of the possible types of structures (e.g., rigid frame or truss) that may be feasible and the types of materials to be used (e.g., structural steel or reinforced concrete). This phase may also involve consideration of nonstructural factors, such as aesthetics, environmental impact of the structure, and so on. The outcome of this phase is usually a structural system that meets the functional requirements and is expected to be the most economical. This phase is perhaps the most crucial one of the entire project and requires experience and knowledge of construction practices in addition to a thorough understanding of the behavior of structures.

2. *Preliminary Structural Design* In the preliminary structural design phase, the sizes of the various members of the structural system selected in the planning phase are estimated based on approximate analysis, past experience, and code requirements. The member sizes thus selected are used in the next phase to estimate the weight of the structure.

3. *Estimation of Loads* Estimation of loads involves determination of all the loads that can be expected to act on the structure.

4. *Structural Analysis* In structural analysis, the values of the loads are used to carry out an analysis of the structure in order to determine the stresses or stress resultants in the members and the deflections at various points of the structure.

5. *Safety and Serviceability Checks* The results of the analysis are used to determine whether or not the structure satisfies the safety and serviceability requirements of the design codes. If these requirements are satisfied, then the design drawings and the construction specifications are prepared, and the construction phase begins.

6. *Revised Structural Design* If the code requirements are not satisfied, then the member sizes are revised, and phases 3 through 5 are repeated until all the safety and serviceability requirements are satisfied.

Except for a discussion of the types of loads that can be expected to act on structures (Chapter 2), our primary focus in this text will be on the analysis of structures.

1.3 CLASSIFICATION OF STRUCTURES

As discussed in the preceding section, perhaps the most important decision made by a structural engineer in implementing an engineering project is the selection of the type of structure to be used for supporting or transmitting loads. Commonly used structures can be classified into five basic categories, depending on the type of primary stresses that may develop in their members under major design loads. However, it should be realized that any two or more of the basic structural types described in the following may be combined in a single structure, such as a building or a bridge, to meet the structure's functional requirements.

Tension Structures

The members of tension structures are subjected to pure tension under the action of external loads. Because the tensile stress is distributed uniformly over the cross-sectional areas of members, the material of such a structure is utilized in the most efficient manner. Tension structures composed of flexible steel cables are frequently employed to support bridges and long-span roofs. Because of their flexibility, cables have negligible bending stiffness and can develop only tension. Thus, under external loads, a cable adopts a shape that enables it to support the load by tensile forces alone. In other words, the shape of a cable changes as the loads acting on it change. As an example, the shapes that a single cable may assume under two different loading conditions are shown in Fig. 1.3.

Figure 1.4 shows a familiar type of cable structure—the *suspension bridge*. In a suspension bridge, the roadway is suspended from two main cables by means of vertical hangers. The main cables pass over a pair of towers and are anchored into solid rock or a concrete foundation at their ends. Because suspension bridges and other cable structures lack stiffness in lateral directions, they are susceptible to wind-induced oscillations (see Fig. 1.5). Bracing or stiffening systems are therefore provided to reduce such oscillations.

Besides cable structures, other examples of tension structures include vertical rods used as hangers (for example, to support balconies or tanks) and membrane structures such as tents.

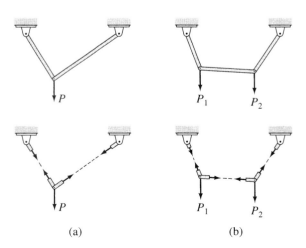

FIG. **1.3**   (a)   (b)

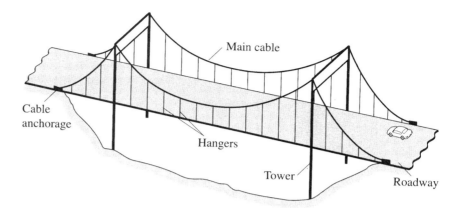

FIG. 1.4 Suspension Bridge

FIG. 1.5 Tacoma Narrows Bridge
Oscillating before Its Collapse in 1940
Smithsonian Institution Photo No. 72-787

Compression Structures

Compression structures develop mainly compressive stresses under the action of external loads. Two common examples of such structures are *columns* and *arches*. Columns are straight members subjected to axially compressive loads, as shown in Fig. 1.6. When a straight member is subjected to lateral loads and/or moments in addition to axial loads, it is called a *beam-column*.

An arch is a curved structure, with a shape similar to that of an inverted cable, as shown in Fig. 1.7. Such structures are frequently used to support bridges and long-span roofs. Arches develop mainly compres-

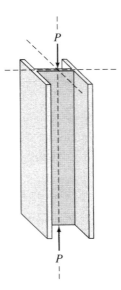

FIG. **1.6** Column

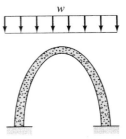

FIG. **1.7** Arch

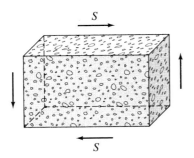

FIG. **1.9** Shear Wall

sive stresses when subjected to loads and are usually designed so that they will develop only compression under a major design loading. However, because arches are rigid and cannot change their shapes as can cables, other loading conditions usually produce secondary bending and shear stresses in these structures, which, if significant, should be considered in their designs.

Because compression structures are susceptible to buckling or instability, the possibility of such a failure should be considered in their designs; if necessary, adequate bracing must be provided to avoid such failures.

Trusses

Trusses are composed of straight members connected at their ends by hinged connections to form a stable configuration (Fig. 1.8). When the loads are applied to a truss only at the joints, its members either elongate or shorten. Thus, the members of an ideal truss are always either in uniform tension or in uniform compression. Real trusses are usually constructed by connecting members to gusset plates by bolted or welded connections. Although the rigid joints thus formed cause some bending in the members of a truss when it is loaded, in most cases such secondary bending stresses are small, and the assumption of hinged joints yields satisfactory designs.

Trusses, because of their light weight and high strength, are among the most commonly used types of structures. Such structures are used in a variety of applications, ranging from supporting roofs of buildings to serving as support structures in space stations.

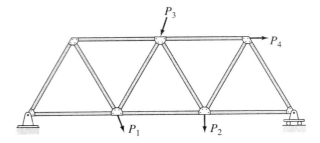

FIG. **1.8** Plane Truss

Shear Structures

Shear structures, such as reinforced concrete *shear walls* (Fig. 1.9), are used in multistory buildings to reduce lateral movements due to wind loads and earthquake excitations. Shear structures develop mainly in-plane shear, with relatively small bending stresses under the action of external loads.

Bending Structures

Bending structures develop mainly bending stresses under the action of external loads. In some structures, the shear stresses associated with the changes in bending moments may also be significant and should be considered in their designs.

Some of the most commonly used structures, such as *beams*, *rigid frames*, *slabs*, and *plates*, can be classified as bending structures. *A beam is a straight member that is loaded perpendicular to its longitudinal axis* (Fig. 1.10). Recall from previous courses on *statics* and *mechanics of materials* that the bending (normal) stress varies linearly over the depth of a beam from the maximum compressive stress at the fiber farthest from the neutral axis on the concave side of the bent beam to the maximum tensile stress at the outermost fiber on the convex side. For example, in the case of a horizontal beam subjected to a vertically downward load, as shown in Fig. 1.10, the bending stress varies from the maximum compressive stress at the top edge to the maximum tensile stress at the bottom edge of the beam. To utilize the material of a beam cross section most efficiently under this varying stress distribution, the cross sections of beams are often I-shaped (see Fig. 1.10), with most of the material in the top and bottom flanges. The I-shaped cross sections are most effective in resisting bending moments.

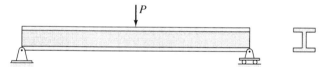

FIG. 1.10 Beam

Rigid frames are composed of straight members connected together either by rigid (moment-resisting) connections or by hinged connections to form stable configurations. Unlike trusses, which are subjected only to joint loads, the external loads on frames may be applied on the members as well as on the joints (see Fig. 1.11). The members of a rigid frame are, in general, subjected to bending moment, shear, and axial compression or tension under the action of external loads. However, the design of horizontal members or beams of rectangular frames is often governed by bending and shear stresses only, since the axial forces in such members are usually small.

Frames, like trusses, are among the most commonly used types of structures. Structural steel and reinforced concrete frames are commonly used in multistory buildings (Fig. 1.12), bridges, and industrial plants. Frames are also used as supporting structures in airplanes, ships, aerospace vehicles, and other aerospace and mechanical applications.

It may be of interest to note that the generic term *framed structure* is frequently used to refer to any structure composed of straight members, including a truss. In that context, this textbook is devoted primarily to the analysis of plane framed structures.

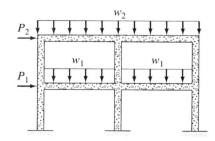

FIG. 1.11 Rigid Frame

FIG. 1.12 Skeleton of a Structural Steel
Frame Building
Photo courtesy of Bethlehem Steel Corporation

1.4 ANALYTICAL MODELS

An analytical model is a simplified representation, or an ideal, of a real structure for the purpose of analysis. The objective of the model is to simplify the analysis of a complicated structure. The analytical model represents, as accurately as practically possible, the behavioral characteristics of the structure of interest to the analyst, while discarding much of the detail about the members, connections, and so on, that is expected to have little effect on the desired characteristics. Establishment of the analytical model is one of the most important steps of the analysis process; it requires experience and knowledge of design practices in addition to a thorough understanding of the behavior of structures. Remember that the structural response predicted from the analysis of the model is valid only to the extent that the model represents the actual structure.

Development of the analytical model generally involves consideration of the following factors.

Plane Versus Space Structure

If all the members of a structure as well as the applied loads lie in a single plane, the structure is called a *plane structure*. The analysis of plane, or two-dimensional, structures is considerably simpler than the analysis of space, or three-dimensional, structures. Fortunately, many actual three-dimensional structures can be subdivided into plane structures for analysis.

As an example, consider the framing system of a bridge shown in Fig. 1.13(a). The main members of the system, designed to support vertical loads, are shown by solid lines, whereas the secondary bracing members, necessary to resist lateral wind loads and to provide stability, are represented by dashed lines. The deck of the bridge rests on beams called *stringers*; these beams are supported by *floor beams*, which, in turn, are connected at their ends to the joints on the bottom panels of the two longitudinal trusses. Thus, the weight of the traffic, deck, stringers, and floor beams is transmitted by the floor beams to the supporting trusses at their joints; the trusses, in turn, transmit the load to the foundation. Because this applied loading acts on each truss in its own plane, the trusses can be treated as plane structures.

As another example, the framing system of a multistory building is shown in Fig. 1.14(a). At each story, the floor slab rests on floor beams, which transfer any load applied to the floor, the weight of the slab, and their own weight to the girders of the supporting rigid frames. This applied loading acts on each frame in its own plane, so each frame can, therefore, be analyzed as a plane structure. The loads thus transferred to each frame are further transmitted from the girders to the columns and then finally to the foundation.

Although a great majority of actual three-dimensional structural systems can be subdivided into plane structures for the purpose of analysis, some structures, such as latticed domes, aerospace structures, and transmission towers, cannot, due to their shape, arrangement of members, or applied loading, be subdivided into planar components. Such structures, called *space structures*, are analyzed as three-dimensional bodies subjected to three-dimensional force systems.

Line Diagram

The analytical model of the two- or three-dimensional body selected for analysis is represented by a *line diagram*. On this diagram, each member of the structure is represented by a line coinciding with its centroidal axis. The dimensions of the members and the size of the connections are not shown on the diagram. The line diagrams of the bridge truss of Fig. 1.13(a), and the rigid frame of Fig. 1.14(a) are shown in Figs. 1.13(b) and 1.14(b), respectively. Note that two lines (⇌) are sometimes used in this text to represent members on the line diagrams. This is done,

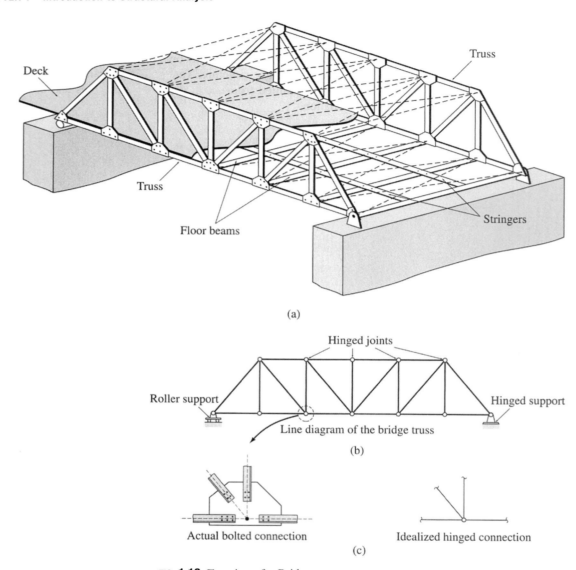

(a)

(b)

(c)

Fig. **1.13** Framing of a Bridge

when necessary, for clarity of presentation; in such cases, the distance between the lines does not represent the member depth.

Connections

Two types of connections are commonly used to join members of structures: (1) *rigid connections* and (2) *flexible*, or *hinged, connections*. (A third type of connection, termed a *semirigid connection*, although rec-

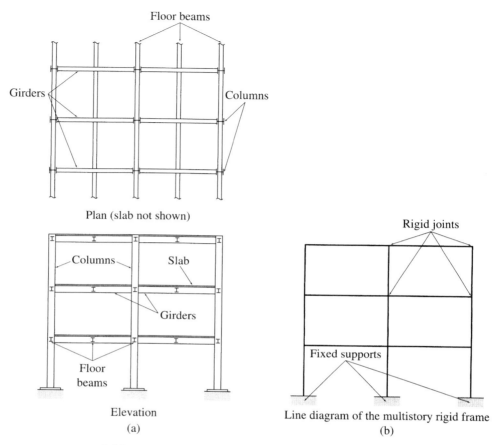

FIG. 1.14 Framing of a Multistory Building

ognized by structural steel design codes, is not commonly used in practice and, therefore, is not considered in this text.)

A rigid connection or joint prevents relative translations and rotations of the member ends connected to it; that is, all member ends connected to a rigid joint have the same translation and rotation. In other words, the original angles between the members intersecting at a rigid joint are maintained after the structure has deformed under the action of loads. Such joints are, therefore, capable of transmitting forces as well as moments between the connected members. Rigid joints are usually represented by points at the intersections of members on the line diagram of the structure, as shown in Fig. 1.14(b).

A hinged connection or joint prevents only relative translations of member ends connected to it; that is, all member ends connected to a hinged joint have the same translation but may have different rotations. Such joints are thus capable of transmitting forces but not moments between the connected members. Hinged joints are usually depicted by small circles at the intersections of members on the line diagram of the structure, as shown in Fig. 1.13(b).

The perfectly rigid connections and the perfectly flexible frictionless hinges used in the analysis are merely idealizations of the actual connections, which are seldom perfectly rigid or perfectly flexible (see Fig. 1.13(c)). However, actual bolted or welded connections are purposely designed to behave like the idealized cases. For example, the connections of trusses are designed with the centroidal axes of the members concurrent at a point, as shown in Fig. 1.13(c), to avoid eccentricities that may cause bending of members. For such cases, the analysis based on the idealized connections and supports (described in the following paragraph) generally yields satisfactory results.

Supports

Supports for plane structures are commonly idealized as either *fixed supports*, which do not allow any movement; *hinged supports*, which can prevent translation but permit rotation; or *roller*, or *link*, *supports*, which can prevent translation in only one direction. A more detailed description of the characteristics of these supports is presented in Chapter 3. The symbols commonly used to represent roller and hinged supports on line diagrams are shown in Fig. 1.13(b), and the symbol for fixed supports is depicted in Fig. 1.14(b).

SUMMARY

In this chapter, we learned about structural analysis and its role in structural engineering. *Structural analysis* is the prediction of the performance of a given structure under prescribed loads. Structural engineering has long been a part of human endeavor, but Galileo is considered to be the originator of the *theory* of structures. Following his pioneering work, many other people have made significant contributions. The availability of computers has revolutionized structural analysis.

Structural engineering is the science of planning, designing, and constructing safe, economical structures. Structural analysis is an integral part of this process.

Structures can be classified into five basic categories, namely, tension structures (e.g., cables and hangers), compression structures (e.g., columns and arches), trusses, shear structures (e.g., shear walls), and bending structures (e.g., beams and rigid frames).

An analytical model is a simplified representation of a real structure for the purpose of analysis. Development of the model generally involves (1) determination of whether or not the structure can be treated as a plane structure, (2) construction of the line diagram of the structure, and (3) idealization of connections and supports.

Earthquake-Damaged Bridge
Photo courtesy of Bethlehem Steel Corporation

2

Loads on Structures

2.1 Dead Loads
2.2 Live Loads
2.3 Impact
2.4 Wind Loads
2.5 Snow Loads
2.6 Earthquake Loads
2.7 Hydrostatic and Soil Pressures
2.8 Thermal and Other Effects
2.9 Load Combinations
Summary
Problems

The objective of a structural engineer is to design a structure that will be able to withstand all the loads to which it is subjected while serving its intended purpose throughout its intended life span. In designing a structure, an engineer must, therefore, consider all the loads that can realistically be expected to act on the structure during its planned life span. The loads that act on common civil engineering structures can be grouped according to their nature and source into three classes: (1) *dead loads* due to the weight of the structural system itself and any other material permanently attached to it; (2) *live loads*, which are movable or moving loads due to the use of the structure; and (3) *environmental loads*, which are caused by environmental effects, such as wind, snow, and earthquakes.

In addition to estimating the magnitudes of the design loads, an engineer must also consider the possibility that some of these loads might act simultaneously on the structure. The structure is finally designed so that it will be able to withstand the most unfavorable combination of loads that is likely to occur in its lifetime.

The minimum design loads and the load combinations for which the structures must be designed are usually specified in building codes. The national codes providing guidance on loads for buildings, bridges, and other structures include *ASCE Standard Minimum Design Loads for*

Buildings and Other Structures (SEI/ASCE 7-02) [1],* *Manual for Railway Engineering* [26], *Standard Specifications for Highway Bridges* [36], and *International Building Code* [15].

Although the load requirements of most local building codes are generally based on those of the national codes listed herein, local codes may contain additional provisions warranted by such regional conditions as earthquakes, tornadoes, hurricanes, heavy snow, and the like. Local building codes are usually legal documents enacted to safeguard public welfare and safety, and the engineer must become thoroughly familiar with the building code for the area in which the structure is to be built.

The loads described in the codes are usually based on past experience and study and are the *minimum* for which the various types of structures must be designed. However, the engineer must decide if the structure is to be subjected to any loads in addition to those considered by the code, and, if so, must design the structure to resist the additional loads. Remember that the engineer is ultimately responsible for the safe design of the structure.

The objective of this chapter is to describe the types of loads commonly encountered in the design of structures and to introduce the basic concepts of load estimation. We first describe dead loads and then discuss live loads for buildings and bridges. We next consider the dynamic effect, or the impact, of live loads. We describe environmental loads, including wind loads, snow loads, and earthquake loads. We give a brief discussion of hydrostatic and soil pressures and thermal effects and conclude with a discussion about the combinations of loads used for design purposes.

The material presented herein is mainly based on the *ASCE Standard Minimum Design Loads for Buildings and Other Structures* (SEI/ASCE 7-02), which is commonly referred to as the *ASCE 7 Standard* and is perhaps the most widely used standard in practice. Since the intent here is to familiarize the reader with the general topic of loads on structures, many of the details have not been included. Needless to say, the complete provisions of the local building codes or the *ASCE 7 Standard*[†] must be followed in designing structures.

2.1 DEAD LOADS

Dead loads are gravity loads of constant magnitudes and fixed positions that act permanently on the structure. Such loads consist of the weights of the structural system itself and of all other material and equipment

*The numbers in brackets refer to items listed in the bibliography.

[†] Copies of this standard may be purchased from the American Society of Civil Engineers, 1801 Alexander Bell Drive, Reston, Virginia 20191-4400.

TABLE 2.1 UNIT WEIGHTS OF CONSTRUCTION MATERIALS

Material	Unit Weight	
	lb/ft³	kN/m³
Aluminum	165	25.9
Brick	120	18.8
Concrete, reinforced	150	23.6
Structural steel	490	77.0
Wood	40	6.3

permanently attached to the structural system. For example, the dead loads for a building structure include the weights of frames, framing and bracing systems, floors, roofs, ceilings, walls, stairways, heating and air-conditioning systems, plumbing, electrical systems, and so forth.

The weight of the structure is not known in advance of design and is usually assumed based on past experience. After the structure has been analyzed and the member sizes determined, the actual weight is computed by using the member sizes and the unit weights of materials. The actual weight is then compared to the assumed weight, and the design is revised if necessary. The unit weights of some common construction materials are given in Table 2.1. The weights of permanent service equipment, such as heating and air-conditioning systems, are usually obtained from the manufacturer.

Example 2.1

The floor system of a building consists of a 5-in.-thick reinforced concrete slab resting on four steel floor beams, which in turn are supported by two steel girders, as shown in Fig. 2.1(a). The cross-sectional areas of the floor beams and the girders are 14.7 in.² and 52.3 in.², respectively. Determine the dead loads acting on the beams CG and DH and the girder AD.

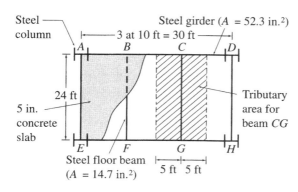

(a) Framing Plan

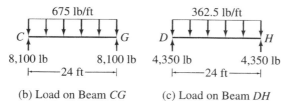

(b) Load on Beam CG

(c) Load on Beam DH

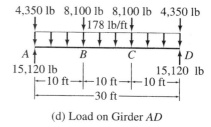

(d) Load on Girder AD

FIG. 2.1

continued

Solution

Beam CG As shown in Fig. 2.1(a), the portion of the slab supported by beam *CG* has a width of 10 ft (i.e., half the distance between beams *CG* and *BF* plus half the distance between beams *CG* and *DH*) and a length of 24 ft. This surface area (24 × 10 = 240 ft²) supported by beam *CG* (the shaded area in Fig. 2.1(a)) is referred to as the *tributary area* for beam *CG*.

We use the unit weights of reinforced concrete and structural steel from Table 2.1 to compute the dead load per foot of length of beam *CG* as follows:

$$\text{Concrete slab:}\quad (150 \text{ lb/ft}^3)(10 \text{ ft})(1 \text{ ft})\left(\frac{5}{12}\right) \text{ ft} = 625 \text{ lb/ft}$$

$$\text{Steel beam:}\quad (490 \text{ lb/ft}^3)\left(\frac{14.7}{144} \text{ ft}^2\right)(1 \text{ ft}) = 50 \text{ lb/ft}$$

$$\text{Total load} = \overline{675 \text{ lb/ft}} \qquad \textbf{Ans.}$$

This load is uniformly distributed on the beam, as shown in Fig. 2.1(b). This figure also shows the reactions exerted by the supporting girders at the ends of the beam. As the beam is symmetrically loaded, the magnitudes of the reactions are equal to half of the total load acting on the beam:

$$R_C = R_G = \tfrac{1}{2}(675 \text{ lb/ft})(24 \text{ ft}) = 8100 \text{ lb}$$

Note that the magnitudes of these end reactions represent the downward loads being transmitted to the supporting girders *AD* and *EH* at points *C* and *G*, respectively.

Beam DH The tributary area for beam *DH* is 5 ft wide and 24 ft long. The dead load per foot of length of this beam is computed as follows:

$$\text{Concrete slab:}\quad (150 \text{ lb/ft}^3)(5 \text{ ft})(1 \text{ ft})\left(\frac{5}{12} \text{ ft}\right) = 312.5 \text{ lb/ft}$$

$$\text{Steel beam:}\qquad (\text{same as for beam } CG) = 50.0 \text{ lb/ft}$$

$$\text{Total load} = \overline{362.5 \text{ lb/ft}} \qquad \textbf{Ans.}$$

As shown in Fig. 2.1(c), the end reactions are

$$R_D = R_H = \tfrac{1}{2}(362.5 \text{ lb/ft})(24 \text{ ft}) = 4350 \text{ lb}$$

Girder AD Because of the symmetry of the framing system and loading, the loads acting on beams *BF* and *AE* are the same as those on beams *CG* and *DH*, respectively. The load on girder *AD* consists of the uniformly distributed load due to its own weight, which has a magnitude of

$$(490 \text{ lb/ft}^3)\left(\frac{52.3}{144} \text{ ft}^2\right)(1 \text{ ft}) = 178 \text{ lb/ft}$$

and the concentrated loads transmitted to it by the beams at points *A*, *B*, *C*, and *D*, as shown in Fig. 2.1(d).   **Ans.**

2.2 LIVE LOADS

Live loads are loads of varying magnitudes and/or positions caused by the use of the structure. Sometimes, the term live loads is used to refer to all loads on the structure that are not dead loads, including environmental loads, such as snow loads or wind loads. However, since the probabilities of occurrence for environmental loads are different from those due to the use of structures, the current codes use the term live loads to refer only to those variable loads caused by the use of the structure. It is in the latter context that this text uses this term.

The magnitudes of design live loads are usually specified in building codes. The position of a live load may change, so each member of the structure must be designed for the position of the load that causes the maximum stress in that member. Different members of a structure may reach their maximum stress levels at different positions of the given load. For example, as a truck moves across a truss bridge, the stresses in the truss members vary as the position of the truck changes. If member *A* is subjected to its maximum stress when the truck is at a certain position *x*, then another member *B* may reach its maximum stress level when the truck is in a different position *y* on the bridge. The procedures for determining the position of a live load at which a particular response characteristic, such as a stress resultant or a deflection, of a structure is maximum (or minimum) are discussed in subsequent chapters.

Live Loads for Buildings

Live loads for buildings are usually specified as uniformly distributed surface loads in pounds per square foot or kilopascals. Minimum floor live loads for some common types of buildings are given in Table 2.2.

TABLE 2.2 MINIMUM FLOOR LIVE LOADS FOR BUILDINGS

Occupancy or Use	Live Load	
	psf	kPa
Hospital private rooms and wards, residential dwellings, apartments, hotel guest rooms, school classrooms	40	1.92
Library reading rooms, hospital operating rooms and laboratories	60	2.87
Dance halls and ballrooms, restaurants, gymnasiums	100	4.79
Light manufacturing, light storage warehouses, wholesale stores	125	5.98
Heavy manufacturing, heavy storage warehouses	250	11.97

Source: Adapted with permission from ASCE 7-88, *Minimum Design Loads for Buildings and Other Structures,* July 1990.

For a comprehensive list of live loads for various types of buildings and for provisions regarding roof live loads, concentrated loads, and reduction in live loads, the reader is referred to the *ASCE 7 Standard*.

Live Loads for Bridges

Live loads due to vehicular traffic on highway bridges are specified by the American Association of State Highway and Transportation Officials in the *Standard Specifications for Highway Bridges* [36], which is commonly referred to as the *AASHTO Specification*.

As the heaviest loading on highway bridges is usually caused by trucks, the *AASHTO Specification* defines two systems of standard trucks, *H trucks* and *HS trucks*, to represent the vehicular loads for design purposes.

The H-truck loadings (or H loadings), representing a two-axle truck, are designated by the letter *H*, followed by the total weight of the truck and load in tons and the year in which the loading was initially specified. For example, the loading H20-44 represents a code for a two-axle truck weighing 20 tons initially instituted in the 1944 edition of the *AASHTO Specification*. The axle spacing, axle loads, and wheel spacing for the H trucks are shown in Fig. 2.2(a).

The HS-truck loadings (or HS loadings) represent a two-axle tractor truck with a single-axle semitrailer. These loadings are designated by the

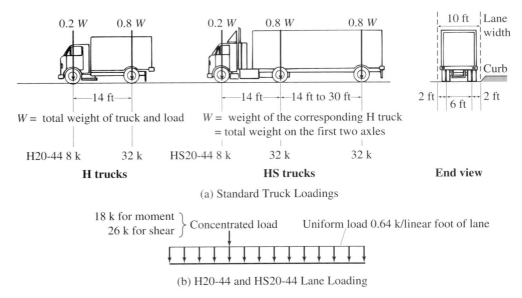

(a) Standard Truck Loadings

(b) H20-44 and HS20-44 Lane Loading

FIG. 2.2 Live Loads for Highway Bridges

Source: Taken from the *Standard Specifications for Highway Bridges*. Copyright 1989. American Association of State Highway and Transportation Officials, Washington, D.C. Used by permission.

letters *HS* followed by the weight of the corresponding *H* truck in tons and the year in which the loading was initially specified. The axle spacing, axle loads, and wheel spacing for the HS trucks are shown in Fig. 2.2(a). Note that the spacing between the rear axle of the tractor truck and the axle of the semitrailer should be varied between 14 ft and 30 ft, and the spacing causing the maximum stress should be used for design.

The particular type of truck loading to be used in design depends on the anticipated traffic on the bridge. The H20-44 and HS20-44 are the most commonly used loadings; the axle loads for these loadings are shown in Fig. 2.2(a).

In addition to the aforementioned single-truck loading, which must be placed to produce the most unfavorable effect on the member being designed, AASHTO specifies that a lane loading, consisting of a uniformly distributed load combined with a single concentrated load, be considered. The lane loading represents the effect of a lane of medium-weight vehicles containing a heavy truck. The lane loading must also be placed on the structure so that it causes maximum stress in the member under consideration. As an example, the lane loading corresponding to the H20-44 and HS20-44 truck loadings is shown in Fig. 2.2(b). The type of loading, either truck loading or lane loading, that causes the maximum stress in a member should be used for the design of that member. Additional information regarding multiple lanes, loadings for continuous spans, reduction in load intensity, and so on, can be found in the *AASHTO Specification*.

Live loads for railroad bridges are specified by the American Railway Engineering and Maintenance of Way Association (AREMA) in the *Manual for Railway Engineering* [26]. These loadings, which are commonly known as *Cooper E loadings*, consist of two sets of nine concentrated loads, each separated by specified distances, representing the two locomotives followed by a uniform loading representing the weight of the freight cars. An example of such a loading, called the E80 loading, is depicted in Fig. 2.3. The design loads for heavier or lighter trains can be obtained from this loading by proportionately increasing or decreasing the magnitudes of the loads while keeping the same distances

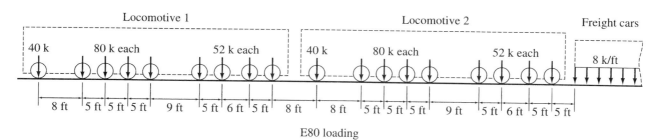

E80 loading

FIG. 2.3 Live Loads for Railroad Bridges

between the concentrated loads. For example, the E40 loading can be obtained from the E80 loading by simply dividing the magnitudes of the loads by 2. As in the case of highway bridges considered previously, live loads on railroad bridges must be placed so that they will cause the most unfavorable effect on the member under consideration.

2.3 IMPACT

When live loads are applied rapidly to a structure, they cause larger stresses than those that would be produced if the same loads would have been applied gradually. The dynamic effect of the load that causes this increase in stress in the structure is referred to as *impact*. To account for the increase in stress due to impact, the live loads expected to cause such a dynamic effect on structures are increased by certain impact percentages, or impact factors. The impact percentages and factors, which are usually based on past experience and/or experimental results, are specified in the building codes. For example, the *ASCE 7 Standard* specifies that all elevator loads for buildings be increased by 100% to account for impact.

For highway bridges, the *AASHTO Specification* gives the expression for the impact factor as

$$I = \frac{50}{L + 125} \le 0.3$$

in which L is the length in feet of the portion of the span loaded to cause the maximum stress in the member under consideration. Similar empirical expressions for impact factors to be used in designing railroad bridges are specified in [26].

2.4 WIND LOADS

Wind loads are produced by the flow of wind around the structure. The magnitudes of wind loads that may act on a structure depend on the geographical location of the structure, obstructions in its surrounding terrain, such as nearby buildings, and the geometry and the vibrational characteristics of the structure itself. Although the procedures described in the various codes for the estimation of wind loads usually vary in detail, most of them are based on the same basic relationship between the wind speed V and the dynamic pressure q induced on a flat surface normal to the wind flow, which can be obtained by applying Bernoulli's principle and is expressed as

$$q = \tfrac{1}{2}\rho V^2 \tag{2.1}$$

in which ρ is the mass density of the air. Using the unit weight of air of 0.0765 lb/ft^3 for the standard atmosphere (at sea level, with a temperature of $59°$F), and expressing the wind speed V in miles per hour, the dynamic pressure q in pounds per square foot is given by

$$q = \frac{1}{2}\left(\frac{0.0765}{32.2}\right)\left(\frac{5280}{3600}\right)^2 V^2 = 0.00256 V^2 \tag{2.2}$$

The wind speed V to be used in the determination of the design loads on a structure depends on its geographical location and can be obtained from meteorological data for the region. The *ASCE 7 Standard* provides a contour map of the basic wind speeds for the United States (Fig. 2.4). This map, which is based on data collected at 485 weather stations, gives the 3-second gust speeds in miles per hour (m/s). These speeds are for open terrain at the heights of 33 ft (10 m) above ground level. To account for the variation in wind speed with the height and the surroundings in which a structure is located and to account for the consequences of the failure of structures, the *ASCE 7 Standard* modifies Eq. (2.2) as

$$q_z = 0.00256 K_z K_{zt} K_d V^2 I \tag{2.3}$$

in which q_z is the velocity pressure at height z in pounds per square foot; V is the basic wind speed in miles per hour (Fig. 2.4); I is the *importance factor*; K_z is the *velocity pressure exposure coefficient*; K_{zt} is the *topographic factor*; and K_d is the *wind directionality factor*. When converted to SI units, Eq. (2.3) becomes

$$q_z = 0.613 K_z K_{zt} K_d V^2 I \quad \text{[SI units]} \tag{2.4}$$

with q_z and V now expressed in units of N/m^2 and m/s, respectively.

The importance factor I accounts for hazard to human life and damage to property in the event of failure of the structure. The values of I to be used for estimating wind loads for the various categories of buildings are listed in Table 2.3.

The velocity pressure exposure coefficient, K_z, is given by

$$K_z = \begin{cases} 2.01(z/z_g)^{2/\alpha} & \text{for } 15 \text{ ft } (4.6 \text{ m}) \leq z \leq z_g \\ 2.01(15/z_g)^{2/\alpha} & \text{for } z < 15 \text{ ft } (4.6 \text{ m}) \end{cases} \tag{2.5}$$

in which $z =$ height above ground in feet (or meters); $z_g =$ gradient height in feet (or meters); and $\alpha =$ power law coefficient. The constants z_g and α depend on the obstructions on the terrain immediately surrounding the structure. The *ASCE 7 Standard* classifies the terrains to which the structures may be exposed into three categories. These three categories are briefly described in Table 2.4, which also provides the

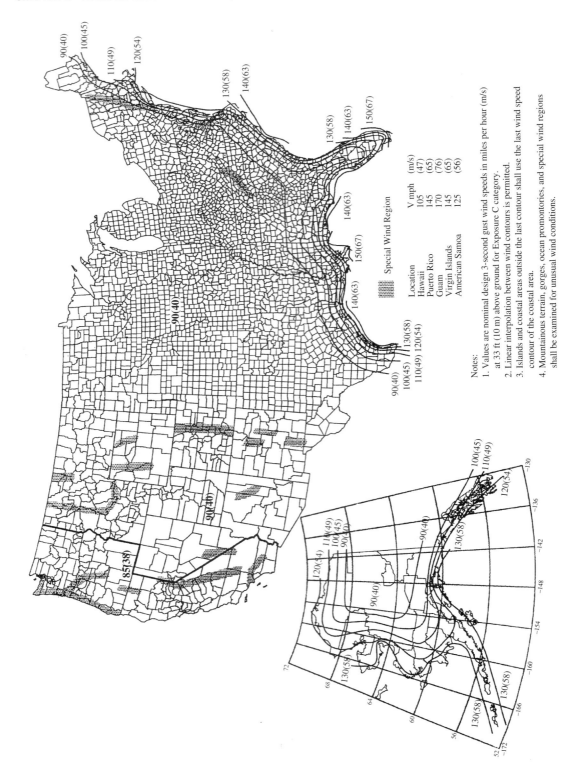

Special Wind Region	Location	V mph	(m/s)
	Hawaii	105	(47)
	Puerto Rico	145	(65)
	Guam	170	(76)
	Virgin Islands	145	(65)
	American Samoa	125	(56)

Notes:
1. Values are nominal design 3-second gust wind speeds in miles per hour (m/s) at 33 ft (10 m) above ground for Exposure C category.
2. Linear interpolation between wind contours is permitted.
3. Islands and coastal areas outside the last contour shall use the last wind speed contour of the coastal area.
4. Mountainous terrain, gorges, ocean promontories, and special wind regions shall be examined for unusual wind conditions.

FIG. 2.4 Basic Wind Speeds for the United States

Source: Reproduced with permission from ASCE 7-02, *Minimum Design Loads for Buildings and Other Structures*. This information is extracted from ASCE 7-02; for further information, the complete text of the manual should be referenced (http://www.pubs.asce.org/ASCE7.html?9991330).

TABLE 2.3 CLASSIFICATION OF BUILDINGS FOR ENVIRONMENTAL LOADS

Occupancy or use	Category	Importance Factor, I		
		Wind loads	Snow loads	Earthquake loads
Buildings representing low hazard to human life in the case of failure, such as agricultural and minor storage facilities	I	0.87 for $V \leq 100$ mph 0.77 for $V > 100$ mph	0.8	1.00
All buildings other than those listed in Categories I, III, and IV	II	1.00	1.0	1.00
Buildings representing a substantial hazard to human life in the case of failure, such as: those where more than 300 people congregate in one area; day-care facilities with capacity greater than 150; schools with capacity greater than 250; colleges with capacity greater than 500; hospitals without emergency treatment or surgery facilities but with patient capacity greater than 50; jails; power stations and utilities not essential in an emergency; and buildings containing hazardous and explosive materials	III	1.15	1.1	1.25
Essential facilities, including hospitals, fire and police stations, national defense facilities and emergency shelters, communication centers, power stations, and utilities required in an emergency	IV	1.15	1.2	1.5

Source: Adapted with permission from ASCE 7-02, *Minimum Design Loads for Buildings and Other Structures.* This information is extracted from ASCE 7-02; for further information, the complete text of the manual should be referenced (http://www.pubs.asce.org/ASCE7.html?9991330).

TABLE 2.4 EXPOSURE CATEGORIES FOR BUILDINGS FOR WIND LOADS

Exposure	Category	Constants	
		z_g ft(m)	α
Urban and suburban areas with closely spaced obstructions of the size of single family houses or larger. This terrain must prevail in the upwind direction for a distance of at least 2,630 ft (800 m) or 10 times the building height, whichever is greater	B	1,200(366)	7.0
Applies to all buildings to which exposures B or D do not apply	C	900(274)	9.5
Flat, unobstructed areas and water surfaces outside hurricane-prone regions. This terrain must prevail in the upwind direction for a distance of at least 5,000 ft (1,524 m) or 10 times the building height, whichever is greater	D	700(213)	11.5

Source: Adapted with permission from ASCE 7-02, *Minimum Design Loads for Buildings and Other Structures.* This information is extracted from ASCE 7-02; for further information, the complete text of the manual should be referenced (http://www.pubs.asce.org/ASCE7.html?9991330).

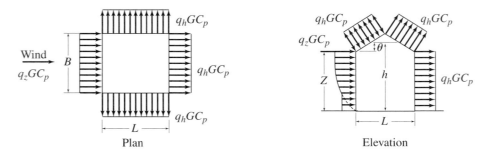

Gable, Hip Roof

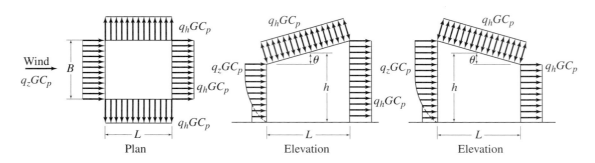

Monoslope Roof (Note 4)

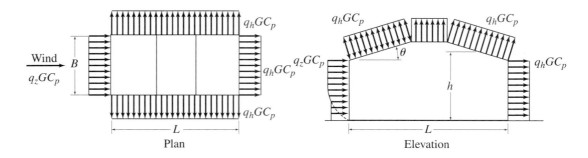

Mansard Roof (Note 6)

FIG. 2.5 *External Pressure Coefficients*, C_p, for Loads on Main Wind-Force Resisting Systems for Enclosed or Partially Enclosed Buildings of All Heights

Source: Reproduced with permission from ASCE 7-02, *Minimum Design Loads for Buildings and Other Structures.* This information is extracted from ASCE 7-02; for further information, the complete text of the manual should be referenced (http://www.pubs.asce.org/ASCE7.html?9991330).

Wall Pressure Coefficients, C_p			
Surface	L/B	C_p	Use with
Windward wall	All values	0.8	q_z
Leeward wall	0–1	−0.5	q_h
	2	−0.3	
	≥4	−0.2	
Side wall	All values	−0.7	q_h

Roof Pressure Coefficients, C_p, for use with q_h												
Wind direction		Windward								Leeward		
		Angle, θ (degrees)								Angle, θ (degrees)		
	h/L	10	15	20	25	30	35	45	≥60#	10	15	≥20
Normal to ridge for $\theta \geq 10°$	≤0.25	−0.7 / −0.18	−0.5 / 0.0*	−0.3 / 0.2	−0.2 / 0.3	−0.2 / 0.3	0.0* / 0.4	0.4	0.01θ	−0.3	−0.5	−0.6
	0.5	−0.9 / −0.18	−0.7 / −0.18	−0.4 / 0.0*	−0.3 / 0.2	−0.2 / 0.2	−0.2 / 0.3	0.0* / 0.4	0.01θ	−0.5	−0.5	−0.6
	≥1.0	−1.3** / −0.18	−1.0 / −0.18	−0.7 / −0.18	−0.5 / 0.0*	−0.3 / 0.2	−0.2 / 0.2	0.0* / 0.3	0.01θ	−0.7	−0.6	−0.6

Normal to ridge for $\theta < 10°$ and Parallel to ridge for all θ	≤0.5	Horiz distance from windward edge	C_p	*Value is provided for interpolation purposes.
		0 to h/2	−0.9, −0.18	
		h/2 to h	−0.9, −0.18	**Value can be reduced linearly with area over which it is applicable as follows.
		h to 2 h	−0.5, −0.18	
		>2 h	−0.3, −0.18	

		Horiz distance	C_p	Area (sq ft)	Reduction factor
	≥1.0	0 to h/2	−1.3**, −0.18	≤100 (9.3 sq m)	1.0
				200 (23.2 sq m)	0.9
		>h/2	−0.7, −0.18	≥1,000 (92.9 sq m)	0.8

Notes:
1. Plus and minus signs signify pressures acting toward and away from the surfaces, respectively.
2. Linear interpolation is permitted for values of $L/B, h/L$, and θ other than shown. Interpolation shall only be carried out between values of the same sign. Where no value of the same sign is given, assume 0.0 for interpolation purposes.
3. Where two values of C_p are listed, this indicates that the windward roof slope is subjected to either positive or negative pressures and the roof structure shall be designed for both conditions. Interpolation for intermediate ratios of h/L in this case shall only be carried out between C_p values of like sign.
4. For monoslope roofs, the entire roof surface is either a windward or leeward surface.
5. Notation:
   B: Horizontal dimension of building, in feet (meters), measured normal to wind direction.
   L: Horizontal dimension of building, in feet (meters), measured parallel to wind direction.
   h: Mean roof height in feet (meters), except that eave height shall be used for $\theta \leq 10$ degrees.
   z: Height above ground, in feet (meters).
   G: Gust effect factor.
   q_z, q_h: Velocity pressure, in pounds per square foot (N/m^2), evaluated at respective height.
   θ: Angle of plane of roof from horizontal, in degrees.
6. For mansard roofs, the top horizontal surface and leeward inclined surface shall be treated as leeward surfaces from the table.
7. Except for MWFRS's at the roof consisting of moment resisting frames, the total horizontal shear shall not be less than that determined by neglecting wind forces on roof surfaces.
#For roof slopes greater than 80°, use $C_p = 0.8$.

FIG. 2.5 (contd.)

values of the constants for each of the categories. A more detailed description of the exposure categories can be found in the *ASCE 7 Standard*. The topographic factor, K_{zt}, takes into account the effect of increase in wind speed due to abrupt changes in topography, such as isolated hills and steep cliffs. For structures located on or near the tops of such hills, the value of K_{zt} should be determined using the procedure specified in the *ASCE 7 Standard*. For other structures, $K_{zt} = 1$. The wind directionality factor, K_d, takes into account the reduced probability of maximum winds coming from the direction that is most unfavorable for the structure. This factor is used only when wind loads are applied in combination with other types of loads (such as dead loads, live loads, etc.). For structures subjected to such load combinations, the values of K_d should be obtained from the *ASCE 7 Standard*. For structures subjected only to wind loads, $K_d = 1$.

The external wind pressures to be used for designing the main framing of structures are given by

$$p_z = q_z G C_p \quad \text{for windward wall}$$
$$p_h = q_h G C_p \quad \text{for leeward wall, sidewalls, and roof}$$
(2.6)

in which h = mean roof height above ground; q_h = velocity pressure at height h (evaluated by substituting $z = h$ in Eq. (2.3) or (2.4)); p_z = design wind pressure at height z above ground; p_h = design wind pressure at mean roof height h; G = *gust effect factor*; and C_p = *external pressure coefficient*.

The gust effect factor, G, is used to consider the loading effect of wind turbulence on the structure. For a rigid structure, whose fundamental frequency is greater than or equal to 1 Hz., $G = 0.85$. For flexible structures, the value of G should be calculated using the equations given in the *ASCE 7 Standard*.

The values of the external pressure coefficients, C_p, based on wind tunnel and full-scale tests, have been provided in the *ASCE 7 Standard* for various types of structures. Figure 2.5 shows the coefficients specified for designing the main framing of structures. We can see from this figure that the external wind pressure varies with height on the windward wall of the structure but is uniform on the leeward wall and the sidewalls. Note that the positive pressures act toward the surfaces, whereas the negative pressures, called suctions, act away from the surfaces of the structures.

Once the external wind pressures have been established, they are combined with the internal pressures to obtain the design wind pressures. With the design wind pressures known, we can determine the corresponding design loads on members of the structures by multiplying the pressures by the appropriate tributary areas of the members.

Example 2.2

Determine the external wind pressure on the roof of the rigid gabled frame of a nonessential industrial building shown in Fig. 2.6(a). The structure is located in a suburb of Boston, Massachusetts, where the terrain is representative of exposure B. The wind direction is normal to the ridge of the frame as shown.

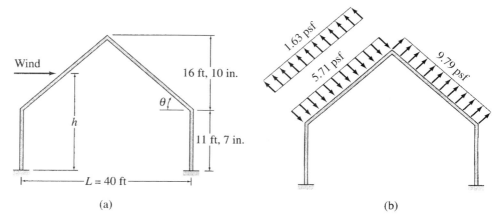

(a)   (b)

FIG. 2.6

Solution

Roof Slope and Mean Roof Height From Fig. 2.6(a), we obtain

$$\tan \theta = \frac{16.83}{20} = 0.842, \quad \text{or} \quad \theta = 40.1°$$

$$h = 11.58 + \frac{16.83}{2} = 20.0$$

$$\frac{h}{L} = \frac{20}{40} = 0.5$$

Velocity Pressure at $z = h = 20'$ From Fig. 2.4, we obtain the basic wind speed for Boston as

$$V = 110 \text{ mph}$$

From Table 2.3, we can see that the importance factor for wind loads for nonessential buildings (category II) is

$$I = 1.0$$

and from Table 2.4, for the exposure category B, we obtain the following values of the constants:

$$z_g = 1,200 \text{ ft} \quad \text{and} \quad \alpha = 7.0$$

By using Eq. (2.5), we determine the velocity pressure exposure coefficient:

$$K_h = 2.01 \left(\frac{h}{z_g}\right)^{2/\alpha} = 2.01 \left(\frac{20}{1,200}\right)^{2/7} = 0.62$$

continued

Using $K_{zt} = 1$ and $K_d = 1$, we apply Eq. (2.3) to obtain the velocity pressure at height h as

$$q_h = 0.00256 K_h K_{zt} K_d V^2 I$$

$$= 0.00256(0.62)(1)(1)(110)^2(1.0)$$

$$= 19.2 \text{ psf}$$

External Wind Pressure on Roof For rigid structures, the gust effect factor is

$$G = 0.85$$

For $\theta \approx 40°$ and $h/L = 0.5$, the values of the external pressure coefficients are (Fig. 2.5):

For windward side: $C_p = 0.35$ and -0.1

For leeward side: $C_p = -0.6$

Finally, by substituting the values of q_h, G, and C_p into Eq. (2.6), we obtain the following wind pressures: for the windward side,

$$p_h = q_h G C_p = (19.2)(0.85)(0.35) = 5.71 \text{ psf} \qquad \textbf{Ans.}$$

and

$$p_h = q_h G C_p = (19.2)(0.85)(-0.1) = -1.63 \text{ psf} \qquad \textbf{Ans.}$$

and for the leeward side

$$p_h = q_h G C_p = (19.2)(0.85)(-0.6) = -9.79 \text{ psf} \qquad \textbf{Ans.}$$

These wind pressures are applied to the roof of the frame, as shown in Fig. 2.6(b). The two wind pressures (positive and negative) on the windward side are treated as separate loading conditions, and the structure is designed for both conditions.

2.5 SNOW LOADS

In many parts of the United States and the world, snow loads must be considered in designing structures. The design snow load for a structure is based on the ground snow load for its geographical location, which can be obtained from building codes or meteorological data for that region. The *ASCE 7 Standard* provides contour maps (similar to Fig. 2.4) of the ground snow loads for various parts of the United States. These maps, which are based on data collected at 204 weather stations and over 9000 other locations, give the snow loads (in pounds per square foot) that have a 2% probability of being exceeded in any given year.

Once the ground snow load has been established, the design snow load for the roof of the structure is determined by considering such factors as the structure's exposure to wind, and its thermal, geometric, and functional characteristics. In most cases, there is less snow on roofs than on the ground. The *ASCE 7 Standard* recommends that the design snow load for flat roofs be expressed as

$$p_f = 0.7 C_e C_t I p_g \tag{2.7}$$

in which p_f = design flat-roof snow load in pounds per square foot (kN/m^2); p_g = ground snow load in pounds per square foot (kN/m^2); C_e = *exposure factor*; C_t = *thermal factor*; and I = importance factor.

In Eq. (2.7), the numerical factor 0.7, which is referred to as the basic exposure factor, accounts for the general effect of wind, which is likely to blow some of the snow off the roofs. The local effects of wind, which depend on the particular terrain surrounding the structure and the exposure of its roof, are accounted for by the exposure factor C_e. The *ASCE 7 Standard* provides the values of C_e, which range from 0.7 for structures in windy areas with exposed roofs to 1.3 for structures exposed to little wind.

The thermal factor, C_t, accounts for the fact that there will be more snow on the roofs of unheated structures than on those of heated ones. The values of C_t are specified as 1.0 and 1.2 for heated and unheated structures, respectively. As in the case of wind loads, the importance factor I in Eq. (2.7) accounts for hazard to human life and damage to property in the case of failure of the structure. The values of I to be used for estimating roof snow loads are given in Table 2.3.

The design snow load for a sloped roof is determined by multiplying the corresponding flat-roof snow load by a *slope factor* C_s. Thus,

$$p_s = C_s p_f \tag{2.8}$$

in which p_s is the design sloped-roof snow load considered to act on the horizontal projection of the roof surface, and the slope factor C_s is given by

For warm roofs $(C_t \leq 1.0)$
$$\begin{cases} C_s = 1 & \text{for } 0 \leq \theta < 30° \\ C_s = 1 - \dfrac{\theta - 30°}{40°} & \text{for } 30° \leq \theta \leq 70° \\ C_s = 0 & \text{for } \theta > 70° \end{cases} \tag{2.9}$$

For cold roofs $(C_t = 1.2)$
$$\begin{cases} C_s = 1 & \text{for } 0 \leq \theta < 45° \\ C_s = 1 - \dfrac{\theta - 45°}{25°} & \text{for } 45° \leq \theta \leq 70° \\ C_s = 0 & \text{for } \theta > 70° \end{cases} \tag{2.10}$$

In Eqs. (2.9) and (2.10), θ denotes the slope of the roof from the horizontal, in degrees. These slope factors are based on the considerations that more snow is likely to slide off of steep roofs, as compared to shallow ones, and that more snow is likely to melt and slide off the roofs of heated structures than those of unheated structures.

The *ASCE 7 Standard* specifies minimum values of snow loads for which structures with low-slope roofs must be designed. For such structures, if $P_g \leq 20$ psf $(0.96 \ kN/m^2)$, then P_f shall not be less than $P_g I$; if $P_g > 20$ psf $(0.96 \ kN/m^2)$, then P_f shall not be less than

$20I$ psf $(0.96I\ \text{kN/m}^2)$. These minimum values of P_f apply to monoslope roofs with $\theta \leq 15°$, and to hip and gable roofs with $\theta \leq (70/W) + 0.5$, where W is the horizontal distance from the eave to the ridge in feet.

In some structures, the snow load acting on only a part of the roof may cause higher stresses than when the entire roof is loaded. To account for such a possibility, the *ASCE 7 Standard* recommends that the effect of unbalanced snow loads also be considered in the design of structures. A detailed description of unbalanced snow load distributions to be considered in the design of various types of roofs can be found in the *ASCE 7 Standard*. For example, for gable roofs with $(70/W) + 0.5 \leq \theta \leq 70°$ and $W \leq 20$ ft, the *ASCE 7 Standard* specifies that the structures be designed to resist an unbalanced uniform load of magnitude $1.5\ P_s/C_e$ applied to the leeward side of the roof, with the windward side free of snow.

Example 2.3

Determine the design snow loads for the roof of the gabled frame of an apartment building shown in Fig. 2.7(a). The building is located in Chicago, Illinois, where the ground snow load is 25 psf. Because of several trees near the structure, assume the exposure factor is $C_e = 1$.

Solution

Flat-Roof Snow Load

$p_g = 25$ psf

$C_e = 1$

$C_t = 1$ (heated structure)

$I = 1$ (from Table 2.3 for nonessential building, category II)

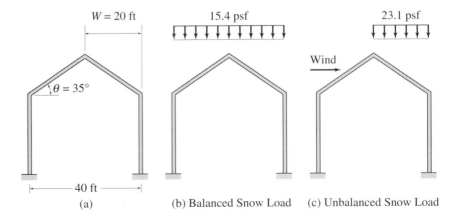

FIG. **2.7**

(a)   (b) Balanced Snow Load   (c) Unbalanced Snow Load

From Eq. (2.7), the flat-roof snow load is obtained as

$$p_f = 0.7C_eC_tIp_g = 0.7(1)(1)(1)(25)$$

$$= 17.5 \text{ psf}$$

From Fig. 2.7(a), we can see that $W = 20$ ft. Thus,

$$\frac{70}{W} + 0.5 = \frac{70}{20} + 0.5 = 4°$$

The slope is $\theta = 35°$, which is greater than $4°$, so the minimum values of p_f need not be considered.

Sloped-Roof Snow Load By applying Eq. (2.9), we compute the slope factor as

$$C_s = 1 - \frac{\theta - 30°}{40°} = 1 - \frac{35° - 30°}{40°} = 0.88$$

From Eq. (2.8), we determine the design sloped-roof snow load:

$$p_s = C_sp_f = 0.88(17.5) = 15.4 \text{ psf} \qquad \textbf{Ans.}$$

This load is called the *balanced design snow load* and is applied to the entire roof of the structure, as shown in Fig. 2.7(b).

$$\textit{Unbalanced Design Snow Load} = \frac{1.5\,p_s}{C_e} = \frac{1.5(15.4)}{1}$$

$$= 23.1 \text{ psf} \qquad \textbf{Ans.}$$

This load is applied only to the leeward side of the roof, as shown in Fig. 2.7(c).

2.6 EARTHQUAKE LOADS

An *earthquake* is a sudden undulation of a portion of the earth's surface. Although the ground surface moves in both horizontal and vertical directions during an earthquake, the magnitude of the vertical component of ground motion is usually small and does not have a significant effect on most structures. It is the horizontal component of ground motion that causes structural damage and that must be considered in designs of structures located in earthquake-prone areas.

During an earthquake, as the foundation of the structure moves with the ground, the above-ground portion of the structure, because of the inertia of its mass, resists the motion, thereby causing the structure to vibrate in the horizontal direction (Fig. 2.8). These vibrations produce horizontal shear forces in the structure. For an accurate prediction of the stresses that may develop in a structure in the case of an earthquake, a dynamic analysis, considering the mass and stiffness characteristics of the structure, must be performed. However, for low- to medium-height rectangular buildings, most codes employ equivalent

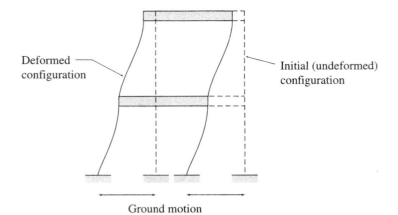

Deformed configuration

Initial (undeformed) configuration

Ground motion

FIG. 2.8 Effect of Earthquake on a Structure

static forces to design for earthquake resistance. In this empirical approach, the dynamic effect of the earthquake is approximated by a set of lateral (horizontal) forces applied to the structure, and static analysis is performed to evaluate stresses in the structure.

The *ASCE 7 Standard* permits the use of this equivalent lateral-force procedure for earthquake design of buildings. According to the *ASCE 7 Standard*, the total lateral seismic force that a building is designed to resist is given by the equation

$$V = C_S W \qquad (2.11)$$

in which V = total lateral force or base shear, W = dead load of the building, and C_S = seismic response coefficient. The latter is defined by the equation

$$C_S = \frac{S_{DS}}{R/I} \qquad (2.12)$$

in which S_{DS} is the design spectral response acceleration in the short period range; R denotes the response modification factor; and I represents the importance factor. The *ASCE 7 Standard* further specifies upper and lower limits for the values of C_S to be used in design.

The design spectral response acceleration (S_{DS}), used in the evaluation of the design base shear, depends on the geographical location of the structure, and can be obtained from the contour maps provided in the *ASCE 7 Standard*. The response modification factor R takes into consideration the energy-dissipation capacity of the structure; its values range from 1.25 to 8. For example, for plain unreinforced masonry shear walls, $R = 1.5$; whereas, for moment resisting frames, $R = 8$. The values of I to be used for estimating earthquake loads are given in Table 2.3.

The total lateral force V thus obtained is then distributed to the various floor levels of the building using the formulas provided in the *ASCE 7 Standard*. For additional details about this equivalent lateral-force procedure, and for limitations on the use of this procedure, the reader is referred to the *ASCE 7 Standard.*

2.7 HYDROSTATIC AND SOIL PRESSURES

Structures used to retain water, such as dams and tanks, as well as coastal structures partially or fully submerged in water must be designed to resist hydrostatic pressure. Hydrostatic pressure acts normal to the submerged surface of the structure, with its magnitude varying linearly with height, as shown in Fig. 2.9. Thus, the pressure at a point located at a distance h below the surface of the liquid can be expressed as

$$p = \gamma h \tag{2.13}$$

in which γ = unit weight of the liquid.

Underground structures, basement walls and floors, and retaining walls must be designed to resist soil pressure. The vertical soil pressure is given by Eq. (2.13), with γ now representing the unit weight of the soil. The lateral soil pressure depends on the type of soil and is usually considerably smaller than the vertical pressure. For the portions of structures below the water table, the combined effect of hydrostatic pressure and soil pressure due to the weight of the soil, reduced for buoyancy, must be considered.

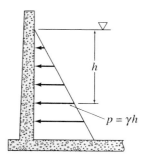

FIG. 2.9 Hydrostatic Pressure

2.8 THERMAL AND OTHER EFFECTS

Statically indeterminate structures may be subjected to stresses due to temperature changes, shrinkage of material, fabrication errors, and differential settlements of supports. Although these effects are usually not addressed in building codes, they may cause significant stresses in structures and should be considered in their designs. The procedures for determining the forces induced in structures due to these effects are considered in Part III.

2.9 LOAD COMBINATIONS

As stated previously, once the magnitudes of the design loads for a structure have been estimated, an engineer must consider all loads that might act simultaneously on the structure at a given time. For example, it is highly unlikely that an earthquake and the maximum wind loads will occur simultaneously. Based on past experience and probability analysis, the *ASCE 7 Standard* specifies various load combinations to be considered when designing structures. It is important to realize that the structure must be designed to have adequate strength to resist the most unfavorable of all the load combinations.

In addition to the aforementioned strength or safety requirements, a structure must also satisfy any serviceability requirements related to its intended use. For example, a high-rise building may be perfectly safe, yet unserviceable if it deflects or vibrates excessively due to wind. The serviceability requirements are specified in building codes for most common types of structures and are usually concerned with deflections, vibrations, cracking, corrosion, and fatigue.

SUMMARY

In this chapter, we learned about the loads that act on common civil engineering structures. These loads can be grouped into three classes: (1) dead loads, (2) live loads, and (3) environmental loads.

Dead loads have constant magnitudes and fixed positions, and they act permanently on the structure. Live loads have varying magnitudes and/or positions and are caused by the use or occupancy of the structure. Each member of the structure must be designed for that position of the live load that produces the most unfavorable effect on that member. For structures subjected to rapidly applied live loads, the dynamic effect, or the impact, of the loads should be considered in design.

The external wind pressures used for designing the main framing of structures are given by

$$p_z = q_z G C_p \quad \text{for windward wall}$$
$$p_h = q_h G C_p \quad \text{for leeward wall, sidewalls, and roof} \tag{2.6}$$

where h is the mean roof height, G is the gust effect factor, C_p is the external pressure coefficient, and q_z is the velocity pressure at height z, which is expressed in psf as

$$q_z = 0.00256 K_z K_{zt} K_d V^2 I \tag{2.3}$$

with K_z = velocity pressure exposure coefficient, K_{zt} = topographic factor, K_d = directionality factor, V = basic wind speed in mph, and I = importance factor.

The design flat-roof snow load for buildings is given by

$$p_f = 0.7 C_e C_t I p_g \tag{2.7}$$

where p_g = ground snow load, C_e = exposure factor, and C_t = thermal factor. The design sloped-roof snow load is expressed as

$$p_s = C_s p_f \tag{2.8}$$

with C_s = slope factor.

The total lateral seismic design force for buildings is given by

$$V = C_S W \qquad (2.11)$$

in which C_S = seismic response coefficient, and W = dead load of the building.

The magnitude of the hydrostatic pressure at a point located at a distance h below the surface of the liquid is given by

$$p = \gamma h \qquad (2.13)$$

in which γ = unit weight of the liquid.

The effects of temperature changes, shrinkage of material, fabrication errors, and support settlements should be considered in designing statically indeterminate structures. The structure must be designed to withstand the most unfavorable combination of loads.

PROBLEMS

Section 2.1

2.1 The floor system of an apartment building consists of a 4-in.-thick reinforced concrete slab resting on three steel floor beams, which in turn are supported by two steel girders, as shown in Fig. P2.1. The areas of cross section of the floor beams and the girders are 18.3 in.^{2} and 32.7 in.^{2}, respectively. Determine the dead loads acting on the beam CD and the girder AE.

2.2 Solve Problem 2.1 if a 6-in.-thick brick wall, which is 7 ft high and 25 ft long, bears directly on the top of beam CD. See Fig. P2.1.

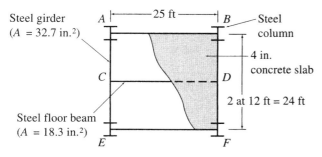

FIG. **P2.1, P2.2, P2.5**

2.3 The floor system of a gymnasium consists of a 130-mm-thick concrete slab resting on four steel beams

(A = 9,100 mm^2) that, in turn, are supported by two steel girders (A = 25,600 mm^2), as shown in Fig. P2.3. Determine the dead loads acting on beam BF and girder AD.

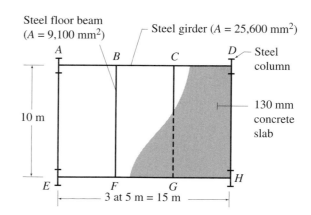

FIG. **P2.3, P2.6**

2.4 The roof system of an office building consists of a 4-in.-thick reinforced concrete slab resting on four steel beams (A = 16.2 in.^{2}), which are supported by two steel girders (A = 42.9 in.^{2}). The girders, in turn, are supported by four columns, as shown in Fig. P2.4. Determine the dead loads acting on the girder AG.

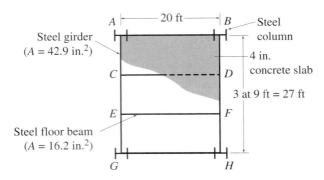

FIG. **P2.4, P2.7**

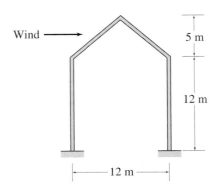

FIG. **P2.9, P2.13**

Section 2.2

2.5 For the apartment building whose floor system was described in Problem 2.1, determine the live loads acting on the beam *CD* and the girder *AE*. See Fig. P2.1.

2.6 For the gymnasium whose floor system was described in Problem 2.3, determine the live loads acting on beam *BF* and girder *AD*. See Fig. P2.3.

2.7 The roof of the office building considered in Problem 2.4 is subjected to a live load of 20 psf. Determine the live loads acting on the beam *EF*, the girder *AG*, and the column *A*. See Fig. P2.4.

Section 2.4

2.8 Determine the external wind pressure on the roof of the rigid-gabled frame of an apartment building shown in Fig. P2.8. The building is located in the Los Angeles area of California, where the terrain is representative of exposure *B*. The wind direction is normal to the ridge as shown.

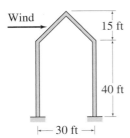

FIG. **P2.8**

2.9 Determine the external wind pressure on the roof of the rigid-gabled frame of a school building shown in Fig. P2.9. The structure is located in a suburb of Chicago, Illinois, where the terrain is representative of exposure *B*. The wind direction is normal to the ridge as shown.

2.10 Determine the external wind pressure on the roof of the rigid-gabled frame of a building for an essential disaster operation center shown in Fig. P2.10. The building is located in Kansas City, Missouri, where the terrain is representative of exposure *C*. The wind direction is normal to the ridge, as shown in the figure.

2.11 Determine the external wind pressures on the windward and leeward walls of the building of Problem 2.10. See Fig. P2.10.

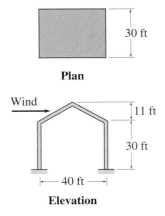

FIG. **P2.10, P2.11, P2.12**

Section 2.5

2.12 Determine the design snow loads for the roof of the disaster operation center building of Problem 2.10. The ground snow load in Kansas City is 20 psf. Because of trees near the building, assume the exposure factor is $C_e = 1$. See Fig. P2.10.

2.13 Determine the design snow loads for the roof of the school building of Problem 2.9. The ground snow load in Chicago is 1.2 kN/m^2. Assume the exposure factor is $C_e = 1$. See Fig. P2.9.

Part Two

Analysis of Statically
Determinate Structures

3

Equilibrium and Support Reactions

Bridge Construction on an Expressway
Photo courtesy of Donovan Reese/Photodisc Green

The objective of this chapter is to review the basic concept of equilibrium of structures under the action of forces and to develop the analysis of reactions exerted by supports on plane (two-dimensional) structures subjected to coplanar force systems.

We first review the concept of equilibrium and develop the equations of equilibrium of structures. Next we discuss the external and internal forces. We then describe the common types of supports used to restrict movements of plane structures. Structures can be classified as externally statically determinate, indeterminate, or unstable. We discuss how this classification can be made for plane structures. We then develop a procedure for determining reactions at supports for plane statically determinate structures. Finally, we define the principle of superposition and show how to use proportions in the computation of reactions of simply supported structures.

3.1 EQUILIBRIUM OF STRUCTURES

A structure is considered to be in equilibrium if, initially at rest, it remains at rest when subjected to a system of forces and couples. If a structure is in equilibrium, then all its members and parts are also in equilibrium.

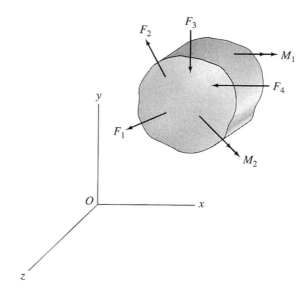

FIG. 3.1

In order for a structure to be in equilibrium, all the forces and couples (including support reactions) acting on it must balance each other, and there must neither be a resultant force nor a resultant couple acting on the structure. Recall from *statics* that for a space (three-dimensional) structure subjected to three-dimensional systems of forces and couples (Fig. 3.1), the conditions of zero resultant force and zero resultant couple can be expressed in a Cartesian (xyz) coordinate system as

$$\sum F_x = 0 \qquad \sum F_y = 0 \qquad \sum F_z = 0$$
$$\sum M_x = 0 \qquad \sum M_y = 0 \qquad \sum M_z = 0 \tag{3.1}$$

These six equations are called the *equations of equilibrium of space structures* and are the necessary and sufficient conditions for equilibrium. The first three equations ensure that there is no resultant force acting on the structure, and the last three equations express the fact that there is no resultant couple acting on the structure.

For a plane structure lying in the xy plane and subjected to a coplanar system of forces and couples (Fig. 3.2), the necessary and sufficient conditions for equilibrium can be expressed as

$$\sum F_x = 0 \qquad \sum F_y = 0 \qquad \sum M_z = 0 \tag{3.2}$$

These three equations are referred to as the *equations of equilibrium of plane structures*. The first two of the three equilibrium equations express, respectively, that the algebraic sums of the x components and y components of all the forces are zero, thereby indicating that the resultant

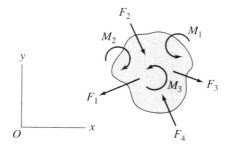

FIG. 3.2

force acting on the structure is zero. The third equation indicates that the algebraic sum of the moments of all the forces about any point in the plane of the structure and the moments of any couples acting on the structure is zero, thereby indicating that the resultant couple acting on the structure is zero. All the equilibrium equations must be satisfied simultaneously for the structure to be in equilibrium.

It should be realized that if a structure (e.g., an aerospace vehicle) initially in motion is subjected to forces that satisfy the equilibrium equations, it will maintain its motion with a constant velocity, since the forces cannot accelerate it. Such structures may also be considered to be in equilibrium. However, the term equilibrium is commonly used to refer to the state of rest of structures and is used in this context herein.

Alternative Forms of Equations of Equilibrium of Plane Structures

Although the equilibrium equations as expressed in Eq. (3.2) provide the most convenient means of analyzing a majority of plane structures, the analysis of some structures can be expedited by employing one of the following two alternative forms of the equations of equilibrium:

$$\sum F_q = 0 \qquad \sum M_A = 0 \qquad \sum M_B = 0 \qquad (3.3)$$

in which A and B are any two points in the plane of the structure, provided that the line connecting A and B is not perpendicular to the q axis, and

$$\sum M_A = 0 \qquad \sum M_B = 0 \qquad \sum M_C = 0 \qquad (3.4)$$

in which A, B, and C are any points in the plane of the structure, provided that these three points do not lie on the same straight line.

Concurrent Force Systems

When a structure is in equilibrium under the action of a concurrent force system—that is, the lines of action of all the forces intersect at a single point—the moment equilibrium equations are automatically satisfied, and only the force equilibrium equations need to be considered.

Therefore, for a space structure subjected to a concurrent three-dimensional force system, the equations of equilibrium are

$$\sum F_x = 0 \qquad \sum F_y = 0 \qquad \sum F_z = 0 \qquad (3.5)$$

Similarly, for a plane structure subjected to a concurrent coplanar force system, the equilibrium equations can be expressed as

$$\sum F_x = 0 \qquad \sum F_y = 0 \qquad (3.6)$$

Two-Force and Three-Force Structures

Throughout this text, we will encounter several structures and structural members that will be in equilibrium under the action of only two, or three, forces. The analysis of such structures and of structures composed of such members can be considerably expedited by recalling from *statics* the following characteristics of such systems:

1. If a structure is in equilibrium under the action of only two forces, the forces must be equal, opposite, and collinear.
2. If a structure is in equilibrium under the action of only three forces, the forces must be either concurrent or parallel.

3.2 EXTERNAL AND INTERNAL FORCES

The forces and couples to which a structure may be subjected can be classified into two types, external forces and internal forces.

External Forces

External forces are the actions of other bodies on the structure under consideration. For the purposes of analysis, it is usually convenient to further classify these forces as applied forces and reaction forces. *Applied forces*, usually referred to as *loads* (e.g., live loads and wind loads), have a tendency to move the structure and are usually *known* in the analysis. Reaction forces, or *reactions*, are the forces exerted by supports on the structure and have a tendency to prevent its motion and keep it in equilibrium. The reactions are usually among the *unknowns* to be determined by the analysis. The state of equilibrium or motion of the structure as a whole is governed solely by the external forces acting on it.

Internal Forces

Internal forces are the forces and couples exerted on a member or portion of the structure by the rest of the structure. These forces develop within the structure and hold the various portions of it together. The internal forces always occur in equal but opposite pairs, because each member or portion exerts back on the rest of the structure the same forces acting upon it but in opposite directions, according to Newton's third law. Because the internal forces cancel each other, they do not appear in the equations of equilibrium of the entire structure. The internal forces are also among the unknowns in the analysis and are determined by applying the equations of equilibrium to the individual members or portions of the structure.

3.3 TYPES OF SUPPORTS FOR PLANE STRUCTURES

Supports are used to attach structures to the ground or other bodies, thereby restricting their movements under the action of applied loads. The loads tend to move the structures; but supports prevent the movements by exerting opposing forces, or reactions, to neutralize the effects of loads, thereby keeping the structures in equilibrium. The type of reaction a support exerts on a structure depends on the type of supporting device used and the type of movement it prevents. A support that prevents translation of the structure in a particular direction exerts a reaction force on the structure in that direction. Similarly, a support that prevents rotation of the structure about a particular axis exerts a reaction couple on the structure about that axis.

The types of supports commonly used for plane structures are depicted in Fig. 3.3. These supports are grouped into three categories, depending on the number of reactions (1, 2, or 3) they exert on the structures. The figure also gives the types of reactions that these supports exert, as well as the number of unknowns that the various supports introduce in the analysis. Figures 3.4 through 3.6 illustrate roller, rocker, and hinged supports.

3.4 STATIC DETERMINACY, INDETERMINACY, AND INSTABILITY

Internal Stability

A structure is considered to be *internally stable, or rigid, if it maintains its shape and remains a rigid body when detached from the supports.* Conversely, a structure is termed *internally unstable* (or nonrigid) if it cannot maintain its shape and may undergo large displacements under

Category	Type of support	Symbolic representation	Reactions	Number of unknowns
I	Roller	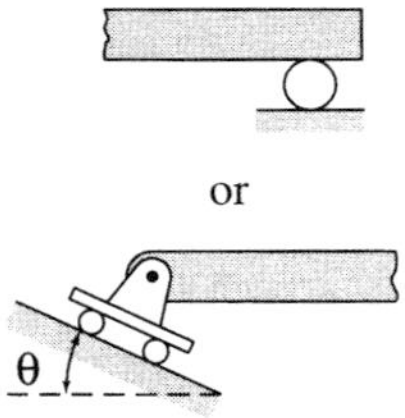 or	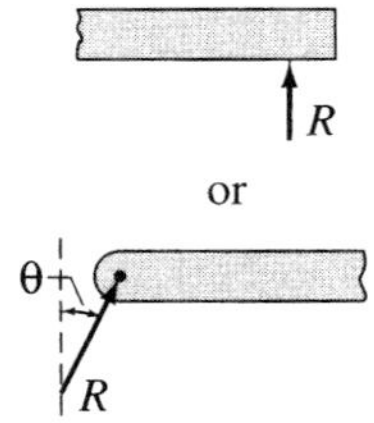 or	**1** The reaction force R acts perpendicular to the supporting surface and may be directed either into or away from the structure. The magnitude of R is the unknown.
	Rocker	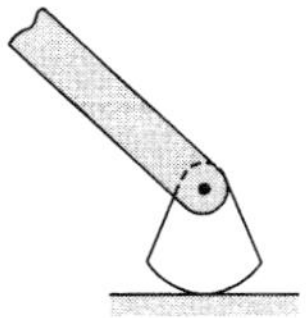	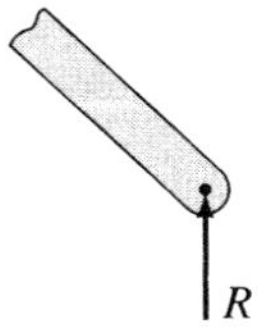	
	Link	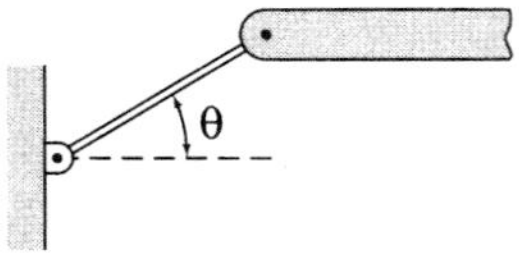	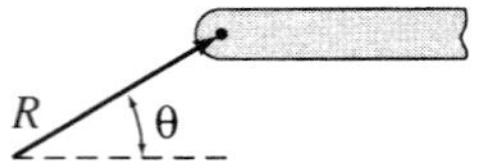	**1** The reaction force R acts in the direction of the link and may be directed either into or away from the structure. The magnitude of R is the unknown.
II	Hinge	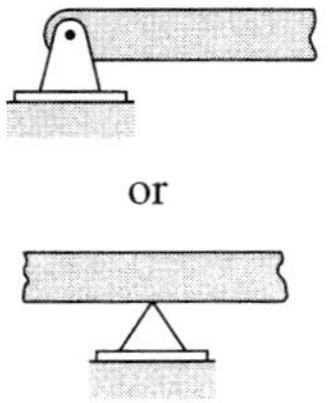 or	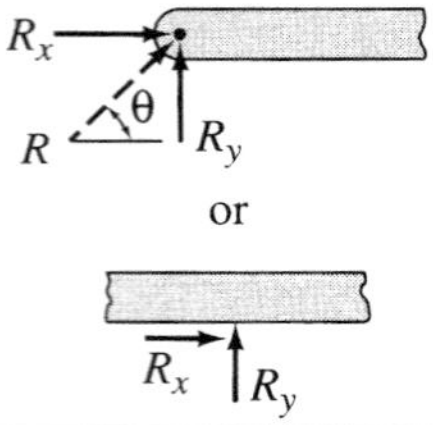 or	**2** The reaction force R may act in any direction. It is usually convenient to represent R by its rectangular components, R_x and R_y. The magnitudes of R_x and R_y are the two unknowns.
III	Fixed	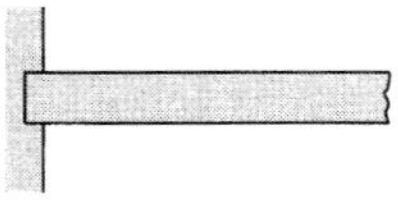	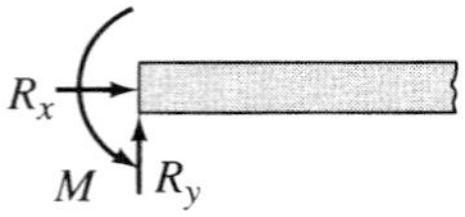	**3** The reactions consist of two force components R_x and R_y and a couple of moment M. The magnitudes of R_x, R_y, and M are the three unknowns.

FIG. 3.3 Types of Supports for Plane Structures

small disturbances when not supported externally. Some examples of internally stable structures are shown in Fig. 3.7. Note that each of the structures shown forms a rigid body, and each can maintain its shape under loads. Figure 3.8 shows some examples of internally unstable structures. A careful look at these structures indicates that each structure is composed of two rigid parts, AB and BC, connected by a hinged joint B, which cannot prevent the rotation of one part with respect to the other.

FIG. 3.4 Roller Support
Photo courtesy of the Illinois Department of Transportation.

FIG. 3.5 Rocker Support

FIG. 3.6 Hinged Support

It should be realized that all physical bodies deform when subjected to loads; the deformations in most engineering structures under service conditions are so small that their effect on the equilibrium state of the structure can be neglected. The term *rigid structure* as used here implies that the structure offers significant resistance to its change of shape,

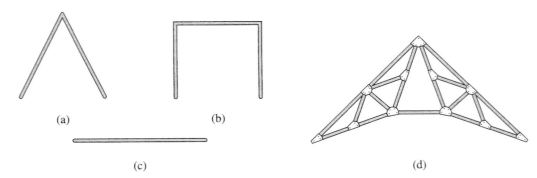

FIG. 3.7 Examples of Internally Stable Structures

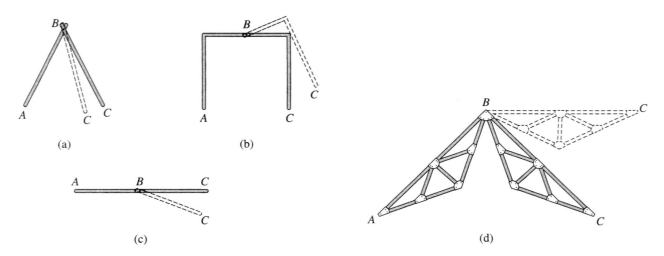

FIG. 3.8 Examples of Internally Unstable Structures

whereas a nonrigid structure offers negligible resistance to its change of shape when detached from the supports and would often collapse under its own weight when not supported externally.

Static Determinacy of Internally Stable Structures

An internally stable structure is considered to be *statically determinate externally if all its support reactions can be determined by solving the equations of equilibrium*. Since a plane internally stable structure can be treated as a plane rigid body, in order for it to be in equilibrium under a general system of coplanar loads, it must be supported by at least three reactions that satisfy the three equations of equilibrium (Eqs. 3.2, 3.3, or 3.4). Also, since there are only three equilibrium equations, they cannot

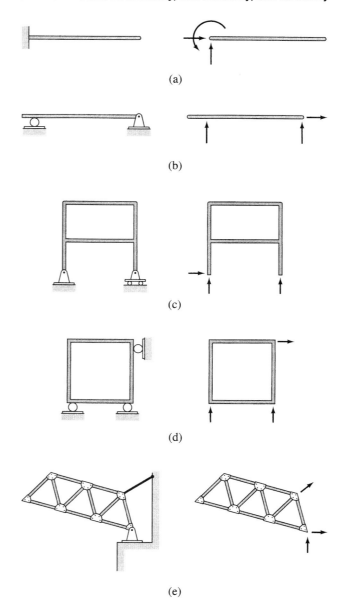

FIG. 3.9 Examples of Externally
Statically Determinate Plane Structures

be used to determine more than three reactions. Thus, a plane structure
that is statically determinate externally must be supported by exactly
three reactions. Some examples of externally statically determinate
plane structures are shown in Fig. 3.9. It should be noted that each of
these structures is supported by three reactions that can be determined
by solving the three equilibrium equations.

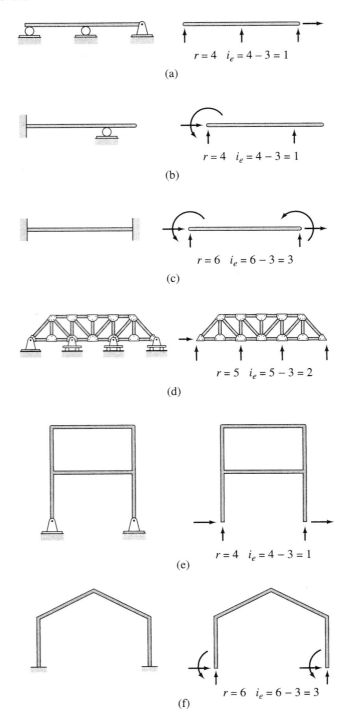

If a structure is supported by more than three reactions, then all the reactions cannot be determined from the three equations of equilibrium. Such structures are termed *statically indeterminate externally*. The reactions in excess of those necessary for equilibrium are called *external redundants*, and the number of external redundants is referred to as the *degree of external indeterminacy*. Thus, if a structure has r reactions $(r > 3)$, then the degree of external indeterminacy can be written as

$$i_e = r - 3 \qquad (3.7)$$

Figure 3.10 shows some examples of externally statically indeterminate plane structures.

If a structure is supported by fewer than three support reactions, the reactions are not sufficient to prevent all possible movements of the structure in its plane. Such a structure cannot remain in equilibrium under a general system of loads and is, therefore, referred to as *statically unstable externally*. An example of such a structure is shown in Fig. 3.11. The truss shown in this figure is supported on only two rollers. It should be obvious that although the two reactions can prevent the truss from rotating and translating in the vertical direction, they cannot prevent its translation in the horizontal direction. Thus, the truss is not fully constrained and is statically unstable.

The conditions of static instability, determinacy, and indeterminacy of plane internally stable structures can be summarized as follows:

$r < 3$ the structure is statically unstable externally

$r = 3$ the structure is statically determinate externally   (3.8)

$r > 3$ the structure is statically indeterminate externally

where r = number of reactions.

It should be realized that the first of three conditions stated in Eq. (3.8) is both necessary and sufficient in the sense that if $r < 3$, the structure is definitely unstable. However, the remaining two conditions, $r = 3$ and $r > 3$, although necessary, are not sufficient for static determinacy and indeterminacy, respectively. In other words, a structure may be supported by a sufficient number of reactions $(r \geq 3)$ but may still be unstable due to improper arrangement of supports. Such structures are

FIG. 3.11 An Example of Externally Statically Unstable Plane Structure

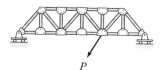

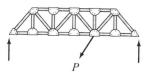

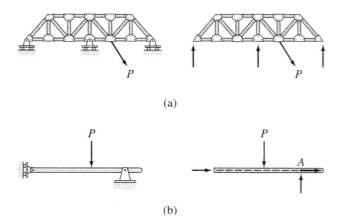

(a)

(b)

FIG. 3.12 Reaction Arrangements Causing External Geometric Instability in Plane Structures

referred to as *geometrically unstable externally*. The two types of reaction arrangements that cause geometric instability in plane structures are shown in Fig. 3.12. The truss in Fig. 3.12(a) is supported by three parallel reactions. It can be seen from this figure that although there is a sufficient number of reactions ($r = 3$), all of them are in the vertical direction, so they cannot prevent translation of the structure in the horizontal direction. The truss is, therefore, geometrically unstable. The other type of reaction arrangement that causes geometric instability is shown in Fig. 3.12(b). In this case, the beam is supported by three nonparallel reactions. However, since the lines of action of all three reaction forces are concurrent at the same point, A, they cannot prevent rotation of the beam about point A. In other words, the moment equilibrium equation $\sum M_A = 0$ cannot be satisfied for a general system of coplanar loads applied to the beam. The beam is, therefore, geometrically unstable.

Based on the preceding discussion, we can conclude that in order for a plane internally stable structure to be geometrically stable externally so that it can remain in equilibrium under the action of any arbitrary coplanar loads, it must be supported by at least three reactions, all of which must be neither parallel nor concurrent.

Static Determinacy of Internally Unstable Structures—Equations of Condition

Consider an internally unstable structure composed of two rigid members AB and BC connected by an internal hinge at B, as shown in Fig. 3.13(a). The structure is supported by a roller support at A and a hinged support at C, which provide three nonparallel nonconcurrent external reactions. As this figure indicates, these reactions, which would have

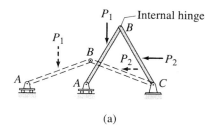

(a)

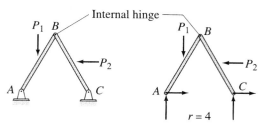

One equation of condition: $\Sigma M_B^{AB} = 0$ or $\Sigma M_B^{BC} = 0$

(b)

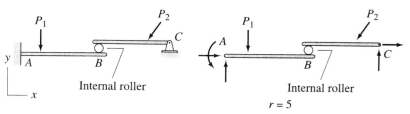

Two equations of condition: $\Sigma F_x^{AB} = 0$ or $\Sigma F_x^{BC} = 0$

$\Sigma M_B^{AB} = 0$ or $\Sigma M_B^{BC} = 0$

FIG. 3.13

(c)

been sufficient to fully constrain an internally stable or rigid structure, are not sufficient for this structure. The structure can, however, be made externally stable by replacing the roller support at A by a hinged support to prevent the horizontal movement of end A of the structure. Thus, as shown in Fig. 3.13(b), the minimum number of external reactions required to fully constrain this structure is four.

Obviously, the three equilibrium equations are not sufficient to determine the four unknown reactions at the supports for this structure. However, the presence of the internal hinge at B yields an additional equation that can be used with the three equilibrium equations to determine the four unknowns. The additional equation is based on the condition that an internal hinge cannot transmit moment; that is, the moments at the ends of the parts of the structure connected to a hinged joint are zero. Therefore, when an internal hinge is used to connect two portions of a structure, the algebraic sum of the moments about the hinge of the loads and reactions acting on each portion of the structure on either side of the hinge must be zero. Thus, for the structure of Fig. 3.13(b), the presence of the internal hinge at B requires that the algebraic sum of moments about B of the loads and reactions acting on the individual members AB and BC must be zero; that is, $\sum M_B^{AB} = 0$ and $\sum M_B^{BC} = 0$. Such equations are commonly referred to as the *equations of condition or construction*. It is important to realize that these two equations are not independent. When one of the two equations—for example, $\sum M_B^{AB} = 0$—is satisfied along with the moment equilibrium equation $\sum M = 0$ for the entire structure, the remaining equation $\sum M_B^{BC} = 0$ is automatically satisfied. Thus, an internal hinge connecting two members or portions of a structure provides one independent equation of condition. (The structures that contain hinged joints connecting more than two members are considered in subsequent chapters.) Because all four unknown reactions for the structure of Fig. 3.13(b) can be determined by solving the three equations of equilibrium plus one equation of condition ($\sum M_B^{AB} = 0$ or $\sum M_B^{BC} = 0$), the structure is considered to be statically determinate externally.

Occasionally, connections are used in structures that permit not only relative rotations of the member ends but also relative translations in certain directions of the ends of the connected members. Such connections are modeled as internal roller joints for the purposes of analysis. Figure 3.13(c) shows a structure consisting of two rigid members AB and BC that are connected by such an internal roller at B. The structure is internally unstable and requires a minimum of five external support reactions to be fully constrained against all possible movements under a general system of coplanar loads. Since an internal roller can transmit neither moment nor force in the direction parallel to the supporting surface, it provides two equations of condition;

$$\sum F_x^{AB} = 0 \quad \text{or} \quad \sum F_x^{BC} = 0$$

and

$$\sum M_B^{AB} = 0 \quad \text{or} \quad \sum M_B^{BC} = 0$$

These two equations of condition can be used in conjunction with the three equilibrium equations to determine the five unknown external reactions. Thus, the structure of Fig. 3.13(c) is statically determinate externally.

From the foregoing discussion, we can conclude that if there are e_c equations of condition (one equation for each internal hinge and two equations for each internal roller) for an internally unstable structure, which is supported by r external reactions, then if

$$r < 3 + e_c \quad \text{the structure is statically unstable externally}$$

$$r = 3 + e_c \quad \text{the structure is statically determinate externally} \qquad (3.9)$$

$$r > 3 + e_c \quad \text{the structure is statically indeterminate externally}$$

For an externally indeterminate structure, the degree of external indeterminacy is expressed as

$$i_e = r - (3 + e_c) \qquad (3.10)$$

Alternative Approach   An alternative approach that can be used for determining the static instability, determinacy, and indeterminacy of internally unstable structures is as follows:

1. Count the total number of support reactions, r.
2. Count the total number of internal forces, f_i, that can be transmitted through the internal hinges and the internal rollers of the structure. Recall that an internal hinge can transmit two force components, and an internal roller can transmit one force component.
3. Determine the total number of unknowns, $r + f_i$.
4. Count the number of rigid members or portions, n_r, contained in the structure.
5. Because each of the individual rigid portions or members of the structure must be in equilibrium under the action of applied loads, reactions, and/or internal forces, each member must satisfy the three equations of equilibrium ($\sum F_x = 0$, $\sum F_y = 0$, and $\sum M = 0$). Thus, the total number of equations available for the entire structure is $3n_r$.
6. Determine whether the structure is statically unstable, determinate, or indeterminate by comparing the total number of unknowns, $r + f_i$, to the total number of equations. If

$$r + f_i < 3n_r \quad \text{the structure is statically unstable externally}$$

$$r + f_i = 3n_r \quad \text{the structure is statically determinate externally} \qquad (3.11)$$

$$r + f_i > 3n_r \quad \text{the structure is statically indeterminate externally}$$

For indeterminate structures, the degree of external indeterminacy is given by

$$i_e = (r + f_i) - 3n_r \qquad (3.12)$$

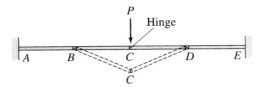

FIG. 3.14

Applying this alternative procedure to the structure of Fig. 3.13(b), we can see that for this structure, $r = 4$, $f_i = 2$, and $n_r = 2$. As the total number of unknowns $(r + f_i = 6)$ is equal to the total number of equations $(3n_r = 6)$, the structure is statically determinate externally. Similarly, for the structure of Fig. 3.13(c), $r = 5$, $f_i = 1$, and $n_r = 2$. Since $r + f_i = 3n_r$, this structure is also statically determinate externally.

The criteria for the static determinacy and indeterminacy as described in Eqs. (3.9) and (3.11), although necessary, are not sufficient because they cannot account for the possibility of geometric instability. To avoid geometric instability, the internally unstable structures, like the internally stable structures considered previously, must be supported by reactions, all of which are neither parallel nor concurrent. An additional type of geometric instability that may arise in internally unstable structures is depicted in Fig. 3.14. For the beam shown, which contains three internal hinges at B, C, and D, $r = 6$ and $e_c = 3$ (i.e., $r = 3 + e_c$); therefore, according to Eq. (3.9), the beam is supported by a sufficient number of reactions, and it should be statically determinate. However, it can be seen from the figure that portion BCD of the beam is unstable because it cannot support the vertical load P applied to it in its undeformed position. Members BC and CD must undergo finite rotations to develop any resistance to the applied load. Such a type of geometric instability can be avoided by externally supporting any portion of the structure that contains three or more internal hinges that are collinear.

Example 3.1

Classify each of the structures shown in Fig. 3.15 as externally unstable, statically determinate, or statically indeterminate. If the structure is statically indeterminate externally, then determine the degree of external indeterminacy.

Solution

   (a) This beam is internally stable with $r = 5 > 3$. Therefore, it is statically indeterminate externally with the degree of external indeterminacy of

$$i_e = r - 3 = 5 - 3 = 2 \qquad \text{Ans.}$$

   (b) This beam is internally unstable. It is composed of two rigid members AB and BC connected by an internal hinge at B. For this beam, $r = 6$ and $e_c = 1$. Since $r > 3 + e_c$, the structure is statically indeterminate externally with the degree of external indeterminacy of

$$i_e = r - (3 + e_c) = 6 - (3 + 1) = 2 \qquad \text{Ans.}$$

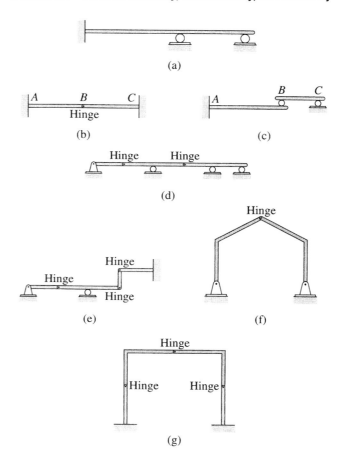

(a)

(b)

(c)

(d)

(e)

(f)

(g)

FIG. 3.15

Alternative Method $f_i = 2$, $n_r = 2$, $r + f_i = 6 + 2 = 8$, and $3n_r = 3(2) = 6$. As $r + f_i > 3n_r$, the beam is statically indeterminate externally, with

$$i_e = (r + f_i) - 3n_r = 8 - 6 = 2 \qquad \text{Checks}$$

(c) This structure is internally unstable with $r = 4$ and $e_c = 2$. Since $r < 3 + e_c$, the structure is statically unstable externally. This can be verified from the figure, which shows that the member BC is not restrained against movement in the horizontal direction. **Ans.**

Alternative Method $f_i = 1$, $n_r = 2$, $r + f_i = 4 + 1 = 5$, and $3n_r = 6$. Since $r + f_i < 3n_r$, the structure is statically unstable externally. **Checks**

(d) This beam is internally unstable with $r = 5$ and $e_c = 2$. Because $r = 3 + e_c$, the beam is statically determinate externally. **Ans.**

Alternative Method $f_i = 4$, $n_r = 3$, $r + f_i = 5 + 4 = 9$, and $3n_r = 3(3) = 9$. Because $r + f_i = 3n_r$, the beam is staticaly determinate externally. **Checks**

continued

(e) This is an internally unstable structure with $r = 6$ and $e_c = 3$. Since $r = 3 + e_c$, the structure is statically determinate externally.   **Ans.**

Alternative Method $f_i = 6$, $n_r = 4$, $r + f_i = 6 + 6 = 12$, and $3n_r = 3(4) = 12$. Because $r + f_i = 3n_r$, the structure is statically determinate externally.
                                                                            Checks

(f) This frame is internally unstable with $r = 4$ and $e_c = 1$. Since $r = 3 + e_c$, the frame is statically determinate externally.   **Ans.**

Alternative Method $f_i = 2$, $n_r = 2$, $r + f_i = 4 + 2 = 6$, and $3n_r = 3(2) = 6$. Since $r + f_i = 3n_r$, the frame is statically determinate externally.   **Checks**

(g) This frame is internally unstable with $r = 6$ and $e_c = 3$. Since $r = 3 + e_c$, the frame is statically determinate externally.   **Ans.**

Alternative Method $f_i = 6$, $n_r = 4$, $r + f_i = 6 + 6 = 12$, and $3n_r = 3(4) = 12$. Because $r + f_i = 3n_r$, the frame is statically determinate externally.
                                                                            Checks

3.5 COMPUTATION OF REACTIONS

The following step-by-step procedure can be used to determine the reactions of plane statically determinate structures subjected to coplanar loads.

1. Draw a free-body diagram (FBD) of the structure.
   a. Show the structure under consideration detached from its supports and disconnected from all other bodies to which it may be connected.
   b. Show each known force or couple on the FBD by an arrow indicating its direction and sense. Write the magnitude of each known force or couple by its arrow.
   c. Show the orientation of the mutually perpendicular xy coordinate system to be used in the analysis. It is usually convenient to orient the x and y axes in the horizontal (positive to the right) and vertical (positive upward) directions, respectively. However, if the dimensions of the structure and/or the lines of action of most of the applied loads are in an inclined direction, selection of the x (or y) axis in that direction may considerably expedite the analysis.
   d. At each point where the structure has been detached from a support, show the unknown external reactions being exerted on the structure. The type of reactions that can be exerted by the various supports are given in Fig. 3.3. The reaction forces are represented on the FBD by arrows in the known directions of their lines of action. The reaction couples are represented by

curved arrows. The senses of the reactions are not known and can be arbitrarily assumed. However, it is usually convenient to assume the senses of the reaction forces in the positive x and y directions and of reaction couples as counterclockwise. The actual senses of the reactions will be known after their magnitudes have been determined by solving the equations of equilibrium and condition (if any). A positive magnitude for a reaction will imply that the sense initially assumed was correct, whereas a negative value of the magnitude will indicate that the actual sense is opposite to the one assumed on the FBD. Since the magnitudes of the reactions are not yet known, they are denoted by appropriate letter symbols on the FBD.

e. To complete the FBD, draw the dimensions of the structure, showing the locations of all the known and unknown external forces.

2. Check for static determinacy. Using the procedure described in Section 3.4, determine whether or not the given structure is statically determinate externally. If the structure is either statically or geometrically unstable or indeterminate externally, end the analysis at this stage.

3. Determine the unknown reactions by applying the equations of equilibrium and condition (if any) to the entire structure. To avoid solving simultaneous equations, write the equilibrium and condition equations so that each equation involves only one unknown. For some internally unstable structures, it may not be possible to write equations containing one unknown each. For such structures, the reactions are determined by solving the equations simultaneously. The analysis of such internally unstable structures can sometimes be expedited and the solution of simultaneous equations avoided by disconnecting the structure into rigid portions and by applying the equations of equilibrium to the individual portions to determine the reactions. In such a case, you must construct the free-body diagrams of the portions of the structure; these diagrams must show, in addition to any applied loads and support reactions, all the internal forces being exerted upon that portion at connections. Remember that the internal forces acting on the adjacent portions of a structure must have the same magnitudes but opposite senses in accordance with Newton's third law.

4. Apply an alternative equilibrium equation that has not been used before to the entire structure to check the computations. This alternative equation should preferably involve all the reactions that were determined in the analysis. You may use a moment equilibrium equation involving a summation of moments about a point that does not lie on lines of action of reaction forces for this purpose. If the analysis has been carried out correctly, then this alternative equilibrium equation must be satisfied.

Example 3.2

Determine the reactions at the supports for the beam shown in Fig. 3.16(a).

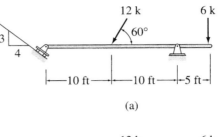

(a)

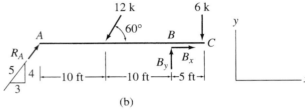

(b)

FIG. 3.16

Solution

Free-Body Diagram The free-body diagram of the beam is shown in Fig. 3.16(b). Note that the roller at A exerts reaction R_A in the direction perpendicular to the inclined supporting surface.

Static Determinacy The beam is internally stable and is supported by three reactions, R_A, B_x, and B_y, all of which are neither parallel nor concurrent. Therefore, the beam is statically determinate.

Support Reactions Since two of the three reactions, namely, B_x and B_y, are concurrent at B, their moments about B are zero. Therefore, the equilibrium equation $\sum M_B = 0$, which involves the summation of moments of all the forces about B, contains only one unknown, R_A. Thus,

$$+\circlearrowleft \sum M_B = 0$$

$$-\frac{4}{5}R_A(20) + 12\sin 60°(10) - 6(5) = 0$$

$$R_A = 4.62 \text{ k}$$

The positive answer for R_A indicates that our initial assumption about the sense of this reaction was correct. Therefore,

$$R_A = 4.62 \text{ k} \nearrow \qquad\qquad \text{Ans.}$$

Next, in order to determine B_x, we apply the equilibrium equation,

$$+ \rightarrow \sum F_x = 0$$

$$\frac{3}{5}(4.62) - 12 \cos 60° + B_x = 0$$

$$B_x = 3.23 \text{ k}$$

$$B_x = 3.23 \text{ k} \rightarrow \qquad \text{Ans.}$$

The only remaining unknown, B_y, can now be determined by applying the remaining equation of equilibrium:

$$+ \uparrow \sum F_y = 0$$

$$\frac{4}{5}(4.62) - 12 \sin 60° + B_y - 6 = 0$$

$$B_y = 12.7 \text{ k}$$

$$B_y = 12.7 \text{ k} \uparrow \qquad \text{Ans.}$$

In order to avoid having to solve simultaneous equations in the preceding computations, we applied the equilibrium equations in such a manner that each equation contained only one unknown.

Checking Computations Finally, to check our computations, we apply an alternative equation of equilibrium (see Fig. 3.16(b)):

$$+ \circlearrowleft \sum M_C = -\frac{4}{5}(4.62)(25) + 12 \sin 60°(15) - 12.7(5)$$

$$= -0.01 \text{ k-ft} \qquad \text{Checks}$$

Example 3.3

Determine the reactions at the supports for the beam shown in Fig. 3.17(a).

Solution

Free-Body Diagram See Fig. 3.17(b).

Static Determinacy The beam is internally stable with $r = 3$. Thus, it is statically determinate.

Support Reactions By applying the three equations of equilibrium, we obtain

$$+ \rightarrow \sum F_x = 0$$

$$B_x = 0 \qquad \text{Ans.}$$

continued

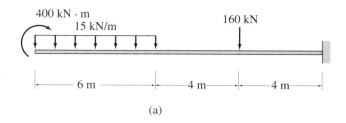

(a)

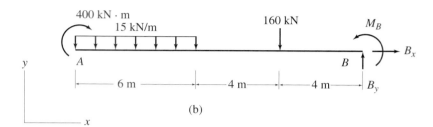

(b)

FIG. 3.17

$$+\uparrow \sum F_y = 0$$

$$-15(6) - 160 + B_y = 0$$

$$B_y = 250 \text{ kN}$$

$$B_y = 250 \text{ kN} \uparrow \qquad \text{Ans.}$$

$$+\zeta \sum M_B = 0$$

$$-400 + 15(6)(3+8) + 160(4) + M_B = 0$$

$$M_B = -1230 \text{ kN} \cdot \text{m}$$

$$M_B = 1230 \text{ kN} \cdot \text{m} \, \rangle \qquad \text{Ans.}$$

Checking Computations

$$+\zeta \sum M_A = -400 - 15(6)(3) - 160(10) + 250(14) - 1230 = 0 \quad \text{Checks}$$

Example 3.4

Determine the reactions at the support for the frame shown in Fig. 3.18(a).

Solution

Free-Body Diagram The free-body diagram of the frame is shown in Fig. 3.18(b). Note that the trapezoidal loading distribution has been divided into two simpler, uniform, and triangular, distributions whose areas and centroids are easier to compute.

Static Determinacy The frame is internally stable with $r = 3$. Therefore, it is statically determinate.

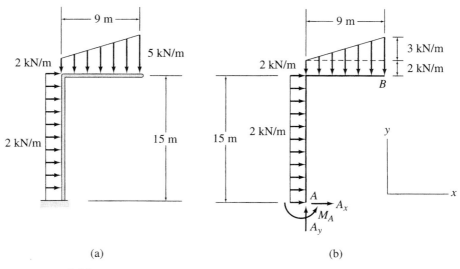

FIG. 3.18

Support Reactions By applying the three equations of equilibrium, we obtain

$$+ \rightarrow \sum F_x = 0$$

$$A_x + 2(15) = 0$$

$$A_x = -30 \text{ kN}$$

$$A_x = 30 \text{ kN} \leftarrow \qquad \text{Ans.}$$

$$+ \uparrow \sum F_y = 0$$

$$A_y - 2(9) - \frac{1}{2}(3)(9) = 0$$

$$A_y = 31.5 \text{ kN}$$

$$A_y = 31.5 \text{ kN} \uparrow \qquad \text{Ans.}$$

$$+ \zeta \sum M_A = 0$$

$$M_A - [2(15)]\left(\frac{15}{2}\right) - [2(9)]\left(\frac{9}{2}\right) - \left[\frac{1}{2}(3)(9)\right]\frac{2}{3}(9) = 0$$

$$M_A = 387 \text{ kN-m}$$

$$M_A = 387 \text{ kN-m} \; \zeta \quad \text{Ans.}$$

Checking Computations

$$+ \zeta \sum M_B = -30(15) - 31.5(9) + 387 + [2(15)]\left(\frac{15}{2}\right)$$

$$+ [2(9)]\left(\frac{9}{2}\right) + \left[\frac{1}{2}(3)(9)\right]\left(\frac{9}{3}\right)$$

$$= 0 \qquad \text{Checks}$$

Example 3.5

Determine the reactions at the supports for the frame shown in Fig. 3.19(a).

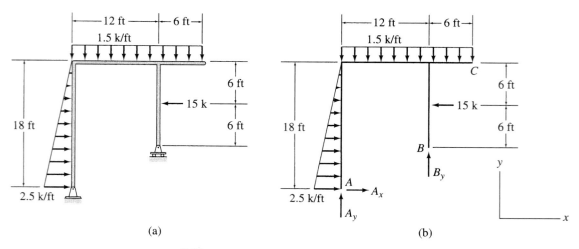

(a)   (b)

FIG. **3.19**

Solution

Free-Body Diagram See Fig. 3.19(b).

Static Determinacy The frame is internally stable with $r = 3$. Thus, it is statically determinate.

Support Reactions

$$+ \rightarrow \sum F_x = 0$$

$$A_x + \frac{1}{2}(2.5)(18) - 15 = 0$$

$$A_x = -7.5 \text{ k}$$

$$A_x = 7.5 \text{ k} \leftarrow \qquad \text{Ans.}$$

$$+ \circlearrowleft \sum M_A = 0$$

$$-\left[\frac{1}{2}(2.5)(18)\right]\left(\frac{18}{3}\right) - [1.5(18)](9) + 15(12) + B_y(12) = 0$$

$$B_y = 16.5 \text{ k}$$

$$B_y = 16.5 \text{ k} \uparrow \quad \text{Ans.}$$

$$+ \uparrow \sum F_y = 0$$

$$A_y - 1.5(18) + 16.5 = 0$$

$$A_y = 10.5 \text{ k}$$

$$A_y = 10.5 \text{ k} \uparrow \qquad \text{Ans.}$$

Checking Computations

$$+\zeta \sum M_C = -7.5(18) - 10.5(18) + \left[\frac{1}{2}(2.5)(18)\right]\frac{2}{3}(18)$$

$$+ 1.5(18)\left(\frac{18}{2}\right) - 15(6) - 16.5(6)$$

$$= 0 \qquad\qquad\qquad \text{Checks}$$

Example 3.6

Determine the reactions at the supports for the frame shown in Fig. 3.20(a).

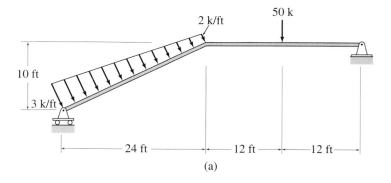

(a)

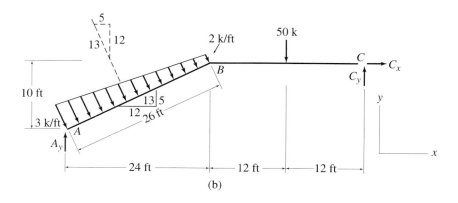

(b)

FIG. 3.20

Solution

Free-Body Diagram See Fig. 3.20(b).

Static Determinacy The frame is internally stable with $r = 3$. Therefore, it is statically determinate.

Support Reactions

$$+ \rightarrow \sum F_x = 0$$

$$\left(\frac{2+3}{2}\right)(26)\left(\frac{5}{13}\right) + C_x = 0$$

$$C_x = -25 \text{ k}$$

$$C_x = 25 \text{ k} \leftarrow \qquad \text{Ans.}$$

$$+ \zeta \sum M_A = 0$$

$$-2(26)(13) - \frac{1}{2}(1)(26)\left(\frac{26}{3}\right) - 50(24 + 12) + 25(10) + C_y(48) = 0$$

$$C_y = 48.72 \text{ k}$$

$$C_y = 48.72 \text{ k} \uparrow$$
$$\text{Ans.}$$

$$+ \uparrow \sum F_y = 0$$

$$A_y - \left(\frac{2+3}{2}\right)(26)\left(\frac{12}{13}\right) - 50 + 48.72 = 0$$

$$A_y = 61.28 \text{ k}$$

$$A_y = 61.28 \text{ k} \uparrow \qquad \text{Ans.}$$

Checking Computations

$$+ \zeta \sum M_B = -61.28(24) + 2(26)(13) + \frac{1}{2}(1)(26)\left(\frac{2}{3}\right)(26) - 50(12) + 48.72(24)$$

$$= -0.107 \text{ k-ft} \approx 0 \qquad \text{Checks}$$

Example 3.7

Determine the reactions at the supports for the beam shown in Fig. 3.21(a).

Solution

Free-Body Diagram See Fig. 3.21(b).

Static Determinacy The beam is internally unstable. It is composed of three rigid members, AB, BE, and EF, connected by two internal hinges at B and E. The structure has $r = 5$ and $e_c = 2$; because $r = 3 + e_c$, the structure is statically determinate.

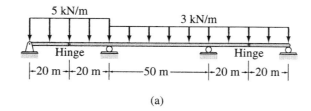

(a)

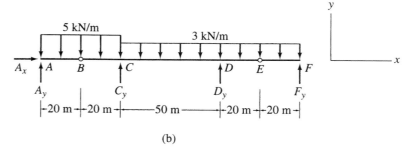

(b)

FIG. 3.21

Support Reactions

$$+ \rightarrow \sum F_x = 0$$

$$A_x = 0 \qquad \text{Ans.}$$

Next, we apply the equation of condition, $\sum M_B^{AB} = 0$, which involves the summation of moments about B of all the forces acting on the portion AB.

$$+ \zeta \sum M_B^{AB} = 0$$

$$-A_y(20) + [5(20)](10) = 0$$

$$A_y = 50 \text{ kN}$$

$$A_y = 50 \text{ kN} \uparrow \qquad \text{Ans.}$$

Similarly, by applying the equation of condition $\sum M_E^{EF} = 0$, we determine the reaction F_y as follows:

$$+ \zeta \sum M_E^{EF} = 0$$

$$-[3(20)](10) + F_y(20) = 0$$

$$F_y = 30 \text{ kN}$$

$$F_y = 30 \text{ kN} \uparrow \qquad \text{Ans.}$$

The remaining two equilibrium equations can now be applied to determine the remaining two unknowns, C_y and D_y:

$$+ \zeta \sum M_D = 0$$

$$-50(90) + [5(40)](70) - C_y(50) + [3(90)](5) + 30(40) = 0$$

$$C_y = 241 \text{ kN}$$

$$C_y = 241 \text{ kN} \uparrow \quad \text{Ans.}$$

continued

It is important to realize that the moment equilibrium equations involve the moments of *all* the forces acting on the entire structure, whereas, the moment equations of condition involve only the moments of those forces that act on the portion of the structure on *one side* of the internal hinge.

Finally, we compute D_y by using the equilibrium equation,

$$+\uparrow \sum F_y = 0$$

$$50 - 5(40) + 241 - 3(90) + D_y + 30 = 0$$

$$D_y = 149 \text{ kN}$$

$$D_y = 149 \text{ kN} \uparrow \qquad \text{Ans.}$$

Checking Computations

$$+\circlearrowleft \sum M_F = -50(130) + [5(40)](110) - 241(90) + [3(90)](45) - 149(40)$$

$$= 0 \qquad \text{Checks}$$

Example 3.8

Determine the reactions at the supports for the three-hinged arch shown in Fig. 3.22(a).

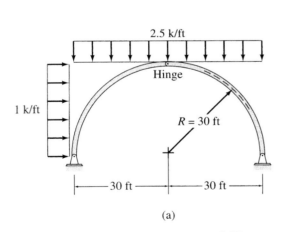

(a)

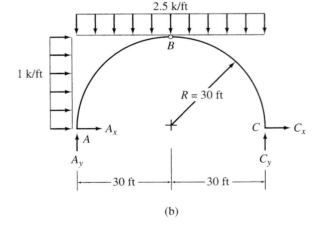

(b)

FIG. 3.22

Solution

Free-Body Diagram See Fig. 3.22(b).

Static Determinacy The arch is internally unstable; it is composed of two rigid portions, AB and BC, connected by an internal hinge at B. The arch has $r = 4$ and $e_c = 1$; since $r = 3 + e_c$, it is statically determinate.

Support Reactions

$$+\circlearrowleft \sum M_C = 0$$

$$-A_y(60) - [1(30)](15) + [2.5(60)](30) = 0$$

$$A_y = 67.5 \text{ k}$$

$$A_y = 67.5 \text{ k} \uparrow \qquad \text{Ans.}$$

$$+\circlearrowleft \sum M_B^{AB} = 0$$

$$A_x(30) - 67.5(30) + [1(30)](15) + [2.5(30)](15) = 0$$

$$A_x = 15 \text{ k}$$

$$A_x = 15 \text{ k} \rightarrow \qquad \text{Ans.}$$

$$+ \rightarrow \sum F_x = 0$$

$$15 + 1(30) + C_x = 0$$

$$C_x = -45 \text{ k}$$

$$C_x = 45 \text{ k} \leftarrow \qquad \text{Ans.}$$

$$+ \uparrow \sum F_y = 0$$

$$67.5 - 2.5(60) + C_y = 0$$

$$C_y = 82.5 \text{ k}$$

$$C_y = 82.5 \uparrow \qquad \text{Ans.}$$

Checking Computations To check our computations, we apply the equilibrium equation $\sum M_B = 0$ for the entire structure:

$$+\circlearrowleft \sum M_B = 15(30) - 67.5(30) + [1(30)](15) + [2.5(60)](0)$$

$$- 45(30) + 82.5(30)$$

$$= 0 \qquad \text{Checks}$$

Example 3.9

Determine the reactions at the supports for the beam shown in Fig. 3.23(a).

Solution

Free-Body Diagram The free-body diagram of the entire structure is shown in Fig. 3.23(b).

Static Determinacy The beam is internally unstable, with $r = 5$ and $e_c = 2$. Since $r = 3 + e_c$, the structure is statically determinate.

Support Reactions Using the free-body diagram of the entire beam shown in Fig. 3.23(b), we determine the reactions as follows:

continued

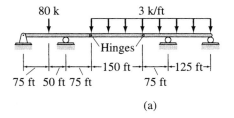

(a)

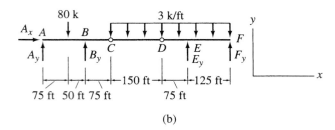

(b)

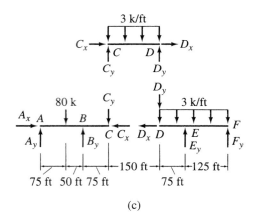

FIG. 3.23

(c)

$$+ \rightarrow \sum F_x = 0$$

$$A_x = 0 \qquad\qquad \text{Ans.}$$

$$+ \circlearrowleft \sum M_C^{AC} = 0$$

$$-A_y(200) + 80(125) - B_y(75) = 0$$

$$8A_y + 3B_y = 400 \qquad (1)$$

In order to obtain another equation containing the same two unknowns, A_y and B_y, we write the second equation of condition as

$$+ \circlearrowleft \sum M_D^{AD} = 0$$

$$-A_y(350) + 80(275) - B_y(225) + [(3)(150)](75) = 0$$

$$14A_y + 9B_y = 2230 \qquad (2)$$

Solving Eqs. (1) and (2) simultaneously, we obtain

$$A_y = -103 \text{ k} \quad \text{and} \quad B_y = 408 \text{ k}$$

$$A_y = 103 \text{ k} \downarrow \qquad \qquad \text{Ans.}$$

$$B_y = 408 \text{ k} \uparrow \qquad \qquad \text{Ans.}$$

The remaining two unknowns, E_y and F_y, are determined from the remaining two equilibrium equations as follows:

$$+\,\zeta \sum M_F = 0$$

$$103(550) + 80(475) - 408(425) + [3(350)](175) - E_y(125) = 0$$

$$E_y = 840 \text{ k}$$

$$E_y = 840 \text{ k} \uparrow \;\; \text{Ans.}$$

$$+\uparrow \sum F_y = 0$$

$$-103 - 80 + 408 - 3(350) + 840 + F_y = 0$$

$$F_y = -15 \text{ k}$$

$$F_y = 15 \text{ k} \downarrow \qquad \qquad \text{Ans.}$$

Alternative Method The reactions of the beam can be determined alternatively by applying the three equations of equilibrium to each of the three rigid portions, AC, CD, and DF, of the beam. The free-body diagrams of these rigid portions are shown in Fig. 3.23(c). These diagrams show, in addition to the applied loads and support reactions, the internal forces being exerted through the internal hinges at C and D. Note that the internal forces acting at the ends C of portions AC and CD and at the ends D of portions CD and DF have the same magnitudes but opposite senses, according to Newton's law of action and reaction.

The total number of unknowns (including the internal forces) is nine. Since there are three equilibrium equations for each of the three rigid portions, the total number of equations available is also nine ($r + f_i = 3n_r = 9$). Therefore, all nine unknowns (reactions plus internal forces) can be determined from the equilibrium equations, and the beam is statically determinate.

Applying the three equations of equilibrium to the portion CD, we obtain the following:

$$+\,\zeta \sum M_C^{CD} = 0$$

$$-[3(150)](75) + D_y(150) = 0$$

$$D_y = 225 \text{ k}$$

$$+\uparrow \sum F_y^{CD} = 0$$

$$C_y - 3(150) + 225 = 0$$

$$C_y = 225 \text{ k}$$

$$+\rightarrow \sum F_x^{CD} = 0$$

$$C_x + D_x = 0 \qquad \qquad (3)$$

continued

Next, we consider the equilibrium of portion DF:

$$+ \rightarrow \sum F_x^{DF} = 0$$

$$-D_x = 0 \quad \text{or} \quad D_x = 0$$

From Eq. (3), we obtain $C_x = 0$

$$+ \zeta \sum M_F^{DF} = 0$$

$$225(200) + [3(200)](100) - E_y(125) = 0$$

$$E_y = 840 \text{ k} \qquad \text{Checks}$$

$$+ \uparrow \sum F_y^{DF} = 0$$

$$-225 - 3(200) + 840 + F_y = 0$$

$$F_y = -15 \text{ k} \qquad \text{Checks}$$

Considering the equilibrium of portion AC, we write

$$+ \rightarrow \sum F_x^{AC} = 0$$

$$A_x - 0 = 0$$

$$A_x = 0 \qquad \text{Checks}$$

$$+ \zeta \sum M_A^{AC} = 0$$

$$-80(75) + B_y(125) - 225(200) = 0$$

$$B_y = 408 \text{ k} \qquad \text{Checks}$$

$$+ \uparrow \sum F_y^{AC} = 0$$

$$A_y - 80 + 408 - 225 = 0$$

$$A_y = -103 \text{ k} \qquad \text{Checks}$$

Example 3.10

A gable frame is subjected to a wind loading, as shown in Fig. 3.24(a). Determine the reactions at its supports due to the loading.

Solution

Free-Body Diagram See Fig. 3.24(b).

Static Determinacy The frame is internally unstable, with $r = 4$ and $e_c = 1$. Since $r = 3 + e_c$, it is statically determinate.

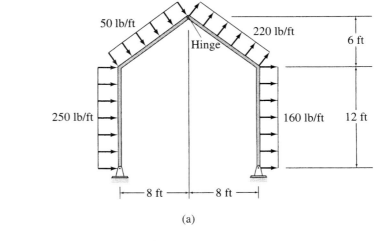

(a)

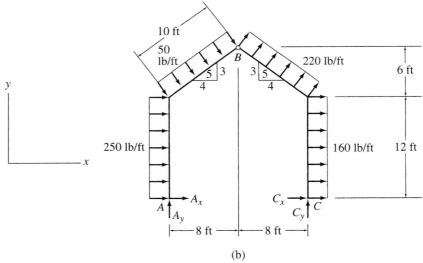

FIG. **3.24**

(b)

Support Reactions

$$+\circlearrowleft \sum M_C = 0$$

$$-A_y(16) - [250(12)](6) - \left[\frac{3}{5}(50)(10)\right](12 + 3)$$

$$+ \left[\frac{4}{5}(50)(10)\right](8 + 4) - \left[\frac{3}{5}(220)(10)\right](12 + 3)$$

$$- \left[\frac{4}{5}(220)(10)\right](4) - [160(12)](6) = 0$$

$$A_y = -3503.75 \text{ lb}$$

$$A_y = 3503.75 \text{ lb} \downarrow \qquad \text{Ans.}$$

continued

$$+\zeta \sum M_B^{AB} = 0$$

$$A_x(18) + 3503.75(8) + [250(12)](6+6) + [50(10)](5) = 0$$

$$A_x = -3696.11 \text{ lb}$$

$$A_x = 3696.11 \text{ lb} \leftarrow \quad \text{Ans.}$$

$$+ \rightarrow \sum F_x = 0$$

$$-3696.11 + 250(12) + \frac{3}{5}(50)(10) + \frac{3}{5}(220)(10) + 160(12) + C_x = 0$$

$$C_x = -2843.89 \text{ lb}$$

$$C_x = 2843.89 \text{ lb} \leftarrow \quad\quad \text{Ans.}$$

$$+ \uparrow \sum F_y = 0$$

$$-3503.75 - \frac{4}{5}(50)(10) + \frac{4}{5}(220)(10) + C_y = 0$$

$$C_y = 2143.75 \text{ lb}$$

$$C_y = 2143.75 \text{ lb} \uparrow \quad \text{Ans.}$$

Checking Computations

$$+\zeta \sum M_B = (-3696.11 - 2843.89)(18)$$
$$+ (3503.75 + 2143.75)(8) + [(250 + 160)(12)](12)$$
$$+ [(50 + 220)(10)](5)$$
$$= 0 \quad\quad \text{Checks}$$

Example 3.11

Determine the reactions at the supports for the frame shown in Fig. 3.25(a).

Solution

Free-Body Diagram See Fig. 3.25(b).

Static Determinacy The frame has $r = 4$ and $e_c = 1$; since $r = 3 + e_c$, it is statically determinate.

Support Reactions

$$+\zeta \sum M_C = 0$$

$$A_x(10) - A_y(40) - 25(20) + 3(40)(20) = 0$$

$$A_x - 4A_y = -190 \quad\quad (1)$$

$$+\zeta \sum M_B^{AB} = 0$$

$$A_x(30) - A_y(20) + 3(20)(10) = 0$$

$$3A_x - 2A_y = -60 \quad\quad (2)$$

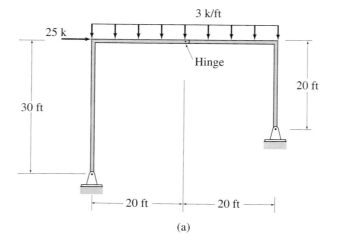

(a)

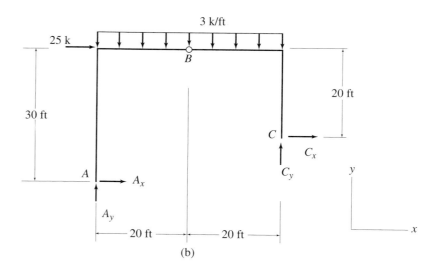

FIG. 3.25

(b)

Solving Eqs. (1) and (2) simultaneously, we obtain $A_x = 14$ k and $A_y = 51$ k

$$A_x = 14 \text{ k} \rightarrow \qquad \text{Ans.}$$

$$A_y = 51 \text{ k} \uparrow \qquad \text{Ans.}$$

$$+ \rightarrow \sum F_x = 0$$

$$14 + 25 + C_x = 0$$

$$C_x = -39 \text{ k}$$

$$C_x = 39 \text{ k} \leftarrow \qquad \text{Ans.}$$

continued

$$+\uparrow \sum F_y = 0$$

$$51 - 3(40) + C_y = 0$$

$$C_y = 69 \text{ k}$$

$$C_y = 69 \text{ k} \uparrow \qquad\qquad \text{Ans.}$$

Checking Computations

$$+\zeta \sum M_B = 14(30) - 51(20) - 39(20) + 69(20) = 0 \qquad \text{Checks}$$

3.6 PRINCIPLE OF SUPERPOSITION

The *principle of superposition* simply states that *on a linear elastic structure, the combined effect of several loads acting simultaneously is equal to the algebraic sum of the effects of each load acting individually.* For example, this principle implies, for the beam of Example 3.2, that the total reactions due to the two loads acting simultaneously could have been obtained by algebraically summing, or *superimposing*, the reactions due to each of the two loads acting individually.

The principle of superposition considerably simplifies the analysis of structures subjected to different types of loads acting simultaneously and is used extensively in structural analysis. The principle is valid for structures that satisfy the following two conditions: (1) the deformations of the structure must be so small that the equations of equilibrium can be based on the undeformed geometry of the structure; and (2) the structure must be composed of linearly elastic material; that is, the stress-strain relationship for the structural material must follow Hooke's law. The structures that satisfy these two conditions respond linearly to applied loads and are referred to as *linear elastic structures*. Engineering structures are generally designed so that under service loads they undergo small deformations with stresses within the initial linear portions of the stress-strain curves of their materials. Thus, most common types of structures under service loads can be classified as linear elastic; therefore, the principle of superposition can be used in their analysis. The principle of superposition is considered valid throughout this text.

3.7 REACTIONS OF SIMPLY SUPPORTED STRUCTURES USING PROPORTIONS

Consider a simply supported beam subjected to a vertical concentrated load P, as shown in Fig. 3.26. By applying the moment equilibrium equations, $\sum M_B = 0$ and $\sum M_A = 0$, we obtain the expressions for the vertical reactions at supports A and B, respectively, as

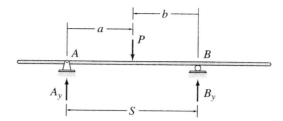

FIG. **3.26**

$$A_y = P\left(\frac{b}{S}\right) \quad \text{and} \quad B_y = P\left(\frac{a}{S}\right) \tag{3.13}$$

where, as shown in Fig. 3.26, a = distance of the load P from support A (measured positive to the right); b = distance of P from support B (measured positive to the left); and S = distance between supports A and B.

The first of the two expressions in Eq. (3.13) indicates that the magnitude of the vertical reaction at A is equal to the magnitude of the load P times the ratio of the distance of P from support B to the distance between the supports A and B. Similarly, the second expression in Eq. (3.13) states that the magnitude of the vertical reaction at B is equal to the magnitude of P times the ratio of the distance of P from A to the distance between A and B. These expressions involving proportions, when used in conjunction with the principle of superposition, make it very convenient to determine reactions of simply supported structures subjected to series of concentrated loads, as illustrated by the following example.

Example 3.12

Determine the reactions at the supports for the truss shown in Fig. 3.27(a).

Solution

Free-Body Diagram See Fig. 3.27(b).

Static Determinacy The truss is internally stable with $r = 3$. Therefore, it is statically determinate.

Support Reactions

$$+ \rightarrow \Sigma F_x = 0$$
$$A_x = 0 \qquad\qquad \text{Ans.}$$

continued

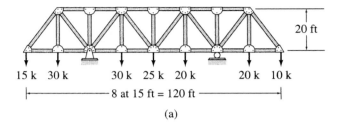

(a)

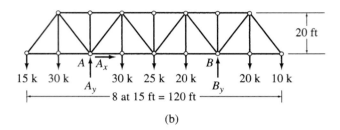

FIG. 3.27

(b)

$$A_y = 15\left(\frac{6}{4}\right) + 30\left(\frac{5}{4} + \frac{3}{4}\right) + 25\left(\frac{2}{4}\right) + 20\left(\frac{1}{4} - \frac{1}{4}\right) + 10\left(\frac{-2}{4}\right)$$

$$= 90 \text{ k}$$

$$A_y = 90 \text{ k} \uparrow \qquad\qquad\qquad \text{Ans.}$$

$$B_y = 15\left(\frac{-2}{4}\right) + 30\left(\frac{-1}{4} + \frac{1}{4}\right) + 25\left(\frac{2}{4}\right) + 20\left(\frac{3}{4} + \frac{5}{4}\right) + 10\left(\frac{6}{4}\right)$$

$$= 60 \text{ k}$$

$$B_y = 60 \text{ k} \uparrow \qquad\qquad\qquad \text{Ans.}$$

Checking Computations

$$+\uparrow \sum F_y = -15 - 2(30) - 25 - 2(20) - 10 + 90 + 60 = 0 \quad \text{Checks}$$

SUMMARY

In this chapter, we have learned that a structure is considered to be in equilibrium if, initially at rest, it remains at rest when subjected to a system of forces and couples. The equations of equilibrium of space structures can be expressed as

$$\sum F_x = 0 \qquad \sum F_y = 0 \qquad \sum F_z = 0$$
$$\sum M_x = 0 \qquad \sum M_y = 0 \qquad \sum M_z = 0 \tag{3.1}$$

For plane structures, the equations of equilibrium are expressed as

$$\sum F_x = 0 \qquad \sum F_y = 0 \qquad \sum M_z = 0 \tag{3.2}$$

Two alternative forms of the equilibrium equations for plane structures are given in Eqs. (3.3) and (3.4).

The common types of supports used for plane structures are summarized in Fig. 3.3. A structure is considered to be internally stable, or rigid, if it maintains its shape and remains a rigid body when detached from the supports.

A structure is called statically determinate externally if all of its support reactions can be determined by solving the equations of equilibrium and condition. For a plane internally stable structure supported by r number of reactions, if

$$r < 3 \quad \text{the structure is statically unstable externally}$$

$$r = 3 \quad \text{the structure is statically determinate externally} \tag{3.8}$$

$$r > 3 \quad \text{the structure is statically indeterminate externally}$$

The degree of external indeterminacy is given by

$$i_e = r - 3 \tag{3.7}$$

For a plane internally unstable structure, which has r number of external reactions and e_c number of equations of condition, if

$$r < 3 + e_c \quad \text{the structure is statically unstable externally}$$

$$r = 3 + e_c \quad \text{the structure is statically determinate externally} \tag{3.9}$$

$$r > 3 + e_c \quad \text{the structure is statically indeterminate externally}$$

The degree of external indeterminacy for such a structure is given by

$$i_e = r - (3 + e_c) \tag{3.10}$$

In order for a plane structure to be geometrically stable, it must be supported by reactions, all of which are neither parallel nor concurrent. A procedure for the determination of reactions at supports for plane structures is presented in Section 3.5.

The principle of superposition states that on a linear elastic structure, the combined effect of several loads acting simultaneously is equal to the algebraic sum of the effects of each load acting individually. The determination of reactions of simply supported structures using proportions is discussed in Section 3.7.

PROBLEMS

Section 3.4

3.1 through 3.4 Classify each of the structures shown as externally unstable, statically determinate, or statically inde-

terminate. If the structure is statically indeterminate externally, then determine the degree of external indeterminacy.

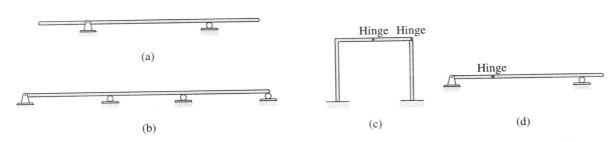

FIG. **P3.1**

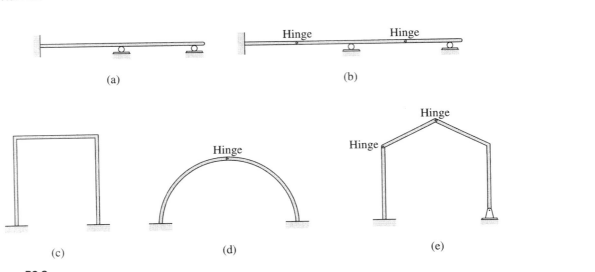

FIG. **P3.2**

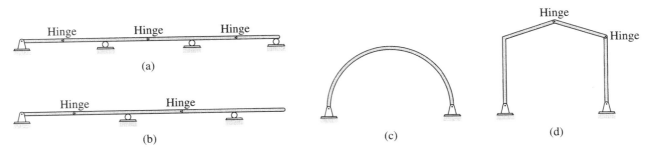

FIG. **P3.3**

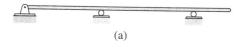

(a)

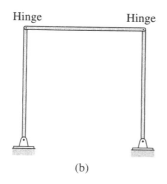

Hinge   Hinge

(b)

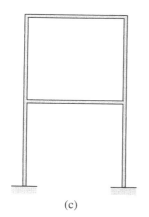

(c)

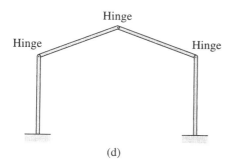

Hinge

Hinge   Hinge

(d)

FIG. P3.4

Sections 3.5 and 3.7

3.5 through 3.13 Determine the reactions at the supports for the beam shown.

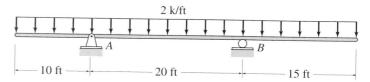

2 k/ft

A

B

10 ft   20 ft   15 ft

FIG. P3.5

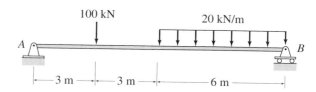

100 kN   20 kN/m

A   B

3 m   3 m   6 m

FIG. P3.6

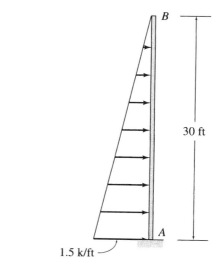

B

30 ft

A

1.5 k/ft

FIG. P3.7

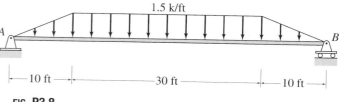

1.5 k/ft

A   B

10 ft   30 ft   10 ft

FIG. P3.8

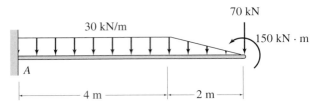

FIG. **P3.9**

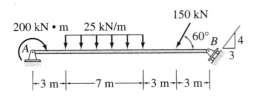

FIG. **P3.10**

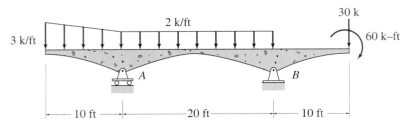

FIG. **P3.11**

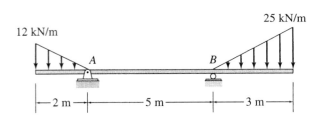

FIG. **P3.12**

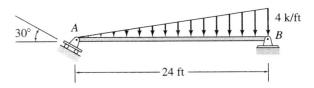

FIG. **P3.13**

3.14 The weight of a car, moving at a constant speed on a beam bridge, is modeled as a single concentrated load, as shown in Fig. P3.14. Determine the expressions for the vertical reactions at the supports in terms of the position of the car as measured by the distance x, and plot the graphs showing the variations of these reactions as functions of x.

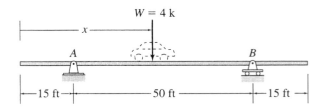

FIG. **P3.14**

3.15 The weight of a 5-m-long trolley, moving at a constant speed on a beam bridge, is modeled as a moving uniformly distributed load, as shown in Fig. P3.15. Determine the expressions for the vertical reactions at the supports in terms of the position of the trolley as measured by the distance x, and plot the graphs showing the variations of these reactions as functions of x.

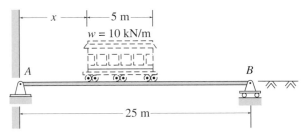

FIG. **P3.15**

3.16 through 3.41 Determine the reactions at the supports for the structures shown.

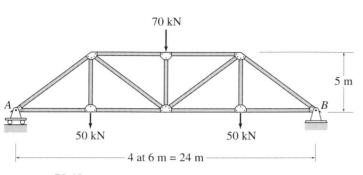

70 kN

5 m

A

B

50 kN

50 kN

4 at 6 m = 24 m

FIG. **P3.16**

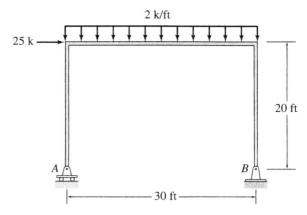

2 k/ft

25 k

20 ft

A

B

30 ft

FIG. **P3.19**

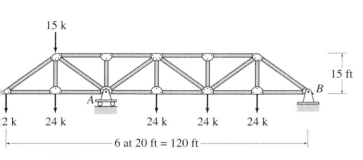

15 k

15 ft

A

B

2 k   24 k   24 k   24 k   24 k

6 at 20 ft = 120 ft

FIG. **P3.17**

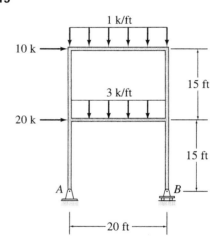

1 k/ft

10 k

15 ft

3 k/ft

20 k

15 ft

A

B

20 ft

FIG. **P3.20**

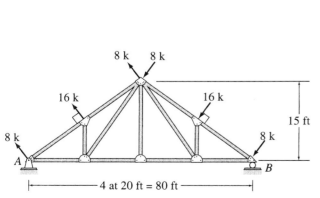

8 k   8 k

16 k   16 k

8 k   8 k

A   B

4 at 20 ft = 80 ft

15 ft

FIG. **P3.18**

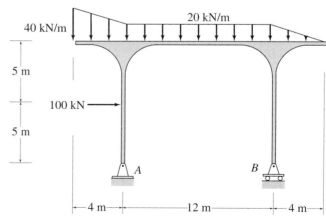

20 kN/m

40 kN/m

5 m

100 kN

5 m

A   B

4 m   12 m   4 m

FIG. **P3.21**

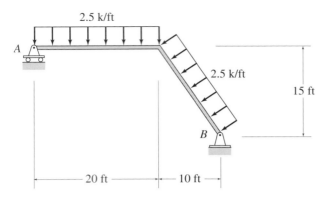

FIG. **P3.22**

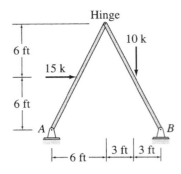

FIG. **P3.26**

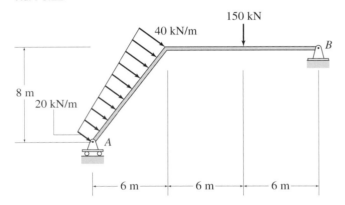

FIG. **P3.23**

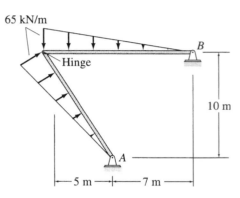

FIG. **P3.27**

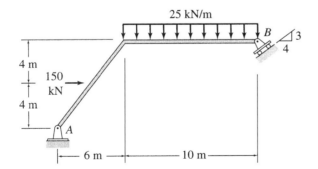

FIG. **P3.24**

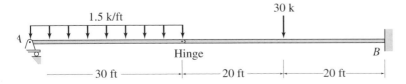

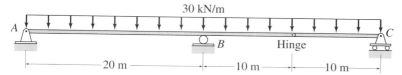

FIG. **P3.28**

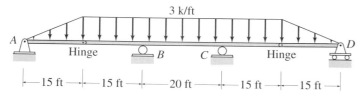

FIG. **P3.29**

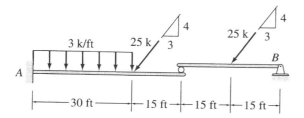

FIG. **P3.30**

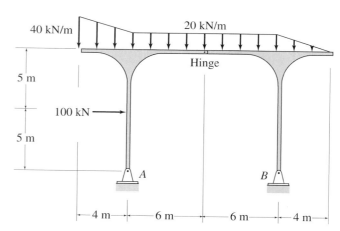

FIG. **P3.33**

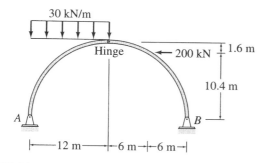

FIG. **P3.31**

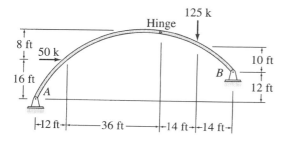

FIG. **P3.32**

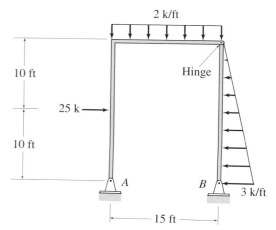

FIG. **P3.34**

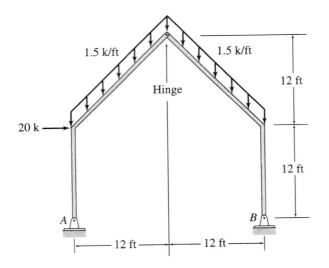

FIG. P3.35

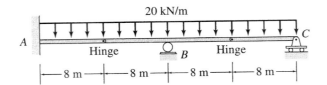

FIG. P3.36

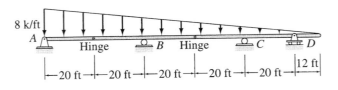

FIG. P3.37

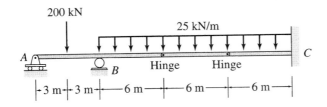

FIG. P3.38

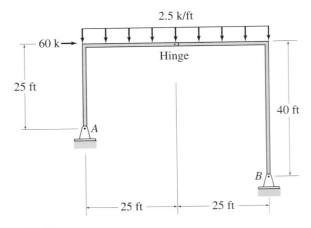

FIG. P3.39

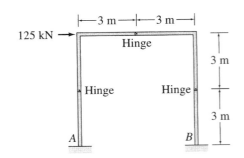

FIG. P3.40

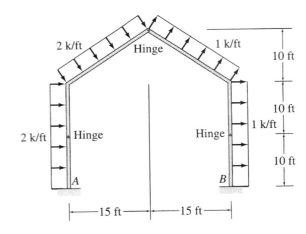

FIG. P3.41

4

Plane and Space Trusses

A Truss Bridge over the Mississippi River
Photo courtesy of the Illinois Department of Transportation

A *truss* is an assemblage of straight members connected at their ends by flexible connections to form a rigid configuration. Because of their light weight and high strength, trusses are widely used, and their applications range from supporting bridges and roofs of buildings (Fig. 4.1) to being support structures in space stations (Fig. 4.2). Modern trusses are constructed by connecting members, which usually consist of structural steel or aluminum shapes or wood struts, to gusset plates by bolted or welded connections.

As discussed in Section 1.4, if all the members of a truss and the applied loads lie in a single plane, the truss is called a *plane truss*.

Plane trusses are commonly used for supporting decks of bridges and roofs of buildings. A typical framing system for truss bridges was described in Section 1.4 (see Fig. 1.13(a)). Figure 4.3 shows a typical framing system for a roof supported by plane trusses. In this case, two or more trusses are connected at their joints by beams, termed *purlins*, to form a three-dimensional framework. The roof is attached to the purlins, which transmit the roof load (weight of the roof plus any other load due to snow, wind, etc.) as well as their own weight to the supporting trusses at the joints. Because this applied loading acts on each truss in its own plane, the trusses can be treated as plane trusses. Some of the

FIG. 4.1 Roof Truss
Photo courtesy of Bethlehem Steel Corporation

FIG. 4.2 Conceptual Model of a Space
Station
Photo courtesy of National Aeronautics and Space
Administration

common configurations of bridge and roof trusses, many of which have
been named after their original designers, are shown in Figs. 4.4 and 4.5
(see pp. 92 and 93), respectively.

Although a great majority of trusses can be analyzed as plane
trusses, there are some truss systems, such as transmission towers and
latticed domes (Fig. 4.6), that cannot be treated as plane trusses because

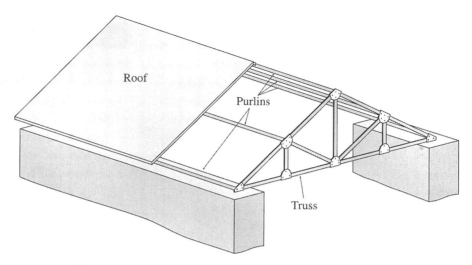

FIG. 4.3 Framing of a Roof Supported by Trusses

of their shape, arrangement of members, or applied loading. Such trusses, which are called *space trusses*, are analyzed as three-dimensional bodies subjected to three-dimensional force systems.

The objective of this chapter is to develop the analysis of member forces of statically determinate plane and space trusses. We begin by discussing the basic assumptions underlying the analysis presented in this chapter, and then we consider the number and arrangement of members needed to form internally stable or rigid plane trusses. As part of this discussion, we define *simple* and *compound* trusses. We also present the equations of condition commonly encountered in plane trusses. We next establish the classification of plane trusses as statically determinate, indeterminate, and unstable and present the procedures for the analysis of simple plane trusses by the methods of joints and sections. We conclude with an analysis of compound plane trusses, a brief discussion of complex trusses, and analysis of space trusses.

4.1 ASSUMPTIONS FOR ANALYSIS OF TRUSSES

The analysis of trusses is usually based on the following simplifying assumptions:

1. All members are connected only at their ends by frictionless hinges in plane trusses and by frictionless ball-and-socket joints in space trusses.
2. All loads and support reactions are applied only at the joints.
3. The centroidal axis of each member coincides with the line connecting the centers of the adjacent joints.

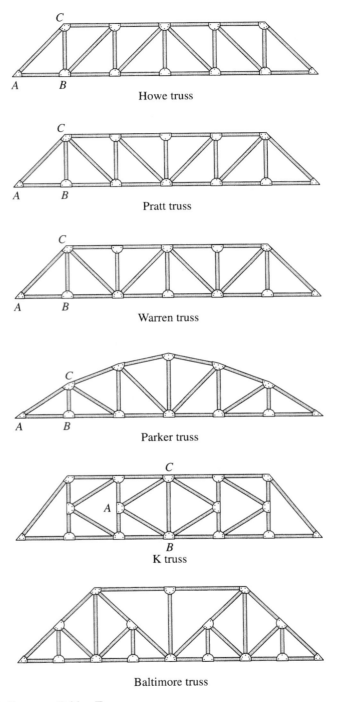

FIG. 4.4 Common Bridge Trusses

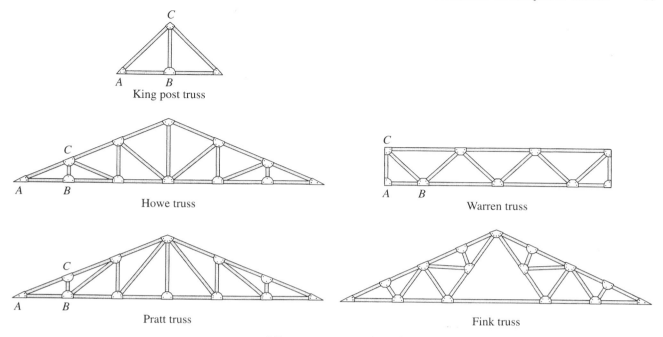

FIG. **4.5** Common Roof Trusses

FIG. **4.6** Lattice Dome
Photo courtesy of Bethlehem Steel Corporation

The reason for making these assumptions is to obtain an *ideal truss*, whose members are subjected only to axial forces. Since each member of an ideal truss is connected at its ends by frictionless hinges (assumption 1) with no loads applied between its ends (assumption 2), the member would be subjected to only two forces at its ends, as shown in Fig. 4.7(a). Since the member is in equilibrium, the resultant force and

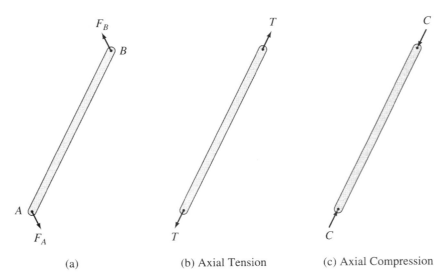

FIG. 4.7

(a)   (b) Axial Tension   (c) Axial Compression

the resultant couple of the two forces F_A and F_B must be zero; that is, the forces must satisfy the three equations of equilibrium. From Fig. 4.7(a), we can see that in order for the resultant force of the two forces to be zero ($\sum F_x = 0$ and $\sum F_y = 0$), the two forces must be equal in magnitude but with opposite senses. For their resultant couple to be also equal to zero ($\sum M = 0$), the two forces must be collinear—that is, they must have the same line of action. Moreover, since the centroidal axis of each truss member is a straight line coinciding with the line connecting the centers of the adjacent joints (assumption 3), the member is not subjected to any bending moment or shear force and is either in axial tension (being elongated, as shown in Fig. 4.7(b)) or in axial compression (being shortened, as shown in Fig. 4.7(c)). Such member axial forces determined from the analysis of an ideal truss are called the *primary forces*.

In real trusses, these idealizations are almost never completely realized. As stated previously, real trusses are constructed by connecting members to gusset plates by welded or bolted connections (Fig. 4.8). Some members of the truss may even be continuous at the joints. Furthermore, although the external loads are indeed transmitted to the trusses at joints by means of floor beams, purlins, and so on, the dead weights of the members are distributed along their lengths. The bending moments and shear and axial forces caused by these and other deviations from the aforementioned idealized conditions are commonly referred to as *secondary forces*. Although secondary forces cannot be eliminated, they can be substantially reduced in most trusses by using relatively slender members and by designing connections so that the centroidal axes of the members meeting at a joint are concurrent at a point (as shown in Fig. 1.13). The secondary forces in such trusses are

FIG. 4.8 Truss Bridge with Bolted
Connections
Photo courtesy of the Illinois Department of Transportation

small compared to the primary forces and are usually not considered in their designs. In this chapter, we focus only on primary forces. If large secondary forces are anticipated, the truss should be analyzed as a rigid frame using the methods presented in subsequent chapters.

4.2 ARRANGEMENT OF MEMBERS OF PLANE TRUSSES—INTERNAL STABILITY

Based on our discussion in Section 3.4, we can define a plane truss as internally stable if the number and geometric arrangement of its members is such that the truss does not change its shape and remains a rigid body when detached from the supports. The term *internal* is used here to refer to the number and arrangement of members contained within the truss. The instability due to insufficient external supports or due to improper arrangement of external supports is referred to as *external*.

Basic Truss Element

The simplest internally stable (or rigid) plane truss can be formed by connecting three members at their ends by hinges to form a triangle, as shown in Fig. 4.9(a). This triangular truss is called the *basic truss element*. Note that this triangular truss is internally stable in the sense that it is a rigid body that will not change its shape under loads. In contrast, a rectangular truss formed by connecting four members at their ends by hinges, as shown in Fig. 4.9(b), is internally unstable because it will

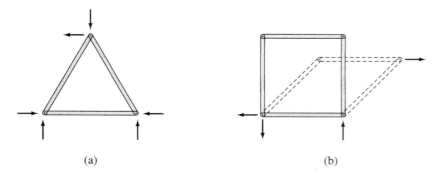

FIG. 4.9

(a)   (b)

change its shape and collapse when subjected to a general system of co-planar forces.

Simple Trusses

The basic truss element ABC of Fig. 4.10(a) can be enlarged by attaching two new members, BD and CD, to two of the existing joints B and C and by connecting them to form a new joint D, as shown in Fig. 4.10(b). As long as the new joint D does not lie on the straight line passing through the existing joints B and C, the new enlarged truss will be internally stable. The truss can be further enlarged by repeating the same procedure (as shown in Fig. 4.10(c)) as many times as desired. Trusses constructed by this procedure are called *simple trusses*. The reader should examine the trusses depicted in Figs. 4.4 and 4.5 to verify that each of them, with the exception of the Baltimore truss (Fig. 4.4) and the Fink truss (Fig. 4.5), is a simple truss. The basic truss element of the simple trusses is identified as ABC in these figures.

A simple truss is formed by enlarging the basic truss element, which contains three members and three joints, by adding two additional members for each additional joint, so the total number of members m in a simple truss is given by

$$m = 3 + 2(j - 3) = 2j - 3 \qquad (4.1)$$

in which j = total number of joints (including those attached to the supports).

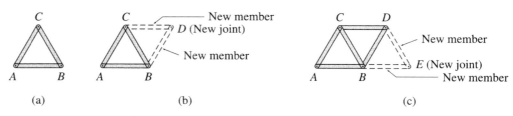

(a)   (b)   (c)

FIG. 4.10 Simple Truss

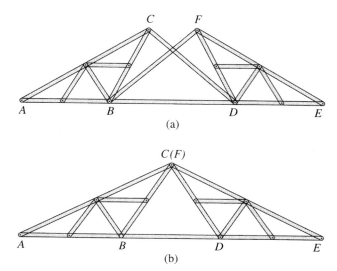

FIG. 4.11 Compound Trusses

Compound Trusses

Compound trusses are constructed by connecting two or more simple trusses to form a single rigid body. To prevent any relative movement between the simple trusses, each truss must be connected to the other(s) by means of connections capable of transmitting at least three force components, all of which are neither parallel nor concurrent. Two examples of connection arrangements used to form compound trusses are shown in Fig. 4.11. In Fig. 4.11(a), two simple trusses ABC and DEF are connected by three members, BD, CD, and BF, which are nonparallel and nonconcurrent. Another type of connection arrangement is shown in Fig. 4.11(b). This involves connecting the two simple trusses ABC and DEF by a common joint C and a member BD. In order for the compound truss to be internally stable, the common joint C and joints B and D must not lie on a straight line. The relationship between the total number of members m and the total number of joints j for an internally stable compound truss remains the same as for the simple trusses. This relationship, which is given by Eq. (4.1), can be easily verified for the compound trusses shown in Fig. 4.11.

Internal Stability

Equation (4.1) expresses the requirement of the minimum number of members that a plane truss of j joints must contain if it is to be internally stable. If a plane truss contains m members and j joints, then if

$$m < 2j - 3 \quad \text{the truss is internally unstable}$$
$$m \geq 2j - 3 \quad \text{the truss is internally stable}$$

(4.2)

It is very important to realize that although the foregoing criterion for internal stability is *necessary*, it is *not sufficient* to ensure internal stability. A truss must not only contain enough members to satisfy the $m \geq 2j - 3$ condition, but the members must also be properly arranged to ensure rigidity of the entire truss. Recall from our discussion of simple and compound trusses that in a stable truss, each joint is connected to the rest of the structure by at least two nonparallel members, and each portion of the truss must be connected to the remainder of the truss by connections capable of transmitting at least three nonparallel and nonconcurrent force components.

Example 4.1

Classify each of the plane trusses shown in Fig. 4.12 as internally stable or unstable.

Solution

(a) The truss shown in Fig. 4.12(a) contains 20 members and 12 joints. Therefore, $m = 20$ and $2j - 3 = 2(12) - 3 = 21$. Since m is less than $2j - 3$, this truss does not have a sufficient number of members to form a rigid body; therefore, it is internally unstable. A careful look at the truss shows that it contains two rigid bodies, $ABCD$ and $EFGH$, connected by two parallel members, BE and DG. These two horizontal members cannot prevent the relative displacement in the vertical direction of one rigid part of the truss with respect to the other. **Ans.**

(b) The truss shown in Fig. 4.12(b) is the same as that of Fig. 4.12(a), except that a diagonal member DE has now been added to prevent the relative displacement between the two portions $ABCD$ and $EFGH$. The entire truss now acts as a single rigid body. Addition of member DE increases the number of members to 21 (while the number of joints remains the same at 12), thereby satisfying the equation $m = 2j - 3$. The truss is now internally stable. **Ans.**

(c) Four more diagonals are added to the truss of Fig. 4.12(b) to obtain the truss shown in Fig. 4.12(c), thereby increasing m to 25, while j remains constant at 12. Because $m > 2j - 3$, the truss is internally stable. Also, since the difference $m - (2j - 3) = 4$, the truss contains four more members than required for internal stability. **Ans.**

(d) The truss shown in Fig. 4.12(d) is obtained from that of Fig. 4.12(c) by removing two diagonals, BG and DE, from panel BE, thereby decreasing m to 23; j remains constant at 12. Although $m - (2j - 3) = 2$—that is, the truss contains two more members than the minimum required for internal stability—its two rigid portions, $ABCD$ and $EFGH$, are not connected properly to form a single rigid body. Therefore, the truss is internally unstable. **Ans.**

(e) The roof truss shown in Fig. 4.12(e) is internally unstable because $m = 26$ and $j = 15$, thereby yielding $m < 2j - 3$. This is also clear from the diagram of the truss which shows that the portions ABE and CDE of the truss can

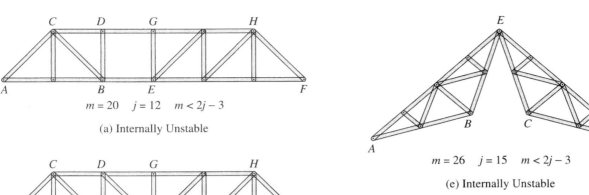

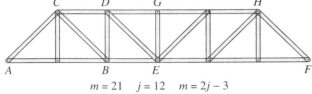

$m = 20$ $j = 12$ $m < 2j - 3$

(a) Internally Unstable

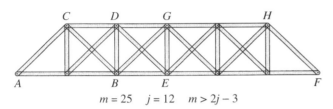

$m = 21$ $j = 12$ $m = 2j - 3$

(b) Internally Stable

$m = 26$ $j = 15$ $m < 2j - 3$

(e) Internally Unstable

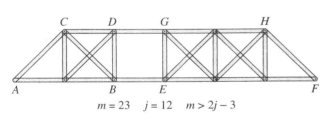

$m = 25$ $j = 12$ $m > 2j - 3$

(c) Internally Stable

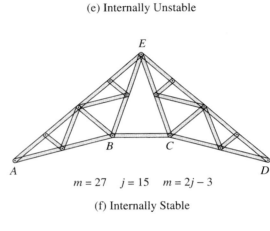

$m = 27$ $j = 15$ $m = 2j - 3$

(f) Internally Stable

$m = 23$ $j = 12$ $m > 2j - 3$

(d) Internally Unstable

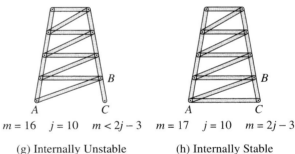

$m = 16$ $j = 10$ $m < 2j - 3$ $m = 17$ $j = 10$ $m = 2j - 3$

(g) Internally Unstable   (h) Internally Stable

FIG. 4.12

rotate with respect to each other. The difference $m - (2j - 3) = -1$ indicates that this truss has one less member than required for internal stability.   **Ans.**

(f) In Fig. 4.12(f), a member BC has been added to the truss of Fig. 4.12(e), which prevents the relative movement of the two portions ABE and CDE, thereby making the truss internally stable. As m has now been increased to 27, it satisfies the equation $m = 2j - 3$ for $j = 15$.   **Ans.**

continued

(g) The tower truss shown in Fig. 4.12(g) has 16 members and 10 joints. Because $m < 2j - 3$, the truss is internally unstable. This is also obvious from Fig. 4.12(g), which shows that member BC can rotate with respect to the rest of the structure. This rotation can occur because joint C is connected by only one member instead of the two required to completely constrain a joint of a plane truss. **Ans.**

(h) In Fig. 4.12(h), a member AC has been added to the truss of Fig. 4.12(g), which makes it internally stable. Here $m = 17$ and $j = 10$, so the equation $m = 2j - 3$ is satisfied. **Ans.**

4.3 EQUATIONS OF CONDITION FOR PLANE TRUSSES

In Section 3.4, we indicated that the types of connections used to connect rigid portions of internally unstable structures provide equations of condition that, along with the three equilibrium equations, can be used to determine the reactions needed to constrain such structures fully.

Three types of connection arrangements commonly used to connect two rigid trusses to form a single (internally unstable) truss are shown in Fig. 4.13. In Fig. 4.13(a), two rigid trusses, AB and BC, are connected together by an internal hinge at B. Because an internal hinge cannot transmit moment, it provides an equation of condition:

$$\sum M_B^{AB} = 0 \quad \text{or} \quad \sum M_B^{BC} = 0$$

Another type of connection arrangement is shown in Fig. 4.13(b). This involves connecting two rigid trusses, AB and CD, by two parallel members. Since these parallel (horizontal) bars cannot transmit force in the direction perpendicular to them, this type of connection provides an equation of condition:

$$\sum F_y^{AB} = 0 \quad \text{or} \quad \sum F_y^{CD} = 0$$

A third type of connection arrangement involves connecting two rigid trusses, AB and CD, by a single link, BC, as shown in Fig. 4.13(c). Since a link can neither transmit moment nor force in the direction perpendicular to it, it provides two equations of condition:

$$\sum F_x^{AB} = 0 \quad \text{or} \quad \sum F_x^{CD} = 0$$

and

$$\sum M_B^{AB} = 0 \quad \text{or} \quad \sum M_C^{CD} = 0$$

As we indicated in the previous chapter, these equations of condition can be used with the three equilibrium equations to determine the unknown reactions of externally statically determinate plane trusses. The reader should verify that all three trusses shown in Fig. 4.13 are statically determinate externally.

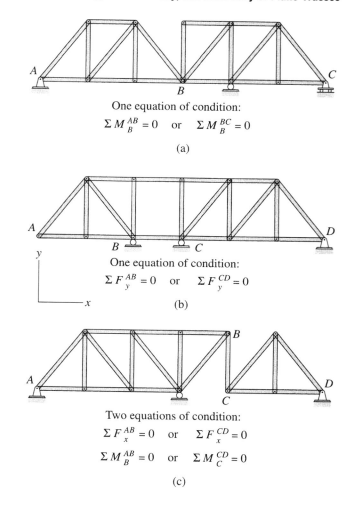

One equation of condition:

$$\Sigma M_B^{AB} = 0 \quad \text{or} \quad \Sigma M_B^{BC} = 0$$

(a)

One equation of condition:

$$\Sigma F_y^{AB} = 0 \quad \text{or} \quad \Sigma F_y^{CD} = 0$$

(b)

Two equations of condition:

$$\Sigma F_x^{AB} = 0 \quad \text{or} \quad \Sigma F_x^{CD} = 0$$

$$\Sigma M_B^{AB} = 0 \quad \text{or} \quad \Sigma M_C^{CD} = 0$$

(c)

FIG. 4.13 Equations of Condition for Plane Trusses

4.4 STATIC DETERMINACY, INDETERMINACY, AND INSTABILITY OF PLANE TRUSSES

We consider a truss to be *statically determinate if the forces in all its members, as well as all the external reactions, can be determined by using the equations of equilibrium.*

Since the two methods of analysis presented in the following sections can be used to analyze only statically determinate trusses, it is important for the student to be able to recognize statically determinate trusses before proceeding with the analysis.

Consider a plane truss subjected to external loads P_1, P_2, and P_3, as shown in Fig. 4.14(a). The free-body diagrams of the five members and

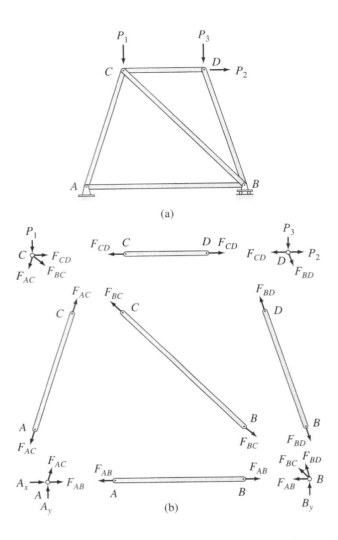

FIG. 4.14

the four joints are shown in Fig. 4.14(b). Each member is subjected to two axial forces at its ends, which are collinear (with the member centroidal axis) and equal in magnitude but opposite in sense. Note that in Fig. 4.14(b), all members are assumed to be in tension; that is, the forces are pulling on the members. The free-body diagrams of the joints show the same member forces but in opposite directions, in accordance with Newton's third law. The analysis of the truss involves the calculation of the magnitudes of the five member forces, $F_{AB}, F_{AC}, F_{BC}, F_{BD}$, and F_{CD} (the lines of action of these forces are known), and the three reactions, A_x, A_y, and B_y. Therefore, the total number of unknown quantities to be determined is eight.

Because the entire truss is in equilibrium, each of its joints must also be in equilibrium. As shown in Fig. 4.14(b), at each joint the internal and external forces form a coplanar and concurrent force system, which must satisfy the two equations of equilibrium, $\sum F_x = 0$ and $\sum F_y = 0$. Since the truss contains four joints, the total number of equations available is $2(4) = 8$. These eight joint equilibrium equations can be solved to calculate the eight unknowns. The plane truss of Fig. 4.14(a) is, therefore, statically determinate.

Three equations of equilibrium of the entire truss as a rigid body could be written and solved for the three unknown reactions (A_x, A_y, and B_y). However, these equilibrium equations (as well as the equations of condition in the case of internally unstable trusses) are *not independent* from the joint equilibrium equations and do not contain any additional information.

Based on the preceding discussion, we can develop the criteria for the static determinacy, indeterminacy, and instability of general plane trusses containing m members and j joints and supported by r (number of) external reactions. For the analysis, we need to determine m member forces and r external reactions; that is, we need to calculate a total of $m + r$ unknown quantities. Since there are j joints and we can write two equations of equilibrium ($\sum F_x = 0$ and $\sum F_y = 0$) for each joint, the total number of equilibrium equations available is $2j$. If the number of unknowns ($m + r$) for a truss is equal to the number of equilibrium equations ($2j$)—that is, $m + r = 2j$—all the unknowns can be determined by solving the equations of equilibrium, and the truss is statically determinate.

If a truss has more unknowns ($m + r$) than the available equilibrium equations ($2j$)—that is, $m + r > 2j$—all the unknowns cannot be determined by solving the available equations of equilibrium. Such a truss is called *statically indeterminate*. Statically indeterminate trusses have more members and/or external reactions than the minimum required for stability. The excess members and reactions are called *redundants*, and the number of excess members and reactions is referred to as the *degree of static indeterminacy*, i, which can be expressed as

$$i = (m + r) - 2j \qquad (4.3)$$

If the number of unknowns ($m + r$) for a truss is less than the number of equations of joint equilibrium ($2j$)—that is, $m + r < 2j$—the truss is called *statically unstable*. The static instability may be due to the truss having fewer members than the minimum required for internal stability or due to an insufficient number of external reactions or both.

The conditions of static instability, determinacy, and indeterminacy of plane trusses can be summarized as follows:

$$
\begin{aligned}
m + r &< 2j \quad \text{statically unstable truss} \\
m + r &= 2j \quad \text{statically determinate truss} \qquad (4.4)\\
m + r &> 2j \quad \text{statically indeterminate truss}
\end{aligned}
$$

The first condition, for the static instability of trusses, is both necessary and sufficient in the sense that if $m < 2j - r$, the truss is definitely statically unstable. However, the remaining two conditions, for static determinacy ($m = 2j - r$) and indeterminacy ($m > 2j - r$), are necessary but not sufficient conditions. In other words, these two equations simply tell us that the *number* of members and reactions is sufficient for stability. They do not provide any information regarding their *arrangement*. A truss may have a sufficient number of members and external reactions but may still be unstable due to improper arrangement of members and/ or external supports.

We emphasize that in order for the criteria for static determinacy and indeterminacy, as given by Eqs. (4.3) and (4.4), to be valid, the truss must be stable and act as a single rigid body under a general system of coplanar loads when attached to the supports. Internally stable trusses must be supported by at least three reactions, all of which must be neither parallel nor concurrent. If a truss is internally unstable, then it must be supported by reactions equal in number to at least three plus the number of equations of condition $(3 + e_c)$, and all the reactions must be neither parallel nor concurrent. In addition, each joint, member, and portion of the truss must be constrained against all possible rigid body movements in the plane of the truss, either by the rest of the truss or by external supports. If a truss contains a sufficient number of members, but they are not properly arranged, the truss is said to have *critical form*. For some trusses, it may not be obvious from the drawings whether or not their members are arranged properly. However, if the member arrangement is improper, it will become evident during the analysis of the truss. The analysis of such unstable trusses will always lead to inconsistent, indeterminate, or infinite results.

Example 4.2

Classify each of the plane trusses shown in Fig. 4.15 as unstable, statically determinate, or statically indeterminate. If the truss is statically indeterminate, then determine the degree of static indeterminacy.

Solution

(a) The truss shown in Fig. 4.15(a) contains 17 members and 10 joints and is supported by 3 reactions. Thus, $m + r = 2j$. Since the three reactions are neither parallel nor concurrent and the members of the truss are properly arranged, it is statically determinate.   Ans.

(b) For this truss, $m = 17$, $j = 10$, and $r = 2$. Because $m + r < 2j$, the truss is unstable.   Ans.

(c) For this truss, $m = 21$, $j = 10$, and $r = 3$. Because $m + r > 2j$, the truss is statically indeterminate, with the degree of static indeterminacy $i = (m + r) - 2j = 4$. It should be obvious from Fig. 4.15(c) that the truss contains four more members than required for stability.   Ans.

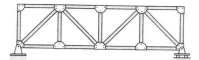

$m = 17 \quad j = 10 \quad r = 3$
$m + r = 2j$

(a) Statically Determinate

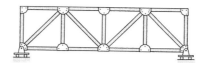

$m = 17 \quad j = 10 \quad r = 2$
$m + r < 2j$

(b) Unstable

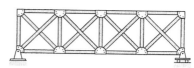

$m = 21 \quad j = 10 \quad r = 3$
$m + r > 2j$

(c) Statically Indeterminate ($i = 4$)

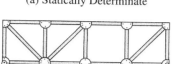

$m = 16 \quad j = 10 \quad r = 3$
$m + r < 2j$

(d) Unstable

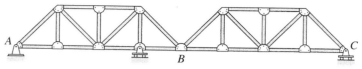

$m = 26 \quad j = 15 \quad r = 4$
$m + r = 2j$

(e) Statically Determinate

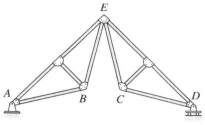

$m = 10 \quad j = 7 \quad r = 3$
$m + r < 2j$

(f) Unstable

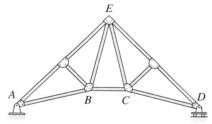

$m = 11 \quad j = 7 \quad r = 3$
$m + r = 2j$

(g) Statically Determinate

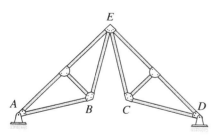

$m = 10 \quad j = 7 \quad r = 4$
$m + r = 2j$

(h) Statically Determinate

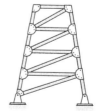

$m = 16 \quad j = 10 \quad r = 4$
$m + r = 2j$

(i) Statically Determinate

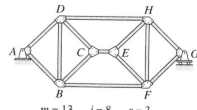

$m = 13 \quad j = 8 \quad r = 3$
$m + r = 2j$

(j) Unstable

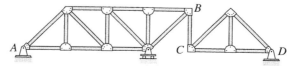

$m = 19 \quad j = 12 \quad r = 5$
$m + r = 2j$

(k) Statically Determinate

FIG. 4.15

continued

(d) This truss has $m = 16$, $j = 10$, and $r = 3$. The truss is unstable, since $m + r < 2j$.   **Ans.**

(e) This truss is composed of two rigid portions, AB and BC, connected by an internal hinge at B. The truss has $m = 26$, $j = 15$, and $r = 4$. Thus, $m + r = 2j$. The four reactions are neither parallel nor concurrent and the entire truss is properly constrained, so the truss is statically determinate.   **Ans.**

(f) For this truss, $m = 10$, $j = 7$, and $r = 3$. Because $m + r < 2j$, the truss is unstable.   **Ans.**

(g) In Fig. 4.15(g), a member BC has been added to the truss of Fig. 4.15(f), which prevents the relative rotation of the two portions ABE and CDE. Since m has now been increased to 11, with j and r kept constant at 7 and 3, respectively, the equation $m + r = 2j$ is satisfied. Thus, the truss of Fig. 4.15(g) is statically determinate.   **Ans.**

(h) The truss of Fig. 4.15(f) is stabilized by replacing the roller support at D by a hinged support, as shown in Fig. 4.15(h). Thus, the number of reactions has been increased to 4, but m and j remain constant at 10 and 7, respectively. With $m + r = 2j$, the truss is now statically determinate.   **Ans.**

(i) For the tower truss shown in Fig. 4.15(i), $m = 16$, $j = 10$, and $r = 4$. Because $m + r = 2j$, the truss is statically determinate.   **Ans.**

(j) This truss has $m = 13$, $j = 8$, and $r = 3$. Although $m + r = 2j$, the truss is unstable, because it contains two rigid portions $ABCD$ and $EFGH$ connected by three parallel members, BF, CE, and DH, which cannot prevent the relative displacement, in the vertical direction, of one rigid part of the truss with respect to the other.   **Ans.**

(k) For the truss shown in Fig. 4.15(k), $m = 19$, $j = 12$, and $r = 5$. Because $m + r = 2j$, the truss is statically determinate.   **Ans.**

4.5 ANALYSIS OF PLANE TRUSSES BY THE METHOD OF JOINTS

In the *method of joints, the axial forces in the members of a statically determinate truss are determined by considering the equilibrium of its joints.* Since the entire truss is in equilibrium, each of its joints must also be in equilibrium. At each joint of the truss, the member forces and any applied loads and reactions form a coplanar concurrent force system (see Fig. 4.14), which must satisfy two equilibrium equations, $\sum F_x = 0$ and $\sum F_y = 0$, in order for the joint to be in equilibrium. These two equilibrium equations must be satisfied at each joint of the truss. There are only two equations of equilibrium at a joint, so they cannot be used to determine more than two unknown forces.

The method of joints consists of selecting a joint with no more than two unknown forces (which must not be collinear) acting on it and applying the two equilibrium equations to determine the unknown forces. The procedure may be repeated until all the desired forces have been obtained. As we discussed in the preceding section, all the unknown member forces and the reactions can be determined from the joint equilibrium equations, but in many trusses it may not be possible to find a joint with two or fewer unknowns to start the analysis unless the reactions are known beforehand. In such cases, the reactions are computed by using the equations of equilibrium and condition (if any) for the entire truss before proceeding with the method of joints to determine member forces.

To illustrate the analysis by this method, consider the truss shown in Fig. 4.16(a). The truss contains five members, four joints, and three reactions. Since $m + r = 2j$, the truss is statically determinate. The free-body diagrams of all the members and the joints are given in Fig. 4.16(b). Because the member forces are not yet known, the sense of axial forces (tension or compression) in the members has been arbitrarily assumed. As shown in Fig. 4.16(b), members AB, BC, and AD are assumed to be in tension, with axial forces tending to elongate the members, whereas members BD and CD are assumed to be in compression, with axial forces tending to shorten them. The free-body diagrams of the joints show the member forces in directions opposite to their directions on the member ends in accordance with Newton's law of action and reaction. Focusing our attention on the free-body diagram of joint C, we observe that the *tensile force* F_{BC} is *pulling away* on the joint, whereas the *compressive force* F_{CD} is *pushing toward* the joint. This effect of members in tension pulling on the joints and members in compression pushing into the joints can be seen on the free-body diagrams of all the joints shown in Fig. 4.16(b). The free-body diagrams of members are usually omitted in the analysis and only those of joints are drawn, so it is important to understand that *a tensile member axial force is always indicated on the joint by an arrow pulling away on the joint, and a compressive member axial force is always indicated by an arrow pushing toward the joint*.

The analysis of the truss by the method of joints is started by selecting a joint that has two or fewer unknown forces (which must not be collinear) acting on it. An examination of the free-body diagrams of the joints in Fig. 4.16(b) indicates that none of the joints satisfies this requirement. We therefore compute reactions by applying the three equilibrium equations to the free body of the entire truss shown in Fig. 4.16(c), as follows:

$$+ \rightarrow \sum F_x = 0 \qquad\qquad A_x - 28 = 0 \qquad A_x = 28 \text{ k} \rightarrow$$

$$+ \circlearrowleft \sum M_C = 0 \qquad -A_y(35) + 28(20) + 42(15) = 0 \qquad A_y = 34 \text{ k} \uparrow$$

$$+ \uparrow \sum F_y = 0 \qquad\qquad 34 - 42 + C_y = 0 \qquad C_y = 8 \text{ k} \uparrow$$

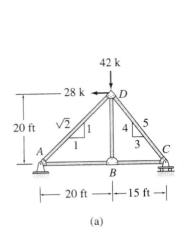

42 k

28 k ← D

20 ft   √2 1   4   5
         1   3   C
A
         B
├— 20 ft —┼—15 ft —┤

(a)

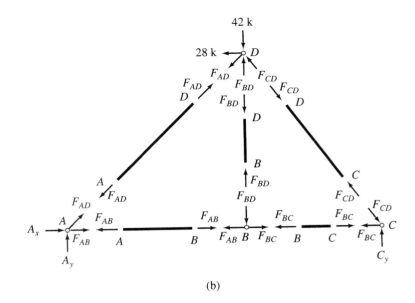

(b)

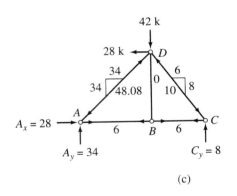

42 k

28 k ← D

34   0   6
34 48.08   10 8
A
A_x = 28 → 6 B 6 C
A_y = 34   C_y = 8

(c)

y

x

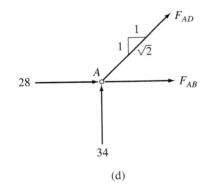

F_AD

1
1 √2
A
28 →   F_AB

34

(d)

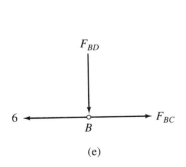

F_BD

6 ←   → F_BC
       B

(e)

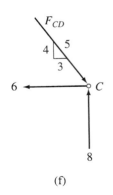

F_CD

4 5
3

6 ←   C

8

(f)

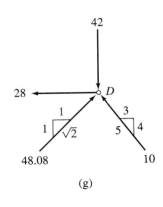

42

28 ←   D
      1   3
    1 √2   5 4
  48.08   10

(g)

FIG. 4.16

Having determined the reactions, we can now begin computing member forces either at joint A, which now has two unknown forces, F_{AB} and F_{AD}, or at joint C, which also has two unknowns, F_{BC} and F_{CD}. Let us start with joint A. The free-body diagram of this joint is shown in Fig. 4.16(d). Although we could use the sines and cosines of the angles of inclination of inclined members in writing the joint equilibrium equations, it is usually more convenient to use the slopes of the inclined members instead. The slope of an inclined member is simply the ratio of the vertical projection of the length of the member to the horizontal projection of its length. For example, from Fig. 4.16(a), we can see that member CD of the truss under consideration rises 20 ft in the vertical direction over a horizontal distance of 15 ft. Therefore, the slope of this member is 20:15, or 4:3. Similarly, we can see that the slope of member AD is 1:1. The slopes of inclined members thus determined from the dimensions of the truss are usually depicted on the diagram of the truss by means of small right-angled triangles drawn on the inclined members, as shown in Fig. 4.16(a).

Refocusing our attention on the free-body diagram of joint A in Fig. 4.16(d), we determine the unknowns F_{AB} and F_{AD} by applying the two equilibrium equations:

$$+\uparrow \sum F_y = 0 \qquad\qquad 34 + \frac{1}{\sqrt{2}} F_{AD} = 0 \qquad F_{AD} = -48.08 \text{ k}$$

$$= 48.08 \text{ k (C)}$$

$$+\rightarrow \sum F_x = 0 \qquad 28 - \frac{1}{\sqrt{2}}(48.08) + F_{AB} = 0 \qquad F_{AB} = +6 \text{ k}$$

$$= 6 \text{ k (T)}$$

Note that the equilibrium equations were applied in such an order so that each equation contains only one unknown. The negative answer for F_{AD} indicates that the member AD is in compression instead of in tension, as initially assumed, whereas the positive answer for F_{AB} indicates that the assumed sense of axial force (tension) in member AB was correct.

Next, we draw the free-body diagram of joint B, as shown in Fig. 4.16(e), and determine F_{BC} and F_{BD} as follows:

$$+\rightarrow \sum F_x = 0 \qquad -6 + F_{BC} = 0 \qquad F_{BC} = +6 \text{ k}, \quad \text{or} \quad F_{BC} = 6 \text{ k (T)}$$

$$+\uparrow \sum F_y = 0 \qquad -F_{BD} = 0 \qquad F_{BD} = 0$$

Applying the equilibrium equation $\sum F_x = 0$ to the free-body diagram of joint C (Fig. 4.16(f)), we obtain

$$+\rightarrow \sum F_x = 0 \qquad -6 + \frac{3}{5} F_{CD} = 0 \qquad F_{CD} = +10 \text{ k}, \quad \text{or}$$

$$F_{CD} = 10 \text{ k (C)}$$

We have determined all the member forces, so the three remaining equilibrium equations, $\sum F_y = 0$ at joint C and $\sum F_x = 0$ and $\sum F_y = 0$ at joint D, can be used to check our calculations. Thus, at joint C,

$$+ \uparrow \sum F_y = 8 - \frac{4}{5}(10) = 0 \qquad \qquad \text{Checks}$$

and at joint D (Fig. 4.16(g)),

$$+ \rightarrow \sum F_x = -28 + \frac{1}{\sqrt{2}}(48.08) - \frac{3}{5}(10) = 0 \qquad \text{Checks}$$

$$+ \uparrow \sum F_y = \frac{1}{\sqrt{2}}(48.08) - 42 + \frac{4}{5}(10) = 0 \qquad \text{Checks}$$

In the preceding paragraphs, the analysis of a truss has been carried out by drawing a free-body diagram and writing the two equilibrium equations for each of its joints. However, the analysis of trusses can be considerably expedited if we can determine some (preferably all) of the member forces by inspection—that is, without drawing the joint free-body diagrams and writing the equations of equilibrium. This approach can be conveniently used for the joints at which at least one of the two unknown forces is acting in the horizontal or vertical direction. When both of the unknown forces at a joint have inclined directions, it usually becomes necessary to draw the free-body diagram of the joint and determine the unknowns by solving the equilibrium equations simultaneously. To illustrate this procedure, consider again the truss of Fig. 4.16(a). The free-body diagram of the entire truss is shown in Fig. 4.16(c), which also shows the support reactions computed previously. Focusing our attention on joint A in this figure, we observe that in order to satisfy the equilibrium equation $\sum F_y = 0$ at joint A, the vertical component of F_{AD} must push downward into the joint with a magnitude of 34 k to balance the vertically upward reaction of 34 k. The fact that member AD is in compression is indicated on the diagram of the truss by drawing arrows near joints A and D pushing into the joints, as shown in Fig. 4.16(c). Because the magnitude of the vertical component of F_{AD} has been found to be 34 k and since the slope of member AD is 1:1, the magnitude of the horizontal component of F_{AD} must also be 34 k; therefore, the magnitude of the resultant force F_{AD} is $F_{AD} = \sqrt{(34)^2 + (34)^2} = 48.08$ k. The components of F_{AD}, as well as F_{AD} itself are shown on the corresponding sides of a right-angled triangle drawn on member AD, as shown in Fig. 4.16(c). With the horizontal component of F_{AD} now known, we observe (from Fig. 4.16(c)) that in order to satisfy the equilibrium equation $\sum F_x = 0$ at joint A, the force in member AB (F_{AB}) must pull to the right on the joint with a magnitude of 6 k to balance the horizontal component of F_{AD} of 34 k acting to the left and the horizontal reaction of 28 k acting to the right. The magnitude of

F_{AB} is now written on member AB, and the arrows, pulling away on the joints, are drawn near joints A and B to indicate that member AB is in tension.

Next, we focus our attention on joint B of the truss. It should be obvious from Fig. 4.16(c) that in order to satisfy $\sum F_y = 0$ at B, the force in member BD must be zero. To satisfy $\sum F_x = 0$, the force in member BC must have a magnitude of 6 k, and it must pull to the right on joint B, indicating tension in member BC. This latest information is recorded in the diagram of the truss in Fig. 4.16(c). Considering now the equilibrium of joint C, we can see from the figure that in order to satisfy $\sum F_y = 0$, the vertical component of F_{CD} must push downward into the joint with a magnitude of 8 k to balance the vertically upward reaction of 8 k. Thus, member CD is in compression. Since the magnitude of the vertical component of F_{CD} is 8 k and since the slope of member CD is 4:3, the magnitude of the horizontal component of F_{CD} is equal to $(3/4)(8) = 6$ k; therefore, the magnitude of F_{CD} itself is $F_{CD} = \sqrt{(6)^2 + (8)^2} = 10$ k. Having determined all the member forces, we check our computations by applying the equilibrium equations $\sum F_x = 0$ at joint C and $\sum F_x = 0$ and $\sum F_y = 0$ at joint D. The horizontal and vertical components of the member forces are already available in Fig. 4.16(c), so we can easily check by inspection to find that these equations of equilibrium are indeed satisfied. We must recognize that all the arrows shown on the diagram of the truss in Fig. 4.16(c) indicate forces acting at the joints (not at the ends of the members).

Identification of Zero-Force Members

Because trusses are usually designed to support several different loading conditions, it is not uncommon to find members with zero forces in them when a truss is being analyzed for a particular loading condition. Zero-force members are also added to trusses to brace compression members against buckling and slender tension members against vibrating. The analysis of trusses can be expedited if we can identify the zero-force members by inspection. Two common types of member arrangements that result in zero-force members are the following:

1. If only two noncollinear members are connected to a joint that has no external loads or reactions applied to it, then the force in both members is zero.
2. If three members, two of which are collinear, are connected to a joint that has no external loads or reactions applied to it, then the force in the member that is not collinear is zero.

The first type of arrangement is shown in Fig. 4.17(a). It consists of two noncollinear members AB and AC connected to a joint A. Note that no external loads or reactions are applied to the joint. From this figure we can see that in order to satisfy the equilibrium equation $\sum F_y = 0$, the y component of F_{AB} must be zero; therefore, $F_{AB} = 0$.

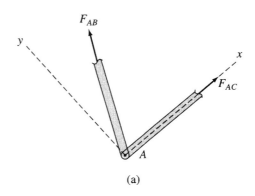

(a)

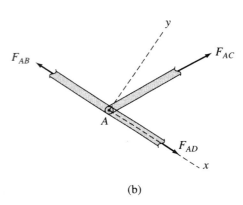

(b)

FIG. 4.17

Because the x component of F_{AB} is zero, the second equilibrium equation, $\sum F_x = 0$, can be satisfied only if F_{AC} is also zero.

The second type of arrangement is shown in Fig. 4.17(b), and it consists of three members, AB, AC, and AD, connected together at a joint A. Note that two of the three members, AB and AD, are collinear. We can see from the figure that since there is no external load or reaction applied to the joint to balance the y component of F_{AC}, the equilibrium equation $\sum F_y = 0$ can be satisfied only if F_{AC} is zero.

Example 4.3

Identify all zero-force members in the Fink roof truss subjected to an unbalanced snow load, as shown in Fig. 4.18.

Solution

It can be seen from the figure that at joint B, three members, AB, BC, and BJ, are connected, of which AB and BC are collinear and BJ is not. Since no external loads are applied at joint B, member BJ is a zero-force member. A similar reasoning can be used for joint D to identify member DN as a zero-force mem-

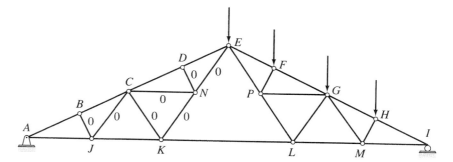

FIG. **4.18**

ber. Next, we focus our attention on joint J, where four members, AJ, BJ, CJ, and JK, are connected and no external loads are applied. We have already identified BJ as a zero-force member. Of the three remaining members, AJ and JK are collinear; therefore, CJ must be a zero-force member. Similarly, at joint N, member CN is identified as a zero-force member; the same type of arguments can be used for joint C to identify member CK as a zero-force member and for joint K to identify member KN as a zero-force member. Finally, we consider joint N, where four members, CN, DN, EN, and KN, are connected, of which three members, CN, DN, and KN, have already been identified as zero-force members. No external loads are applied at joint N, so the force in the remaining member, EN, must also be zero.

Procedure for Analysis

The following step-by-step procedure can be used for the analysis of statically determinate simple plane trusses by the method of joints.

1. Check the truss for static determinacy, as discussed in the preceding section. If the truss is found to be statically determinate and stable, proceed to step 2. Otherwise, end the analysis at this stage. (The analysis of statically indeterminate trusses is considered in Part Three of this text.)
2. Identify by inspection any zero-force members of the truss.
3. Determine the slopes of the inclined members (except the zero-force members) of the truss.
4. Draw a free-body diagram of the whole truss, showing all external loads and reactions. Write zeros by the members that have been identified as zero-force members.
5. Examine the free-body diagram of the truss to select a joint that has no more than two unknown forces (which must not be collinear) acting on it. If such a joint is found, then go directly to the next step. Otherwise, determine reactions by applying the three equations of equilibrium and the equations of condition (if any) to the free body of the whole truss; then select a joint with two or fewer unknowns, and go to the next step.

6. **a.** Draw a free-body diagram of the selected joint, showing tensile forces by arrows pulling away from the joint and compressive forces by arrows pushing into the joint. It is usually convenient to assume the unknown member forces to be tensile.

    b. Determine the unknown forces by applying the two equilibrium equations $\sum F_x = 0$ and $\sum F_y = 0$. A positive answer for a member force means that the member is in tension, as initially assumed, whereas a negative answer indicates that the member is in compression.

    If at least one of the unknown forces acting at the selected joint is in the horizontal or vertical direction, the unknowns can be conveniently determined by satisfying the two equilibrium equations by inspection of the joint on the free-body diagram of the truss.

7. If all the desired member forces and reactions have been determined, then go to the next step. Otherwise, select another joint with no more than two unknowns, and return to step 6.

8. If the reactions were determined in step 5 by using the equations of equilibrium and condition of the whole truss, then apply the remaining joint equilibrium equations that have not been utilized so far to check the calculations. If the reactions were computed by applying the joint equilibrium equations, then use the equilibrium equations of the entire truss to check the calculations. If the analysis has been performed correctly, then these extra equilibrium equations must be satisfied.

Example 4.4

Determine the force in each member of the Warren truss shown in Fig. 4.19(a) by the method of joints.

Solution

Static Determinacy The truss has 13 members and 8 joints and is supported by 3 reactions. Because $m + r = 2j$ and the reactions and the members of the truss are properly arranged, it is statically determinate.

Zero-Force Members It can be seen from Fig. 4.19(a) that at joint G, three members, CG, FG, and GH, are connected, of which FG and GH are collinear and CG is not. Since no external load is applied at joint G, member CG is a zero-force member.

$$F_{CG} = 0 \qquad\qquad \text{Ans.}$$

From the dimensions of the truss, we find that all inclined members have slopes of 3:4, as shown in Fig. 4.19(a). The free-body diagram of the entire truss is shown in Fig. 4.19(b). As a joint with two or fewer unknowns—which should not be collinear—cannot be found, we calculate the support reactions. (Although joint G has only two unknown forces, F_{FG} and F_{GH}, acting on it, these forces are

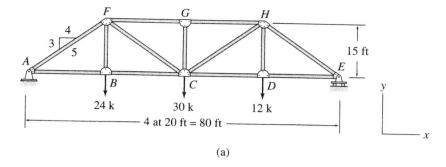

(a)

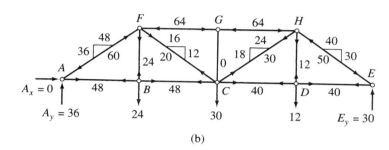

(b)

FIG. 4.19

collinear, so they cannot be determined from the joint equilibrium equation, $\sum F_x = 0$.)

Reactions By using proportions,

$$A_y = 24\left(\frac{3}{4}\right) + 30\left(\frac{1}{2}\right) + 12\left(\frac{1}{4}\right) = 36$$

$$\sum F_y = 0 \qquad E_y = (24 + 30 + 12) - 36 = 30 \text{ k}$$

$$\sum F_x = 0 \qquad A_x = 0$$

Joint A Focusing our attention on joint A in Fig. 4.19(b), we observe that in order to satisfy $\sum F_y = 0$, the vertical component of F_{AF} must push downward into the joint with a magnitude of 36 k to balance the upward reaction of 36 k. The slope of member AF is 3:4, so the magnitude of the horizontal component of F_{AF} is $(4/3)(36)$, or 48 k. Thus, the force in member AF is compressive, with a magnitude of $F_{AF} = \sqrt{(48)^2 + (36)^2} = 60$ k.

$$F_{AF} = 60 \text{ k (C)} \qquad \text{Ans.}$$

With the horizontal component of F_{AF} now known, we can see from the figure that in order for $\sum F_x = 0$ to be satisfied, F_{AB} must pull to the right with a magnitude of 48 k to balance the horizontal component of F_{AF} of 48 k acting to the left. Therefore, member AB is in tension with a force of 48 k.

$$F_{AB} = 48 \text{ k (T)} \qquad \text{Ans.}$$

Joint B Next, we consider the equilibrium of joint B. Applying $\sum F_x = 0$, we obtain F_{BC}.

$$F_{BC} = 48 \text{ k (T)} \qquad \text{Ans.}$$

From $\sum F_y = 0$, we obtain F_{BF}.

$$F_{BF} = 24 \text{ k (T)} \qquad \text{Ans.}$$

Joint F This joint now has two unknowns, F_{CF} and F_{FG}, so they can be determined by applying the equations of equilibrium as follows. We can see from Fig. 4.19(b) that in order to satisfy $\sum F_y = 0$, the vertical component of F_{CF} must pull downward on joint F with a magnitude of $36 - 24 = 12$ k. Using the 3:4 slope of member CF, we obtain the magnitude of the horizontal component as $(4/3)(12) = 16$ k and the magnitude of F_{CF} itself as 20 k.

$$F_{CF} = 20 \text{ k (T)} \qquad \text{Ans.}$$

Considering the equilibrium of joint F in the horizontal direction ($\sum F_x = 0$), it should be obvious from Fig. 4.19(b) that F_{FG} must push to the left on the joint with a magnitude of $48 + 16 = 64$ k.

$$F_{FG} = 64 \text{ k (C)} \qquad \text{Ans.}$$

Joint G Similarly, by applying $\sum F_x = 0$, we obtain F_{GH}.

$$F_{GH} = 64 \text{ k (C)} \qquad \text{Ans.}$$

Note that the second equilibrium equation, $\sum F_y = 0$, at this joint has already been utilized in the identification of member CG as a zero-force member.

Joint C By considering equilibrium in the vertical direction, $\sum F_y = 0$, we observe (from Fig. 4.19(b)) that member CH should be in tension and that the magnitude of the vertical component of its force must be equal to $30 - 12 = 18$ k. Therefore, the magnitudes of the horizontal component of F_{CH} and of F_{CH} itself are 24 k and 30 k, respectively, as shown in Fig. 4.19(b).

$$F_{CH} = 30 \text{ k (T)} \qquad \text{Ans.}$$

By considering equilibrium in the horizontal direction, $\sum F_x = 0$, we observe that member CD must be in tension and that the magnitude of its force should be equal to $48 + 16 - 24 = 40$ k.

$$F_{CD} = 40 \text{ k (T)} \qquad \text{Ans.}$$

Joint D By applying $\sum F_x = 0$, we obtain F_{DE}.

$$F_{DE} = 40 \text{ k (T)} \qquad \text{Ans.}$$

From $\sum F_y = 0$, we determine F_{DH}.

$$F_{DH} = 12 \text{ k (T)} \qquad \text{Ans.}$$

Joint E Considering the vertical components of all the forces acting at joint E, we find that in order to satisfy $\sum F_y = 0$, the vertical component of F_{EH} must push downward into joint E with a magnitude of 30 k to balance the upward reaction $E_y = 30$ k. The magnitude of the horizontal component of F_{EH} is equal to $(4/3)(30)$, or 40 k. Thus, F_{EH} is a compressive force with a magnitude of 50 k.

$$F_{EH} = 50 \text{ k (C)} \qquad \text{Ans.}$$

Checking Computations To check our computations, we apply the following remaining joint equilibrium equations (see Fig. 4.19(b)). At joint E,

$$+ \rightarrow \sum F_x = -40 + 40 = 0 \qquad \text{Checks}$$

At joint H,

$$+ \rightarrow \sum F_x = 64 - 24 - 40 = 0 \qquad \text{Checks}$$

$$+ \uparrow \sum F_y = -18 - 12 + 30 = 0 \qquad \text{Checks}$$

Example 4.5

Determine the force in each member of the truss shown in Fig. 4.20(a) by the method of joints.

Solution

Static Determinacy The truss is composed of 7 members and 5 joints and is supported by 3 reactions. Thus, $m + r = 2j$. Since the reactions and the members of the truss are properly arranged, it is statically determinate.

From the dimensions of the truss given in Fig. 4.20(a), we find that all inclined members have slopes of 12:5. Since joint E has two unknown non-collinear forces, F_{CE} and F_{DE}, acting on it, we can begin the method of joints without first calculating the support reactions.

Joint E Focusing our attention on joint E in Fig. 4.20(b), we observe that in order to satisfy $\sum F_x = 0$, the horizontal component of F_{DE} must push to the left into the joint with a magnitude of 25 kN to balance the 25 kN external load acting to the right. The slope of member DE is 12:5, so the magnitude of the vertical component of F_{DE} is $(12/5)(25)$, or 60 kN. Thus, the force in member DE is compressive, with a magnitude of

$$F_{DE} = \sqrt{(25)^2 + (60)^2} = 65 \text{ kN}$$

$$F_{DE} = 65 \text{ kN (C)} \qquad \text{Ans.}$$

With the vertical component of F_{DE} now known, we can see from the figure that in order for $\sum F_y = 0$ to be satisfied, F_{CE} must pull downward on joint E with a magnitude of $60 - 30 = 30$ kN.

$$F_{CE} = 30 \text{ kN (T)} \qquad \text{Ans.}$$

Joint C Next, we consider the equilibrium of joint C. Applying $\sum F_x = 0$, we obtain F_{CD}.

$$F_{CD} = 50 \text{ kN (C)} \qquad \text{Ans.}$$

From $\sum F_y = 0$, we obtain F_{AC}.

$$F_{AC} = 30 \text{ kN (T)} \qquad \text{Ans.}$$

continued

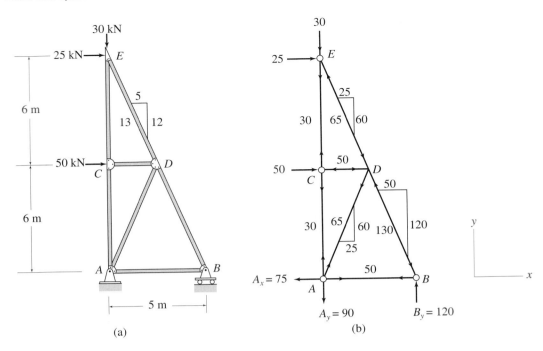

(a)   (b)

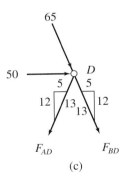

(c)

FIG. **4.20**

Joint D Both of the unknown forces, F_{AD} and F_{BD}, acting at this joint have inclined directions, so we draw the free-body diagram of this joint as shown in Fig. 4.20(c) and determine the unknowns by solving the equilibrium equations simultaneously:

$$+ \rightarrow \sum F_x = 0 \qquad 50 + \frac{5}{13}(65) - \frac{5}{13}F_{AD} + \frac{5}{13}F_{BD} = 0$$

$$+ \uparrow \sum F_y = 0 \qquad -\frac{12}{13}(65) - \frac{12}{13}F_{AD} - \frac{12}{13}F_{BD} = 0$$

Solving these equations simultaneously, we obtain

$$F_{AD} = 65 \text{ kN} \quad \text{and} \quad F_{BD} = -130 \text{ kN}$$

$$F_{AD} = 65 \text{ kN (T)} \qquad \text{Ans.}$$

$$F_{BD} = 130 \text{ kN (C)} \qquad \text{Ans.}$$

Joint B (See Fig. 4.20(b).) By considering the equilibrium of joint B in the horizontal direction ($\sum F_x = 0$), we obtain F_{AB}.

$$F_{AB} = 50 \text{ kN (T)} \qquad \text{Ans.}$$

Having determined all the member forces, we apply the remaining equilibrium equation ($\sum F_y = 0$) at joint B to calculate the support reaction B_y.

$$B_y = 120 \text{ kN} \uparrow \qquad \text{Ans.}$$

Joint A By applying $\sum F_x = 0$, we obtain A_x.

$$A_x = 75 \text{ kN} \leftarrow \qquad \text{Ans.}$$

From $\sum F_y = 0$, we obtain A_y.

$$A_y = 90 \text{ kN} \downarrow \qquad \text{Ans.}$$

Checking Computations To check our computations, we consider the equilibrium of the entire truss. Applying the three equilibrium equations to the free body of the entire truss shown in Fig. 4.20(b), we obtain

$$+ \rightarrow \sum F_x = 25 + 50 - 75 = 0 \qquad \text{Checks}$$

$$+ \uparrow \sum F_y = -30 - 90 + 120 = 0 \qquad \text{Checks}$$

$$+ \zeta \sum M_B = 30(5) - 25(12) - 50(6) + 90(5) = 0 \qquad \text{Checks}$$

Example 4.6

Determine the force in each member of the three-hinged trussed arch shown in Fig. 4.21(a) by the method of joints.

Solution

Static Determinacy The truss contains 10 members and 7 joints and is supported by 4 reactions. Since $m + r = 2j$ and the reactions and the members of the truss are properly arranged, it is statically determinate. Note that since $m < 2j - 3$, the truss is not internally stable, and it will not remain a rigid body when it is detached from its supports. However, when attached to the supports, the truss will maintain its shape and can be treated as a rigid body.

Zero-Force Members It can be seen from Fig. 4.21(a) that at joint C, three members, $AC, CE,$ and CF, are connected, of which members AC and CF are collinear. Since joint C does not have any external load applied to it, the non-collinear member CE is a zero-force member.

$$F_{CE} = 0 \qquad \text{Ans.}$$

continued

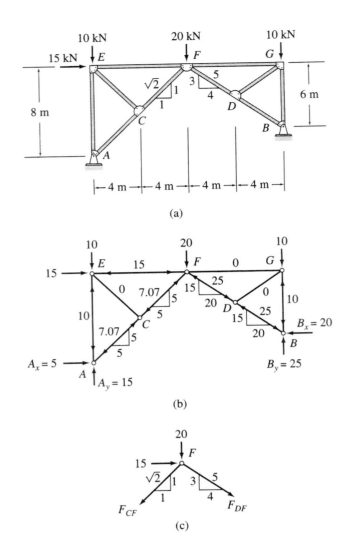

FIG. **4.21**

Similar reasoning can be used for joint D to identify member DG as a zero-force member.

$$F_{DG} = 0$$ Ans.

The slopes of the non-zero-force inclined members are shown in Fig. 4.21(a). The free-body diagram of the entire truss is shown in Fig. 4.21(b). The method of joints can be started either at joint E, or at joint G, since both of these joints have only two unknowns each.

Joint E Beginning with joint E, we observe from Fig. 4.21(b) that in order for $\sum F_x = 0$ to be satisfied, the force in member EF must be compressive with a magnitude of 15 kN.

$$F_{EF} = 15 \text{ kN (C)} \qquad \text{Ans.}$$

Similarly, from $\sum F_y = 0$, we obtain F_{AE}.

$$F_{AE} = 10 \text{ kN (C)} \qquad \text{Ans.}$$

Joint G By considering the equilibrium of joint G in the horizontal direction ($\sum F_x = 0$), we observe that the force in member FG is zero.

$$F_{FG} = 0 \qquad \text{Ans.}$$

Similarly, by applying $\sum F_y = 0$, we obtain F_{BG}.

$$F_{BG} = 10 \text{ kN (C)} \qquad \text{Ans.}$$

Joint F Next, we consider joint F. Both of the unknown forces, F_{CF} and F_{DF}, acting at this joint have inclined directions, so we draw the free-body diagram of this joint as shown in Fig. 4.21(c) and determine the unknowns by solving the equilibrium equations simultaneously:

$$+ \rightarrow \sum F_x = 0 \qquad 15 - \frac{1}{\sqrt{2}} F_{CF} + \frac{4}{5} F_{DF} = 0$$

$$+ \uparrow \sum F_y = 0 \qquad -20 - \frac{1}{\sqrt{2}} F_{CF} - \frac{3}{5} F_{DF} = 0$$

Solving these equations, we obtain

$$F_{DF} = -25 \text{ kN} \quad \text{and} \quad F_{CF} = -7.07 \text{ kN}$$

$$F_{DF} = 25 \text{ kN (C)} \qquad \text{Ans.}$$

$$F_{CF} = 7.07 \text{ kN (C)} \qquad \text{Ans.}$$

Joint C (See Fig. 4.21(b).) In order for joint C to be in equilibrium, the two nonzero collinear forces acting at it must be equal and opposite.

$$F_{AC} = 7.07 \text{ kN (C)} \qquad \text{Ans.}$$

Joint D Using a similar reasoning at joint D, we obtain F_{BD}.

$$F_{BD} = 25 \text{ kN (C)} \qquad \text{Ans.}$$

Joint A Having determined all the member forces, we apply the two equilibrium equations at joint A to calculate the support reactions, A_x and A_y. By applying $\sum F_x = 0$, we obtain A_x.

$$A_x = 5 \text{ kN} \rightarrow \qquad \text{Ans.}$$

By applying $\sum F_y = 0$, we find that A_y is equal to $10 + 5 = 15$ kN.

$$A_y = 15 \text{ kN} \uparrow \qquad\qquad \text{Ans.}$$

Joint B By applying $\sum F_x = 0$, we obtain B_x.

$$B_x = 20 \text{ kN} \leftarrow \qquad\qquad \text{Ans.}$$

From $\sum F_y = 0$, we find that $B_y = 15 + 10 = 25$ kN.

$$B_y = 25 \text{ kN} \uparrow \qquad\qquad \text{Ans.}$$

Equilibrium Check of Entire Truss Finally, to check our computations, we consider the equilibrium of the entire truss. Applying the three equations of equilibrium to the free body of the entire truss shown in Fig. 4.21(b), we have

$$+ \rightarrow \sum F_x = 5 + 15 - 20 = 0 \qquad\qquad \text{Checks}$$

$$+ \uparrow \sum F_y = 15 - 10 - 20 - 10 + 25 = 0 \qquad\qquad \text{Checks}$$

$$+ \zeta \sum M_B = 5(2) - 15(16) - 15(6) + 10(16) + 20(8) = 0 \quad \text{Checks}$$

4.6 ANALYSIS OF PLANE TRUSSES BY THE METHOD OF SECTIONS

The method of joints, presented in the preceding section, proves to be very efficient when forces in all the members of a truss are to be determined. However, if the forces in only certain members of a truss are desired, the method of joints may not prove to be efficient, because it may involve calculation of forces in several other members of the truss before a joint is reached that can be analyzed for a desired member force. The *method of sections* enables us to determine forces in the specific members of trusses directly, without first calculating many unnecessary member forces, as may be required by the method of joints.

The method of sections involves cutting the truss into two portions by passing an imaginary section through the members whose forces are desired. The desired member forces are then determined by considering the equilibrium of one of the two portions of the truss. Each portion of the truss is treated as a rigid body in equilibrium, under the action of any applied loads and reactions and the forces in the members that have been cut by the section. The unknown member forces are determined by applying the three equations of equilibrium to one of the two portions of the truss. There are only three equilibrium equations available, so they cannot be used to determine more than three unknown forces. Thus, in general, *sections should be chosen that do not pass through more than three members with unknown forces.* In some trusses, the arrangement of members may be such that by using sections that pass through more than three members with unknown forces, we can determine one or, at most, two unknown forces. Such sections are, however, employed in the analysis of only certain types of trusses (see Example 4.9).

Procedure for Analysis

The following step-by-step procedure can be used for determining the member forces of statically determinate plane trusses by the method of sections.

1. Select a section that passes through as many members as possible whose forces are desired, but not more than three members with unknown forces. The section should cut the truss into two parts.
2. Although either of the two portions of the truss can be used for computing the member forces, we should select the portion that will require the least amount of computational effort in determining the unknown forces. To avoid the necessity for the calculation of reactions, if one of the two portions of the truss does not have any reactions acting on it, then select this portion for the analysis of member forces and go to the next step. If both portions of the truss are attached to external supports, then calculate reactions by applying the equations of equilibrium and condition (if any) to the free body of the entire truss. Next, select the portion of the truss for analysis of member forces that has the least number of external loads and reactions applied to it.
3. Draw the free-body diagram of the portion of the truss selected, showing all external loads and reactions applied to it and the forces in the members that have been cut by the section. The unknown member forces are usually assumed to be *tensile* and are, therefore, shown on the free-body diagram by arrows *pulling away* from the joints.
4. Determine the unknown forces by applying the three equations of equilibrium. To avoid solving simultaneous equations, try to apply the equilibrium equations in such a manner that each equation involves only one unknown. This can sometimes be achieved by using the alternative systems of equilibrium equations $(\sum F_q = 0, \sum M_A = 0, \sum M_B = 0$ or $\sum M_A = 0, \sum M_B = 0, \sum M_C = 0)$ described in Section 3.1 instead of the usual two-force summations and a moment summation $(\sum F_x = 0, \sum F_y = 0, \sum M = 0)$ system of equations.
5. Apply an alternative equilibrium equation, which was not used to compute member forces, to check the calculations. This alternative equation should preferably involve all three member forces determined by the analysis. If the analysis has been performed correctly, then this alternative equilibrium equation must be satisfied.

Example 4.7

Determine the forces in members CD, DG, and GH of the truss shown in Fig. 4.22(a) by the method of sections.

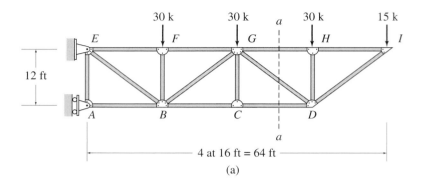

(a)

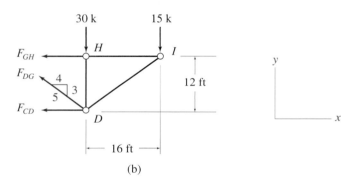

(b)

FIG. 4.22

Solution

Section aa As shown in Fig. 4.22(a), a section *aa* is passed through the three members of interest, CD, DG, and GH, cutting the truss into two portions, $ACGE$ and DHI. To avoid the calculation of support reactions, we will use the right-hand portion, DHI, to calculate the member forces.

Member Forces The free-body diagram of the portion DHI of the truss is shown in Fig. 4.22(b). All three unknown forces F_{CD}, F_{DG}, and F_{GH}, are assumed to be tensile and are indicated by arrows pulling away from the corresponding joints on the diagram. The slope of the inclined force, F_{DG}, is also shown on the free-body diagram. The desired member forces are calculated by applying the equilibrium equations as follows (see Fig. 4.22(b)).

$$+\circlearrowleft \sum M_D = 0 \qquad -15(16) + F_{GH}(12) = 0$$

$$F_{GH} = 20 \text{ k (T)} \qquad\qquad \text{Ans.}$$

$$+\uparrow \sum F_y = 0 \qquad -30 - 15 + \frac{3}{5}F_{DG} = 0$$

$$F_{DG} = 75 \text{ k (T)} \qquad\qquad \text{Ans.}$$

$$+\rightarrow \sum F_x = 0 \qquad -20 - \frac{4}{5}(75) - F_{CD} = 0$$

$$F_{CD} = -80 \text{ k}$$

The negative answer for F_{CD} indicates that our initial assumption about this force being tensile was incorrect, and F_{CD} is actually a compressive force.

$$F_{CD} = 80 \text{ k (C)} \qquad\qquad \text{Ans.}$$

Checking Computations (See Fig. 4.22(b).)

$$+\circlearrowleft \sum M_I = 30(16) - (-80)12 - \frac{4}{5}(75)(12) - \frac{3}{5}(75)(16) = 0 \quad \text{Checks}$$

Example 4.8

Determine the forces in members CJ and IJ of the truss shown in Fig. 4.23(a) by the method of sections.

Solution

Section aa As shown in Fig. 4.23(a), a section *aa* is passed through members IJ, CJ, and CD, cutting the truss into two portions, ACI and DGJ. The left-hand portion, ACI, will be used to analyze the member forces.

Reactions Before proceeding with the calculation of member forces, we need to determine reactions at support A. By considering the equilibrium of the entire truss (Fig. 4.23(b)), we determine the reactions to be $A_x = 0$, $A_y = 50$ k $\uparrow$, and $G_y = 50$ k $\uparrow$.

Member Forces The free-body diagram of the portion ACI of the truss is shown in Fig. 4.23(c). The slopes of the inclined forces, F_{IJ} and F_{CJ}, are obtained from the dimensions of the truss given in Fig. 4.23(a) and are shown on the free-body diagram. The unknown member forces are determined by applying the equations of equilibrium, as follows.

Because F_{CJ} and F_{CD} pass through point C, by summing moments about C, we obtain an equation containing only F_{IJ}:

$$+\circlearrowleft \sum M_C = 0 \qquad -50(40) + 20(20) - \frac{4}{\sqrt{17}}F_{IJ}(25) = 0$$

$$F_{IJ} = -65.97 \text{ k}$$

continued

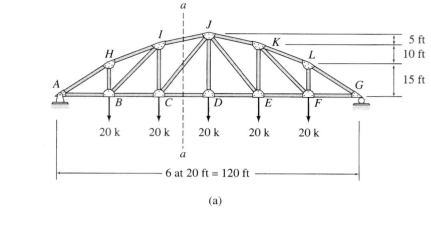

(a)

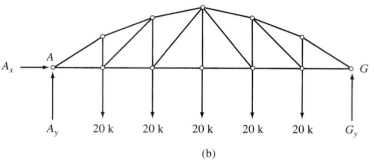

(b)

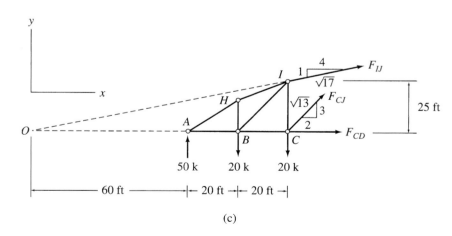

FIG. 4.23

(c)

The negative answer for F_{IJ} indicates that our initial assumption about this force being tensile was incorrect. Force F_{IJ} is actually a compressive force.

$$F_{IJ} = 65.97 \text{ k (C)}$$ **Ans.**

Next, we calculate F_{CJ} by summing moments about point O, which is the point of intersection of the lines of action of F_{IJ} and F_{CD}. Because the slope of member IJ is 1:4, the distance $OC = 4(IC) = 4(25) = 100$ ft (see Fig. 4.23(c)). Equilibrium of moments about O yields

$$+\zeta \sum M_O = 0 \qquad 50(60) - 20(80) - 20(100) + \frac{3}{\sqrt{13}} F_{CJ}(100) = 0$$

$$F_{CJ} = 7.21 \text{ k (T)} \qquad\qquad\qquad \text{Ans.}$$

Checking Computations To check our computations, we apply an alternative equation of equilibrium, which involves the two member forces just determined.

$$+\uparrow \sum F_y = 50 - 20 - 20 - \frac{1}{\sqrt{17}}(65.97) + \frac{3}{\sqrt{13}}(7.21) = 0 \quad \text{Checks}$$

Example 4.9

Determine the forces in members FJ, HJ, and HK of the K truss shown in Fig. 4.24(a) by the method of sections.

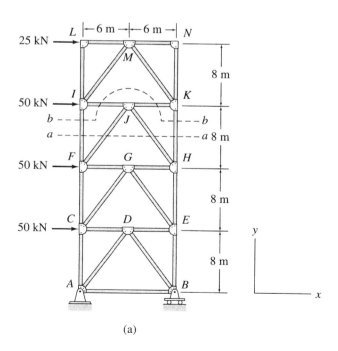

FIG. **4.24**

(a)

continued

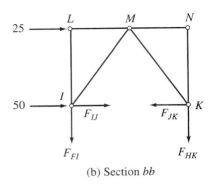

(b) Section bb

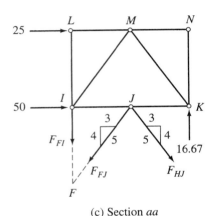

(c) Section aa

FIG. 4.24 (contd.)

Solution

From Fig. 4.24(a), we can observe that the horizontal section *aa* passing through the three members of interest, FJ, HJ, and HK, also cuts an additional member FI, thereby releasing four unknowns, which cannot be determined by three equations of equilibrium. Trusses such as the one being considered here with the members arranged in the form of the letter K can be analyzed by a section curved around the middle joint, like section *bb* shown in Fig. 4.24(a). To avoid the calculation of support reactions, we will use the upper portion *IKNL* of the truss above section *bb* for analysis. The free-body diagram of this portion is shown in Fig. 4.24(b). It can be seen that although section *bb* has cut four members, FI, IJ, JK, and HK, forces in members FI and HK can be determined by summing moments about points K and I, respectively, because the lines of action of three of the four unknowns pass through these points. We will, therefore, first compute F_{HK} by considering section *bb* and then use section *aa* to determine F_{FJ} and F_{HJ}.

Section bb Using Fig. 4.24(b), we write

$$+\circlearrowleft \sum M_I = 0 \qquad -25(8) - F_{HK}(12) = 0$$

$$F_{HK} = -16.67 \text{ kN}$$

$$F_{HK} = 16.67 \text{ kN (C)} \qquad \text{Ans.}$$

Section aa The free-body diagram of the portion *IKNL* of the truss above section *aa* is shown in Fig. 4.24(c). To determine F_{HJ}, we sum moments about F, which is the point of intersection of the lines of action of F_{FI} and F_{FJ}. Thus,

$$+\circlearrowleft \sum M_F = 0 \qquad -25(16) - 50(8) + 16.67(12) - \frac{3}{5}F_{HJ}(8) - \frac{4}{5}F_{HJ}(6) = 0$$

$$F_{HJ} = -62.5 \text{ kN}$$

$$F_{HJ} = 62.5 \text{ kN (C)} \qquad \text{Ans.}$$

By summing forces in the horizontal direction, we obtain

$$+\rightarrow \sum F_x = 0 \qquad 25 + 50 - \frac{3}{5}F_{FJ} - \frac{3}{5}(62.5) = 0$$

$$F_{FJ} = 62.5 \text{ kN (T)} \qquad \text{Ans.}$$

Checking Computations Finally, to check our calculations, we apply an alternative equilibrium equation, which involves the three member forces determined by the analysis. Using Fig. 4.24(c), we write

$$+\circlearrowleft \sum M_I = -25(8) - \frac{4}{5}(62.5)(6) + \frac{4}{5}(62.5)(6) + 16.67(12) = 0 \quad \text{Checks}$$

4.7 ANALYSIS OF COMPOUND TRUSSES

Although the method of joints and the method of sections described in the preceding sections can be used individually for the analysis of compound trusses, the analysis of such trusses can sometimes be expedited by using a combination of the two methods. For some types of compound trusses, the sequential analysis of joints breaks down when a joint with two or fewer unknown forces cannot be found. In such a case, the method of sections is then employed to calculate some of the member forces, thereby yielding a joint with two or fewer unknowns, from which the method of joints may be continued. This approach is illustrated by the following examples.

Example 4.10

Determine the force in each member of the compound truss shown in Fig. 4.25(a).

Solution

Static Determinacy The truss has 11 members and 7 joints and is supported by 3 reactions. Since $m + r = 2j$ and the reactions and the members of the truss are properly arranged, it is statically determinate.

The slopes of the inclined members, as determined from the dimensions of the truss, are shown in Fig. 4.25(a).

Reactions The reactions at supports A and B, as computed by applying the three equilibrium equations to the free-body diagram of the entire truss (Fig. 4.25(b)), are

$$A_x = 25 \text{ k} \leftarrow \qquad A_y = 5 \text{ k} \uparrow \qquad B_y = 35 \text{ k} \uparrow$$

Section aa Since a joint with two or fewer unknown forces cannot be found to start the method of joints, we first calculate F_{AB} by using section aa, as shown in Fig. 4.25(a).

The free-body diagram of the portion of the truss on the left side of section aa is shown in Fig. 4.25(c). We determine F_{AB} by summing moments about point G, the point of intersection of the lines of action of F_{CG} and F_{DG}.

continued

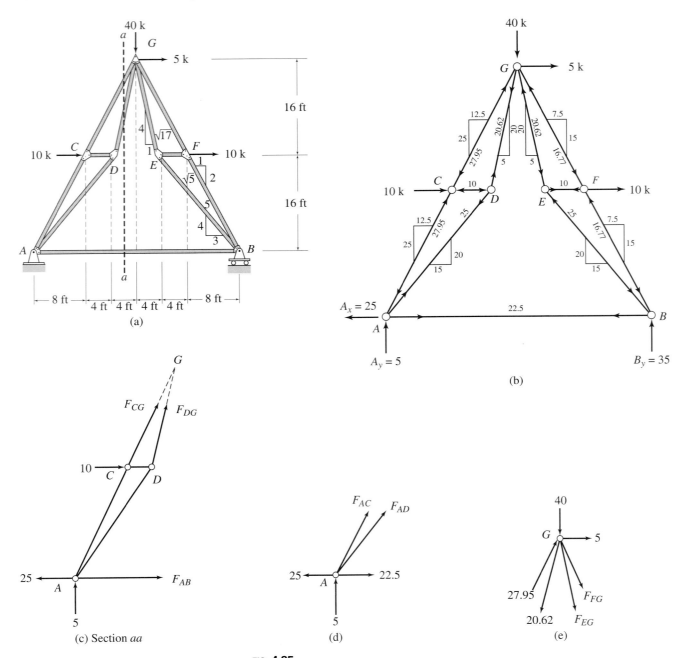

FIG. **4.25**

$$+ \circlearrowleft \sum M_G = 0 \qquad -25(32) - 5(16) + 10(16) + F_{AB}(32) = 0$$

$$F_{AB} = 22.5 \text{ k (T)} \qquad \text{Ans.}$$

With F_{AB} now known, the method of joints can be started either at joint A, or at joint B, since both of these joints have only two unknowns each. We begin with joint A.

Joint A The free-body diagram of joint A is shown in Fig. 4.25(d).

$$+ \rightarrow \sum F_x = 0 \qquad -25 + 22.5 + \frac{1}{\sqrt{5}} F_{AC} + \frac{3}{5} F_{AD} = 0$$

$$+ \uparrow \sum F_y = 0 \qquad 5 + \frac{2}{\sqrt{5}} F_{AC} + \frac{4}{5} F_{AD} = 0$$

Solving these equations simultaneously, we obtain

$$F_{AC} = -27.95 \text{ k} \quad \text{and} \quad F_{AD} = 25 \text{ k}$$

$$F_{AC} = 27.95 \text{ k (C)} \qquad \text{Ans.}$$

$$F_{AD} = 25 \text{ k (T)} \qquad \text{Ans.}$$

Joints C and D Focusing our attention on joints C and D in Fig. 4.25(b), and by satisfying the two equilibrium equations by inspection at each of these joints, we determine

$$F_{CG} = 27.95 \text{ k (C)} \qquad \text{Ans.}$$

$$F_{CD} = 10 \text{ k (C)} \qquad \text{Ans.}$$

$$F_{DG} = 20.62 \text{ k (T)} \qquad \text{Ans.}$$

Joint G Next, we consider the equilibrium of joint G (see Fig. 4.25(e)).

$$+ \rightarrow \sum F_x = 0 \qquad 5 + \frac{1}{\sqrt{5}}(27.95) - \frac{1}{\sqrt{17}}(20.62) + \frac{1}{\sqrt{17}} F_{EG} + \frac{1}{\sqrt{5}} F_{FG} = 0$$

$$+ \uparrow \sum F_y = 0 \qquad -40 + \frac{2}{\sqrt{5}}(27.95) - \frac{4}{\sqrt{17}}(20.62) - \frac{4}{\sqrt{17}} F_{EG} - \frac{2}{\sqrt{5}} F_{FG} = 0$$

Solving these equations, we obtain

$$F_{EG} = -20.62 \text{ k} \quad \text{and} \quad F_{FG} = -16.77 \text{ k}$$

$$F_{EG} = 20.62 \text{ k (C)} \qquad \text{Ans.}$$

$$F_{FG} = 16.77 \text{ k (C)} \qquad \text{Ans.}$$

Joints E and F Finally, by considering the equilibrium, by inspection, of joints E and F (see Fig. 4.25(b)), we obtain

$$F_{BE} = 25 \text{ k (C)} \qquad \text{Ans.}$$

$$F_{EF} = 10 \text{ k (T)} \qquad \text{Ans.}$$

$$F_{BF} = 16.77 \text{ k (C)} \qquad \text{Ans.}$$

Example 4.11

Determine the force in each member of the Fink truss shown in Fig. 4.26(a).

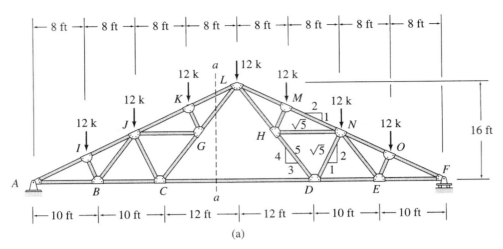

(a)

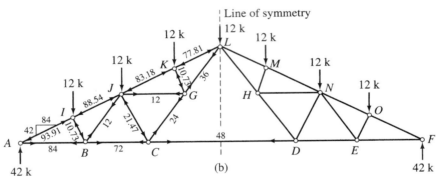

(b)

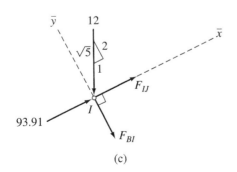

(c)

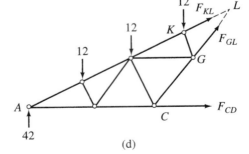

(d)

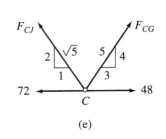

(e)

FIG. 4.26

Solution

The Fink truss shown in Fig. 4.26(a) is a compound truss formed by connecting two simple trusses, ACL and DFL, by a common joint L and a member CD.

Static Determinacy The truss contains 27 members and 15 joints and is supported by 3 reactions. Because $m + r = 2j$ and the reactions and the members of the truss are properly arranged, it is statically determinate.

Reactions The reactions at supports A and F of the truss, as computed by applying the three equations of equilibrium to the free-body diagram of the entire truss (Fig. 4.26(b)), are

$$A_x = 0 \qquad A_y = 42 \text{ k} \uparrow \qquad F_y = 42 \text{ k} \uparrow$$

Joint A The method of joints can now be started at joint A, which has only two unknown forces, F_{AB} and F_{AI}, acting on it. By inspection of the forces acting at this joint (see Fig. 4.26(b)), we obtain the following:

$$F_{AI} = 93.91 \text{ k } (C) \qquad\qquad \text{Ans.}$$

$$F_{AB} = 84 \text{ k } (T) \qquad\qquad \text{Ans.}$$

Joint I The free-body diagram of joint I is shown in Fig. 4.26(c). Member BI is perpendicular to members AI and IJ, which are collinear, so the computation of member forces can be simplified by using an $\bar{x}$ axis in the direction of the collinear members, as shown in Fig. 4.26(c).

$$+\nwarrow \sum F_{\bar{y}} = 0 \qquad -\frac{2}{\sqrt{5}}(12) - F_{BI} = 0$$

$$F_{BI} = -10.73 \text{ k}$$

$$F_{BI} = 10.73 \text{ k } (C) \qquad\qquad \text{Ans.}$$

$$+\nearrow \sum F_{\bar{x}} = 0 \qquad 93.91 - \frac{1}{\sqrt{5}}(12) + F_{IJ} = 0$$

$$F_{IJ} = -88.54 \text{ k}$$

$$F_{IJ} = 88.54 \text{ k } (C) \qquad\qquad \text{Ans.}$$

Joint B Considering the equilibrium of joint B, we obtain (see Fig. 4.26(b)) the following:

$$+\uparrow \sum F_y = 0 \qquad -\frac{2}{\sqrt{5}}(10.73) + \frac{4}{5}F_{BJ} = 0$$

$$F_{BJ} = 12 \text{ k } (T) \qquad\qquad \text{Ans.}$$

$$+\rightarrow \sum F_x = 0 \qquad -84 + \frac{1}{\sqrt{5}}(10.73) + \frac{3}{5}(12) + F_{BC} = 0$$

$$F_{BC} = 72 \text{ k } (T) \qquad\qquad \text{Ans.}$$

Section aa Since at each of the next two joints, C and J, there are three unknowns (F_{CD}, F_{CG}, and F_{CJ} at joint C and F_{CJ}, F_{GJ}, and F_{JK} at joint J), we calculate F_{CD} by using section aa, as shown in Fig. 4.26(a). (If we moved to joint F and started computing member forces from that end of the truss, we would encounter similar difficulties at joints D and N.)

continued

The free-body diagram of the portion of the truss on the left side of section *aa* is shown in Fig. 4.26(d). We determine F_{CD} by summing moments about point L, the point of intersection of the lines of action of F_{GL} and F_{KL}.

$$+\circlearrowleft \sum M_L = 0 \qquad -42(32) + 12(24) + 12(16) + 12(8) + F_{CD}(16) = 0$$

$$F_{CD} = 48 \text{ k (T)} \qquad \text{Ans.}$$

Joint C With F_{CD} now known, there are only two unknowns, F_{CG} and F_{CJ}, at joint C. These forces can be determined by applying the two equations of equilibrium to the free body of joint C, as shown in Fig. 4.26(e).

$$+\uparrow \sum F_y = 0 \qquad \frac{2}{\sqrt{5}}F_{CJ} + \frac{4}{5}F_{CG} = 0$$

$$+\rightarrow \sum F_x = 0 \qquad -72 + 48 - \frac{1}{\sqrt{5}}F_{CJ} + \frac{3}{5}F_{CG} = 0$$

Solving these equations simultaneously, we obtain

$$F_{CJ} = -21.47 \text{ k} \quad \text{and} \quad F_{CG} = 24 \text{ k}$$

$$F_{CJ} = 21.47 \text{ k (C)} \qquad \text{Ans.}$$

$$F_{CG} = 24 \text{ k (T)} \qquad \text{Ans.}$$

Joints J, K, and G Similarly, by successively considering the equilibrium of joints J, K, and G, in that order, we determine the following:

$$F_{JK} = 83.18 \text{ k (C)} \qquad \text{Ans.}$$

$$F_{GJ} = 12 \text{ k (T)} \qquad \text{Ans.}$$

$$F_{KL} = 77.81 \text{ k (C)} \qquad \text{Ans.}$$

$$F_{GK} = 10.73 \text{ k (C)} \qquad \text{Ans.}$$

$$F_{GL} = 36 \text{ k (T)} \qquad \text{Ans.}$$

Symmetry Since the geometry of the truss and the applied loading are symmetrical about the center line of the truss (shown in Fig. 4.26(b)), its member forces will also be symmetrical with respect to the line of symmetry. It is, therefore, sufficient to determine member forces in only one-half of the truss. The member forces determined here for the left half of the truss are shown in Fig. 4.26(b). The forces in the right half can be obtained from the consideration of symmetry; for example, the force in member MN is equal to that in member JK, and so forth. The reader is urged to verify this by computing a few member forces in the right half of the truss.   **Ans.**

4.8 COMPLEX TRUSSES

Trusses that can be classified neither as simple trusses nor as compound trusses are referred to as *complex trusses*. Two examples of complex trusses are shown in Fig. 4.27. From an analytical viewpoint, the main difference between simple or compound trusses and complex trusses

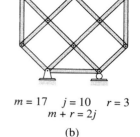

$$m = 9 \quad j = 6 \quad r = 3$$
$$m + r = 2j$$

(a)

$$m = 17 \quad j = 10 \quad r = 3$$
$$m + r = 2j$$

(b)

FIG. **4.27** Complex Trusses

stems from the fact that the methods of joints and sections, as described previously, cannot be used for the analysis of complex trusses. We can see from Fig. 4.27 that although the two complex trusses shown are statically determinate, after the computation of reactions the method of joints cannot be applied because we cannot find a joint at which there are two or fewer unknown member forces. Likewise, the method of sections cannot be employed, because every section would pass through more than three members with unknown forces. The member forces in such trusses can be determined by writing two equilibrium equations in terms of unknown member forces for each joint of the truss and then solving the system of $2j$ equations simultaneously. Today, complex trusses are usually analyzed on computers using the matrix formulation presented in Chapter 18.

4.9 SPACE TRUSSES

Space trusses, because of their shape, arrangement of members, or applied loading, cannot be subdivided into plane trusses for the purposes of analysis and must, therefore, be analyzed as three-dimensional structures subjected to three-dimensional force systems. As stated in Section 4.1, to simplify the analysis of space trusses, it is assumed that the truss members are connected at their ends by frictionless ball-and-socket joints, all external loads and reactions are applied only at the joints, and the centroidal axis of each member coincides with the line connecting the centers of the adjacent joints. Because of these simplifying assumptions, the members of space trusses can be treated as axial force members.

The simplest internally stable (or rigid) space truss can be formed by connecting six members at their ends by four ball-and-socket joints to form a *tetrahedron*, as shown in Fig. 4.28(a). This tetrahedron truss may be considered as the *basic space truss element*. It should be realized that this basic space truss is internally stable in the sense that it is a three-dimensional rigid body that will not change its shape under a general

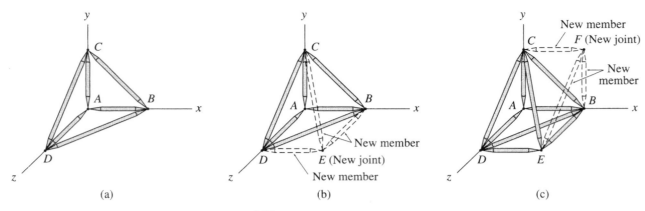

FIG. 4.28 Simple Space Truss

three-dimensional loading applied at its joints. The basic truss *ABCD* of Fig. 4.28(a) can be enlarged by attaching three new members, *BE, CE,* and *DE,* to three of the existing joints *B, C,* and *D,* and by connecting them to form a new joint *E,* as depicted in Fig. 4.28(b). As long as the new joint *E* does not lie in the plane containing the existing joints *B, C,* and *D,* the new enlarged truss will be internally stable. The truss can be further enlarged by repeating the same procedure (as shown in Fig. 4.28(c)) as many times as desired. Trusses constructed by this procedure are termed *simple space trusses.*

A simple space truss is formed by enlarging the basic tetrahedron element containing six members and four joints by adding three additional members for each additional joint, so the total number of members *m* in a simple space truss is given by

$$m = 6 + 3(j - 4) = 3j - 6 \qquad (4.5)$$

in which j = total number of joints (including those attached to the supports).

Reactions

The types of supports commonly used for space trusses are depicted in Fig. 4.29. The number and directions of the reaction forces that a support may exert on the truss depend on the number and directions of the translations it prevents.

As suggested in Section 3.1, in order for an internally stable space structure to be in equilibrium under a general system of three-dimensional forces, it must be supported by at least six reactions that satisfy the six equations of equilibrium (Eq. (3.1)):

$$\sum F_x = 0 \qquad \sum F_y = 0 \qquad \sum F_z = 0$$
$$\sum M_x = 0 \qquad \sum M_y = 0 \qquad \sum M_z = 0$$

Category	Type of support	Symbolic representation	Reactions	Number of unknowns
I	Ball	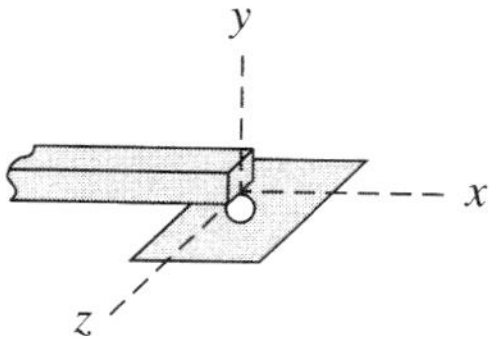	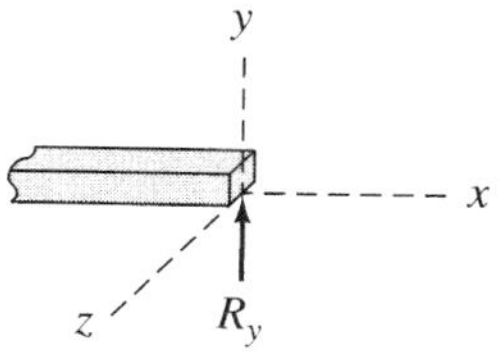	**1** The reaction force R_y acts perpendicular to the supporting surface and may be directed either into or away from the structure. The magnitude of R_y is the unknown.
	Link	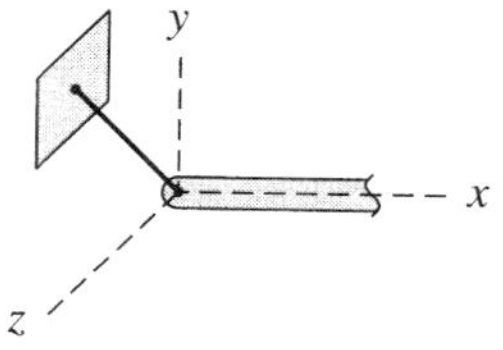	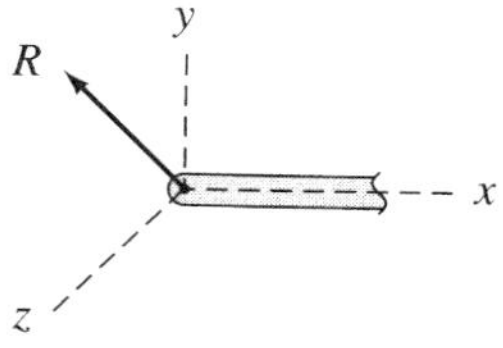	**1** The reaction force R acts in the direction of the link and may be directed either into or away from the structure. The magnitude of R is the unknown.
II	Roller	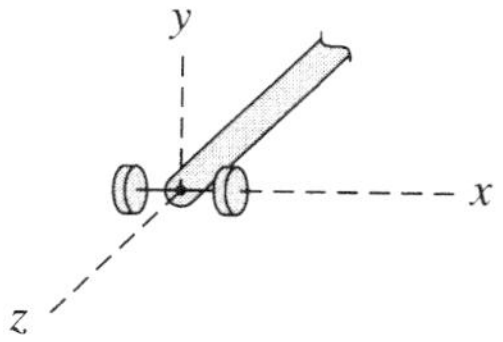	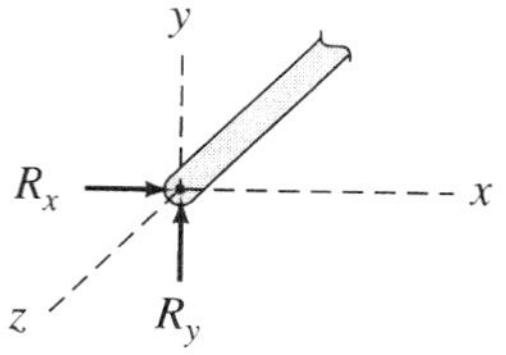	**2** Two reaction force components R_x and R_y act in a plane perpendicular to the direction in which the roller is free to roll. The magnitudes of R_x and R_y are the two unknowns.
III	Ball and socket	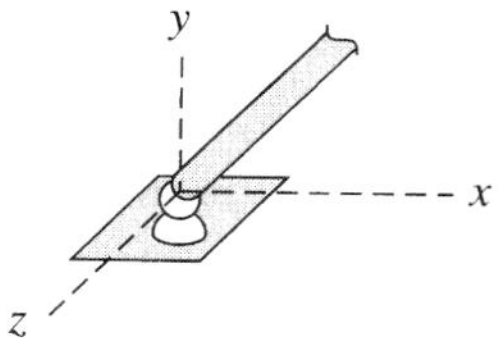	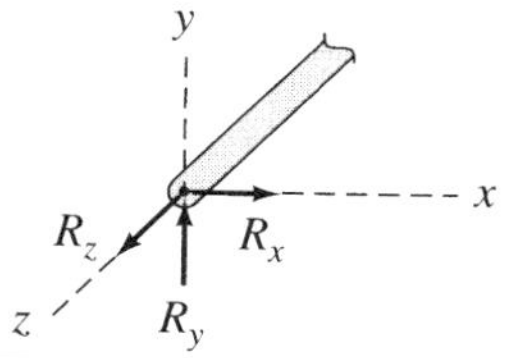	**3** The reaction force R may act in any direction. It is usually represented by its rectangular components, R_x, R_y, and R_z. The magnitudes of R_x, R_y, and R_z are the three unknowns.

FIG. 4.29 Types of Supports for Space Trusses

Because there are only six equilibrium equations, they cannot be used to determine more than six reactions. Thus, an internally stable space structure that is statically determinate externally must be supported by exactly six reactions. If a space structure is supported by more than six reactions, then all the reactions cannot be determined from the six equilibrium equations, and such a structure is termed statically indeterminate externally. Conversely, if a space structure is supported by fewer than six reactions, the reactions are not sufficient to prevent all possible movements of the structure in three-dimensional space, and such a structure is referred to as statically unstable externally. Thus, if

$r < 6$ the space structure is statically unstable externally

$r = 6$ the space structure is statically determinate externally   (4.6)

$r > 6$ the space structure is statically indeterminate externally

where r = number of reactions.

As in the case of plane structures discussed in the previous chapter, the conditions for static determinacy and indeterminacy, as given in Eq. (4.6), are necessary but not sufficient. In order for a space structure to be geometrically stable externally, the reactions must be properly arranged so that they can prevent translations in the directions of, as well as rotations about, each of the three coordinate axes. For example, if the lines of action of all the reactions of a space structure are either parallel or intersect a common axis, the structure would be geometrically unstable.

Static Determinacy, Indeterminacy, and Instability

If a space truss contains m members and is supported by r external reactions, then for its analysis we need to determine a total of $m + r$ unknown forces. Since the truss is in equilibrium, each of its joints must also be in equilibrium. At each joint, the internal and external forces form a three-dimensional concurrent force system that must satisfy the three equations of equilibrium, $\sum F_x = 0$, $\sum F_y = 0$, and $\sum F_z = 0$. Therefore, if the truss contains j joints, the total number of equilibrium equations available is $3j$. If $m + r = 3j$, all the unknowns can be determined by solving the $3j$ equations of equilibrium, and the truss is statically determinate.

Space trusses containing more unknowns than the available equilibrium equations $(m + r > 3j)$ are statically indeterminate, and those with fewer unknowns than the equilibrium equations $(m + r < 3j)$ are statically unstable. Thus, the conditions of static instability, determinacy, and indeterminacy of space trusses can be summarized as follows:

$$m + r < 3j \quad \text{statically unstable space truss}$$
$$m + r = 3j \quad \text{statically determinate space truss} \qquad (4.7)$$
$$m + r > 3j \quad \text{statically indeterminate space truss}$$

In order for the criteria for static determinacy and indeterminacy, as given by Eq. (4.7), to be valid, the truss must be stable and act as a single rigid body, under a general three-dimensional system of loads, when attached to the supports.

Analysis of Member Forces

The two methods for analysis of plane trusses discussed in Sections 4.5 and 4.6 can be extended to the analysis of space trusses. The *method of joints* essentially remains the same, except that three equilibrium equations $(\sum F_x = 0, \sum F_y = 0, \text{and} \sum F_z = 0)$ must now be satisfied at each joint of the space truss. Since the three equilibrium equations cannot be used to determine more than three unknown forces, the analysis is

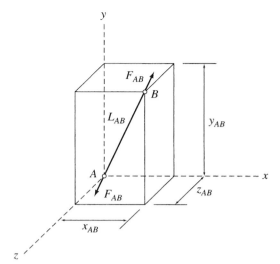

FIG. 4.30

started at a joint that has a maximum of three unknown forces (which must not be coplanar) acting on it. The three unknowns are determined by applying the three equations of equilibrium. We then proceed from joint to joint, computing three or fewer unknown forces at each subsequent joint, until all the desired forces have been determined.

Since it is difficult to visualize the orientations of inclined members in three-dimensional space, it is usually convenient to express the rectangular components of forces in such members in terms of the projections of member lengths in the $x, y,$ and z directions. Consider a member AB of a space truss, as shown in Fig. 4.30. The projections of its length L_{AB} in the $x, y,$ and z directions are $x_{AB}, y_{AB},$ and $z_{AB},$ respectively, as shown, with

$$L_{AB} = \sqrt{(x_{AB})^2 + (y_{AB})^2 + (z_{AB})^2}$$

Because the force F_{AB} acts in the direction of the member, its components $F_{xAB}, F_{yAB},$ and F_{zAB} in the $x, y,$ and z directions, respectively, can be expressed as

$$F_{xAB} = F_{AB}\left(\frac{x_{AB}}{L_{AB}}\right)$$

$$F_{yAB} = F_{AB}\left(\frac{y_{AB}}{L_{AB}}\right)$$

$$F_{zAB} = F_{AB}\left(\frac{z_{AB}}{L_{AB}}\right)$$

and the resultant force F_{AB} is given by

$$F_{AB} = \sqrt{(F_{xAB})^2 + (F_{yAB})^2 + (F_{zAB})^2}$$

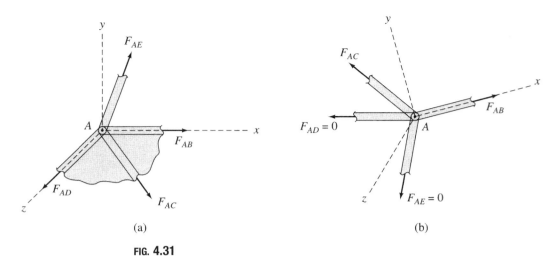

(a)   (b)

FIG. 4.31

The analysis of space trusses can be expedited by identifying the zero-force members by inspection. Two common types of member arrangements that result in zero-force members are the following:

1.   If all but one of the members connected to a joint lie in a single plane and no external loads or reactions are applied to the joint, then the force in the member that is not coplanar is zero.
2.   If all but two of the members connected to a joint have zero force and no external loads or reactions are applied to the joint, then unless the two remaining members are collinear, the force in each of them is also zero.

The first type of arrangement is shown in Fig. 4.31(a). It consists of four members AB, AC, AD, and AE connected to a joint A. Of these, AB, AC, and AD lie in the xz plane, whereas member AE does not. Note that no external loads or reactions are applied to joint A. It should be obvious that in order to satisfy the equilibrium equation $\sum F_y = 0$, the y component of F_{AE} must be zero, and therefore $F_{AE} = 0$.

The second type of arrangement is shown in Fig. 4.31(b). It consists of four members AB, AC, AD, and AE connected to a joint A, of which AD and AE are zero-force members, as shown. Note that no external loads or reactions are applied to the joint. By choosing the orientation of the x axis in the direction of member AB, we can see that the equilibrium equations $\sum F_y = 0$ and $\sum F_z = 0$ can be satisfied only if $F_{AC} = 0$. Because the x component of F_{AC} is zero, the equation $\sum F_x = 0$ is satisfied only if F_{AB} is also zero.

As in the case of plane trusses, the *method of sections* can be employed for determining forces in specific members of space trusses. An imaginary section is passed through the truss, cutting the members whose forces are desired. The desired member forces are then calculated by applying the six equations of equilibrium (Eq. (3.1)) to one of the two portions of the truss. No more than six unknown forces can be determined from the six equilibrium equations, so a section is generally chosen that does not pass through more than six members with unknown forces.

Because of the considerable amount of computational effort involved, the analysis of space trusses is performed today on computers. However, it is important to analyze at least a few relatively small space trusses manually to gain an understanding of the basic concepts involved in the analysis of such structures.

Example 4.12

Determine the reactions at the supports and the force in each member of the space truss shown in Fig. 4.32(a).

Solution

Static Determinacy The truss contains 9 members and 5 joints and is supported by 6 reactions. Because $m + r = 3j$ and the reactions and the members of the truss are properly arranged, it is statically determinate.

Member Projections The projections of the truss members in the x, y, and z directions, as obtained from Fig. 4.32(a), as well as their lengths computed from these projections, are tabulated in Table 4.1.

Zero-Force Members It can be seen from Fig. 4.32(a) that at joint D, three members, AD, CD, and DE, are connected. Of these members, AD and CD lie in the same (xz) plane, whereas DE does not. Since no external loads or reactions are applied at the joint, member DE is a zero-force member.

$$F_{DE} = 0$$ Ans.

Having identified DE as a zero-force member, we can see that since the two remaining members AD and CD are not collinear, they must also be zero-force members.

$$F_{AD} = 0$$ Ans.

$$F_{CD} = 0$$ Ans.

Reactions See Fig. 4.32(a).

$$+\nearrow \sum F_z = 0$$
$$B_z + 15 = 0$$
$$B_z = -15 \text{ k}$$
$$B_z = 15 \text{ k} \nearrow$$ Ans.

continued

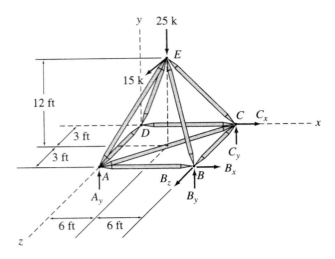

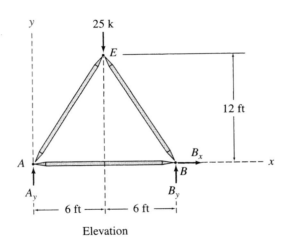

Elevation

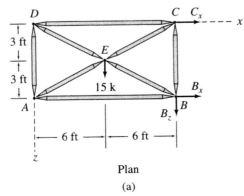

Plan

(a)

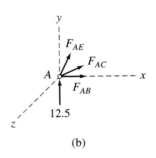

(b)

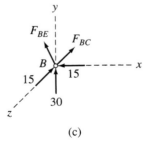

(c)

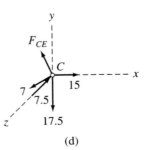

(d)

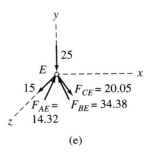

(e)

FIG. 4.32

TABLE 4.1

Member	Projection			Length (ft)
	x (ft)	y (ft)	z (ft)	
AB	12	0	0	12.0
BC	0	0	6	6.0
CD	12	0	0	12.0
AD	0	0	6	6.0
AC	12	0	6	13.42
AE	6	12	3	13.75
BE	6	12	3	13.75
CE	6	12	3	13.75
DE	6	12	3	13.75

$$+\zeta \sum M_y = 0$$

$$B_x(6) + 15(12) - 15(6) = 0$$

$$B_x = -15 \text{ k}$$

$$B_x = 15 \text{ k} \leftarrow \qquad \text{Ans.}$$

$$+ \rightarrow \sum F_x = 0$$

$$-15 + C_x = 0$$

$$C_x = 15 \text{ k} \rightarrow \qquad \text{Ans.}$$

$$+\zeta \sum M_x = 0$$

$$-A_y(6) - B_y(6) + 25(3) + 15(12) = 0$$

$$A_y + B_y = 42.5 \qquad (1)$$

$$+ \uparrow \sum F_y = 0$$

$$A_y + B_y + C_y - 25 = 0 \qquad (2)$$

By substituting Eq. (1) into Eq. (2), we obtain

$$C_y = -17.5 \text{ k}$$

$$C_y = 17.5 \text{ k} \downarrow \qquad \text{Ans.}$$

$$+\zeta \sum M_z = 0$$

$$B_y(12) - 17.5(12) - 25(6) = 0$$

$$B_y = 30 \text{ k} \uparrow \qquad \text{Ans.}$$

By substituting $B_y = 30$ into Eq. (1), we obtain A_y.

$$A_y = 12.5 \text{ k} \uparrow \qquad \text{Ans.}$$

continued

Joint A See Fig. 4.32(b).

$$+\uparrow \sum F_y = 0 \qquad 12.5 + \left(\frac{y_{AE}}{L_{AE}}\right) F_{AE} = 0$$

in which the second term on the left-hand side represents the y component of F_{AE}. Substituting the values of y and L for member AE from Table 4.1, we write

$$12.5 + \left(\frac{12}{13.75}\right) F_{AE} = 0$$

$$F_{AE} = -14.32 \text{ k}$$

$$F_{AE} = 14.32 \text{ k (C)} \qquad \text{Ans.}$$

Similarly, we apply the remaining equilibrium equations:

$$+\swarrow \sum F_z = 0 \qquad -\left(\frac{6}{13.42}\right) F_{AC} + \left(\frac{3}{13.75}\right)(14.32) = 0$$

$$F_{AC} = 7.0 \text{ k (T)} \qquad \text{Ans.}$$

$$+\rightarrow \sum F_x = 0 \qquad F_{AB} + \left(\frac{12}{13.42}\right)(7) - \left(\frac{6}{13.75}\right)(14.32) = 0$$

$$F_{AB} = 0 \qquad \text{Ans.}$$

Joint B (See Fig. 4.32(c).)

$$+\rightarrow \sum F_x = 0 \qquad -\left(\frac{6}{13.75}\right) F_{BE} - 15 = 0$$

$$F_{BE} = -34.38 \text{ k}$$

$$F_{BE} = 34.38 \text{ k (C)} \qquad \text{Ans.}$$

$$+\swarrow \sum F_z = 0 \qquad -15 - F_{BC} + \left(\frac{3}{13.75}\right)(34.38) = 0$$

$$F_{BC} = -7.5 \text{ k}$$

$$F_{BC} = 7.5 \text{ k (C)} \qquad \text{Ans.}$$

As all the unknown forces at joint B have been determined, we will use the remaining equilibrium equation to check our computations:

$$+\uparrow \sum F_y = 30 - \left(\frac{12}{13.75}\right)(34.38) = 0 \qquad \text{Checks}$$

Joint C See Fig. 4.32(d).

$$+\uparrow \sum F_y = 0 \qquad -17.5 + \left(\frac{12}{13.75}\right) F_{CE} = 0$$

$$F_{CE} = 20.05 \text{ k (T)} \qquad \text{Ans.}$$

Checking Computations At joint C (Fig. 4.32(d)),

$$+\rightarrow \Sigma F_x = 15 - \left(\frac{6}{13.75}\right)(20.05) - \left(\frac{12}{13.42}\right)(7) = 0 \qquad \text{Checks}$$

$$+\nearrow \Sigma F_z = -7.5 + \left(\frac{6}{13.42}\right)(7) + \left(\frac{3}{13.75}\right)(20.05) = 0 \qquad \text{Checks}$$

At joint E (Fig. 4.32(e)),

$$+\rightarrow \Sigma F_x = \frac{6}{13.75}(14.32 - 34.38 + 20.05) = 0 \qquad \text{Checks}$$

$$+\uparrow \Sigma F_y = -25 + \left(\frac{12}{13.75}\right)(14.32 + 34.38 - 20.05) = 0 \qquad \text{Checks}$$

$$+\nearrow \Sigma F_z = 15 - \left(\frac{3}{13.75}\right)(14.32 + 34.38 + 20.05) = 0 \qquad \text{Checks}$$

SUMMARY

A truss is defined as a structure that is composed of straight members connected at their ends by flexible connections to form a rigid configuration. The analysis of trusses is based on three simplifying assumptions:

1. All members are connected only at their ends by frictionless hinges in plane trusses and by frictionless ball-and-socket joints in space trusses.
2. All loads and reactions are applied only at the joints.
3. The centroidal axis of each member coincides with the line connecting the centers of the adjacent joints. The effect of these assumptions is that all the members of the truss can be treated as axial force members.

A truss is considered to be internally stable if the number and arrangement of its members is such that it does not change its shape and remains a rigid body when detached from its supports. The common types of equations of condition for plane trusses are described in Section 4.3.

A truss is considered to be statically determinate if all of its member forces and reactions can be determined by using the equations of equilibrium. If a plane truss contains m members, j joints, and is supported by r reactions, then if

$$m + r < 2j \quad \text{the truss is statically unstable}$$

$$m + r = 2j \quad \text{the truss is statically determinate} \qquad (4.4)$$

$$m + r > 2j \quad \text{the truss is statically indeterminate}$$

The degree of static indeterminacy is given by

$$i = (m + r) - 2j \tag{4.3}$$

The foregoing conditions for static determinacy and indeterminacy are necessary but not sufficient conditions. In order for these criteria to be valid, the truss must be stable and act as a single rigid body under a general system of coplanar loads when it is attached to the supports.

To analyze statically determinate plane trusses, we can use the method of joints, which essentially consists of selecting a joint with no more than two unknown forces acting on it and applying the two equilibrium equations to determine the unknown forces. We repeat the procedure until we obtain all desired forces. This method is most efficient when forces in all or most of the members of a truss are desired.

The method of sections usually proves to be more convenient when forces in only a few specific members of the truss are desired. This method essentially involves cutting the truss into two portions by passing an imaginary section through the members whose forces are desired and determining the desired forces by applying the three equations of equilibrium to the free body of one of the two portions of the truss.

The analysis of compound trusses can usually be expedited by using a combination of the method of joints and the method of sections. A procedure for the determination of reactions and member forces in space trusses is also presented.

PROBLEMS

Section 4.4

4.1 through 4.5 Classify each of the plane trusses shown as unstable, statically determinate, or statically indeterminate. If the truss is statically indeterminate, then determine the degree of static indeterminacy.

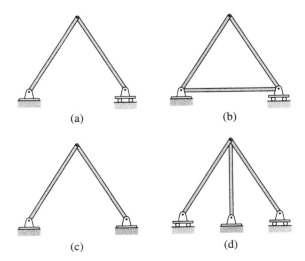

FIG. **P4.1**

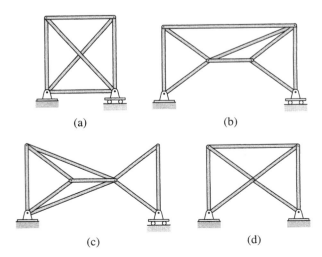

FIG. **P4.2**

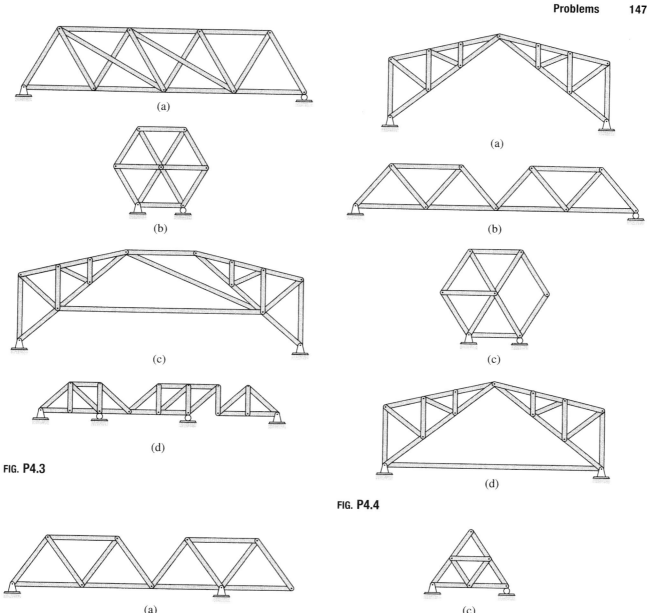

(a)

(b)

(c)

(d)

FIG. P4.3

(a)

(b)

(c)

(d)

FIG. P4.4

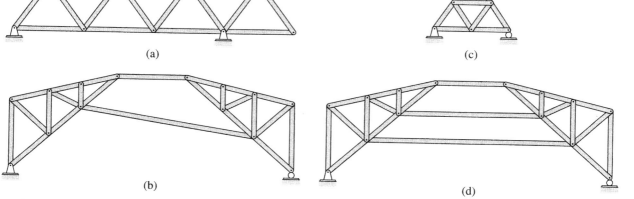

(a)

(c)

(b)

(d)

FIG. P4.5

Section 4.5

4.6 through 4.27 Determine the force in each member of the truss shown by the method of joints.

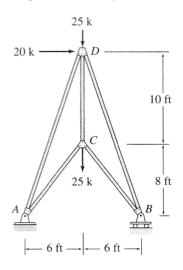

FIG. **P4.6**

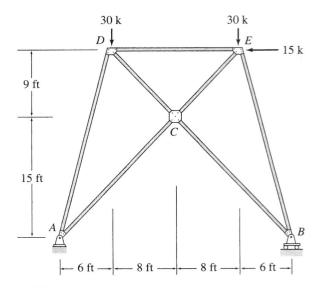

FIG. **P4.9**

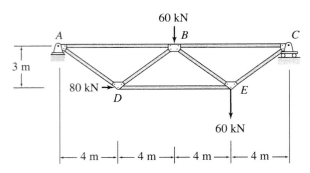

FIG. **P4.7**

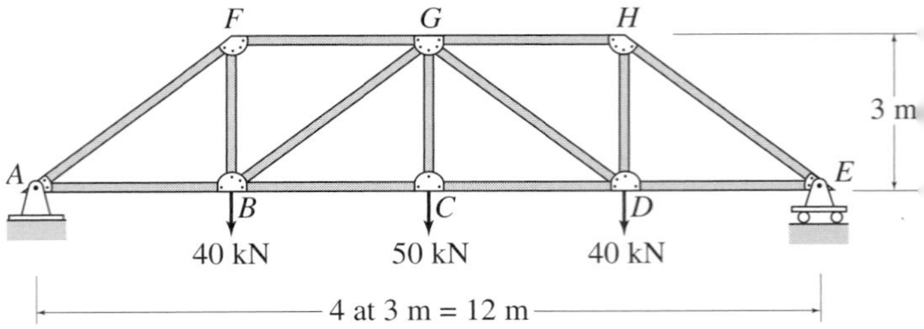

FIG. **P4.10**

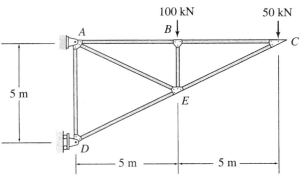

FIG. **P4.8**

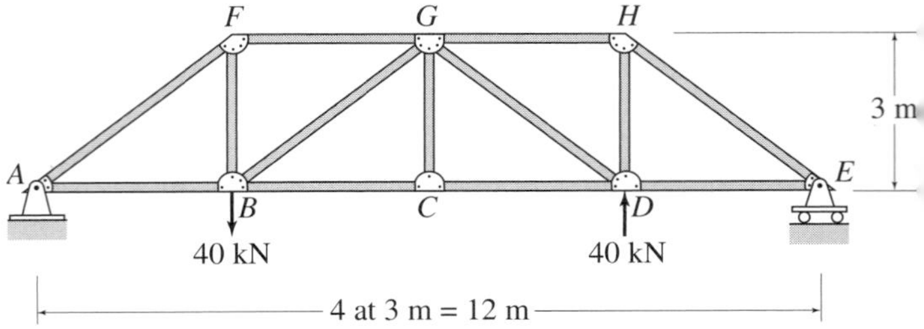

FIG. **P4.11**

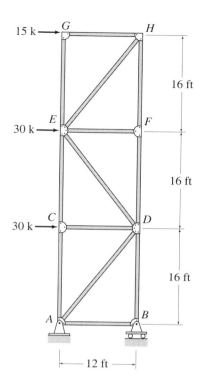

FIG. **P4.12**

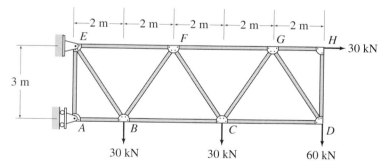

FIG. **P4.13**

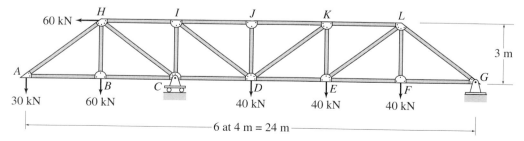

FIG. **P4.14**

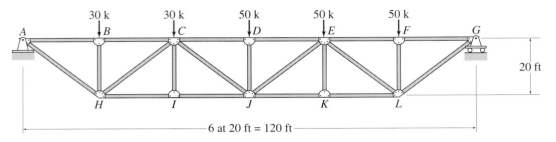

FIG. **P4.15**

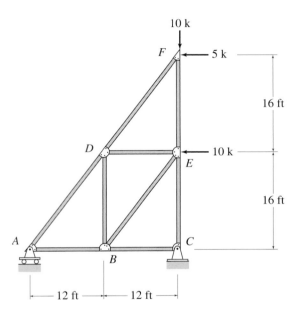

FIG. **P4.16**

FIG. **P4.18**

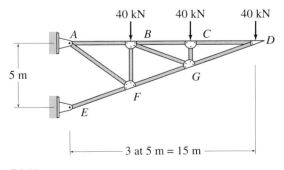

FIG. **P4.17**

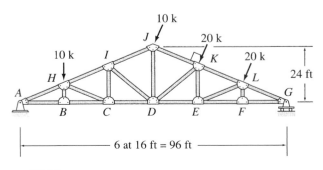

FIG. **P4.19**

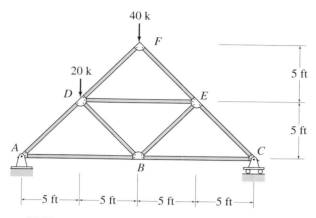

FIG. **P4.20**

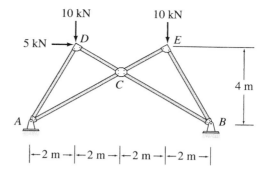

FIG. **P4.23**

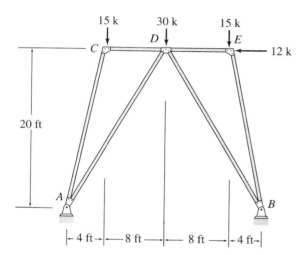

FIG. **P4.21**

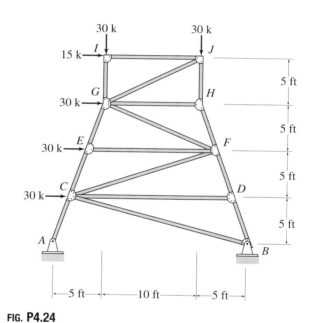

FIG. **P4.24**

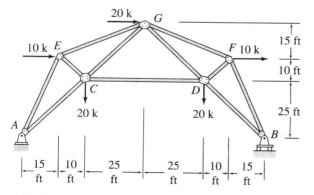

FIG. **P4.22**

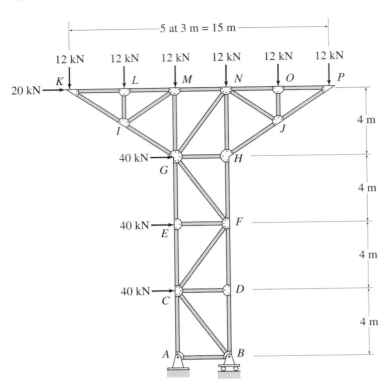

FIG. **P4.25**

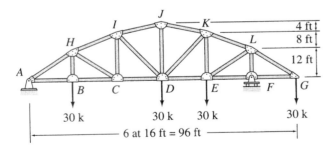

FIG. **P4.26**

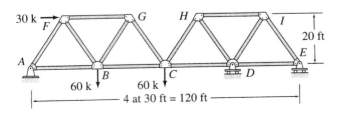

FIG. **P4.27**

4.28 Determine the force in each member of the truss supporting a floor deck as shown in Fig. P4.28. The deck is simply supported on floor beams which, in turn, are connected to the joints of the truss. Thus, the uniformly distributed loading on the deck is transmitted by the floor beams as concentrated loads to the top joints of the truss.

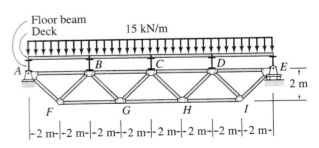

FIG. **P4.28**

4.29 and 4.30 Determine the force in each member of the roof truss shown. The roof is simply supported on purlins which, in turn, are attached to the joints of the top chord of the truss. Thus, the uniformly distributed loading on the roof is transmitted by the purlins as concentrated loads to the truss joints.

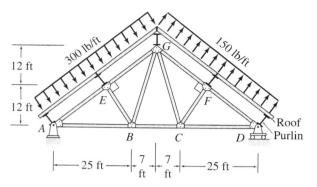

FIG. P4.29

Section 4.6

4.31 Determine the forces in the top chord member GH and the bottom chord member BC of the truss, if $h = 3$ ft. How would the forces in these members change if the height h of the truss was doubled to 6 ft?

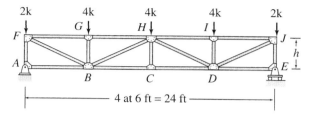

FIG. P4.31

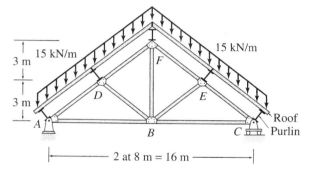

FIG. P4.30

4.32 through 4.45 Determine the forces in the members identified by "x" of the truss shown by the method of sections.

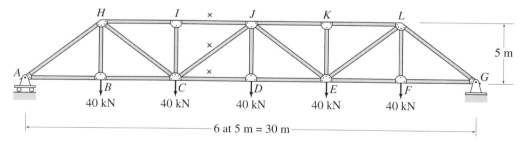

FIG. P4.32

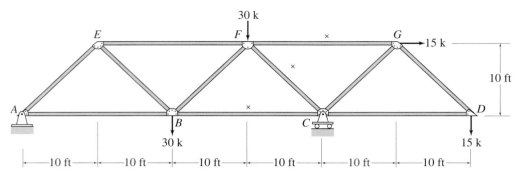

FIG. **P4.33**

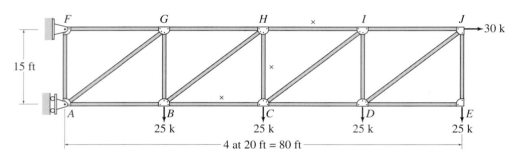

FIG. **P4.34**

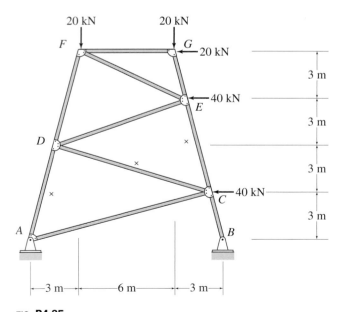

FIG. **P4.35**

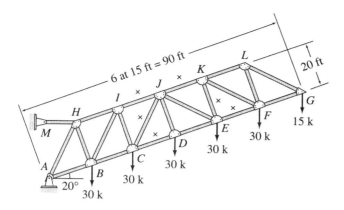

FIG. **P4.36**

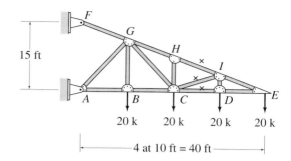

FIG. **P4.38**

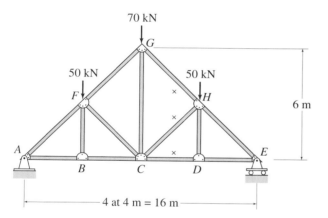

FIG. **P4.37**

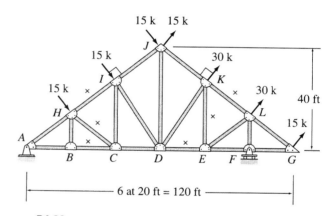

FIG. **P4.39**

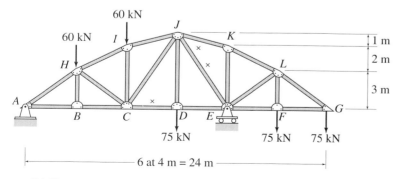

FIG. **P4.40**

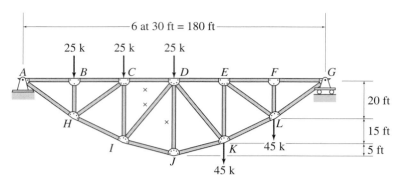

FIG. **P4.41**

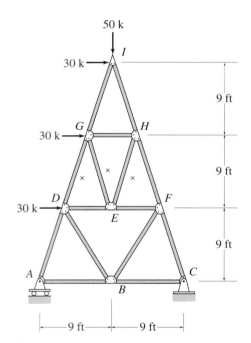

FIG. **P4.42**

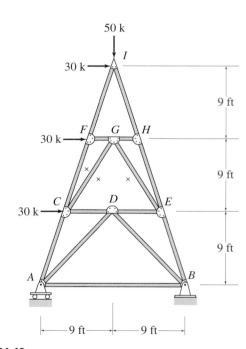

FIG. **P4.43**

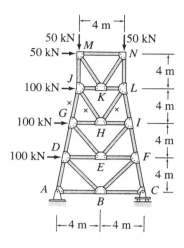

FIG. **P4.44**

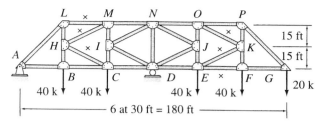

FIG. **P4.45**

Section 4.7

4.46 through 4.50 Determine the force in each member of the truss shown.

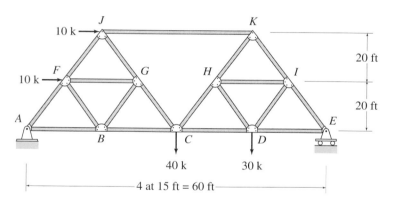

FIG. **P4.46**

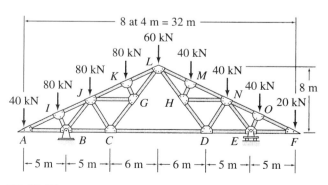

FIG. **P4.47**

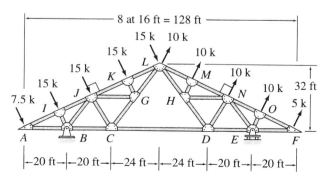

FIG. **P4.48**

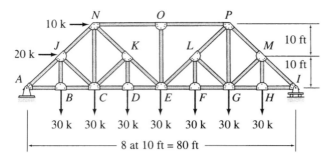

FIG. **P4.49**

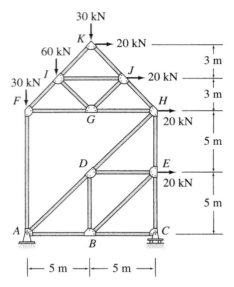

FIG. **P4.50**

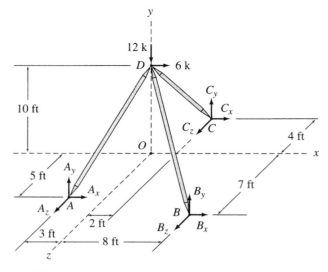

FIG. **P4.51**

Section 4.9

4.51 through 4.55 Determine the force in each member of the space truss shown.

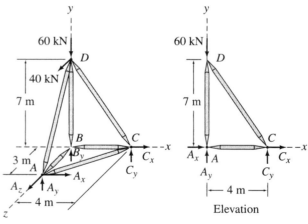

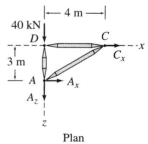

FIG. **P4.52**

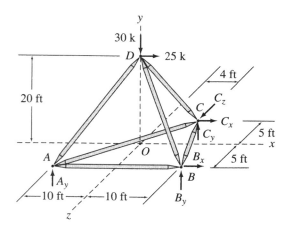

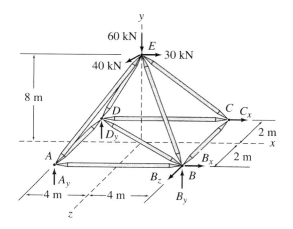

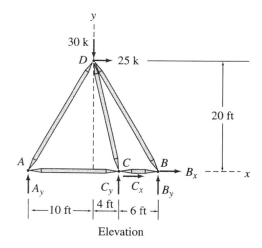

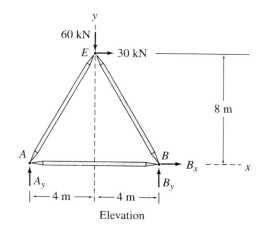

Elevation

Elevation

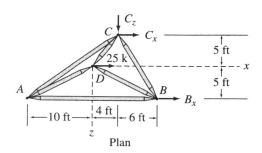

Plan

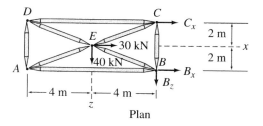

Plan

FIG. **P4.54**

FIG. **P4.53**

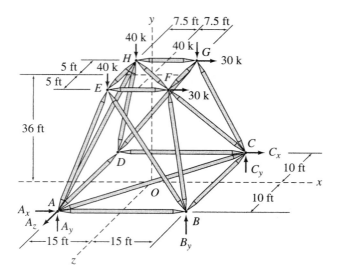

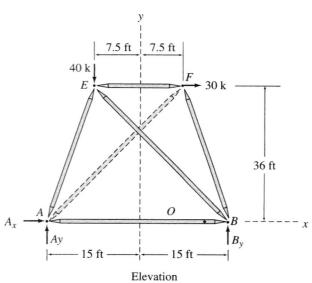

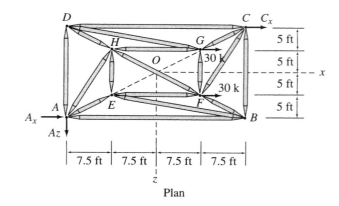

Elevation

Plan

FIG. P4.55

Steel Girder
Photo courtesy of Bethlehem Steel Corporation

5

Beams and Frames: Shear and Bending Moment

5.1 Axial Force, Shear, and Bending Moment
5.2 Shear and Bending Moment Diagrams
5.3 Qualitative Deflected Shapes
5.4 Relationships between Loads, Shears, and Bending Moments
5.5 Static Determinacy, Indeterminacy, and Instability of Plane Frames
5.6 Analysis of Plane Frames
Summary
Problems

Unlike trusses, considered in the preceding chapter, whose members are always subjected to only axial forces, the members of rigid frames and beams may be subjected to shear forces and bending moments as well as axial forces under the action of external loads. The determination of these internal forces and moments (stress resultants) is necessary for the design of such structures. The objective of this chapter is to present the analysis of internal forces and moments that may develop in beams, and the members of plane frames, under the action of coplanar systems of external forces and couples.

We begin by defining the three types of stress resultants—axial forces, shear forces, and bending moments—that may act on the cross sections of beams and the members of plane frames. We next discuss construction of the shear and bending moment diagrams by the method of sections. We also consider qualitative deflected shapes of beams and the relationships between loads, shears, and bending moments. In addition, we develop the procedures for constructing the shear and bending moment diagrams using these relationships. Finally we present the classification of plane frames as statically determinate, indeterminate, and unstable; and the analysis of statically determinate plane frames.

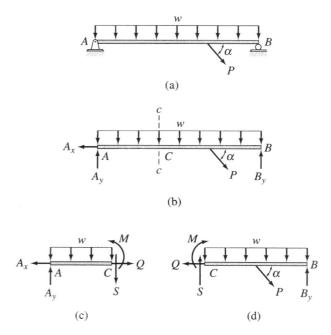

FIG. 5.1

5.1 AXIAL FORCE, SHEAR, AND BENDING MOMENT

Internal forces were defined in Section 3.2 as the forces and couples exerted on a portion of the structure by the rest of the structure. Consider, for example, the simply supported beam shown in Fig. 5.1(a). The free-body diagram of the entire beam is depicted in Fig. 5.1(b), which shows the external loads, as well as the reactions A_x and A_y, and B_y at supports A and B, respectively. As discussed in Chapter 3, the support reactions can be computed by applying the equations of equilibrium to the free body of the entire beam. In order to determine the internal forces acting on the cross section of the beam at a point C, we pass an imaginary section cc through C, thereby cutting the beam into two parts, AC and CB, as shown in Figs. 5.1(c) and 5.1(d). The free-body diagram of the portion AC (Fig. 5.1(c)) shows, in addition to the external loads and support reactions acting on the portion AC, the internal forces, Q, S, and M exerted upon portion AC at C by the removed portion of the structure. Note that without these internal forces, portion AC is not in equilibrium. Also, under a general coplanar system of external loads and reactions, three internal forces (two perpendicular force components and a couple) are necessary at a section to maintain a portion of the beam in equilibrium. The two internal force components are usually oriented in the direction of, and perpendicular to, the centroidal axis of the beam at the section under consideration, as shown in Fig. 5.1(c). The internal force Q in the direction of the centroidal axis of the beam is called the *axial force*, and the internal force S in the direction perpendicular to the centroidal axis is referred to as the *shear force* (or, simply, *shear*). The moment M of the internal couple is termed the *bending moment*. Recall from *mechanics of materials* that these internal forces, Q, S, and M, represent the resultants of the stress distribution acting on the cross section of the beam.

The free-body diagram of the portion CB of the beam is shown in Fig. 5.1(d). Note that this diagram shows the same internal forces, Q, S, and M, but in opposite directions, being exerted upon portion CB at C by the removed portion AC in accordance with Newton's third law. The magnitudes and the correct senses of the internal forces can be determined by simply applying the three equations of equilibrium, $\sum F_x = 0$, $\sum F_y = 0$, and $\sum M = 0$, to one of the two portions (AC or CB) of the beam.

It can be seen from Figs. 5.1(c) and 5.1(d), that in order for the equilibrium equation $\sum F_x = 0$ to be satisfied for a portion of the beam, the internal axial force Q must be equal in magnitude (but opposite in direction) to the algebraic sum (resultant) of the components in the direction parallel to the axis of the beam of all the external forces acting on that portion. Since the entire beam is in equilibrium—that is, $\sum F_x = 0$ for the entire beam—the application of $\sum F_x = 0$ individually to its two portions will yield the same magnitude of the axial force Q. Thus, we can state the following:

> The internal axial force Q at any section of a beam is equal in magnitude but opposite in direction to the algebraic sum (resultant) of the components in the direction parallel to the axis of the beam of all the external loads and support reactions acting on either side of the section under consideration.

Using similar reasoning, we can define the shear and bending moment as follows:

> The shear S at any section of a beam is equal in magnitude but opposite in direction to the algebraic sum (resultant) of the components in the direction perpendicular to the axis of the beam of all the external loads and support reactions acting on either side of the section under consideration.
>
> The bending moment M at any section of a beam is equal in magnitude but opposite in direction to the algebraic sum of the moments about (the centroid of the cross section of the beam at) the section under consideration of all the external loads and support reactions acting on either side of the section.

Sign Convention

The sign convention commonly used for the axial forces, shears, and bending moments is depicted in Fig. 5.2. An important feature of this sign convention, which is often referred to as the *beam convention*, is that it yields the same (positive or negative) results regardless of which

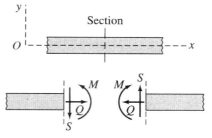

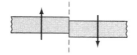

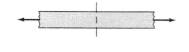

(a) Positive Internal Axial Force, Shear and Bending Moment at a Section

(b) External Forces Causing Positive Axial Force

(c) External Forces Causing Positive Shear

(d) External Forces Causing Positive Bending Moment

FIG. 5.2 Beam Convention

side of the section is considered for computing the internal forces. The positive directions of the internal forces acting on the portions of the member on each side of the section are shown in Fig. 5.2(a).

From a computational viewpoint, however, it is usually more convenient to express this sign convention in terms of the external loads and reactions acting on the beam or frame member, as shown in Fig. 5.2(b) to 5.2(d). As indicated in Fig. 5.2(b), the *internal axial force Q is considered to be positive when the external forces acting on the member produce tension or have the tendency to pull the member apart at the section.*

As shown in Fig. 5.2(c), *the shear S is considered to be positive when the external forces tend to push the portion of the member on the left of the section upward with respect to the portion on the right of the section.* It can be seen from this figure that an external force that acts upward on the left portion or downward on the right portion causes positive shear. Alternatively, this sign convention for shear can be remembered by realizing that any force that produces clockwise moment about a section causes positive shear at that section and vice versa.

The positive bending moment is shown in Fig. 5.2(d). *The bending moment M is considered to be positive when the external forces and couples tend to bend the beam concave upward, causing compression in the upper fibers and tension in the lower fibers of the beam at the section.* When the left portion is used for computing the bending moment, the forces acting on the portion that produce clockwise moments about the section, as well as clockwise couples, cause positive bending moment at the section. When the right portion is considered, however, the forces producing counterclockwise moments about the section, and counter-clockwise couples, cause positive bending moment and vice versa.

In our discussion thus far, the beam or frame member has been as-sumed to be horizontal, but the foregoing sign convention can be used for inclined and vertical members by employing an xy coordinate sys-tem, as shown in Fig. 5.2(a). The x axis of the coordinate system is ori-ented in the direction of the centroidal axis of the member, and the positive direction of the y axis is chosen so that the coordinate system is right-handed, with the z axis always pointing out of the plane of the paper. The sign convention can now be used for an inclined or a vertical member by considering the positive y direction as the upward direction and the portion of the member near the origin O as the portion to the left of the section.

Procedure for Analysis

The procedure for determining internal forces at a specified location on a beam can be summarized as follows:

1. Compute the support reactions by applying the equations of equi-librium and condition (if any) to the free body of the entire beam. (In cantilever beams, this step can be avoided by selecting the free, or externally unsupported, portion of the beam for analysis; see Example 5.2.)

2. Pass a section perpendicular to the centroidal axis of the beam at the point where the internal forces are desired, thereby cutting the beam into two portions.

3. Although either of the two portions of the beam can be used for computing internal forces, we should select the portion that will re-quire the least amount of computational effort, such as the portion that does not have any reactions acting on it or that has the least number of external loads and reactions applied to it.

4. Determine the axial force at the section by algebraically summing the components in the direction parallel to the axis of the beam of all the external loads and support reactions acting on the selected portion. According to the sign convention adopted in the preceding paragraphs, if the portion of the beam to the left of the section is being used for computing the axial force, then the external forces acting to the left are considered positive, whereas the external forces acting to the right are considered to be negative (see Fig. 5.2(b)). If the right portion is being used for analysis, then the external forces acting to the right are considered to be positive and vice versa.

5. Determine the shear at the section by algebraically summing the components in the direction perpendicular to the axis of the beam of all the external loads and reactions acting on the selected por-tion. If the left portion of the beam is being used for analysis, then

the external forces acting upward are considered positive, whereas the external forces acting downward are considered to be negative (see Fig. 5.2(c)). If the right portion has been selected for analysis, then the downward external forces are considered positive and vice versa.

6. Determine the bending moment at the section by algebraically summing the moments about the section of all the external forces plus the moments of any external couples acting on the selected portion. If the left portion is being used for analysis, then the clockwise moments are considered to be positive, and the counterclockwise moments are considered negative (see Fig. 5.2(d)). If the right portion has been selected for analysis, then the counterclockwise moments are considered positive and vice versa.

7. To check the calculations, values of some or all of the internal forces may be computed by using the portion of the beam not utilized in steps 4 through 6. If the analysis has been performed correctly, then the results based on both left and right portions must be identical.

Example 5.1

Determine the axial force, shear, and bending moment at point B of the beam shown in Fig. 5.3(a).

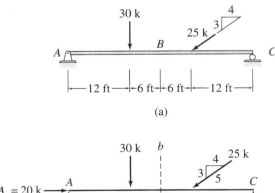

(a)

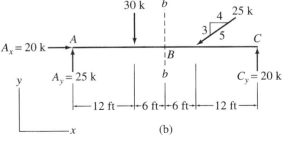

(b)

FIG. **5.3**

Solution

Reactions Considering the equilibrium of the free body of the entire beam (Fig. 5.3(b)), we write

$$+ \rightarrow \sum F_x = 0 \qquad\qquad A_x - \left(\frac{4}{5}\right)(25) = 0 \qquad A_x = 20 \text{ k} \rightarrow$$

$$+ \circlearrowleft \sum M_c = 0 \qquad -A_y(36) + 30(24) + \left(\frac{3}{5}\right)(25)(12) = 0 \qquad A_y = 25 \text{ k} \uparrow$$

$$+ \uparrow \sum F_y = 0 \qquad\qquad 25 - 30 - \left(\frac{3}{5}\right)(25) + C_y = 0 \qquad C_y = 20 \text{ k} \uparrow$$

Section bb A section *bb* is passed through point *B*, cutting the beam into two portions, *AB* and *BC* (see Fig. 5.3(b)). The portion *AB*, which is to the left of the section, is used here to compute the internal forces.

Axial Force Considering the external forces acting to the left as positive, we write

$$Q = -20 \text{ k} \qquad\qquad \text{Ans.}$$

Shear Considering the external forces acting upward as positive, we write

$$S = 25 - 30 = -5$$
$$S = -5 \text{ k} \qquad\qquad \text{Ans.}$$

Bending Moment Considering the clockwise moments of the external forces about *B* as positive, we write

$$M = 25(18) - 30(6) = 270$$
$$M = 270 \text{ k-ft} \qquad\qquad \text{Ans.}$$

Checking Computations To check our calculations, we compute the internal forces using portion *BC*, which is to the right of the section under consideration.

By considering the horizontal components of the external forces acting to the right on portion *BC* as positive, we obtain

$$Q = -\left(\frac{4}{5}\right)(25) = -20 \text{ k} \qquad\qquad \text{Checks}$$

By considering the external forces acting downward as positive, we obtain

$$S = -20 + \left(\frac{3}{5}\right)(25) = -5 \text{ k} \qquad\qquad \text{Checks}$$

Finally, by considering the counterclockwise moments of the external forces about *B* as positive, we obtain

$$M = 20(18) - \left(\frac{3}{5}\right)(25)(6) = 270 \text{ k-ft} \qquad\qquad \text{Checks}$$

Example 5.2

Determine the shear and bending moment at point B of the beam shown in Fig. 5.4.

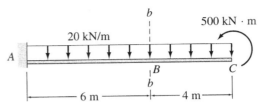

FIG. 5.4

Solution

Section bb (See Fig. 5.4.) To avoid computing reactions, we select externally unsupported portion BC, which is to the right of the section bb, for computing the internal forces.

Shear Considering the external forces acting downward as positive, we write

$$S = +20(4) = +80 \text{ kN}$$

$$S = 80 \text{ kN} \qquad \text{Ans.}$$

Note that the 500 kN · m couple does not have any effect on shear.

Bending Moment Considering the counterclockwise moments as positive, we write

$$M = 500 - 20(4)(2) = 340 \text{ kN} \cdot \text{m}$$

$$M = 340 \text{ kN} \cdot \text{m} \qquad \text{Ans.}$$

The reader may check the results by summing forces and moments on portion AB of the beam after computing the reactions at support A.

5.2 SHEAR AND BENDING MOMENT DIAGRAMS

Shear and bending moment diagrams depict the variations of these quantities along the length of the member. Such diagrams can be constructed by using the method of sections described in the preceding section. Proceeding from one end of the member to the other (usually from left to right), sections are passed, after each successive change in loading, along the length of the member to determine the equations expressing the shear and bending moment in terms of the distance of the section from a fixed origin. The values of shear and bending moments determined from these equations are then plotted as ordinates against the position with respect to a member end as abscissa to obtain the shear and bending moment diagrams. This procedure is illustrated by the following examples.

Example 5.3

Draw the shear and bending moment diagrams for the beam shown in Fig. 5.5(a).

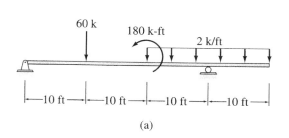

(a)

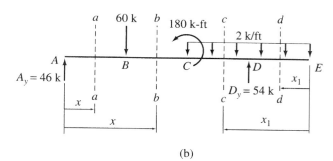

(b)

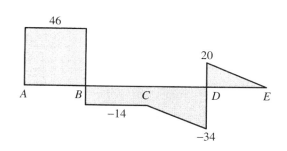

(c) Shear Diagram (k)

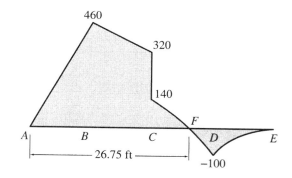

(d) Bending Moment Diagram (k-ft)

FIG. 5.5

Solution

Reactions See Fig. 5.5(b).

$$+ \rightarrow \sum F_x = 0 \qquad A_x = 0$$

$$+ \circlearrowleft \sum M_D = 0$$

$$-A_y(30) + 60(20) + 180 + 2(20)(0) = 0$$

$$A_y = 46 \text{ k} \uparrow$$

$$+ \uparrow \sum F_y = 0$$

$$46 - 60 - 2(20) + D_y = 0$$

$$D_y = 54 \text{ k} \uparrow$$

continued

Shear Diagram To determine the equation for shear in segment AB of the beam, we pass a section aa at a distance x from support A, as shown in Fig. 5.5(b). Considering the free body to the left of this section, we obtain

$$S = 46 \text{ k} \quad \text{for } 0 < x < 10 \text{ ft}$$

As this equation indicates, the shear is constant at 46 k from an infinitesimal distance to the right of point A to an infinitesimal distance to the left of point B. At point A, the shear increases abruptly from 0 to 46 k, so a vertical line is drawn from 0 to 46 on the shear diagram (Fig. 5.5(c)) at A to indicate this change. This is followed by a horizontal line from A to B to indicate that the shear remains constant in this segment.

Next, by using section bb (Fig. 5.5(b)), we determine the equation for shear in segment BC as

$$S = 46 - 60 = -14 \text{ k} \quad \text{for } 10 \text{ ft} < x \leq 20 \text{ ft}$$

The abrupt change in shear from 46 k at an infinitesimal distance to the left of B to -14 k at an infinitesimal distance to the right of B is shown on the shear diagram (Fig. 5.5(c)) by a vertical line from $+46$ to -14. A horizontal line at -14 is then drawn from B to C to indicate that the shear remains constant at this value throughout this segment.

To determine the equations for shear in the right half of the beam, it is convenient to use another coordinate, x_1, directed to the left from the end E of the beam, as shown in Fig. 5.5(b). The equations for shear in segments ED and DC are obtained by considering the free bodies to the right of sections dd and cc, respectively. Thus,

$$S = 2x_1 \quad \text{for } 0 \leq x_1 < 10 \text{ ft}$$

and

$$S = 2x_1 - 54 \quad \text{for } 10 \text{ ft} < x_1 \leq 20 \text{ ft}$$

These equations indicate that the shear increases linearly from zero at E to $+20$ k at an infinitesimal distance to the right of D; it then drops abruptly to -34 k at an infinitesimal distance to the left of D; and from there it increases linearly to -14 k at C. This information is plotted on the shear diagram, as shown in Fig. 5.5(c).   **Ans.**

Bending Moment Diagram Using the same sections and coordinates employed previously for computing shear, we determine the following equations for bending moment in the four segments of the beam. For segment AB:

$$M = 46x \quad \text{for } 0 \leq x \leq 10 \text{ ft}$$

For segment BC:

$$M = 46x - 60(x - 10) = -14x + 600 \quad \text{for } 10 \text{ ft} \leq x < 20 \text{ ft}$$

For segment ED:

$$M = -2x_1 \left(\frac{x_1}{2}\right) = -x_1^2 \quad \text{for } 0 \leq x_1 \leq 10 \text{ ft}$$

For segment DC:

$$M = -x_1^2 + 54(x_1 - 10) = -x_1^2 + 54x_1 - 540 \quad \text{for } 10 \text{ ft} \leq x_1 < 20 \text{ ft}$$

The first two equations, for the left half of the beam, indicate that the bending moment increases linearly from 0 at A to 460 k-ft at B; it then decreases linearly to 320 k-ft at C, as shown on the bending moment diagram in Fig. 5.5(d). The last two equations for the right half of the beam are quadratic in x_1. The values of M computed from these equations are plotted on the bending moment diagram shown in Fig. 5.5(d). It can be seen that M decreases from 0 at E to -100 k-ft at D, and it then increases to $+140$ k-ft at an infinitesimal distance to the right of C. Note that at C, the bending moment drops abruptly by an amount $320 - 140 = 180$ k-ft, which is equal to the magnitude of the moment of the counterclockwise external couple acting at this point.

A point at which the bending moment is zero is termed the *point of inflection*. To determine the location of the point of inflection F (Fig. 5.5(d)), we set $M = 0$ in the equation for bending moment in segment DC to obtain

$$M = -x_1^2 + 54x_1 - 540 = 0$$

from which $x_1 = 13.25$ ft; that is, point F is located at a distance of 13.25 ft from end E, or $40 - 13.25 = 26.75$ ft from support A of the beam, as shown in Fig. 5.5(d).   Ans.

Example 5.4

Draw the shear and bending moment diagrams for the beam shown in Fig. 5.6(a).

Solution

Reactions See Fig. 5.6(b).

$$+ \rightarrow \sum F_x = 0 \qquad B_x = 0$$

$$+ \circlearrowleft \sum M_c = 0$$

$$\left(\frac{1}{2}\right)(9)(27)\left(\frac{9}{3}\right) - B_y(6) = 0 \qquad B_y = 60.75 \text{ kN} \uparrow$$

$$+ \uparrow \sum F_y = 0$$

$$-\left(\frac{1}{2}\right)(9)(27) + 60.75 + C_y = 0 \qquad C_y = 60.75 \text{ kN} \uparrow$$

Shear Diagram To determine the equations for shear in segments AB and BC of the beam, we pass sections aa and bb through the beam, as shown in Fig. 5.6(b). Considering the free bodies to the left of these sections and realizing that the load intensity, $w(x)$, at a point at a distance x from end A is $w(x) = \left(\frac{27}{9}\right)x = 3x$ kN/m, we obtain the following equations for shear in segments AB and BC, respectively:

$$S = -\left(\frac{1}{2}\right)(x)(3x) = -\frac{3x^2}{2} \quad \text{for } 0 \leq x < 3 \text{ m}$$

$$S = -\left(\frac{3x^2}{2}\right) + 60.75 \qquad \text{for } 3 \text{ m} < x < 9 \text{ m}$$

continued

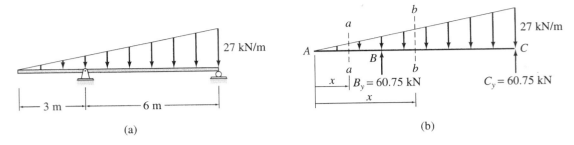

(a)   (b)

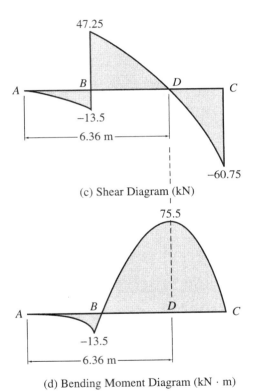

(c) Shear Diagram (kN)

(d) Bending Moment Diagram (kN · m)

FIG. **5.6**

The values of S computed from these equations are plotted to obtain the shear diagram shown in Fig. 5.6(c). The point D at which the shear is zero is obtained from the equation

$$S = -\left(\frac{3x^2}{2}\right) + 60.75 = 0$$

from which $x = 6.36$ m.   Ans.

Bending Moment Diagram Using the same sections employed previously for computing shear, we determine the following equations for bending moment in segments AB and BC, respectively:

$$M = -\left(\frac{1}{2}\right)(x)(3x)\left(\frac{x}{3}\right) = -\frac{x^3}{2} \quad \text{for } 0 \leq x \leq 3 \text{ m}$$

$$M = -\left(\frac{x^3}{2}\right) + 60.75(x-3) \quad \text{for } 3 \text{ m} \leq x \leq 9 \text{ m}$$

The values of M computed from these equations are plotted to obtain the bending moment diagram shown in Fig. 5.6(d). To locate the point at which the bending moment is maximum, we differentiate the equation for M in segment BC with respect to x and set the derivative dM/dx equal to zero; that is,

$$\frac{dM}{dx} = \left(-\frac{3x^2}{2}\right) + 60.75 = 0$$

from which $x = 6.36$ m. This indicates that the maximum bending moment occurs at the same point at which the shear is zero. Also, a comparison of the expressions for dM/dx and S in segment BC indicates that the two equations are identical; that is, the slope of the bending moment diagram at a point is equal to the shear at that point. (This relationship, which is generally valid, is discussed in detail in a subsequent section.)

Finally, the magnitude of the maximum moment is determined by substituting $x = 6.36$ m into the equation for M in segment BC:

$$M_{\max} = -\left[\frac{(6.36)^3}{2}\right] + 60.75(6.36 - 3) = 75.5 \text{ kN} \cdot \text{m} \qquad \text{Ans.}$$

5.3 QUALITATIVE DEFLECTED SHAPES

A *qualitative deflected shape* (*elastic curve*) of a structure is simply a rough (usually exaggerated) sketch of the neutral surface of the structure, in the deformed position, under the action of a given loading condition. Such sketches, which can be constructed without any knowledge of the numerical values of deflections, provide valuable insights into the behavior of structures and are often useful in computing the numerical values of deflections. (Procedures for the quantitative analysis of deflections are presented in the following chapters.)

According to the sign convention adopted in Section 5.1, a positive bending moment bends a beam concave upward (or toward the positive y direction), whereas a negative bending moment bends a beam concave downward (or toward the negative y direction). Thus, the sign (positive or negative) of the curvature at any point along the axis of a beam can be obtained from the bending moment diagram. Using the

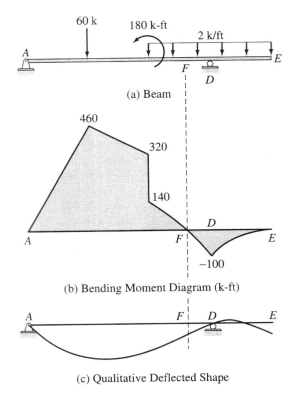

(a) Beam

(b) Bending Moment Diagram (k-ft)

(c) Qualitative Deflected Shape

FIG. 5.7

signs of curvatures, a qualitative deflected shape (elastic curve) of the beam, which is consistent with its support conditions, can be easily sketched (see Fig. 5.7).

For example, consider the beam analyzed in Example 5.3. The beam and its bending moment diagram are redrawn in Fig. 5.7(a) and (b), respectively. A qualitative deflected shape of the beam is shown in Fig. 5.7(c). Because the bending moment is positive in segment AF, the beam is bent concave upward in this region. Conversely, the bending moment is negative in segment FE; therefore, in this region, the beam is bent concave downward. Regarding the support conditions, note that at both supports A and D the deflection of the beam is zero, but its slope (rotation) is not zero at these points.

It is important to realize that a qualitative deflected shape is approximate, because it is based solely on the signs of curvatures; the numerical values of deflections along the axis of the beam are not known (except at supports). For example, numerical computations could possibly indicate that the end E of the beam actually deflects upward, instead of downward as assumed in Fig. 5.7(c).

5.4 RELATIONSHIPS BETWEEN LOADS, SHEARS, AND BENDING MOMENTS

The construction of shear and bending moment diagrams can be considerably expedited by using the basic differential relationships that exist between the loads, the shears, and the bending moments.

To derive these relationships, consider a beam subjected to an arbitrary loading, as shown in Fig. 5.8(a). All the external loads shown in this figure are assumed to be acting in their positive directions. As indicated in this figure, the external distributed and concentrated loads acting upward (in the positive y direction) are considered positive; the external couples acting clockwise are also considered to be positive and vice versa. Next, we consider the equilibrium of a differential element of length dx, isolated from the beam by passing imaginary sections at distances x and $x + dx$ from the origin O, as shown in Fig. 5.8(a). The free-body diagram of the element is shown in Fig. 5.8(b), in which S and M represent the shear and bending moment, respectively, acting on the left face of the element (i.e., at distance x from the origin O), and dS and dM denote the changes in shear and bending moment, respectively, over the distance dx. As the distance dx is infinitesimally small, the distributed load w acting on the element can be considered to be uniform of magnitude $w(x)$. In order for the element to be in equilibrium, the forces and couples acting on it must satisfy the two equations of equilibrium, $\sum F_y = 0$ and $\sum M = 0$. The third equilibrium equation, $\sum F_x = 0$, is

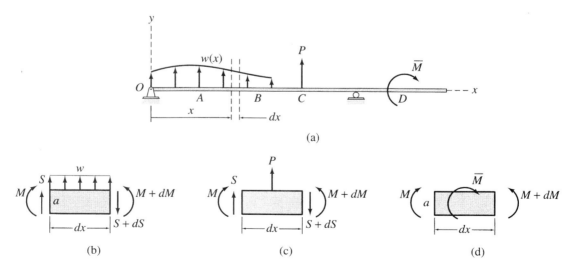

FIG. 5.8

automatically satisfied, since no horizontal forces are acting on the element. Applying the equilibrium equation $\sum F_y = 0$, we obtain

$$+\uparrow \sum F_y = 0$$

$$S + w\,dx - (S + dS) = 0$$

$$dS = w\,dx \tag{5.1}$$

Dividing by dx, we write Eq. (5.1) as

$$\frac{dS}{dx} = w \tag{5.2}$$

in which dS/dx represents the slope of the shear diagram. Thus, Eq. (5.2) can be expressed as

$$\begin{matrix}\text{slope of shear diagram} \\ \text{at a point}\end{matrix} = \begin{matrix}\text{intensity of distributed} \\ \text{load at that point}\end{matrix} \tag{5.3}$$

To determine the change in shear between points A and B along the axis of the member (see Fig. 5.8(a)), we integrate Eq. (5.1) from A to B to obtain

$$\int_A^B dS = S_B - S_A = \int_A^B w\,dx \tag{5.4}$$

in which $(S_B - S_A)$ represents the change in shear between points A and B and $\int_A^B w\,dx$ represents the area under the distributed load diagram between points A and B. Thus, Eq. (5.4) can be stated as

$$\begin{matrix}\text{change in shear between} \\ \text{points } A \text{ and } B\end{matrix} = \begin{matrix}\text{area under the distributed load} \\ \text{diagram between points } A \text{ and } B\end{matrix} \tag{5.5}$$

Applying the moment equilibrium equation to the free body of the beam element shown in Fig. 5.8(b), we write

$$+\curvearrowleft \sum M_a = 0 \qquad -M + w(dx)(dx/2) - (S + dS)\,dx + (M + dM) = 0$$

By neglecting the terms containing second-order differentials, we obtain

$$dM = S\,dx \tag{5.6}$$

which can also be written as

$$\frac{dM}{dx} = S \tag{5.7}$$

in which dM/dx represents the slope of the bending moment diagram. Thus, Eq. (5.7) can be stated as

$$\frac{\text{slope of bending moment}}{\text{diagram at a point}} = \text{shear at that point} \qquad (5.8)$$

To obtain the change in bending moment between points A and B (see Fig. 5.8(a)), we integrate Eq. (5.6) to obtain

$$\int_A^B dM = M_B - M_A = \int_A^B S\, dx \qquad (5.9)$$

in which $(M_B - M_A)$ represents the change in bending moment between points A and B and $\int_A^B S\, dx$ represents the area under the shear diagram between points A and B. Thus, Eq. (5.9) can be stated as

$$\frac{\text{change in bending moment}}{\text{between points } A \text{ and } B} = \frac{\text{area under the shear diagram}}{\text{between points } A \text{ and } B} \qquad (5.10)$$

Concentrated Loads

The relationships between the loads and shears derived thus far (Eqs. (5.1) through (5.5)) are not valid at the point of application of concentrated loads. As we illustrated in Example 5.3, at such a point the shear changes abruptly by an amount equal to the magnitude of the concentrated load. To verify this relationship, we consider the equilibrium of a differential element that is isolated from the beam of Fig. 5.8(a) by passing imaginary sections at infinitesimal distances to the left and to the right of the point of application C of the concentrated load P. The free-body diagram of this element is shown in Fig. 5.8(c). Applying the equilibrium equation $\sum F_y = 0$, we obtain

$$+\uparrow \sum F_y = 0$$
$$S + P - (S + dS) = 0$$
$$dS = P \qquad (5.11)$$

which can be stated as

$$\frac{\text{change in shear at the point of}}{\text{application of a concentrated load}} = \frac{\text{magnitude of}}{\text{the load}} \qquad (5.12)$$

The relationships between the shears and bending moments (Eqs. (5.6) through (5.10)) derived previously remain valid at the points of application of concentrated loads. Note that because of the abrupt change in the shear diagram at such a point, there will be an abrupt change in the slope of the bending moment diagram at that point.

Couples or Concentrated Moments

Although the relationships between the loads and shears derived thus far (Eqs. (5.1) through (5.5), (5.11), and (5.12)) are valid at the points of application of couples or concentrated moments, the relationships between the shears and bending moments as given by Eqs. (5.6) through (5.10) are not valid at such points. As illustrated in Example 5.3, at the point of application of a couple, the bending moment changes abruptly by an amount equal to the magnitude of the moment of the couple. To derive this relationship, we consider the equilibrium of a differential element that is isolated from the beam of Fig. 5.8(a) by passing imaginary sections at infinitesimal distances to the left and to the right of the point of application D of the couple $\overline{M}$. The free-body diagram of this element is shown in Fig. 5.8(d). Applying the moment equilibrium equation, we write

$$+\zeta \sum M_a = 0$$
$$-M - \overline{M} + (M + dM) = 0$$
$$dM = \overline{M} \qquad (5.13)$$

which can be stated as

$$\begin{array}{c} \text{change in bending moment at the} \\ \text{point of application of a couple} \end{array} = \begin{array}{c} \text{magnitude of the} \\ \text{moment of the couple} \end{array} \qquad (5.14)$$

Procedure for Analysis

The following step-by-step procedure can be used for constructing the shear and bending moment diagrams for beams by applying the foregoing relationships between the loads, the shears, and the bending moments.

1. Calculate the support reactions.
2. Construct the shear diagram as follows:
    a. Determine the shear at the left end of the beam. If no concentrated load is applied at this point, the shear is zero at this point; go to step 2(b). Otherwise, the ordinate of the shear diagram at this point changes abruptly from zero to the magnitude of the concentrated force. Recall that an upward force causes the shear to increase, whereas a downward force causes the shear to decrease.
    b. Proceeding from the point at which the shear was computed in the previous step toward the right along the length of the beam, identify the next point at which the numerical value of the or-

dinate of the shear diagram is to be determined. Usually, it is necessary to determine such values only at the ends of the beam and at points at which the concentrated forces are applied and where the load distributions change.

   c. Determine the ordinate of the shear diagram at the point selected in step 2(b) (or just to the left of it, if a concentrated load acts at the point) by adding algebraically the area under the load diagram between the previous point and the point currently under consideration to the shear at the previous point (or just to the right of it, if a concentrated force acts at the point). The formulas for the areas of common geometric shapes are listed in Appendix A.

   d. Determine the shape of the shear diagram between the previous point and the point currently under consideration by applying Eq. (5.3), which states that the slope of the shear diagram at a point is equal to the load intensity at that point.

   e. If no concentrated force is acting at the point under consideration, then proceed to step 2(f). Otherwise, determine the ordinate of the shear diagram just to the right of the point by adding algebraically the magnitude of the concentrated load to the shear just to the left of the point. Thus, the shear diagram at this point changes abruptly by an amount equal to the magnitude of the concentrated force.

   f. If the point under consideration is not located at the right end of the beam, then return to step 2(b). Otherwise, the shear diagram has been completed. If the analysis has been carried out correctly, then the value of shear just to the right of the right end of the beam must be zero, except for the round-off errors.

3. Construct the bending moment diagram as follows:

   a. Determine the bending moment at the left end of the beam. If no couple is applied at this point, the bending moment is zero at this point; go to step 3(b). Otherwise, the ordinate of the bending moment diagram at this point changes abruptly from zero to the magnitude of the moment of the couple. Recall that a clockwise couple causes the bending moment to increase, whereas a counterclockwise couple causes the bending moment to decrease at its point of application.

   b. Proceeding from the point at which the bending moment was computed in the previous step toward the right along the length of the beam, identify the next point at which the numerical value of the ordinate of the bending moment diagram is to be determined. It is usually necessary to determine such values only at the points where the numerical values of shear were computed in step 2, where the couples are applied, and where the maximum and minimum values of bending moment occur. In addition to the points of application of couples, the maximum and minimum values of bending moment occur at points where the shear is zero. At a point of zero shear, if the shear changes from positive to the left to negative to the right, the slope of the bending moment diagram will change from positive to the left of the point to negative to the right of it; that is, the bending moment will be maximum at this point. Conversely, at a point of zero shear, where the shear changes from negative to the left to positive to the right, the bending moment will be minimum. For most common loading conditions, such as concentrated loads and uniformly and linearly distributed loads, the points of zero shear can be located by considering the geometry of the shear diagram. However, for some cases of linearly distributed loads, as well as for nonlinearly distributed loads, it becomes necessary to locate the points of zero shear by solving the expressions for shear, as illustrated in Example 5.4.

c. Determine the ordinate of the bending moment diagram at the point selected in step 3(b) (or just to the left of it, if a couple acts at the point) by adding algebraically the area under the shear diagram between the previous point and the point currently under consideration to the bending moment at the previous point (or just to the right of it, if a couple acts at the point).

d. Determine the shape of the bending moment diagram between the previous point and the point currently under consideration by applying Eq. (5.8), which states that the slope of the bending moment diagram at a point is equal to the shear at that point.

e. If no couple is acting at the point under consideration, then proceed to step 3(f). Otherwise, determine the ordinate of the bending moment diagram just to the right of the point by adding algebraically the magnitude of the moment of the couple to the bending moment just to the left of the point. Thus, the bending moment diagram at this point changes abruptly by an amount equal to the magnitude of the moment of the couple.

f. If the point under consideration is not located at the right end of the beam, then return to step 3(b). Otherwise, the bending moment diagram has been completed. If the analysis has been carried out correctly, then the value of bending moment just to the right of the right end of the beam must be zero, except for the round-off errors.

The foregoing procedure can be used for constructing the shear and bending moment diagrams by proceeding from the left end of the beam to its right end, as is currently the common practice. However, if we wish to construct these diagrams by proceeding from the right end of the beam toward the left, the procedure essentially remains the same except that downward forces must now be considered to cause increase in shear, counterclockwise couples are now considered to cause increase in bending moment, and vice versa.

Example 5.5

Draw the shear and bending moment diagrams and the qualitative deflected shape for the beam shown in Fig. 5.9(a).

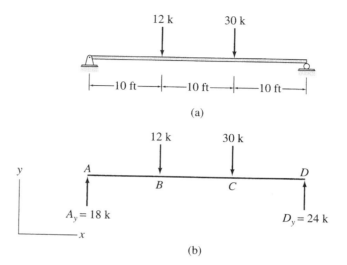

(a)

(b)

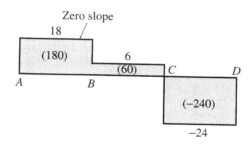

(c) Shear Diagram (k)

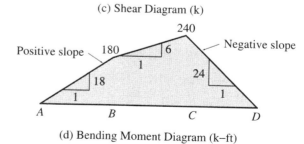

(d) Bending Moment Diagram (k–ft)

(e) Qualitative Deflected Shape

FIG. 5.9

continued

Solution

Reactions (See Fig. 5.9(b).)

$$+ \rightarrow \sum F_x = 0 \qquad A_x = 0$$

By proportions,

$$A_y = 12\left(\frac{20}{30}\right) + 30\left(\frac{10}{30}\right) = 18 \text{ k} \qquad A_y = 18 \text{ k} \uparrow$$

$$+ \uparrow \sum F_y = 0$$

$$18 - 12 - 30 + D_y = 0$$

$$D_y = 24 \text{ k} \qquad\qquad D_y = 24 \text{ k} \uparrow$$

Shear Diagram

Point A Since a positive (upward) concentrated force of 18-k magnitude acts at point A, the shear diagram increases abruptly from 0 to +18 k at this point.

Point B The shear just to the left of point B is given by

$$S_{B,L} = S_{A,R} + \text{area under the load diagram between just to the right of } A \text{ to just to the left of } B$$

in which the subscripts ", L" and ", R" are used to denote "just to the left" and "just to the right," respectively. As no load is applied to this segment of the beam,

$$S_{B,L} = 18 + 0 = 18 \text{ k}$$

Because a negative (downward) concentrated load of 12-k magnitude acts at point B, the shear just to the right of B is

$$S_{B,R} = 18 - 12 = 6 \text{ k}$$

Point C

$$S_{C,L} = S_{B,R} + \text{area under the load diagram between just to the right of } B \text{ to just to the left of } C$$

$$S_{C,L} = 6 + 0 = 6 \text{ k}$$

$$S_{C,R} = 6 - 30 = -24 \text{ k}$$

Point D $\qquad\qquad S_{D,L} = -24 + 0 = -24 \text{ k}$

$$S_{D,R} = -24 + 24 = 0 \qquad\qquad \text{Checks}$$

The numerical values of shear computed at points A, B, C, and D are used to construct the shear diagram as shown in Fig. 5.9(c). The shape of the diagram between these ordinates has been established by applying Eq. (5.3), which states that the slope of the shear diagram at a point is equal to the load intensity at that point. Because no load is applied to the beam between these points, the slope of the shear diagram is zero between these points, and the shear diagram consists of a series of horizontal lines, as shown in the figure. Note that the shear diagram closes (i.e., returns to zero) just to the right of the right end D of the beam, indicating that the analysis has been carried out correctly. **Ans.**

To facilitate the construction of the bending moment diagram, the areas of the various segments of the shear diagram have been computed and are shown in parentheses on the shear diagram (Fig. 5.9(c)).

Bending Moment Diagram

Point A Because no couple is applied at end A, $M_A = 0$.

Point B $M_B = M_A + $ area under the shear diagram
                                        between A and B

$$M_B = 0 + 180 = 180 \text{ k-ft}$$

Point C $M_C = 180 + 60 = 240 \text{ k-ft}$

Point D $M_D = 240 - 240 = 0$ Checks

The numerical values of bending moment computed at points A, B, C, and D are used to construct the bending moment diagram shown in Fig. 5.9(d). The shape of the diagram between these ordinates has been established by applying Eq. (5.8), which states that the slope of the bending moment diagram at a point is equal to the shear at that point. As the shear between these points is constant, the slope of the bending moment diagram must be constant between these points. Therefore, the ordinates of the bending moment diagram are connected by straight, sloping lines. In segment AB, the shear is $+18$ k. Therefore, the slope of the bending moment diagram in this segment is 18:1, and it is positive—that is, *upward to the right* (/). In segment BC, the shear drops to $+6$ k; therefore, the slope of the bending moment diagram reduces to 6:1 but remains positive. In segment CD, the shear becomes -24; consequently, the slope of the bending moment diagram becomes negative—that is, *downward to the right* (\\), as shown in Fig. 5.9(d). Note that the maximum bending moment occurs at point C, where the shear changes from positive to the left to negative to the right.

Ans.

Qualitative Deflected Shape A qualitative deflected shape of the beam is shown in Fig. 5.9(e). As the bending moment is positive over its entire length, the beam bends concave upward, as shown.

Ans.

Example 5.6

Draw the shear and bending moment diagrams and the qualitative deflected shape for the beam shown in Fig. 5.10(a).

Solution

Reactions (See Fig. 5.10(b).)

$$+ \rightarrow \textstyle\sum F_x = 0 \qquad\qquad A_x = 0$$

$$+ \uparrow \textstyle\sum F_y = 0$$

$$A_y - 70 = 0$$

$$A_y = 70 \text{ kN} \qquad\qquad A_y = 70 \text{ kN} \uparrow$$

continued

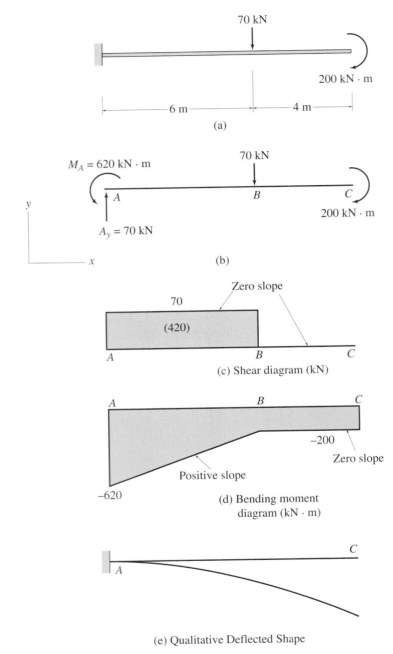

(a)

(b)

(c) Shear diagram (kN)

(d) Bending moment
diagram (kN · m)

(e) Qualitative Deflected Shape

FIG. 5.10

$$+\circlearrowleft \sum M_A = 0$$

$$M_A - 70(6) - 200 = 0$$

$$M_A = 620 \text{ kN} \cdot \text{m} \qquad M_A = 620 \text{ kN} \cdot \text{m} \circlearrowright$$

Shear Diagram

Point A $S_{A,R} = 70$ kN

Point B $S_{B,L} = 70 + 0 = 70$ kN

                 $S_{B,R} = 70 - 70 = 0$

Point C $S_{C,L} = 0 + 0 = 0$

                 $S_{C,R} = 0 + 0 = 0$ Checks

The numerical values of shear evaluated at points A, B, and C are used to construct the shear diagram as shown in Fig. 5.10(c). Because no load is applied to the beam between these points, the slope of the shear diagram is zero between these points. To facilitate the construction of the bending moment diagram, the area of the segment AB of the shear diagram has been computed and is shown in parentheses on the shear diagram (Fig. 5.10(c)).   Ans.

Bending Moment Diagram

Point A Since a negative (counterclockwise) couple of 620 kN · m moment acts at point A, the bending moment diagram decreases abruptly from 0 to -620 kN · m at this point; that is,

$$M_{A,R} = -620 \text{ kN} \cdot \text{m}$$

Point B $M_B = -620 + 420 = -200$ kN · m

Point C $M_{C,L} = -200 + 0 = -200$ kN · m

                 $M_{C,R} = -200 + 200 = 0$ Checks

The bending moment diagram is shown in Fig. 5.10(d). The shape of this diagram between the ordinates just computed is based on the condition that the slope of the bending moment diagram at a point is equal to shear at that point. As the shear in the segments AB and BC is constant, the slope of the bending moment diagram must be constant in these segments. Therefore, the ordinates of the bending moment diagram are connected by straight lines. In segment AB, the shear is positive, and so is the slope of the bending moment diagram in this segment. In segment BC, the shear becomes zero; consequently, the slope of the bending moment diagram becomes zero, as shown in Fig. 5.10(d).   Ans.

Qualitative Deflected Shape A qualitative deflected shape of the beam is shown in Fig. 5.10(e). As the bending moment is negative over its entire length, the beam bends concave downward, as shown.   Ans.

Example 5.7

Draw the shear and bending moment diagrams and the qualitative deflected shape for the beam shown in Fig. 5.11(a).

Solution

Reactions (See Fig. 5.11(b).)

continued

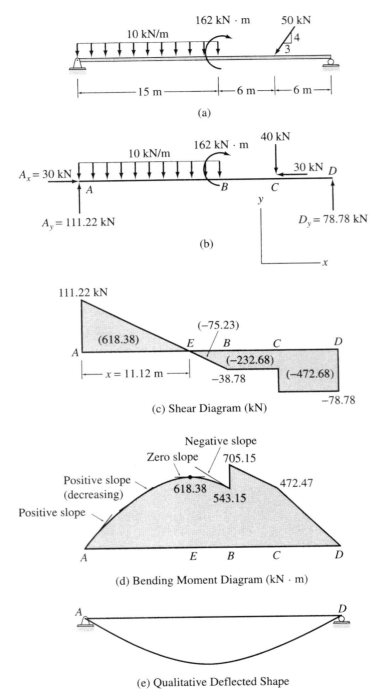

FIG. **5.11**

$$+ \rightarrow \sum F_x = 0$$

$$A_x - 30 = 0$$

$$A_x = 30 \text{ kN} \qquad A_x = 30 \text{ kN} \rightarrow$$

$$+ \zeta \sum M_D = 0$$

$$-A_y(27) + 10(15)(19.5) - 162 + 40(6) = 0$$

$$A_y = 111.22 \text{ kN} \qquad A_y = 111.22 \text{ kN} \uparrow$$

$$+ \uparrow \sum F_y = 0$$

$$111.22 - 10(15) - 40 + D_y = 0$$

$$D_y = 78.78 \text{ kN} \qquad D_y = 78.78 \text{ kN} \uparrow$$

Shear Diagram

Point A $\qquad S_{A,R} = 111.22 \text{ kN}$

Point B $\qquad S_B = 111.22 - 10(15) = -38.78 \text{ kN}$

Point C $\qquad S_{C,L} = -38.78 + 0 = -38.78 \text{ kN}$

$\qquad\qquad S_{C,R} = -38.78 - 40 = -78.78 \text{ kN}$

Point D $\qquad S_{D,L} = -78.78 + 0 = -78.78 \text{ kN}$

$\qquad\qquad S_{D,R} = -78.78 + 78.78 = 0 \qquad\qquad$ Checks

The shear diagram is shown in Fig. 5.11(c). In segment *AB*, the beam is subjected to a downward (negative) uniformly distributed load of 10 kN/m. Because the load intensity is constant and negative in segment *AB*, the shear diagram in this segment is a straight line with negative slope. No distributed load is applied to the beam in segments *BC* and *CD*, so the shear diagram in these segments consists of horizontal lines, indicating zero slopes. **Ans.**

The point of zero shear, *E*, can be located by using the similar triangles forming the shear diagram between *A* and *B*. Thus,

$$\frac{x}{111.22} = \frac{15}{(111.22 + 38.78)}$$

$$x = 11.12 \text{ m}$$

To facilitate the construction of the bending moment diagram, the areas of the various segments of the shear diagram have been computed; they are shown in parentheses on the shear diagram (Fig. 5.11(c)).

Bending Moment Diagram

Point A $\qquad M_A = 0$

Point E $\qquad M_E = 0 + 618.38 = 618.38 \text{ kN} \cdot \text{m}$

Point B $\qquad M_{B,L} = 618.38 - 75.23 = 543.15 \text{ kN} \cdot \text{m}$

$\qquad\qquad M_{B,R} = 543.15 + 162 = 705.15 \text{ kN} \cdot \text{m}$

continued

Point C $M_C = 705.15 - 232.68 = 472.47 \text{ kN} \cdot \text{m}$

Point D $M_D = 472.47 - 472.68 = -0.21 \approx 0$ Checks

The bending moment diagram is shown in Fig. 5.11(d). The shape of this diagram between the ordinates just computed has been based on the condition that the slope of the bending moment diagram at any point is equal to the shear at that point. Just to the right of A, the shear is positive, and so is the slope of the bending moment diagram at this point. As we move to the right from A, the shear decreases linearly (but remains positive), until it becomes zero at E. Therefore, the slope of the bending moment diagram gradually decreases, or becomes less steep (but remains positive), as we move to the right from A, until it becomes zero at E. Note that the shear diagram in segment AE is linear, but the bending moment diagram in this segment is parabolic, or a second-degree curve, because the bending moment diagram is obtained by integrating the shear diagram (Eq. 5.11). Therefore, the bending moment curve will always be one degree higher than the corresponding shear curve.

We can see from Fig. 5.11(d) that the bending moment becomes locally maximum at point E, where the shear changes from positive to the left to negative to the right. As we move to the right from E, the shear becomes negative, and it decreases linearly between E and B. Accordingly, the slope of the bending moment diagram becomes negative to the right of E, and it decreases continuously (becomes more steep downward to the right) between E and just to the left of B. A positive (clockwise) couple acts at B, so the bending moment increases abruptly at this point by an amount equal to the magnitude of the moment of the couple. The largest value (global maximum) of the bending moment over the entire length of the beam occurs at just to the right of B. (Note that no abrupt change, or discontinuity, occurs in the shear diagram at this point.) Finally, as the shear in segments BC and CD is constant and negative, the bending moment diagram in these segments consists of straight lines with negative slopes.   Ans.

Qualitative Deflected Shape See Fig. 5.11(e).   Ans.

Example 5.8

Draw the shear and bending moment diagrams and the qualitative deflected shape for the beam shown in Fig. 5.12(a).

Solution

Reactions (See Fig. 5.12(b).)

$$+ \rightarrow \sum F_x = 0 \qquad\qquad B_x = 0$$

$$+ \circlearrowleft \sum M_C = 0$$

$$\frac{1}{2}(3)(12)(24) - B_y(20) + 3(20)(10) - \frac{1}{2}(3)(6)(2) = 0$$

$$B_y = 50.7 \text{ k} \qquad B_y = 50.7 \text{ k} \uparrow$$

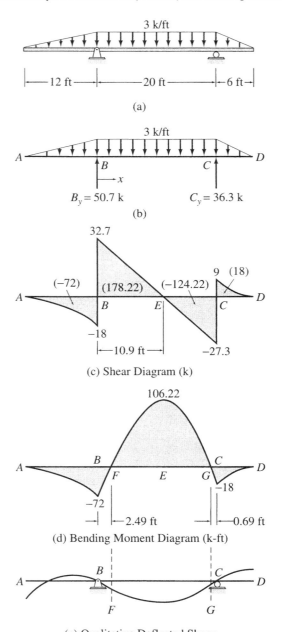

(a)

(b)

(c) Shear Diagram (k)

(d) Bending Moment Diagram (k-ft)

(e) Qualitative Deflected Shape

FIG. 5.12

continued

$$+\uparrow \sum F_y = 0$$

$$-\frac{1}{2}(3)(12) + 50.7 - 3(20) - \frac{1}{2}(3)(6) + C_y = 0$$

$$C_y = 36.3 \text{ k} \qquad C_y = 36.3 \text{ k} \uparrow$$

Shear Diagram

Point A	$S_A = 0$
Point B	$S_{B,L} = 0 - \frac{1}{2}(3)(12) = -18$ k
	$S_{B,R} = -18 + 50.7 = 32.7$ k
Point C	$S_{C,L} = 32.7 - 3(20) = -27.3$ k
	$S_{C,R} = -27.3 + 36.3 = 9$ k
Point D	$S_D = 9 - \frac{1}{2}(3)(6) = 0$ Checks

The shear diagram is shown in Fig. 5.12(c). The shape of the diagram be-tween the ordinates just computed is obtained by applying the condition that the slope of the shear diagram at any point is equal to the load intensity at that point. For example, as the load intensity at A is zero, so is the slope of the shear diagram at A. Between A and B, the load intensity is negative and it decreases linearly from zero at A to -3 k/ft at B. Thus, the slope of the shear diagram is negative in this segment, and it decreases (becomes more steep) continuously from A to just to the left of B. The rest of the shear diagram is constructed by using similar reasoning.   **Ans.**

The point of zero shear, E, is located by using the similar triangles forming the shear diagram between B and C.

To facilitate the construction of the bending moment diagram, the areas of the various segments of the shear diagram have been computed and are shown in parentheses on the shear diagram (Fig. 5.12(c)). It should be noted that the areas of the parabolic spandrels, AB and CD, can be obtained by using the for-mula for the area of this shape given in Appendix A.

Bending Moment Diagram

Point A	$M_A = 0$
Point B	$M_B = 0 - 72 = -72$ k-ft
Point E	$M_E = -72 + 178.22 = 106.22$ k-ft
Point C	$M_C = 106.22 - 124.22 = -18$ k-ft
Point D	$M_D = -18 + 18 = 0$ Checks

The shape of the bending moment diagram between these ordinates is ob-tained by using the condition that the slope of the bending moment diagram at any point is equal to the shear at that point. The bending moment diagram thus constructed is shown in Fig. 5.12(d).

It can be seen from this figure that the maximum negative bending mo-ment occurs at point B, whereas the maximum positive bending moment, which has the largest absolute value over the entire length of the beam, occurs at point E. **Ans.**

To locate the points of inflection, F and G, we set equal to zero the equation for bending moment in segment BC, in terms of the distance x from the left support point B (Fig. 5.12(b)):

$$M = -\left(\frac{1}{2}\right)(3)(12)(4+x) + 50.7x - 3(x)\left(\frac{x}{2}\right) = 0$$

or

$$-1.5x^2 + 32.7x - 72 = 0$$

from which $x = 2.49$ ft and $x = 19.31$ ft from B.

Qualitative Deflected Shape A qualitative deflected shape of the beam is shown in Fig. 5.12(e). The bending moment is positive in segment FG, so the beam is bent concave upward in this region. Conversely, since the bending moment is negative in segments AF and GD, the beam is bent concave downward in these segments. **Ans.**

Example 5.9

Draw the shear and bending moment diagrams and the qualitative deflected shape for the beam shown in Fig. 5.13(a).

Solution

Reactions (See Fig. 5.13(b).)

$$+\zeta \sum M_B^{BD} = 0$$
$$-20(10)(5) + C_y(10) - 100(15) = 0$$
$$C_y = 250 \text{ kN} \qquad C_y = 250 \text{ kN} \uparrow$$
$$+\uparrow \sum F_y = 0$$
$$A_y - 20(10) + 250 - 100 = 0$$
$$A_y = 50 \text{ kN} \qquad A_y = 50 \text{ kN} \uparrow$$
$$+\zeta \sum M_A = 0$$
$$M_A - 20(10)(15) + 250(20) - 100(25) = 0$$
$$M_A = 500 \text{ kN} \cdot \text{m} \qquad M_A = 500 \text{ kN} \cdot \text{m} \, \zeta$$

Shear Diagram

Point A \qquad $S_{A,R} = 50$ kN

Point B \qquad $S_B = 50 + 0 = 50$ kN

continued

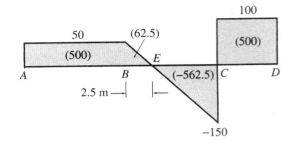

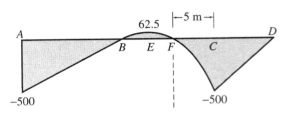

(c) Shear Diagram (kN)

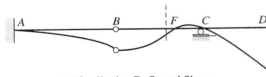

(d) Bending Moment Diagram (kN · m)

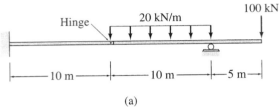

(a)

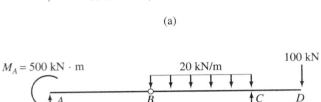

(b)

(e) Qualitative Deflected Shape

FIG. 5.13

Point C $S_{C,L} = 50 - 20(10) = -150$ kN

$S_{C,R} = -150 + 250 = 100$ kN

Point D $S_{D,L} = 100 + 0 = 100$ kN

$S_{D,R} = 100 - 100 = 0$ Checks

The shear diagram is shown in Fig. 5.13(c).   Ans.

Bending Moment Diagram

Point A $M_{A,R} = -500$ kN · m

Point B $M_B = -500 + 500 = 0$

Point E $M_E = 0 + 62.5 = 62.5$ kN · m

Point C $M_C = 62.5 - 562.5 = -500$ kN · m

Point D $M_D = -500 + 500 = 0$ Checks

The bending moment diagram is shown in Fig. 5.13(d). The point of inflection F can be located by setting equal to zero the equation for bending moment in segment BC, in terms of the distance x_1 from the right support point C (Fig. 5.13(b)):

$$M = -100(5 + x_1) + 250x_1 - 20(x_1)\left(\frac{x_1}{2}\right) = 0$$

or

$$-10x_1^2 + 150x_1 - 500 = 0$$

from which $x_1 = 5$ m and $x_1 = 10$ m from C. Note that the solution $x_1 = 10$ m represents the location of the internal hinge at B, at which the bending moment is zero. Thus, the point of inflection F is located at a distance of 5 m to the left of C, as shown in Fig. 5.13(d). **Ans.**

Qualitative Deflected Shape A qualitative deflected shape of the beam is shown in Fig. 5.13(e). Note that at the fixed support A, both the deflection and the slope of the beam are zero, whereas at the roller support C, only the deflection is zero, but the slope is not. The internal hinge B does not provide any rotational restraint, so the slope at B can be discontinuous. **Ans.**

Example 5.10

Draw the shear and bending moment diagrams and the qualitative deflected shape for the beam shown in Fig. 5.14(a).

Solution

Reactions (See Fig. 5.14(b).)

$$+ \circlearrowleft \sum M_C^{CD} = 0$$

$$D_y(24) - 2(24)(12) = 0$$

$$D_y = 24 \text{ k} \qquad D_y = 24 \text{ k} \uparrow$$

$$+ \circlearrowleft \sum M_A = 0$$

$$24(60) + B_y(30) - 2(60)(30) = 0$$

$$B_y = 72 \text{ k} \qquad B_y = 72 \text{ k} \uparrow$$

$$+ \uparrow \sum F_y = 0$$

$$A_y - 2(60) + 72 + 24 = 0$$

$$A_y = 24 \text{ k} \qquad A_y = 24 \text{ k} \uparrow$$

Shear Diagram

Point A $S_{A,R} = 24$ k

Point B $S_{B,L} = 24 - 2(30) = -36$ k

$$S_{B,R} = -36 + 72 = 36 \text{ k}$$

continued

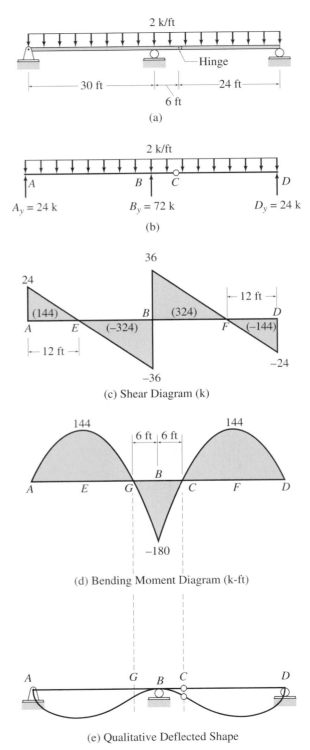

(a)

(b)

(c) Shear Diagram (k)

(d) Bending Moment Diagram (k-ft)

(e) Qualitative Deflected Shape

FIG. 5.14

Point D $S_{D,L} = 36 - 2(30) = -24$ k

                    $S_{D,R} = -24 + 24 = 0$ Checks

The shear diagram is shown in Fig. 5.14(c).   Ans.

Bending Moment Diagram

Point A $M_A = 0$

Point E $M_E = 0 + 144 = 144$ k-ft

Point B $M_B = 144 - 324 = -180$ k-ft

Point F $M_F = -180 + 324 = 144$ k-ft

Point D $M_D = 144 - 144 = 0$ Checks

The bending moment diagram is shown in Fig. 5.14(d).   Ans.

Qualitative Deflected Shape See Fig. 5.14(e).   Ans.

Example 5.11

Draw the shear and bending moment diagrams and the qualitative deflected shape for the statically indeterminate beam shown in Fig. 5.15. The support reactions, determined by using the procedures for the analysis of statically indeterminate beams (presented in Part Three of this text), are given in Fig. 5.15(a).

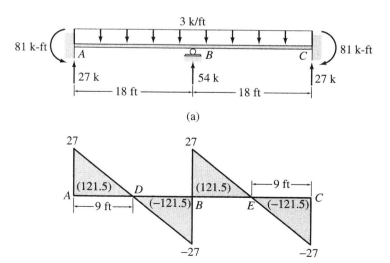

(a)

(b) Shear Diagram (k)

FIG. 5.15

continued

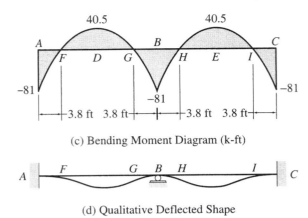

(c) Bending Moment Diagram (k-ft)

(d) Qualitative Deflected Shape

FIG. **5.15** (contd.)

Solution

Regardless of whether a beam is statically determinate or indeterminate, once the support reactions have been determined, the procedure for constructing the shear and bending moment diagrams remains the same. The shear and bending moment diagrams for the given statically indeterminate beam are shown in Fig. 5.15(b) and (c), respectively, and a qualitative deflected shape of the beam is shown in Fig. 5.15(d).

5.5 STATIC DETERMINACY, INDETERMINACY, AND INSTABILITY OF PLANE FRAMES

As defined in Section 1.3, rigid frames, usually referred to simply as *frames*, are composed of straight members connected either by rigid (moment-resisting) connections or by hinged connections to form stable configurations. The members of frames are usually connected by rigid joints, although hinged connections are sometimes used. A rigid joint prevents relative translations and rotations of the member ends connected to it, so the joint is capable of transmitting two rectangular force components and a couple between the connected members. Under the action of external loads, the members of a frame may be, in general, subjected to bending moment, shear, and axial tension or compression.

A frame is considered to be *statically determinate if the bending moments, shears, and axial forces in all its members, as well as all the external reactions, can be determined by using the equations of equilibrium and condition.*

Since the method of analysis presented in the following section can be used only to analyze statically determinate frames, it is important for the student to be able to recognize statically determinate frames before proceeding with the analysis.

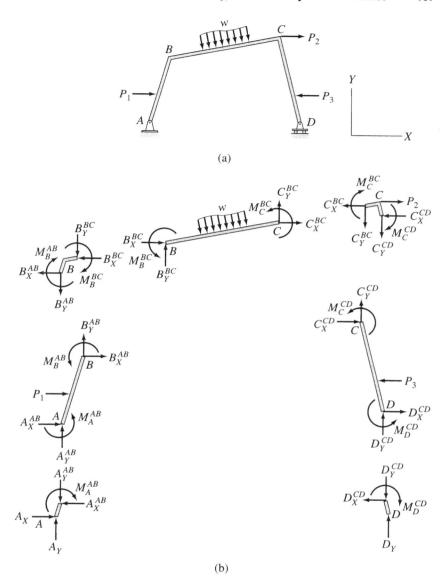

(a)

(b)

FIG. 5.16

Consider a plane frame subjected to an arbitrary loading, as shown in Fig. 5.16(a). The free-body diagrams of the three members and the four joints of the frame are shown in Fig. 5.16(b). Each member is subjected to, in addition to the external forces, two internal force components and an internal couple at each of its ends. Of course, the correct senses of the internal forces and couples, which are commonly referred to as the *member end forces*, are not known before the analysis and are chosen arbitrarily. The free-body diagrams of the joints show the same member end forces but in opposite directions, in accordance with Newton's third law. The analysis of the frame involves the determination of the magnitudes of the 18 member end forces (six per member), and the three support reactions, $A_X, A_Y,$ and D_Y. Therefore, the total number of unknown quantities to be determined is 21.

Because the entire frame is in equilibrium, each of its members and joints must also be in equilibrium. As shown in Fig. 5.16(b), each member and each joint are subjected to a general coplanar system of forces and couples, which must satisfy the three equations of equilibrium, $\sum F_X = 0$, $\sum F_Y = 0$, and $\sum M = 0$. Since the frame contains three members and four joints (including the two joints connected to supports), the total number of equations available is $3(3) + 3(4) = 21$. These 21 equilibrium equations can be solved to calculate the 21 unknowns. The member end forces thus obtained can then be used to determine axial forces, shears, and bending moments at various points along the lengths of members. The frame of Fig. 5.16(a) is, therefore, statically determinate.

Three equations of equilibrium of the entire frame as a rigid body could be written and solved for the three unknown reactions $(A_X, A_Y,$ and $D_Y)$. However, these equilibrium equations are not independent from the member and joint equilibrium equations and do not contain any additional information.

Based on the foregoing discussion, we can develop the criteria for the static determinacy, indeterminacy, and instability of general plane frames containing m members and j joints and supported by r (number of) external reactions. For the analysis, we need to determine $6m$ member forces and r external reactions; that is, we need to calculate a total of $6m + r$ unknown quantities. Since there are m members and j joints and we can write three equations of equilibrium for each member and each joint, the number of equilibrium equations available is $3(m + j)$. Furthermore, if a frame contains internal hinges and/or internal rollers, these internal conditions provide additional equations, which can be used in conjunction with the equilibrium equations to determine the unknowns. Thus, if there are e_c equations of condition for a frame, the total number of equations (equilibrium equations plus equations of condition) available is $3(m + j) + e_c$. For a frame, if the number of unknowns is equal to the number of equations, that is,

$$6m + r = 3(m + j) + e_c$$

or

$$3m + r = 3j + e_c$$

then all the unknowns can be determined by solving the equations of equilibrium and condition, and the frame is statically determinate. If a frame has more unknowns than the available equations—that is, $3m + r > 3j + e_c$—all the unknowns cannot be determined by solving the available equations, and the frame is called statically indeterminate. Statically indeterminate frames have more members and/or external reactions than the minimum required for stability. The excess members and reactions are called *redundants,* and the number of excess member forces and reactions is referred to as the degree of static indeterminacy, i, which can be expressed as

$$i = (3m + r) - (3j + e_c) \tag{5.15}$$

For a frame, if the number of unknowns is less than the number of available equations—that is, $3m + r < 3j + e_c$—the frame is called statically unstable. The conditions for static instability, determinacy, and indeterminacy of plane frames can be summarized as follows:

$$
\begin{aligned}
3m + r < 3j + e_c &\quad \text{statically unstable frame} \\
3m + r = 3j + e_c &\quad \text{statically determinate frame} \\
3m + r > 3j + e_c &\quad \text{statically indeterminate frame}
\end{aligned}
\tag{5.16}
$$

In applying Eq. (5.16), the ends of the frame attached to supports as well as any free ends are treated as joints. The conditions for static determinacy and indeterminacy, as given by Eq. (5.16), are necessary but not sufficient conditions. In order for these criteria for static determinacy and indeterminacy to be valid, the arrangement of the members, support reactions, and internal hinges and rollers (if any) must be such that the frame will remain geometrically stable under a general system of coplanar loads.

The procedure for determining the number of equations of condition remains the same as discussed in Chapter 3. Recall that an internal hinge provides one equation of condition, and an internal roller provides two such equations. When several members of a frame are

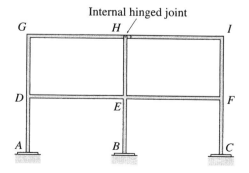

Internal hinged joint

Any two of the following may be considered as equations of condition:

$$M_H^{EH} = 0, \quad M_H^{GH} = 0, \quad M_H^{HI} = 0$$

FIG. 5.17

connected at a hinged joint, the number of equations of condition at the joint is equal to the number of members meeting at the joint minus one. For example, consider the hinged joint H of the frame shown in Fig. 5.17. As a hinge cannot transmit moment, the moments at the ends H of the three members EH, GH, and HI meeting at the joint must be zero; that is, $M_H^{EH} = 0$, $M_H^{GH} = 0$, and $M_H^{HI} = 0$. However, these three equations are not independent in the sense that if any two of these three equations are satisfied along with the moment equilibrium equation for the joint H, the remaining equation will automatically be satisfied. Thus, the hinged joint H provides two independent equations of condition. Using a similar reasoning, it can be shown that an internal roller joint provides the equations of condition whose number is equal to $2 \times$ (number of members meeting at the joint -1).

Alternative Approach

An alternative approach that can be used for determining the degree of static indeterminacy of a frame is to cut enough members of the frame by passing imaginary sections and/or to remove enough supports to render the structure statically determinate. The total number of internal and external restraints thus removed equals the degree of static indeterminacy. As an example, consider the frame shown in Fig. 5.18(a).

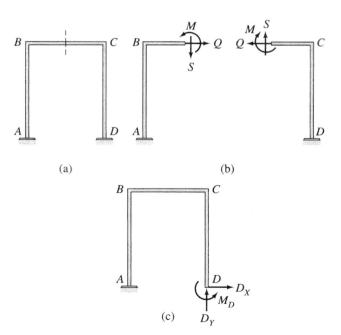

(a) (b)

(c)

FIG. 5.18

The frame can be made statically determinate by passing an imaginary section through the girder BC, thereby removing three internal restraints (the axial force Q, the shear S, and the bending moment M), as shown in Fig. 5.18(b). Note that the two cantilever structures thus produced are both statically determinate and geometrically stable. Because three restraints $(Q, S,$ and $M)$ had to be removed from the original statically indeterminate frame of Fig. 5.18(a) to obtain the statically determinate frames of Fig. 5.18(b), the degree of static indeterminacy of the original frame is three. There are many possible choices regarding the restraints that can be removed from a statically indeterminate structure to render it statically determinate. For example, the frame of Fig. 5.18(a) could alternatively be rendered statically determinate by disconnecting it from the fixed support at D, as shown in Fig. 5.18(c). Since three external restraints or reactions, $D_X, D_Y,$ and M_D, must be removed in this process, the degree of static indeterminacy of the frame is three, as concluded previously.

This alternative approach of establishing the degree of indeterminacy (instead of applying Eq. (5.15)) provides the most convenient means of determining the degrees of static indeterminacy of multistory building frames. An example of such a frame is shown in Fig. 5.19(a). The structure can be made statically determinate by passing an imaginary section through each of the girders, as shown in Fig. 5.19(b). Because each cut removes three restraints, the total number of restraints that must be removed to render the structure statically determinate is equal to three times the number of girders in the frame. Thus, the degree of static indeterminacy of a multistory frame with fixed supports is equal to three times the number of girders, provided that the frame does not contain any internal hinges or rollers.

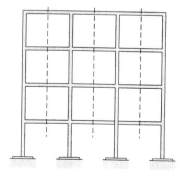

(a)

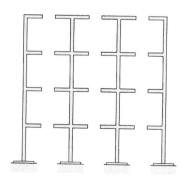

$i = 3(\text{number of girders}) = 3(12) = 36$

(b)

FIG. 5.19

Example 5.12

Verify that each of the plane frames shown in Fig. 5.20 is statically indeterminate and determine its degree of static indeterminacy.

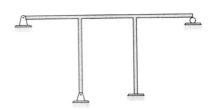

$m = 5$ $j = 6$ $r = 8$ $e_c = 0$
$3m + r > 3j + e_c$
(a) Statically Indeterminate ($i = 5$)

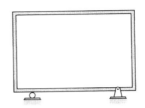

$m = 4$ $j = 4$ $r = 3$ $e_c = 0$
$3m + r > 3j + e_c$
(b) Statically Indeterminate ($i = 3$)

$m = 6$ $j = 6$ $r = 4$ $e_c = 0$
$3m + r > 3j + e_c$
(c) Statically Indeterminate ($i = 4$)

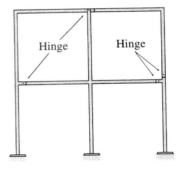

$m = 10$ $j = 9$ $r = 9$ $e_c = 5$
$3m + r > 3j + e_c$
(d) Statically Indeterminate ($i = 7$)

$i = 3$ (number of girders) $= 3(4) = 12$
(e)

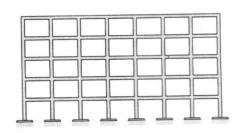

$i = 3$ (number of girders) $= 3(35) = 105$
(f)

FIG. 5.20

Solution
See Fig. 5.20(a) through (f).

5.6 ANALYSIS OF PLANE FRAMES

The following step-by-step procedure can be used for determining the member end forces as well as the shears, bending moments, and axial forces in members of plane statically determinate frames.

1. Check for static determinacy. Using the procedure described in the preceding section, determine whether or not the given frame is stat-

ically determinate. If the frame is found to be statically determinate and stable, proceed to step 2. Otherwise, end the analysis at this stage. (The analysis of statically indeterminate frames is considered in Part Three of this text.)

2. Determine the support reactions. Draw a free-body diagram of the entire frame, and determine reactions by applying the equations of equilibrium and any equations of condition that can be written in terms of external reactions only (without involving any internal member forces). For some internally unstable frames, it may not be possible to express all the necessary equations of condition exclusively in terms of external reactions; therefore, it may not be possible to determine all the reactions. However, some of the reactions for such structures can usually be calculated from the available equations.

3. Determine member end forces. It is usually convenient to specify the directions of the unknown forces at the ends of the members of the frame by using a common structural (or global) XY coordinate system, with the X and Y axes oriented in the horizontal (positive to the right) and vertical (positive upward) directions, respectively. Draw free-body diagrams of all the members and joints of the structure. These free-body diagrams must show, in addition to any external loads and support reactions, all the internal forces being exerted upon the member or the joint. Remember that a rigid joint is capable of transmitting two force components and a couple, a hinged joint can transmit two force components, and a roller joint can transmit only one force component. If there is a hinge at an end of a member, the internal moment at that end should be set equal to zero. Any load acting at a joint should be shown on the free-body diagrams of the joint, not at the ends of the members connected to the joint. The senses of the member end forces are not known and can be arbitrarily assumed. However, it is usually convenient to assume the senses of the unknown forces at member ends in the positive X and Y directions and of the unknown couples as counterclockwise. The senses of the internal forces and couples on the free-body diagrams of joints must be in directions opposite to those assumed on the member ends in accordance with Newton's third law. Compute the member end forces as follows:
   a. Select a member or a joint with three or fewer unknowns.
   b. Determine the unknown forces and moments by applying the three equations of equilibrium ($\sum F_X = 0$, $\sum F_Y = 0$, and $\sum M = 0$) to the free body of the member or joint selected in step 3(a).
   c. If all the unknown forces, moments, and reactions have been determined, then proceed to step 3(d). Otherwise, return to step 3(a).
   d. Since the support reactions were calculated in step 2 by using the equations of equilibrium and condition of the entire

structure, there should be some equations remaining that have not been utilized so far. The number of leftover equations should be equal to the number of reactions computed in step 2. Use these remaining equations to check the calculations. If the analysis has been carried out correctly, then the remaining equations must be satisfied.

For some types of frames, a member or a joint that has a number of unknowns less than or equal to the number of equilibrium equations may not be found to start or continue the analysis. In such a case, it may be necessary to write equilibrium equations in terms of unknowns for two or more free bodies and solve the equations simultaneously to determine the unknown forces and moments.

4. For each member of the frame, construct the shear, bending moment, and axial force diagrams as follows:

a. Select a member (local) xy coordinate system with origin at either end of the member and x axis directed along the centroidal axis of the member. The positive direction of the y axis is chosen so that the coordinate system is right-handed, with the z axis pointing out of the plane of the paper.

b. Resolve all the external loads, reactions, and end forces acting on the member into components in the x and y directions (i.e., in the directions parallel and perpendicular to the centroidal axis of the member). Determine the total (resultant) axial force and shear at each end of the member by algebraically adding the x components and y components, respectively, of the forces acting at each end of the member.

c. Construct the shear and bending moment diagrams for the member by using the procedure described in Section 5.4. The procedure can be applied to nonhorizontal members by considering the member end at which the origin of the xy coordinate system is located as the left end of the member (with x axis pointing toward the right) and the positive y direction as the upward direction.

d. Construct the axial force diagram showing the variation of axial force along the length of the member. Such a diagram can be constructed by using the method of sections. Proceeding in the positive x direction from the member end at which the origin of the xy coordinate system is located, sections are passed after each successive change in loading along the length of the member to determine the equations for the axial force in terms of x. According to the sign convention adopted in Section 5.1, the external forces acting in the negative x direction (causing tension at the section) are considered to be positive. The values of axial forces determined from these equations are plotted as ordinates against x to obtain the axial force diagram.

5. Draw a qualitative deflected shape of the frame. Using the bending moment diagrams constructed in step 4, draw a qualitative deflected

shape for each member of the frame. The deflected shape of the entire frame is then obtained by connecting the deflected shapes of the individual members at joints so that the original angles between the members at the rigid joints are maintained and the support conditions are satisfied. The axial and shear deformations, which are usually negligibly small as compared to the bending deformations, are neglected in sketching the deflected shapes of frames.

It should be noted that the bending moment diagrams constructed by using the procedure described in step 4(c) will always show moments on the *compression sides* of the members. For example, at a point along a vertical member, if the left side of the member is in compression, then the value of the moment at that point will appear on the left side. Since the side of the member on which a moment appears indicates the direction of the moment, it is not necessary to use plus and minus signs on the moment diagrams. When designing reinforced concrete frames, the moment diagrams are sometimes drawn on the tension sides of the members to facilitate the placement of steel bars used to reinforce concrete that is weak in tension. A tension-side moment diagram can be obtained by simply inverting (i.e., rotating through 180° about the member's axis) the corresponding compression-side moment diagram. Only compression-side moment diagrams are considered in this text.

Example 5.13

Draw the shear, bending moment, and axial force diagrams and the qualitative deflected shape for the frame shown in Fig. 5.21(a).

Solution

Static Determinacy $m = 3$, $j = 4$, $r = 3$, and $e_c = 0$. Because $3m + r = 3j + e_c$ and the frame is geometrically stable, it is statically determinate.

Reactions Considering the equilibrium of the entire frame (Fig. 5.21(b)), we observe that in order to satisfy $\sum F_X = 0$, the reaction component A_X must act to the left with a magnitude of 18 k to balance the horizontal load of 18 k to the right. Thus,

$$A_X = -18 \text{ k} \qquad A_X = 18 \text{ k} \leftarrow$$

We compute the remaining two reactions by applying the two equilibrium equations as follows:

$$+\circlearrowleft \sum M_A = 0 \qquad -18(20) - 2(30)(15) + D_Y(30) = 0 \qquad D_Y = 42 \text{ k} \uparrow$$

$$+\uparrow \sum F_Y = 0 \qquad A_Y - 2(30) + 42 = 0 \qquad A_Y = 18 \text{ k} \uparrow$$

continued

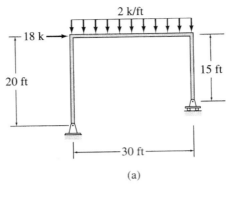

(a)

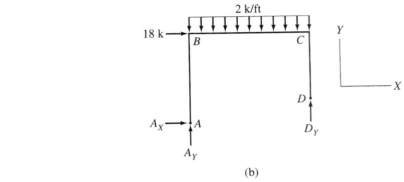

(b)

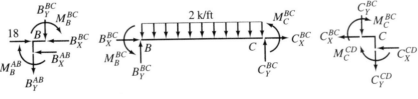

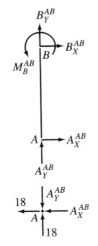

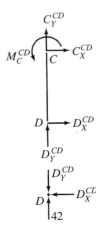

(c)

FIG. 5.21

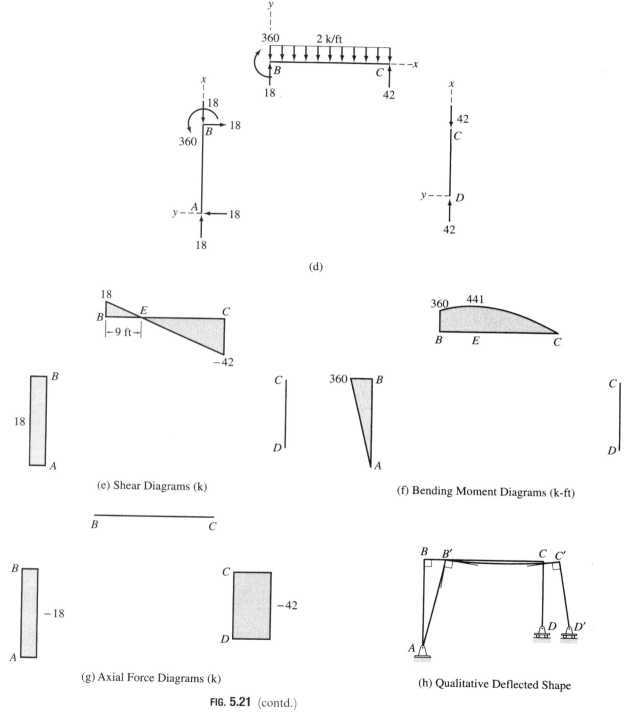

(d)

(e) Shear Diagrams (k)

(f) Bending Moment Diagrams (k-ft)

(g) Axial Force Diagrams (k)

(h) Qualitative Deflected Shape

FIG. 5.21 (contd.)

continued

Member End Forces The free-body diagrams of all the members and joints of the frame are shown in Fig. 5.21(c). We can begin the computation of internal forces either at joint A or at joint D, both of which have only three unknowns.

Joint A Beginning with joint A, we can see from its free-body diagram that in order to satisfy $\sum F_X = 0$, A_X^{AB} must act to the right with a magnitude of 18 k to balance the horizontal reaction of 18 k to the left. Thus,

$$A_X^{AB} = -18 \text{ k}$$

Similarly, by applying $\sum F_Y = 0$, we obtain

$$A_Y^{AB} = 18 \text{ k}$$

Member AB With the magnitudes of A_X^{AB} and A_Y^{AB} now known, member AB has three unknowns, B_X^{AB}, B_Y^{AB}, and M_B^{AB}, which can be determined by applying $\sum F_X = 0$, $\sum F_Y = 0$, and $\sum M_A = 0$. Thus,

$$B_X^{AB} = 18 \text{ k} \qquad B_Y^{AB} = -18 \text{ k} \qquad M_B^{AB} = 360 \text{ k-ft}$$

Joint B Proceeding next to joint B and considering its equilibrium, we obtain

$$B_X^{BC} = 0 \qquad B_Y^{BC} = 18 \text{ k} \qquad M_B^{BC} = -360 \text{ k-ft}$$

Member BC Next, considering the equilibrium of member BC, we write

$+ \rightarrow \sum F_X = 0$ $\qquad\qquad\qquad\qquad\qquad\qquad C_X^{BC} = 0$

$+ \uparrow \sum F_Y = 0$ $\qquad\qquad 18 - 2(30) + C_Y^{BC} = 0 \qquad C_Y^{BC} = 42 \text{ k}$

$+ \circlearrowleft \sum M_B = 0 \quad -360 - 2(30)(15) + 42(30) + M_C^{BC} = 0 \quad M_C^{BC} = 0$

Joint C Applying the three equilibrium equations, we obtain

$$C_X^{CD} = 0 \qquad C_Y^{CD} = -42 \text{ k} \qquad M_C^{CD} = 0$$

Member CD Applying $\sum F_X = 0$ and $\sum F_Y = 0$ in order, we obtain

$$D_X^{CD} = 0 \qquad D_Y^{CD} = 42 \text{ k}$$

Since all unknown forces and moments have been determined, we check our computations by using the third equilibrium equations for member CD.

$$+ \circlearrowleft \sum M_D = 0 \qquad\qquad\qquad \text{Checks}$$

Joint D (Checking computations)

$$+ \rightarrow \sum F_X = 0 \qquad\qquad\qquad \text{Checks}$$

$$+ \uparrow \sum F_Y = 0 \qquad 42 - 42 = 0 \qquad \text{Checks}$$

Shear Diagrams The xy coordinate systems selected for the three members of the frame are shown in Fig. 5.21(d), and the shear diagrams for the members constructed by using the procedure described in Section 5.4 are depicted in Fig. 5.21(e).   Ans.

Bending Moment Diagrams The bending moment diagrams for the three members of the frame are shown in Fig. 5.21(f).

Axial Force Diagrams From the free-body diagram of member *AB* in Fig. 5.21(d), we observe that the axial force throughout the length of this member is compressive, with a constant magnitude of 18 k. Therefore, the axial force diagram for this member is a straight line parallel to the *x* axis at a value of -18 k, as shown in Fig. 5.21(g). Similarly, it can be seen from Fig. 5.21(d) that the axial forces in members *BC* and *CD* are also constant, with magnitudes of 0 and -42 k, respectively. The axial force diagrams thus constructed for these members are shown in Fig. 5.21(g). **Ans.**

Qualitative Deflected Shape From the bending moment diagrams of the members of the frame (Fig. 5.21(f)), we observe that the members *AB* and *BC* bend concave to the left and concave upward, respectively. As no bending moment develops in member *CD*, it does not bend but remains straight. A qualitative deflected shape of the frame obtained by connecting the deflected shapes of the three members at the joints is shown in Fig. 5.21(h). As this figure indicates, the deflection of the frame at support *A* is zero. Due to the horizontal load at *B*, joint *B* deflects to the right to *B'*. Since the axial deformations of members are neglected and bending deformations are assumed to be small, joint *B* deflects only in the horizontal direction, and joint *C* deflects by the same amount as joint *B*; that is, $BB' = CC'$. Note that the curvatures of the members are consistent with their bending moment diagrams and that the original 90° angles between members at the rigid joints *B* and *C* have been maintained. **Ans.**

Example 5.14

Draw the shear, bending moment, and axial force diagrams and the qualitative deflected shape for the frame shown in Fig. 5.22(a).

Solution

Static Determinacy $m = 2$, $j = 3$, $r = 3$, and $e_c = 0$. Because $3m + r = 3j + e_c$ and the frame is geometrically stable, it is statically determinate.

Reactions (See Fig. 5.22(b).)

$$+ \rightarrow \sum F_x = 0$$

$$-A_x + 25 = 0 \qquad A_x = 25 \text{ k} \leftarrow$$

$$+ \uparrow \sum F_y = 0$$

$$A_y - 1.6(15) = 0 \qquad A_y = 24 \text{ k} \uparrow$$

$$+ \circlearrowleft \sum M_A = 0$$

$$M_A - 25(10) - 1.6(15)(7.5) = 0 \qquad M_A = 430 \text{ k-ft} \circlearrowright$$

Member End Forces (See Fig. 5.22(c).)

Joint A By applying the equilibrium equations $\sum F_X = 0$, $\sum F_Y = 0$, and $\sum M_A = 0$, we obtain

$$A_X^{AB} = -25 \text{ k} \qquad A_Y^{AB} = 24 \text{ k} \qquad M_A^{AB} = 430 \text{ k-ft}$$

continued

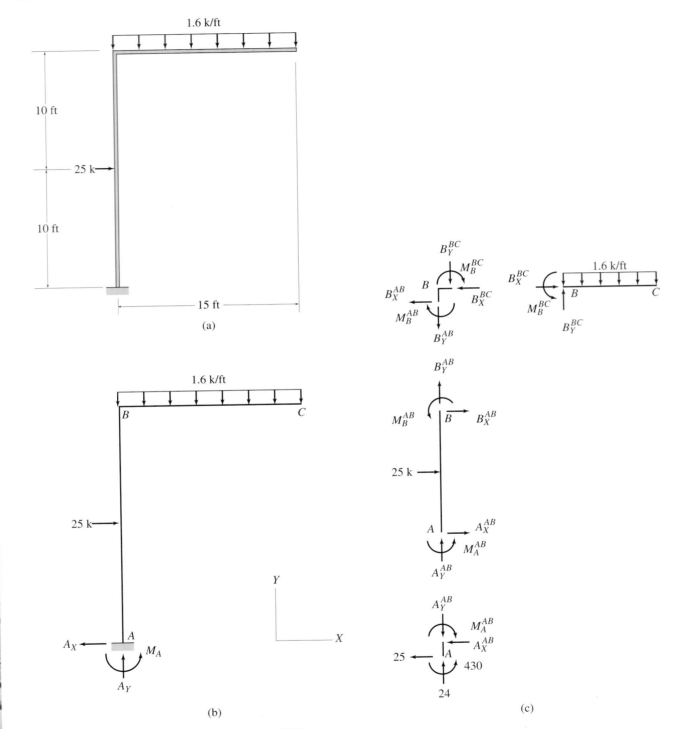

FIG. **5.22**

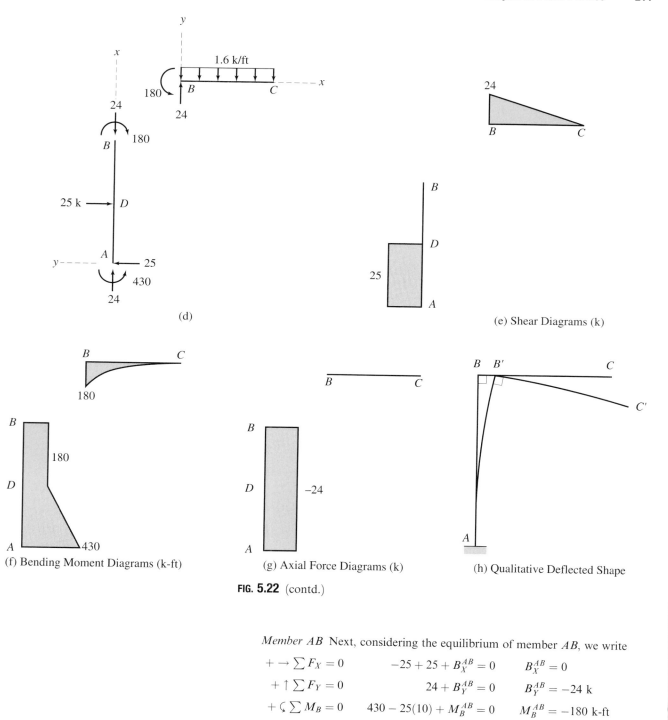

(d)

(e) Shear Diagrams (k)

(f) Bending Moment Diagrams (k-ft)

(g) Axial Force Diagrams (k)

(h) Qualitative Deflected Shape

FIG. 5.22 (contd.)

Member AB Next, considering the equilibrium of member AB, we write

$$+ \rightarrow \sum F_X = 0 \qquad -25 + 25 + B_X^{AB} = 0 \qquad B_X^{AB} = 0$$

$$+ \uparrow \sum F_Y = 0 \qquad 24 + B_Y^{AB} = 0 \qquad B_Y^{AB} = -24 \text{ k}$$

$$+ \zeta \sum M_B = 0 \qquad 430 - 25(10) + M_B^{AB} = 0 \qquad M_B^{AB} = -180 \text{ k-ft}$$

continued

Joint B Applying the three equations of equilibrium, we obtain

$$B_X^{BC} = 0 \qquad B_Y^{BC} = 24 \text{ k} \qquad M_B^{BC} = 180 \text{ k-ft}$$

Member BC (Checking computations.)

$+ \rightarrow \sum F_X = 0$		Checks
$+ \uparrow \sum F_Y = 0$	$24 - 1.6(15) = 0$	Checks
$+ \zeta \sum M_B = 0$	$180 - 1.6(15)(7.5) = 0$	Checks

The member end forces are shown in Fig. 5.22(d).

Shear Diagrams See Fig. 5.22(e).   Ans.

Bending Moment Diagrams See Fig. 5.22(f).   Ans.

Axial Force Diagrams See Fig. 5.22(g).   Ans.

Qualitative Deflected Shape See Fig. 5.22(h).   Ans.

Example 5.15

A gable frame is subjected to a snow loading, as shown in Fig. 5.23(a). Draw the shear, bending moment, and axial force diagrams and the qualitative deflected shape for the frame.

Solution

Static Determinacy $m = 4$, $j = 5$, $r = 4$, and $e_c = 1$. Because $3m + r = 3j + e_c$ and the frame is geometrically stable, it is statically determinate.

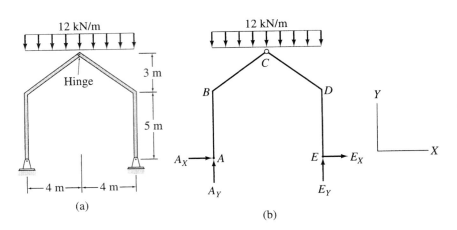

(a)

(b)

FIG. 5.23

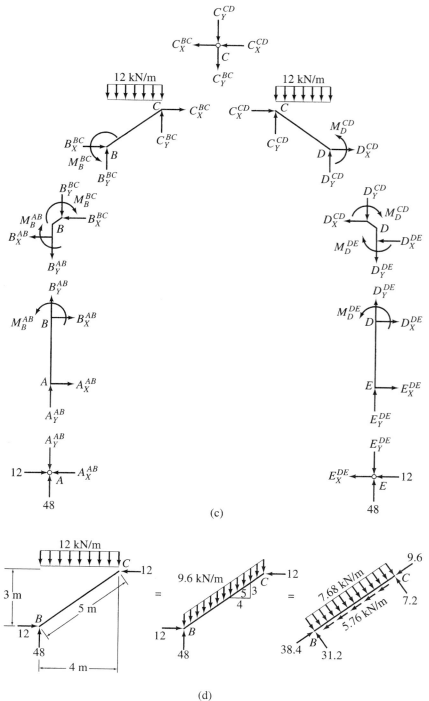

FIG. **5.23** (contd.)

(c)

(d)

continued

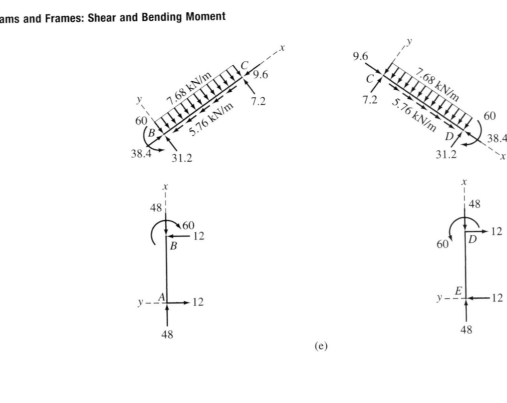

(e)

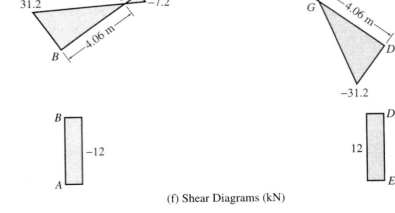

(f) Shear Diagrams (kN)

FIG. 5.23 (contd.)

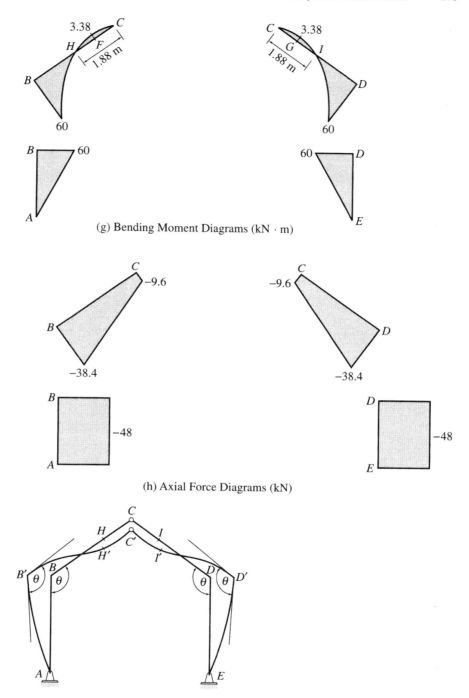

(g) Bending Moment Diagrams (kN · m)

(h) Axial Force Diagrams (kN)

(i) Qualitative Deflected Shape

FIG. 5.23 (contd.)

continued

Reactions (See Fig. 5.23(b).)

$$+\circlearrowleft \sum M_E = 0$$

$$-A_Y(8) + 12(8)(4) = 0 \qquad\qquad A_Y = 48 \text{ kN } \uparrow$$

$$+\uparrow \sum F_Y = 0$$

$$48 - 12(8) + E_Y = 0 \qquad\qquad E_Y = 48 \text{ kN } \uparrow$$

$$+\circlearrowleft \sum M_C^{AC} = 0$$

$$A_X(8) - 48(4) + 12(4)(2) = 0 \qquad\qquad A_X = 12 \text{ kN } \rightarrow$$

$$+\rightarrow \sum F_X = 0$$

$$12 + E_X = 0$$

$$E_X = -12 \text{ kN} \qquad E_X = 12 \text{ kN } \leftarrow$$

Member End Forces (See Fig. 5.23(c).)

Joint A By applying the equations of equilibrium $\sum F_X = 0$ and $\sum F_Y = 0$, we obtain

$$A_X^{AB} = 12 \text{ kN} \qquad A_Y^{AB} = 48 \text{ kN}$$

Member AB Considering the equilibrium of member AB, we obtain

$$B_X^{AB} = -12 \text{ kN} \qquad B_Y^{AB} = -48 \text{ kN} \qquad M_B^{AB} = -60 \text{ kN} \cdot \text{m}$$

Joint B Applying the three equilibrium equations, we obtain

$$B_X^{BC} = 12 \text{ kN} \qquad B_Y^{BC} = 48 \text{ kN} \qquad M_B^{BC} = 60 \text{ kN} \cdot \text{m}$$

Member BC

$$+\rightarrow \sum F_X = 0 \qquad C_X^{BC} = -12 \text{ kN}$$

$$+\uparrow \sum F_Y = 0$$

$$48 - 12(4) + C_Y^{BC} = 0 \qquad C_Y^{BC} = 0$$

$$+\circlearrowleft \sum M_B = 0$$

$$60 - 12(4)(2) + 12(3) = 0 \qquad\qquad\qquad\qquad \text{Checks}$$

Joint C Considering the equilibrium of joint C, we determine

$$C_X^{CD} = 12 \text{ kN} \qquad C_Y^{CD} = 0$$

Member CD

$$+\rightarrow \sum F_X = 0 \qquad D_X^{CD} = -12 \text{ kN}$$

$$+\uparrow \sum F_Y = 0$$

$$-12(4) + D_Y^{CD} = 0 \qquad D_Y^{CD} = 48 \text{ kN}$$

$$+\circlearrowleft \sum M_D = 0$$

$$-12(3) + 12(4)(2) + M_D^{CD} = 0 \qquad M_D^{CD} = -60 \text{ kN} \cdot \text{m}$$

Joint D Applying the three equilibrium equations, we obtain

$$D_X^{DE} = 12 \text{ kN} \qquad D_Y^{DE} = -48 \text{ kN} \qquad M_D^{DE} = 60 \text{ kN} \cdot \text{m}$$

Member DE

$$+ \rightarrow \sum F_X = 0 \qquad E_X^{DE} = -12 \text{ kN}$$

$$+ \uparrow \sum F_Y = 0 \qquad E_Y^{DE} = 48 \text{ kN}$$

$$+ \circlearrowleft \sum M_E = 0$$

$$60 - 12(5) = 0 \qquad\qquad\qquad\qquad\qquad \text{Checks}$$

Joint E

$$+ \rightarrow \sum F_X = 0 \qquad -12 + 12 = 0 \qquad \text{Checks}$$

$$+ \uparrow \sum F_Y = 0 \qquad 48 - 48 = 0 \qquad \text{Checks}$$

Distributed Loads on Inclined Members BC and CD As the 12-kN/m snow loading is specified per horizontal meter, it is necessary to resolve it into components parallel and perpendicular to the directions of members *BC* and *CD*. Consider, for example, member *BC*, as shown in Fig. 5.23(d). The total vertical load acting on this member is $(12 \text{ kN/m})(4 \text{ m}) = 48 \text{ kN}$. Dividing this total vertical load by the length of the member, we obtain the intensity of the vertical distributed load per meter along the inclined length of the member as $48/5 = 9.6 \text{ kN/m}$. The components of this vertical distributed load in the directions parallel and perpendicular to the axis of the member are $(3/5)(9.6) = 5.76 \text{ kN/m}$ and $(4/5)(9.6) = 7.68 \text{ kN/m}$, respectively, as shown in Fig. 5.23(d). The distributed loading for member *CD* is computed similarly and is shown in Fig. 5.23(e).

Shear and Bending Moment Diagrams See Fig. 5.23(f) and (g). **Ans.**

Axial Force Diagrams The equations for axial force in the members of the frame are:

$$\text{Member } AB \quad Q = -48$$

$$\text{Member } BC \quad Q = -38.4 + 5.76x$$

$$\text{Member } CD \quad Q = -9.6 - 5.76x$$

$$\text{Member } DE \quad Q = -48$$

The axial force diagrams are shown in Fig. 5.23(h). **Ans.**

Qualitative Deflected Shape See Fig. 5.23(i). **Ans.**

SUMMARY

In this chapter, we have learned that the internal axial force at any section of a member is equal in magnitude, but opposite in direction, to the algebraic sum of the components in the direction parallel to the axis of the member of all the external loads and reactions acting on either side of the section. We consider it to be positive when the external forces tend to produce tension. The shear at any section of a member is equal in magnitude, but opposite in direction, to the algebraic sum of the components in the direction perpendicular to the axis of the member of all the external loads and reactions acting on either side of the section. We consider it to be positive when the external forces tend to push the portion of the member on the left of the section upward with respect to the portion on the right of the section. The bending moment at any section of a member is equal in magnitude, but opposite in direction, to the algebraic sum of the moments about the section of all the external loads and reactions acting on either side of the section. We consider it to be positive when the external forces and couples tend to bend the member concave upward, causing compression in the upper fibers and tension in the lower fibers at the section.

Shear, bending moment, and axial force diagrams depict the variations of these quantities along the length of the member. Such diagrams can be constructed by determining and plotting the equations expressing these stress resultants in terms of the distance of the section from an end of the member. The construction of shear and bending moment diagrams can be considerably expedited by applying the following relationships that exist between the loads, shears, and bending moments:

slope of shear diagram at a point	= intensity of distributed load at that point	(5.3)
change in shear between points A and B	= area under the distributed load diagram between points A and B	(5.5)
change in shear at the point of application of a concentrated load	= magnitude of the load	(5.12)
slope of bending moment diagram at a point	= shear at that point	(5.8)
change in bending moment between points A and B	= area under the shear diagram between points A and B	(5.10)
change in bending moment at the point of application of a couple	= magnitude of the moment of the couple	(5.14)

A frame is considered to be statically determinate if the shears, bending moments, and axial forces in all its members as well as all the external reactions can be determined by using the equations of equilibrium and condition. If a plane frame contains m members and j joints, is supported by r reactions, and has e_c equations of condition, then if

$$3m + r < 3j + e_c \quad \text{the frame is statically unstable}$$

$$3m + r = 3j + e_c \quad \text{the frame is statically determinate} \quad (5.16)$$

$$3m + r > 3j + e_c \quad \text{the frame is statically indeterminate}$$

The degree of static indeterminacy is given by

$$i = (3m + r) - (3j + e_c) \quad (5.15)$$

A procedure for the determination of member end forces, shears, bending moments, and axial forces in the members of plane statically determinate frames is presented in Section 5.6.

PROBLEMS

Section 5.1

5.1 through 5.11 Determine the axial forces, shears, and bending moments at points A and B of the structure shown.

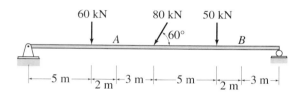

FIG. **P5.1**

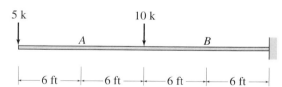

FIG. **P5.2**

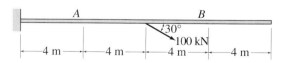

FIG. **P5.3**

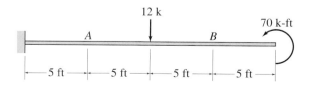

FIG. **P5.4**

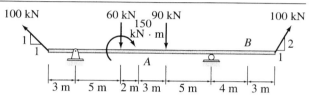

FIG. **P5.5**

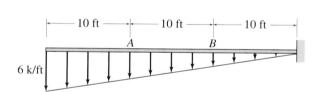

FIG. **P5.6**

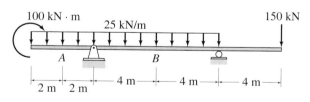

FIG. **P5.7**

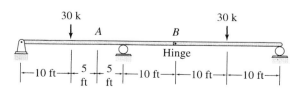

FIG. **P5.8**

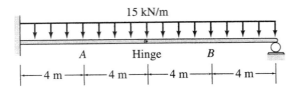

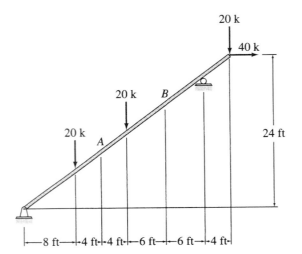

FIG. **P5.9**

FIG. **P5.10**

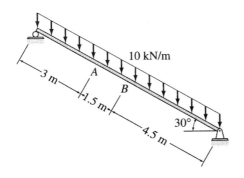

FIG. **P5.11**

Section 5.2

5.12 through 5.28 Determine the equations for shear and bending moment for the beam shown. Use the resulting equations to draw the shear and bending moment diagrams.

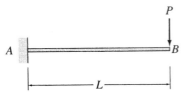

FIG. **P5.12**

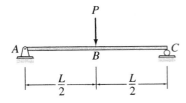

FIG. **P5.13**

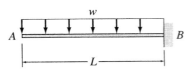

FIG. **P5.14**

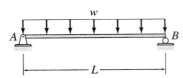

FIG. **P5.15**

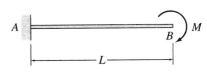

FIG. **P5.16**

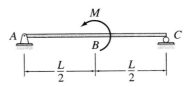

FIG. **P5.17**

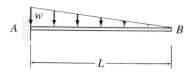

FIG. **P5.18**

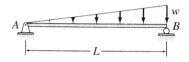

FIG. **P5.19**

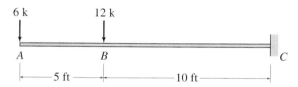

FIG. **P5.20**

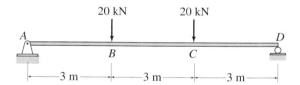

FIG. **P5.21**

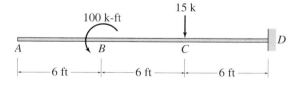

FIG. **P5.22**

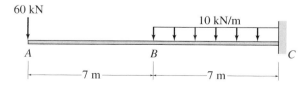

FIG. **P5.23**

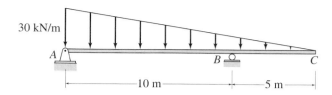

FIG. **P5.24**

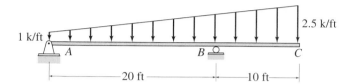

FIG. **P5.25**

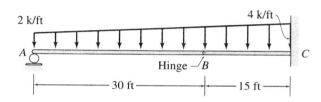

FIG. **P5.26**

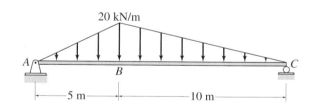

FIG. **P5.27**

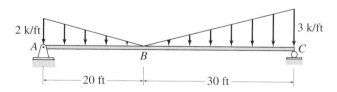

FIG. **P5.28**

Section 5.4

5.29 through 5.51 Draw the shear and bending moment diagrams and the qualitative deflected shape for the beam shown.

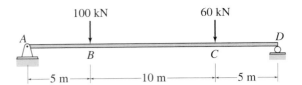

FIG. **P5.29**

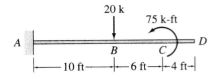

FIG. **P5.35**

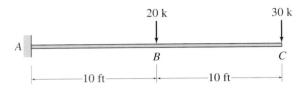

FIG. **P5.30**

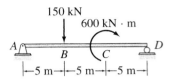

FIG. **P5.36**

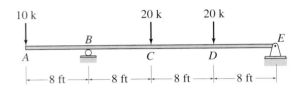

FIG. **P5.31**

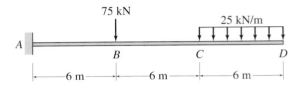

FIG. **P5.37**

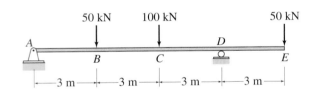

FIG. **P5.32**

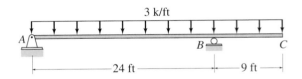

FIG. **P5.38**

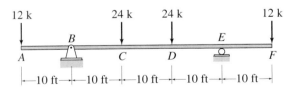

FIG. **P5.33**

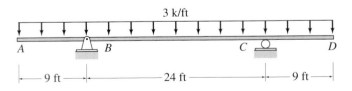

FIG. **P5.39**

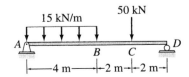

FIG. **P5.34**

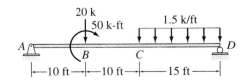

FIG. **P5.40**

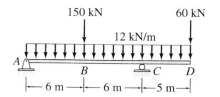

FIG. **P5.41**

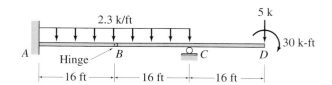

FIG. **P5.46**

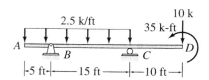

FIG. **P5.42**

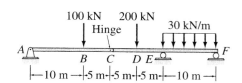

FIG. **P5.47**

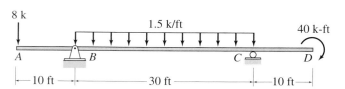

FIG. **P5.43**

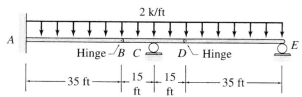

FIG. **P5.48**

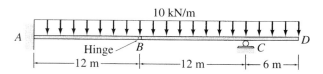

FIG. **P5.44**

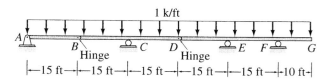

FIG. **P5.49**

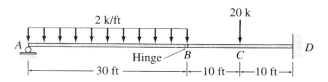

FIG. **P5.45**

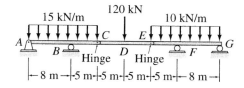

FIG. **P5.50**

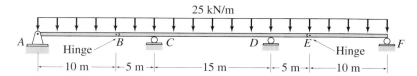

FIG. **P5.51**

5.52 Draw the shear and bending moment diagrams for the reinforced concrete footing subjected to the downward column loading of 1.5 k/ft and the upward soil reaction of 0.5 k/ft, as shown in the figure.

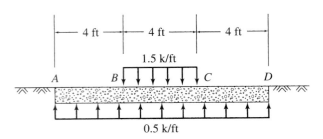

FIG. P5.52

5.53 and 5.54 For the beam shown: (a) determine the distance a for which the maximum positive and negative bending moments in the beam are equal; and (b) draw the corresponding shear and bending moment diagrams for the beam.

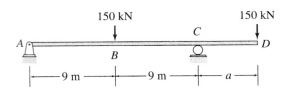

FIG. P5.53

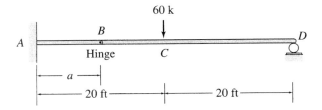

FIG. P5.54

Section 5.5

5.55 and 5.56 Classify each of the plane frames shown as unstable, statically determinate, or statically indeterminate. If statically indeterminate, then determine the degree of static indeterminacy.

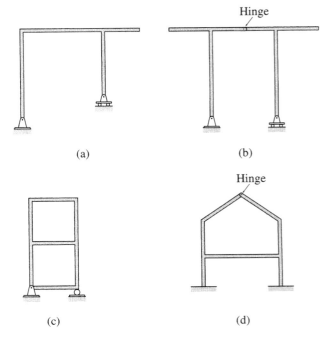

(a)

(b)

(c)

(d)

FIG. P5.55

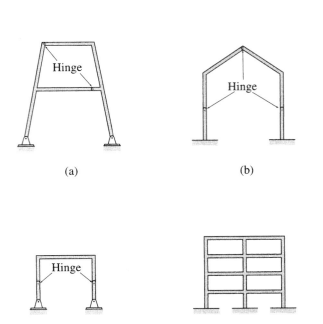

(a)

(b)

(c)

(d)

FIG. P5.56

Section 5.6

5.57 through 5.71 Draw the shear, bending moment, and axial force diagrams and the qualitative deflected shape for the frame shown.

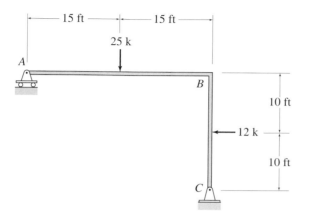

FIG. **P5.57**

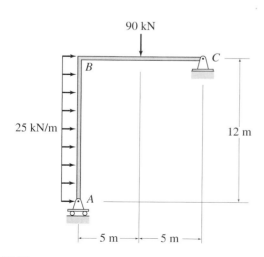

FIG. **P5.58**

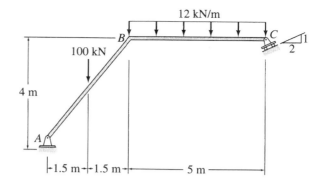

FIG. **P5.59**

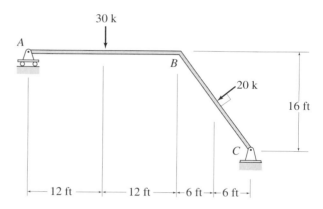

FIG. **P5.60**

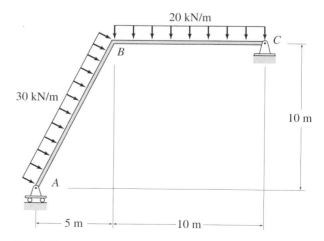

FIG. **P5.61**

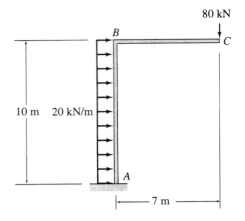

FIG. **P5.62**

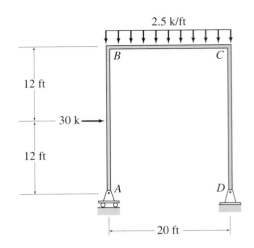

FIG. **P5.64**

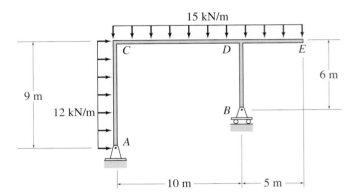

FIG. **P5.65**

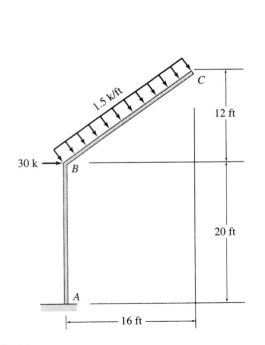

FIG. **P5.63**

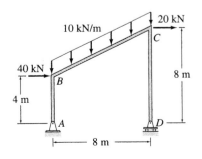

FIG. **P5.66**

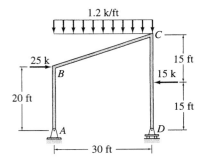

FIG. **P5.67**

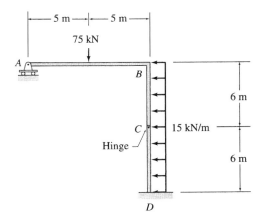

FIG. **P5.68**

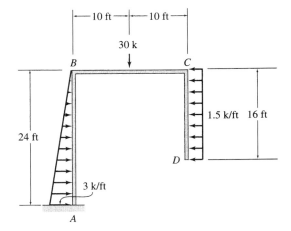

FIG. **P5.69**

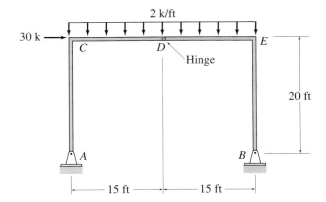

FIG. **P5.70**

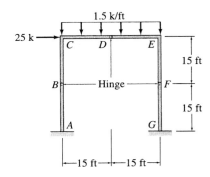

FIG. **P5.71**

6

Deflections of Beams: Geometric Methods

The John Hancock Building,
Chicago
Photo courtesy of Bethlehem Steel Corporation

Structures, like all other physical bodies, deform and change shape when subjected to forces. Other common causes of deformations of structures include temperature changes and support settlements. If the deformations disappear and the structure regains its original shape when the actions causing the deformations are removed, the deformations are termed *elastic deformations*. The permanent deformations of structures are referred to as *inelastic, or plastic, deformations*. In this text, we will focus our attention on *linear elastic deformations*. Such deformations vary linearly with applied loads (for instance, if the magnitudes of the loads acting on the structure are doubled, its deformations are also doubled, and so forth). Recall from Section 3.6 that in order for a structure to respond linearly to applied loads, it must be composed of linear elastic material, and it must undergo small deformations. The principle of superposition is valid for such structures.

For most structures, excessive deformations are undesirable, as they may impair the structure's ability to serve its intended purpose. For example, a high-rise building may be perfectly safe in the sense that the allowable stresses are not exceeded, yet useless (unoccupied) if it deflects excessively due to wind, causing cracks in the walls and windows. Structures are usually designed so that their deflections under normal service conditions will not exceed the allowable values specified in building codes.

From the foregoing discussion, we can see that the computation of deflections forms an essential part of structural analysis. Deflection calculations are also required in the determination of the reactions and stress resultants for statically indeterminate structures, to be considered in Part Three of this text.

The methods that have been developed for computing deflections can be broadly classified into two categories, (1) geometric methods and (2) work-energy methods. As these names imply, geometric methods are based on a consideration of the geometry of the deflected shapes of structures, whereas the work-energy methods are based on the basic principles of work and energy.

In this chapter, we study geometric methods commonly used for determining the slopes and deflections of statically determinate beams. We discuss work-energy methods in the following chapter. First, we derive the differential equation for the deflection of beams; we follow this derivation with brief reviews of the direct (double) integration and superposition methods of computing deflections. (We assume here that the reader is familiar with these methods from a previous course in *mechanics of materials*.) Next, we present the moment-area method for calculating slopes and deflections of beams, the construction of bending moment diagrams by parts, and finally the conjugate-beam method for computing slopes and deflections of beams.

6.1 DIFFERENTIAL EQUATION FOR BEAM DEFLECTION

Consider an initially straight elastic beam subjected to an arbitrary loading acting perpendicular to its centroidal axis and in the plane of symmetry of its cross section, as shown in Fig. 6.1(a). The neutral surface of the beam in the deformed state is referred to as the *elastic curve*. To derive the differential equation defining the elastic curve, we focus our attention on a differential element dx of the beam. The element in the deformed position is shown in Fig. 6.1(b). As this figure indicates, we assume that the plane sections perpendicular to the neutral surface of the beam before bending remain plane and perpendicular to the neutral surface after bending. The sign convention for bending moment M remains the same as established in Chapter 5; that is, a positive bending moment causes compression in the fibers above the neutral surface (in the positive y direction). Tensile strains and stresses are considered to be positive. The slope of the elastic curve, $\theta = dy/dx$, is assumed to be so small that θ^2 is negligible compared to unity; $\sin \theta \approx \theta$ and $\cos \theta \approx 1$. Note that $d\theta$ represents the change in slope over the differential length dx. It can be seen from Fig. 6.1(b) that the deformation of an arbitrary fiber ab located at a distance y from the neutral surface can be expressed as

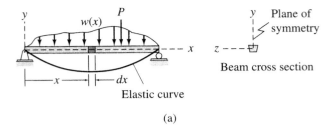

(a)

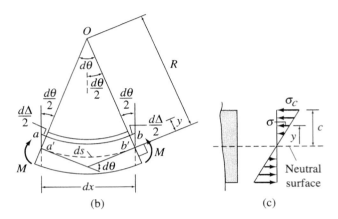

FIG. 6.1

(b)   (c)

$$dΔ = a'b' - ab = -2y\left(\frac{dθ}{2}\right) = -y\, dθ$$

Thus, the strain in fiber ab is equal to

$$ε = \frac{dΔ}{dx} = \frac{dΔ}{ds} = -\frac{y\, dθ}{R\, dθ} = -\frac{y}{R} \qquad (6.1)$$

in which R is the radius of curvature. By substituting the linear stress-strain relationship $ε = σ/E$ into Eq. (6.1), we obtain

$$σ = -\frac{Ey}{R} \qquad (6.2)$$

in which $σ$ is the stress in fiber ab and E represents Young's modulus of elasticity. Equation (6.2) indicates that the stress varies linearly with the distance y from the neutral surface, as shown in Fig. 6.1(c). If $σ_c$ denotes the stress at the uppermost fiber located at a distance c from the neutral surface (Fig. 6.1(c)), then the stress $σ$ at a distance y from the neutral surface can be written as

$$σ = \frac{y}{c}σ_c \qquad (6.3)$$

Since the bending moment M is equal to the sum of the moments about the neutral axis of the forces acting at all the fibers of the beam cross section, we write

$$M = \int_A -\sigma y \, dA \tag{6.4}$$

Substituting Eq. (6.3) into Eq. (6.4), we obtain

$$M = -\frac{\sigma_c}{c} \int_A y^2 \, dA = -\frac{\sigma_c}{c} I$$

or

$$\sigma_c = -\frac{Mc}{I}$$

Using Eq. (6.3), we obtain

$$\sigma = -\frac{My}{I} \tag{6.5}$$

where I is the moment of inertia of the beam cross section.

Next, by combining Eqs. (6.2) and (6.5), we obtain the moment-curvature relationship

$$\frac{1}{R} = \frac{M}{EI} \tag{6.6}$$

in which the product EI is commonly referred to as the *flexural rigidity* of the beam. To express Eq. (6.6) in Cartesian coordinates, we recall (from *calculus*) the relationship

$$\frac{1}{R} = \frac{d^2y/dx^2}{[1 + (dy/dx)^2]^{3/2}} \tag{6.7}$$

in which y represents the vertical deflection. As stated previously, for small slopes the square of the slope, $(dy/dx)^2$, is negligible in comparison with unity. Thus, Eq. (6.7) reduces to

$$\frac{1}{R} \approx \frac{d^2y}{dx^2} \tag{6.8}$$

By substituting Eq. (6.8) into Eq. (6.6), we obtain the following differential equation for the deflection of beams:

$$\frac{d^2y}{dx^2} = \frac{M}{EI} \tag{6.9}$$

This equation is also referred to as the Bernoulli-Euler beam equation. Because $\theta = dy/dx$, Eq. (6.9) can also be expressed as

$$\frac{d\theta}{dx} = \frac{M}{EI} \tag{6.10}$$

6.2 DIRECT INTEGRATION METHOD

The direct integration method essentially involves writing the expression for M/EI (bending moment divided by flexural rigidity of the beam) in terms of the distance x along the axis of the beam and integrating this expression successively to obtain equations for the slope and deflection of the elastic curve. The constants of integration are determined from the boundary conditions. The direct integration method proves to be most convenient for computing slopes and deflections of beams for which M/EI can be expressed as a single continuous function of x over the entire length of the beam. However, the application of the method to structures for which the M/EI function is not continuous can become quite complicated. This problem occurs because each discontinuity, due to a change in loading and/or the flexural rigidity (EI), introduces two additional constants of integration in the analysis, which must be evaluated by applying the conditions of continuity of the elastic curve, a process that can be quite tedious. The difficulty can, however, be circumvented, and the analysis can be somewhat simplified by employing the *singularity functions* defined in most textbooks on *mechanics of materials*.

Example 6.1

Determine the equations for the slope and deflection of the beam shown in Fig. 6.2(a) by the direct integration method. Also, compute the slope at each end and the deflection at the midspan of the beam. EI is constant.

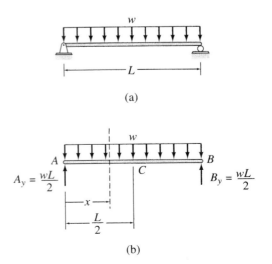

FIG. **6.2**

Solution

Reactions See Fig. 6.2(b).

$$+ \to \sum F_x = 0 \qquad A_x = 0$$

$$+ \circlearrowleft \sum M_B = 0$$

$$-A_y(L) + w(L)\left(\frac{L}{2}\right) = 0 \qquad A_y = \frac{wL}{2} \uparrow$$

$$+ \uparrow \sum F_y = 0$$

$$\left(\frac{wL}{2}\right) - (wL) + B_y = 0 \qquad B_y = \frac{wL}{2} \uparrow$$

Equation for Bending Moment To determine the equation for bending moment for the beam, we pass a section at a distance x from support A, as shown in Fig. 6.2(b). Considering the free body to the left of this section, we obtain

$$M = \frac{wL}{2}(x) - (wx)\left(\frac{x}{2}\right) = \frac{w}{2}(Lx - x^2)$$

Equation for M/EI The flexural rigidity, EI, of the beam is constant, so the equation for M/EI can be written as

$$\frac{d^2y}{dx^2} = \frac{M}{EI} = \frac{w}{2EI}(Lx - x^2)$$

Equations for Slope and Deflection The equation for the slope of the elastic curve of the beam can be obtained by integrating the equation for M/EI as

$$\theta = \frac{dy}{dx} = \frac{w}{2EI}\left(\frac{Lx^2}{2} - \frac{x^3}{3}\right) + C_1$$

Integrating once more, we obtain the equation for deflection as

$$y = \frac{w}{2EI}\left(\frac{Lx^3}{6} - \frac{x^4}{12}\right) + C_1 x + C_2$$

The constants of integration, C_1 and C_2, are evaluated by applying the following boundary conditions:

$$\text{At end } A, \qquad x = 0, \qquad y = 0$$

$$\text{At end } B, \qquad x = L, \qquad y = 0$$

By applying the first boundary condition—that is, by setting $x = 0$ and $y = 0$ in the equation for y—we obtain $C_2 = 0$. Next, by using the second boundary condition—that is, by setting $x = L$ and $y = 0$ in the equation for y—we obtain

$$0 = \frac{w}{2EI}\left(\frac{L^4}{6} - \frac{L^4}{12}\right) + C_1 L$$

from which

$$C_1 = -\frac{wL^3}{24EI}$$

continued

Thus, the equations for slope and deflection of the beam are

$$\theta = \frac{w}{2EI}\left(\frac{Lx^2}{2} - \frac{x^3}{3} - \frac{L^3}{12}\right) \qquad (1) \qquad \text{Ans.}$$

$$y = \frac{wx}{12EI}\left(Lx^2 - \frac{x^3}{2} - \frac{L^3}{2}\right) \qquad (2) \qquad \text{Ans.}$$

Slopes at Ends A and B By substituting $x = 0$ and L, respectively, into Eq. (1), we obtain

$$\theta_A = -\frac{wL^3}{24EI} \quad \text{or} \quad \theta_A = \frac{wL^3}{24EI} \qquad \qquad \text{Ans.}$$

$$\theta_B = \frac{wL^3}{24EI} \quad \text{or} \quad \theta_B = \frac{wL^3}{24EI} \qquad \qquad \text{Ans.}$$

Deflection at Midspan By substituting $x = L/2$ into Eq. (2), we obtain

$$y_C = -\frac{5wL^4}{384EI} \quad \text{or} \quad y_C = \frac{5wL^4}{384EI} \downarrow \qquad \text{Ans.}$$

Example 6.2

Determine the slope and deflection at point B of the cantilever beam shown in Fig. 6.3(a) by the direct integration method.

Solution

Equation for Bending Moment We pass a section at a distance x from support A, as shown in Fig. 6.3(b). Considering the free body to the right of this section, we write the equation for bending moment as

$$M = -15(20 - x)$$

Equation for M/EI

$$\frac{d^2y}{dx^2} = \frac{M}{EI} = -\frac{15}{EI}(20 - x)$$

Equations for Slope and Deflection By integrating the equation for M/EI, we determine the equation for slope as

$$\theta = \frac{dy}{dx} = -\frac{15}{EI}\left(20x - \frac{x^2}{2}\right) + C_1$$

Integrating once more, we obtain the equation for deflection as

$$y = -\frac{15}{EI}\left(10x^2 - \frac{x^3}{6}\right) + C_1x + C_2$$

The constants of integration, C_1 and C_2, are evaluated by using the boundary conditions that $\theta = 0$ at $x = 0$, and $y = 0$ at $x = 0$. By applying the first boundary condition—that is, by setting $\theta = 0$ and $x = 0$ in the equation for θ— we obtain $C_1 = 0$. Similarly, by applying the second boundary condition—that

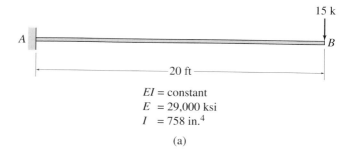

EI = constant
E = 29,000 ksi
I = 758 in.^{4}

(a)

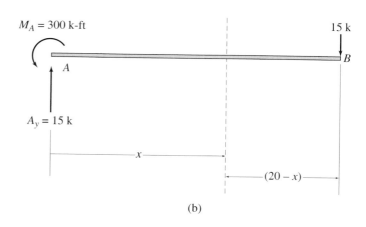

FIG. 6.3

(b)

is, by setting $y = 0$ and $x = 0$ in the equation for y—we obtain $C_2 = 0$. Thus, the equations for slope and deflection of the beam are

$$\theta = -\frac{15}{EI}\left(20x - \frac{x^2}{2}\right)$$

$$y = -\frac{15}{EI}\left(10x^2 - \frac{x^3}{6}\right)$$

Slope and Deflection at End B By substituting $x = 20$ ft, $E = 29,000(12^2)$ ksf, and $I = 758/(12^4)$ ft^4 into the foregoing equations for slope and deflection, we obtain

$$\theta_B = -0.0197 \text{ rad} \qquad \text{or} \quad \theta_B = 0.0197 \text{ rad} \qquad \text{Ans.}$$

$$y_B = -0.262 \text{ ft} = -3.14 \text{ in.} \quad \text{or} \quad y_B = 3.14 \text{ in.} \downarrow \qquad \text{Ans.}$$

6.3 SUPERPOSITION METHOD

When a beam is subjected to several loads, it is usually convenient to determine slope or deflection caused by the combined effect of loads by superimposing (algebraically adding) the slopes or deflections due to each of the loads acting individually on the beam. The slope and deflection due to each individual load can be computed by using either the direct integration method described previously or one of the other methods discussed in subsequent sections. Also, many structural engineering handbooks (e.g., *Manual of Steel Construction* published by the *American Institute of Steel Construction*) contain deflection formulas for beams for various types of loads and support conditions, which can be used for this purpose. Such formulas for slopes and deflections of beams for some common types of loads and support conditions are given inside the front cover of this book for convenient reference.

6.4 MOMENT-AREA METHOD

The moment-area method for computing slopes and deflections of beams was developed by Charles E. Greene in 1873. The method is based on two theorems, called the *moment-area theorems*, relating the geometry of the elastic curve of a beam to its M/EI diagram, which is constructed by dividing the ordinates of the bending moment diagram by the flexural rigidity EI. The method utilizes graphical interpretations of integrals involved in the solution of the deflection differential equation (Eq. (6.9)) in terms of the areas and the moments of areas of the M/EI diagram. Therefore, it is more convenient to use for beams with loading discontinuities and the variable EI, as compared to the direct integration method described previously.

To derive the moment-area theorems, consider a beam subjected to an arbitrary loading as shown in Fig. 6.4. The elastic curve and the M/EI diagram for the beam are also shown in the figure. Focusing our attention on a differential element dx of the beam, we recall from the previous section (Eq. (6.10)) that $d\theta$, which represents the change in slope of the elastic curve over the differential length dx, is given by

$$d\theta = \frac{M}{EI}dx \qquad (6.11)$$

Note that the term $(M/EI)\,dx$ represents an infinitesimal area under the M/EI diagram, as shown in Fig. 6.4. To determine the change in slope between two arbitrary points A and B on the beam, we integrate Eq. (6.11) from A to B to obtain

$$\int_A^B d\theta = \int_A^B \frac{M}{EI}dx$$

or

$$\theta_{BA} = \theta_B - \theta_A = \int_A^B \frac{M}{EI}dx \qquad (6.12)$$

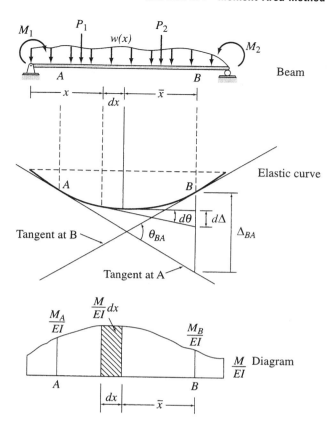

FIG. **6.4**

in which θ_A and θ_B are the slopes of the elastic curve at points A and B, respectively, with respect to the axis of the beam in the undeformed (horizontal) state, θ_{BA} denotes the angle between the tangents to the elastic curve at A and B, and $\int_A^B (M/EI)\, dx$ represents the area under the M/EI diagram between points A and B.

Equation (6.12) represents the mathematical expression of the *first moment-area theorem*, which can be stated as follows:

> The change in slope between the tangents to the elastic curve at any two points is equal to the area under the M/EI diagram between the two points, provided that the elastic curve is continuous between the two points.

As noted, this theorem applies only to those portions of the elastic curve in which there are no discontinuities due to the presence of internal hinges. In applying the first moment-area theorem, if the area of the M/EI diagram between any two points is positive, then the angle from the tangent at the point to the left to the tangent at the point to the right will be counterclockwise, and this change in slope is considered to be positive; and vice versa.

Considering again the beam shown in Fig. 6.4, we observe that the deviation $d\Delta$ between the tangents drawn at the ends of the differential element dx on a line perpendicular to the undeformed axis of the beam from a point B is given by

$$d\Delta = \bar{x}(d\theta) \tag{6.13}$$

where $\bar{x}$ is the distance from B to the element dx. Substitution of Eq. (6.11) into Eq. (6.13) yields

$$d\Delta = \left(\frac{M}{EI}\right)\bar{x}\,dx \tag{6.14}$$

Note that the term on the right-hand side of Eq. (6.14) represents the moment of the infinitesimal area corresponding to dx about B. Integrating Eq. (6.14) between any two arbitrary points A and B on the beam, we obtain

$$\int_A^B d\Delta = \int_A^B \frac{M}{EI}\bar{x}\,dx$$

or

$$\Delta_{BA} = \int_A^B \frac{M}{EI}\bar{x}\,dx \tag{6.15}$$

in which Δ_{BA} represents the *tangential deviation of B* from the tangent at A, which is the deflection of point B in the direction perpendicular to the undeformed axis of the beam from the tangent at point A, and $\int_A^B (M/EI)\bar{x}\,dx$ represents the moment of the area under the M/EI diagram between points A and B about point B.

Equation (6.15) represents the mathematical expression of the *second moment-area theorem*, which can be stated as follows:

> The tangential deviation in the direction perpendicular to the undeformed axis of the beam of a point on the elastic curve from the tangent to the elastic curve at another point is equal to the moment of the area under the M/EI diagram between the two points about the point at which the deviation is desired, provided that the elastic curve is continuous between the two points.

It is important to note the order of the subscripts used for Δ in Eq. (6.15). The first subscript denotes the point where the deviation is determined and about which the moments are evaluated, whereas the second subscript denotes the point where the tangent to the elastic curve is drawn. Also, since the distance $\bar{x}$ in Eq. (6.15) is always taken as positive, the sign of Δ_{BA} is the same as that of the area of the M/EI diagram between A and B. If the area of the M/EI diagram between A and B is positive, then Δ_{BA} is also positive, and point B lies above (in the positive y direction) the tangent to the elastic curve at point A and vice versa.

Procedure for Analysis

In order to apply the moment-area theorems to compute the slopes and deflections of a beam, it is necessary to draw a qualitative deflected shape of the beam using its bending moment diagram. In this regard, recall from Section 5.3 that a positive bending moment bends the beam concave upward, whereas a negative bending moment bends it concave downward. Also, at a fixed support, both the slope and the deflection of the beam must be zero; therefore, the tangent to the elastic curve at this point is in the direction of the undeformed axis, whereas at a hinged or a roller support, the deflection is zero, but the slope may not be zero. To facilitate the computation of areas and moments of areas of the M/EI diagrams, the formulas for the areas and centroids of common geometric shapes are listed in Appendix A.

Instead of adopting a formal sign convention, it is common practice to use an intuitive approach in solving problems using the moment-area method. In this approach, the slopes and deflections at the various points are assumed to be positive in the directions shown on the sketch of the deflected shape or elastic curve of the structure. Any area of the M/EI diagram that tends to increase the quantity under consideration is considered to be positive and vice versa. A positive answer for a slope or deflection indicates that the sense of that quantity as assumed on the elastic curve is correct. Conversely, a negative answer indicates that the correct sense is opposite to that initially assumed on the elastic curve.

In applying the moment-area theorems, it is important to realize that these theorems in general do not directly provide the slope and deflection at a point with respect to the undeformed axis of the beam (which are usually of practical interest); instead, they provide the slope and deflection of a point relative to the tangent to the elastic curve at another point. Therefore, before the slope or deflection at an arbitrary point on the beam can be computed, a point must be identified where the slope of the tangent to the elastic curve is either initially known or can be determined by using the support conditions. Once this *reference tangent* has been established, the slope and deflection at any point on the beam can be computed by applying the moment-area theorems. In cantilever beams, since the slope of the tangent to the elastic curve at the fixed support is zero, this tangent can be used as the reference tangent. In the case of beams for which a tangent with zero slope cannot be located by inspection, it is usually convenient to use the tangent at one of the supports as the reference tangent. The slope of this reference tangent can be determined by using the conditions of zero deflections at the reference support and an adjacent support.

The magnitudes of the slopes and deflections of structures are usually very small, so from a computational viewpoint it is usually convenient to determine the solution in terms of EI and then substitute the numerical values of E and I at the final stage of the analysis to obtain the numerical magnitudes of the slopes and deflections. When the moment of inertia varies along the length of a beam, it is convenient to express the moments of inertia of the various segments of the beam in terms of a single *reference moment of inertia*, which is then carried symbolically through the analysis.

Example 6.3

Determine the slopes and deflections at points B and C of the cantilever beam shown in Fig. 6.5(a) by the moment-area method.

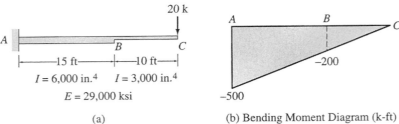

(a)

(b) Bending Moment Diagram (k-ft)

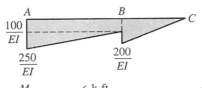

(c) $\dfrac{M}{EI}$ Diagram $\left(\dfrac{\text{k-ft}}{EI}\ \text{with}\ I = 3{,}000\ \text{in.}^4\right)$

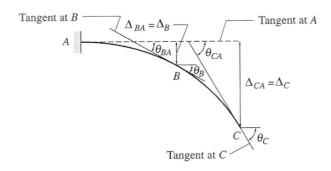

(d) Elastic Curve

FIG. **6.5**

Solution

Bending Moment Diagram The bending moment diagram for the beam is shown in Fig. 6.5(b).

M/EI Diagram As indicated in Fig. 6.5(a), the values of the moment of inertia of the segments AB and BC of the beam are 6,000 in.[4] and 3,000 in.[4], respectively. Using $I = I_{BC} = 3{,}000$ in.[4] as the reference moment of inertia, we express I_{AB} in terms of I as

$$I_{AB} = 6{,}000 = 2(3{,}000) = 2I$$

which indicates that in order to obtain the M/EI diagram in terms of EI, we must divide the bending moment diagram for segment AB by 2, as shown in Fig. 6.5(c).

Elastic Curve The elastic curve for the beam is shown in Fig. 6.5(d). Note that because the M/EI diagram is negative, the beam bends concave downward. Since the support at A is fixed, the slope at A is zero ($\theta_A = 0$); that is, the tangent to the elastic curve at A is horizontal, as shown in the figure.

Slope at B With the slope at A known, we can determine the slope at B by evaluating the change in slope θ_{BA} between A and B (which is the angle between the tangents to the elastic curve at points A and B, as shown in Fig. 6.5(d)). According to the first moment-area theorem, θ_{BA} = area of the M/EI diagram between A and B. This area can be conveniently evaluated by dividing the M/EI diagram into triangular and rectangular parts, as shown in Fig. 6.5(c). Thus,

$$\theta_{BA} = \frac{1}{EI}\left[(100)(15) + \frac{1}{2}(150)(15)\right] = \frac{2{,}625 \text{ k-ft}^2}{EI}$$

From Fig. 6.5(d), we can see that because the tangent at A is horizontal (in the direction of the undeformed axis of the beam), the slope at $B(\theta_B)$ is equal to the angle θ_{BA} between the tangents at A and B; that is,

$$\theta_B = \theta_{BA} = \frac{2{,}625 \text{ k-ft}^2}{EI} = \frac{2{,}625(12)^2 \text{ k-in.}^2}{EI}$$

Substituting the numerical values of $E = 29{,}000$ ksi and $I = 3{,}000$ in.^{4}, we obtain

$$\theta_B = \frac{2{,}625(12)^2}{(29{,}000)(3{,}000)} \text{ rad} = 0.0043 \text{ rad}$$

$$\theta_B = 0.0043 \text{ rad} \qquad\qquad\qquad \text{Ans.}$$

Deflection at B From Fig. 6.5(d), it can be seen that the deflection of B with respect to the undeformed axis of the beam is equal to the tangential deviation of B from the tangent at A; that is,

$$\Delta_B = \Delta_{BA}$$

According to the second moment-area theorem,

Δ_{BA} = moment of the area of the M/EI diagram between A and B about B

$$= \frac{1}{EI}\left[(100)(15)(7.5) + \frac{1}{2}(150)(15)(10)\right] = \frac{22{,}500 \text{ k-ft}^3}{EI}$$

Therefore,

$$\Delta_B = \Delta_{BA} = \frac{22{,}500 \text{ k-ft}^3}{EI}$$

$$= \frac{22{,}500(12)^3}{(29{,}000)(3{,}000)} = 0.45 \text{ in.}$$

$$\Delta_B = 0.45 \text{ in.} \downarrow \qquad\qquad\qquad \text{Ans.}$$

Slope at C From Fig. 6.5(d), we can see that

$$\theta_C = \theta_{CA}$$

continued

where

$$\theta_{CA} = \text{area of the } M/EI \text{ diagram between } A \text{ and } C$$

$$= \frac{1}{EI}\left[(100)(15) + \frac{1}{2}(150)(15) + \frac{1}{2}(200)(10)\right] = \frac{3,625 \text{ k-ft}^2}{EI}$$

Therefore,

$$\theta_C = \theta_{CA} = \frac{3,625 \text{ k-ft}^2}{EI}$$

$$= \frac{3,625(12)^2}{(29,000)(3,000)} = 0.006 \text{ rad}$$

$$\theta_C = 0.006 \text{ rad} \qquad \qquad \text{Ans.}$$

Deflection at C It can be seen from Fig. 6.5(d) that

$$\Delta_C = \Delta_{CA}$$

where

$$\Delta_{CA} = \text{moment of the area of the } M/EI \text{ diagram between } A \text{ and } C \text{ about } C$$

$$= \frac{1}{EI}\left[(100)(15)(7.5 + 10) + \frac{1}{2}(150)(15)(10 + 10) + \frac{1}{2}(200)(10)(6.67)\right]$$

$$= \frac{55,420 \text{ k-ft}^3}{EI}$$

Therefore,

$$\Delta_C = \Delta_{CA} = \frac{55,420 \text{ k-ft}^3}{EI}$$

$$= \frac{55,420(12)^3}{(29,000)(3,000)} = 1.1 \text{ in.}$$

$$\Delta_C = 1.1 \text{ in.} \downarrow \qquad \qquad \text{Ans.}$$

Example 6.4

Use the moment-area method to determine the slopes at ends A and D and the deflections at points B and C of the beam shown in Fig. 6.6(a).

Solution

M/EI Diagram Because EI is constant along the length of the beam, the shape of the M/EI diagram is the same as that of the bending moment diagram. The M/EI diagram is shown in Fig. 6.6(b).

Elastic Curve The elastic curve for the beam is shown in Fig. 6.6(c).

Slope at A The slope of the elastic curve is not known at any point on the beam, so we will use the tangent at support A as the reference tangent and determine its slope, θ_A, from the conditions that the deflections at the support points A and D are zero. From Fig. 6.6(c), we can see that

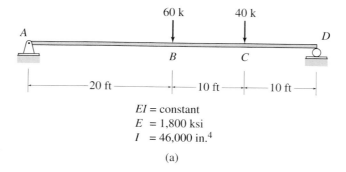

EI = constant
E = 1,800 ksi
I = 46,000 in.^{4}

(a)

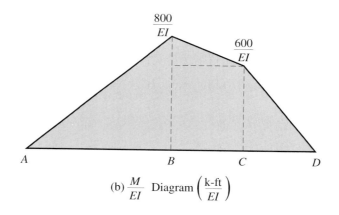

(b) $\dfrac{M}{EI}$ Diagram $\left(\dfrac{\text{k-ft}}{EI}\right)$

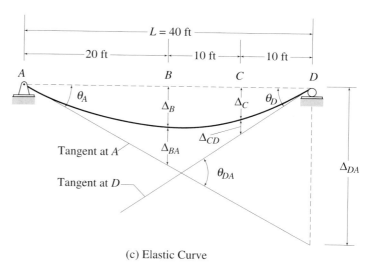

(c) Elastic Curve

FIG. 6.6

continued

$$\theta_A = \frac{\Delta_{DA}}{L}$$

in which θ_A is assumed to be so small that $\tan \theta_A \approx \theta_A$. To evaluate the tangential deviation Δ_{DA}, we apply the second moment-area theorem:

Δ_{DA} = moment of the area of the M/EI diagram between A and D about D

$$\Delta_{DA} = \frac{1}{EI}\left[\frac{1}{2}(800)(20)\left(\frac{20}{3}+20\right)+\frac{1}{2}(200)(10)\left(\frac{20}{3}+10\right)\right.$$

$$\left.+ 600(10)(15)+\frac{1}{2}(600)(10)\left(\frac{20}{3}\right)\right]$$

$$= \frac{340{,}000 \text{ k-ft}^3}{EI}$$

Therefore, the slope at A is

$$\theta_A = \frac{\Delta_{DA}}{L} = \frac{340{,}000/EI}{40} = \frac{8{,}500 \text{ k-ft}^2}{EI}$$

Substituting the numerical values of E and I, we obtain

$$\theta_A = \frac{8{,}500(12)^2}{(1{,}800)(46{,}000)} = 0.015 \text{ rad}$$

$$\theta_A = 0.015 \text{ rad} \quad \diagdown \qquad \qquad \text{Ans.}$$

Slope at D From Fig. 6.6(c), we can see that

$$\theta_D = \theta_{DA} - \theta_A$$

in which, according to the first moment-area theorem,

θ_{DA} = area of the M/EI diagram between A and D

$$= \frac{1}{EI}\left[\frac{1}{2}(800)(20)+\frac{1}{2}(200)(10)+600(10)+\frac{1}{2}(600)(10)\right]$$

$$= \frac{18{,}000 \text{ k-ft}^2}{EI}$$

Therefore,

$$\theta_D = \frac{18{,}000}{EI}-\frac{8{,}500}{EI} = \frac{9{,}500 \text{ k-ft}^2}{EI}$$

$$\theta_D = \frac{9{,}500(12)^2}{(1{,}800)(46{,}000)} = 0.017 \text{ rad}$$

$$\theta_D = 0.017 \text{ rad} \quad \diagup \qquad \qquad \text{Ans.}$$

Deflection at B Considering the portion AB of the elastic curve in Fig. 6.6(c), and realizing that θ_A is so small that $\tan \theta_A \approx \theta_A$, we write

$$\theta_A = \frac{\Delta_B + \Delta_{BA}}{20}$$

from which

$$\Delta_B = 20\theta_A - \Delta_{BA}$$

where

Δ_{BA} = moment of the area of the M/EI diagram between A and B about B

$$= \frac{1}{EI}\left[\frac{1}{2}(800)(20)\left(\frac{20}{3}\right)\right]$$

$$= \frac{53,333.33 \text{ k-ft}^3}{EI}$$

Therefore,

$$\Delta_B = 20\left(\frac{8,500}{EI}\right) - \frac{53,333.33}{EI} = \frac{116,666.67 \text{ k-ft}^3}{EI}$$

$$\Delta_B = \frac{116,666.67(12)^3}{(1,800)(46,000)} = 2.43 \text{ in.}$$

$$\Delta_B = 2.43 \text{ in.} \downarrow \qquad\qquad\qquad \text{Ans.}$$

Deflection at C Finally, considering the portion CD of the elastic curve in Fig. 6.6(c) and assuming θ_D to be small (so that $\tan\theta_D \approx \theta_D$), we write

$$\theta_D = \frac{\Delta_C + \Delta_{CD}}{10}$$

or

$$\Delta_C = 10\theta_D - \Delta_{CD}$$

where

$$\Delta_{CD} = \frac{1}{EI}\left[\frac{1}{2}(600)(10)\left(\frac{10}{3}\right)\right] = \frac{10,000 \text{ k-ft}^3}{EI}$$

Therefore,

$$\Delta_C = 10\left(\frac{9,500}{EI}\right) - \frac{10,000}{EI} = \frac{85,000 \text{ k-ft}^3}{EI}$$

$$\Delta_C = \frac{85,000(12)^3}{(1,800)(46,000)} = 1.77 \text{ in.}$$

$$\Delta_C = 1.77 \text{ in.} \downarrow \qquad\qquad\qquad \text{Ans.}$$

Example 6.5

Determine the maximum deflection for the beam shown in Fig. 6.7(a) by the moment-area method.

Solution

M/EI Diagram The M/EI diagram is shown in Fig. 6.7(b).

Elastic Curve The elastic curve for the beam is shown in Fig. 6.7(c).

continued

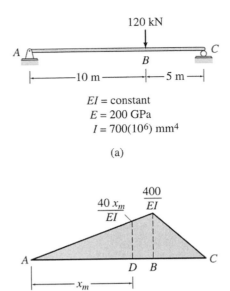

$EI = $ constant
$E = 200$ GPa
$I = 700(10^6)$ mm^4

(a)

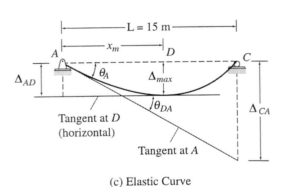

(b) $\dfrac{M}{EI}$ Diagram $\left(\dfrac{\text{kN} \cdot \text{m}}{EI}\right)$

(c) Elastic Curve

FIG. **6.7**

Slope at A   The slope of the elastic curve is not known at any point on the beam, so we will use the tangent at support A as the reference tangent and determine its slope, θ_A, from the conditions that the deflections at the support points A and C are zero. From Fig. 6.7(c), we can see that

$$\theta_A = \frac{\Delta_{CA}}{15}$$

To evaluate the tangential deviation Δ_{CA}, we apply the second moment-area theorem:

Δ_{CA} = moment of the area of the M/EI diagram between A and C about C

$$\Delta_{CA} = \frac{1}{EI}\left[\frac{1}{2}(400)(10)\left(\frac{10}{3}+5\right)+\frac{1}{2}(400)(5)\left(\frac{10}{3}\right)\right]$$

$$= \frac{20{,}000 \text{ kN} \cdot \text{m}^3}{EI}$$

Therefore, the slope at A is

$$\theta_A = \frac{20{,}000/EI}{15} = \frac{1{,}333.33 \text{ kN} \cdot \text{m}^2}{EI}$$

Location of the Maximum Deflection If the maximum deflection occurs at point D, located at a distance x_m from the left support A (see Fig. 6.7(c)), then the slope at D must be zero; therefore,

$$\theta_{DA} = \theta_A = \frac{1{,}333.33 \text{ kN} \cdot \text{m}^2}{EI}$$

which indicates that in order for the slope at D to be zero (i.e., the maximum deflection occurs at D), the area of the M/EI diagram between A and D must be equal to $1{,}333.33/EI$. We use this condition to determine the location of point D:

$$\theta_{DA} = \text{area of the } \frac{M}{EI} \text{ diagram between } A \text{ and } D = \frac{1{,}333.33}{EI}$$

or

$$\frac{1}{2}\left(\frac{40x_m}{EI}\right)x_m = \frac{1{,}333.33}{EI}$$

from which

$$x_m = 8.16 \text{ m}$$

Maximum Deflection From Fig. 6.7(c), we can see that

$$\Delta_{\max} = \Delta_{AD}$$

where

Δ_{AD} = moment of the area of the M/EI diagram between A and D about A

$$= \frac{1}{2}\frac{(40)(8.16)}{EI}(8.16)\left(\frac{2}{3}\right)(8.16)$$

$$= \frac{7{,}244.51 \text{ kN} \cdot \text{m}^3}{EI}$$

Therefore,

$$\Delta_{\max} = \frac{7{,}244.51 \text{ kN} \cdot \text{m}^3}{EI}$$

Substituting $E = 200$ GPa $= 200(10^6)$ kN/m^2 and $I = 700(10^6)$ mm$^4 =$ $700(10^{-6})$ m^4, we obtain

$$\Delta_{\max} = \frac{7{,}244.51}{200(10^6)(700)(10^{-6})} = 0.0517 \text{ m}$$

$$\Delta_{\max} = 51.7 \text{ mm} \downarrow \qquad\qquad \textbf{Ans.}$$

Example 6.6

Use the moment-area method to determine the slope at point A and the deflection at point C of the beam shown in Fig. 6.8(a).

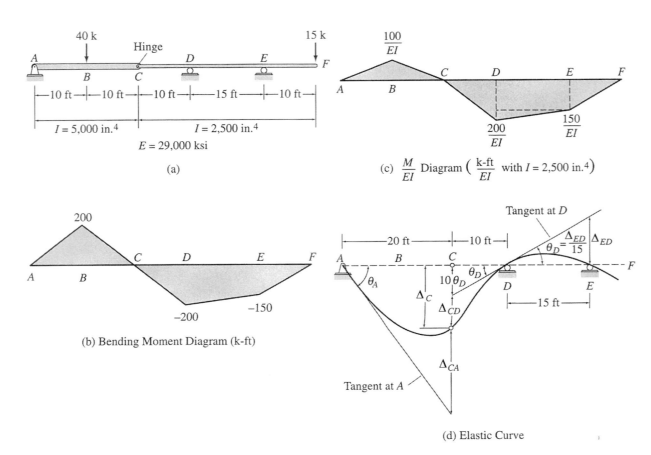

(a)

(c) $\dfrac{M}{EI}$ Diagram $\left(\dfrac{\text{k-ft}}{EI}\text{ with }I = 2{,}500\text{ in.}^4\right)$

(b) Bending Moment Diagram (k-ft)

(d) Elastic Curve

FIG. 6.8

Solution

M/EI Diagram The bending moment diagram is shown in Fig. 6.8(b), and the M/EI diagram for a reference moment of inertia $I = 2{,}500$ in.^{4} is shown in Fig. 6.8(c).

Elastic Curve The elastic curve for the beam is shown in Fig. 6.8(d). Note that the elastic curve is discontinuous at the internal hinge C. Therefore, the moment-area theorems must be applied separately over the portions AC and CF of the curve on each side of the hinge.

Slope at D The tangent at support D is selected as the reference tangent. From Fig. 6.8(d), we can see that the slope of this tangent is given by the relationship

$$\theta_D = \frac{\Delta_{ED}}{15}$$

where, from the second moment-area theorem,

$$\Delta_{ED} = \frac{1}{EI}\left[150(15)(7.5) + \frac{1}{2}(50)(15)(10)\right] = \frac{20{,}625 \text{ k-ft}^3}{EI}$$

Therefore,

$$\theta_D = \frac{20{,}625}{15(EI)} = \frac{1{,}375 \text{ k-ft}^2}{EI}$$

Deflection at C From Fig. 6.8(d), we can see that

$$\Delta_C = 10\theta_D + \Delta_{CD}$$

in which

$$\Delta_{CD} = \frac{1}{2}\left(\frac{200}{EI}\right)(10)\left(\frac{20}{3}\right) = \frac{6{,}666.67 \text{ k-ft}^3}{EI}$$

Therefore,

$$\Delta_C = 10\left(\frac{1{,}375}{EI}\right) + \frac{6{,}666.67}{EI} = \frac{20{,}416.67 \text{ k-ft}^3}{EI}$$

Substituting the numerical values of E and I, we obtain

$$\Delta_C = \frac{20{,}416.67(12)^3}{(29{,}000)(2{,}500)} = 0.487 \text{ in.}$$

$$\Delta_C = 0.487 \text{ in.} \downarrow \qquad\qquad\qquad \text{Ans.}$$

Slope at A Considering the portion AC of the elastic curve, we can see from Fig. 6.8(d) that

$$\theta_A = \frac{\Delta_C + \Delta_{CA}}{20}$$

where

$$\Delta_{CA} = \frac{1}{2}\left(\frac{100}{EI}\right)(20)(10) = \frac{10{,}000 \text{ k-ft}^3}{EI}$$

Therefore,

$$\theta_A = \frac{1}{20}\left(\frac{20{,}416.67}{EI} + \frac{10{,}000}{EI}\right) = \frac{1{,}520.83 \text{ k-ft}^2}{EI}$$

$$\theta_A = \frac{1{,}520.83(12)^2}{(29{,}000)(2{,}500)} = 0.003 \text{ rad}$$

$$\theta_A = 0.003 \text{ rad} \qquad\qquad\qquad \text{Ans.}$$

6.5 BENDING MOMENT DIAGRAMS BY PARTS

As illustrated in the preceding section, application of the moment-area method involves computation of the areas and moments of areas of various portions of the M/EI diagram. It will be shown in the following section that the conjugate-beam method for determining deflections of beams also requires computation of these quantities. When a beam is subjected to different types of loads, such as a combination of distributed and concentrated loads, determination of the properties of the resultant M/EI diagram, due to the combined effect of all the loads, can become a formidable task. This difficulty can be avoided by construct-

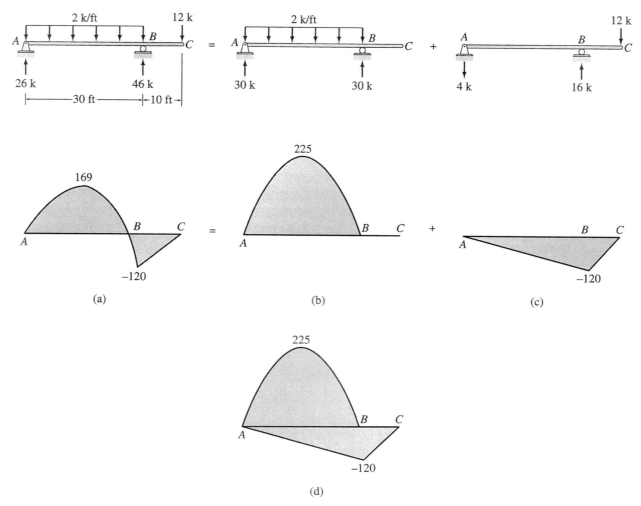

FIG. **6.9**

ing the bending moment diagram in *parts*—that is, constructing a separate bending moment diagram for each of the loads. The ordinates of the bending moment diagrams thus obtained are then divided by *EI* to obtain the M/EI diagrams. These diagrams usually consist of simple geometric shapes, so their areas and moments of areas can be easily computed. The required areas and moments of areas of the resultant M/EI diagram are then obtained by algebraically adding (superimposing) the corresponding areas and moments of areas, respectively, of the bending moment diagrams for the individual loads.

Two procedures are commonly used for constructing bending moment diagrams by parts. The first procedure simply involves applying each of the loads separately on the beam and constructing the corresponding bending moment diagrams. Consider, for example, a beam subjected to a combination of a uniformly distributed load and a concentrated load, as shown in Fig. 6.9(a). To construct the bending moment diagram by parts, we apply the two types of loads separately on the beam, as shown in Fig. 6.9(b) and (c), and draw the corresponding bending moment diagrams. It is usually convenient to draw the parts of the bending moment diagram together, as shown in Fig. 6.9(d). Although it is not necessary for the application of the moment-area and conjugate-beam methods, if so desired, the resultant bending moment diagram, as shown in Fig. 6.9(a), can be obtained by superimposing the two parts shown in Fig. 6.9(b) and (c).

An alternative procedure for constructing bending moment diagrams by parts consists of selecting a point on the beam (usually a support point or an end of the beam) at which the beam is assumed to

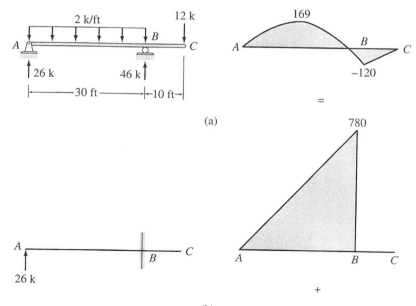

(a)

(b)

FIG. 6.10

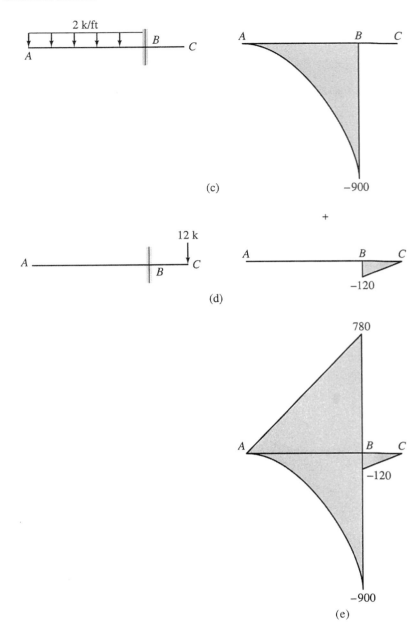

FIG. **6.10** (contd.)

be fixed, applying each of the loads and support reactions separately on this imaginary cantilever beam, and constructing the corresponding bending moment diagrams. This procedure is commonly referred to as constructing the *bending moment diagram by cantilever parts*. To illustrate this procedure, consider again the beam examined in Fig. 6.9. The beam is redrawn in Fig. 6.10(a), which also shows the external loads as well as the support reactions determined from the equations of equilibrium. To construct the bending moment diagram by cantilever parts with respect to the support point B, we imagine the beam to be a cantilever beam with fixed support at point B. Then we apply the two loads and the reaction at support A separately on this imaginary cantilever beam, as shown in Fig. 6.10(b)–(d), and draw the corresponding bending moment diagrams, as shown in these figures. The parts of the bending moment diagram are often drawn together, as shown in Fig. 6.10(e). The resultant bending moment diagram, as depicted in Fig. 6.10(a), can be obtained, if desired, by superimposing the three parts shown in Fig. 6.10(b)–(d).

Example 6.7

Determine the deflection at point C of the beam shown in Fig. 6.11(a) by the moment-area method.

Solution

M/EI Diagram The bending moment diagram for this beam by cantilever parts with respect to the support point B was determined in Fig. 6.10. The ordinates of the bending moment diagram are divided by EI to obtain the M/EI diagram shown in Fig. 6.11(b).

Elastic Curve See Fig. 6.11(c).

Slope at B Selecting the tangent at B as the reference tangent, it can be seen from Fig. 6.11(c) that

$$\theta_B = \frac{\Delta_{AB}}{30}$$

By using the M/EI diagram (Fig. 6.11(b)) and the properties of geometric shapes given in Appendix A, we compute

$$\Delta_{AB} = \frac{1}{EI}\left[\frac{1}{2}(780)(30)(20) - \frac{1}{3}(900)(30)\left(\frac{3}{4}\right)(30)\right]$$

$$= \frac{31{,}500 \text{ k-ft}^3}{EI}$$

Therefore,

$$\theta_B = \frac{31{,}500}{30EI} = \frac{1{,}050 \text{ k-ft}^2}{EI}$$

Deflection at C From Fig. 6.11(c), we can see that

$$\Delta_C = 10\theta_B - \Delta_{CB}$$

continued

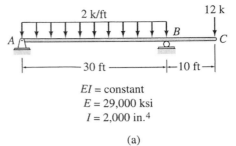

EI = constant
E = 29,000 ksi
I = 2,000 in.^{4}

(a)

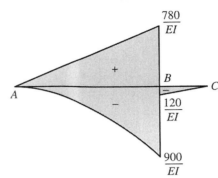

(b) $\dfrac{M}{EI}$ Diagram $\left(\dfrac{\text{k-ft}}{EI}\right)$

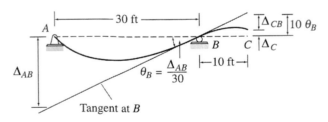

FIG. 6.11

(c) Elastic Curve

where

$$\Delta_{CB} = \frac{1}{2}\left(\frac{120}{EI}\right)(10)\left(\frac{20}{3}\right) = \frac{4,000 \text{ k-ft}^3}{EI}$$

Therefore,

$$\Delta_C = 10\left(\frac{1,050}{EI}\right) - \frac{4,000}{EI} = \frac{6,500 \text{ k-ft}^3}{EI}$$

Substituting the numerical values of E and I, we obtain

$$\Delta_C = \frac{6,500(12)^3}{(29,000)(2,000)} = 0.194 \text{ in.}$$

$$\Delta_C = 0.194 \text{ in.} \uparrow$$

Ans.

6.6 CONJUGATE-BEAM METHOD

The conjugate-beam method, developed by Otto Mohr in 1868, generally provides a more convenient means of computing slopes and deflections of beams than the moment-area method. Although the amount of computational effort required by the two methods is essentially the same, the conjugate-beam method is preferred by many engineers because of its systematic sign convention and straightforward application, which does not require sketching the elastic curve of the structure.

The conjugate-beam method is based on the analogy between the relationships among load, shear, and bending moment and the relationships among M/EI, slope, and deflection. These two types of relationships were derived in Sections 5.4 and 6.1, respectively, and are repeated in Table 6.1 for comparison purposes. As this table indicates, the relationships between M/EI, slope, and deflection have the same form as that of the relationships between load, shear, and bending moment. Therefore, the slope and deflection can be determined from M/EI by the same operations as those performed to compute shear and bending moment, respectively, from the load. Furthermore, if the M/EI diagram for a beam is applied as the load on a fictitious analogous beam, then the shear and bending moment at any point on the fictitious beam will be equal to the slope and deflection, respectively, at the corresponding point on the original real beam. The fictitious beam is referred to as the *conjugate beam*, and it is defined as follows:

> A conjugate beam corresponding to a real beam is a fictitious beam of the same length as the real beam, but it is externally supported and internally connected such that if the conjugate beam is loaded with the M/EI diagram of the real beam, the shear and bending moment at any point on the conjugate beam are equal, respectively, to the slope and deflection at the corresponding point on the real beam.

As the foregoing discussion indicates, the conjugate-beam method essentially involves computing the slopes and deflections of beams by computing the shears and bending moments in the corresponding conjugate beams.

TABLE 6.1

Load–Shear–Bending Moment Relationships	M/EI–Slope–Deflection Relationships
$\dfrac{dS}{dx} = w$	$\dfrac{d\theta}{dx} = \dfrac{M}{EI}$
$\dfrac{dM}{dx} = S$ or $\dfrac{d^2M}{dx^2} = w$	$\dfrac{dy}{dx} = \theta$ or $\dfrac{d^2y}{dx^2} = \dfrac{M}{EI}$

Supports for Conjugate Beams

External supports and internal connections for conjugate beams are determined from the analogous relationships between conjugate beams and the corresponding real beams; that is, the shear and bending moment at any point on the conjugate beam must be consistent with the slope and deflection at that point on the real beam. The conjugate counterparts of the various types of real supports thus determined are shown in Fig. 6.12. As this figure indicates, a hinged or a roller support at an end of the real beam remains the same in the conjugate beam. This is because at such a support there may be slope, but no deflection, of the real beam. Therefore, at the corresponding end of the conjugate beam there must be shear but no bending moment; and a hinged or a roller support at that end would satisfy these conditions. Since at a fixed support of the real beam there is neither slope nor deflection, both shear and bending moment at that end of the conjugate beam must be zero; therefore, the conjugate of a fixed real support is a free end, as shown in Fig. 6.12. Conversely, a free end of a real beam becomes a fixed support

Real Beam			Conjugate Beam
Type of Support	Slope and Deflection	Shear and Bending Moment	Type of Support
Simple end support 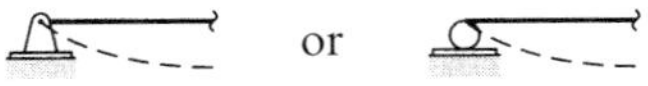	$\theta \neq 0$ $\Delta = 0$	$S \neq 0$ $M = 0$	Simple end support 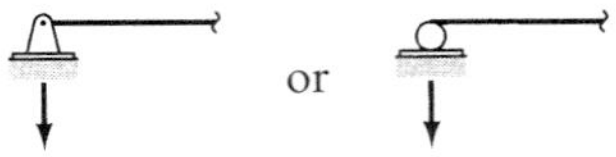
Fixed support 	$\theta = 0$ $\Delta = 0$	$S = 0$ $M = 0$	Free end 
Free end 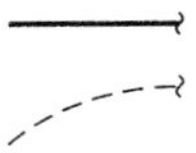	$\theta \neq 0$ $\Delta \neq 0$	$S \neq 0$ $M \neq 0$	Fixed support 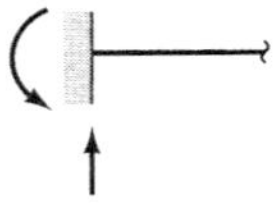
Simple interior support 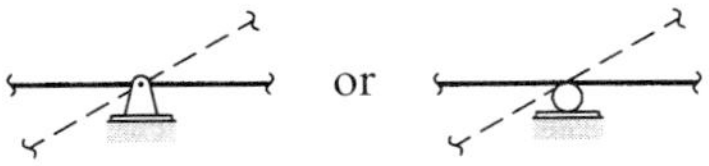	$\theta \neq 0$ and continuous $\Delta = 0$	$S \neq 0$ and continuous $M = 0$	Internal hinge 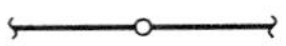
Internal hinge 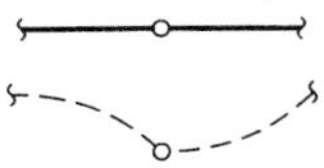	$\theta \neq 0$ and discontinuous $\Delta \neq 0$	$S \neq 0$ and discontinuous $M \neq 0$	Simple interior support 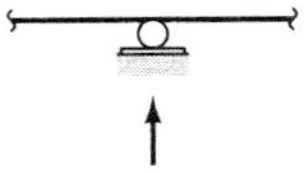

FIG. 6.12 Supports for Conjugate Beams

in the conjugate beam because there may be slope as well as deflection at that end of the real beam; therefore, the conjugate beam must develop both shear and bending moment at that point. At an interior support of a real beam there is no deflection, but the slope is continuous (i.e., there is no abrupt change of slope from one side of the support to the other), so the corresponding point on the conjugate beam becomes an internal hinge at which the bending moment is zero and the shear is continuous. Finally, at an internal hinge in the real beam there may be deflection as well as discontinuous slope of the real beam. Therefore, the conjugate beam must have bending moment and abrupt change of shear at that point. Because an interior support satisfies both of these requirements, an internal hinge in the real beam becomes an interior support in the conjugate beam, as shown in Fig. 6.12.

The conjugates of some common types of (real) beams are depicted in Fig. 6.13. As Fig. 6.13(a)–(e) indicates, the conjugate beams corresponding to statically determinate real beams are always statically determinate, whereas statically indeterminate beams have unstable conjugate beams, as shown in Fig. 6.13(f)–(h). However, since these unstable conjugate beams will be loaded with the M/EI diagrams of statically indeterminate real beams, which are self-balancing, the unstable conjugate beams will be in equilibrium. As the last two examples in Fig. 6.13 illustrate, statically unstable real beams have statically indeterminate conjugate beams.

Sign Convention

If the positive ordinates of the M/EI diagram are applied to the conjugate beam as upward loads (in the positive y direction) and vice versa, then a positive shear in the conjugate beam denotes a positive (counterclockwise) slope of the real beam with respect to the undeformed axis of the real beam; also, a positive bending moment in the conjugate beam denotes a positive (upward or in the positive y direction) deflection of the real beam with respect to the undeformed axis of the real beam and vice versa.

Procedure for Analysis

The following step-by-step procedure can be used for determining the slopes and deflections of beams by the conjugate-beam method.

1. Construct the M/EI diagram for the given (real) beam subjected to the specified (real) loading. If the beam is subjected to a combination of different types of loads (e.g., concentrated loads and distributed loads), the analysis can be considerably expedited by constructing the M/EI diagram by parts, as discussed in the preceding section.

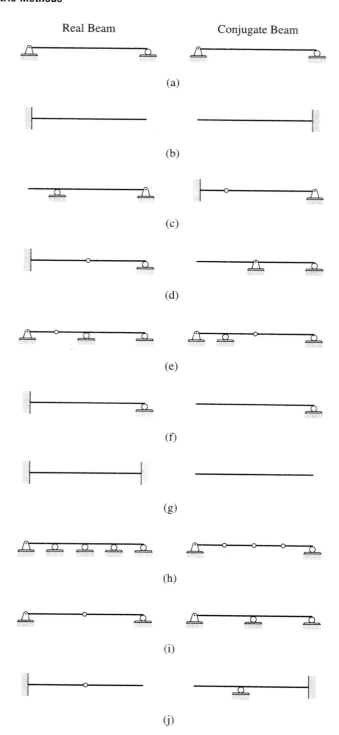

FIG. **6.13**

2. Determine the conjugate beam corresponding to the given real beam. The external supports and internal connections for the conjugate beam must be selected so that the shear and bending moment at any point on the conjugate beam are consistent with the slope and deflection, respectively, at that point on the real beam. The conjugates of various types of real supports are given in Fig. 6.12.

3. Apply the M/EI diagram (from step 1) as the load on the conjugate beam. The positive ordinates of the M/EI diagram are applied as upward loads on the conjugate beam and vice versa.

4. Calculate the reactions at the supports of the conjugate beam by applying the equations of equilibrium and condition (if any).

5. Determine the shears at those points on the conjugate beam where slopes are desired on the real beam. Determine the bending moments at those points on the conjugate beam where deflections are desired on the real beam. The shears and bending moments in conjugate beams are considered to be positive or negative in accordance with the beam sign convention (Fig. 5.2).

6. The slope at a point on the real beam with respect to the undeformed axis of the real beam is equal to the shear at that point on the conjugate beam. A positive shear in the conjugate beam denotes a positive or counterclockwise slope of the real beam and vice versa.

7. The deflection at a point on the real beam with respect to the undeformed axis of the real beam is equal to the bending moment at that point on the conjugate beam. A positive bending moment in the conjugate beam denotes a positive or upward deflection of the real beam and vice versa.

Example 6.8

Determine the slopes and deflections at points B and C of the cantilever beam shown in Fig. 6.14(a) by the conjugate-beam method.

Solution

M/EI Diagram This beam was analyzed in Example 6.3 by the moment-area method. The M/EI diagram for a reference moment of inertia $I = 3,000$ in.^{4} is shown in Fig. 6.14(b).

Conjugate Beam Fig. 6.14(c) shows the conjugate beam, loaded with the M/EI diagram of the real beam. Note that point A, which is fixed on the real beam, becomes free on the conjugate beam, whereas point C, which is free on the real beam, becomes fixed on the conjugate beam. Because the M/EI diagram is negative, it is applied as a downward load on the conjugate beam.

continued

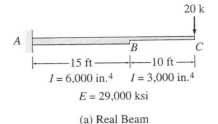

(a) Real Beam

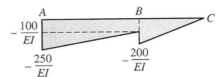

(b) $\dfrac{M}{EI}$ Diagram $\left(\dfrac{\text{k-ft}}{EI}\text{ with } I = 3{,}000 \text{ in.}^4\right)$

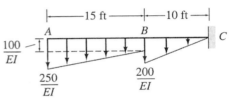

(c) Conjugate Beam

FIG. **6.14**

Slope at B The slope at B on the real beam is equal to the shear at B in the conjugate beam. Using the free body of the conjugate beam to the left of B and considering the external forces acting upward on the free body as positive, in accordance with the beam sign convention (see Fig. 5.2), we compute the shear at B in the conjugate beam as

$$+\uparrow S_B = \frac{1}{EI}\left[-100(15) - \frac{1}{2}(150)(15)\right] = -\frac{2{,}625 \text{ k-ft}^2}{EI}$$

Therefore, the slope at B on the real beam is

$$\theta_B = -\frac{2{,}625 \text{ k-ft}^2}{EI}$$

Substituting the numerical values of E and I, we obtain

$$\theta_B = -\frac{2{,}625(12)^2}{(29{,}000)(3{,}000)} = -0.0043 \text{ rad}$$

$$\theta_B = 0.0043 \text{ rad} \qquad\qquad \textbf{Ans.}$$

Deflection at B The deflection at B on the real beam is equal to the bending moment at B in the conjugate beam. Using the free body of the conjugate beam to the left of B and considering the clockwise moments of the external forces about B as positive, in accordance with the beam sign convention (Fig. 5.2), we compute the bending moment at B on the conjugate beam as

$$+\circlearrowleft M_B = \frac{1}{EI}\left[-100(15)(7.5) - \frac{1}{2}(150)(15)(10)\right] = -\frac{22{,}500 \text{ k-ft}^3}{EI}$$

Therefore, the deflection at B on the real beam is

$$\Delta_B = -\frac{22{,}500 \text{ k-ft}^3}{EI} = -\frac{22{,}500(12)^3}{(29{,}000)(3{,}000)} = -0.45 \text{ in.}$$

$$\Delta_B = 0.45 \text{ in.} \downarrow \qquad\qquad \text{Ans.}$$

Slope at C Using the free body of the conjugate beam to the left of C, we determine the shear at C as

$$+\uparrow S_C = \frac{1}{EI}\left[-100(15) - \frac{1}{2}(150)(15) - \frac{1}{2}(200)(10)\right] = -\frac{3{,}625 \text{ k-ft}^2}{EI}$$

Therefore, the slope at C on the real beam is

$$\theta_C = -\frac{3{,}625 \text{ k-ft}^2}{EI} = -\frac{3{,}625(12)^2}{(29{,}000)(3{,}000)} = -0.006 \text{ rad}$$

$$\theta_C = 0.006 \text{ rad} \quad\searrow \qquad\qquad \text{Ans.}$$

Deflection at C Considering the free body of the conjugate beam to the left of C, we obtain

$$+\circlearrowleft M_C = \frac{1}{EI}\left[-100(15)(17.5) - \frac{1}{2}(150)(15)(20) - \frac{1}{2}(200)(10)(6.67)\right]$$

$$= -\frac{55{,}420 \text{ k-ft}^3}{EI}$$

Therefore, the deflection at C on the real beam is

$$\Delta_C = -\frac{55{,}420 \text{ k-ft}^3}{EI} = -\frac{55{,}420(12)^3}{(29{,}000)(3{,}000)} = -1.1 \text{ in.}$$

$$\Delta_C = 1.1 \text{ in.} \downarrow \qquad\qquad \text{Ans.}$$

Example 6.9

Determine the slope and deflection at point B of the beam shown in Fig. 6.15(a) by the conjugate-beam method.

Solution

M/EI Diagram See Fig. 6.15(b).

Conjugate Beam The conjugate beam, loaded with the M/EI diagram of the real beam, is shown in Fig. 6.15(c).

continued

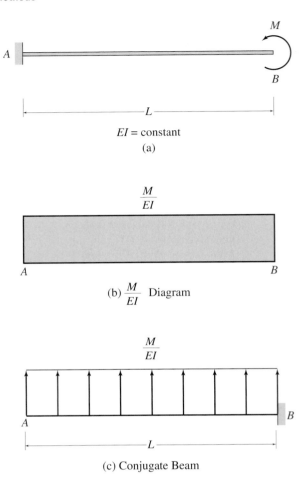

FIG. 6.15

Slope at B Considering the free body of the conjugate beam to the left of B, we determine the shear at B as

$$+\uparrow S_B = \frac{M}{EI}(L) = \frac{ML}{EI}$$

Therefore, the slope at B on the real beam is

$$\theta_B = \frac{ML}{EI}$$

$$\theta_B = \frac{ML}{EI} \quad \diagup \qquad \qquad \textbf{Ans.}$$

Deflection at B Using the free body of the conjugate beam to the left of B, we determine the bending moment at B as

$$+\circlearrowleft M_B = \frac{M}{EI}(L)\left(\frac{L}{2}\right) = \frac{ML^2}{2EI}$$

Therefore, the deflection at B on the real beam is

$$\Delta_B = \frac{ML^2}{2EI}$$

$$\Delta_B = \frac{ML^2}{2EI} \uparrow \qquad\qquad \text{Ans.}$$

Example 6.10

Use the conjugate-beam method to determine the slopes at ends A and D and the deflections at points B and C of the beam shown in Fig. 6.16(a).

Solution

M/EI Diagram This beam was analyzed in Example 6.4 by the moment-area method. The M/EI diagram for this beam is shown in Fig. 6.16(b).

Conjugate Beam Fig. 6.16(c) shows the conjugate beam loaded with the M/EI diagram of the real beam. Points A and D, which are simple end supports on the real beam, remain the same on the conjugate beam. Because the M/EI diagram is positive, it is applied as an upward load on the conjugate beam.

Reactions for Conjugate Beam By applying the equations of equilibrium to the free body of the entire conjugate beam, we obtain the following:

$$+\circlearrowleft \sum M_D = 0$$

$$A_y(40) - \frac{1}{EI}\left[\frac{1}{2}(800)(20)\left(\frac{20}{3}+20\right) + 600(10)(15)\right.$$

$$\left. + \frac{1}{2}(200)(10)\left(\frac{20}{3}+10\right) + \frac{1}{2}(600)(10)\left(\frac{20}{3}\right)\right] = 0$$

$$A_y = \frac{8,500 \text{ k-ft}^2}{EI}$$

$$+\uparrow \sum F_y = 0$$

$$\frac{1}{EI}\left[-8,500 + \frac{1}{2}(800)(20) + 600(10) + \frac{1}{2}(200)(10)\right.$$

$$\left. + \frac{1}{2}(600)(10)\right] - D_y = 0$$

$$D_y = \frac{9,500 \text{ k-ft}^2}{EI}$$

Slope at A The slope at A on the real beam is equal to the shear just to the right of A in the conjugate beam, which is

$$+\uparrow S_{A,R} = -A_y = -\frac{8,500 \text{ k-ft}^2}{EI}$$

continued

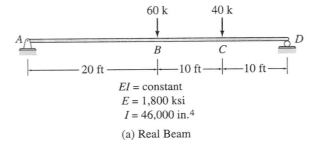

(a) Real Beam

EI = constant
$E = 1,800$ ksi
$I = 46,000$ in.^{4}

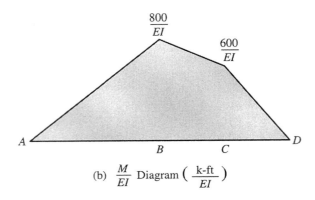

(b) $\dfrac{M}{EI}$ Diagram $\left(\dfrac{\text{k-ft}}{EI} \right)$

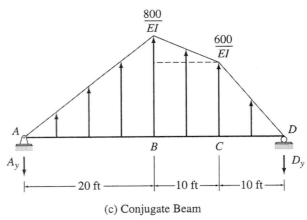

(c) Conjugate Beam

FIG. 6.16

Therefore, the slope at A on the real beam is

$$\theta_A = -\frac{8,500 \text{ k-ft}^2}{EI} = -\frac{8,500(12)^2}{(1,800)(46,000)} = -0.015 \text{ rad}$$

$$\theta_A = 0.015 \text{ rad} \quad \searrow \text{)}$$ **Ans.**

Slope at D The slope at D on the real beam is equal to the shear just to the left of D in the conjugate beam, which is

$$+\downarrow S_{D,L} = +D_y = \frac{+9,500 \text{ k-ft}^2}{EI}$$

Therefore, the slope at D on the real beam is

$$\theta_D = \frac{9{,}500 \text{ k-ft}^2}{EI} = \frac{9{,}500(12)^2}{(1{,}800)(46{,}000)} = 0.017 \text{ rad}$$

$$\theta_D = 0.017 \text{ rad} \quad \text{⟍} \qquad \qquad \textbf{Ans.}$$

Deflection at B   The deflection at B on the real beam is equal to the bending moment at B in the conjugate beam. Using the free body of the conjugate beam to the left of B, we compute

$$+ \circlearrowleft M_B = \frac{1}{EI}\left[-8{,}500(20) + \frac{1}{2}(800)(20)\left(\frac{20}{3}\right)\right] = -\frac{116{,}666.67 \text{ k-ft}^3}{EI}$$

Therefore, the deflection at B on the real beam is

$$\Delta_B = -\frac{116{,}666.67 \text{ k-ft}^3}{EI} = -\frac{116{,}666.67(12)^3}{(1{,}800)(46{,}000)} = -2.43 \text{ in.}$$

$$\Delta_B = 2.43 \text{ in.} \downarrow \qquad \qquad \textbf{Ans.}$$

Deflection at C   The deflection at C on the real beam is equal to the bending moment at C in the conjugate beam. Using the free body of the conjugate beam to the right of C, we determine

$$+ \circlearrowright M_C = \frac{1}{EI}\left[-9{,}500(10) + \frac{1}{2}(600)(10)\left(\frac{10}{3}\right)\right] = -\frac{85{,}000 \text{ k-ft}^3}{EI}$$

Therefore, the deflection at C on the real beam is

$$\Delta_C = -\frac{85{,}000 \text{ k-ft}^3}{EI} = -\frac{85{,}000(12)^3}{(1{,}800)(46{,}000)} = -1.77 \text{ in.}$$

$$\Delta_C = 1.77 \text{ in.} \downarrow \qquad \qquad \textbf{Ans.}$$

Example 6.11

Determine the maximum deflection for the beam shown in Fig. 6.17(a) by the conjugate-beam method.

Solution

M/EI Diagram   This beam was previously analyzed in Example 6.5 by the moment-area method. The M/EI diagram for the beam is shown in Fig. 6.17(b).

Conjugate Beam   The simply supported conjugate beam, loaded with the M/EI diagram of the real beam, is shown in Fig. 6.17(c).

Reaction at Support A of the Conjugate Beam   By applying the moment equilibrium equation $\sum M_C = 0$ to the free body of the entire conjugate beam, we determine

continued

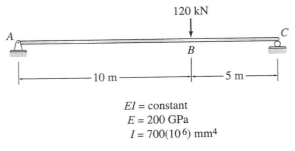

$EI = \text{constant}$
$E = 200 \text{ GPa}$
$I = 700(10^6) \text{ mm}^4$

(a) Real Beam

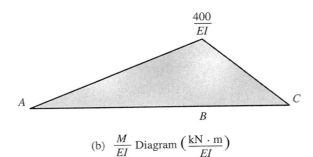

(b) $\dfrac{M}{EI}$ Diagram $\left(\dfrac{\text{kN} \cdot \text{m}}{EI}\right)$

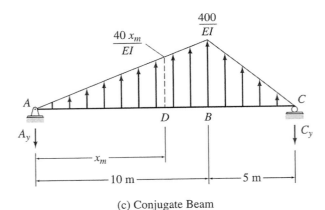

(c) Conjugate Beam

FIG. 6.17

$$+ \circlearrowleft M_C = 0$$

$$A_y(15) - \frac{1}{EI}\left[\frac{1}{2}(400)(10)\left(\frac{10}{3}+5\right) + \frac{1}{2}(400)(5)\left(\frac{10}{3}\right)\right] = 0$$

$$A_y = \frac{1{,}333.33 \text{ kN} \cdot \text{m}^2}{EI}$$

Location of the Maximum Bending Moment in Conjugate Beam If the maximum bending moment in the conjugate beam (or the maximum deflection on the real beam) occurs at point D, located at a distance x_m from the left support A (see Fig. 6.17(c)), then the shear in the conjugate beam at D must be zero. Considering the free body of the conjugate beam to the left of D, we write

$$+\uparrow S_D = \frac{1}{EI}\left[-1{,}333.33 + \frac{1}{2}(40x_m)(x_m)\right] = 0$$

from which

$$x_m = 8.16 \text{ m}$$

Maximum Deflection of the Real Beam The maximum deflection of the real beam is equal to the maximum bending moment in the conjugate beam, which can be determined by considering the free body of the conjugate beam to the left of D, with $x_m = 8.16$ m. Thus,

$$+\circlearrowleft M_{\max} = M_D = \frac{1}{EI}\left[-1{,}333.33(8.16) + \frac{1}{2}(40)(8.16)^2\left(\frac{8.16}{3}\right)\right]$$

$$= -\frac{7{,}244.51 \text{ kN} \cdot \text{m}^3}{EI}$$

Therefore, the maximum deflection of the real beam is

$$\Delta_{\max} = -\frac{7{,}244.51 \text{ kN} \cdot \text{m}^3}{EI} = -\frac{7{,}244.51}{(200)(700)} = -0.0517 \text{ m} = -51.7 \text{ mm}$$

$$\Delta_{\max} = 51.7 \text{ mm} \downarrow \qquad \qquad \textbf{Ans.}$$

Example 6.12

Determine the slope at point A and the deflection at point C of the beam shown in Fig. 6.18(a) by the conjugate-beam method.

Solution

M/EI Diagram This beam was analyzed in Example 6.6 by the moment-area method. The M/EI diagram for a reference moment of inertia $I = 2{,}500$ in.[4] is shown in Fig. 6.18(b).

Conjugate Beam Figure 6.18(c) shows the conjugate beam loaded with the M/EI diagram of the real beam. Note that points D and E, which are simple interior supports on the real beam, become internal hinges on the conjugate beam; point C, which is an internal hinge on the real beam, becomes a simple interior support on the conjugate beam. Also note that the positive part of the M/EI diagram is applied as upward loading on the conjugate beam, whereas the negative part of the M/EI diagram is applied as downward loading.

Reaction at Support A of the Conjugate Beam We determine the reaction A_y of the conjugate beam by applying the equations of condition as follows:

continued

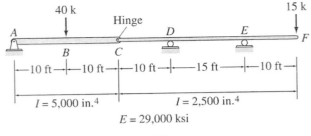

(a) Real Beam

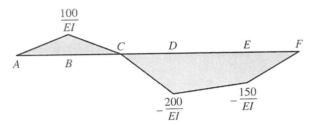

(b) $\dfrac{M}{EI}$ Diagram $\left(\dfrac{\text{k-ft}}{EI}\ \text{with } I = 2{,}500 \text{ in.}^4\right)$

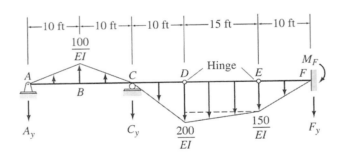

(c) Conjugate Beam

FIG. 6.18

$$+\circlearrowleft\sum M_D^{AD} = 0$$

$$A_y(30) - \frac{1}{2}\left(\frac{100}{EI}\right)(20)(20) + C_y(10) + \frac{1}{2}\left(\frac{200}{EI}\right)(10)\left(\frac{10}{3}\right) = 0$$

or

$$C_y = -3A_y + \frac{1{,}666.67}{EI} \qquad (1)$$

$$+\circlearrowleft\sum M_E^{AE} = 0$$

$$A_y(45) - \frac{1}{2}\left(\frac{100}{EI}\right)(20)(35) + C_y(25) + \frac{1}{2}\left(\frac{200}{EI}\right)(10)\left(\frac{10}{3} + 15\right)$$

$$+ \frac{150}{EI}(15)(7.5) + \frac{1}{2}\left(\frac{50}{EI}\right)(15)(10) = 0$$

or

$$45A_y + 25C_y = -\frac{3,958.33}{EI} \qquad (2)$$

Substituting Eq. (1) into Eq. (2) and solving for A_y, we obtain

$$A_y = \frac{1,520.83 \text{ k-ft}^2}{EI}$$

Slope at A The slope at A on the real beam is equal to the shear just to the right of A in the conjugate beam, which is

$$+\uparrow S_{A,R} = -A_y = -\frac{1,520.83 \text{ k-ft}^2}{EI}$$

Therefore, the slope at A on the real beam is

$$\theta_A = -\frac{1,520.83}{EI} = -\frac{1,520.83(12)^2}{(29,000)(2,500)} = -0.003 \text{ rad}$$

$$\theta_A = 0.003 \text{ rad} \qquad \qquad \text{Ans.}$$

Deflection at C The deflection at C on the real beam is equal to the bending moment at C in the conjugate beam. Considering the free body of the conjugate beam to the left of C, we obtain

$$+\circlearrowleft M_C = \frac{1}{EI}\left[-1,520.83(20) + \frac{1}{2}(100)(20)(10)\right] = -\frac{20,416.67 \text{ k-ft}^3}{EI}$$

Therefore, the deflection at C on the real beam is

$$\Delta_C = -\frac{20,416.67 \text{ k-ft}^3}{EI} = -\frac{20,416.67(12)^3}{(29,000)(2,500)} = -0.487 \text{ in.}$$

$$\Delta_C = 0.487 \text{ in.} \downarrow \qquad \qquad \text{Ans.}$$

Example 6.13

Use the conjugate-beam method to determine the deflection at point C of the beam shown in Fig. 6.19(a).

Solution

M/EI Diagram This beam was previously analyzed in Example 6.7 by the moment-area method. The M/EI diagram by cantilever parts with respect to point B is shown in Fig. 6.19(b).

Conjugate Beam See Fig. 6.19(c).

continued

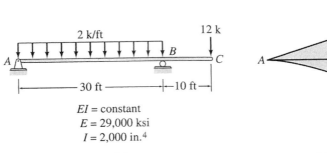

EI = constant
E = 29,000 ksi
I = 2,000 in.^{4}

(a) Real Beam

(b) $\dfrac{M}{EI}$ Diagram $\left(\dfrac{\text{k-ft}}{EI}\right)$

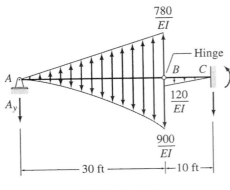

(c) Conjugate Beam

FIG. 6.19

Reaction at Support A of the Conjugate Beam

$$+\circlearrowleft \sum M_B^{AB} = 0$$

$$A_y(30) + \frac{1}{EI}\left[\frac{1}{3}(900)(30)\left(\frac{30}{4}\right) - \frac{1}{2}(780)(30)\left(\frac{30}{3}\right)\right] = 0$$

$$A_y = \frac{1{,}650 \text{ k-ft}^2}{EI}$$

Deflection at C The deflection at C on the real beam is equal to the bending moment at C in the conjugate beam. Considering the free body of the conjugate beam to the left of C, we obtain

$$+\circlearrowleft M_C = \frac{1}{EI}\left[-1{,}650(40) - \frac{1}{3}(900)(30)\left(\frac{30}{4}+10\right) + \frac{1}{2}(780)(30)(20)\right.$$

$$\left. - \frac{1}{2}(120)(10)\left(\frac{20}{3}\right)\right] = \frac{6{,}500 \text{ k-ft}^3}{EI}$$

Therefore, the deflection at C on the real beam is

$$\Delta_C = \frac{6{,}500 \text{ k-ft}^3}{EI} = \frac{6{,}500(12)^3}{(29{,}000)(2{,}000)} = 0.194 \text{ in.}$$

$$\Delta_C = 0.194 \text{ in.} \uparrow$$ Ans.

SUMMARY

In this chapter we have discussed the geometric methods for determining the slopes and deflections of statically determinate beams. The differential equation for the deflection of beams can be expressed as

$$\frac{d^2y}{dx^2} = \frac{M}{EI} \qquad (6.9)$$

The direct integration method essentially involves writing expression(s) for M/EI for the beam in terms of x and integrating the expression(s) successively to obtain equations for the slope and deflection of the elastic curve. The constants of integration are determined from the boundary conditions and the conditions of continuity of the elastic curve. If a beam is subjected to several loads, the slope or deflection due to the combined effects of the loads can be determined by algebraically adding the slopes or deflections due to each of the loads acting individually on the beam.

The moment-area method is based on two theorems, which can be mathematically expressed as follows:

$$\text{First moment-area theorem:} \qquad \theta_{BA} = \int_A^B \frac{M}{EI} dx \qquad (6.12)$$

$$\text{Second moment-area theorem:} \qquad \Delta_{BA} = \int_A^B \frac{M}{EI} \bar{x} dx \qquad (6.15)$$

Two procedures for constructing bending moment diagrams by parts are presented in Section 6.5.

A conjugate beam is a fictitious beam of the same length as the corresponding real beam; but it is externally supported and internally connected such that, if the conjugate beam is loaded with the M/EI diagram of the real beam, the shear and bending moment at any point on the conjugate beam are equal, respectively, to the slope and deflection at the corresponding point on the real beam. The conjugate-beam method essentially involves determining the slopes and deflections of beams by computing the shears and bending moments in the corresponding conjugate beams.

PROBLEMS

Section 6.2

6.1 through 6.6 Determine the equations for slope and deflection of the beam shown by the direct integration method. EI = constant.

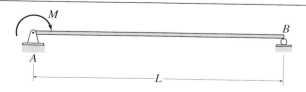

FIG. **P6.1**

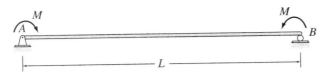

FIG. **P6.2**

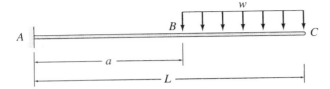

FIG. **P6.3**

FIG. **P6.4**

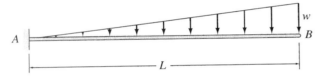

FIG. **P6.5**

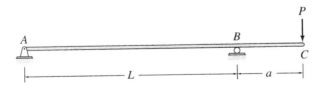

FIG. **P6.6**

6.7 and 6.8 Determine the slope and deflection at point B of the beam shown by the direct integration method.

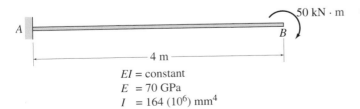

EI = constant
E = 70 GPa
I = 164 (10^6) mm^4

FIG. **P6.7**

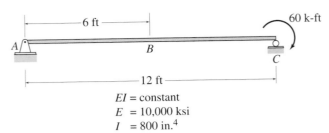

EI = constant
E = 10,000 ksi
I = 800 in.^{4}

FIG. **P6.8**

Sections 6.4 and 6.5

6.9 through 6.12 Determine the slope and deflection at point B of the beam shown by the moment-area method.

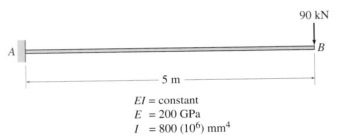

EI = constant
E = 200 GPa
I = 800 (10^6) mm^4

FIG. **P6.9, P6.35**

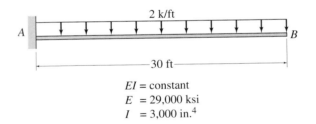

EI = constant
E = 29,000 ksi
I = 3,000 in.^{4}

FIG. **P6.10, P6.36**

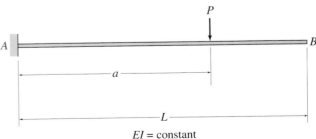

EI = constant

FIG. **P6.11, P6.37**

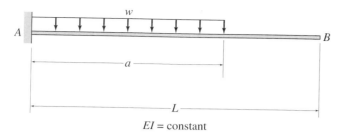

FIG. P6.12, P6.38

6.13 Determine the slope and deflection at point A of the beam shown by the moment-area method.

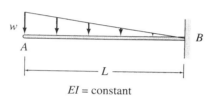

EI = constant

FIG. P6.13, P6.39

6.14 through 6.17 Use the moment-area method to determine the slopes and deflections at points B and C of the beam shown.

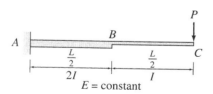

FIG. P6.14, P6.40

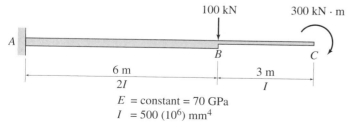

FIG. P6.15, P6.41

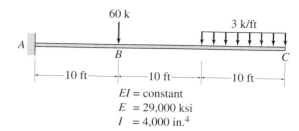

FIG. P6.16, P6.42

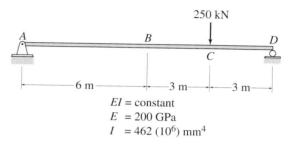

FIG. P6.17, P6.43

6.18 through 6.22 Determine the smallest moment of inertia I required for the beam shown, so that its maximum deflection does not exceed the limit of $1/360$ of the span length (i.e., $\Delta_{max} \leq L/360$). Use the moment-area method.

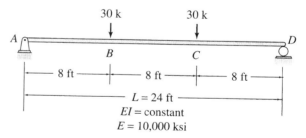

FIG. P6.18, P6.44

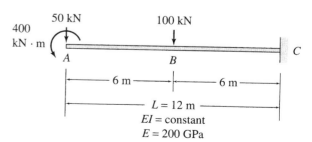

FIG. P6.19, P6.45

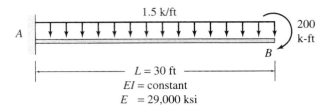

FIG. **P6.20, P6.46**

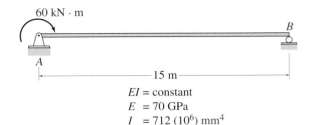

FIG. **P6.24, P6.50**

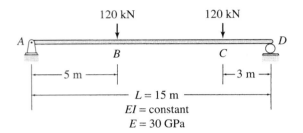

FIG. **P6.21, P6.47**

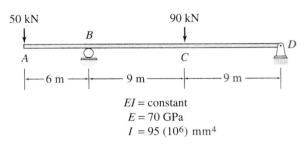

FIG. **P6.25, P6.51**

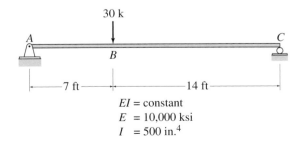

FIG. **P6.22, P6.48**

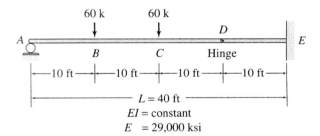

FIG. **P6.26, P6.52**

6.23 through 6.30 Determine the maximum deflection for the beam shown by the moment-area method.

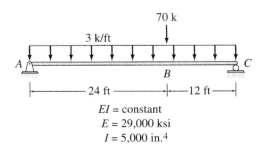

FIG. **P6.23, P6.49**

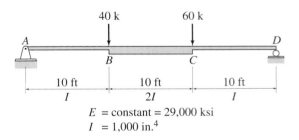

FIG. **P6.27, P6.53**

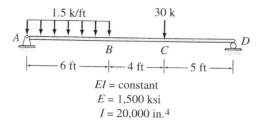

FIG. P6.28, P6.54

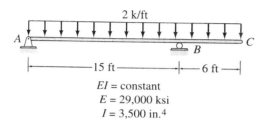

FIG. P6.29, P6.55

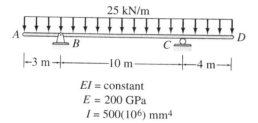

FIG. P6.30, P6.56

6.31 and 6.32 Use the moment-area method to determine the slope and deflection at point D of the beam shown.

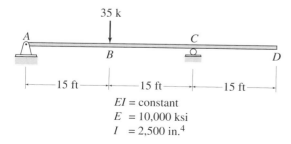

FIG. P6.31, P6.57

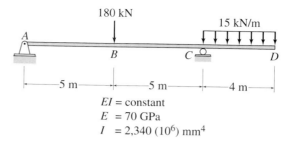

FIG. P6.32, P6.58

6.33 and 6.34 Use the moment-area method to determine the slopes and deflections at points B and D of the beam shown.

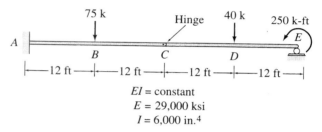

FIG. P6.33, P6.59

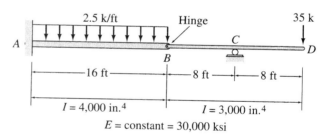

FIG. P6.34, P6.60

Section 6.6

6.35 through 6.38 Use the conjugate-beam method to determine the slope and deflection at point B of the beams shown in Figs. P6.9 through P6.12.

6.39 Determine the slope and deflection at point A of the beam shown in Fig. P6.13 by the conjugate-beam method.

6.40 through 6.43 Use the conjugate-beam method to determine the slopes and deflections at points B and C of the beams shown in Figs. P6.14 through P6.17.

6.44 through 6.48 Using the conjugate-beam method, determine the smallest moments of inertia I required for the beams shown in Figs. P6.18 through P6.22, so that the maximum beam deflection does not exceed the limit of $1/360$ of the span length (i.e., $\Delta_{max} \leq L/360$).

6.49 through 6.56 Determine the maximum deflection for the beams shown in Figs. P6.23 through P6.30 by the conjugate-beam method.

6.57 and 6.58 Use the conjugate-beam method to determine the slope and deflection at point D of the beam shown in Figs. P6.31 and P6.32.

6.59 and 6.60 Use the conjugate-beam method to determine the slopes and deflections at points B and D of the beams shown in Figs. P6.33 and P6.34.

7

Deflections of Trusses, Beams, and Frames: Work–Energy Methods

Bridge Collapse Due to Impact of
Wide-Load Truck
Photo courtesy of the Illinois Department of Transportation

In this chapter, we develop methods for the analysis of deflections of statically determinate structures by using some basic principles of work and energy. Work–energy methods are more general than the geometric methods considered in the previous chapter in the sense that they can be applied to various types of structures, such as trusses, beams, and frames. A disadvantage of these methods is that with each application, only one deflection component, or slope, at one point of the structure can be computed.

We begin by reviewing the basic concept of work performed by forces and couples during a deformation of the structure and then discuss the principle of virtual work. This principle is used to formulate the method of virtual work for the deflections of trusses, beams, and frames. We derive the expressions for strain energy of trusses, beams, and frames and then consider Castigliano's second theorem for computing deflections. Finally, we present Betti's law and Maxwell's law of reciprocal deflections.

7.1 WORK

The work done by a force acting on a structure is simply defined as the force times the displacement of its point of application in the direction of the force. Work is considered to be positive when the force and the displacement in the direction of the force have the same sense and negative when the force and the displacement have opposite sense.

Let us consider the work done by a force P during the deformation of a structure under the action of a system of forces (which includes P), as shown in Fig. 7.1(a). The magnitude of P may vary as its point of application displaces from A in the undeformed position of the structure to A' in the final deformed position. The work dW that P performs as its point of application undergoes an infinitesimal displacement, $d\Delta$ (Fig. 7.1(a)), can be written as

$$dW = P(d\Delta)$$

The total work W that the force P performs over the entire displacement Δ is obtained by integrating the expression of dW as

$$W = \int_0^\Delta P \, d\Delta \tag{7.1}$$

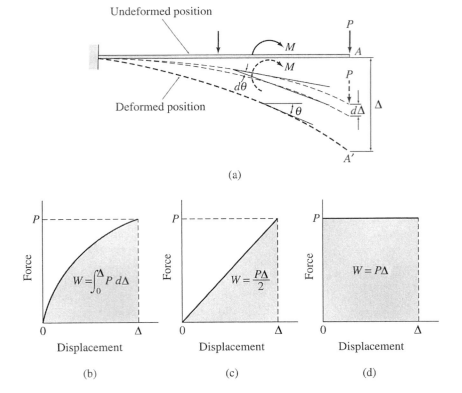

(a)

(b)   (c)   (d)

FIG. **7.1**

As Eq. (7.1) indicates, the work is equal to the area under the force-displacement diagram as shown in Fig. 7.1(b). In this text, we are focusing our attention on the analysis of linear elastic structures, so an expression for work of special interest is for the case when the force varies linearly with displacement from zero to its final value, as shown in Fig. 7.1(c). The work for such a case is given by the triangular area under the force-displacement diagram and is expressed as

$$W = \frac{1}{2}P\Delta \tag{7.2}$$

Another special case of interest is depicted in Fig. 7.1(d). In this case, the force remains constant at P while its point of application undergoes a displacement Δ caused by some other action independent of P. The work done by the force P in this case is equal to the rectangular area under the force-displacement diagram and is expressed as

$$W = P\Delta \tag{7.3}$$

It is important to distinguish between the two expressions for work as given by Eqs. (7.2) and (7.3). Note that the expression for work for the case when the force varies linearly with displacement (Eq. 7.2) contains a factor of $\frac{1}{2}$, whereas the expression for work for the case of a constant force (Eq. 7.3) does not contain this factor. These two expressions for work will be used subsequently in developing different methods for computing deflections of structures.

The expressions for the work of couples are similar in form to those for the work of forces. The work done by a couple acting on a structure is defined as the moment of the couple times the angle through which the couple rotates. The work dW that a couple of moment M performs through an infinitesimal rotation $d\theta$ (see Fig. 7.1(a)) is given by

$$dW = M(d\theta)$$

Therefore, the total work W of a couple with variable moment M over the entire rotation θ can be expressed as

$$W = \int_0^\theta M \, d\theta \tag{7.4}$$

When the moment of the couple varies linearly with rotation from zero to its final value, the work can be expressed as

$$W = \frac{1}{2}M\theta \tag{7.5}$$

and, if M remains constant during a rotation θ, then the work is given by

$$W = M\theta \tag{7.6}$$

7.2 PRINCIPLE OF VIRTUAL WORK

The *principle of virtual work*, which was introduced by John Bernoulli in 1717, provides a powerful analytical tool for many problems of structural mechanics. In this section, we study two formulations of this principle, namely, the *principle of virtual displacements for rigid bodies* and the *principle of virtual forces for deformable bodies*. The latter formulation is used in the following sections to develop the *method of virtual work*, which is considered to be one of the most general methods for determining deflections of structures.

Principle of Virtual Displacements for Rigid Bodies

The principle of virtual displacements for rigid bodies can be stated as follows:

> If a rigid body is in equilibrium under a system of forces and if it is subjected to any small virtual rigid-body displacement, the virtual work done by the external forces is zero.

The term *virtual* simply means imaginary, not real. Consider the beam shown in Fig. 7.2(a). The free-body diagram of the beam is shown in Fig. 7.2(b), in which P_x and P_y represent the components of the external load P in the x and y directions, respectively.

Now, suppose that the beam is given an arbitrary small virtual rigid-body displacement from its initial equilibrium position ABC to another position $A'B'C'$, as shown in Fig. 7.2(c). As shown in this figure, the total virtual rigid-body displacement of the beam can be decomposed into translations Δ_{vx} and Δ_{vy} in the x and y directions, respectively, and a rotation θ_v about point A. Note that the subscript v is used here to identify the displacements as virtual quantities. As the beam undergoes the virtual displacement from position ABC to position $A'B'C'$, the forces acting on it perform work, which is called *virtual work*. The total virtual work, W_{ve}, performed by the external forces acting on the beam can be expressed as the sum of the virtual work W_{vx} and W_{vy} done during translations in the x and y directions, respectively, and the virtual work W_{vr}, done during the rotation; that is,

$$W_{ve} = W_{vx} + W_{vy} + W_{vr} \qquad (7.7)$$

During the virtual translations Δ_{vx} and Δ_{vy} of the beam, the virtual work done by the forces is given by

$$W_{vx} = A_x\Delta_{vx} - P_x\Delta_{vx} = (A_x - P_x)\Delta_{vx} = \left(\sum F_x\right)\Delta_{vx} \qquad (7.8)$$

and

$$W_{vy} = A_y\Delta_{vy} - P_y\Delta_{vy} + C_y\Delta_{vy} = (A_y - P_y + C_y)\Delta_{vy} = \left(\sum F_y\right)\Delta_{vy} \qquad (7.9)$$

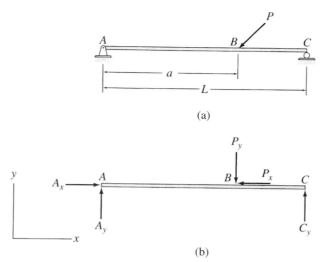

(a)

(b)

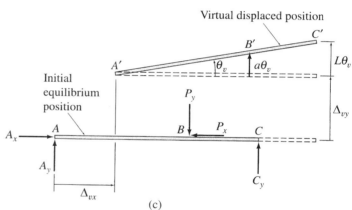

(c)

FIG. 7.2

(see Fig. 7.2(c)). The virtual work done by the forces during the small virtual rotation θ_v can be expressed as

$$W_{vr} = -P_y(a\theta_v) + C_y(L\theta_v) = (-aP_y + LC_y)\theta_v = (\textstyle\sum M_A)\theta_v \qquad (7.10)$$

By substituting Eqs. (7.8) through (7.10) into Eq. (7.7), we write the total virtual work done as

$$W_{ve} = (\textstyle\sum F_x)\Delta_{vx} + (\textstyle\sum F_y)\Delta_{vy} + (\textstyle\sum M_A)\theta_v \qquad (7.11)$$

Because the beam is in equilibrium, $\sum F_x = 0$, $\sum F_y = 0$, and $\sum M_A = 0$; therefore, Eq. (7.11) becomes

$$W_{ve} = 0 \qquad (7.12)$$

which is the mathematical statement of the principle of virtual displacements for rigid bodies.

Principle of Virtual Forces for Deformable Bodies

The principle of virtual forces for deformable bodies can be stated as follows:

> If a deformable structure is in equilibrium under a virtual system of forces (and couples) and if it is subjected to any small real deformation consistent with the support and continuity conditions of the structure, then the virtual external work done by the virtual external forces (and couples) acting through the real external displacements (and rotations) is equal to the virtual internal work done by the virtual internal forces (and couples) acting through the real internal displacements (and rotations).

In this statement, the term *virtual* is associated with the forces to indicate that the force system is arbitrary and does not depend on the action causing the real deformation.

To demonstrate the validity of this principle, consider the two-member truss shown in Fig. 7.3(a). The truss is in equilibrium under the action of a virtual external force P_v as shown. The free-body diagram of joint C of the truss is shown in Fig. 7.3(b). Since joint C is in equilibrium, the virtual external and internal forces acting on it must satisfy the following two equilibrium equations:

$$\sum F_x = 0 \qquad P_v - F_{vAC} \cos \theta_1 - F_{vBC} \cos \theta_2 = 0$$
$$\sum F_y = 0 \qquad -F_{vAC} \sin \theta_1 + F_{vBC} \sin \theta_2 = 0 \tag{7.13}$$

in which F_{vAC} and F_{vBC} represent the virtual internal forces in members AC and BC, respectively, and θ_1 and θ_2 denote, respectively, the angles of inclination of these members with respect to the horizontal (Fig. 7.3(a)).

Now, let us assume that joint C of the truss is given a small real displacement, Δ, to the right from its equilibrium position, as shown in Fig. 7.3(a). Note that the deformation is consistent with the support

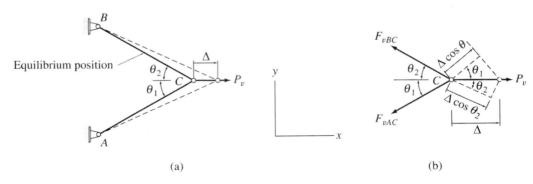

(a)   (b)

FIG. 7.3

conditions of the truss; that is, joints A and B, which are attached to supports, are not displaced. Because the virtual forces acting at joints A and B do not perform any work, the total virtual work for the truss (W_v) is equal to the algebraic sum of the work of the virtual forces acting at joint C; that is,

$$W_v = P_v\Delta - F_{vAC}(\Delta \cos \theta_1) - F_{vBC}(\Delta \cos \theta_2)$$

or

$$W_v = (P_v - F_{vAC} \cos \theta_1 - F_{vBC} \cos \theta_2)\Delta \qquad (7.14)$$

As indicated by Eq. (7.13), the term in the parentheses on the right-hand side of Eq. (7.14) is zero; therefore, the total virtual work is $W_v = 0$. Thus, Eq. (7.14) can be expressed as

$$P_v\Delta = F_{vAC}(\Delta \cos \theta_1) + F_{vBC}(\Delta \cos \theta_2) \qquad (7.15)$$

in which the quantity on the left-hand side represents the virtual external work (W_{ve}) done by the virtual external force, P_v, acting through the real external displacement, Δ. Also, realizing that the terms $\Delta \cos \theta_1$ and $\Delta \cos \theta_2$ are equal to the real internal displacements (elongations) of members AC and BC, respectively, we can conclude that the right-hand side of Eq. (7.15) represents the virtual internal work (W_{vi}) done by the virtual internal forces acting through the real internal displacements; that is

$$W_{ve} = W_{vi} \qquad (7.16)$$

which is the mathematical statement of the principle of virtual forces for deformable bodies.

It should be realized that the principle of virtual forces as described here is applicable regardless of the cause of real deformations; that is, deformations due to loads, temperature changes, or any other effect can be determined by the application of the principle. However, the deformations must be small enough so that the virtual forces remain constant in magnitude and direction while performing the virtual work. Also, although the application of this principle in this text is limited to elastic structures, the principle is valid regardless of whether the structure is elastic or not.

The method of virtual work is based on the principle of virtual forces for deformable bodies as expressed by Eq. (7.16), which can be rewritten as

$$\text{virtual external work} = \text{virtual internal work} \qquad (7.17)$$

or, more specifically, as

Virtual system

$$\sum \left(\begin{array}{c} \text{virtual external force} \times \\ \text{real external displacement} \end{array} \right) = \sum \left(\begin{array}{c} \text{virtual internal force} \times \\ \text{real internal displacement} \end{array} \right)$$

Real system (7.18)

in which the terms *forces* and *displacements* are used in a general sense and include moments and rotations, respectively. Note that because the virtual forces are independent of the actions causing the real deformation and remain constant during the real deformation, the expressions of the external and internal virtual work in Eq. (7.18) do not contain the factor 1/2.

As Eq. (7.18) indicates, the method of virtual work employs two separate systems: a virtual force system and the real system of loads (or other effects) that cause the deformation to be determined. To determine the deflection (or slope) at any point of a structure, a virtual force system is selected so that the desired deflection (or rotation) will be the only unknown in Eq. (7.18). The explicit expressions of the virtual work method to be used for computing deflections of trusses, beams, and frames are developed in the following three sections.

7.3 DEFLECTIONS OF TRUSSES BY THE VIRTUAL WORK METHOD

To develop the expression of the virtual work method that can be used to determine the deflections of trusses, consider an arbitrary statically determinate truss, as shown in Fig. 7.4(a). Let us assume that we want to determine the vertical deflection, Δ, at joint B of the truss due to the given external loads P_1 and P_2. The truss is statically determinate, so the axial forces in its members can be determined from the method of joints described previously in Chapter 4. If F represents the axial force in an arbitrary member j (e.g., member CD in Fig. 7.4(a)) of the truss, then (from *mechanics of materials*) the axial deformation, δ, of this member is given by

$$\delta = \frac{FL}{AE} \qquad (7.19)$$

in which $L, A,$ and E denote, respectively, the length, cross-sectional area, and modulus of elasticity of member j.

To determine the vertical deflection, Δ, at joint B of the truss, we select a virtual system consisting of a unit load acting at the joint and in the direction of the desired deflection, as shown in Fig. 7.4(b). Note

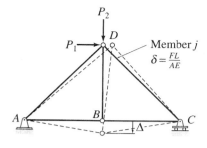

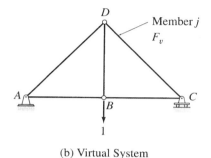

(a) Real System

(b) Virtual System

FIG. 7.4

that the (downward) sense of the unit load in Fig. 7.4(b) is the same as the assumed sense of the desired deflection Δ in Fig. 7.4(a). The forces in the truss members due to the virtual unit load can be determined from the method of joints. Let F_v denote the virtual force in member j. Next, we subject the truss with the virtual unit load acting on it (Fig. 7.4(b)) to the deformations of the real loads (Fig. 7.4(a)). The virtual external work performed by the virtual unit load as it goes through the real deflection Δ is equal to

$$W_{ve} = 1(\Delta) \tag{7.20}$$

To determine the virtual internal work, let us focus our attention on member j (member CD in Fig. 7.4). The virtual internal work done on member j by the virtual axial force F_v, acting through the real axial deformation δ, is equal to $F_v\delta$. Therefore, the total virtual internal work done on all the members of the truss can be written as

$$W_{vi} = \sum F_v(\delta) \tag{7.21}$$

By equating the virtual external work (Eq. (7.20)) to the virtual internal work (Eq. (7.21)) in accordance with the principle of virtual forces for deformable bodies, we obtain the following expression for the method of virtual work for truss deflections:

$$1(\Delta) = \sum F_v(\delta) \tag{7.22}$$

When the deformations are caused by external loads, Eq. (7.19) can be substituted into Eq. (7.22) to obtain

$$1(\Delta) = \sum F_v\left(\frac{FL}{AE}\right) \tag{7.23}$$

Because the desired deflection, Δ, is the only unknown in Eq. (7.23), its value can be determined by solving this equation.

Temperature Changes and Fabrication Errors

The expression of the virtual work method as given by Eq. (7.22) is quite general in the sense that it can be used to determine truss deflections due to temperature changes, fabrication errors, and any other effect for which the member axial deformations, δ, are either known or can be evaluated beforehand.

The axial deformation of a truss member j of length L due to a change in temperature (ΔT) is given by

$$\delta = \alpha(\Delta T)L \tag{7.24}$$

in which α denotes the coefficient of thermal expansion of member j. Substituting Eq. (7.24) into Eq. (7.22), we obtain the following expression:

$$1(\Delta) = \sum F_v \alpha (\Delta T) L \qquad (7.25)$$

which can be used to compute truss deflections due to the changes in temperature.

Truss deflections due to fabrication errors can be determined by simply substituting changes in member lengths due to fabrication errors for δ in Eq. (7.22).

Procedure for Analysis

The following step-by-step procedure can be used to determine the deflections of trusses by the virtual work method.

1. **Real System** If the deflection of the truss to be determined is caused by external loads, then apply the method of joints and/or the method of sections to compute the (real) axial forces (F) in all the members of the truss. In the examples given at the end of this section, tensile member forces are considered to be positive and vice versa. Similarly, increases in temperature and increases in member lengths due to fabrication errors are considered to be positive and vice versa.

2. **Virtual System** Remove all the given (real) loads from the truss; then apply a unit load at the joint where the deflection is desired and in the direction of the desired deflection to form the virtual force system. By using the method of joints and/or the method of sections, compute the virtual axial forces (F_v) in all the members of the truss. The sign convention used for the virtual forces must be the same as that adopted for the real forces in step 1; that is, if real tensile forces, temperature increases, or member elongations due to fabrication errors were considered as positive in step 1, then the virtual tensile forces must also be considered to be positive and vice versa.

3. The desired deflection of the truss can now be determined by applying Eq. (7.23) if the deflection is due to external loads, Eq. (7.25) if the deflection is caused by temperature changes, or Eq. (7.22) in the case of the deflection due to fabrication errors. The application of these virtual work expressions can be facilitated by arranging the real and virtual quantities, computed in steps 1 and 2, in a tabular form, as illustrated in the following examples. A positive answer for the desired deflection means that the deflection occurs in the same direction as the unit load, whereas a negative answer indicates that the deflection occurs in the direction opposite to that of the unit load.

Example 7.1

Determine the horizontal deflection at joint C of the truss shown in Fig. 7.5(a) by the virtual work method.

Solution

Real System The real system consists of the loading given in the problem, as shown in Fig. 7.5(b). The member axial forces due to the real loads (F) obtained by using the method of joints are also depicted in Fig. 7.5(b).

Virtual System The virtual system consists of a unit (1-k) load applied in the horizontal direction at joint C, as shown in Fig. 7.5(c). The member axial forces due to the 1-k virtual load (F_v) are determined by applying the method of joints. These member forces are also shown in Fig. 7.5(c).

Horizontal Deflection at C, Δ_C To facilitate the computation of the desired deflection, the real and virtual member forces are tabulated along with the member lengths (L), as shown in Table 7.1. As the values of the cross-sectional

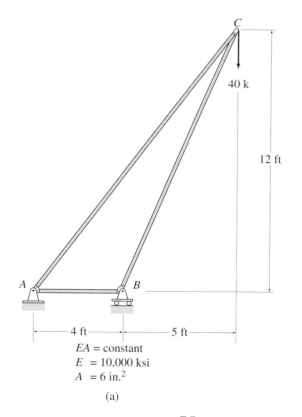

EA = constant
E = 10,000 ksi
A = 6 in.^{2}

(a)

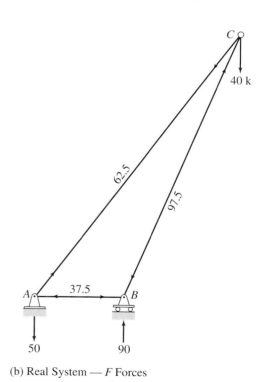

(b) Real System — F Forces

FIG. 7.5

continued

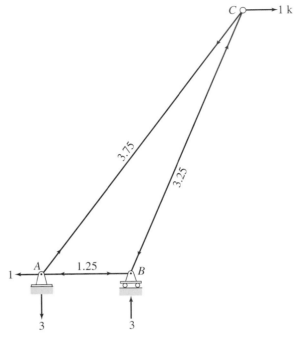

FIG. **7.5** (contd.)

(c) Virtual System — F_v Forces

TABLE 7.1

Member	L (in.)	F (k)	F_v (k)	$F_v(FL)$ (k² · in.)
AB	48	−37.5	−1.25	2,250
AC	180	62.5	3.75	42,187.5
BC	156	−97.5	−3.25	49,432.5

$$\sum F_v(FL) = 93,870$$

$$1(\Delta_C) = \frac{1}{EA}\sum F_v(FL)$$

$$(1\text{ k})\Delta_C = \frac{93,870\text{ k}^2 \cdot \text{in.}}{(10,000\text{ k/in.}^2)(6\text{ in.}^2)}$$

$$\Delta_C = 1.56\text{ in.}$$

$$\Delta_C = 1.56\text{ in.} \rightarrow \qquad \text{Ans.}$$

area, A, and modulus of elasticity, E, are the same for all the members, these are not included in the table. Note that the same sign convention is used for both real and virtual systems; that is, in both the third and the fourth columns of the table, tensile forces are entered as positive numbers and compressive forces as negative numbers. Then, for each member, the quantity $F_v(FL)$ is computed, and its value is entered in the fifth column of the table.

The algebraic sum of all of the entries in the fifth column, $\sum F_v(FL)$, is then determined, and its value is recorded at the bottom of the fifth column, as shown. The total virtual internal work done on all of the members of the truss is given by

$$W_{vi} = \frac{1}{EA}\sum F_v(FL)$$

The virtual external work done by the 1-k load acting through the desired horizontal deflection at C, Δ_C, is

$$W_{ve} = (1\ \text{k})\Delta_C$$

Finally, we determine the desired deflection Δ_C by equating the virtual external work to the virtual internal work and solving the resulting equation for Δ_C as shown in Table 7.1. Note that the positive answer for Δ_C indicates that joint C deflects to the right, in the direction of the unit load.

Example 7.2

Determine the horizontal deflection at joint G of the truss shown in Fig. 7.6(a) by the virtual work method.

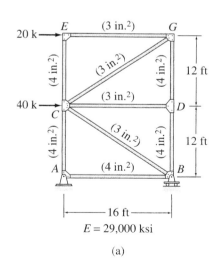

(a)

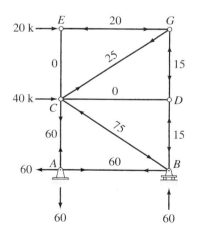

(b) Real System — F Forces

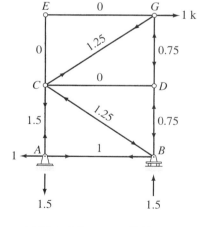

(c) Virtual System — F_v Forces

FIG. 7.6

continued

Solution

Real System The real system consists of the loading given in the problem, as shown in Fig. 7.6(b). The member axial forces due to the real loads (F) obtained by using the method of joints are also shown in Fig. 7.6(b).

Virtual System The virtual system consists of a unit (1-k) load applied in the horizontal direction at joint G, as shown in Fig. 7.6(c). The member axial forces due to the 1-k virtual load (F_v) are also depicted in Fig. 7.6(c).

Horizontal Deflection at G, Δ_G To facilitate the computation of the desired deflection, the real and virtual member forces are tabulated along with the lengths (L) and the cross-sectional areas (A) of the members, as shown in Table 7.2. The modulus of elasticity, E, is the same for all the members, so its value is not included in the table. Note that the same sign convention is used for both real and virtual systems; that is, in both the fourth and the fifth columns of the table, tensile forces are entered as positive numbers, and compressive forces as negative numbers. Then, for each member the quantity $F_v(FL/A)$ is computed, and its value is entered in the sixth column of the table. The algebraic sum of all the entries in the sixth column, $\sum F_v(FL/A)$, is then determined, and its value is recorded at the bottom of the sixth column, as shown. Finally, the desired deflection Δ_G is determined by applying the virtual work expression (Eq. (7.23)) as shown in Table 7.2. Note that the positive answer for Δ_G indicates that joint G deflects to the right, in the direction of the unit load.

TABLE 7.2

Member	L (in.)	A (in.^{2})	F (k)	F_v (k)	$F_v(FL/A)$ (k^2/in.)
AB	192	4	60	1	2,880
CD	192	3	0	0	0
EG	192	3	-20	0	0
AC	144	4	60	1.5	3,240
CE	144	4	0	0	0
BD	144	4	-15	-0.75	405
DG	144	4	-15	-0.75	405
BC	240	3	-75	-1.25	7,500
CG	240	3	25	1.25	2,500

$$\sum F_v\left(\frac{FL}{A}\right) = 16{,}930$$

$$1(\Delta_G) = \frac{1}{E}\sum F_v\left(\frac{FL}{A}\right)$$

$$(1\ \text{k})\Delta_G = \frac{16{,}930\ \text{k}^2/\text{in.}}{29{,}000\ \text{k}/\text{in.}^2}$$

$$\Delta_G = 0.584\ \text{in.}$$

$$\Delta_G = 0.584\ \text{in.} \rightarrow$$

Ans.

Example 7.3

Determine the horizontal and vertical components of the deflection at joint B of the truss shown in Fig. 7.7(a) by the virtual work method.

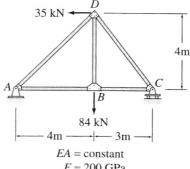

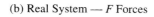

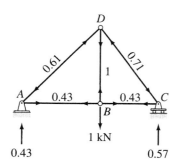

EA = constant
E = 200 GPa
A = 1,200 mm^2

(a)

(b) Real System — F Forces

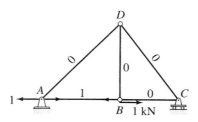

(c) Virtual System for
Determining Δ_{BH} (F_{v1} Forces)

(d) Virtual System for
Determining Δ_{BV} (F_{v2} Forces)

FIG. 7.7

Solution

Real System The real system and the corresponding member axial forces (F) are shown in Fig. 7.7(b).

Horizontal Deflection at B, Δ_{BH} The virtual system used for determining the horizontal deflection at B consists of a 1-kN load applied in the horizontal direction at joint B, as shown in Fig. 7.7(c). The member axial forces (F_{v1}) due to this virtual load are also shown in this figure. The member axial forces due to the real system (F) and this virtual system (F_{v1}) are then tabulated, and the virtual work expression given by Eq. (7.23) is applied to determine Δ_{BH}, as shown in Table 7.3.

continued

TABLE 7.3

Member	L (m)	F (kN)	F_{v1} (kN)	$F_{v1}(FL)$ $(kN^2 \cdot m)$	F_{v2} (kN)	$F_{v2}(FL)$ $(kN^2 \cdot m)$
AB	4	21	1	84	0.43	36.12
BC	3	21	0	0	0.43	27.09
AD	5.66	−79.2	0	0	−0.61	273.45
BD	4	84	0	0	1	336.00
CD	5	−35	0	0	−0.71	124.25
		$\sum F_v(FL)$		84		796.91

$$1(\Delta_{BH}) = \frac{1}{EA}\sum F_{v1}(FL) \qquad\qquad 1(\Delta_{BV}) = \frac{1}{EA}\sum F_{v2}(FL)$$

$$(1 \text{ kN})\Delta_{BH} = \frac{84}{200(10^6)(0.0012)} \text{ kN} \cdot \text{m} \qquad (1 \text{ kN})\Delta_{BV} = \frac{796.91}{200(10^6)(0.0012)} \text{ kN} \cdot \text{m}$$

$$\Delta_{BH} = 0.00035 \text{ m} \qquad\qquad\qquad \Delta_{BV} = 0.00332 \text{ m}$$

$$\Delta_{BH} = 0.35 \text{ mm} \rightarrow \qquad \text{Ans.} \qquad \Delta_{BV} = 3.32 \text{ mm} \downarrow \qquad \text{Ans.}$$

Vertical Deflection at B, Δ_{BV} The virtual system used for determining the vertical deflection at B consists of a 1-kN load applied in the vertical direction at joint B, as shown in Fig. 7.7(d). The member axial forces (F_{v2}) due to this virtual load are also shown in this figure. These member forces are tabulated in the sixth column of Table 7.3, and Δ_{BV} is computed by applying the virtual work expression (Eq. (7.23)), as shown in the table.

Example 7.4

Determine the vertical deflection at joint C of the truss shown in Fig. 7.8(a) due to a temperature drop of 15°F in members AB and BC and a temperature increase of 60°F in members AF, FG, GH, and EH. Use the virtual work method.

Solution

Real System The real system consists of the temperature changes (ΔT) given in the problem, as shown in Fig. 7.8(b).

Virtual System The virtual system consists of a 1-k load applied in the vertical direction at joint C, as shown in Fig. 7.8(c). Note that the virtual axial forces (F_v) are computed for only those members that are subjected to temperature changes. Because the temperature changes in the remaining members of the truss are zero, their axial deformations are zero; therefore, no internal virtual work is done on those members.

Vertical Deflection at C, Δ_C The temperature changes (ΔT) and the virtual member forces (F_v) are tabulated along with the lengths (L) of the members, in

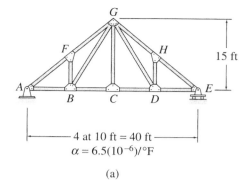

(a)

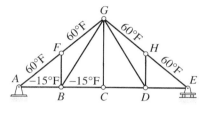

(b) Real System — ΔT

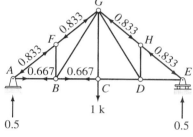

(c) Virtual System — F_v Forces

FIG. 7.8

TABLE 7.4

Member	L (ft)	ΔT (°F)	F_v (k)	$F_v(\Delta T)L$ (k-°F-ft)
AB	10	−15	0.667	−100
BC	10	−15	0.667	−100
AF	12.5	60	−0.833	−625
FG	12.5	60	−0.833	−625
GH	12.5	60	−0.833	−625
EH	12.5	60	−0.833	−625

$$\sum F_v(\Delta T)L = -2{,}700$$

$$1(\Delta_C) = \alpha \sum F_v(\Delta T)L$$
$$(1 \text{ k})\Delta_C = 6.5(10^{-6})(-2{,}700) \text{ k-ft}$$
$$\Delta_C = -0.0176 \text{ ft} = -0.211 \text{ in.}$$
$$\Delta_C = 0.211 \text{ in.} \uparrow \qquad\qquad\qquad \textbf{Ans.}$$

Table 7.4. The coefficient of thermal expansion, α, is the same for all the members, so its value is not included in the table. The desired deflection Δ_C is determined by applying the virtual work expression given by Eq. (7.25), as shown in the table. Note that the negative answer for Δ_C indicates that joint C deflects upward, in the direction opposite to that of the unit load.

Example 7.5

Determine the vertical deflection at joint D of the truss shown in Fig. 7.9(a) if member CF is 0.6 in. too long and member EF is 0.4 in. too short. Use the method of virtual work.

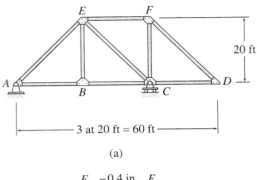

20 ft

3 at 20 ft = 60 ft

(a)

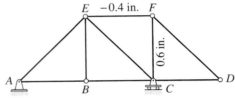

E −0.4 in. F

0.6 in.

(b) Real System — δ

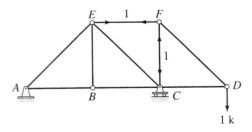

E 1 F

1

1 k

(c) Virtual System — F_v Forces

FIG. 7.9

Solution

Real System The real system consists of the changes in the lengths (δ) of members CF and EF of the truss, as shown in Fig. 7.9(b).

Virtual System The virtual system consists of a 1-k load applied in the vertical direction at joint D, as shown in Fig. 7.9(c). The necessary virtual forces (F_v) in members CF and EF can be easily computed by using the method of sections.

Vertical Deflection at D, Δ_D The desired deflection is determined by applying the virtual work expression given by Eq. (7.22), as shown in Table 7.5.

TABLE 7.5

Member	δ (in.)	F_v (k)	$F_v(\delta)$ (k-in.)
CF	0.6	-1	-0.6
EF	-0.4	1	-0.4
			$\sum F_v(\delta) = -1.0$

$$1(\Delta_D) = \sum F_v(\delta)$$
$$(1\text{ k})\Delta_D = -1.0 \text{ k-in.}$$
$$\Delta_D = -1.0 \text{ in.}$$
$$\Delta_D = 1.0 \text{ in. } \uparrow$$

Ans.

7.4 DEFLECTIONS OF BEAMS BY THE VIRTUAL WORK METHOD

To develop an expression for the virtual work method for determining the deflections of beams, consider a beam subjected to an arbitrary loading, as shown in Fig. 7.10(a). Let us assume that the vertical deflection, Δ, at a point B of the beam is desired. To determine this deflection, we select a virtual system consisting of a unit load acting at the point and in the direction of the desired deflection, as shown in Fig. 7.10(b). Now, if we subject the beam with the virtual unit load acting on it (Fig. 7.10(b)), to the deformations due to the real loads (Fig. 7.10(a)), the virtual external work performed by the virtual unit load as it goes through the real deflection Δ is $W_{ve} = 1(\Delta)$.

To obtain the virtual internal work, we focus our attention on a differential element dx of the beam located at a distance x from the left support A, as shown in Fig. 7.10(a) and (b). Because the beam with the virtual load (Fig. 7.10(b)) is subjected to the deformation due to the real loading (Fig. 7.10(a)), the virtual internal bending moment, M_v, acting on the element dx performs virtual internal work as it undergoes the real rotation $d\theta$, as shown in Fig. 7.10(c). Thus, the virtual internal work done on the element dx is given by

$$dW_{vi} = M_v(d\theta) \tag{7.26}$$

Note that because the virtual moment M_v remains constant during the real rotation $d\theta$, Eq. (7.26) does not contain a factor of 1/2. Recall from Eq. (6.10) that the change of slope $d\theta$ over the differential length dx can be expressed as

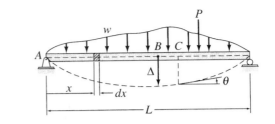

(a) Real System

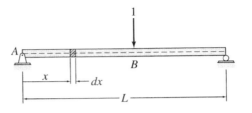

(b) Virtual System for Determining Δ

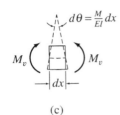

(c)

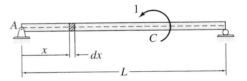

(d) Virtual System for Determining θ

FIG. **7.10**

$$d\theta = \frac{M}{EI}dx \qquad (7.27)$$

in which M = bending moment due to the real loading causing the rotation $d\theta$. By substituting Eq. (7.27) into Eq. (7.26), we write

$$dW_{vi} = M_v\left(\frac{M}{EI}\right)dx \qquad (7.28)$$

The total virtual internal work done on the entire beam can now be determined by integrating Eq. (7.28) over the length L of the beam as

$$W_{vi} = \int_0^L \frac{M_v M}{EI}dx \qquad (7.29)$$

By equating the virtual external work, $W_{ve} = 1(\Delta)$, to the virtual internal work (Eq. (7.29)), we obtain the following expression for the method of virtual work for beam deflections:

$$1(\Delta) = \int_0^L \frac{M_v M}{EI}dx \qquad (7.30)$$

If we want the slope θ at a point C of the beam (Fig. 7.10(a)), then we use a virtual system consisting of a unit couple acting at the point, as shown in Fig. 7.10(d). When the beam with the virtual unit couple is subjected to the deformations due to the real loading, the virtual external work performed by the virtual unit couple, as it undergoes the real rotation θ, is $W_{ve} = 1(\theta)$. The expression for the internal virtual work remains the same as given in Eq. (7.29), except that M_v now denotes the bending moment due to the virtual unit couple. By setting $W_{ve} = W_{vi}$, we obtain the following expression for the method of virtual work for beam slopes:

$$1(\theta) = \int_0^L \frac{M_v M}{EI} dx \qquad (7.31)$$

In the derivation of Eq. (7.29) for virtual internal work, we have neglected the internal work performed by the virtual shear forces acting through the real shear deformations. Therefore, the expressions of the virtual work method as given by Eqs. (7.30) and (7.31) do not account for the shear deformations of beams. However, for most beams (except for very deep beams), shear deformations are so small as compared to the bending deformations that their effect can be neglected in the analysis.

Procedure for Analysis

The following step-by-step procedure can be used to determine the slopes and deflections of beams by the virtual work method.

1. **Real System** Draw a diagram of the beam showing all the real (given) loads acting on it.
2. **Virtual System** Draw a diagram of the beam without the real loads. If deflection is to be determined, then apply a unit load at the point and in the direction of the desired deflection. If the slope is to be calculated, then apply a unit couple at the point on the beam where the slope is desired.
3. By examining the real and virtual systems and the variation of the flexural rigidity EI specified along the length of the beam, divide the beam into segments so that the real and virtual loadings as well as EI are continuous in each segment.
4. For each segment of the beam, determine an equation expressing the variation of the bending moment due to real loading (M) along the length of the segment in terms of a position coordinate x. The origin for x may be located anywhere on the beam and should be chosen so that the number of terms in the equation for M is minimum. It is usually convenient to consider the bending moments as positive or negative in accordance with the *beam sign convention* (Fig. 5.2).

5. For each segment of the beam, determine the equation for the bending moment due to virtual load or couple (M_v) using the same x coordinate that was used for this segment in step 4 to establish the expression for the real bending moment, M. The sign convention for the virtual bending moment (M_v) must be the same as that adopted for the real bending moment in step 4.
6. Determine the desired deflection or slope of the beam by applying the appropriate virtual work expression, Eq. (7.30) or Eq. (7.31). If the beam has been divided into segments, then the integral on the right-hand side of Eq. (7.30) or (7.31) can be evaluated by algebraically adding the integrals for all the segments of the beam.

Example 7.6

Determine the slope and deflection at point A of the beam shown in Fig. 7.11(a) by the virtual work method.

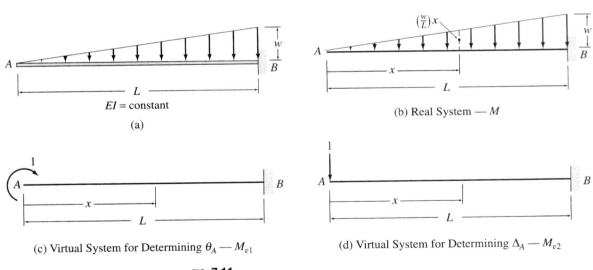

(a)
$EI = $ constant

(b) Real System — M

(c) Virtual System for Determining θ_A — M_{v1}

(d) Virtual System for Determining Δ_A — M_{v2}

FIG. 7.11

Solution

Real System See Fig. 7.11(b).

Slope at A, θ_A The virtual system consists of a unit couple applied at A, as shown in Fig. 7.11(c). From Fig. 7.11(a) through (c), we can see that there are no discontinuities of the real and virtual loadings or of EI along the length of the beam. Therefore, there is no need to subdivide the beam into segments. To determine the equation for the bending moment M due to real loading, we select an x coordinate with its origin at end A of the beam, as shown in Fig. 7.11(b). By applying the method of sections described in Section 5.2, we determine the equation for M as

$$0 < x < L \qquad M = -\frac{1}{2}(x)\left(\frac{wx}{L}\right)\left(\frac{x}{3}\right) = -\frac{wx^3}{6L}$$

Similarly, the equation for the bending moment M_{v1} due to virtual unit moment in terms of the same x coordinate is

$$0 < x < L \qquad M_{v1} = 1$$

To determine the desired slope θ_A, we apply the virtual work expression given by Eq. (7.31):

$$1(\theta_A) = \int_0^L \frac{M_{v1}M}{EI}\,dx = \int_0^L 1\left(-\frac{wx^3}{6LEI}\right)dx$$

$$\theta_A = -\frac{w}{6EIL}\left[\frac{x^4}{4}\right]_0^L = -\frac{wL^3}{24EI}$$

The negative answer for θ_A indicates that point A rotates counterclockwise, in the direction opposite to that of the unit moment.

$$\theta_A = \frac{wL^3}{24EI} \qquad\qquad \text{Ans.}$$

Deflection at A, Δ_A   The virtual system consists of a unit load applied at A, as shown in Fig. 7.11(d). If we use the same x coordinate as we used for computing θ_A, then the equation for M remains the same as before, and the equation for bending moment M_{v2} due to virtual unit load (Fig. 7.11(d)) is given by

$$0 < x < L \qquad M_{v2} = -1(x) = -x$$

By applying the virtual work expression given by Eq. (7.30), we determine the desired deflection Δ_A as

$$1(\Delta_A) = \int_0^L \frac{M_{v2}M}{EI}\,dx = \int_0^L (-x)\left(-\frac{wx^3}{6LEI}\right)dx$$

$$\Delta_A = \frac{w}{6EIL}\left[\frac{x^5}{5}\right]_0^L = \frac{wL^4}{30EI}$$

The positive answer for Δ_A indicates that point A deflects downward, in the direction of the unit load.

$$\Delta_A = \frac{wL^4}{30EI} \downarrow \qquad\qquad \text{Ans.}$$

Example 7.7

Determine the slope at point B of the cantilever beam shown in Fig. 7.12(a) by the virtual work method.

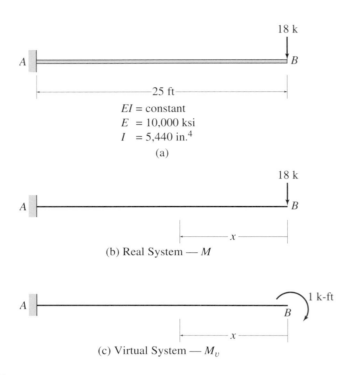

18 k

A ————————————————— B

—25 ft—

EI = constant
E = 10,000 ksi
I = 5,440 in.^{4}

(a)

18 k

A ————————————————— B

x

(b) Real System — M

A ————————————————— B 1 k-ft

x

(c) Virtual System — M_v

FIG. 7.12

Solution

The real and virtual systems are shown in Figs. 7.12(b) and (c), respectively. As shown in these figures, an x coordinate with its origin at end B of the beam is selected to obtain the bending moment equations. From Fig. 7.12(b), we can see that the equation for M in terms of the x coordinate is

$$0 < x < 25 \text{ ft} \qquad M = -18x$$

Similarly, from Fig. 7.12(c), we obtain the equation for M_v to be

$$0 < x < 25 \text{ ft} \qquad M_v = -1$$

The slope at B can now be computed by applying the virtual work expression given by Eq. (7.31), as follows:

$$1(\theta_B) = \int_0^L \frac{M_v M}{EI} dx$$

$$1(\theta_B) = \frac{1}{EI} \int_0^{25} -1(-18x) \, dx$$

$$(1 \text{ k-ft})\theta_B = \frac{5,625 \text{ k}^2\text{-ft}^3}{EI}$$

Therefore,

$$\theta_B = \frac{5,625 \text{ k-ft}^2}{EI} = \frac{5,625(12)^2}{(10,000)(5,440)} = 0.0149 \text{ rad.}$$

The positive answer for θ_B indicates that point B rotates clockwise, in the direction of the unit moment.

$$\theta_B = 0.0149 \text{ rad.}$$ Ans.

Example 7.8

Determine the deflection at point D of the beam shown in Fig. 7.13(a) by the virtual work method.

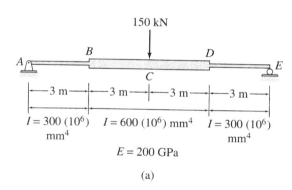

(a)

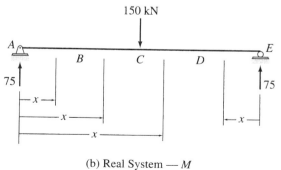

(b) Real System — M

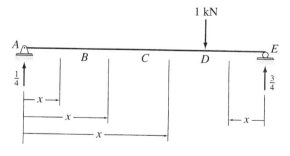

(c) Virtual System — M_v

FIG. **7.13**

continued

Solution

The real and virtual systems are shown in Fig. 7.13(b) and (c), respectively. It can be seen from Fig. 7.13(a) that the flexural rigidity EI of the beam changes abruptly at points B and D. Also, Fig. 7.13(b) and (c) indicates that the real and virtual loadings are discontinuous at points C and D, respectively. Consequently, the variation of the quantity $(M_v M/EI)$ will be discontinuous at points $B, C,$ and D. Thus, the beam must be divided into four segments, $AB, BC, CD,$ and DE; in each segment the quantity $(M_v M/EI)$ will be continuous and, therefore, can be integrated.

The x coordinates selected for determining the bending moment equations are shown in Fig. 7.13(b) and (c). Note that in any particular segment of the beam, the same x coordinate must be used to write both equations—that is, the equation for the real bending moment (M) and the equation for the virtual bending moment (M_v). The equations for M and M_v for the four segments of the beam, determined by using the method of sections, are tabulated in Table 7.6. The deflection at D can now be computed by applying the virtual work expression given by Eq. (7.30).

$$1(\Delta_D) = \int_0^L \frac{M_v M}{EI} dx$$

$$1(\Delta_D) = \frac{1}{EI}\left[\int_0^3 \left(\frac{x}{4}\right)(75x)\,dx + \frac{1}{2}\int_3^6 \left(\frac{x}{4}\right)(75x)\,dx \right.$$

$$\left. + \frac{1}{2}\int_6^9 \left(\frac{x}{4}\right)(-75x + 900)\,dx + \int_0^3 \left(\frac{3}{4}x\right)(75x)\,dx \right]$$

$$(1\text{ kN})\Delta_D = \frac{2{,}193.75 \text{ kN}^2 \cdot \text{m}^3}{EI}$$

Therefore,

$$\Delta_D = \frac{2{,}193.75 \text{ kN} \cdot \text{m}^3}{EI} = \frac{2{,}193.75}{200(300)} = 0.0366 \text{ m} = 36.6 \text{ mm}$$

$$\Delta_D = 36.6 \text{ mm} \downarrow \qquad\qquad\qquad \text{Ans.}$$

TABLE 7.6

Segment	x Coordinate Origin	x Coordinate Limits (m)	EI ($I = 300 \times 10^6$ mm^4)	M (kN · m)	M_v (kN · m)
AB	A	0–3	EI	$75x$	$\dfrac{x}{4}$
BC	A	3–6	$2EI$	$75x$	$\dfrac{x}{4}$
CD	A	6–9	$2EI$	$75x - 150(x-6)$	$\dfrac{x}{4}$
ED	E	0–3	EI	$75x$	$\dfrac{3}{4}x$

Example 7.9

Determine the deflection at point C of the beam shown in Fig. 7.14(a) by the virtual work method.

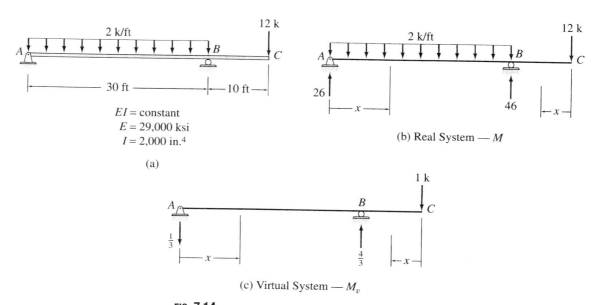

EI = constant
E = 29,000 ksi
I = 2,000 in.^{4}

(a)

(b) Real System — M

(c) Virtual System — M_v

FIG. 7.14

Solution

This beam was previously analyzed by the moment-area and the conjugate-beam methods in Examples 6.7 and 6.13, respectively.

The real and virtual systems for this problem are shown in Fig. 7.14(b) and (c), respectively. The real and virtual loadings are discontinuous at point B, so the beam is divided into two segments, AB and BC. The x coordinates used for determining the bending moment equations are shown in Fig. 7.14(b) and (c), and the equations for M and M_v obtained for each of the two segments of the beam are tabulated in Table 7.7. The deflection at C can now be determined by applying the virtual work expression given by Eq. (7.30), as follows:

TABLE 7.7

| Segment | x Coordinate | | M (k-ft) | M_v (k-ft) |
	Origin	Limits (ft)		
AB	A	0–30	$26x - x^2$	$-\dfrac{x}{3}$
CB	C	0–10	$-12x$	$-x$

continued

$$1(\Delta_C) = \int_0^L \frac{M_v M}{EI} dx$$

$$1(\Delta_C) = \frac{1}{EI}\left[\int_0^{30} \left(-\frac{x}{3}\right)(26x - x^2)\, dx + \int_0^{10} (-x)(-12x)\, dx\right]$$

$$(1\ k)\Delta_C = -\frac{6{,}500\ k^2\text{-ft}^3}{EI}$$

Therefore,

$$\Delta_C = -\frac{6{,}500\ \text{k-ft}^3}{EI} = -\frac{6{,}500(12)^3}{(29{,}000)(2{,}000)} = -0.194\ \text{in.}$$

$$\Delta_C = 0.194\ \text{in.} \uparrow \qquad\qquad \text{Ans.}$$

7.5 DEFLECTIONS OF FRAMES BY THE VIRTUAL WORK METHOD

Application of the virtual work method to determine the slopes and deflections of frames is similar to that for beams. To determine the deflection, Δ, or rotation, θ, at a point of a frame, a virtual unit load or unit couple is applied at that point. When the virtual system is subjected to the deformations of the frame due to real loads, the virtual external work performed by the unit load or the unit couple is $W_{ve} = 1(\Delta)$, or $W_{ve} = 1(\theta)$. As portions of the frame may undergo axial deformations in addition to the bending deformations, the total virtual internal work done on the frame is equal to the sum of the internal virtual work due to bending and that due to axial deformations. As discussed in the preceding section, when the real and virtual loadings and the flexural rigidity EI are continuous over a segment of the frame, the virtual internal work due to bending for that segment can be obtained by integrating the quantity $M_v M/EI$ over the length of the segment. The virtual internal work due to bending for the entire frame can then be obtained by summing the work for the individual segments; that is,

$$W_{vib} = \sum \int \frac{M_v M}{EI} dx \qquad\qquad (7.32)$$

Similarly, if the axial forces F and F_v due to the real and virtual loads, respectively, and the axial rigidity AE are constant over the length L of a segment of the frame, then, as discussed in Section 7.3, the virtual internal work for that segment due to axial deformation is equal to $F_v(FL/AE)$. Thus, the virtual internal work due to axial deformations for the entire frame can be expressed as

$$W_{via} = \sum F_v \left(\frac{FL}{AE}\right) \qquad\qquad (7.33)$$

By adding Eqs. (7.32) and (7.33), we obtain the total internal virtual work for the frame due to both bending and axial deformations as

$$W_{vi} = \sum F_v \left(\frac{FL}{AE} \right) + \sum \int \frac{M_v M}{EI} dx \qquad (7.34)$$

By equating the virtual external work to the virtual internal work, we obtain the expressions for the method of virtual work for deflections and rotations of frames, respectively, as

$$1(\Delta) = \sum F_v \left(\frac{FL}{AE} \right) + \sum \int \frac{M_v M}{EI} dx \qquad (7.35)$$

and

$$1(\theta) = \sum F_v \left(\frac{FL}{AE} \right) + \sum \int \frac{M_v M}{EI} dx \qquad (7.36)$$

The axial deformations in the members of frames composed of common engineering materials are generally much smaller than the bending deformations and are, therefore, usually neglected in the analysis. In this text, unless stated otherwise, we will neglect the effect of axial deformations in the analysis of frames. The virtual work expressions considering only the bending deformations of frames can be obtained by simply omitting the first term on the right-hand sides of Eqs. (7.35) and (7.36), which are thus reduced to

$$1(\Delta) = \sum \int \frac{M_v M}{EI} dx \qquad (7.37)$$

and

$$1(\theta) = \sum \int \frac{M_v M}{EI} dx \qquad (7.38)$$

Procedure for Analysis

The following step-by-step procedure can be used to determine the slopes and deflections of frames by the virtual work method.

1. **Real System** Determine the internal forces at the ends of the members of the frame due to the real loading by using the procedure described in Section 5.6.

2. ***Virtual System*** If the deflection of the frame is to be determined, then apply a unit load at the point and in the direction of the desired deflection. If the rotation is to be calculated, then apply a unit couple at the point on the frame where the rotation is desired. Determine the member end forces due to the virtual loading.

3. If necessary, divide the members of the frame into segments so that the real and virtual loads and EI are continuous in each segment.

4. For each segment of the frame, determine an equation expressing the variation of the bending moment due to real loading (M) along the length of the segment in terms of a position coordinate x.

5. For each segment of the frame, determine the equation for the bending moment due to virtual load or couple (M_v) using the same x coordinate that was used for this segment in step 4 to establish the expression for the real bending moment, M. Any convenient sign convention can be used for M and M_v. However, it is important that the sign convention be the same for both M and M_v in a particular segment.

6. If the effect of axial deformations is to be included in the analysis, then go to step 7. Otherwise, determine the desired deflection or rotation of the frame by applying the appropriate virtual work expression, Eq. (7.37) or Eq. (7.38). End the analysis at this stage.

7. If necessary, divide the members of the frame into segments so that the real and virtual axial forces and AE are constant in each segment. It is not necessary that these segments be the same as those used in step 3 for evaluating the virtual internal work due to bending. It is important, however, that the same sign convention be used for both the real axial force, F, and the virtual axial force, F_v, in a particular segment.

8. Determine the desired deflection or rotation of the frame by applying the appropriate virtual work expression, Eq. (7.35) or Eq. (7.36).

Example 7.10

Determine the rotation of joint C of the frame shown in Fig. 7.15(a) by the virtual work method.

Solution

The real and virtual systems are shown in Fig. 7.15(b) and (c), respectively. The x coordinates used for determining the bending moment equations for the three segments of the frame, AB, BC, and CD, are also shown in these figures. The equations for M and M_v obtained for the three segments are tabulated in Table 7.8. The rotation of joint C of the frame can now be determined by applying the virtual work expression given by Eq. (7.38).

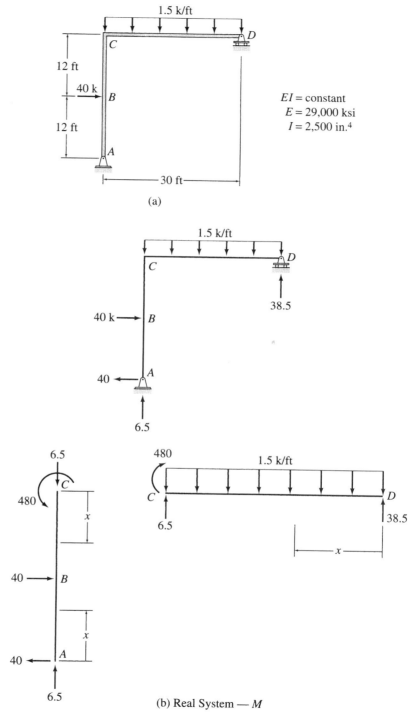

FIG. **7.15**

(a)

EI = constant
E = 29,000 ksi
I = 2,500 in.^{4}

(b) Real System — M

continued

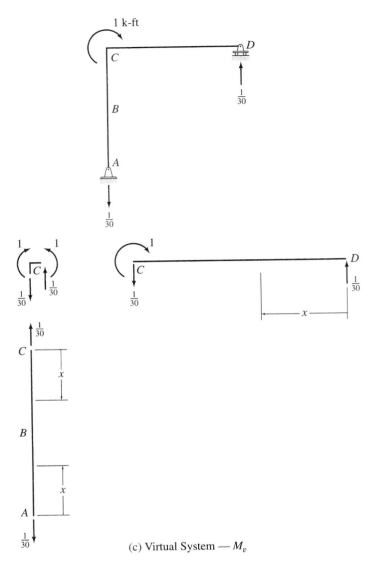

FIG. **7.15** (contd.)

(c) Virtual System — M_v

TABLE 7.8

| Segment | x Coordinate | | M (k-ft) | M_v (k-ft) |
	Origin	Limits (ft)		
AB	A	0–12	40x	0
CB	C	0–12	480	0
DC	D	0–30	$38.5x - 1.5\dfrac{x^2}{2}$	$\dfrac{x}{30}$

$$1(\theta_C) = \sum \int \frac{M_v M}{EI} dx$$

$$= \frac{1}{EI} \int_0^{30} \left(\frac{x}{30}\right)\left(38.5x - 1.5\frac{x^2}{2}\right) dx$$

$$(1 \text{ k-ft})\theta_C = \frac{6{,}487.5 \text{ k}^2\text{-ft}^3}{EI}$$

Therefore,

$$\theta_C = \frac{6{,}487.5 \text{ k-ft}^2}{EI} = \frac{6{,}487.5(12)^2}{(29{,}000)(2{,}500)} = 0.0129 \text{ rad.}$$

$$\theta_C = 0.0129 \text{ rad.} \qquad\qquad \text{Ans.}$$

Example 7.11

Use the virtual work method to determine the vertical deflection at joint C of the frame shown in Fig. 7.16(a).

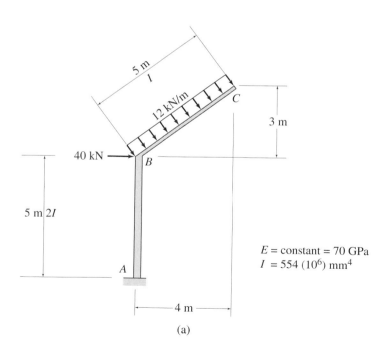

FIG. 7.16

(a)

continued

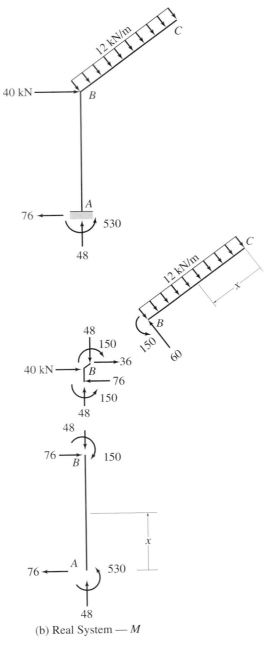

(b) Real System — M

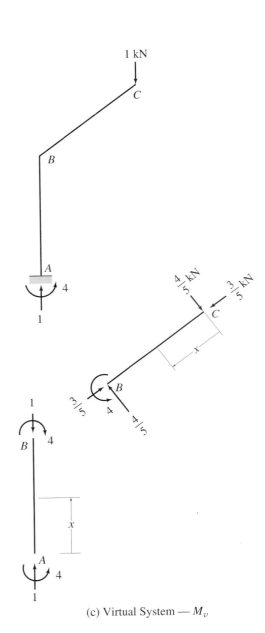

(c) Virtual System — M_v

FIG. 7.16 (contd.)

TABLE 7.9

Segment	x Coordinate Origin	x Coordinate Limits (m)	EI ($I = 554 \times 10^6$ mm^4)	M (kN · m)	M_v (kN · m)
AB	A	0–5	$2EI$	$76x - 530$	-4
CB	C	0–5	EI	$-12\dfrac{x^2}{2}$	$-\dfrac{4}{5}x$

Solution

The real and virtual systems are shown in Figs. 7.16(b) and (c), respectively. The x coordinates used for determining the bending moment equations for the two members of the frame, AB and BC, are also shown in the figures. The equations for M and M_v obtained for the two members are tabulated in Table 7.9. The vertical deflection at joint C of the frame can now be calculated by applying the virtual work expression given by Eq. (7.37):

$$1(\Delta_C) = \sum \int \frac{M_v M}{EI}\, dx$$

$$1(\Delta_C) = \frac{1}{EI}\left[\frac{1}{2}\int_0^5 (-4)(76x - 530)\, dx + \int_0^5 \left(-\frac{4}{5}x\right)(-6x^2)\, dx \right]$$

$$(1 \text{ kN})\Delta_C = \frac{4{,}150 \text{ kN}^2 \cdot \text{m}^3}{EI}$$

Therefore,

$$\Delta_C = \frac{4{,}150 \text{ kN} \cdot \text{m}^3}{EI} = \frac{4{,}150}{70(554)} = 0.107 \text{ m} = 107 \text{ mm}$$

$$\Delta_C = 107 \text{ mm} \downarrow$$

Ans.

Example 7.12

Determine the horizontal deflection at joint C of the frame shown in Fig. 7.17(a) including the effect of axial deformations, by the virtual work method.

Solution

The real and virtual systems are shown in Fig. 7.17(b) and (c), respectively. The x coordinates used for determining the bending moment equations for the three members of the frame, AB, BC, and CD, are also shown in the figures. The equations for M and M_v obtained for the three members are tabulated in Table 7.10 along with the axial forces F and F_v of the members. The horizontal de-

continued

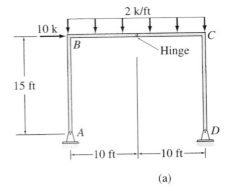

$E = 29{,}000$ ksi
$I = 1{,}000$ in.^{4}
$A = 35$ in.^{2}

(a)

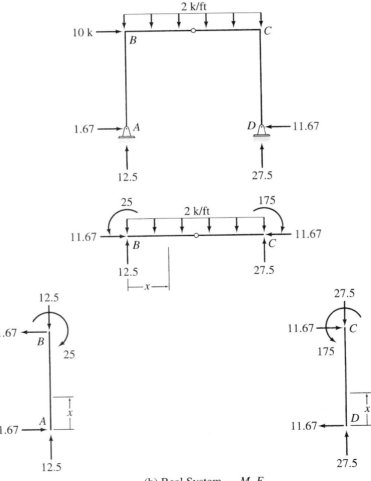

(b) Real System — M, F

FIG. **7.17**

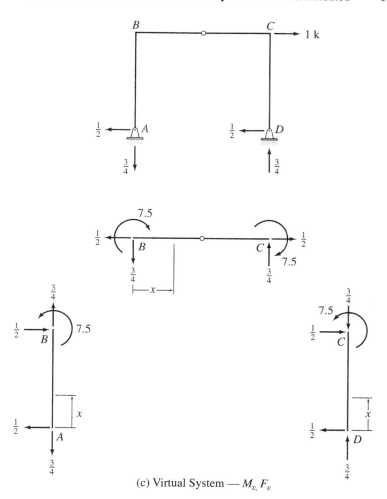

(c) Virtual System — M_v, F_v

FIG. 7.17 (contd.)

TABLE 7.10

Segment	Origin	x Coordinate Limits (ft)	M (k-ft)	F (k)	M_v (k-ft)	F_v (k)
AB	A	0–15	$-1.67x$	-12.50	$\dfrac{x}{2}$	$\dfrac{3}{4}$
BC	B	0–20	$-25 + 12.5x - x^2$	-11.67	$7.5 - \dfrac{3}{4}x$	$\dfrac{1}{2}$
DC	D	0–15	$11.67x$	-27.50	$\dfrac{x}{2}$	$-\dfrac{3}{4}$

continued

flection at joint C of the frame can be determined by applying the virtual work expression given by Eq. (7.35):

$$1(\Delta_C) = \sum F_v\left(\frac{FL}{AE}\right) + \sum \int \frac{M_v M}{EI}\, dx$$

$$1(\Delta_C) = \frac{1}{AE}\left[\frac{3}{4}(-12.5)(15) + \frac{1}{2}(-11.67)(20) - \frac{3}{4}(-27.5)(15)\right]$$

$$+ \frac{1}{EI}\left[\int_0^{15} \frac{x}{2}(-1.67x)\, dx\right.$$

$$+ \left.\int_0^{20}\left(7.5 - \frac{3}{4}x\right)(-25 + 12.5x - x^2)\, dx + \int_0^{15}\frac{x}{2}(11.67x)\, dx\right]$$

$$(1\ \text{k})\Delta_C = \frac{52.08\ \text{k}^2\text{-ft}}{AE} + \frac{9,375\ \text{k}^2\text{-ft}^3}{EI}$$

Therefore,

$$\Delta_C = \frac{52.08\ \text{k-ft}}{AE} + \frac{9,375\ \text{k-ft}^3}{EI}$$

$$= \frac{52.08}{(35)(29,000)} + \frac{9,375(12)^2}{(29,000)(1,000)}$$

$$= 0.00005 + 0.04655$$

$$= 0.0466\ \text{ft} = 0.559\ \text{in.}$$

$$\Delta_C = 0.559\ \text{in.} \rightarrow \qquad\qquad \text{Ans.}$$

Note that the magnitude of the axial deformation term is negligibly small as compared to that of the bending deformation term.

7.6 CONSERVATION OF ENERGY AND STRAIN ENERGY

Before we can develop the next method for computing deflections of structures, it is necessary to understand the concepts of conservation of energy and strain energy.

The *energy* of a structure can be simply defined as its *capacity for doing work*. The term *strain energy* is attributed to the *energy that a structure has because of its deformation*. The relationship between the work and strain energy of a structure is based on the *principle of conservation of energy*, which can be stated as follows:

The work performed on an elastic structure in equilibrium by statically (gradually) applied external forces is equal to the work done by internal forces, or the strain energy stored in the structure.

This principle can be mathematically expressed as

$$W_e = W_i \qquad\qquad (7.39)$$

or

$$W_e = U \tag{7.40}$$

In these equations, W_e and W_i represent the work done by the external and internal forces, respectively, and U denotes the strain energy of the structure. The explicit expression for the strain energy of a structure depends on the types of internal forces that can develop in the members of the structure. Such expressions for the strain energy of trusses, beams, and frames are derived in the following.

Strain Energy of Trusses

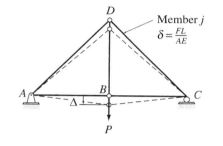

Member j
$\delta = \frac{FL}{AE}$

FIG. 7.18

Consider the arbitrary truss shown in Fig. 7.18. The truss is subjected to a load P, which increases gradually from zero to its final value, causing the structure to deform as shown in the figure. Because we are considering linearly elastic structures, the deflection of the truss Δ at the point of application of P increases linearly with the load; therefore, as discussed in Section 7.1 (see Fig. 7.1(c)), the external work performed by P during the deformation Δ can be expressed as

$$W_e = \frac{1}{2}P\Delta$$

To develop the expression for internal work or strain energy of the truss, let us focus our attention on an arbitrary member j (e.g., member CD in Fig. 7.18) of the truss. If F represents the axial force in this member due to the external load P, then as discussed in Section 7.3, the axial deformation of this member is given by $\delta = (FL)/(AE)$. Therefore, internal work or strain energy stored in member j, U_j, is given by

$$U_j = \frac{1}{2}F\delta = \frac{F^2 L}{2AE}$$

The strain energy of the entire truss is simply equal to the sum of the strain energies of all of its members and can be expressed as

$$U = \sum \frac{F^2 L}{2AE} \tag{7.41}$$

Note that a factor of $\frac{1}{2}$ appears in the expression for strain energy because the axial force F and the axial deformation δ caused by F in each member of the truss are related by the linear relationship $\delta = (FL)/(AE)$.

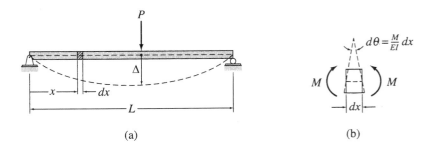

FIG. 7.19

(a)   (b)

Strain Energy of Beams

To develop the expression for the strain energy of beams, consider an arbitrary beam, as shown in Fig. 7.19(a). As the external load P acting on the beam increases gradually from zero to its final value, the internal bending moment M acting on a differential element dx of the beam (Fig. 7.19(a) and (b)) also increases gradually from zero to its final value, while the cross sections of element dx rotate by an angle $d\theta$ with respect to each other. The internal work or the strain energy stored in the element dx is, therefore, given by

$$dU = \frac{1}{2}M(d\theta) \tag{7.42}$$

Recalling from Section 7.4 (Eq. (7.27)) that the change in slope, $d\theta$, can be expressed in terms of the bending moment, M, by the relationship $d\theta = (M/EI)\,dx$, we write Eq. (7.42) as

$$dU = \frac{M^2}{2EI}dx \tag{7.43}$$

The expression for the strain energy of the entire beam can now be obtained by integrating Eq. (7.43) over the length L of the beam:

$$U = \int_0^L \frac{M^2}{2EI}dx \tag{7.44}$$

When the quantity M/EI is not a continuous function of x over the entire length of the beam, then the beam must be divided into segments so that M/EI is continuous in each segment. The integral on the right-hand side of Eq. (7.44) is then evaluated by summing the integrals for all the segments of the beam. We must realize that Eq. (7.44) is based on the consideration of bending deformations of beams and does not include the effect of shear deformations, which, as stated previously, are negligibly small as compared to the bending deformations for most beams.

Strain Energy of Frames

The portions of frames may be subjected to axial forces as well as bending moments, so the total strain energy (U) of frames is expressed as the sum of the strain energy due to axial forces (U_a) and the strain energy due to bending (U_b); that is,

$$U = U_a + U_b \qquad (7.45)$$

If a frame is divided into segments so that the quantity F/AE is constant over the length L of each segment, then—as shown previously in the case of trusses—the strain energy stored in each segment due to the axial force F is equal to $(F^2L)/(2AE)$. Therefore, the strain energy due to axial forces for the entire frame can be expressed as

$$U_a = \sum \frac{F^2L}{2AE} \qquad (7.46)$$

Similarly, if the frame is divided into segments so that the quantity M/EI is continuous over each segment, then the strain energy stored in each segment due to bending can be obtained by integrating the quantity M/EI over the length of the segment (Eq. (7.44)). The strain energy due to bending for the entire frame is equal to the sum of strain energies of bending of all the segments of the frame and can be expressed as

$$U_b = \sum \int \frac{M^2}{2EI} dx \qquad (7.47)$$

By substituting Eqs. (7.46) and (7.47) into Eq. (7.45), we obtain the following expression for the strain energy of frames due to both the axial forces and bending:

$$U = \sum \frac{F^2L}{2AE} + \sum \int \frac{M^2}{2EI} dx \qquad (7.48)$$

As stated previously, the axial deformations of frames are generally much smaller than the bending deformations and are usually neglected in the analysis. The strain energy of frames due only to bending is expressed as

$$U = \sum \int \frac{M^2}{2EI} dx \qquad (7.49)$$

7.7 CASTIGLIANO'S SECOND THEOREM

In this section, we consider another energy method for determining deflections of structures. This method, which can be applied only to linearly elastic structures, was initially presented by Alberto Castigliano in 1873 and is commonly known as *Castigliano's second theorem*. (Castigliano's first theorem, which can be used to establish equations of equilibrium of structures, is not considered in this text.) Castigliano's second theorem can be stated as follows:

> For linearly elastic structures, the partial derivative of the strain energy with respect to an applied force (or couple) is equal to the displacement (or rotation) of the force (or couple) along its line of action.

In mathematical form, this theorem can be stated as:

$$\frac{\partial U}{\partial P_i} = \Delta_i \quad \text{or} \quad \frac{\partial U}{\partial \overline{M}_i} = \theta_i \tag{7.50}$$

in which U = strain energy; Δ_i = deflection of the point of application of the force P_i in the direction of P_i; and θ_i = rotation of the point of application of the couple $\overline{M}_i$ in the direction of $\overline{M}_i$.

To prove this theorem, consider the beam shown in Fig. 7.20. The beam is subjected to external loads P_1, P_2, and P_3, which increase gradually from zero to their final values, causing the beam to deflect, as shown in the figure. The strain energy (U) stored in the beam due to the external work (W_e) performed by these forces is given by

$$U = W_e = \frac{1}{2}P_1\Delta_1 + \frac{1}{2}P_2\Delta_2 + \frac{1}{2}P_3\Delta_3 \tag{7.51}$$

in which Δ_1, Δ_2, and Δ_3 are the deflections of the beam at the points of application of P_1, P_2, and P_3, respectively, as shown in the figure. As Eq. (7.51) indicates, the strain energy U is a function of the external loads and can be expressed as

$$U = f(P_1, P_2, P_3) \tag{7.52}$$

Now, assume that the deflection Δ_2 of the beam at the point of application of P_2 is to be determined. If P_2 is increased by an infinitesimal

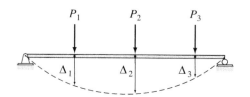

FIG. **7.20**

amount dP_2, then the increase in strain energy of the beam due to the application of dP_2 can be written as

$$dU = \frac{\partial U}{\partial P_2} dP_2 \tag{7.53}$$

and the total strain energy, U_T, now stored in the beam is given by

$$U_T = U + dU = U + \frac{\partial U}{\partial P_2} dP_2 \tag{7.54}$$

The beam is assumed to be composed of linearly elastic material, so regardless of the sequence in which the loads $P_1, (P_2 + dP_2)$, and P_3 are applied, the total strain energy stored in the beam should be the same.

Consider, for example, the sequence in which dP_2 is applied to the beam before the application of P_1, P_2, and P_3. If $d\Delta_2$ is the deflection of the beam at the point of application of dP_2 due to dP_2, then the strain energy stored in the beam is given by $(1/2)(dP_2)(d\Delta_2)$. The loads P_1, P_2, and P_3 are then applied to the beam, causing the additional deflections Δ_1, Δ_2, and Δ_3, respectively, at their points of application. Note that since the beam is linearly elastic, the loads P_1, P_2, and P_3 cause the same deflections, Δ_1, Δ_2, and Δ_3, respectively, and perform the same amount of external work on the beam regardless of whether any other load is acting on the beam or not. The total strain energy stored in the beam during the application of dP_2 followed by P_1, P_2, and P_3 is given by

$$U_T = \frac{1}{2}(dP_2)(d\Delta_2) + dP_2(\Delta_2) + \frac{1}{2}P_1\Delta_1 + \frac{1}{2}P_2\Delta_2 + \frac{1}{2}P_3\Delta_3 \tag{7.55}$$

Since dP_2 remains constant during the additional deflection, Δ_2, of its point of application, the term $dP_2(\Delta_2)$ on the right-hand side of Eq. (7.55) does not contain the factor $1/2$. The term $(1/2)(dP_2)(d\Delta_2)$ represents a small quantity of second order, so it can be neglected, and Eq. (7.55) can be written as

$$U_T = dP_2(\Delta_2) + \frac{1}{2}P_1\Delta_1 + \frac{1}{2}P_2\Delta_2 + \frac{1}{2}P_3\Delta_3 \tag{7.56}$$

By substituting Eq. (7.51) into Eq. (7.56) we obtain

$$U_T = dP_2(\Delta_2) + U \tag{7.57}$$

and by equating Eqs. (7.54) and (7.57), we write

$$U + \frac{\partial U}{\partial P_2} dP_2 = dP_2(\Delta_2) + U$$

or

$$\frac{\partial U}{\partial P_2} = \Delta_2$$

which is the mathematical statement of Castigliano's second theorem.

Application to Trusses

To develop the expression of Castigliano's second theorem, which can be used to determine the deflections of trusses, we substitute Eq. (7.41) for the strain energy (U) of trusses into the general expression of Castigliano's second theorem for deflections as given by Eq. (7.50) to obtain

$$\Delta = \frac{\partial}{\partial P} \sum \frac{F^2 L}{2AE} \qquad (7.58)$$

As the partial derivative $\partial F^2 / \partial P = 2F(\partial F / \partial P)$, the expression of Castigliano's second theorem for trusses can be written as

$$\Delta = \sum \left(\frac{\partial F}{\partial P} \right) \frac{FL}{AE} \qquad (7.59)$$

The foregoing expression is similar in form to the expression of the method of virtual work for trusses (Eq. (7.23)). As illustrated by the solved examples at the end of this section, the procedure for computing deflections by Castigliano's second theorem is also similar to that of the virtual work method.

Application to Beams

By substituting Eq. (7.44) for the strain energy (U) of beams into the general expressions of Castigliano's second theorem (Eq. (7.50)), we obtain the following expressions for the deflections and rotations, respectively, of beams:

$$\Delta = \frac{\partial}{\partial P} \int_0^L \frac{M^2}{2EI} dx \quad \text{and} \quad \theta = \frac{\partial}{\partial \overline{M}} \int_0^L \frac{M^2}{2EI} dx$$

or

$$\Delta = \int_0^L \left(\frac{\partial M}{\partial P} \right) \frac{M}{EI} dx \qquad (7.60)$$

and

$$\theta = \int_0^L \left(\frac{\partial M}{\partial \overline{M}} \right) \frac{M}{EI} dx \qquad (7.61)$$

Application to Frames

Similarly, by substituting Eq. (7.48) for the strain energy (U) of frames due to the axial forces and bending into the general expressions of Castigliano's second theorem (Eq. (7.50)), we obtain the following expressions for the deflections and rotations, respectively, of frames:

$$\Delta = \sum \left(\frac{\partial F}{\partial P}\right)\frac{FL}{AE} + \sum \int \left(\frac{\partial M}{\partial P}\right)\frac{M}{EI}\,dx \qquad (7.62)$$

and

$$\theta = \sum \left(\frac{\partial F}{\partial \overline{M}}\right)\frac{FL}{AE} + \sum \int \left(\frac{\partial M}{\partial \overline{M}}\right)\frac{M}{EI}\,dx \qquad (7.63)$$

When the effect of axial deformations of the members of frames is neglected in the analysis, Eqs. (7.62) and (7.63) reduce to

$$\Delta = \sum \int \left(\frac{\partial M}{\partial P}\right)\frac{M}{EI}\,dx \qquad (7.64)$$

and

$$\theta = \sum \int \left(\frac{\partial M}{\partial \overline{M}}\right)\frac{M}{EI}\,dx \qquad (7.65)$$

Procedure for Analysis

As stated previously, the procedure for computing deflections of structures by Castigliano's second theorem is similar to that of the virtual work method. The procedure essentially involves the following steps.

1. If an external load (or couple) is acting on the given structure at the point and in the direction of the desired deflection (or rotation), then designate that load (or couple) as the variable P (or $\overline{M}$) and go to step 2. Otherwise, apply a fictitious load P (or couple $\overline{M}$) at the point and in the direction of the desired deflection (or rotation).
2. Determine the axial force F and/or the equation(s) for bending moment $M(x)$ in each member of the structure in terms of P (or $\overline{M}$).
3. Differentiate the member axial forces F and/or the bending moments $M(x)$ obtained in step 2 with respect to the variable P (or $\overline{M}$) to compute $\partial F/\partial P$ and/or $\partial M/\partial P$ (or $\partial F/\partial \overline{M}$ and/or $\partial M/\partial \overline{M}$).

4. Substitute the numerical value of P (or $\bar{M}$) into the expressions of F and/or $M(x)$ and their partial derivatives. If P (or $\bar{M}$) represents a fictitious load (or couple), its numerical value is zero.
5. Apply the appropriate expression of Castigliano's second theorem (Eqs. (7.59) through (7.65)) to determine the desired deflection or rotation of the structure. A positive answer for the desired deflection (or rotation) indicates that the deflection (or rotation) occurs in the same direction as P (or $\bar{M}$) and vice versa.

Example 7.13

Determine the deflection at point C of the beam shown in Fig. 7.21(a) by Castigliano's second theorem.

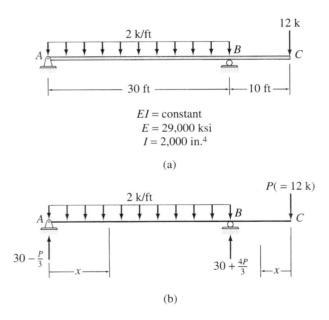

$EI = \text{constant}$
$E = 29{,}000 \text{ ksi}$
$I = 2{,}000 \text{ in.}^4$

(a)

(b)

FIG. 7.21

Solution

This beam was previously analyzed by the moment-area, the conjugate-beam, and the virtual work methods in Examples 6.7, 6.13, and 7.9, respectively.

The 12-k external load is already acting at point C, where the deflection is to be determined, so we designate this load as the variable P, as shown in Fig. 7.21(b). Next, we compute the reactions of the beam in terms of P. These are also shown in Fig. 7.21(b). Since the loading is discontinuous at point B, the beam is divided into two segments, AB and BC. The x coordinates used for determining the equations for the bending moment in the two segments of the beam are shown in Fig. 7.21(b). The equations for M (in terms of P) obtained for the segments of the beam are tabulated in Table 7.11, along with the partial derivatives of M with respect to P.

TABLE 7.11

| Segment | x Coordinate | | M (k-ft) | $\dfrac{\partial M}{\partial P}$ (k-ft/k) |
	Origin	Limits (ft)		
AB	A	0–30	$\left(30 - \dfrac{P}{3}\right)x - x^2$	$-\dfrac{x}{3}$
CB	C	0–10	$-Px$	$-x$

The deflection at C can now be determined by substituting $P = 12$ k into the equations for M and $\partial M/\partial P$ and by applying the expression of Castigliano's second theorem as given by Eq. (7.60):

$$\Delta_C = \int_0^L \left(\frac{\partial M}{\partial P}\right)\left(\frac{M}{EI}\right) dx$$

$$\Delta_C = \frac{1}{EI}\left[\int_0^{30}\left(-\frac{x}{3}\right)\left(30x - \frac{12x}{3} - x^2\right) dx + \int_0^{10}(-x)(-12x)\,dx\right]$$

$$= \frac{1}{EI}\left[\int_0^{30}\left(-\frac{x}{3}\right)(26x - x^2)\,dx + \int_0^{10}(-x)(-12x)\,dx\right]$$

$$= -\frac{6{,}500 \text{ k-ft}^3}{EI} = -\frac{6{,}500(12)^3}{(29{,}000)(2{,}000)} = -0.194 \text{ in.}$$

The negative answer for Δ_C indicates that point C deflects upward in the direction opposite to that of P.

$$\Delta_C = 0.194 \text{ in. } \uparrow \qquad\qquad \text{Ans.}$$

Example 7.14

Use Castigliano's second theorem to determine the deflection at point B of the beam shown in Fig. 7.22(a).

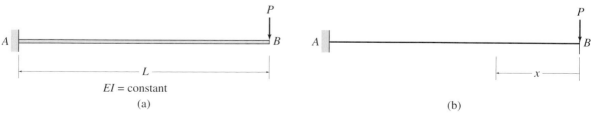

$EI = $ constant

(a)

(b)

FIG. 7.22

continued

Solution

Using the x coordinate shown in Fig. 7.22(b), we write the equation for the bending moment in the beam as

$$M = -Px$$

The partial derivative of M with respect to P is given by

$$\frac{\partial M}{\partial P} = -x$$

The deflection at B can now be obtained by applying the expression of Castigliano's second theorem, as given by Eq. (7.60), as follows:

$$\Delta_B = \int_0^L \left(\frac{\partial M}{\partial P}\right)\left(\frac{M}{EI}\right) dx$$

$$\Delta_B = \int_0^L (-x)\left(-\frac{Px}{EI}\right) dx$$

$$= \frac{P}{EI}\int_0^L x^2\, dx = \frac{PL^3}{3EI}$$

$$\Delta_B = \frac{PL^3}{3EI} \downarrow \qquad\qquad \text{Ans.}$$

Example 7.15

Determine the rotation of joint C of the frame shown in Fig. 7.23(a) by Castigliano's second theorem.

Solution

This frame was previously analyzed by the virtual work method in Example 7.10.

No external couple is acting at joint C, where the rotation is desired, so we apply a fictitious couple $\bar{M}\ (= 0)$ at C, as shown in Fig. 7.23(b). The x coordinates used for determining the bending moment equations for the three segments of the frame are also shown in Fig. 7.23(b), and the equations for M in terms of $\bar{M}$ and $\partial M/\partial\bar{M}$ obtained for the three segments are tabulated in Table 7.12. The rotation of joint C of the frame can now be determined by setting $\bar{M} = 0$ in the equations for M and $\partial M/\partial\bar{M}$ and by applying the expression of Castigliano's second theorem as given by Eq. (7.65):

$$\theta_C = \sum \int \left(\frac{\partial M}{\partial\bar{M}}\right) \frac{M}{EI}\, dx$$

$$= \int_0^{30} \left(\frac{x}{30}\right)\left(38.5x - 1.5\frac{x^2}{2}\right) dx$$

$$= \frac{6{,}487.5 \text{ k-ft}^2}{EI} = \frac{6{,}487.5(12)^2}{(29{,}000)(2{,}500)} = 0.0129 \text{ rad}$$

$$\theta_C = 0.0129 \text{ rad} \qquad\qquad \text{Ans.}$$

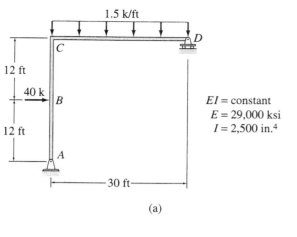

EI = constant
E = 29,000 ksi
I = 2,500 in.[4]

(a)

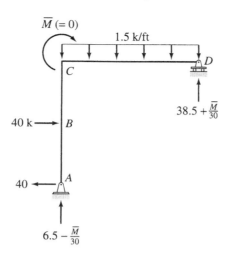

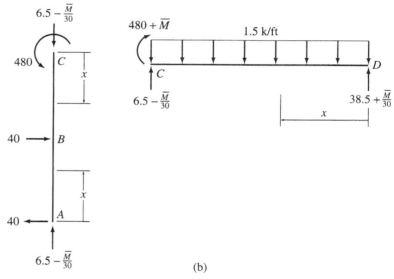

(b)

FIG. 7.23

TABLE 7.12

| Segment | x Coordinate | | M (k-ft) | $\dfrac{\partial M}{\partial \overline{M}}$ (k-ft/k-ft) |
	Origin	Limits (ft)		
AB	A	0–12	$40x$	0
CB	C	0–12	480	0
DC	D	0–30	$\left(38.5 + \dfrac{\overline{M}}{30}\right)x - 1.5\dfrac{x^2}{2}$	$\dfrac{x}{30}$

Example 7.16

Use Castigliano's second theorem to determine the horizontal and vertical components of the deflection at joint B of the truss shown in Fig. 7.24(a).

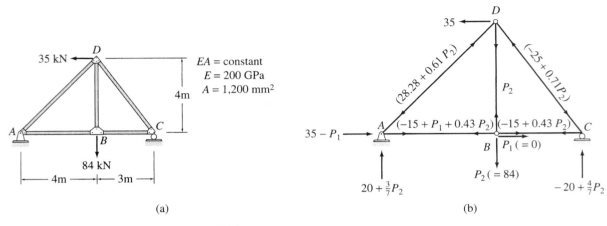

(a)   (b)

FIG. 7.24

Solution

This truss was previously analyzed by the virtual work method in Example 7.3.

TABLE 7.13

Member	L (m)	F (kN)	$\dfrac{\partial F}{\partial P_1}$ (kN/kN)	$\dfrac{\partial F}{\partial P_2}$ (kN/kN)	For $P_1 = 0$ and $P_2 = 84$ kN	
					$(\partial F/\partial P_1)FL$ (kN · m)	$(\partial F/\partial P_2)FL$ (kN · m)
AB	4	$-15 + P_1 + 0.43P_2$	1	0.43	84.48	36.32
BC	3	$-15 + 0.43P_2$	0	0.43	0	27.24
AD	5.66	$-28.28 - 0.61P_2$	0	-0.61	0	274.55
BD	4	P_2	0	1	0	336.00
CD	5	$25 - 0.71P_2$	0	-0.71	0	122.97
				$\Sigma \left(\dfrac{\partial F}{\partial P} \right) FL$	84.48	797.08

$$\Delta_{BH} = \frac{1}{EA} \sum \left(\frac{\partial F}{\partial P_1} \right) FL$$

$$= \frac{84.48}{EA} \text{ kN} \cdot \text{m}$$

$$= \frac{84.48}{200(10^6)(0.0012)} = 0.00035 \text{ m}$$

$$\Delta_{BH} = 0.35 \text{ mm} \rightarrow \qquad \textbf{Ans.}$$

$$\Delta_{BV} = \frac{1}{EA} \sum \left(\frac{\partial F}{\partial P_2} \right) FL$$

$$= \frac{797.08}{EA} \text{ kN} \cdot \text{m}$$

$$= \frac{797.08}{200(10^6)(0.0012)} = 0.00332 \text{ m}$$

$$\Delta_{BV} = 3.32 \text{ mm} \downarrow \qquad \textbf{Ans.}$$

As shown in Fig. 7.24(b), a fictitious horizontal force P_1 ($= 0$) is applied at joint B to determine the horizontal component of deflection, whereas the 84-kN vertical load is designated as the variable P_2 to be used for computing the vertical component of deflection at joint B. The member axial forces, in terms of P_1 and P_2, are then determined by applying the method of joints. These member forces F, along with their partial derivatives with respect to P_1 and P_2, are tabulated in Table 7.13. Note that the tensile axial forces are considered as positive and the compressive forces are negative. Numerical values of $P_1 = 0$ and $P_2 = 84$ kN are then substituted in the equations for F, and the expression of Castigliano's second theorem, as given by Eq. (7.59) is applied, as shown in the table, to determine the horizontal and vertical components of the deflection at joint B of the truss.

7.8 BETTI'S LAW AND MAXWELL'S LAW OF RECIPROCAL DEFLECTIONS

Maxwell's law of reciprocal deflections, initially developed by James C. Maxwell in 1864, plays an important role in the analysis of statically indeterminate structures to be considered in Part Three of this text. Maxwell's law will be derived here as a special case of the more general *Betti's law*, which was presented by E. Betti in 1872. Betti's law can be stated as follows:

> For a linearly elastic structure, the virtual work done by a P system of forces and couples acting through the deformation caused by a Q system of forces and couples is equal to the virtual work of the Q system acting through the deformation due to the P system.

To show the validity of this law, consider the beam shown in Fig. 7.25. The beam is subjected to two different systems of forces, P and Q systems, as shown in Fig. 7.25(a) and (b), respectively. Now, let us assume that we subject the beam that has the P forces already acting on it (Fig. 7.25(a)) to the deflections caused by the Q system of forces (Fig. 7.25(b)). The virtual external work (W_{ve}) done can be written as

$$W_{ve} = P_1\Delta_{Q1} + P_2\Delta_{Q2} + \cdots + P_n\Delta_{Qn}$$

or

$$W_{ve} = \sum_{i=1}^{n} P_i\Delta_{Qi}$$

By applying the principle of virtual forces for deformable bodies, $W_{ve} = W_{vi}$, and using the expression for the virtual internal work done in beams (Eq. (7.29)), we obtain

$$\sum_{i=1}^{n} P_i\Delta_{Qi} = \int_0^L \frac{M_P M_Q}{EI} dx \qquad (7.66)$$

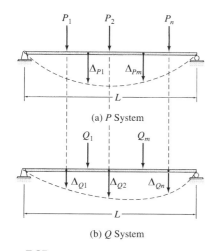

(a) P System

(b) Q System

FIG. 7.25

Next, we assume that the beam with the Q forces acting on it (Fig. 7.25(b)) is subjected to the deflections caused by the P forces (Fig. 7.25(a)). By equating the virtual external work to the virtual internal work, we obtain

$$\sum_{j=1}^{m} Q_j \Delta_{pj} = \int_0^L \frac{M_Q M_P}{EI} dx \tag{7.67}$$

Noting that the right-hand sides of Eqs. (7.66) and (7.67) are identical, we equate the left-hand sides to obtain

$$\sum_{i=1}^{n} P_i \Delta_{Qi} = \sum_{j=1}^{m} Q_j \Delta_{Pj} \tag{7.68}$$

Equation (7.68) represents the mathematical statement of Betti's law.

Maxwell's law of reciprocal deflections states that *for a linearly elastic structure, the deflection at a point i due to a unit load applied at a point j is equal to the deflection at j due to a unit load at i.*

In this statement, the terms *deflection* and *load* are used in the general sense to include rotation and couple, respectively. As mentioned previously, Maxwell's law can be considered as a special case of Betti's law. To prove Maxwell's law, consider the beam shown in Fig. 7.26. The beam is separately subjected to the P and Q systems, consisting of the unit loads at points i and j, respectively, as shown in Fig. 7.26(a) and (b). As the figure indicates, f_{ij} represents the deflection at i due to the unit load at j, whereas f_{ji} denotes the deflection at j due to the unit load at i. These deflections per unit load are referred to as *flexibility coefficients*. By applying Betti's law (Eq. (7.68)), we obtain

$$1(f_{ij}) = 1(f_{ji})$$

or

$$f_{ij} = f_{ji} \tag{7.69}$$

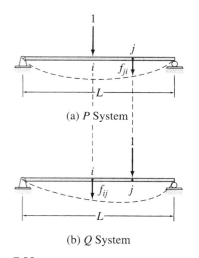

(a) P System

(b) Q System

FIG. 7.26

which is the mathematical statement of Maxwell's law.

The reciprocal relationship remains valid between the rotations caused by two unit couples as well as between the deflection and the rotation caused by a unit couple and a unit force, respectively.

SUMMARY

In this chapter we have learned that the work done by a force P (or couple M) during a displacement Δ (or rotation θ) of its point of application in the direction of its line of action is given by

$$W = \int_0^\Delta P \, d\Delta \tag{7.1}$$

or

$$W = \int_0^\theta M \, d\theta \tag{7.4}$$

The principle of virtual work for rigid bodies states that if a rigid body is in equilibrium under a system of forces and if it is subjected to any small virtual rigid-body displacement, the virtual work done by the external forces is zero.

The principle of virtual forces for deformable bodies can be mathematically stated as

$$W_{ve} = W_{vi} \tag{7.16}$$

in which W_{ve} = virtual external work done by virtual external forces (and couples) acting through the real external displacements (and rotations) of the structure; and W_{vi} = virtual internal work done by the virtual internal forces (and couples) acting through the real internal displacements (and rotations) of the structure.

The method of virtual work for determining the deformations of structures is based on the principle of virtual forces for deformable bodies. The method employs two separate systems: (1) a real system of loads (or other effects) causing the deformation to be determined and (2) a virtual system consisting of a unit load (or unit couple) applied at the point and in the direction of the desired deflection (or rotation). The explicit expressions of the virtual work method to be used to determine the deflections of trusses, beams, and frames are as follows:

$$\text{Trusses} \quad 1(\Delta) = \sum F_v \left(\frac{FL}{AE} \right) \tag{7.23}$$

$$\text{Beams} \quad 1(\Delta) = \int_0^L \frac{M_v M}{EI} \, dx \tag{7.30}$$

$$\text{Frames} \quad 1(\Delta) = \sum F_v \left(\frac{FL}{AE} \right) + \sum \int \frac{M_v M}{EI} \, dx \tag{7.35}$$

The principle of conservation of energy states that the work performed by statically applied external forces on an elastic structure in equilibrium is equal to the work done by internal forces or the strain energy stored in the structure. The expressions for the strain energy of trusses, beams and frames are

$$\text{Trusses} \quad U = \sum \frac{F^2 L}{2AE} \tag{7.41}$$

$$\text{Beams} \quad U = \int_0^L \frac{M^2}{2EI} \, dx \tag{7.44}$$

$$\text{Frames} \quad U = \sum \frac{F^2 L}{2AE} + \sum \int \frac{M^2}{2EI} \, dx \tag{7.48}$$

Castigliano's second theorem for linearly elastic structures can be mathematically expressed as

$$\frac{\partial U}{\partial P_i} = \Delta_i \quad \text{or} \quad \frac{\partial U}{\partial \overline{M}_i} = \theta_i \tag{7.50}$$

The expressions of Castigliano's second theorem, which can be used to determine deflections, are as follows:

$$\text{Trusses} \quad \Delta = \sum \left(\frac{\partial F}{\partial P}\right) \frac{FL}{AE} \tag{7.59}$$

$$\text{Beams} \quad \Delta = \int_0^L \left(\frac{\partial M}{\partial P}\right) \frac{M}{EI} dx \tag{7.60}$$

$$\text{Frames} \quad \Delta = \sum \left(\frac{\partial F}{\partial P}\right) \frac{FL}{AE} + \sum \int \left(\frac{\partial M}{\partial P}\right) \frac{M}{EI} dx \tag{7.62}$$

Maxwell's law of reciprocal deflections states that, for a linearly elastic structure, the deflection at a point i due to a unit load applied at a point j is equal to the deflection at j due to a unit load at i.

PROBLEMS

Section 7.3

7.1 through 7.5 Use the virtual work method to determine the horizontal and vertical components of the deflection at joint B of the truss shown in Figs. P7.1–P7.5.

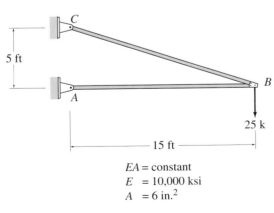

$EA = \text{constant}$
$E = 10,000 \text{ ksi}$
$A = 6 \text{ in.}^2$

FIG. P7.1, P7.45

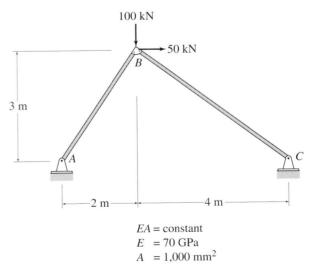

$EA = \text{constant}$
$E = 70 \text{ GPa}$
$A = 1,000 \text{ mm}^2$

FIG. P7.2, P7.46

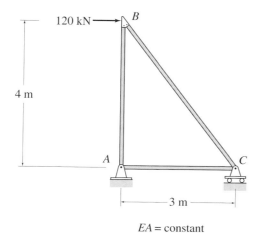

FIG. **P7.3, P7.47**

EA = constant
E = 200 GPa
A = 1,500 mm²

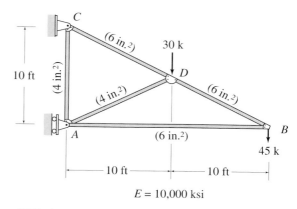

E = 10,000 ksi

FIG. **P7.5, P7.49**

7.6 and 7.7 Use the virtual work method to determine the vertical deflection at joint C of the truss shown in Figs. P7.6 and P7.7.

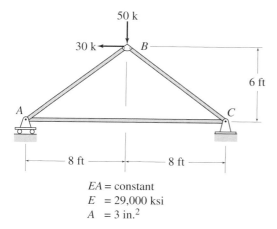

EA = constant
E = 29,000 ksi
A = 3 in.²

FIG. **P7.4, P7.48**

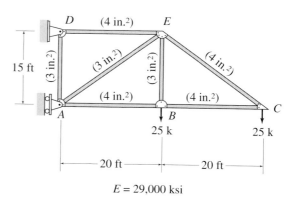

E = 29,000 ksi

FIG. **P7.6**

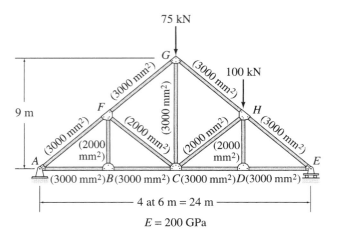

FIG. **P7.7**

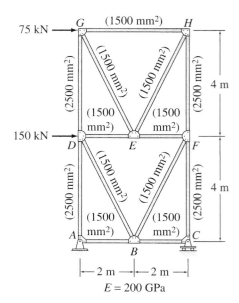

FIG. **P7.9**

7.8 Use the virtual work method to determine the horizontal deflection at joint E of the truss shown in Fig. P7.8.

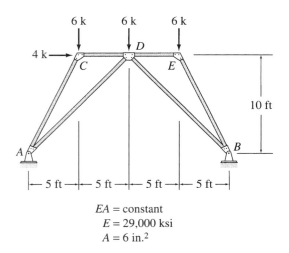

FIG. **P7.8, P7.50**

7.9 Use the virtual work method to determine the horizontal deflection at joint H of the truss shown in Fig. P7.9.

7.10 through 7.12 Determine the smallest cross-sectional area A required for the members of the truss shown, so that the horizontal deflection at joint D does not exceed 10 mm. Use the virtual work method.

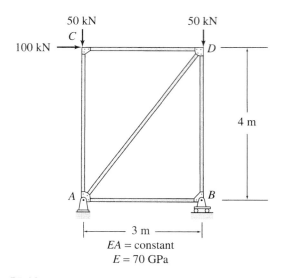

FIG. **P7.10**

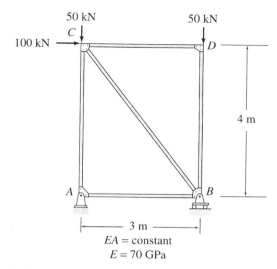

50 kN   50 kN

100 kN

4 m

3 m
EA = constant
E = 70 GPa

FIG. **P7.11**

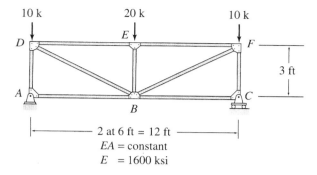

10 k   20 k   10 k

3 ft

2 at 6 ft = 12 ft
EA = constant
E = 1600 ksi

FIG. **P7.13**

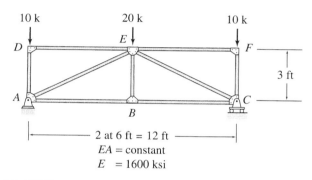

10 k   20 k   10 k

3 ft

2 at 6 ft = 12 ft
EA = constant
E = 1600 ksi

FIG. **P7.14**

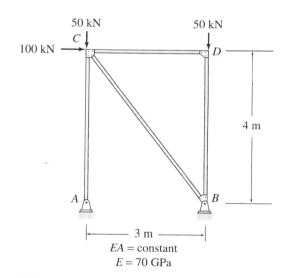

50 kN   50 kN

100 kN

4 m

3 m
EA = constant
E = 70 GPa

FIG. **P7.12**

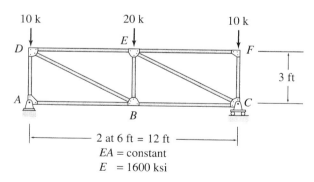

10 k   20 k   10 k

3 ft

2 at 6 ft = 12 ft
EA = constant
E = 1600 ksi

FIG. **P7.15**

7.13 through 7.15 Determine the smallest cross-sectional area A for the members of the truss shown, so that the vertical deflection at joint B does not exceed 0.4 inches. Use the method of virtual work.

7.16 Determine the vertical deflection at joint G of the truss shown in Fig. P7.16 due to a temperature increase of 65°F in members AB, BC, CD, and DE and a temperature drop of 20°F in members FG and GH. Use the method of virtual work.

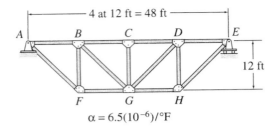

FIG. **P7.16, P7.18**

7.17 Determine the horizontal deflection at joint H of the truss shown in Fig. P7.17 due to a temperature increase of 70°F in members BD, DF, and FH. Use the method of virtual work.

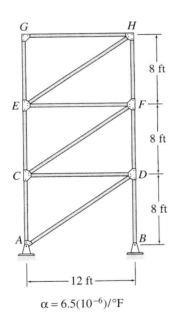

FIG. **P7.17, P7.19**

7.18 Determine the vertical deflection at joint G of the truss shown in Fig. P7.16 if members BC and CG are 0.5 in. too short. Use the method of virtual work.

7.19 Determine the horizontal deflection at joint H of the truss shown in Fig. P7.17 if member AC is 0.5 in. too short and member BD is 0.7 in. too long. Use the method of virtual work.

Section 7.4

7.20 and 7.21 Use the virtual work method to determine the slope and deflection at point B of the beam shown.

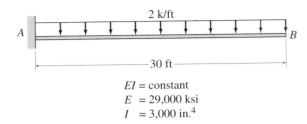

EI = constant
E = 29,000 ksi
I = 3,000 in.^{4}

FIG. **P7.20, P7.51**

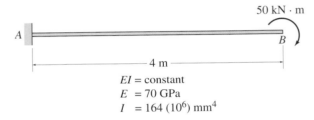

EI = constant
E = 70 GPa
I = 164 (10^6) mm^4

FIG. **P7.21, P7.52**

7.22 through 7.25 Use the virtual work method to determine the deflection at point C of the beam shown.

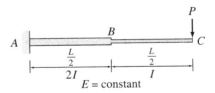

FIG. **P7.22, P7.53**

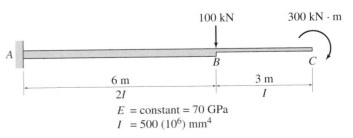

E = constant = 70 GPa
I = 500 (10^6) mm^4

FIG. **P7.23, P7.54**

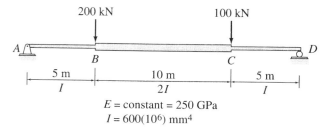

200 kN   100 kN

FIG. **P7.24, P7.55**

E = constant = 250 GPa
I = 600(10⁶) mm⁴

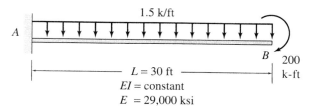

1.5 k/ft

L = 30 ft
EI = constant
E = 29,000 ksi

FIG. **P7.28**

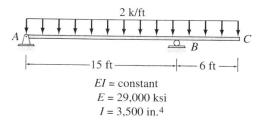

2 k/ft

15 ft ———— 6 ft

EI = constant
E = 29,000 ksi
I = 3,500 in.⁴

FIG. **P7.25, P7.56**

7.29 and 7.30 Use the virtual work method to determine the slope and deflection at point D of the beam shown.

7.26 through 7.28 Determine the smallest moment of inertia I required for the beam shown, so that its maximum deflection does not exceed the limit of 1/360 of the span length (i.e., Δ_max ≤ L/360). Use the method of virtual work.

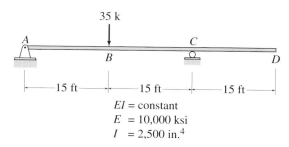

35 k

15 ft ——— 15 ft ——— 15 ft

EI = constant
E = 10,000 ksi
I = 2,500 in.⁴

FIG. **P7.29, P7.57**

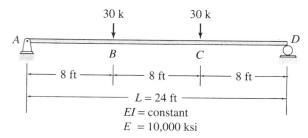

30 k   30 k

8 ft — 8 ft — 8 ft
L = 24 ft
EI = constant
E = 10,000 ksi

FIG. **P7.26**

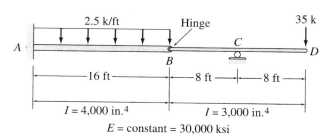

2.5 k/ft   Hinge   35 k

16 ft ——— 8 ft —— 8 ft
I = 4,000 in.⁴   I = 3,000 in.⁴
E = constant = 30,000 ksi

FIG. **P7.30, P7.58**

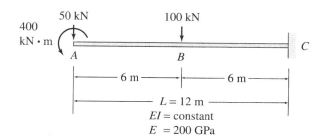

50 kN   100 kN
400 kN · m

6 m —— 6 m
L = 12 m
EI = constant
E = 200 GPa

FIG. **P7.27**

Section 7.5

7.31 and 7.32 Use the virtual work method to determine the vertical deflection at joint C of the frame shown.

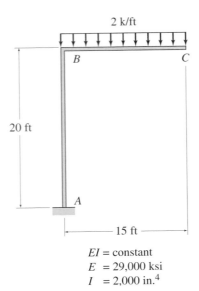

FIG. **P7.31, P7.59**

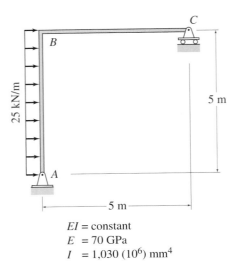

FIG. **P7.33, P7.60**

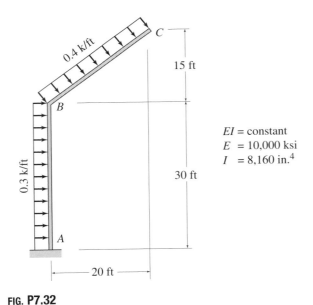

FIG. **P7.32**

7.33 Use the virtual work method to determine the horizontal deflection at joint C of the frame shown.

7.34 Use the virtual work method to determine the rotation of joint D of the frame shown.

7.35 Use the virtual work method to determine the horizontal deflection at joint E of the frame shown in Fig. P7.34.

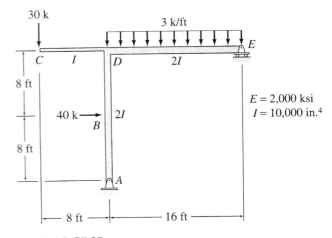

FIG. **P7.34, P7.35**

7.36 Use the virtual work method to determine the rotation of joint B of the frame shown.

7.37 Use the virtual work method to determine the vertical deflection at joint B of the frame shown in Fig. P7.36.

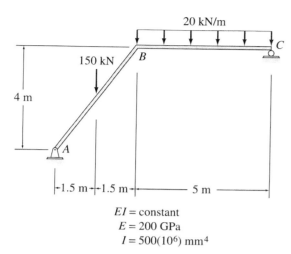

EI = constant
E = 200 GPa
I = 500(10⁶) mm⁴

FIG. P7.36, P7.37

7.38 Use the virtual work method to determine the rotation of joint D of the frame shown.

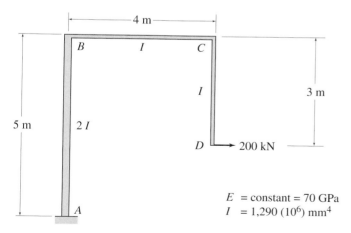

E = constant = 70 GPa
I = 1,290 (10⁶) mm⁴

FIG. P7.38 and P7.61

7.39 and 7.40 Use the virtual work method to determine the horizontal deflection at joint C of the frame shown.

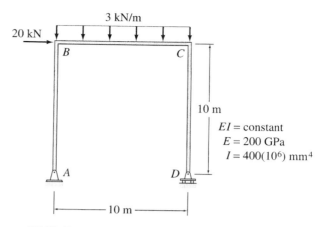

EI = constant
E = 200 GPa
I = 400(10⁶) mm⁴

FIG. P7.39, P7.62

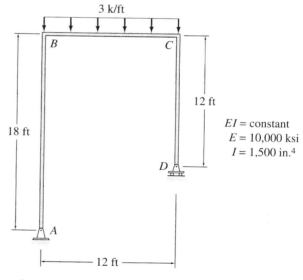

EI = constant
E = 10,000 ksi
I = 1,500 in.⁴

FIG. P7.40, P7.63

7.41 and 7.42 Determine the smallest moment of inertia I required for the members of the frame shown, so that the horizontal deflection at joint C does not exceed 1 inch. Use the virtual work method.

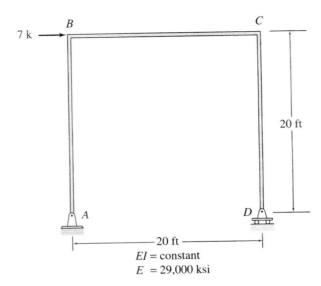

FIG. P7.41

20 ft

EI = constant
E = 29,000 ksi

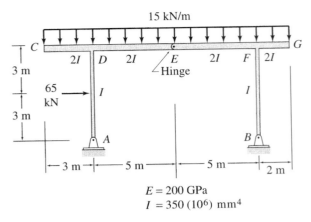

$E = 200$ GPa
$I = 350 (10^6)$ mm^4

FIG. P7.43, P7.44

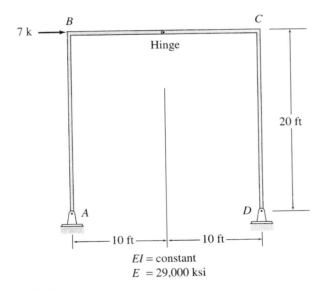

FIG. P7.42

EI = constant
E = 29,000 ksi

7.43 Use the virtual work method to determine the rotation of joint D of the frame shown.

7.44 Using the method of virtual work, determine the vertical deflection at joint E of the frame shown in Fig. P7.43.

Section 7.7

7.45 through 7.49 Use Castigliano's second theorem to determine the horizontal and vertical components of the deflection at joint B of the trusses shown in Figs. P7.1–P7.5.

7.50 Use Castigliano's second theorem to determine the horizontal deflection at joint E of the truss shown in Fig. P7.8.

7.51 and 7.52 Use Castigliano's second theorem to determine the slope and deflection at point B of the beam shown in Figs. P7.20 and P7.21.

7.53 through 7.56 Use Castigliano's second theorem to determine the deflection at point C of the beams shown in Figs. P7.22–P7.25.

7.57 and 7.58 Use Castigliano's second theorem to determine the slope and deflection at point D of the beam shown in Figs. P7.29 and P7.30.

7.59 Use Castigliano's second theorem to determine the vertical deflection at joint C of the frame shown in Fig. P7.31.

7.60 Use Castigliano's second theorem to determine the horizontal deflection at joint C of the frame shown in Fig. P7.33.

7.61 Use Castigliano's second theorem to determine the rotation of joint D of the frame shown in Fig. P7.38.

7.62 and 7.63 Use Castigliano's second theorem to determine the horizontal deflection at joint C of the frames shown in Figs. P7.39 and P7.40.

8

Influence Lines

8.1 Influence Lines for Beams and Frames by Equilibrium Method
8.2 Müller-Breslau's Principle and Qualitative Influence Lines
8.3 Influence Lines for Girders with Floor Systems
8.4 Influence Lines for Trusses
8.5 Influence Lines for Deflections
Summary
Problems

A Bridge Subjected to Variable
Loads Due to Traffic
Photo courtesy of the Illinois Department of Transportation

In the previous chapters, we considered the analysis of structures subjected to loads whose positions were fixed on the structures. An example of such stationary loading is the dead load due to the weight of the structure itself and of other material and equipment permanently attached to the structure. However, structures generally are also subjected to loads (such as live loads and environmental loads) whose positions may vary on the structure. In this chapter, we study the analysis of statically determinate structures subjected to variable loads.

Consider, as an example, the bridge truss shown in Fig. 8.1. As a car moves across the bridge, the forces in the members of the truss will vary with the position x of the car. It should be realized that the forces in different truss members will become maximum at different positions of the car. For example, if the force in member AB becomes maximum when the car is at a certain position $x = x_1$, then the force in another member—for example, member CH—may become maximum when the car is at a different position $x = x_2$. The design of each member of the truss must be based on the maximum force that develops in that member as the car moves across the bridge. Therefore, the analysis of the truss would involve, for each member, determining the position of the car at which the force in the member becomes maximum and then computing the value of the maximum member force.

339

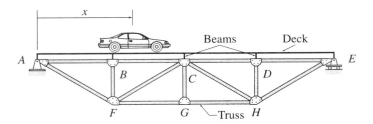

FIG. **8.1**

From the foregoing discussion, we can see that the analysis of structures for variable loads consists of two steps: (1) determining the position(s) of the load(s) at which the response function of interest (e.g., a reaction, shear or bending moment at a section of a beam, or force in a truss member) becomes maximum, and (2) computing the maximum value of the response function.

An important concept used in the analysis of structures subjected to variable loads is that of the *influence lines*, initially introduced by E. Winkler in 1867. *An influence line is a graph of a response function of a structure as a function of the position of a downward unit load moving across the structure.*

We begin this chapter by describing the procedure for constructing influence lines for the reactions, shears, and bending moments of beams and frames by using the equations of equilibrium. We next discuss the *Müller-Breslau principle* and its application for determining influence lines. We also consider the influence lines for the force response functions of girders with floor systems and of trusses and, finally, the influence lines for deflections. The application of influence lines in determining the maximum values of response functions of structures due to variable loads is considered in the next chapter.

8.1 INFLUENCE LINES FOR BEAMS AND FRAMES BY EQUILIBRIUM METHOD

Consider the simply supported beam shown in Fig. 8.2(a). The beam is subjected to a downward concentrated load of unit magnitude, which moves from the left end A of the beam to the right end C. The position of the unit load is defined by the coordinate x measured from the left end A of the beam, as shown in the figure. Suppose that we wish to draw the influence lines for the vertical reactions at supports A and C and the shear and bending moment at point B, which is located at a distance a from the left end of the beam, as shown in the figure.

Influence Lines for Reactions

To develop the influence line for the vertical reaction A_y of the beam, we determine the expression for A_y in terms of the variable position of the unit load, x, by applying the equilibrium equation:

$$+\circlearrowleft \sum M_C = 0$$

$$-A_y(L) + 1(L - x) = 0$$

$$A_y = \frac{1(L - x)}{L} = 1 - \frac{x}{L} \tag{8.1}$$

Equation (8.1) indicates that A_y is a linear function of x, with $A_y = 1$ at $x = 0$ and $A_y = 0$ at $x = L$.

Equation (8.1) represents the equation of the influence line for A_y, which is constructed by plotting this equation with A_y as ordinate

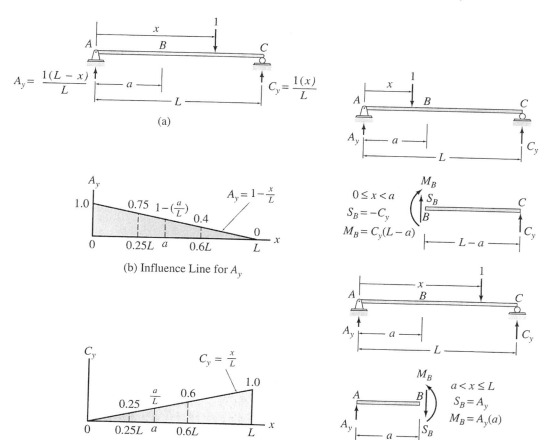

(a)

(b) Influence Line for A_y

(c) Influence Line for C_y

(d)

FIG. 8.2

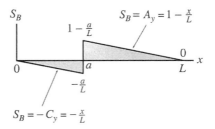

$$S_B = A_y = 1 - \frac{x}{L}$$

$$1 - \frac{a}{L}$$

$$-\frac{a}{L}$$

$$S_B = -C_y = -\frac{x}{L}$$

(e) Influence Line for S_B

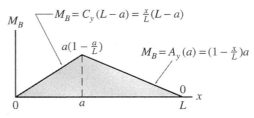

$$M_B = C_y(L - a) = \frac{x}{L}(L - a)$$

$$a\left(1 - \frac{a}{L}\right)$$

$$M_B = A_y(a) = \left(1 - \frac{x}{L}\right)a$$

(f) Influence Line for M_B

FIG. **8.2** (contd.)

against the position of the unit load, x, as abscissa, as shown in Fig. 8.2(b). Note that this influence line (Fig. 8.2(b)) shows graphically how the movement of a unit load across the length of the beam influences the magnitude of the reaction A_y. As this influence line indicates, $A_y = 1$ when the unit load is located at the left support A of the beam (i.e., when $x = 0$). As the unit load moves from A to C, the magnitude of A_y decreases linearly until it becomes zero when the unit load reaches the right support C (i.e., when $x = L$). It is important to realize that the ordinate of the influence line at any position x is equal to the magnitude of A_y due to a unit load acting at the position x on the beam. For example, from the influence line for A_y (Fig. 8.2(b)), we can determine that when a unit load is applied at a distance of $0.25L$ from the end A of the beam, the magnitude of the reaction A_y will be 0.75. Similarly, when the unit load is acting at $x = 0.6L$, the magnitude of A_y will be 0.4, and so on.

The influence line for the vertical reaction C_y of the beam (Fig. 8.2(a)) can be developed by using the procedure just outlined. To determine the expression for C_y in terms of x, we apply the equilibrium equation:

$$+\circlearrowleft \sum M_A = 0$$

$$-1(x) + C_y(L) = 0$$

$$C_y = \frac{1(x)}{L} = \frac{x}{L} \tag{8.2}$$

Equation (8.2) represents the equation of the influence line for C_y, which is constructed by plotting this equation, as shown in Fig. 8.2(c). It can be seen from Fig. 8.2(b) and (c) that the sum of the ordinates of the influence lines for the reactions A_y and C_y at any position of the unit load, x, is equal to 1, indicating that the equilibrium equation $\sum F_y = 0$ is satisfied.

Influence Line for Shear at *B*

The influence lines for shears and bending moments can be developed by employing a procedure similar to that used for constructing the influence lines for reactions. To develop the influence line for the shear at point *B* of the beam (Fig. 8.2(d)), we determine the expressions for S_B. It can be seen from Fig. 8.2(d) that when the unit load is located to the left of point *B*—that is, on segment *AB* of the beam $(0 \leq x < a)$— the shear at *B* can be conveniently obtained by using the free body of the portion *BC* of the beam that is to the right of *B*. Considering the downward external forces and reactions acting on the portion *BC* as positive in accordance with the *beam sign convention* (Section 5.1), we determine the shear at *B* as

$$S_B = -C_y \qquad 0 \leq x < a$$

When the unit load is located to the right of point *B*—that is, on segment *BC* of the beam $(a < x \leq L)$—it is simpler to determine S_B by using the free body of the portion *AB*, which is to the left of *B*. Considering the upward external forces and reactions acting on the portion *AB* as positive, we determine the shear at *B* as

$$S_B = A_y \qquad a < x \leq L$$

Thus the equations of the influence line for S_B can be written as

$$S_B = \begin{cases} -C_y & 0 \leq x < a \\ A_y & a < x \leq L \end{cases} \qquad (8.3)$$

Note that Eq. (8.3) expresses the influence line for S_B in terms of the influence lines for the reactions A_y and C_y. This equation indicates that the segment of the influence line for S_B between points *A* and *B* $(0 \leq x < a)$ can be obtained by multiplying the ordinates of the segment of the influence line for C_y between *A* and *B* by -1. Also, according to this equation, the segment of the influence line for S_B between points *B* and *C* $(a < x \leq L)$ is the same as the segment of the influence line for A_y between the same two points. The influence line for S_B thus constructed from the influence lines for A_y and C_y is shown in Fig. 8.2(e). It is usually more convenient to construct the influence lines for shears and bending moments (to be discussed subsequently) from the influence lines for reactions instead of from the equations expressing the shear or bending moment explicitly in terms of the position of the unit load, x. If desired, such equations for the influence line for S_B in terms of x can be obtained by simply substituting Eqs. (8.1) and (8.2) into Eq. (8.3); that is,

$$
S_B = \begin{cases} -C_y = -\dfrac{x}{L} & 0 \le x < a \\ A_y = 1 - \dfrac{x}{L} & a < x \le L \end{cases} \tag{8.4}
$$

The influence line for S_B (Fig. 8.2(e)) shows that the shear at B is zero when the unit load is located at the left support A of the beam. As the unit load moves from A to B, the shear at B decreases linearly until it becomes $-a/L$ when the unit load reaches just to the left of point B. As the unit load crosses point B, the shear at B increases abruptly to $1 - (a/L)$. It then decreases linearly as the unit load moves toward C until it becomes zero when the unit load reaches the right support C.

Influence Line for Bending Moment at *B*

When the unit load is located to the left of point B (Fig. 8.2(d)), the expression for the bending moment at B can be conveniently obtained by using the free body of the portion BC of the beam to the right of B. Considering the counterclockwise moments of the external forces and reactions acting on the portion BC as positive in accordance with the *beam sign convention* (Section 5.1), we determine the bending moment at B as

$$M_B = C_y(L - a) \qquad 0 \le x \le a$$

When the unit load is located to the right of point B, we use the free body of the portion AB to the left of B to determine M_B. Considering the clockwise moments of the external forces and reactions acting on the portion AB as positive, we determine the bending moment at B as

$$M_B = A_y(a) \qquad a \le x \le L$$

Thus the equations of the influence line for M_B can be written as

$$
M_B = \begin{cases} C_y(L - a) & 0 \le x \le a \\ A_y(a) & a \le x \le L \end{cases} \tag{8.5}
$$

Equation (8.5) indicates that the segment of the influence line for M_B between points A and B ($0 \le x \le a$) can be obtained by multiplying the ordinates of the segment of the influence line for C_y between A and B by $(L - a)$. Also, according to this equation the segment of the influence line for M_B between points B and C ($a \le x \le L$) can be obtained by multiplying the ordinates of the segment of the influence line for A_y between B and C by a. The influence line for M_B thus constructed from the influence lines for A_y and C_y is shown in Fig. 8.2(f). The equations of this influence line in terms of the position of the unit load, x, can be obtained by substituting Eqs. (8.1) and (8.2) into Eq. (8.5); that is,

$$M_B = \begin{cases} C_y(L-a) = \dfrac{x}{L}(L-a) & 0 \le x \le a \\[2mm] A_y(a) = \left(1 - \dfrac{x}{L}\right)a & a \le x \le L \end{cases} \qquad (8.6)$$

Although the influence line for M_B (Fig. 8.2(f)) resembles, in shape, the bending moment diagram of the beam for a concentrated load applied at point B, the influence line for bending moment has an entirely different meaning than the bending moment diagram, and it is essential that we clearly understand the difference between the two. A bending moment diagram shows how the bending moment varies at *all sections* along the length of a member for a loading condition whose position is fixed on the member, whereas an influence line for bending moment shows how the bending moment varies at *one particular section* as a unit load moves across the length of the member.

Note from Fig. 8.2 that the influence lines for the reactions, shear, and bending moment of the simply supported beam consist of straight-line segments. We show in the following section that this is true for the influence lines for all response functions involving forces and moments (e.g., reactions, shears, bending moments, and forces in truss members) for all statically determinate structures. However, influence lines for the deflections of statically determinate structures (discussed in Section 8.5) are composed of curved lines.

Procedure for Analysis

The procedure for constructing influence lines for the reactions, shears, and bending moments of beams and frames by using the equilibrium method can be summarized as follows:

1. Select an origin from which the position of a moving downward concentrated unit load will be measured. It is usually convenient to assume that the unit load moves from the left end of the structure to the right end, with its position defined by a coordinate x measured from the left end of the structure.
2. To construct an influence line for a support reaction:
    a. Place the unit load at a distance x from the left end of the structure, and determine the expression for the reaction in terms of x by applying an equation of equilibrium or condition. If the structure is composed of two or more rigid parts connected together by internal hinges and/or rollers, the expression for the reaction may change as the unit load moves from one rigid part to the next by crossing an internal hinge or roller. Therefore, for such structures, when applying the equations of condition the unit load must be placed successively on each rigid part of the structure in the path of the unit load, and an expression for the reaction must be determined for each position of the load.

b.   Once the expression(s) for the reaction for all the positions of the unit load has been determined, construct the influence line by plotting the expression(s) with the magnitude of the reaction as ordinate against the position x of the unit load as abscissa. A positive ordinate of the influence line indicates that the unit load applied at that point causes the reaction to act in the positive direction (i.e., the direction of the reaction initially used in deriving the equation of the influence line) and vice versa.

c.   Repeat step 2 until all the desired influence lines for reactions have been determined.

3.   It is generally convenient to construct the influence lines for shears and bending moments by using the influence lines for support reactions. Thus, before proceeding with the construction of an influence line for shear or bending moment at a point on the structure, make sure that the influence lines for all the reactions, on either the left or right side of the point under consideration, are available. Otherwise, draw the required influence lines for reactions by using the procedure described in the previous step. An influence line for the shear (or bending moment) at a point on the structure can be constructed as follows:

a.   Place the unit load on the structure at a variable position x to the *left* of the point under consideration, and determine the expression for the shear (or bending moment). If the influence lines for all the reactions are known, then it is usually convenient to use the portion of the structure to the *right* of the point for determining the expression for shear (or bending moment), which will contain terms involving only reactions. The shear (or bending moment) is considered to be positive or negative in accordance with the *beam sign convention* established in Section 5.1 (see Fig. 5.2).

b.   Next, place the unit load to the *right* of the point under consideration, and determine the expression for the shear (or bending moment). If the influence lines for all the reactions are known, then it is usually convenient to use the portion of the structure to the *left* of the point for determining the desired expression, which will contain terms involving only reactions.

c.   If the expressions for the shear (or bending moment) contain terms involving only reactions, then it is generally simpler to construct the influence line for shear (or bending moment) by combining the segments of the reaction influence lines in accordance with these expressions. Otherwise, substitute the expressions for the reactions into the expressions for the shear (or bending moment), and plot the resulting expressions, which will now be in terms only of x, to obtain the influence line.

d.   Repeat step 3 until all the desired influence lines for shears and bending moments have been determined.

Example 8.1

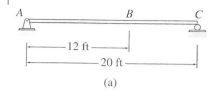

(a)

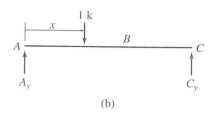

(b)

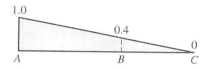

(c) Influence Line for A_y (k/k)

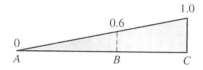

(d) Influence Line for C_y (k/k)

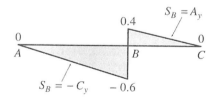

(e) Influence Line for S_B (k/k)

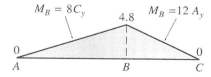

(f) Influence Line for M_B (k-ft/k)

FIG. **8.3**

Draw the influence lines for the vertical reactions at supports A and C, and the shear and bending moment at point B, of the simply supported beam shown in Fig. 8.3(a).

Solution

The free-body diagram of the beam is shown in Fig. 8.3(b). This diagram shows the beam subjected to a moving 1-k load, whose position is defined by the co-ordinate x measured from the left end A of the beam. The two vertical reactions, A_y and C_y, are assumed to be positive in the upward direction, as indicated on the free-body diagram.

Influence Line for A_y To determine the expression for A_y, we apply the equilibrium equation:

$$+\circlearrowleft \sum M_C = 0$$

$$-A_y(20) + 1(20 - x) = 0$$

$$A_y = \frac{1(20 - x)}{20} = 1 - \frac{x}{20}$$

The influence line for A_y, which is obtained by plotting this equation, is shown in Fig. 8.3(c). Note that the ordinates of the influence line are expressed in the units obtained by dividing the units of the response function, A_y, by the units of the unit load—that is, k/k.   **Ans.**

Influence Line for C_y

$$+\circlearrowleft \sum M_A = 0$$

$$-1(x) + C_y(20) = 0$$

$$C_y = \frac{1(x)}{20} = \frac{x}{20}$$

The influence line for C_y, which is obtained by plotting this equation, is shown in Fig. 8.3(d).   **Ans.**

Influence Line for S_B First, we place the unit load at a variable position x to the left of point B—that is, on the segment AB of the beam—and determine the shear at B by using the free body of the portion BC of the beam, which is to the right of B:

$$S_B = -C_y \qquad 0 \le x < 12 \text{ ft}$$

Next, the unit load is located to the right of B—that is, on the segment BC of the beam—and we use the free body of the portion AB, which is to the left of B, to determine S_B:

$$S_B = A_y \qquad 12 \text{ ft} < x \le 20 \text{ ft}$$

Thus, the equations of the influence line for S_B are

continued

$$S_B = \begin{cases} -C_y = -\dfrac{x}{20} & 0 \le x < 12 \text{ ft} \\[2mm] A_y = 1 - \dfrac{x}{20} & 12 \text{ ft} < x \le 20 \text{ ft} \end{cases}$$

The influence line for S_B is shown in Fig. 8.3(e).   Ans.

Influence Line for M_B First, we place the unit load at a position x to the left of B and determine the bending moment at B by using the free body of the portion of the beam to the right of B:

$$M_B = 8C_y \qquad 0 \le x \le 12 \text{ ft}$$

Next, the unit load is located to the right of B, and we use the free body of the portion of the beam to the left of B to determine M_B:

$$M_B = 12A_y \qquad 12 \text{ ft} \le x \le 20 \text{ ft}$$

Thus the equations of the influence line for M_B are

$$M_B = \begin{cases} 8C_y = \dfrac{2x}{5} & 0 \le x \le 12 \text{ ft} \\[2mm] 12A_y = 12 - \dfrac{3x}{5} & 12 \text{ ft} \le x \le 20 \text{ ft} \end{cases}$$

The influence line for M_B is shown in Fig. 8.3(f).   Ans.

Example 8.2

Draw the influence lines for the vertical reaction and the reaction moment at support A and the shear and bending moment at point B of the cantilever beam shown in Fig. 8.4(a).

Solution

Influence Line for A_y

$$+\uparrow \sum F_y = 0$$
$$A_y - 1 = 0$$
$$A_y = 1$$

The influence line for A_y is shown in Fig. 8.4(c).   Ans.

Influence Line for M_A

$$+\zeta \sum M_A = 0$$
$$-M_A - 1(x) = 0$$
$$M_A = -1(x) = -x$$

The influence line for M_A, which is obtained by plotting this equation, is shown in Fig. 8.4(d). As all the ordinates of the influence line are negative, it indicates that the sense of M_A for all the positions of the unit load on the beam is actually counterclockwise, instead of clockwise as initially assumed (see Fig. 8.4(b)) in deriving the equation of the influence line.   Ans.

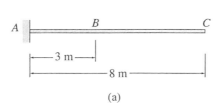

(a)

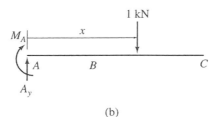

(b)

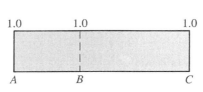

(c) Influence Line for A_y (kN/kN)

FIG. **8.4**

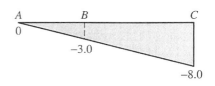

(d) Influence Line for M_A (kN · m/kN)

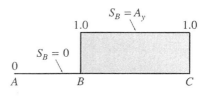

(e) Influence Line for S_B (kN/kN)

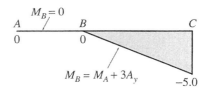

(f) Influence Line for M_B (kN · m/kN)

Influence Line for S_B

$$S_B = \begin{cases} 0 & 0 \le x < 3 \text{ m} \\ A_y = 1 & 3 \text{ m} < x \le 8 \text{ m} \end{cases}$$

The influence line for S_B is shown in Fig. 8.4(e).   Ans.

Influence Line for M_B

$$M_B = \begin{cases} 0 & 0 \le x \le 3 \text{ m} \\ M_A + 3A_y = -x + 3(1) = -x + 3 & 3 \text{ m} \le x \le 8 \text{ m} \end{cases}$$

The influence line for M_B is shown in Fig. 8.4(f).   Ans.

Example 8.3

Draw the influence lines for the vertical reactions at supports $A, C,$ and $E,$ the shear just to the right of support $C,$ and the bending moment at point B of the beam shown in Fig. 8.5(a).

Solution

The beam is composed of two rigid parts, AD and $DE,$ connected by an internal hinge at $D.$ To avoid solving simultaneous equations in determining the expressions for the reactions, we will apply the equations of equilibrium and condition in such an order that each equation involves only one unknown.

continued

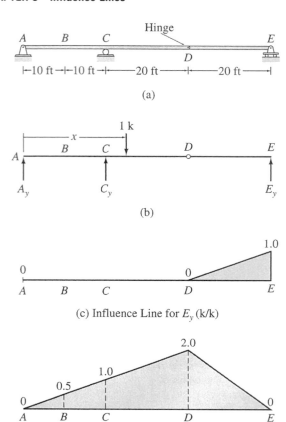

(a)

(b)

(c) Influence Line for E_y (k/k)

(d) Influence Line for C_y (k/k)

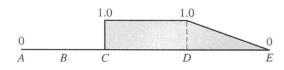

(e) Influence Line for A_y (k/k)

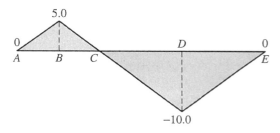

(f) Influence Line for $S_{C,R}$ (k/k)

(g) Influence Line for M_B (k-ft/k)

FIG. 8.5

Influence Line for E_y We will apply the equation of condition, $\sum M_D^{DE} = 0$, to determine the expression for E_y. First, we place the unit load at a variable position x to the left of the hinge D—that is, on the rigid part AD of the beam—to obtain

$$+\circlearrowleft \sum M_D^{DE} = 0$$

$$E_y(20) = 0$$

$$E_y = 0 \qquad 0 \le x \le 40 \text{ ft}$$

Next, the unit load is located to the right of hinge D—that is, on the rigid part DE of the beam—to obtain

$$+\circlearrowleft \sum M_D^{DE} = 0$$

$$-1(x - 40) + E_y(20) = 0$$

$$E_y = \frac{1(x - 40)}{20} = \frac{x}{20} - 2 \qquad 40 \text{ ft} \le x \le 60 \text{ ft}$$

Thus, the equations of the influence line for E_y are

$$E_y = \begin{cases} 0 & 0 \le x \le 40 \text{ ft} \\ \dfrac{x}{20} - 2 & 40 \text{ ft} \le x \le 60 \text{ ft} \end{cases}$$

The influence line for E_y is shown in Fig. 8.5(c). **Ans.**

Influence Line for C_y Applying the equilibrium equation:

$$+\circlearrowleft \sum M_A = 0$$

$$-1(x) + C_y(20) + E_y(60) = 0$$

$$C_y = \frac{x}{20} - 3E_y$$

By substituting the expressions for E_y, we obtain

$$C_y = \begin{cases} \dfrac{x}{20} - 0 = \dfrac{x}{20} & 0 \le x \le 40 \text{ ft} \\ \dfrac{x}{20} - 3\left(\dfrac{x}{20} - 2\right) = 6 - \dfrac{x}{10} & 40 \text{ ft} \le x \le 60 \text{ ft} \end{cases}$$

The influence line for C_y, which is obtained by plotting these equations, is shown in Fig. 8.5(d). **Ans.**

Influence Line for A_y

$$+\uparrow \sum F_y = 0$$

$$A_y - 1 + C_y + E_y = 0$$

$$A_y = 1 - C_y - E_y$$

By substituting the expressions for C_y and E_y, we obtain the following equations of the influence line for A_y:

$$A_y = \begin{cases} 1 - \dfrac{x}{20} - 0 = 1 - \dfrac{x}{20} & 0 \le x \le 40 \text{ ft} \\ 1 - \left(6 - \dfrac{x}{10}\right) - \left(\dfrac{x}{20} - 2\right) = \dfrac{x}{20} - 3 & 40 \text{ ft} \le x \le 60 \text{ ft} \end{cases}$$

The influence line for A_y is shown in Fig. 8.5(e). **Ans.**

Influence Line for Shear at Just to the Right of C, $S_{C,R}$

$$S_{C,R} = \begin{cases} -E_y & 0 \le x < 20 \text{ ft} \\ 1 - E_y & 20 \text{ ft} < x \le 60 \text{ ft} \end{cases}$$

By substituting the expressions for E_y, we obtain

$$S_{C,R} = \begin{cases} 0 & 0 \le x < 20 \text{ ft} \\ 1 - 0 = 1 & 20 \text{ ft} < x \le 40 \text{ ft} \\ 1 - \left(\dfrac{x}{20} - 2\right) = 3 - \dfrac{x}{20} & 40 \text{ ft} \le x \le 60 \text{ ft} \end{cases}$$

The influence line for $S_{C,R}$ is shown in Fig. 8.5(f). **Ans.**

continued

Influence Line for M_B

$$M_B = \begin{cases} 10A_y - 1(10 - x) & 0 \le x \le 10 \text{ ft} \\ 10A_y & 10 \text{ ft} \le x \le 60 \text{ ft} \end{cases}$$

By substituting the expressions for A_y, we obtain

$$M_B = \begin{cases} 10\left(1 - \dfrac{x}{20}\right) - 1(10 - x) = \dfrac{x}{2} & 0 \le x \le 10 \text{ ft} \\ 10\left(1 - \dfrac{x}{20}\right) = 10 - \dfrac{x}{2} & 10 \text{ ft} \le x \le 40 \text{ ft} \\ 10\left(\dfrac{x}{20} - 3\right) = \dfrac{x}{2} - 30 & 40 \text{ ft} \le x \le 60 \text{ ft} \end{cases}$$

The influence line for M_B is shown in Fig. 8.5(g).   Ans.

Example 8.4

Draw the influence lines for the vertical reaction and the reaction moment at support A of the frame shown in Fig. 8.6(a).

Solution

Influence Line for A_y

$$+\uparrow \sum F_y = 0$$
$$A_y - 1 = 0$$
$$A_y = 1$$

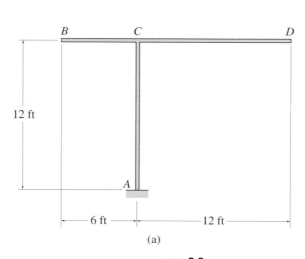

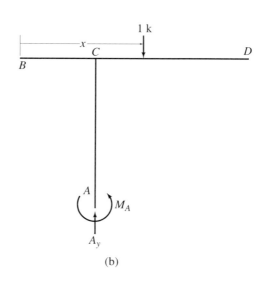

(a)   (b)

FIG. **8.6**

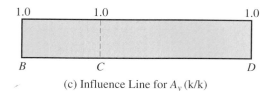

(c) Influence Line for A_y (k/k)

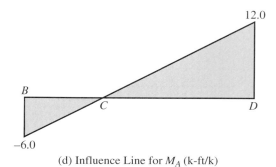

(d) Influence Line for M_A (k-ft/k)

FIG. 8.6 (contd.)

The influence line for A_y is shown in Fig. 8.6(c). Ans.

Influence Line for M_A

$$+\circlearrowleft\sum M_A = 0$$

$$M_A - 1(x - 6) = 0$$

$$M_A = x - 6$$

The influence line for M_A is shown in Fig. 8.6(d). Ans.

Example 8.5

Draw the influence lines for the horizontal and vertical reactions at supports A and B and the shear at hinge E of the three-hinged bridge frame shown in Fig. 8.7(a).

Solution

Influence Line for A_y

$$+\circlearrowleft\sum M_B = 0$$

$$-A_y(10) + 1(15 - x) = 0$$

$$A_y = \frac{1(15 - x)}{10} = 1.5 - \frac{x}{10}$$

The influence line for A_y is shown in Fig. 8.7(c). Ans.

continued

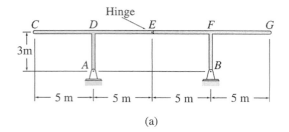

(a)

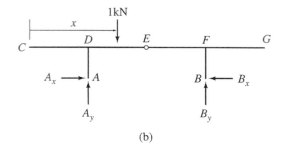

(b)

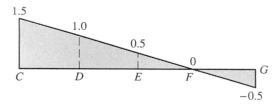

(c) Influence Line for A_y (kN/kN)

FIG. 8.7

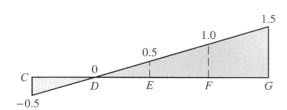

(d) Influence Line for B_y (kN/kN)

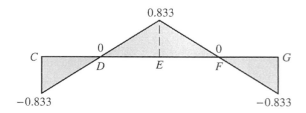

(e) Influence Line for A_x and B_x (kN/kN)

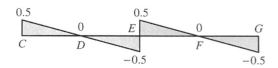

(f) Influence Line for S_E (kN/kN)

Influence Line for B_y

$$+\uparrow \sum F_y = 0$$

$$A_y - 1 + B_y = 0$$

$$B_y = 1 - A_y = 1 - \left(1.5 - \frac{x}{10}\right) = \frac{x}{10} - 0.5$$

The influence line for B_y is shown in Fig. 8.7(d).   Ans.

Influence Line for A_x We will use the equation of condition $\sum M_E^{CE} = 0$ to determine the expressions for A_x. First, we place the unit load to the left of hinge E—that is, on the rigid part CE of the frame—to obtain

$$+\circlearrowleft \sum M_E^{CE} = 0$$

$$A_x(3) - A_y(5) + 1(10 - x) = 0$$

$$A_x = \frac{5}{3}A_y - \frac{1}{3}(10 - x) = \frac{5}{3}\left(1.5 - \frac{x}{10}\right) - \frac{1}{3}(10 - x)$$

$$= \frac{x - 5}{6} \qquad 0 \le x \le 10 \text{ m}$$

Next, the unit load is located to the right of hinge E—that is, on the rigid part EG of the frame—to obtain

$$+\circlearrowleft \sum M_E^{CE} = 0$$

$$A_x(3) - A_y(5) = 0$$

$$A_x = \frac{5}{3}A_y = \frac{5}{3}\left(1.5 - \frac{x}{10}\right) = \frac{15-x}{6} \qquad 10\text{ m} \leq x \leq 20\text{ m}$$

Thus, the equations of the influence line for A_x are

$$A_x = \begin{cases} \dfrac{x-5}{6} & 0 \leq x \leq 10\text{ m} \\[2mm] \dfrac{15-x}{6} & 10\text{ m} \leq x \leq 20\text{ m} \end{cases}$$

The influence line for A_x is shown in Fig. 8.7(e).   Ans.

Influence Line for B_x

$$+\rightarrow \sum F_x = 0$$

$$A_x - B_x = 0$$

$$B_x = A_x$$

which indicates that the influence line for B_x is the same as that for A_x (Fig. 8.7(e)).   Ans.

Influence Line for S_E

$$S_E = \begin{cases} -B_y = -\dfrac{x}{10} + 0.5 & 0 \leq x < 10\text{ m} \\[2mm] A_y = 1.5 - \dfrac{x}{10} & 10\text{ m} < x \leq 20\text{ m} \end{cases}$$

The influence line for S_E is shown in Fig. 8.7(f).   Ans.

8.2 MÜLLER-BRESLAU'S PRINCIPLE AND QUALITATIVE INFLUENCE LINES

The construction of influence lines for the response functions involving forces and moments can be considerably expedited by applying a procedure developed by Heinrich Müller-Breslau in 1886. The procedure, which is commonly known as *Müller-Breslau's principle*, can be stated as follows:

> The influence line for a force (or moment) response function is given by the deflected shape of the released structure obtained by removing the restraint corresponding to the response function from the original structure and by giving the released structure a unit displacement (or rotation) at the location and in the direction of the response function, so that only the response function and the unit load perform external work.

This principle is valid only for the influence lines for response functions involving forces and moments (e.g., reactions, shears, bending moments, or forces in truss members), and it does not apply to the influence lines for deflections.

To prove the validity of Müller-Breslau's principle, consider the simply supported beam subjected to a moving unit load, as shown in Fig. 8.8(a). The influence lines for the vertical reactions at supports A and C and the shear and bending moment at point B of this beam were developed in the previous section by applying the equations of equilib-

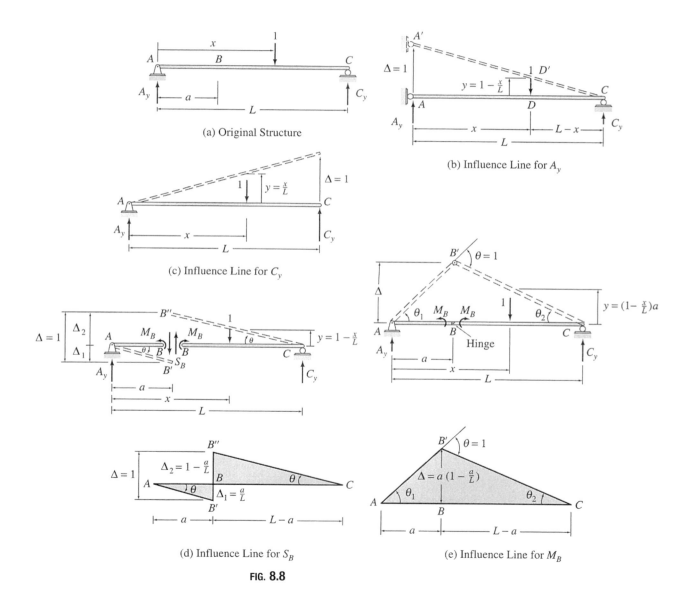

(a) Original Structure

(b) Influence Line for A_y

(c) Influence Line for C_y

(d) Influence Line for S_B

(e) Influence Line for M_B

FIG. 8.8

rium (see Fig. 8.2). Suppose that we now wish to draw the influence lines for the same four response functions by using Müller-Breslau's principle.

To construct the influence line for the vertical reaction A_y, we remove the restraint corresponding to A_y by replacing the hinged support at A by a roller support, which can exert only a horizontal reaction, as shown in Fig. 8.8(b). Note that point A of the beam is now free to displace in the direction of A_y. Although the restraint corresponding to A_y has been removed, the reaction A_y still acts on the beam, which remains in equilibrium in the horizontal position (shown by solid lines in the figure) under the action of the unit load and the reactions A_y and C_y. Next, point A of the released beam is given a virtual unit displacement, $\Delta = 1$, in the positive direction of A_y, causing it to displace, as shown by the dashed lines in Fig. 8.8(b). Note that the pattern of virtual displacement applied is consistent with the support conditions of the released beam; that is, points A and C cannot move in the horizontal and vertical directions, respectively. Also, since the original beam is statically determinate, removal of one restraint from it reduces it to a statically unstable beam. Thus, the released beam remains straight (i.e., it does not bend) during the virtual displacement. Since the beam is in equilibrium, according to the principle of virtual displacements for rigid bodies (Section 7.2), the virtual work done by the real external forces acting through the virtual external displacements must be zero; that is,

$$W_{ve} = A_y(1) - 1(y) = 0$$

from which

$$A_y = y \qquad (8.7)$$

where y represents the displacement of the point of application of the unit load, as shown in Fig. 8.8(b). Equation (8.7) indicates that the displacement y of the beam at any position x is equal to the magnitude of A_y due to a unit load acting at the position x on the beam. Thus, the displacement y at any position x is equal to the ordinate of the influence line for A_y at that position, as stated by Müller-Breslau's principle. Equation (8.7) can be expressed in terms of x by considering the geometry of the deflected shape of the beam. From Fig. 8.8(b), we observe that the triangles $A'AC$ and $D'DC$ are similar. Therefore,

$$\frac{y}{(L-x)} = \frac{1}{L} \quad \text{or} \quad y = 1 - \frac{x}{L}$$

By substituting this expression into Eq. (8.7), we obtain the equation of the influence line for A_y in terms of x as

$$A_y = 1 - \frac{x}{L}$$

which is the same as Eq. (8.1), which was derived by equilibrium consideration.

The influence line for the vertical reaction C_y is determined in a similar manner, as shown in Fig. 8.8(c). Note that this influence line is identical to that constructed previously by equilibrium consideration (Fig. 8.2(c)).

To construct the influence line for the shear S_B at point B of the beam, we remove the restraint corresponding to S_B by cutting the beam at B, as shown in Fig. 8.8(d). Note that points B of the portions AB and BC of the released beam are now free to displace vertically relative to each other. To keep the released beam in equilibrium, we apply at B the shear forces, S_B, and the bending moments, M_B, as shown in the figure. Note that S_B and M_B are assumed to act in their positive directions in accordance with the *beam sign convention*. Next, at B the released beam is given a virtual unit relative displacement, $\Delta = 1$, in the positive direction of S_B (Fig. 8.8(d)) by moving the end B of portion AB downward by Δ_1 and the end B of portion BC upward by Δ_2, so that $\Delta_1 + \Delta_2 = \Delta = 1$. The values of Δ_1 and Δ_2 depend on the requirement that the rotations, θ, of the two portions AB and BC be the same (i.e., the segments AB' and $B''C$ in the displaced position must be parallel to each other), so that the net work done by the two moments M_B is zero, and only the shear forces S_B and the unit load perform work. Applying the principle of virtual displacements, we write

$$W_{ve} = S_B(\Delta_1) + S_B(\Delta_2) - M_B(\theta) + M_B(\theta) - 1(y)$$
$$= S_B(\Delta_1 + \Delta_2) - 1(y)$$
$$= S_B(\Delta) - 1(y)$$
$$= S_B(1) - 1(y) = 0$$

from which

$$S_B = y$$

which indicates that the deflected shape of the beam (Fig. 8.8(d)) is the influence line for S_B, as stated by Müller-Breslau's principle. The values of the ordinates Δ_1 and Δ_2 can be established from the geometry of the deflected shape of the beam. From Fig. 8.8(d), we observe that the triangles ABB' and BCB'' are similar. Therefore,

$$\frac{\Delta_1}{a} = \frac{\Delta_2}{L - a}, \quad \text{or} \quad \Delta_2 = \left(\frac{L - a}{a}\right)\Delta_1 \tag{8.8}$$

Also,

$$\Delta_1 + \Delta_2 = 1, \quad \text{or} \quad \Delta_2 = 1 - \Delta_1 \tag{8.9}$$

By equating Eqs. (8.8) and (8.9) and solving for Δ_1, we obtain

$$\Delta_1 = \frac{a}{L}$$

By substituting the expression for Δ_1 into Eq. (8.9), we obtain

$$\Delta_2 = 1 - \frac{a}{L}$$

These ordinates are the same as determined previously by the equilibrium method (Fig. 8.2(e)).

To construct the influence line for the bending moment M_B, we remove the restraint corresponding to M_B by inserting a hinge at B, as shown in Fig. 8.8(e). The portions AB and BC of the released beam are now free to rotate relative to each other. To keep the released beam in equilibrium, we apply the moments M_B at B, as shown in the figure. The bending moment is assumed to be positive in accordance with the *beam sign convention*. Next, a virtual unit rotation, $\theta = 1$, is introduced at B (Fig. 8.8(e)) by rotating portion AB by θ_1 counterclockwise and portion BC by θ_2 clockwise, so that $\theta_1 + \theta_2 = \theta = 1$. Applying the principle of virtual displacements, we write

$$\begin{aligned} W_{ve} &= M_B(\theta_1) + M_B(\theta_2) - 1(y) \\ &= M_B(\theta_1 + \theta_2) - 1(y) \\ &= M_B(\theta) - 1(y) \\ &= M_B(1) - 1(y) = 0 \end{aligned}$$

from which

$$M_B = y$$

which indicates that the deflected shape of the beam (Fig. 8.8(e)) is the influence line for M_B, as stated by Müller-Breslau's principle. The value of the ordinate Δ can be established from the geometry of the deflected shape of the beam. From Fig. 8.8(e), we can see that

$$\Delta = a\theta_1 = (L - a)\theta_2 \tag{8.10}$$

or

$$\theta_1 = \left(\frac{L-a}{a}\right)\theta_2 \tag{8.11}$$

Also,

$$\theta_1 + \theta_2 = 1, \quad \text{or} \quad \theta_1 = 1 - \theta_2 \tag{8.12}$$

By equating Eqs. (8.11) and (8.12) and solving for θ_2, we obtain

$$\theta_2 = \frac{a}{L}$$

By substituting the expression for θ_2 into Eq. (8.10), we obtain

$$\Delta = (L - a)\frac{a}{L} = a\left(1 - \frac{a}{L}\right)$$

which is the same as obtained previously by the equilibrium method (Fig. 8.2(f)).

In the preceding section, we stated that the influence lines for the force and moment response functions of all statically determinate structures consist of straight-line segments. We can explain this by means of Müller-Breslau's principle. In implementing this principle in constructing an influence line, the restraint corresponding to the force or moment response function of interest needs to be removed from the structure. In the case of a statically determinate structure, removal of any such restraint from the structure reduces it to a statically unstable structure, or a *mechanism*. When this statically unstable released structure is subjected to the unit displacement (or rotation), no stresses are induced in the members of the structure, which remain straight and translate and/or rotate as rigid bodies, thereby forming a deflected shape (and thus an influence line) that consists of straight-line segments. Because the removal of a force or moment restraint from a statically indeterminate structure for the purpose of constructing an influence line does not render it statically unstable, the influence lines for such structures consist of curved lines.

Qualitative Influence Lines

In many practical applications, it is necessary to determine only the general shape of the influence lines but not the numerical values of the ordinates. *A diagram showing the general shape of an influence line without the numerical values of its ordinates is called a qualitative influence line.* In contrast, an influence line with the numerical values of its ordinates known is referred to as a *quantitative influence line*.

Although Müller-Breslau's principle can be used to determine the quantitative influence lines as discussed previously, it is more commonly used to construct qualitative influence lines. The numerical values of the influence-line ordinates, if desired, are then computed by using the equilibrium method.

Procedure for Analysis

A procedure for determining the force and moment influence lines for beams and frames by using the equilibrium method was presented in Section 8.1. The following alternative procedure, which is based on a combination of Müller-Breslau's principle and the equilibrium method, may considerably expedite the construction of such influence lines.

1. Draw the general shape of the influence line by applying Müller-Breslau's principle:
    a. From the given structure remove the restraint corresponding to the response function whose influence line is desired to obtain the released structure.
    b. Apply a small displacement (or rotation) to the released structure at the location and in the positive direction of the response

function. Draw a deflected shape of the released structure that is consistent with the support and continuity conditions of the released structure to obtain the general shape of the influence line. (Remember that the influence lines for statically determinate structures consist only of straight-line segments.) If only a qualitative influence line is desired, then end the analysis at this stage. Otherwise, proceed to the next step.

2. Determine the numerical values of the influence-line ordinates by using the equilibrium method and the geometry of the influence line.

   a. Place a unit load on the given (i.e., not released) structure at the location of the response function, and determine the numerical value of the influence-line ordinate at that location by applying the equation(s) of equilibrium and/or condition. If the response function of interest is a shear, then the unit load must be placed successively at two locations, just to the left and just to the right of the point where the shear is desired, and values of the influence-line ordinates at these locations must be computed. If the influence-line ordinate at the location of the response function is zero, then place the unit load at the location of the maximum or minimum ordinate, and determine the numerical value of that ordinate by equilibrium consideration.

   b. By using the geometry of the influence line, determine the numerical values of all the remaining ordinates where the changes in slope occur in the influence line.

An advantage of the foregoing procedure is that it enables us to construct the influence line for any force or moment response function of interest directly, without having to determine beforehand the influence lines for other functions, which may or may not be needed. For example, the construction of influence lines for shears and bending moments by this procedure does not require the use of influence lines for reactions. The procedure is illustrated by the following examples. The reader is also encouraged to check the influence lines developed in Examples 8.1 through 8.3 by applying this procedure.

Example 8.6

Draw the influence lines for the vertical reactions at supports B and D and the shear and bending moment at point C of the beam shown in Fig. 8.9(a).

Solution

Influence Line for B_y To determine the general shape of the influence line for B_y, we remove the roller support at B from the given beam (Fig. 8.9(a)) to

continued

obtain the released beam shown in Fig. 8.9(b). Next, point B of the released beam is given a small displacement, Δ, in the positive direction of B_y, and a deflected shape of the beam is drawn, as shown by the dashed line in the figure. Note that the deflected shape is consistent with the support conditions of the released structure; that is, the right end of the released beam, which is attached to the hinged support D, does not displace. The shape of the influence line is the same as the deflected shape of the released structure, as shown in Fig. 8.9(b).

To obtain the numerical value of the influence-line ordinate at B, we place a 1-kN load at point B on the original beam (Fig. 8.9(b)) and apply an equilibrium equation to obtain B_y,

$$+\circlearrowleft \sum M_D = 0 \qquad 1(9) - B_y(9) = 0 \qquad B_y = 1 \text{ kN}$$

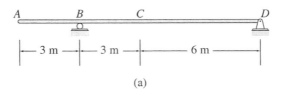

(a)

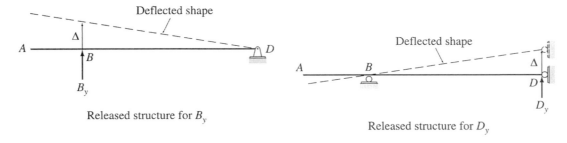

Released structure for B_y

Released structure for D_y

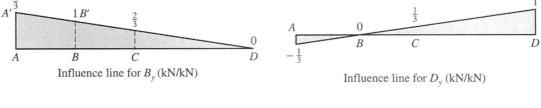

Influence line for B_y (kN/kN)

Influence line for D_y (kN/kN)

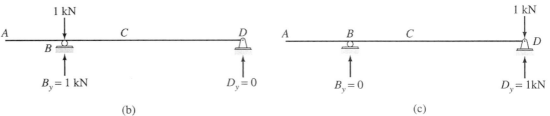

(b)

(c)

FIG. 8.9

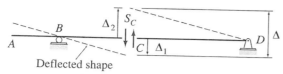

Deflected shape

Released structure for S_C

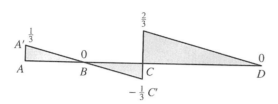

Influence line for S_C (kN/kN)

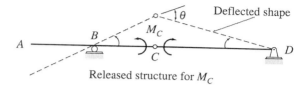

Deflected shape

Released structure for M_C

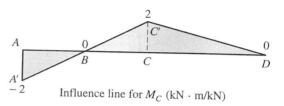

Influence line for M_C (kN · m/kN)

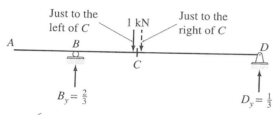

Just to the
left of C 1 kN   Just to the
                               right of C

$B_y = \frac{2}{3}$ $D_y = \frac{1}{3}$

$$S_C = \begin{cases} -\frac{1}{3}\ \text{kN}\quad\text{when 1kN is at just to the left of } C \\ +\frac{2}{3}\ \text{kN}\quad\text{when 1kN is at just to the right of } C \end{cases}$$

(d)

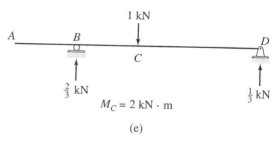

1 kN

$\frac{2}{3}$ kN $\frac{1}{3}$ kN

$M_C = 2\ \text{kN} \cdot \text{m}$

(e)

FIG. 8.9 (contd.)

Thus the value of the influence-line ordinate at B is 1 kN/kN. The value of the ordinate at A can now be determined from the geometry of the influence line (Fig. 8.9(b)). Observing that the triangles $AA'D$ and $BB'D$ are similar, we write

$$AA' = \left(\frac{1}{9}\right)(12) = \frac{4}{3}\ \text{kN/kN}$$

The influence line for B_y thus obtained is shown in Fig. 8.9(b).   Ans.

Influence Line for D_y The influence line for D_y is constructed in a similar manner and is shown in Fig. 8.9(c).   Ans.

Influence Line for S_C To determine the general shape of the influence line for the shear at point C, we cut the given beam at C to obtain the released structure shown in Fig. 8.9(d). Next, the released structure is given a small relative displacement in the positive direction of S_C by moving end C of the portion AC downward by Δ_1 and end C of the portion CD upward by Δ_2 to obtain the

continued

deflected shape shown in Fig. 8.9(d). The shape of the influence line is the same as the deflected shape of the released structure, as shown in the figure.

To obtain the numerical values of the influence-line ordinates at C, we place the 1-kN load first just to the left of C and then just to the right of C, as shown by the solid and dashed arrows, respectively, in Fig. 8.9(d). The reactions B_y and D_y are then determined by applying the equilibrium equations:

$$+\circlearrowleft \sum M_D = 0 \qquad -B_y(9) + 1(6) = 0 \qquad B_y = \frac{2}{3} \text{ kN} \uparrow$$

$$+\uparrow \sum F_y = 0 \qquad \left(\frac{2}{3}\right) - 1 + D_y = 0 \qquad D_y = \frac{1}{3} \text{ kN} \uparrow$$

Note that the magnitudes of B_y and D_y could, alternatively, have been obtained from the influence lines for these reactions constructed previously. It can be seen from Fig. 8.9(b) and (c) that the ordinates at C (or just to the left or right of C) of the influence lines for B_y and D_y are indeed 2/3 and 1/3, respectively. When the unit load is at just to the left of C (see Fig. 8.9(d)), the shear at C is

$$S_C = -D_y = -\frac{1}{3} \text{ kN}$$

When the unit load is at just to the right of C, the shear at C is

$$S_C = B_y = \frac{2}{3} \text{ kN}$$

Thus, the values of the influence-line ordinates at C are $-1/3$ kN/kN (just to the left of C), and $2/3$ kN/kN (just to the right of C), as shown in the figure. The ordinate of the influence line at A can now be obtained from the geometry of the influence line (Fig. 8.9(d)). Observing that the triangles $AA'B$ and BCC' are similar, we obtain the ordinate at A, $AA' = 1/3$ kN/kN. The influence line for S_C thus obtained is shown in Fig. 8.9(d).   Ans.

Influence Line for M_C To obtain the general shape of the influence line for the bending moment at C, we insert a hinge at C in the given beam to obtain the released structure shown in Fig. 8.9(e). Next, a small rotation θ, in the positive direction of M_C, is introduced at C in the released structure by rotating the portion AC counterclockwise and the portion CD clockwise to obtain the deflected shape shown in Fig. 8.9(e). The shape of the influence line is the same as the deflected shape of the released structure, as shown in the figure.

To obtain the numerical value of the influence-line ordinate at C, we place a 1-kN load at C on the original beam (Fig. 8.9(e)). By applying, in order, the equilibrium equations $\sum M_D = 0$ and $\sum F_y = 0$, we compute the reactions $B_y = 2/3$ kN and $D_y = 1/3$ kN, after which the bending moment at C is determined as

$$M_C = \left(\frac{2}{3}\right)(3) = 2 \text{ kN} \cdot \text{m}$$

Thus, the value of the influence-line ordinate at C is 2 kN $\cdot$ m/kN. Finally, to complete the influence line, we determine the ordinate at A by considering the geometry of the influence line. From Fig. 8.9(e), we observe that because the triangles $AA'B$ and BCC' are similar, the ordinate at A is $AA' = -2$ kN $\cdot$ m/kN. The influence line for M_C thus obtained is shown in Fig. 8.9(e).   Ans.

Example 8.7

Draw the influence lines for the vertical reactions at supports A and E, the reaction moment at support A, the shear at point B, and the bending moment at point D of the beam shown in Fig. 8.10(a).

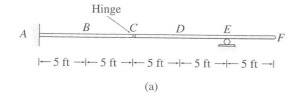

(a)

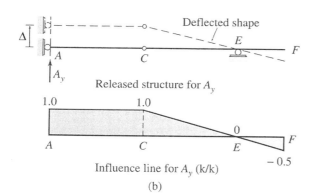

Influence line for A_y (k/k)

(b)

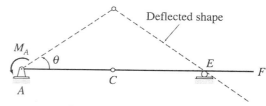

Released structure for M_A

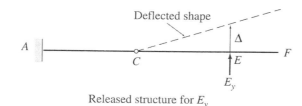

Released structure for E_y

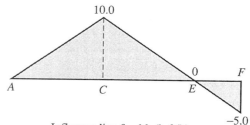

Influence line for M_A (k-ft/k)

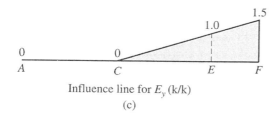

Influence line for E_y (k/k)

(c)

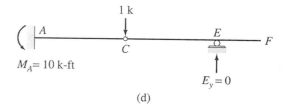

(d)

FIG. **8.10**

continued

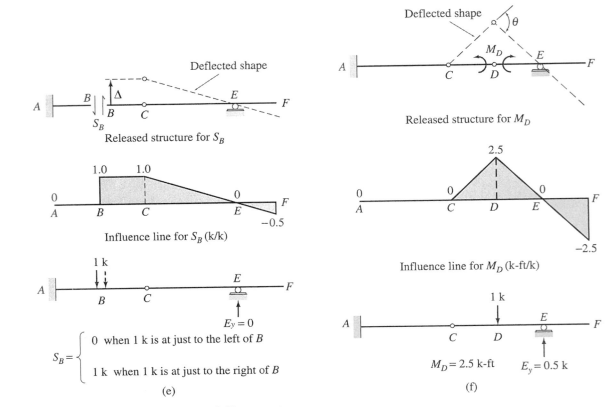

FIG. **8.10** (contd.)

Solution

Influence Line for A_y To determine the general shape of the influence line for A_y, we remove the restraint corresponding to A_y by replacing the fixed support at A by a roller guide that prevents the horizontal displacement and rotation at A but not the vertical displacement. Next, point A of the released structure is given a small displacement Δ, and a deflected shape of the beam is drawn as shown in Fig. 8.10(b). Note that the deflected shape is consistent with the support and continuity conditions of the released structure. The end A of the beam, which is attached to the roller guide, cannot rotate, so the portion AC must remain horizontal in the displaced configuration. Also, point E is attached to the roller support; therefore, it cannot displace in the vertical direction. Thus, the portion CF rotates about E, as shown in the figure. The two rigid portions, AC and CF, of the beam remain straight in the displaced configuration and rotate relative to each other at the internal hinge at C, which permits such a rotation. The shape of the influence line is the same as the deflected shape of the released structure, as shown in Fig. 8.10(b).

By recognizing that $A_y = 1$ k when a 1-k load is placed at A, we obtain the value of 1 k/k for the influence-line ordinate at A. The ordinates at points C and F are then determined from the geometry of the influence line. The influence line for A_y thus obtained is shown in Fig. 8.10(b).   Ans.

Influence Line for E_y   The roller support at E is removed from the given structure, and a small displacement, Δ, is applied at E to obtain the deflected shape shown in Fig. 8.10(c). Because of the fixed support at A, the portion AC of the released beam can neither translate nor rotate as a rigid body. The shape of the influence line is the same as the deflected shape of the released structure, as shown in the figure.

By realizing that $E_y = 1$ k when the 1-k load is placed at E, we obtain the value of 1 k/k for the influence-line ordinate at E. The ordinate at F is then determined from the geometry of the influence line. The influence line thus obtained is shown in Fig. 8.10(c).   **Ans.**

Influence Line for M_A   To remove the restraint corresponding to the reaction moment M_A, we replace the fixed support at A by a hinged support, as shown in Fig. 8.10(d). Next, a small rotation θ in the positive (counterclockwise) direction of M_A is introduced at A in the released structure to obtain the deflected shape shown in the figure. The shape of the influence line is the same as the deflected shape of the released structure.

Because the ordinate of the influence line at A is zero, we determine the ordinate at C by placing the 1-k load at C on the original beam (Fig. 8.10(d)). After computing the reaction $E_y = 0$ by applying the equation of condition $\sum M_C^{CF} = 0$, we determine the moment at A from the equilibrium equation:

$$+\circlearrowleft \sum M_A = 0 \qquad M_A - 1(10) = 0 \qquad M_A = 10 \text{ k-ft}$$

Thus, the value of the influence-line ordinate at C is 10 k-ft/k. The ordinate at F is then determined by considering the geometry of the influence line. The influence line thus obtained is shown in Fig. 8.10(d).   **Ans.**

Influence Line for S_B   To remove the restraint corresponding to the shear at B, we cut the given beam at B to obtain the released structure shown in Fig. 8.10(e). Next, the released structure is given a small relative displacement, Δ, to obtain the deflected shape shown in the figure. Support A is fixed, so portion AB can neither translate nor rotate as a rigid body. Also, the rigid portions AB and BC must remain parallel to each other in the displaced configuration. The shape of the influence line is the same as the deflected shape of the released structure, as shown in the figure.

The numerical values of the influence-line ordinates at B are determined by placing the 1-k load successively just to the left and just to the right of B (Fig. 8.10(e)) and by computing the shears at B for the two positions of the unit load. The ordinates at C and F are then determined from the geometry of the influence line. The influence line thus obtained is shown in Fig. 8.10(e).   **Ans.**

Influence Line for M_D   An internal hinge is inserted in the given beam at point D, and a small rotation θ is applied at D to obtain the deflected shape shown in Fig. 8.10(f). The shape of the influence line is the same as the deflected shape of the released structure, as shown in the figure.

The value of the influence-line ordinate at D is determined by placing the 1-k load at D and by computing the bending moment at D for this position of the unit load (Fig. 8.10(f)). The ordinate at F is then determined from the geometry of the influence line. The influence line thus obtained is shown in Fig. 8.10(f).   **Ans.**

Example 8.8

Draw the influence lines for the vertical reactions at supports A and C of the beam shown in Fig. 8.11(a).

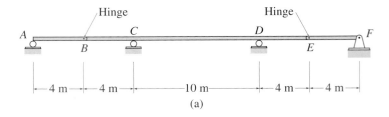

(a)

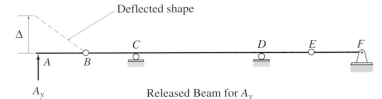

Released Beam for A_y

Influence Line for A_y (kN/kN)

(b)

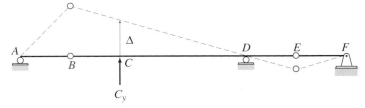

Released Beam for C_y

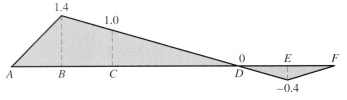

Influence Line for C_y (kN/kN)

(c)

FIG. 8.11

Solution

Influence Line for A_y   To obtain the general shape of the influence line for A_y, the roller support at A is removed from the given beam, and a small displacement, Δ, is given at point A of the released beam as shown in Fig. 8.11(b). The shape of the influence line is the same as the deflected shape of the released beam, as shown in the figure. By realizing that $A_y = 1$ kN when the 1-kN load is placed at A, we obtain the value of 1 kN/kN for the influence-line ordinate at A. The influence line thus obtained is shown in Fig. 8.11(b).   **Ans.**

Influence Line for C_y   The roller support at C is removed from the given beam, and a small displacement, Δ, is applied at C to obtain the deflected shape shown in Fig. 8.11(c). Note that the deflected shape is consistent with the support conditions of the released beam. The shape of the influence line is the same as the deflected shape of the released beam, as shown in the figure. By recognizing that $C_y = 1$ kN when the 1-kN load is placed at C, we obtain the value of 1 kN/kN for the influence-line ordinate at C. The ordinates at B and E are then determined from the geometry of the influence line. The influence line for C_y thus obtained is shown in Fig. 8.11(c).   **Ans.**

8.3 INFLUENCE LINES FOR GIRDERS WITH FLOOR SYSTEMS

In the previous sections, we considered the influence lines for beams that were subjected to a moving unit load applied directly to the beams. In most bridges and buildings, there are some structural members that are not subjected to live loads directly but to which the live loads are transmitted via floor framing systems. Typical framing systems used in bridges and buildings were described in Section 1.4 (Figs. 1.13 and 1.14, respectively). Another example of the framing system of a bridge is shown in Fig. 8.12. The deck of the bridge rests on beams called *stringers*, which are supported by *floor beams*, which, in turn, are supported by the *girders*. Thus, any live loads (e.g., the weight of the traffic), regardless of where they are located on the deck and whether they are concentrated or distributed loads, are always transmitted to the girders as concentrated loads applied at the points where the girders support the floor beams.

To illustrate the procedure for constructing influence lines for shears and bending moments in the girders supporting bridge or building floor systems, consider the simply supported girder shown in Fig. 8.13(a). As shown, a unit load moves from left to right on the stringers, which are assumed to be simply supported on the floor beams. The effect of the unit load is transmitted to the girder at points A through F, at which the girder supports the floor beams. The points A through F are commonly referred to as *panel points*, and the portions of the girder between the panel points (e.g., AB or BC) are called *panels*. Figure 8.13(a) shows

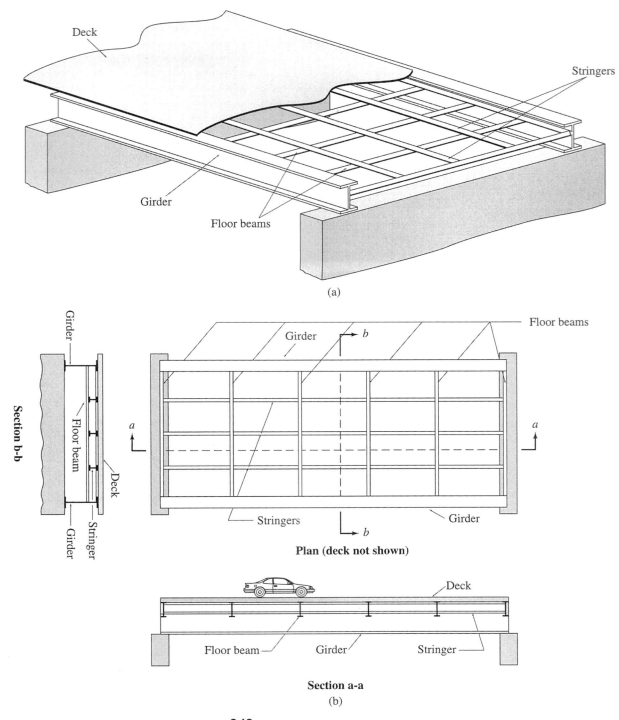

Deck

Stringers

Girder

Floor beams

(a)

Girder

Section b-b

Floor beam

Deck

Stringer

Girder

Girder

Floor beams

b

Stringers

Girder

a

a

b

Plan (deck not shown)

Deck

Floor beam

Girder

Stringer

Section a-a

(b)

FIG. 8.12

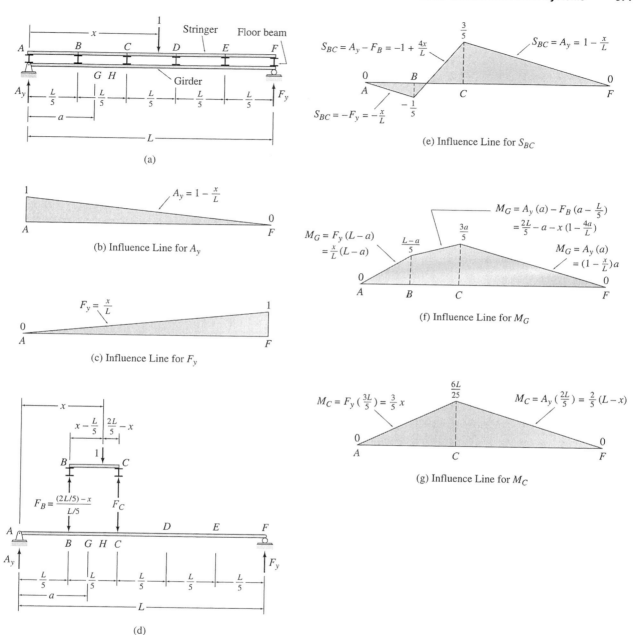

(a)

(b) Influence Line for A_y

$A_y = 1 - \frac{x}{L}$

(c) Influence Line for F_y

$F_y = \frac{x}{L}$

(d)

$F_B = \frac{(2L/5) - x}{L/5}$

(e) Influence Line for S_{BC}

$S_{BC} = A_y - F_B = -1 + \frac{4x}{L}$

$S_{BC} = A_y = 1 - \frac{x}{L}$

$S_{BC} = -F_y = -\frac{x}{L}$

(f) Influence Line for M_G

$M_G = F_y (L - a) = \frac{x}{L}(L - a)$

$M_G = A_y (a) - F_B (a - \frac{L}{5}) = \frac{2L}{5} - a - x(1 - \frac{4a}{L})$

$M_G = A_y (a) = (1 - \frac{x}{L})a$

(g) Influence Line for M_C

$M_C = F_y (\frac{3L}{5}) = \frac{3}{5}x$

$M_C = A_y (\frac{2L}{5}) = \frac{2}{5}(L - x)$

FIG. 8.13

the stringers resting on top of the floor beams, which rest on top of the girder. Although such sketches are used herein to show the manner in which the load is transmitted from one structural member to the other, in actual floor systems, members are seldom supported on top of each other, as depicted in Fig. 8.13(a). Instead, the stringers and the floor beams are usually positioned so that their top edges are even with each other and are either lower than or at the same level as the top edges of the girders (see Fig. 8.12).

Influence Lines for Reactions

The equations of the influence lines for the vertical reactions A_y and F_y can be determined by applying the equilibrium equations (Fig. 8.13(a)):

$$+\circlearrowleft \sum M_F = 0 \qquad -A_y(L) + 1(L - x) = 0 \qquad A_y = 1 - \frac{x}{L}$$

$$+\circlearrowleft \sum M_A = 0 \qquad -1(x) + F_y(L) = 0 \qquad F_y = \frac{x}{L}$$

The influence lines obtained by plotting these equations are shown in Fig. 8.13(b) and (c). Note that these influence lines are identical to those for the reactions of a simply supported beam to which the unit load is applied directly.

Influence Line for Shear in Panel *BC*

Next, suppose that we wish to construct the influence lines for shears at points G and H, which are located in the panel BC, as shown in Fig. 8.13(a). When the unit load is located to the left of the panel point B, the shear at any point within the panel BC (e.g., the points G and H) can be expressed as

$$S_{BC} = -F_y = -\frac{x}{L} \qquad 0 \le x \le \frac{L}{5}$$

Similarly, when the unit load is located to the right of the panel point C, the shear at any point within the panel BC is given by

$$S_{BC} = A_y = 1 - \frac{x}{L} \qquad \frac{2L}{5} \le x \le L$$

When the unit load is located within the panel BC, as shown in Fig. 8.13(d), the force F_B exerted on the girder by the floor beam at B must be included in the expression for shear in panel BC:

$$S_{BC} = A_y - F_B = \left(1 - \frac{x}{L}\right) - \left(2 - \frac{5x}{L}\right) = -1 + \frac{4x}{L} \qquad \frac{L}{5} \le x \le \frac{2L}{5}$$

Thus, the equations of the influence line for S_{BC} can be written as

$$S_{BC} = \begin{cases} -F_y = -\dfrac{x}{L} & 0 \le x \le \dfrac{L}{5} \\[2mm] A_y - F_B = -1 + \dfrac{4x}{L} & \dfrac{L}{5} \le x \le \dfrac{2L}{5} \\[2mm] A_y = 1 - \dfrac{x}{L} & \dfrac{2L}{5} \le x \le L \end{cases} \qquad (8.13)$$

These expressions for shear do not depend on the exact location of a point within the panel; that is, these expressions remain the same for all points located within the panel BC. The expressions do not change because the loads are transmitted to the girder at the panel points only; therefore, *the shear in any panel of the girder remains constant throughout the length of that panel.* Thus for girders with floor systems, the influence lines for shears are usually constructed for panels rather than for specific points along the girders. The influence line for the shear in panel BC, obtained by plotting Eq. (8.13), is shown in Fig. 8.13(e).

Influence Line for Bending Moment at G

The influence line for the bending moment at point G, which is located in the panel BC (Fig. 8.13(a)), can be constructed by using a similar procedure. When the unit load is located to the left of the panel point B, the bending moment at G can be expressed as

$$M_G = F_y(L - a) = \frac{x}{L}(L - a) \qquad 0 \le x \le \frac{L}{5}$$

When the unit load is located to the right of the panel point C, the bending moment at G is given by

$$M_G = A_y(a) = \left(1 - \frac{x}{L}\right)a \qquad \frac{2L}{5} \le x \le L$$

When the unit load is located within the panel BC, as shown in Fig. 8.13(d), the moment of the force F_B exerted on the girder by the floor beam at B, about G, must be included in the expression for bending moment at G:

$$M_G = A_y(a) - F_B\left(a - \frac{L}{5}\right) = \left(1 - \frac{x}{L}\right)a - \left(2 - \frac{5x}{L}\right)\left(a - \frac{L}{5}\right)$$
$$= \frac{2L}{5} - a - x\left(1 - \frac{4a}{L}\right) \qquad \frac{L}{5} \le x \le \frac{2L}{5}$$

Thus, the equations of the influence line for M_G can be written as

$$M_G = \begin{cases} F_y(L-a) = \dfrac{x}{L}(L-a) & 0 \le x \le \dfrac{L}{5} \\[2mm] A_y(a) - F_B\left(a - \dfrac{L}{5}\right) = \dfrac{2L}{5} - a - x\left(1 - \dfrac{4a}{L}\right) & \dfrac{L}{5} \le x \le \dfrac{2L}{5} \\[2mm] A_y(a) = \left(1 - \dfrac{x}{L}\right)a & \dfrac{2L}{5} \le x \le L \end{cases}$$

(8.14)

Equation (8.14) indicates that unlike shear, which remains constant throughout a panel, the expressions for the bending moment depend on the specific location of the point G within the panel BC. The influence line for M_G, obtained by plotting Eq. (8.14), is shown in Fig. 8.13(f). It can be seen from this figure that the influence line for M_G, like the influence line for shear constructed previously (Fig. 8.13(e)), consists of three straight-line segments, with discontinuities at the ends of the panel containing the response function under consideration.

Influence Line for Bending Moment at Panel Point C

When the unit load is located to the left of C (Fig. 8.13(a)), the bending moment at C is given by

$$M_C = F_y\left(\frac{3L}{5}\right) = \frac{x}{L}\left(\frac{3L}{5}\right) = \frac{3}{5}x \qquad 0 \le x \le \frac{2L}{5}$$

When the unit load is located to the right of C,

$$M_C = A_y\left(\frac{2L}{5}\right) = \left(1 - \frac{x}{L}\right)\frac{2L}{5} = \frac{2}{5}(L-x) \qquad \frac{2L}{5} \le x \le L$$

Thus, the equations of the influence line for M_C can be written as

$$M_C = \begin{cases} F_y\left(\dfrac{3L}{5}\right) = \dfrac{3}{5}x & 0 \le x \le \dfrac{2L}{5} \\[2mm] A_y\left(\dfrac{2L}{5}\right) = \dfrac{2}{5}(L-x) & \dfrac{2L}{5} \le x \le L \end{cases}$$

(8.15)

The influence line obtained by plotting these equations is shown in Fig. 8.13(g). Note that this influence line is identical to that for the bending moment of a corresponding beam without the floor system.

Procedure for Analysis

As the foregoing example indicates, the influence lines for the girders supporting floor systems with simply supported stringers consist of straight-line segments with discontinuities or changes in slopes occurring only at the panel points. In the influence lines for shears and for bending moments at points located within panels, the changes in slope occur at the panel points at the ends of the panel containing the response function (Fig. 8.13(e) and (f)), whereas in the influence lines for bending moments at panel points, the change in slope occurs at the panel point where the bending moment is evaluated. The influence lines for the girders can, therefore, be conveniently constructed as follows.

Determine the influence-line ordinates at the support points and at the panel points where the changes in slope occur by placing a unit load successively at each of these points and by applying the equilibrium equations. In the case of an influence line for bending moment at a panel point of a cantilever girder, the influence-line ordinate at the location of the bending moment will be zero. In such a case, it becomes necessary to determine an additional influence-line ordinate (usually at the free end of the cantilever girder) that is not zero to complete the influence line.

If the girder contains internal hinges, its influence lines will be discontinuous at the panel points, where such hinges are located. If an internal hinge is located within a panel, then the discontinuities will occur at the panel points at the ends of that panel. Determine the influence-line ordinates at the panel points where discontinuities occur due to the presence of internal hinges by placing the unit load at these points and by applying the equations of equilibrium and/or condition.

Complete the influence line by connecting the previously computed ordinates by straight lines and by determining any remaining ordinates by using the geometry of the influence line.

Example 8.9

Draw the influence lines for the shear in panel BC and the bending moment at B of the girder with floor system shown in Fig. 8.14(a).

Solution

Influence Line for S_{BC} To determine the influence line for the shear in panel BC, we place a 1-k load successively at the panel points A, B, C, and D. For each position of the unit load, the appropriate support reaction is first determined by proportions, and the shear in panel BC is computed. Thus, when

$$\text{1 k is at } A, \qquad D_y = 0 \qquad S_{BC} = 0$$

$$\text{1 k is at } B, \qquad D_y = \frac{1}{3}\text{ k} \qquad S_{BC} = -\frac{1}{3}\text{ k}$$

$$\text{1 k is at } C, \qquad A_y = \frac{1}{3}\text{ k} \qquad S_{BC} = \frac{1}{3}\text{ k}$$

$$\text{1 k is at } D, \qquad A_y = 0 \qquad S_{BC} = 0$$

continued

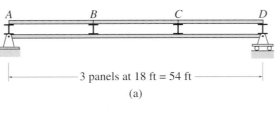

3 panels at 18 ft = 54 ft

(a)

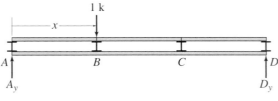

(b)

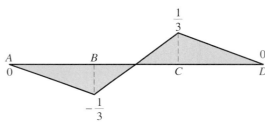

(c) Influence Line for S_{BC} (k/k)

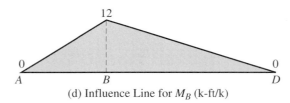

(d) Influence Line for M_B (k-ft/k)

The influence line for S_{BC} is constructed by plotting these ordinates and by connecting them with straight lines, as shown in Fig. 8.14(c).   **Ans.**

Influence Line for M_B To determine the influence line for the bending moment at panel point B, we place the 1-k load successively at the panel points A, B, and D. For each position of the unit load, the bending moment at B is determined as follows: When

$$1 \text{ k is at } A, \quad D_y = 0 \quad\quad M_B = 0$$

$$1 \text{ k is at } B, \quad A_y = \frac{2}{3} \text{ k} \quad\quad M_B = \left(\frac{2}{3}\right)18 = 12 \text{ k-ft}$$

$$1 \text{ k is at } D, \quad A_y = 0 \quad\quad M_B = 0$$

The influence line for M_B thus obtained is shown in Fig. 8.14(d).   **Ans.**

Example 8.10

Draw the influence lines for the shear in panel CD and the bending moment at D of the girder with floor system shown in Fig. 8.15(a).

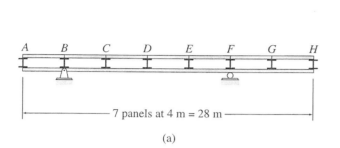

A B C D E F G H

7 panels at 4 m = 28 m

(a)

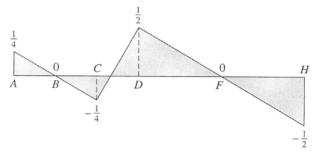

(c) Influence Line for S_{CD} (kN/kN)

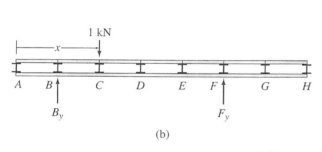

(b)

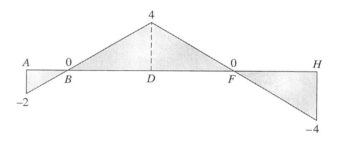

(d) Influence Line for M_D (kN · m/kN)

FIG. 8.15

Solution

Influence Line for S_{CD} To determine the influence line for the shear in panel CD, we place a 1-kN load successively at the panel points B, C, D, and F. For each position of the unit load, the appropriate support reaction is first determined by proportions, and the shear in panel CD is computed. Thus, when

$$
\begin{array}{lll}
1 \text{ kN is at } B, & F_y = 0 & S_{CD} = 0 \\[2mm]
1 \text{ kN is at } C, & F_y = \dfrac{1}{4} \text{ kN} & S_{CD} = -\dfrac{1}{4} \text{ kN} \\[2mm]
1 \text{ kN is at } D, & B_y = \dfrac{2}{4} = \dfrac{1}{2} \text{ kN} & S_{CD} = \dfrac{1}{2} \text{ kN} \\[2mm]
1 \text{ kN is at } F, & B_y = 0 & S_{CD} = 0
\end{array}
$$

The influence line for S_{CD} is constructed by plotting these ordinates and by connecting them with straight lines, as shown in Fig. 8.15(c). The ordinates at

continued

the ends A and H of the girder are then determined from the geometry of the influence line. Ans.

Influence Line for M_D To determine the influence line for the bending moment at panel point D, we place the 1-kN load successively at the panel points B, D, and F. For each position of the unit load, the bending moment at D is determined as follows: When

$$1 \text{ kN is at } B, \qquad F_y = 0 \qquad\qquad M_D = 0$$

$$1 \text{ kN is at } D, \qquad B_y = \frac{1}{2} \text{ kN} \qquad M_D = \left(\frac{1}{2}\right)8 = 4 \text{ kN} \cdot \text{m}$$

$$1 \text{ kN is at } F, \qquad B_y = 0 \qquad\qquad M_D = 0$$

The influence line for M_D thus obtained is shown in Fig. 8.15(d). Ans.

Example 8.11

Draw the influence lines for the reaction at support A, the shear in panel CD, and the bending moment at D of the girder with floor system shown in Fig. 8.16(a).

Solution

Influence Line for A_y To determine the influence line for the reaction A_y, we place a 1-k load successively at the panel points A, B, and C. For each posi-

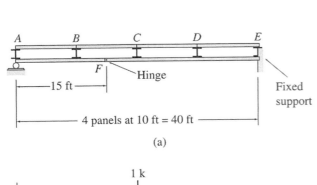

(a)

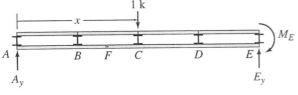

(b)

FIG. **8.16**

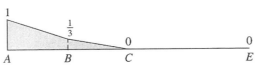

(c) Influence Line for A_y (k/k)

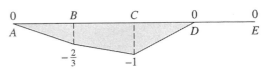

(d) Influence Line for S_{CD} (k/k)

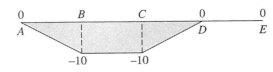

(e) Influence Line for M_D (k-ft/k)

tion of the unit load, the magnitude of A_y is computed by applying the equation of condition $\sum M_F^{AF} = 0$. Thus, when

$$1 \text{ k is at } A, \qquad\qquad A_y = 1 \text{ k}$$

$$1 \text{ k is at } B, \qquad + \zeta \sum M_F^{AF} = 0$$

$$-A_y(15) + 1(5) = 0 \qquad A_y = \frac{1}{3} \text{ k}$$

$$1 \text{ k is at } C, \qquad + \zeta \sum M_F^{AF} = 0$$

$$-A_y(15) = 0$$

$$A_y = 0$$

The influence line for A_y thus obtained is shown in Fig. 8.16(c).   **Ans.**

Influence Line for S_{CD} We place the 1-k load successively at each of the five panel points and determine the influence-line ordinates as follows. When

$$1 \text{ k is at } A, \qquad A_y = 1 \text{ k} \qquad S_{CD} = 0$$

$$1 \text{ k is at } B, \qquad A_y = \frac{1}{3} \text{ k} \qquad S_{CD} = \left(\frac{1}{3}\right) - 1 = -\frac{2}{3} \text{ k}$$

$$1 \text{ k is at } C, \qquad A_y = 0 \qquad S_{CD} = -1 \text{ k}$$

$$1 \text{ k is at } D, \qquad A_y = 0 \qquad S_{CD} = 0$$

$$1 \text{ k is at } E, \qquad A_y = 0 \qquad S_{CD} = 0$$

The influence line for S_{CD} thus obtained is shown in Fig. 8.16(d).   **Ans.**

Influence Line for M_D We place the 1-k load successively at each of the five panel points and determine the influence-line ordinates as follows. When

$$1 \text{ k is at } A, \qquad A_y = 1 \text{ k} \qquad M_D = 0$$

$$1 \text{ k is at } B, \qquad A_y = \frac{1}{3} \text{ k} \qquad M_D = \left(\frac{1}{3}\right)30 - 1(20) = -10 \text{ k-ft}$$

$$1 \text{ k is at } C, \qquad A_y = 0 \qquad M_D = -1(10) = -10 \text{ k-ft}$$

$$1 \text{ k is at } D, \qquad A_y = 0 \qquad M_D = 0$$

$$1 \text{ k is at } E, \qquad A_y = 0 \qquad M_D = 0$$

The influence line for M_D is shown in Fig. 8.16(e).   **Ans.**

8.4 INFLUENCE LINES FOR TRUSSES

The floor framing systems commonly used to transmit live loads to trusses are similar to those used for the girders discussed in the preceding section. Figure 8.17 shows a typical floor system of a truss bridge, described previously in Section 1.4 (Fig. 1.13). The deck of the bridge rests on stringers that are supported by floor beams, which, in turn, are

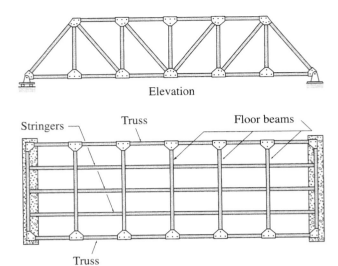

Elevation

Stringers

Truss

Floor beams

Truss

Plan (deck not shown)

FIG. 8.17

connected at their ends to the joints on the bottom chords of the two longitudinal trusses. Thus, any live loads (e.g., the weight of the traffic), regardless of where they are located on the deck and whether they are concentrated or distributed loads, are always transmitted to the trusses as concentrated loads applied at the joints. Live loads are transmitted to the roof trusses in a similar manner. As in the case of the girder floor systems, the stringers of the floor systems of trusses are assumed to be simply supported at their ends on the adjacent floor beams. Thus, the influence lines for trusses also contain straight-line segments between panel points.

To illustrate the construction of influence lines for trusses, consider the Pratt bridge truss shown in Fig. 8.18(a). A unit (1-k) load moves from left to right on the stringers of a floor system attached to the bottom chord AG of the truss. The effect of the unit load is transmitted to the truss at joints (or panel points) A through G, where the floor beams are connected to the truss. Suppose that we wish to draw the influence lines for the vertical reactions at supports A and E and for the axial forces in members CI, CD, DI, IJ, and FL of the truss.

Influence Lines for Reactions

The equations of the influence lines for the vertical reactions, A_y and E_y, can be determined by applying the equilibrium equations (Fig. 8.18(b)):

$$+\circlearrowleft \sum M_E = 0 \qquad -A_y(60) + 1(60 - x) = 0 \qquad A_y = 1 - \frac{x}{60}$$

$$+\circlearrowleft \sum M_A = 0 \qquad -1(x) + E_y(60) = 0 \qquad E_y = \frac{x}{60}$$

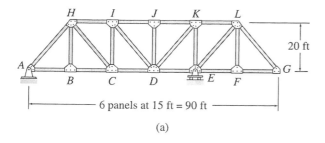

6 panels at 15 ft = 90 ft

(a)

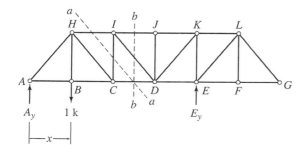

(b)

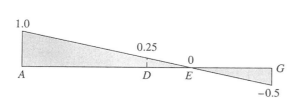

(c) Influence Line for A_y (k/k)

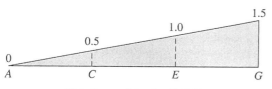

(d) Influence Line for E_y (k/k)

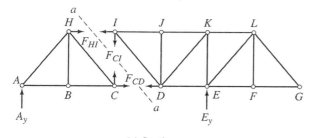

(e) Section aa

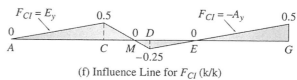

(f) Influence Line for F_{CI} (k/k)

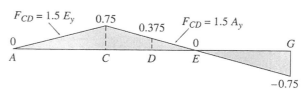

(g) Influence Line for F_{CD} (k/k)

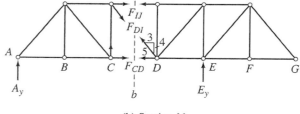

(h) Section bb

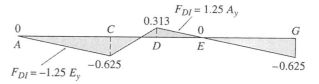

(i) Influence Line for F_{DI} (k/k)

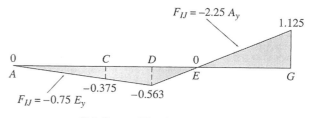

(j) Influence Line for F_{IJ} (k/k)

FIG. **8.18**

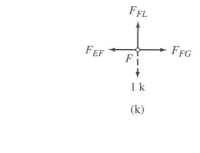

(k)

FIG. 8.18 (contd.)   (1) Influence Line for F_{FL} (k/k)

The influence lines obtained by plotting these equations are shown in Fig. 8.18(c) and (d). Note that these influence lines are identical to those for the reactions of a corresponding beam to which the unit load is applied directly.

Influence Line for Force in Vertical Member *CI*

The expressions for the member force F_{CI} can be determined by passing an imaginary section *aa* through the members CD, CI, and HI, as shown in Fig. 8.18(e), and by applying the equilibrium equation $\sum F_y = 0$ to one of the two portions of the truss. It can be seen from Fig. 8.18(e) that when the 1-k load is located to the left of joint C—that is, on the portion AC of the truss—then F_{CI} can be conveniently determined by considering the equilibrium of the free body of the right portion DG as

$$+\uparrow \sum F_y = 0 \qquad -F_{CI} + E_y = 0 \qquad F_{CI} = E_y \qquad 0 \leq x \leq 30 \text{ ft}$$

which indicates that the segment of the influence line for F_{CI} between A and C is identical to the corresponding segment of the influence line for E_y. When the 1-k load is located to the right of joint D, it is convenient to determine F_{CI} by using the free body of the left portion AC:

$$+\uparrow \sum F_y = 0 \qquad A_y + F_{CI} = 0 \qquad F_{CI} = -A_y \qquad 45 \text{ ft} \leq x \leq 90 \text{ ft}$$

which indicates that the segment of the influence line for F_{CI} between D and G can be obtained by multiplying the corresponding segment of the influence line for A_y by -1. The segments of the influence line for F_{CI} between A and C and between D and G thus constructed from the influence lines for E_y and A_y, respectively, by using the preceding expressions are shown in Fig. 8.18(f). When the 1-k load is located between C and D, the part of the load transmitted to the truss by the floor beam at C, $F_C = (45 - x)/15$, must be included in the equilibrium equation $\sum F_y = 0$ for the left portion AC to obtain F_{CI}:

$$+ \uparrow \sum F_y = 0 \qquad A_y - \left(\frac{45 - x}{15}\right) + F_{CI} = 0$$

$$F_{CI} = -A_y + \left(\frac{45 - x}{15}\right)$$

$$30 \text{ ft} \le x \le 45 \text{ ft}$$

Thus the influence line for F_{CI} is composed of three straight-line segments, as shown in Fig. 8.18(f). Since the member force F_{CI} was assumed to be tensile (Fig. 8.18(e)) in the derivation of the influence-line equations, a positive ordinate of the influence line indicates that the 1-k load applied at that point causes a tensile force in the member CI and vice versa. Thus, the influence line for F_{CI} (Fig. 8.18(f)) indicates that member CI will be in tension when the 1-k load is located between A and M and between E and G, whereas it will be in compression when the unit load is placed between M and E.

Influence Line for Force in Bottom Chord Member *CD*

The expressions for the member force F_{CD} can be determined by considering the same section aa used for F_{CI} but by applying the moment equilibrium equation, $\sum M_I = 0$. It can be seen from Fig. 8.18(e) that when the 1-k load is located to the left of joint C, then F_{CD} can be conveniently determined by considering the equilibrium of the free body of the right portion DG of the truss:

$$+ \zeta \sum M_I = 0 \qquad -F_{CD}(20) + E_y(30) = 0$$

$$F_{CD} = 1.5 E_y \qquad 0 \le x \le 30 \text{ ft}$$

which indicates that the segment of the influence line for F_{CD} between A and C can be obtained by multiplying the corresponding segment of the influence line for E_y by 1.5. When the 1-k load is located to the right of C, it is convenient to determine F_{CD} by using the free body of the left portion AC:

$$+ \zeta \sum M_I = 0 \qquad -A_y(30) + F_{CD}(20) = 0$$

$$F_{CD} = 1.5 A_y \qquad 30 \text{ ft} \le x \le 90 \text{ ft}$$

which indicates that the segment of the influence line for F_{CD} between C and G can be obtained by multiplying the corresponding segment of the influence line for A_y by 1.5. The influence line for F_{CD} thus constructed from the influence lines for A_y and E_y is shown in Fig. 8.18(g).

The influence line for F_{CD} could alternatively have been determined by considering the vertical section bb passing through the members CD, DI, and IJ, as shown in Fig. 8.18(h), instead of the inclined section aa.

Influence Line for Force in Diagonal Member *DI*

The expressions for F_{DI} can be determined by considering section *bb* (Fig. 8.18(h)) and by applying the equilibrium equation $\sum F_y = 0$ to one of the two portions of the truss. When the unit load is located to the left of joint *C*, application of the equilibrium equation $\sum F_y = 0$ to the right portion *DG* of the truss yields

$$+\uparrow \sum F_y = 0 \qquad \frac{4}{5}F_{DI} + E_y = 0$$

$$F_{DI} = -1.25E_y \qquad 0 \le x \le 30 \text{ ft}$$

When the 1-k load is located to the right of joint *D*, we write

$$+\uparrow \sum F_y = 0 \qquad A_y - \frac{4}{5}F_{DI} = 0$$

$$F_{DI} = 1.25A_y \qquad 45 \text{ ft} \le x \le 90 \text{ ft}$$

The segments of the influence line for F_{DI} between *A* and *C* and between *D* and *G* thus constructed from the influence lines for E_y and A_y, respectively, are shown in Fig. 8.18(i). The ordinates at *C* and *D* are then connected by a straight line to complete the influence line for F_{DI}, as shown in the figure.

Influence Line for Force in Top Chord Member *IJ*

By considering section *bb* (Fig. 8.18(h)), and by placing the unit load first to the left and then to the right of joint *D*, we obtain the following expressions for F_{IJ}:

$$+\circlearrowleft \sum M_D = 0$$
$$F_{IJ}(20) + E_y(15) = 0$$
$$F_{IJ} = -0.75E_y \qquad 0 \le x \le 45 \text{ ft}$$
$$+\circlearrowleft \sum M_D = 0$$
$$-A_y(45) - F_{IJ}(20) = 0$$
$$F_{IJ} = -2.25A_y \qquad 45 \text{ ft} \le x \le 90 \text{ ft}$$

The influence line for F_{IJ} thus obtained is shown in Fig. 8.18(j).

Influence Line for Force in Vertical Member *FL*

The influence line for F_{FL} can be constructed by considering the equilibrium of joint *F*. The free-body diagram of this joint is shown in Fig. 8.18(k). By applying the equilibrium equation $\sum F_y = 0$ to the free body of joint *F*, we determine that F_{FL} is zero when the 1-k load is located at joints *A* through *E* and at joint *G* and that $F_{FL} = 1$ k when the unit load is applied to the joint F. Thus, the influence-line ordinate at F is equal to 1, whereas the ordinates at A through E and G are zero. The influence line for F_{FL}, obtained by connecting these ordinates by straight lines, is shown in Fig. 8.18(l). As this influence line indicates, the force in member FL will be nonzero only when the unit load is located in the panels EF and FG of the truss.

Procedure for Analysis

The influence lines for the reactions of trusses can be constructed by using the same procedure used for the reactions of beams described in Sections 8.1 and 8.2.

Perhaps the most straightforward procedure for constructing the influence lines for axial forces in the members of trusses is to apply a unit load successively at each joint on the loaded chord of the truss and for each position of the unit load, determine the magnitude of the member force under consideration by using the method of joints and/or the method of sections. The influence-line ordinates thus computed are then connected by straight lines to obtain the desired influence line. This procedure generally proves to be very time-consuming for constructing influence lines for most truss members, except for the vertical members that are connected at an end to two horizontal members (e.g., members BH, DJ, and FL of the truss shown in Fig. 8.18(a)), whose forces can be determined by inspection.

The following alternative procedure may considerably expedite the construction of influence lines for axial forces in members of most common types of trusses:

1. Draw the influence lines for the reactions of the given truss.
2. By using the method of sections or the method of joints, obtain the equilibrium equation that will be used to determine the expression(s) of the member force whose influence line is desired. The desired member force must be the only unknown in the equilibrium equation. If such an equilibrium equation cannot be found, then it becomes necessary to construct the influence lines for the other member forces that appear in the equation before the desired influence line can be constructed (see Examples 8.12 and 8.13).
3. If using the method of sections, then apply a unit load to the left of the left end of the panel through which the section passes, and determine the expression for the member force by applying the equilibrium equation to the free body of the truss to the right of the section. Next, apply the unit load to the right of the right end of the sectioned panel, and determine the member force expression by applying the equilibrium equation to the free body to the left of the section. Construct the influence line by plotting the member force expressions and by connecting the ordinates at the ends of the sectioned panel by a straight line.

4. When using the method of joints, if the joint being considered is not located on the loaded chord of the truss, then determine the expression of the desired member force directly by applying the equation of equilibrium to the free body of the joint. Otherwise, apply a unit load at the joint under consideration, and determine the magnitude of the member force by considering the equilibrium of the joint. Next, determine the expression for the member force for a position of the unit load outside the panels adjacent to the joint under consideration. Finally, connect the influence-line segments and ordinates thus obtained by straight lines to complete the influence line.

If the member force was initially assumed to be tensile in deriving the equations of the influence line, then a positive ordinate of the influence line indicates that the unit load applied at that point causes a tensile force in the member and vice versa.

Example 8.12

Draw the influence lines for the forces in members AF, CF, and CG of the Parker truss shown in Fig. 8.19(a). Live loads are transmitted to the bottom chord of the truss.

Solution

Influence Lines for Reactions The influence lines for the reactions A_y and E_y (Fig. 8.19(b)), obtained by applying the equilibrium equations, $\sum M_E = 0$ and $\sum M_A = 0$, respectively, to the free body of the entire truss, are shown in Fig. 8.19(c) and (d).

Influence Line for F_{AF} The expressions for F_{AF} can be determined by applying the equilibrium equation $\sum F_y = 0$ to the free-body diagram of joint A shown in Fig. 8.19(e). When the 1-k load is located at joint A, we write

$$+\uparrow \sum F_y = 0 \qquad A_y - 1 + \frac{3}{5}F_{AF} = 0$$

Because $A_y = 1$ k (see Fig. 8.19(c)), we obtain

$$F_{AF} = 0 \quad \text{for } x = 0$$

When the 1-k load is located to the right of joint B, we write

$$+\uparrow \sum F_y = 0 \qquad A_y + \frac{3}{5}F_{AF} = 0$$

$$F_{AF} = -1.667A_y \qquad 20 \text{ ft} \le x \le 80 \text{ ft}$$

Thus, the segment of the influence line for F_{AF} between B and E is obtained by multiplying the corresponding segment of the influence line for A_y by -1.667, as shown in Fig. 8.19(f). The ordinates at A and B are then connected by a straight line to complete the influence line as shown in the figure.   Ans.

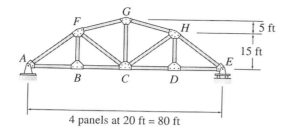

(a)

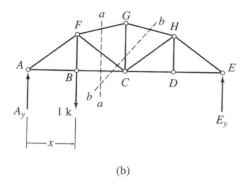

(b)

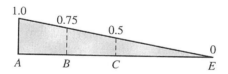

(c) Influence Line for A_y (k/k)

(d) Influence Line for E_y (k/k)

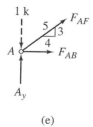

(e)

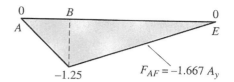

(f) Influence Line for F_{AF} (k/k)

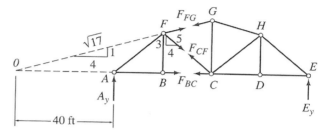

(g) Section aa

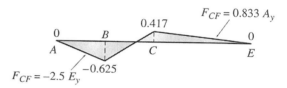

(h) Influence Line for F_{CF} (k/k)

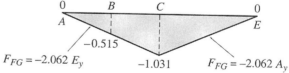

(i)

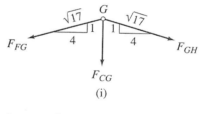

(j) Influence Line for F_{FG} (k/k)

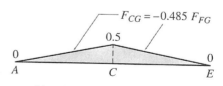

(k) Influence Line for F_{CG} (k/k)

FIG. 8.19

continued

Influence Line for F_{CF} The expressions for F_{CF} can be determined by passing a section aa through the members BC, CF, and FG as shown in Fig. 8.19(b). The free-body diagrams of the two portions of the truss thus obtained are shown in Fig. 8.19(g). The lines of action of F_{FG} and F_{BC} intersect at point O, so the equilibrium equation $\sum M_O = 0$ will contain only one unknown, namely, F_{CF}. Because the slope of member FG is 1:4, the distance $OB = 4(FB) = 4(15) = 60$ ft. Thus, the distance $OA = OB - AB = 60 - 20 = 40$ ft as shown in Fig. 8.19(g). When the 1-k load is located to the left of B, we apply the equilibrium equation $\sum M_O = 0$ to the free body of the right portion CE of the truss to obtain

$$+\zeta \sum M_O = 0$$

$$\frac{3}{5}F_{CF}(80) + E_y(120) = 0$$

$$F_{CF} = -2.5E_y \qquad 0 \le x \le 20 \text{ ft}$$

When the 1-k load is located to the right of C, we consider the equilibrium of the left portion AB to obtain

$$+\zeta \sum M_O = 0$$

$$A_y(40) - \frac{4}{5}F_{CF}(15) - \frac{3}{5}F_{CF}(60) = 0$$

$$F_{CF} = 0.833A_y \qquad 40 \text{ ft} \le x \le 80 \text{ ft}$$

The segments of the influence line for F_{CF} between A and B and between C and E are constructed by using the influence lines for E_y and A_y, respectively, in accordance with the preceding expressions. The ordinates at B and C are then connected by a straight line to complete the influence line, as shown in Fig. 8.19(h).   **Ans.**

Influence Line for F_{CG} We will determine the influence line for F_{CG} by considering the equilibrium of joint G. By applying the equations of equilibrium to the free-body diagram of joint G (Fig. 8.19(i)), we write

$$+\uparrow \sum F_y = 0$$

$$-F_{CG} - \left(\frac{1}{\sqrt{17}}\right)F_{FG} - \left(\frac{1}{\sqrt{17}}\right)F_{GH} = 0$$

$$F_{CG} = -\left(\frac{1}{\sqrt{17}}\right)(F_{FG} + F_{GH}) \qquad (1)$$

$$+\rightarrow \sum F_x = 0$$

$$-\left(\frac{4}{\sqrt{17}}\right)F_{FG} + \left(\frac{4}{\sqrt{17}}\right)F_{GH} = 0$$

$$F_{GH} = F_{FG} \qquad (2)$$

By substituting Eq. (2) into Eq. (1), we obtain

$$F_{CG} = -\left(\frac{2}{\sqrt{17}}\right)F_{FG} = -0.485F_{FG} \qquad (3)$$

Note that Eq. (3), which is valid for any position of the unit load, indicates that the influence line for F_{CG} can be obtained by multiplying the influence line for F_{FG} by -0.485. Thus we will first construct the influence line for F_{FG} by using section aa (Fig. 8.19(g)) and then apply Eq. (3) to obtain the desired influence line for F_{CG}.

It can be seen from Fig. 8.19(g) that when the 1-k load is located to the left of B, the expression for F_{FG} can be determined by applying the equilibrium equation $\sum M_C = 0$ to the free body of the right portion CE of the truss. Thus,

$$+ \zeta \sum M_C = 0$$

$$\left(\frac{4}{\sqrt{17}}\right) F_{FG}(20) + E_y(40) = 0$$

$$F_{FG} = -2.062 E_y \qquad 0 \le x \le 20 \text{ ft}$$

When the 1-k load is located to the right of C, we consider the equilibrium of the left portion AB to obtain

$$+ \zeta \sum M_C = 0$$

$$-\left(\frac{1}{\sqrt{17}}\right) F_{FG}(20) - \left(\frac{4}{\sqrt{17}}\right) F_{FG}(15) - A_y(40) = 0$$

$$F_{FG} = -2.062 A_y \qquad 40 \text{ ft} \le x \le 80 \text{ ft}$$

The influence line for F_{FG}, constructed by using the preceding expressions, is shown in Fig. 8.19(j).

The desired influence line for F_{CG} can now be obtained by multiplying the influence line for F_{FG} by -0.485, in accordance with Eq. (3). The influence line for F_{CG} thus obtained is shown in Fig. 8.19(k).   **Ans.**

The influence line for F_{CG} can also be constructed by considering the section bb shown in Fig. 8.19(b). By summing moments about the point of intersection of the axes of members BC and GH, we can determine the expressions for F_{CG} in terms of F_{CF} and A_y or E_y, whose influence lines are known. The influence line for F_{CG} can then be constructed by plotting these expressions. The reader is encouraged to check the influence line for F_{CG} shown in Fig. 8.19(k) by employing this alternative approach.

Example 8.13

Draw the influence line for the force in member HL of the K truss shown in Fig. 8.20(a). Live loads are transmitted to the bottom chord of the truss.

Solution

Influence Lines for Reactions See Fig. 8.20(c) and (d).

Influence Lines for F_{HL} From Fig. 8.20(b), we can observe that any section, such as section aa, passing through the member HL cuts three or more additional members, thereby releasing four or more unknowns, which cannot be determined by the three equations of equilibrium. We will, therefore, first construct

continued

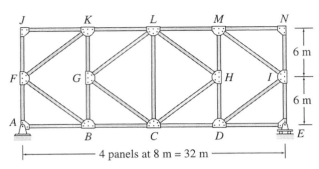

(a)

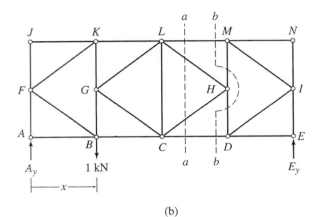

(b)

(c) Influence Line for A_y (kN/kN)

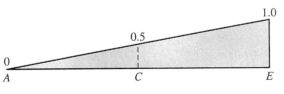

(d) Influence Line for E_y (kN/kN)

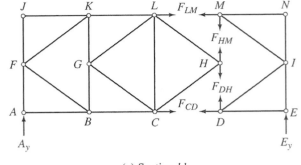

(e) Section *bb*

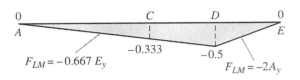

(f) Influence Line for F_{LM} (kN/kN)

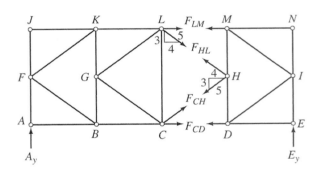

(g) Section *aa*

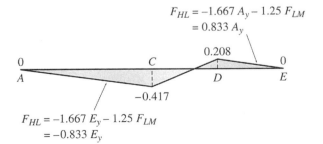

(h) Influence Line for F_{HL} (kN/kN)

FIG. 8.20

the influence line for F_{LM} by considering the curved section bb, as shown in Fig. 8.20(b), and then use section aa to determine the desired influence line for F_{HL}.

The free-body diagrams of the two portions of the truss, obtained by passing section bb, are shown in Fig. 8.20(e). It can be seen that although section bb has cut four members, CD, DH, HM, and LM, the force in member LM can be determined by summing moments about point D, because the lines of action of three remaining unknowns pass through this point. When the 1-kN load is located to the left of C, the expression for F_{LM} can be obtained as

$$+\circlearrowleft \sum M_D = 0$$

$$F_{LM}(12) + E_y(8) = 0$$

$$F_{LM} = -0.667E_y \qquad 0 \leq x \leq 16 \text{ m} \qquad (1)$$

When the unit load is located to the right of D, we obtain

$$+\circlearrowleft \sum M_D = 0$$

$$-F_{LM}(12) - A_y(24) = 0$$

$$F_{LM} = -2A_y \qquad 24 \text{ m} \leq x \leq 32 \text{ m} \qquad (2)$$

The influence line for F_{LM} thus obtained is shown in Fig. 8.20(f).

The desired influence line for F_{HL} can now be constructed by considering section aa. The free-body diagrams of the two portions of the truss, obtained by passing section aa, are shown in Fig. 8.20(g). When the 1-kN load is located to the left of C, the expression for F_{HL} can be determined by applying the equilibrium equation $\sum M_C = 0$:

$$+\circlearrowleft \sum M_C = 0$$

$$F_{LM}(12) + \frac{4}{5}F_{HL}(6) + \frac{3}{5}F_{HL}(8) + E_y(16) = 0$$

$$F_{HL} = -1.667E_y - 1.25F_{LM} \qquad 0 \leq x \leq 16 \text{ m} \quad (3)$$

When the 1-kN load is to the right of D, we obtain

$$+\circlearrowleft \sum M_C = 0$$

$$-A_y(16) - F_{LM}(12) - \frac{4}{5}F_{HL}(12) = 0$$

$$F_{HL} = -1.667A_y - 1.25F_{LM} \qquad 24 \text{ m} \leq x \leq 32 \text{ m} \quad (4)$$

To obtain the expressions for F_{HL} in terms of the reactions only, we substitute Eqs. (1) and (2) into Eqs. (3) and (4), respectively, to obtain

$$F_{HL} = -0.833E_y \qquad 0 \leq x \leq 16 \text{ m} \qquad (5)$$

$$F_{HL} = 0.833A_y \qquad 24 \text{ m} \leq x \leq 32 \text{ m} \qquad (6)$$

The influence line for F_{HL} can now be constructed by using either Eqs. (3) and (4) or Eqs. (5) and (6). The influence line thus obtained is shown in Fig. 8.20(h).

Ans.

8.5 INFLUENCE LINES FOR DEFLECTIONS

A deflection influence line depicts the variation of a deflection of a structure as a concentrated load of unit magnitude moves across the structure. Let us assume that it is desired to construct the influence line for the vertical deflection at point B of the simply supported beam shown in Fig. 8.21(a). We can construct the influence line by placing a unit load successively at arbitrary points to the left and to the right of B; determining an expression for the vertical deflection at B for each position of the unit load by using one of the methods for computing deflections described in Chapters 6 and 7; and plotting the expressions.

A more efficient procedure for constructing the deflection influence lines can be devised by the application of *Maxwell's law of reciprocal deflections* (Section 7.8). Considering again the beam of Fig. 8.21(a), if f_{BX} is the vertical deflection at B when the unit load is placed at an arbitrary point X, then f_{BX} represents the ordinate at X of the influence line for the vertical deflection at B. Now, suppose that we place the unit load at B, as shown in Fig. 8.21(b), and compute the vertical deflection at point X, f_{XB}. According to *Maxwell's law of reciprocal deflections,*

$$f_{XB} = f_{BX}$$

which indicates that the deflection at X due to the unit load at B, f_{XB}, also represents the ordinate at X of the influence line for the vertical deflection at B. Because the point X was arbitrarily chosen, we can con-

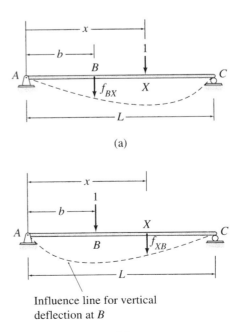

(a)

Influence line for vertical deflection at B

FIG. **8.21**

(b)

clude that *the deflected shape (elastic curve) of a structure due to a unit load applied at a point represents the influence line for deflection at the point where the unit load is applied.* Thus, an influence line for deflection at a point of a structure can be constructed by placing a unit load at the point where the deflection is desired; determining the corresponding deflected shape (elastic curve) of the structure by using one of the methods for computing deflections described in Chapters 6 and 7; and plotting the deflected shape. The procedure is illustrated by the following example.

Example 8.14

Draw the influence line for the vertical deflection at end B of the cantilever beam shown in Fig. 8.22(a).

Solution

To determine the influence line for the vertical deflection at B, we place a 1-k load at B, as shown in Fig. 8.22(b), and determine the expression for the deflected shape of the beam by using the conjugate-beam method described in Section 6.6. The M/EI diagram of the real beam due to the 1-k load applied at B is shown in Fig. 8.22(c), and the conjugate beam, loaded with the M/EI diagram of the real beam, is shown in Fig. 8.22(d). The deflection at an arbitrary point X located at a distance x from A in the real beam is equal to the bending moment at X in the conjugate beam. From Fig. 8.22(d), we can see that the bending moment at X in the conjugate beam is given by

$$M_X = \frac{1}{EI}\left\{-15\left(1-\frac{x}{15}\right)x\left(\frac{x}{2}\right) - \left(\frac{1}{2}\right)\left[15 - 15\left(1-\frac{x}{15}\right)\right]x\left(\frac{2x}{3}\right)\right\}$$

$$= \frac{1}{6EI}(x^3 - 45x^2)$$

Thus, the deflection at X on the real beam is

$$f_{XB} = \frac{1}{6EI}(x^3 - 45x^2)$$

which represents the expression for the deflected shape of the beam due to the 1-k load at B (Fig. 8.22(b)). By applying *Maxwell's law of reciprocal deflections*, $f_{BX} = f_{XB}$, we obtain the equation of the influence line for the vertical deflection at B as

$$f_{BX} = \frac{1}{6EI}(x^3 - 45x^2)$$

By substituting the numerical values of E and I, we get

$$f_{BX} = \frac{x^3 - 45x^2}{604{,}167}$$

The influence line for vertical deflection at B, obtained by plotting the preceding equation, is shown in Fig. 8.22(e).   Ans.

continued

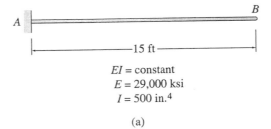

EI = constant
E = 29,000 ksi
I = 500 in.^{4}

(a)

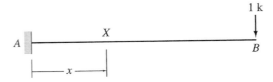

(b) Real Beam

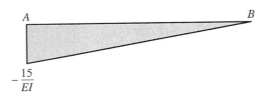

(c) $\dfrac{M}{EI}$ Diagram $\left(\dfrac{\text{k-ft/k}}{EI}\right)$

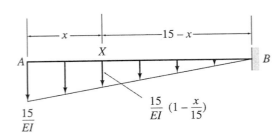

(d) Conjugate Beam

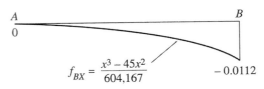

$$f_{BX} = \dfrac{x^3 - 45x^2}{604,167}$$

(e) Influence Line for Vertical
Deflection at B (ft/k)

FIG. 8.22

SUMMARY

In this chapter we have learned that an influence line is a graph of a response function of a structure as a function of the position of a downward unit load moving across the structure. The influence lines for the force and moment response functions of all statically determinate structures consist of straight-line segments.

The influence line for a reaction can be constructed by placing a unit load at a variable position x on the structure, applying an equilibrium equation to determine the expression for the reaction in terms of x, and plotting the expression. The influence line for shear (or bending moment) at a point of a beam can be constructed by placing a unit load successively to the left and to the right of the point under consideration, determining the expressions for shear (or bending moment) for the two positions of the unit load, and plotting the expressions.

Müller-Breslau's principle states that the influence line for a force (or moment) response function is given by the deflected shape of the released structure obtained by removing the restraint corresponding to the response function from the original structure and by giving the released structure a unit displacement (or rotation) at the location and in the direction of the response function, so that only the response function and the unit load perform external work. This principle is commonly used to construct qualitative influence lines (i.e., the general shape of influence lines). The numerical values of the influence-line ordinates, if desired, are then computed by applying the equations of equilibrium. Procedures for constructing influence lines for girders with floor systems and trusses were presented in Sections 8.3 and 8.4, respectively.

The deflected shape (elastic curve) of a structure, due to a unit load applied at a point, represents the influence line for deflection at the point where the unit load is applied.

PROBLEMS

Sections 8.1 and 8.2

8.1 through 8.4 Draw the influence lines for vertical reactions at supports A and C and the shear and bending moment at point B of the beams shown in Figs. P8.1 through P8.4.

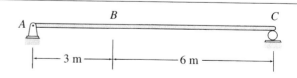

FIG. **P8.2, P8.60**

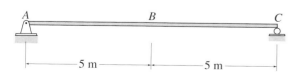

FIG. **P8.1, P8.59**

FIG. **P8.3**

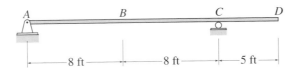

FIG. **P8.4, P8.61**

8.5 and 8.6 Draw the influence lines for vertical reactions at supports B and D and the shear and bending moment at point C of the beams shown in Figs. P8.5 and P8.6.

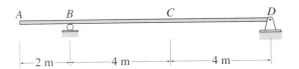

FIG. **P8.5**

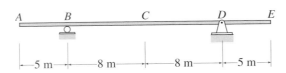

FIG. **P8.6**

8.7 Draw the influence lines for the vertical reactions at supports A and C, the shear at just to the right of A, and the bending moment at point B of the beam shown in Fig. P8.7.

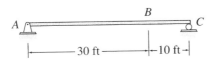

FIG. **P8.7**

8.8 Draw the influence lines for the shear and bending moment at point B of the cantilever beam shown in Fig. P8.8.

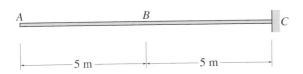

FIG. **P8.8**

8.9 Draw the influence lines for the vertical reaction and the reaction moment at support A and the shear and bending moment at point B of the cantilever beam shown in Fig. P8.9.

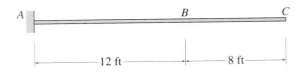

FIG. **P8.9, P8.58**

8.10 Draw the influence lines for the shear and bending moment at point C and the shears just to the left and just to the right of support D of the beam shown in Fig. P8.10.

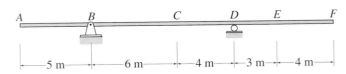

FIG. **P8.10, P8.11**

8.11 Draw the influence lines for the shear and bending moment at point E of the beam shown in Fig. P8.10.

8.12 Draw the influence lines for the shear and bending moment at point B and the shears just to the left and just to the right of support C of the beam shown in Fig. P8.12.

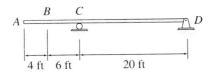

FIG. **P8.12**

8.13 Draw the influence lines for the vertical reactions at supports A and E and the reaction moment at support E of the beam shown in Fig. P8.13.

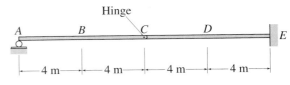

FIG. **P8.13, P8.14, P8.15**

8.14 Draw the influence lines for the shear and bending moment at point B of the beam shown in Fig. P8.13.

8.15 Draw the influence lines for the shear and bending moment at point D of the beam shown in Fig. P8.13.

8.16 Draw the influence lines for the vertical reactions at supports A and E and the shear and bending moment at point D of the frame shown in Fig. P8.16.

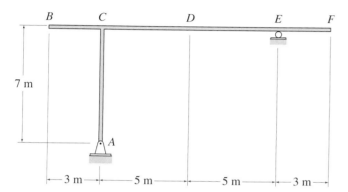

FIG. **P8.16**

8.17 Draw the influence lines for the vertical reactions at supports A and B and the shear and bending moment at point D of the frame shown in Fig. P8.17.

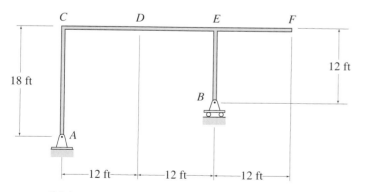

FIG. **P8.17**

8.18 Draw the influence lines for the vertical reaction and reaction moment at support A and the shear and bending moment at point C of the frame shown in Fig. P8.18.

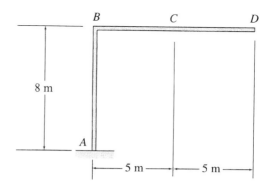

FIG. **P8.18**

8.19 Draw the influence lines for the vertical reactions at supports A, B, and E and the shear at internal hinge D of the frame shown in Fig. P8.19.

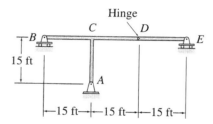

FIG. **P8.19**

8.20 Draw the influence lines for the vertical reactions at supports A, D, and F of the beam shown in Fig. P8.20.

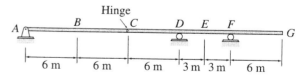

FIG. **P8.20, P8.21, P8.22**

8.21 Draw the influence lines for the shear and bending moment at point B and the shear at internal hinge C of the beam shown in Fig. P8.20.

8.22 Draw the influence lines for the shear and bending moment at point E of the beam shown in Fig. P8.20.

8.23 and 8.24 Draw the influence lines for the vertical reactions at supports $A, C, E,$ and G of the beams shown in Figs. P8.23 and P8.24.

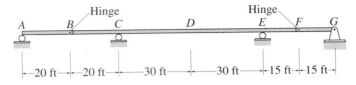

FIG. **P8.23, P8.25**

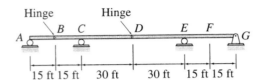

FIG. **P8.24, P8.26**

8.25 Draw the influence lines for the shear and bending moment at point D of the beam shown in Fig. P8.23.

8.26 Draw the influence lines for the shear and bending moment at point F of the beam shown in Fig. P8.24.

8.27 Draw the influence lines for the reaction moment at support A and the vertical reactions at supports $A, D,$ and F of the beam shown in Fig. P8.27.

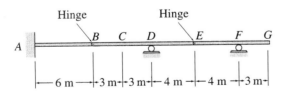

FIG. **P8.27, P8.28**

8.28 Draw the influence lines for the shear and bending moment at point C of the beam shown in Fig. P8.27.

8.29 Draw the influence lines for the reaction moment at support A and the vertical reactions at supports $A, E,$ and G of the beam shown in Fig. P8.29.

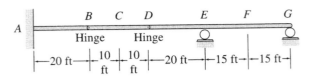

FIG. **P8.29, P8.30**

8.30 Draw the influence lines for the shears and bending moments at points C and F of the beam shown in Fig. P8.29.

8.31 Draw the influence lines for the reaction moments and the vertical reactions at supports A and F of the beam shown in Fig. P8.31.

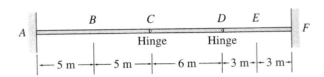

FIG. **P8.31, P8.32**

8.32 Draw the influence lines for the shears and bending moments at points B and E of the beam shown in Fig. P8.31.

8.33 Draw the influence lines for the vertical reactions at supports $A, B, G,$ and H of the beam shown in Fig. P8.33.

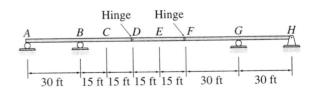

FIG. **P8.33, P8.34**

8.34 Draw the influence lines for the shears and bending moments at points C and E of the beam shown in Fig. P8.33.

8.35 and 8.36 Draw the influence lines for the horizontal and vertical reactions at supports A and B of the frames shown in Figs. P8.35 and P8.36.

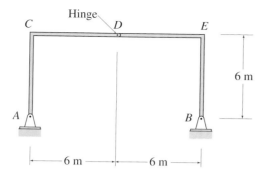

FIG. **P8.35**

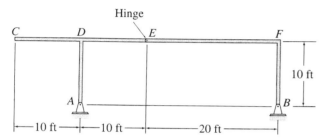

FIG. **P8.36**

8.37 Draw the influence lines for the reaction moment at support A, the vertical reactions at supports A and F, and the shear and bending moment at point E of the frame shown in Fig. P8.37.

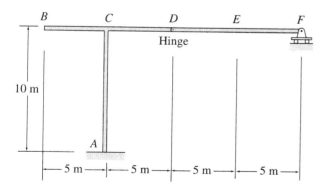

FIG. **P8.37**

8.38 Draw the influence lines for the reaction moment at support A, the vertical reactions at supports A and B, and the shear at the internal hinge C of the frame shown in Fig. P8.38.

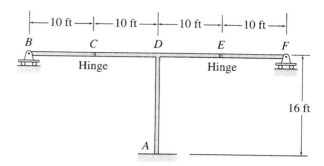

FIG. **P8.38**

8.39 Draw the influence lines for the vertical reactions at supports A, B, C, and the shear and bending moment at point E of the frame shown in Fig. P8.39.

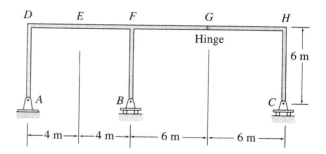

FIG. **P8.39**

Section 8.3

8.40 Draw the influence lines for the shear in panel CD and the bending moment at D of the girder with the floor system shown in Fig. P8.40.

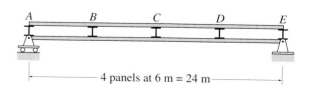

FIG. **P8.40**

8.41 Draw the influence lines for the shear in panel *CD* and the bending moment at *C* of the girder with the floor system shown in Fig. P8.41.

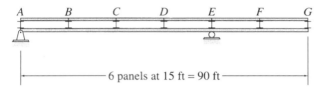

6 panels at 15 ft = 90 ft

FIG. **P8.41, P8.42**

8.42 Draw the influence lines for the shear in panel *EF* and the bending moment at *F* of the girder with the floor system shown in Fig. P8.41.

8.43 Draw the influence lines for the shear in panel *BC* and the bending moment at *C* of the girder with the floor system shown in Fig. P8.43.

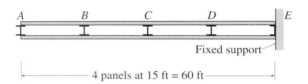

Fixed support

4 panels at 15 ft = 60 ft

FIG. **P8.43**

8.44 Draw the influence line for the shear in panel *CD* and the bending moment at *D* of the girder with the floor system shown in Fig. P8.44.

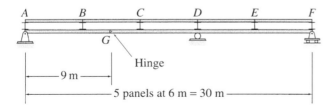

Hinge

9 m

5 panels at 6 m = 30 m

FIG. **P8.44**

Section 8.4

8.45 through 8.52 Draw the influence lines for the forces in the members identified by an "×" of the trusses shown in Figs. P8.45–P8.52. Live loads are transmitted to the bottom chords of the trusses.

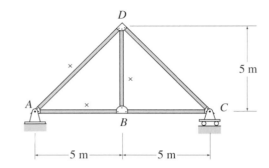

5 m

5 m   5 m

FIG. **P8.45**

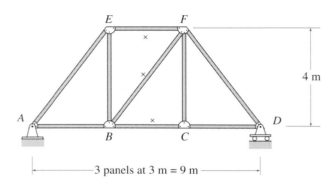

4 m

3 panels at 3 m = 9 m

FIG. **P8.46**

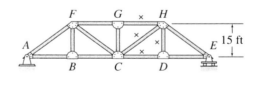

15 ft

4 panels at 20 ft = 80 ft

FIG. **P8.47**

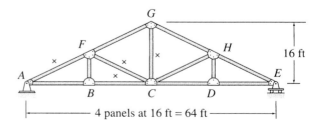

16 ft

4 panels at 16 ft = 64 ft

FIG. **P8.48**

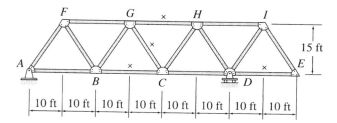

FIG. **P8.49**

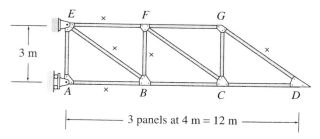

FIG. **P8.50**

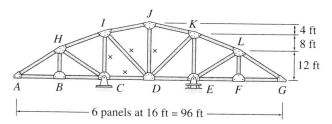

FIG. **P8.51**

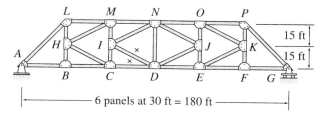

FIG. **P8.52**

8.53 through 8.57 Draw the influence lines for the forces
in the members identified by an "×" of the trusses shown
in Figs. P8.53–P8.57. Live loads are transmitted to the top
chords of the trusses.

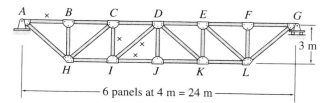

FIG. **P8.53**

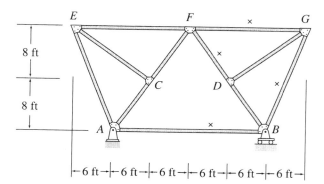

FIG. **P8.54**

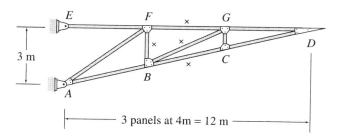

FIG. **P8.55**

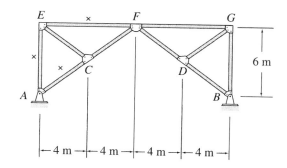

FIG. **P8.56**

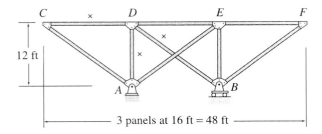

12 ft

3 panels at 16 ft = 48 ft

FIG. **P8.57**

Section 8.5

8.58 Draw the influence line for the vertical deflection at point B of the cantilever beam of Problem 8.9. $EI =$ constant. See Fig. P8.9.

8.59 and 8.60 Draw the influence line for the vertical deflection at point B of the simply supported beams of Problems 8.1 and 8.2. $EI =$ constant. See Figs. P8.1 and P8.2.

8.61 Draw the influence line for the vertical deflection at point D of the beam of Problem 8.4. $EI =$ constant. See Fig. P8.4.

9

Application of Influence Lines

A Highway Bridge Subjected to Moving Loads
Photo courtesy of the Illinois Department of Transportation

In the preceding chapter, we learned how to construct influence lines for various response functions of structures. In this chapter, we consider the application of influence lines in determining the maximum values of response functions at particular locations in structures due to variable loads. We also discuss the procedures for evaluating the absolute maximum value of a response function that may occur anywhere in a structure.

9.1 RESPONSE AT A PARTICULAR LOCATION DUE TO A SINGLE MOVING CONCENTRATED LOAD

As discussed in the preceding chapter, each ordinate of an influence line gives the value of the response function due to a single concentrated load of unit magnitude placed on the structure at the location of that ordinate. Thus, we can state the following.

1. The value of a response function due to any single concentrated load can be obtained by multiplying the magnitude of the load by the ordinate of the response function influence line at the position of the load.

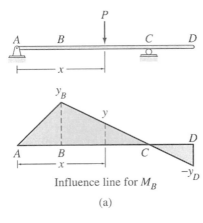

Influence line for M_B

(a)

(b) Position of Load P for Maximum
    Positive M_B

(c) Position of Load P for Maximum
    Negative M_B

FIG. 9.1

2. To determine the maximum positive value of a response function due to a single moving concentrated load, the load must be placed at the location of the maximum positive ordinate of the response function influence line, whereas to determine the maximum negative value of the response function, the load must be placed at the location of the maximum negative ordinate of the influence line.

Consider, for example, a beam subjected to a moving concentrated load of magnitude P, as shown in Fig. 9.1(a). Suppose that we wish to determine the bending moment at B when the load P is located at a distance x from the left support A. The influence line for M_B, given in Fig. 9.1(a), has an ordinate y at the position of the load P, indicating that a unit load placed at the position of P causes a bending moment $M_B = y$. Because the principle of superposition is valid, the load of magnitude P must cause a bending moment at B, which is P times as large as that caused by the load of unit magnitude. Thus, the bending moment at B due to the load P is $M_B = Py$.

Next, suppose that our objective is to determine the maximum positive and the maximum negative bending moments at B due to the load P. From the influence line for M_B (Fig. 9.1(a)), we observe that the maximum positive and the maximum negative influence-line ordinates occur at points B and D, respectively. Therefore, to obtain the maximum positive bending moment at B, we place the load P at point B, as shown in Fig. 9.1(b), and compute the magnitude of the maximum positive bending moment as $M_B = Py_B$, where y_B is the influence-line ordinate at B (Fig. 9.1(a)). Similarly, to obtain the maximum negative bending moment at B, we place the load P at point D, as shown in Fig. 9.1(c), and compute the magnitude of the maximum negative bending moment as $M_B = -Py_D$.

Example 9.1

For the beam shown in Fig. 9.2(a), determine the maximum upward reaction at support C due to a 50-kN concentrated live load.

Solution

Influence Line The influence line for the vertical reaction at support C of this beam was previously constructed in Example 8.8 and is shown in Fig. 9.2(b). Recall that C_y was assumed to be positive in the upward direction in the construction of this influence line.

Maximum Upward Reaction at C To obtain the maximum positive value of C_y due to the 50-kN concentrated live load, we place the load at B (Fig. 9.2(c)), where the maximum positive ordinate (1.4 kN/kN) of the influence line occurs. By multiplying the magnitude of the load by the value of this ordinate, we determine the maximum upward reaction at C as

$$C_y = 50(+1.4) = +70 \text{ kN} = 70 \text{ kN} \uparrow \qquad \textbf{Ans.}$$

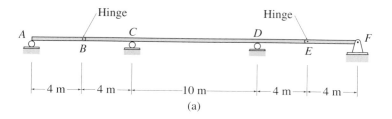

(a)

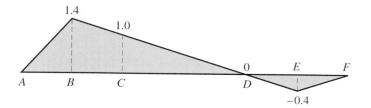

(b) Influence Line for C_y (kN/kN)

(c) Position of 50-kN Load for
Maximum Upward C_y

FIG. 9.2

9.2 RESPONSE AT A PARTICULAR LOCATION DUE TO A UNIFORMLY DISTRIBUTED LIVE LOAD

Influence lines can also be employed to determine the values of response functions of structures due to distributed loads. Consider, for example, a beam subjected to a uniformly distributed live load of intensity w_ℓ, as shown in Fig. 9.3(a). Suppose that we wish to determine the bending moment at B when the load is placed on the beam, from $x = a$ to $x = b$, as shown in the figure. The influence line for M_B is also given in the figure. By treating the distributed load applied over a differential length dx of the beam as a concentrated load of magnitude $dP = w_\ell \, dx$, as shown in the figure, we can express the bending moment at B due to the load dP as

$$dM_B = dP \, y = w_\ell \, dx \, y \qquad (9.1)$$

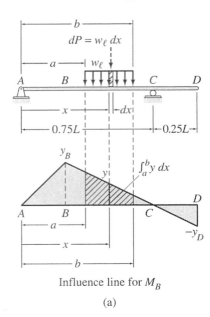

Influence line for M_B

(a)

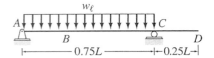

(b) Arrangement of Uniformly
Distributed Live Load w_ℓ for
Maximum Positive M_B

(c) Arrangement of Uniformly
Distributed Live Load w_ℓ for
Maximum Negative M_B

FIG. 9.3

where y is the influence line ordinate at x, which is the point of application of dP, as shown in the figure. To determine the total bending moment at B due to the distributed load from $x = a$ to $x = b$, we integrate Eq. (9.1) between these limits to obtain

$$M_B = \int_a^b w_\ell \, y \, dx = w_\ell \int_a^b y \, dx \qquad (9.2)$$

in which the integral $\int_a^b y \, dx$ represents the area under the segment of the influence line, which corresponds to the loaded portion of the beam. This area is shown as a shaded area on the influence line for M_B in Fig. 9.3(a).

Equation (9.2) also indicates that the bending moment at B will be maximum positive if the uniformly distributed load is placed over all those portions of the beam where the influence-line ordinates are positive and vice versa. From Fig. 9.3(a), we can see that the ordinates of the influence line for M_B are positive between the points A and C and negative between C and D. Therefore, to obtain the maximum positive bending moment at B, we place the uniformly distributed load w_ℓ from A to C, as shown in Fig. 9.3(b), and compute the magnitude of the maximum positive bending moment as

$$M_B = w_\ell(\text{area under the influence line between } A \text{ and } C)$$

$$= w_\ell\left(\frac{1}{2}\right)(0.75L)(y_B) = 0.375 w_\ell \, y_B L$$

Similarly, to obtain the maximum negative bending moment at B, we place the load from C to D, as shown in Fig. 9.3(c), and compute the magnitude of the maximum negative bending moment as

$$M_B = w_\ell(\text{area under the influence line between } C \text{ and } D)$$

$$= w_\ell\left(\frac{1}{2}\right)(0.25L)(-y_D) = -0.125 w_\ell \, y_D L$$

Based on the foregoing discussion, we can state the following.

1. The value of a response function due to a uniformly distributed load applied over a portion of the structure can be obtained by multiplying the load intensity by the net area under the corresponding portion of the response function influence line.

2. To determine the maximum positive (or negative) value of a response function due to a uniformly distributed live load, the load must be placed over those portions of the structure where the ordinates of the response function influence line are positive (or negative).

Example 9.2

For the beam shown in Fig. 9.4(a), determine the maximum upward reaction at support C due to a 15-kN/m uniformly distributed live load.

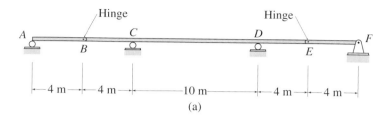

(a)

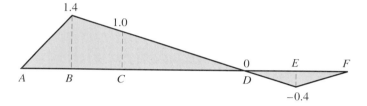

(b) Influence Line for C_y (kN/kN)

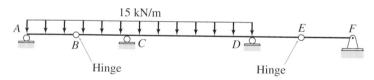

(c) Arrangement of 15-kN/m
Load for Maximum Upward C_y

FIG. 9.4

Solution

Influence Line The influence line for the vertical reaction at support C of this beam was previously constructed in Example 8.8 and is shown in Fig. 9.4(b). Recall that C_y was assumed to be positive in the upward direction in the construction of this influence line.

Maximum Upward Reaction at C From Fig. 9.4(b), we observe that the ordinates of the influence line for C_y are positive between points A and D. Therefore, to obtain the maximum positive value of C_y, we place the 15-kN/m uniformly distributed live load over the portion AD of the beam, as shown in Fig. 9.4(c). By multiplying the load intensity by the area under the portion AD of the influence line, we determine the maximum upward reaction at C as

$$C_y = 15\left[\frac{1}{2}(+1.4)(18)\right] = +189 \text{ kN} = 189 \text{ kN} \uparrow \qquad \textbf{Ans.}$$

Example 9.3

For the beam shown in Fig. 9.5(a), determine the maximum positive and nega-tive shears and the maximum positive and negative bending moments at point C due to a concentrated live load of 90 kN, a uniformly distributed live load of 40 kN/m, and a uniformly distributed dead load of 20 kN/m.

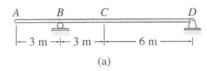

(a)

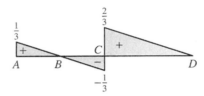

(b) Influence Line for S_C (kN/kN)

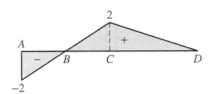

(e) Influence Line for M_C (kN·m/kN)

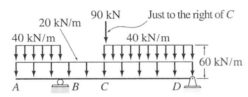

(c) Loading Arrangement for Maximum
    Positive S_C

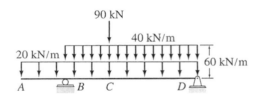

(f) Loading Arrangement for Maximum
    Positive M_C

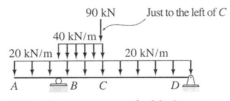

(d) Loading Arrangement for Maximum
    Negative S_C

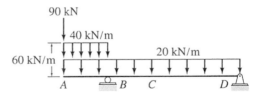

(g) Loading Arrangement for Maximum
    Negative M_C

FIG. 9.5

Solution

Influence Lines The influence lines for the shear and bending moment at point C of this beam were previously constructed in Example 8.6 and are shown in Fig. 9.5(b) and (e), respectively.

Maximum Positive Shear at C To obtain the maximum positive shear at C due to the 90-kN concentrated live load, we place the load just to the right of C (Fig. 9.5(c)), where the maximum positive ordinate (2/3 kN/kN) of the influence line for S_C occurs. By multiplying the magnitude of the load by the value of this ordinate, we determine the maximum positive value of S_C due to the concentrated live load as

$$S_C = 90\left(\frac{2}{3}\right) = 60 \text{ kN}$$

From Fig. 9.5(b), we observe that the ordinates of the influence line for S_C are positive between the points A and B and between the points C and D. Therefore, to obtain the maximum positive shear at C due to the 40-kN/m uniformly distributed live load, we place the load over the portions AB and CD of the beam, as shown in Fig. 9.5(c), and compute the maximum positive value of S_C due to this load by multiplying the load intensity by the area under the portions AB and CD of the influence line. Thus

$$S_C = 40\left[\left(\frac{1}{2}\right)(3)\left(\frac{1}{3}\right) + \left(\frac{1}{2}\right)(6)\left(\frac{2}{3}\right)\right] = 100 \text{ kN}$$

Unlike live loads, the dead loads always act at fixed positions on structures; that is, their positions cannot be varied to maximize response functions. Therefore, the 20-kN/m uniformly distributed dead load is placed over the entire length of the beam, as shown in Fig. 9.5(c), and the corresponding shear at C is determined by multiplying the dead-load intensity by the net area under the entire influence line as

$$S_C = 20\left[\left(\frac{1}{2}\right)(3)\left(\frac{1}{3}\right) + \left(\frac{1}{2}\right)(3)\left(-\frac{1}{3}\right) + \left(\frac{1}{2}\right)(6)\left(\frac{2}{3}\right)\right] = 40 \text{ kN}$$

The total maximum positive shear at C can now be obtained by algebraically adding the values of S_C determined for the three types of loads.

$$\text{Maximum positive } S_C = 60 + 100 + 40 = 200 \text{ kN} \qquad \text{Ans.}$$

Maximum Negative Shear at C The arrangement of the loads to obtain the maximum negative shear at C is shown in Fig. 9.5(d). The maximum negative shear at C is given by

$$\text{Maximum negative } S_C = 90\left(-\frac{1}{3}\right) + 40\left(\frac{1}{2}\right)(3)\left(-\frac{1}{3}\right) + 20\left[\left(\frac{1}{2}\right)(3)\left(\frac{1}{3}\right)\right.$$
$$\left. + \left(\frac{1}{2}\right)(3)\left(-\frac{1}{3}\right) + \left(\frac{1}{2}\right)(6)\left(\frac{2}{3}\right)\right]$$
$$= -10 \text{ kN} \qquad \text{Ans.}$$

Maximum Positive Bending Moment at C The arrangement of the loads to obtain the maximum positive bending moment at C is shown in Fig. 9.5(f). Note that the 90-kN concentrated live load is placed at the location of the maximum positive ordinate of the influence line for M_C (Fig. 9.5(e)); the 40-kN/m uniformly distributed live load is placed over the portion BD of the beam, where

continued

the ordinates of the influence line are positive; whereas the 20-kN/m uniformly distributed dead load is placed over the entire length of the beam. The maximum positive bending moment at C is given by

$$\text{Maximum positive } M_C = 90(2) + 40\left(\frac{1}{2}\right)(9)(2)$$

$$+ 20\left[\left(\frac{1}{2}\right)(3)(-2) + \left(\frac{1}{2}\right)(9)(2)\right]$$

$$= 660 \text{ kN} \cdot \text{m} \qquad \text{Ans.}$$

Maximum Negative Bending Moment at C The loading arrangement to obtain the maximum negative bending moment at C is shown in Fig. 9.5(g). The maximum negative M_C is given by

$$\text{Maximum negative } M_C = 90(-2) + 40\left(\frac{1}{2}\right)(3)(-2)$$

$$+ 20\left[\left(\frac{1}{2}\right)(3)(-2) + \left(\frac{1}{2}\right)(9)(2)\right]$$

$$= -180 \text{ kN} \cdot \text{m} \qquad \text{Ans.}$$

9.3 RESPONSE AT A PARTICULAR LOCATION DUE TO A SERIES OF MOVING CONCENTRATED LOADS

As discussed in Section 2.2, live loads due to vehicular traffic on highway and railway bridges are represented by a series of moving concentrated loads with specified spacing between the loads (see Figs. 2.2 and 2.3). Influence lines provide a convenient means of analyzing structures subjected to such moving loads. In this section, we discuss how the influence line for a response function can be used to determine (1) the value of the response function for a given position of a series of concentrated loads and (2) the maximum value of the response function due to a series of moving concentrated loads.

Consider, for example, the bridge beam shown in Fig. 9.6. Suppose that we wish to determine the shear at point B of the beam due to the wheel loads of an HS20-44 truck when the front axle of the truck is located at a distance of 16 ft from the left support A, as shown in the figure. The influence line for the shear at B is also shown in the figure. The distances between the three loads as well as the location of the 4-k load are known, so the locations of the other two loads can be easily established. Although the influence-line ordinates corresponding to the loads can be obtained by using the properties of the similar triangles formed by the influence line, it is usually convenient to evaluate such an ordinate by multiplying the slope of the segment of the influence line where the load is located by the distance of the load from the point at which

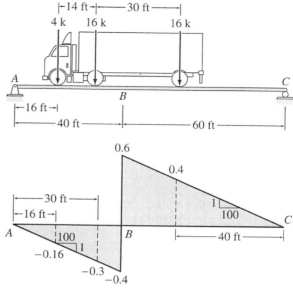

FIG. **9.6**

Influence Line for S_B (k/k)

the influence line segment intersects the horizontal axis (i.e., becomes zero). The sign (plus or minus) of the ordinate is obtained by inspection. For example, the influence-line ordinate corresponding to the 4-k load (Fig. 9.6), can be computed by multiplying the slope (1:100) of the influence-line segment for the portion AB by the distance (16 ft) of the load from point A. Thus the ordinate of the influence line for S_B corresponding to the 4-k load equals $-(1/100)(16) = -0.16$ k/k. The ordinates corresponding to the three loads thus obtained are shown in Fig. 9.6.

It may be recalled that the shear at B due to a single concentrated load is given by the product of the magnitude of the load and the influence-line ordinate at the location of the load. Because superposition is valid, the total shear at B caused by the three concentrated loads can be determined by algebraically summing the shears at B due to the individual loads, that is, by summing the products of the load magnitudes and the respective influence-line ordinates. Thus

$$S_B = -4(0.16) - 16(0.3) + 16(0.4) = 0.96 \text{ k}$$

The foregoing procedure can be employed to determine the value of any force or moment response function of a structure for a given position of a series of concentrated loads.

Influence lines can also be used for determining the maximum values of response functions at particular locations of structures due to a series of concentrated loads. Consider the beam shown in Fig. 9.7(a), and suppose that our objective is to determine the maximum positive

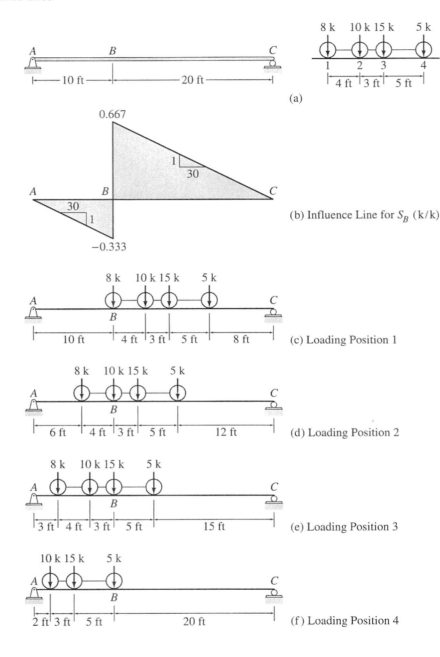

(a)

(b) Influence Line for S_B (k/k)

(c) Loading Position 1

(d) Loading Position 2

(e) Loading Position 3

(f) Loading Position 4

FIG. 9.7

shear at point B due to the series of four concentrated loads shown in the figure. The influence line for S_B is shown in Fig. 9.7(b). Assuming that the load series moves from right to left on the beam, we can observe from these figures that as the series moves from the end C of the beam toward point B, the shear at B increases continuously as the ordinates of the influence line under the loads increase. The shear at B reaches a relative maximum when the first load of the series, the 8-k load, reaches just to the right of B, where the maximum positive ordinate of the influence line is located. As the 8-k load crosses point B, the shear at B decreases abruptly by an amount equal to $-8(0.667 + 0.333) = -8$ k. With the series of loads continuing to move toward the left, S_B increases again, and it reaches another relative maximum when the second load of the series, the 10-k load, reaches just to the right of B, and so on. Because S_B becomes a relative maximum whenever one of the loads of the series reaches the maximum positive influence-line ordinate, we can conclude that during the movement of the series of loads across the entire length of the beam, the (absolute) maximum shear at B occurs when one of the loads of the series is at the location of the maximum positive ordinate of the influence line for S_B. Since it is not possible to identify by inspection the load that will cause the maximum positive S_B when placed at the maximum influence-line ordinate, we use a trial-and-error procedure to determine the value of the maximum positive shear at B. As shown in Fig. 9.7(c), the series of loads is initially positioned on the beam with its first load, the 8-k load, placed just to the right of B, where the maximum positive ordinate of the influence line is located. Noting that the slope of the influence-line segment for the portion BC is 1:30 (Fig. 9.7(b)), we compute the value of S_B for this loading position as

$$S_B = 8(20)\left(\frac{1}{30}\right) + 10(16)\left(\frac{1}{30}\right) + 15(13)\left(\frac{1}{30}\right) + 5(8)\left(\frac{1}{30}\right)$$

$$= 18.5 \text{ k}$$

Next, the entire series of loads is moved to the left by 4 ft to place the second load of the series, the 10-k load, at the location of the maximum positive ordinate of the influence line, as shown in Fig. 9.7(d). The shear at B for this loading position is given by

$$S_B = -8(6)\left(\frac{1}{30}\right) + 10(20)\left(\frac{1}{30}\right) + 15(17)\left(\frac{1}{30}\right) + 5(12)\left(\frac{1}{30}\right)$$

$$= 15.567 \text{ k}$$

The series of loads is then moved further to the left by 3 ft to place the third load of the series, the 15-k load, just to the right of B (Fig. 9.7(e)). The shear at B is now given by

$$S_B = -8(3)\left(\frac{1}{30}\right) - 10(7)\left(\frac{1}{30}\right) + 15(20)\left(\frac{1}{30}\right) + 5(15)\left(\frac{1}{30}\right)$$

$$= 9.367 \text{ k}$$

Finally, the series is positioned so that its last load, the 5-k load, is just to the right of B, as shown in Fig. 9.7(f). Note that the 8-k load has moved off the span of the beam; therefore, it does not contribute to the shear at B, which is given by

$$S_B = -10(2)\left(\frac{1}{30}\right) - 15(5)\left(\frac{1}{30}\right) + 5(20)\left(\frac{1}{30}\right) = 0.167 \text{ k}$$

By comparing the values of S_B determined for the four loading positions, we conclude that the maximum positive shear at B occurs for the first loading position—that is, when the 8-k load is placed just to the right of B (Fig. 9.7(c)):

$$\text{Maximum positive } S_B = 18.5 \text{ k}$$

Procedure for Analysis

The procedure for determining the maximum value of a force or moment response function at a particular location in a structure due to a series of moving concentrated loads can be summarized as follows.

1. Construct an influence line for the response function whose maximum value is desired, and locate its maximum positive or negative ordinate, depending on whether the maximum positive or negative value of the response function is desired. (This ordinate is referred to simply as the *maximum ordinate* in the following.)
2. Select the direction (either from right to left or vice versa) in which the load series will be moved on the structure. If the series is to move from right to left, then the load at the left end of the series is considered to be the first load, whereas if the series is to move from left to right, then the load at the right end is considered to be the first load. Beginning with the first load, sequentially number (as 1, 2, 3, ...) all the loads of the series. The position of the entire load series is referred to by the number of the load, which is placed at the location of the maximum influence line ordinate; for example, when the third load of the series is placed at the location of the maximum influence line ordinate, then the position of the load series is referred to as the loading position 3, and so on (for an example, see Fig. 9.7).
3. Position the given series of concentrated loads on the structure, with the first load of the series at the location of the maximum ordinate of the influence line. Establish the locations of the rest of the loads of the series.
4. Evaluate the influence-line ordinates corresponding to the loads of the series, and determine the value of the response function by algebraically summing the products of the load magnitudes and the respective influence-line ordinates. If the value of the response function determined herein is for the last loading position (with the last load of the series placed at the location of the maximum influence-line ordinate), then go to step 6. Otherwise, continue to the next step.
5. Move the load series in the direction selected in step 2 until the next load of the series reaches the location of the maximum influence-

line ordinate. Establish the positions of the rest of the loads of the series, and return to step 4.

6. By comparing the magnitudes of the response function determined for all the loading positions considered, obtain the maximum value of the response function.

If the arrangement of loads is such that all or most of the heavier loads are located near one of the ends of the series, then the analysis can be expedited by selecting a direction of movement for the series, so that the heavier loads will reach the maximum influence-line ordinate before the lighter loads of the series. For example, a load series in which the heavier loads are to the left should be moved on the structure from right to left and vice versa. In such a case, it may not be necessary to examine all the loading positions obtained by successively placing each load of the series at the location of the maximum influence-line ordinate. Instead, the analysis can be ended when the value of the response function begins to decrease; that is, if the value of the response function for a loading position is found to be less than that for the preceding loading position, then the value of the response function for the preceding loading position is considered to be the maximum value. Although this criterion may also work for series with heavier loads near the middle of the group, it is not valid for any general series of loads. In general, depending on the load magnitudes and spacing, and the shape of the influence line, the value of the response function, after declining for some loading positions, may start increasing again for subsequent loading positions and may attain a higher maximum.

Example 9.4

Determine the maximum axial force in member BC of the Warren truss due to the series of four moving concentrated loads shown in Fig. 9.8(a).

Solution

Influence Line for F_{BC} See Fig. 9.8(b).

Maximum Force in Member BC To determine the maximum value of F_{BC}, we move the load series from right to left, successively placing each load of the series at point B, where the maximum ordinate of the influence line for F_{BC} is located (see Fig. 9.8(c) through (f)). The value of F_{BC} is then computed for each loading position as follows.

*   For loading position 1 (Fig. 9.8(c)):

$$F_{BC} = [16(60) + 32(50) + 8(35) + 32(15)]\left(\frac{1}{80}\right) = 41.5 \text{ k (T)}$$

continued

- For loading position 2 (Fig. 9.8(d)):

$$F_{BC} = 16(10)\left(\frac{3}{80}\right) + [32(60) + 8(45) + 32(25)]\left(\frac{1}{80}\right) = 44.5 \text{ k (T)}$$

- For loading position 3 (Fig. 9.8(e)):

$$F_{BC} = 32(5)\left(\frac{3}{80}\right) + [8(60) + 32(40)]\left(\frac{1}{80}\right) = 28.0 \text{ k (T)}$$

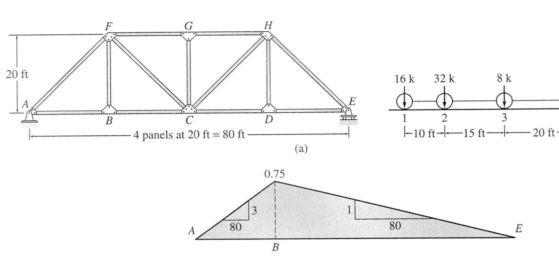

(a)

(b) Influence Line for F_{BC} (k/k)

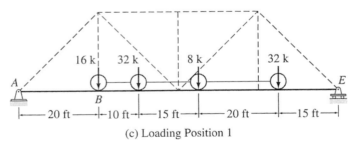

(c) Loading Position 1

(d) Loading Position 2

FIG. 9.8

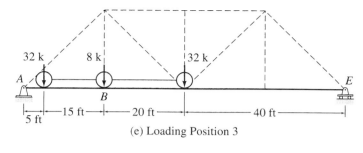

(e) Loading Position 3

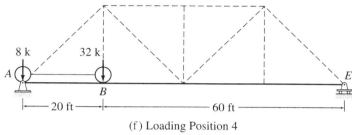

FIG. 9.8 (contd.)

(f) Loading Position 4

- For loading position 4 (Fig. 9.8(f)):

$$F_{BC} = 32(60)\left(\frac{1}{80}\right) = 24.0 \text{ k (T)}$$

By comparing the values of F_{BC} for the four loading positions, we conclude that the magnitude of the maximum axial force that develops in member BC is $F_{BC} = 44.5$ k tension. This maximum force occurs when the second load of the series is placed at joint B of the truss, as shown in Fig. 9.8(d).

Maximum $F_{BC} = 44.5$ k (T)   Ans.

9.4 ABSOLUTE MAXIMUM RESPONSE

Thus far, we have considered the maximum response that may occur at a *particular location* in a structure. In this section, we discuss how to determine the *absolute maximum* value of a response function that may occur at any location throughout a structure. Although only simply supported beams are considered in this section, the concepts presented herein can be used to develop procedures for the analysis of absolute maximum responses of other types of structures.

Single Concentrated Load

Consider the simply supported beam shown in Fig. 9.9(a). The influence lines for the shear and bending moment at an arbitrary section $a'a'$ located at a distance a from the left support A are shown in Fig. 9.9(b)

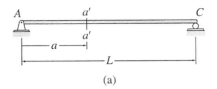

(a)

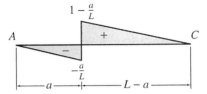

(b) Influence Line for Shear at
Section $a'a'$

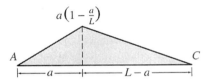

(c) Influence Line for Bending Moment
at Section $a'a'$

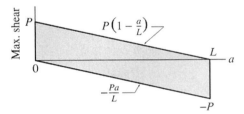

(d) Envelope of Maximum Shears — Single
Concentrated Load

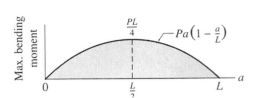

(e) Envelope of Maximum Bending Moments
— Single Concentrated Load

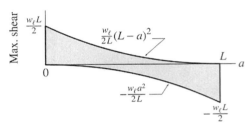

(f) Envelope of Maximum Shears — Uniformly
Distributed Load

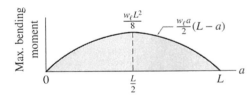

(g) Envelope of Maximum Bending Moments
— Uniformly Distributed Load

FIG. 9.9

and (c), respectively. Recall that these influence lines were initially developed in Section 8.1 (Fig. 8.2(e) and (f)).

Suppose that we wish to determine the absolute maximum shear in the beam due to a single moving concentrated load of magnitude P. As discussed in Section 9.1, the maximum positive shear at the section $a'a'$ is given by the product of the load magnitude, P, and the maximum positive ordinate, $1 - (a/L)$, of the influence line for shear at section $a'a'$ (Fig. 9.9(b)). Thus,

$$\text{maximum positive shear} = P\left(1 - \frac{a}{L}\right) \tag{9.3}$$

Similarly, the maximum negative shear at section $a'a'$ is given by

$$\text{maximum negative shear} = -\frac{Pa}{L} \tag{9.4}$$

These equations indicate that the maximum positive and maximum negative shears at a section due to a single moving concentrated load vary linearly with the distance a of the section from the left support A of the beam. A plot of Eqs. (9.3) and (9.4), with maximum shear as ordinate, against the location a of the section as abscissa is shown in Fig. 9.9(d). Such a graph, depicting the variation of the maximum value of a response function as a function of the location of the section, is referred to as the *envelope of the maximum values of a response function*. An envelope of maximum values of a response function provides a convenient means of determining the absolute maximum value of the response function as well as its location. It can be seen from the envelope of maximum shears (Fig. 9.9(d)) that in a simply supported beam subjected to a moving concentrated load P, the absolute maximum shear develops at sections just inside the supports and has the magnitude of P.

The envelope of maximum bending moments due to a single moving concentrated load P can be generated in a similar manner. By using the influence line for bending moment at the arbitrary section $a'a'$ given in Fig. 9.9(c), we determine the expression for the maximum bending moment at the section $a'a'$ as

$$\text{maximum bending moment} = Pa\left(1 - \frac{a}{L}\right) \tag{9.5}$$

The envelope of maximum bending moments constructed by plotting Eq. (9.5) is shown in Fig. 9.9(e). It can be seen that the absolute maximum bending moment occurs at midspan of the beam and has magnitude $PL/4$.

Uniformly Distributed Load

Next, let us determine the absolute maximum shear and bending moment in the simply supported beam of Fig. 9.9(a) due to a uniformly distributed live load of intensity w_ℓ. As discussed in Section 9.2, the maximum positive (or negative) shear at the section $a'a'$ can be obtained by placing the load over the portion of the beam where the ordinates of the shear influence line (Fig. 9.9(b)) are positive (or negative), and by multiplying the load intensity by the area of the influence line under the loaded portion of the beam. Thus,

$$\text{maximum positive shear} = \frac{w_\ell}{2L}(L - a)^2 \tag{9.6}$$

$$\text{maximum negative shear} = -\frac{w_\ell a^2}{2L} \tag{9.7}$$

The envelope of maximum shears due to a uniformly distributed live load, constructed by plotting Eqs. (9.6) and (9.7), is shown in Fig. 9.9(f). It can be seen that the absolute maximum shear develops at sections just inside the supports and has magnitude $w_\ell L/2$.

To determine the expression for the maximum bending moment at section $a'a'$, we multiply the load intensity, w_ℓ, by the area of the bending moment influence line (Fig. 9.9(c)), to obtain

$$\text{maximum bending moment} = \frac{w_\ell a}{2}(L - a) \tag{9.8}$$

The envelope of maximum bending moments due to a uniformly distributed live load, constructed by plotting Eq. (9.8), is shown in Fig. 9.9(g). It can be seen from this envelope that the absolute maximum bending moment occurs at midspan of the beam and has magnitude $w_\ell L^2/8$.

Series of Concentrated Loads

The absolute maximum value of a response function in any structure subjected to a series of moving concentrated loads or any other live loading condition can be determined from the envelope of maximum values of the response function. Such an envelope can be constructed by evaluating the maximum values of the response function at a number of points along the length of the structure by using the procedures described in Sections 9.1 through 9.3, and by plotting the maximum values. Because of the considerable amount of computational effort involved, except for some simple structures, the analysis of absolute maximum response is usually performed using computers. In the following section, we discuss the direct methods that are commonly employed to determine the absolute maximum shears and bending moments in simply supported beams subjected to a series of moving concentrated loads.

As in the case of single concentrated and uniformly distributed loads, the absolute maximum shear in a simply supported beam due to a series of moving concentrated loads always occurs at sections just inside the supports. From the influence line for shear at an arbitrary section $a'a'$ of a simply supported beam shown in Fig. 9.9(b), we can see that in order to develop the maximum positive shear at the section, we must place as many loads of the series as possible on the portion of the beam for which the influence line is positive and as few loads as possible on the portion where the influence line is negative. Moreover, as section $a'a'$ is shifted toward the left support of the beam, the value of the maximum positive shear will continuously increase, because the length and the maximum ordinate of the positive portion of the influence line increase, whereas those of the negative portion decrease. Thus, the absolute maximum positive shear will occur when the section $a'a'$ is located just to the right of the left support A. Using a similar reasoning,

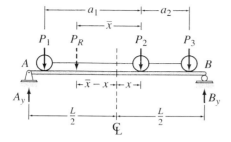

FIG. 9.10

it can be shown that the absolute maximum negative shear occurs at a section located just to the left of the right support C of the simply supported beam. Since the location of the absolute maximum shear is known, the procedure for computing maximum response at a section due to a series of concentrated loads, developed in Section 9.3, can be employed to determine the magnitude of the absolute maximum shear. Because the influence line for shear just inside the left support is identical to the influence line for reaction at the left support, the latter can be conveniently used for determining the magnitude of the absolute maximum shear.

To determine the location of the absolute maximum bending moment, consider the simply supported beam subjected to an arbitrary series of moving concentrated loads P_1, P_2, and P_3, as shown in Fig. 9.10. The resultant of the loads P_1, P_2, and P_3 is denoted by P_R, which is located at the distance $\bar{x}$ from the load P_2, as shown in the figure. The bending moment diagram of the beam consists of straight-line segments between the load points regardless of the position of the loads, so the absolute maximum bending moment occurs under one of the loads. Assuming that the absolute maximum bending moment occurs under the load P_2, our objective is to determine its position x from the midspan of the beam, as shown in the figure. By applying the equilibrium equation $\sum M_B = 0$ and using the resultant P_R instead of the individual loads in the equilibrium equation, we determine the vertical reaction A_y to be

$$+ \circlearrowleft \sum M_B = 0$$

$$-A_y(L) + P_R\left(\frac{L}{2} + \bar{x} - x\right) = 0$$

$$A_y = P_R\left(\frac{1}{2} + \frac{\bar{x}}{L} - \frac{x}{L}\right)$$

Thus the bending moment under the load P_2 is given by

$$M_2 = A_y\left(\frac{L}{2} + x\right) - P_1 a_1$$

$$= P_R\left(\frac{1}{2} + \frac{\bar{x}}{L} - \frac{x}{L}\right)\left(\frac{L}{2} + x\right) - P_1 a_1$$

$$= P_R\left(\frac{L}{4} + \frac{\bar{x}}{2} + \frac{x\bar{x}}{L} - \frac{x^2}{L}\right) - P_1 a_1$$

For M_2 to be maximum, its derivative with respect to x must be zero; that is,

$$\frac{dM_2}{dx} = P_R\left(\frac{\bar{x}}{L} - \frac{2x}{L}\right) = 0$$

from which we obtain

$$x = \frac{\bar{x}}{2} \tag{9.9}$$

Based on Eq. (9.9), we can conclude that *in a simply supported beam subjected to a series of moving concentrated loads, the maximum bending moment develops under a load when the midspan of the beam is located halfway between the load and the resultant of all the loads on the beam.* By applying this criterion, a maximum bending moment can be computed for each load acting on the beam. The largest of the maximum bending moments thus obtained is the absolute maximum bending moment. However, in general it is not necessary to examine all the loads acting on the beam, since the absolute maximum bending moment usually occurs under the load closest to the resultant, provided that it is of equal or larger magnitude than the next adjacent load. Otherwise, the maximum bending moments should be computed for the two loads adjacent to the resultant and compared to obtain the absolute maximum bending moment.

Example 9.5

Determine the absolute maximum bending moment in the simply supported beam due to the wheel loads of the HS20-44 truck shown in Fig. 9.11(a).

Solution

Resultant of Load Series The magnitude of the resultant is obtained by summing the magnitudes of the loads of the series. Thus

$$P_R = \sum P_i = 4 + 16 + 16 = 36 \text{ k}$$

The location of the resultant can be determined by using the condition that the moment of the resultant about a point equals the sum of the moments of the individual loads about the same point. Thus, by summing moments about the 16-k trailer-wheel load, we obtain

$$P_R(\bar{x}) = \sum P_i x_i$$

$$36(\bar{x}) = 4(28) + 16(14)$$

$$\bar{x} = 9.33 \text{ ft}$$

Absolute Maximum Bending Moment From Fig. 9.11(b), we observe that the second load of the series (the 16-k rear-wheel load) is located closest to the resultant. Thus the absolute maximum bending moment occurs under the second load when the series is positioned on the beam so that the midspan of the beam is located halfway between the load and the resultant. The resultant is located 4.67 ft to the right of the second load (Fig. 9.11(b)), so we position this load at a distance of $4.67/2 = 2.33$ ft to the left of the beam midspan, as shown in Fig. 9.11(c). Next we compute the vertical reaction at A to be

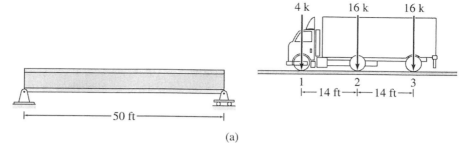

(a)

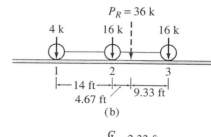

(b)

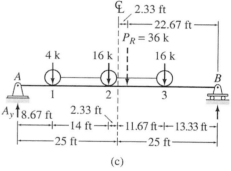

(c)

FIG. **9.11**

$$A_y = 36\left(\frac{22.67}{50}\right) = 16.32 \text{ k}$$

Thus the absolute maximum bending moment, which occurs under the second load of the series, is

Absolute maximum bending moment $= M_2 = 16.32(8.67 + 14) - 4(14)$

$$= 313.97 \text{ k-ft} \qquad \text{Ans.}$$

SUMMARY

In this chapter we have learned that the value of a response function due to a single concentrated load can be obtained by multiplying the magnitude of the load by the ordinate of the response function influence line at the position of the load. To determine the maximum positive (or negative) value of a response function due to a single moving concentrated load, the load must be placed at the location of the maximum positive (or negative) ordinate of the response function influence line.

The value of a response function due to a uniformly distributed load applied over a portion of the structure can be obtained by multiplying the load intensity by the net area under the corresponding portion of the response function influence line. To determine the maximum positive (or negative) value of a response function due to a uniformly distributed live load, the load must be placed over those portions of the structure where the ordinates of the response function influence line are positive (or negative).

The maximum value of a response function at a particular location in a structure due to a series of moving concentrated loads can be determined by successively placing each load of the series on the structure at the location of the maximum ordinate of the response function influence line, by computing the value of the response function for each position of the series through algebraically summing the products of the load magnitudes and the respective influence-line ordinates, and by comparing the values of the response function thus obtained to determine the maximum value of the response function.

In simply supported beams (a) the absolute maximum shear develops at sections just inside the supports, (b) the absolute maximum bending moment due to a single concentrated, or a uniformly distributed, live load occurs at the beam midspan, and (c) the absolute maximum bending moment due to a series of moving concentrated loads occurs under one of the loads near the resultant of the loads when the midspan of the beam is located halfway between the load and the resultant.

PROBLEMS

Sections 9.1 and 9.2

9.1 For the beam of Problem 8.4, determine the maximum negative bending moment at point B due to a 15-k concentrated live load.

9.2 For the beam of Problem 8.4, determine the maximum upward reaction at support A due to a 3-k/ft uniformly distributed live load.

9.3 For the beam of Problem 8.4, determine the maximum negative shear at point B due to a 3-k/ft uniformly distributed live load.

9.4 For the beam of Problem 8.5, determine the maximum positive and negative shears and the maximum positive and negative bending moments at point C due to a concentrated live load of 100 kN, a uniformly distributed live load of 50 kN/m, and a uniformly distributed dead load of 20 kN/m.

9.5 For the cantilever beam of Problem 8.9, determine the maximum upward vertical reaction and the maximum counterclockwise reaction moment at support A due to a concentrated live load of 25 k, a uniformly distributed live load of 2 k/ft, and a uniformly distributed dead load of 0.5 k/ft.

9.6 For the beam of Problem 8.10, determine the maximum positive and negative shears and the maximum positive and negative bending moments at point C due to a concentrated live load of 150 kN, a uniformly distributed live load of 50 kN/m, and a uniformly distributed dead load of 25 kN/m.

9.7 For the beam of Problem 8.23, determine the maximum positive and negative shears and the maximum positive and negative bending moments at point D due to a concentrated live load of 30 k, a uniformly distributed live load of 3 k/ft, and a uniformly distributed dead load of 1 k/ft.

9.8 For the beam of Problem 8.29, determine the maximum positive and negative shears and the maximum positive and negative bending moments at point F due to a concentrated live load of 40 k, a uniformly distributed live load of 2 k/ft, and a uniformly distributed dead load of 1 k/ft.

9.9 For the truss of Problem 8.47, determine the maximum tensile and compressive axial forces in member CH due to a concentrated live load of 30 k, a uniformly distributed live load of 2 k/ft, and a uniformly distributed dead load of 1 k/ft.

9.10 For the truss of Problem 8.50, determine the maximum compressive axial force in member AB and the maximum tensile axial force in member EF due to a concentrated live load of 120 kN, a uniformly distributed live load of 40 kN/m, and a uniformly distributed dead load of 20 kN/m.

9.11 For the truss of Problem 8.51, determine the maximum tensile and compressive axial forces in member IJ due to a concentrated live load of 40 k, a uniformly distributed live load of 4 k/ft, and a uniformly distributed dead load of 2 k/ft.

Section 9.3

9.12 For the beam of Problem 8.2, determine the maximum positive shear and bending moment at point B due to the wheel loads of the moving H20-44 truck shown in Fig. P9.12.

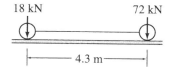

FIG. **P9.12, P9.20**

9.13 For the beam of Problem 8.1, determine the maximum positive shear and bending moment at point B due to the series of three moving concentrated loads shown in Fig. P9.13.

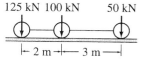

FIG. **P9.13, P9.17, P9.18, P9.22**

9.14 For the beam of Problem 8.7, determine the maximum positive bending moment at point B due to the series of four moving concentrated loads shown in Fig. P9.14.

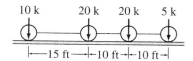

FIG. **P9.14, P9.16, P9.19, P9.23**

9.15 For the beam of Problem 8.23, determine the maximum positive bending moment at point D due to the wheel loads of the moving HS15-44 truck shown in Fig. P9.15.

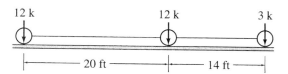

FIG. **P9.15, P9.21**

9.16 For the truss of Problem 8.49, determine the maximum compressive axial force in member GH due to the series of four moving concentrated loads shown in Fig. P9.14.

9.17 For the truss of Problem 8.53, determine the maximum tensile axial force in member DI due to the series of three moving concentrated loads shown in Fig. P9.13.

Section 9.4

9.18 Determine the absolute maximum shear in a 15-m-long simply supported beam due to the series of three moving concentrated loads shown in Fig. P9.13.

9.19 Determine the absolute maximum shear in a 60-ft-long simply supported beam due to the series of four moving concentrated loads shown in Fig. P9.14.

9.20 Determine the absolute maximum bending moment in a 12-m-long simply supported beam due to the wheel loads of the moving H20-44 truck shown in Fig. P9.12.

9.21 Determine the absolute maximum bending moment in a 75-ft-long simply supported beam due to the wheel loads of the moving HS15-44 truck shown in Fig. P9.15.

9.22 Determine the absolute maximum bending moment in a 15-m-long simply supported beam due to the series of three moving concentrated loads shown in Fig. P9.13.

9.23 Determine the absolute maximum bending moment in a 60-ft-long simply supported beam due to the series of four moving concentrated loads shown in Fig. P9.14.

10

Analysis of Symmetric Structures

10.1 Symmetric Structures
10.2 Symmetric and Antisymmetric Components of Loadings
10.3 Behavior of Symmetric Structures under Symmetric and Antisymmetric Loadings
10.4 Procedure for Analysis of Symmetric Structures
Summary
Problems

Taj Mahal, Built in the Seventeenth Century in Agra, India
Photo courtesy of the Indian Government Tourist Office, New York

Many structures, because of aesthetic and/or functional considerations, are arranged in symmetric forms. Provided a symmetric structure is linearly elastic, the response (i.e., member forces and deformations) of the entire structure under any general loading can be obtained from the response of one of its portions separated by the axes of symmetry. Thus only a portion (usually half) of the symmetric structure needs to be analyzed. In this chapter we discuss how to recognize structural symmetry and how to utilize it to reduce the computational effort required in the analysis of symmetric structures.

We first define symmetric structures and then discuss symmetric and antisymmetric loadings. In this presentation, we develop a procedure for decomposing a general loading into symmetric and antisymmetric components. Next we examine the behavior of symmetric structures under the symmetric and antisymmetric loadings; finally, we present a step-by-step procedure for the analysis of symmetric structures.

Although the discussion in this chapter is confined to structures with a single axis of symmetry, the concepts developed herein can be extended to the analysis of structures with multiple axes of symmetry.

10.1 SYMMETRIC STRUCTURES

Reflection

The definition of symmetry can be developed by using the concept of *reflection*, or mirror image. Consider a structure located in the xy plane, as shown in Fig. 10.1(a). The reflection of the structure about the y axis is obtained by rotating the structure through 180° about the y axis, as shown in Fig. 10.1(b). It can be seen from Fig. 10.1(a) and (b) that if the coordinates of a point D of the structure are x_1 and y_1, then the coordinates of that point on the reflection of the structure about the y axis become $-x_1$ and y_1. The reflection of the structure about the x axis can be obtained in a similar manner—that is, by rotating the structure through 180° about the x axis, as shown in Fig. 10.1(c). Note that the coordinates of point D on the reflection of the structure about the x axis become x_1 and $-y_1$.

Based on the foregoing discussion, we realize that the reflection of a structure about an arbitrary s axis can be obtained by rotating the structure through 180° about the s axis. Alternatively, the structure's reflection can be obtained by joining the reflections of its various joints

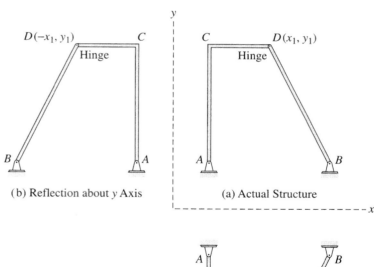

(b) Reflection about y Axis   (a) Actual Structure

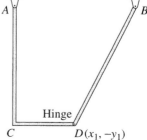

FIG. **10.1**   (c) Reflection about x Axis

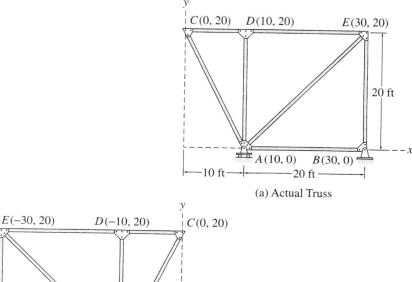

(a) Actual Truss

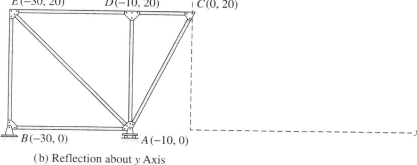

(b) Reflection about y Axis

FIG. **10.2**

and/or ends, which are determined by changing the signs of their coordinates in the direction perpendicular to the s axis. To illustrate the latter approach, consider the truss shown in Fig. 10.2(a). Suppose that we wish to determine its reflection about the y axis. As shown in Fig. 10.2(b), the reflections of the five joints of the truss are first determined by changing the signs of the x coordinates of the joints. The reflections of the joints are then connected by straight lines to obtain the reflection of the entire truss. Note that the reflection of joint C, which is located on the y axis, is in the same position as joint C itself.

Symmetric Structures

A plane structure is considered to be symmetric with respect to an axis of symmetry in its plane if the reflection of the structure about the axis is identical in geometry, supports, and material properties to the structure itself.

Some examples of symmetric structures are shown in Fig. 10.3. For each structure, the axis of symmetry is identified as the s axis. Note that the

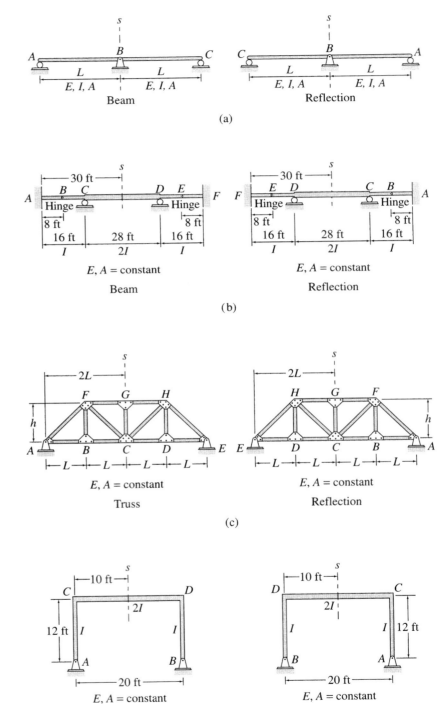

FIG. 10.3 Examples of Symmetric
Structures

Frame   Reflection

(e)

Frame   Reflection

FIG. 10.3 (contd.)

(f)

reflection of each structure about its axis of symmetry is identical in geometry, supports, and material properties to the structure itself.

Although the concept of reflection provides a mathematically
precise means of defining symmetry, it is usually not necessary to draw
the reflection of a structure to determine whether or not the structure
is symmetric. Instead, most symmetric structures can be identified by
inspection—that is, by simply comparing the geometry, supports, and
material properties of the two halves of the structure on each side of the
axis of symmetry. Considering any of the structures of Fig. 10.3, if we
imagine that a half of the structure on either side of the axis of symmetry is rotated through 180° about the axis of symmetry, it will exactly
overlay the other half of the structure, indicating that the structure is
symmetric.

As stated previously, a structure, in general, is considered to be
symmetric if its geometry, supports, and material properties are symmetric with respect to the axis of symmetry. However, when examining

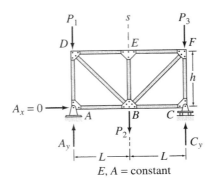

$A_x = 0$

$E, A = $ constant

FIG. 10.4

structural symmetry for the purpose of an analysis, it is necessary to consider the symmetry of only those structural properties that have an effect on results of that particular type of analysis. In other words, a structure can be considered to be symmetric for the purpose of an analysis if its structural properties that have an effect on the results of the analysis are symmetric.

Consider, for example, the statically determinate truss subjected to vertical loads, as shown in Fig. 10.4. We can see from the figure that the geometry of the truss (i.e., the dimensions of the truss and the arrangement of truss members) and its material and cross-sectional properties (E and A) are symmetric with respect to the s axis, but the supports violate symmetry because the hinged support at A can exert both horizontal and vertical reactions, whereas the roller support at C can exert only a vertical reaction. However, the truss can be considered to be symmetric when subjected to vertical loads only because under such loads, the horizontal reaction at the hinged support will be zero ($A_x = 0$); therefore, it will not have any effect on the response (e.g., member axial forces and deflections) of the truss. This truss cannot be considered to be symmetric when subjected to any horizontal loads, however.

Example 10.1

The truss shown in Fig. 10.5(a) is to be analyzed to determine its member axial forces and deflections due to a general system of loads acting at the joints. Can the truss be considered to be symmetric for such an analysis?

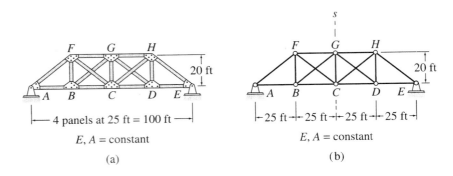

(a)   (b)

FIG. 10.5

Solution
We can see from Fig. 10.5(b) that the dimensions, the arrangement of members, the material and cross-sectional properties (E and A), and the supports of the given truss are all symmetric with respect to the vertical s axis passing through the member CG of the truss. Thus the truss is symmetric with respect to the s axis.   Ans.

Example 10.2

The beam shown in Fig. 10.6(a) is to be analyzed to determine the member end forces and deflections due to the vertical loading shown. Can the beam be considered to be symmetric for the analysis?

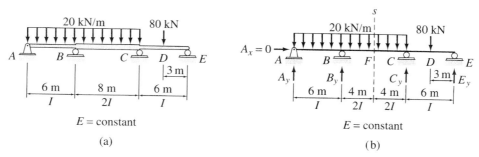

FIG. **10.6**

Solution

We can see from Fig. 10.6(b) that the dimensions and properties (E and I) of the beam are symmetric with respect to the vertical s axis passing through the mid-point F of the beam, but the supports are not symmetric because the hinged support at A can develop both horizontal and vertical reactions, whereas the roller supports at B, C, and E can develop only vertical reactions. However, the beam can be considered to be symmetric under the vertical loads because the horizontal reaction at A is zero ($A_x = 0$); therefore, it does not have any effect on the member end forces and deflections of the beam.   Ans.

Example 10.3

The frame shown in Fig. 10.7(a) is to be analyzed to determine its member end forces and deflections due to a general system of loads. Can the frame be considered to be symmetric?

Solution

From Fig. 10.7(b) we can see that although the frame's geometry and supports are symmetric with respect to the vertical s axis passing through the internal hinge D, its moment of inertia (I) is not symmetric. Since the frame is statically determinate, its member end forces are independent of the material and cross-sectional properties (E, I, and A); therefore, the frame can be considered to be symmetric for the purpose of analysis of its member forces. However, this frame cannot be considered to be symmetric for the analysis of deflections, which depend on the moments of inertia of the members of the frame.   Ans.

continued

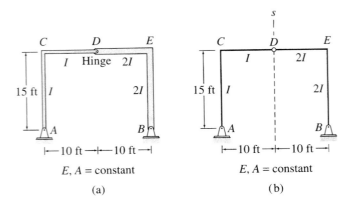

E, A = constant

(a)

E, A = constant

(b)

FIG. 10.7

10.2 SYMMETRIC AND ANTISYMMETRIC COMPONENTS OF LOADINGS

As discussed in the preceding section for structures, the reflection of a system of forces (or deflections) about an axis can be obtained by rotating the force system (or deflections) through 180° about the axis. Consider a system of forces and moments, $F_x, F_y,$ and M, acting at a point A in the xy plane, as shown in Fig. 10.8(a). The reflections of the force

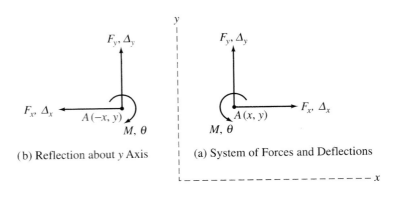

(b) Reflection about y Axis

(a) System of Forces and Deflections

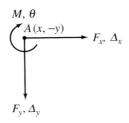

(c) Reflection about x Axis

FIG. 10.8

system about the y and x axes are shown in Fig. 10.8(b) and (c), respectively. As shown in these figures, the reflections of the counterclockwise moment M are clockwise. Conversely, the reflections of a clockwise moment will always be counterclockwise. The reflections of the deflections Δ_x and Δ_y and the rotation θ of point A (Fig. 10.8(a)) can be obtained in a similar manner and are also shown in Fig. 10.8(b) and (c).

Symmetric Loadings

A loading is considered to be symmetric with respect to an axis in its plane if the reflection of the loading about the axis is identical to the loading itself.

Some examples of symmetric loadings are shown in Fig. 10.9. The reflection of each loading about its axis of symmetry is also shown in the figure for verification. However, it is usually not necessary to draw the reflections, since most loadings can be identified as symmetric, or not, by inspection.

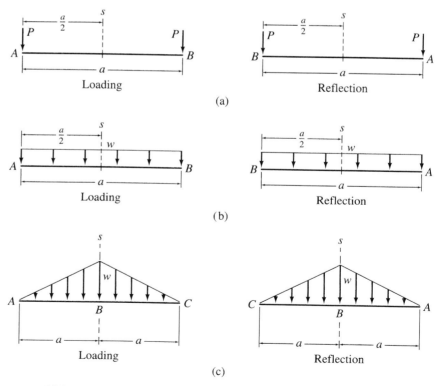

FIG. 10.9 Examples of Symmetric Loadings

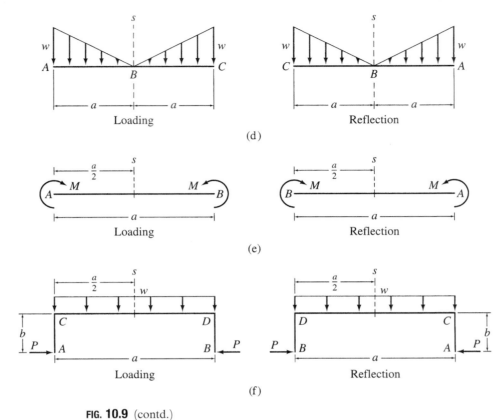

FIG. 10.9 (contd.)

Antisymmetric Loadings

A loading is considered to be antisymmetric with respect to an axis in its plane if the negative of the reflection of the loading about the axis is identical to the loading itself.

Some examples of antisymmetric loadings are shown in Fig. 10.10. For each loading case, the reflection and the negative of reflection are also

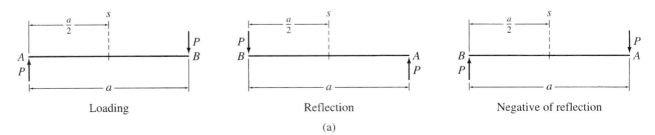

FIG. 10.10 Examples of Antisymmetric Loadings

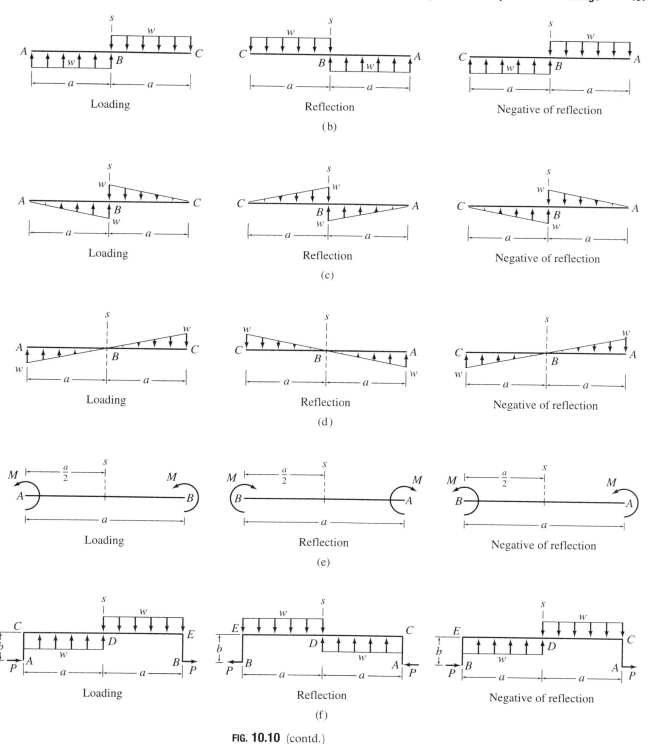

Loading   Reflection   Negative of reflection

(b)

Loading   Reflection   Negative of reflection

(c)

Loading   Reflection   Negative of reflection

(d)

Loading   Reflection   Negative of reflection

(e)

Loading   Reflection   Negative of reflection

(f)

FIG. 10.10 (contd.)

shown in the figure. The negative of a reflection is obtained by simply reversing the directions of all the forces and moments on the reflection. It can be seen from the figure that the negative of reflection of each loading about its *s* axis is identical to the loading itself.

Decomposition of a General Loading into Symmetric and Antisymmetric Components

Any general loading can be decomposed into symmetric and anti-symmetric components with respect to an axis by applying the following procedure:

1. Divide the magnitudes of the forces and/or moments of the given loading by 2 to obtain the half loading.
2. Draw a reflection of the half loading about the specified axis.
3. Determine the symmetric component of the given loading by adding the half loading to its reflection.
4. Determine the antisymmetric component of the given loading by subtracting the symmetric loading component from the given loading.

To illustrate this procedure, consider the unsymmetric loading shown in Fig. 10.11(a). Suppose that we wish to determine the com-

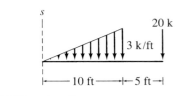

(a) Given Loading

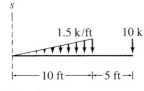

(b) Half Loading

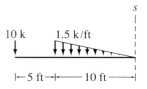

(c) Reflection of Half Loading

FIG. 10.11

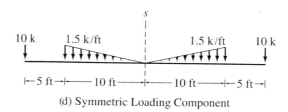

(d) Symmetric Loading Component

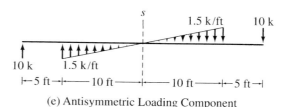

(e) Antisymmetric Loading Component

FIG. 10.11 (contd.)

ponents of this loading, which are symmetric and antisymmetric with respect to an arbitrarily located s axis shown in the figure. We first compute the half loading by dividing the magnitudes of the distributed and the concentrated loads by 2 (Fig. 10.11(b)). The reflection of this half loading about the s axis is then drawn, as shown in Fig. 10.11(c). The symmetric component of the given loading is determined by adding the half loading (Fig. 10.11(b)) to its reflection (Fig. 10.11(c)). The symmetric loading component thus obtained is shown in Fig. 10.11(d). Finally, the antisymmetric component is computed by subtracting the symmetric component (Fig. 10.11(d)) from the given loading (Fig. 10.11(a)). The antisymmetric loading component thus obtained is shown in Fig. 10.11(e). Note that the sum of the symmetric and antisymmetric components is equal to the given loading.

Example 10.4

A Pratt bridge truss is subjected to the loading shown in Fig. 10.12(a). Determine the symmetric and antisymmetric components of the loading with respect to the axis of symmetry of the truss.

Solution

Symmetric Loading Component The axis of symmetry (s axis) of the truss and the half loading are shown in Fig. 10.12(b); the reflection of the half loading about the s axis is drawn in Fig. 10.12(c). The symmetric component of the given loading is determined by adding the half loading (Fig. 10.12(b)) to its reflection (Fig. 10.12(c)), as shown in Fig. 10.12(d).   Ans.

continued

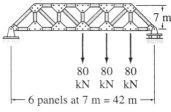

7 m

80 80 80
kN kN kN

── 6 panels at 7 m = 42 m ──

(a) Given Loading

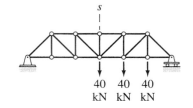

s

40   40   40
kN   kN   kN

(b) Half Loading

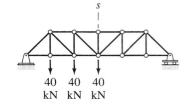

s

40   40   40
kN   kN   kN

(c) Reflection of Half Loading

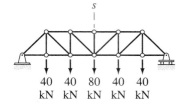

s

40   40   80   40   40
kN   kN   kN   kN   kN

(d) Symmetric Loading Component

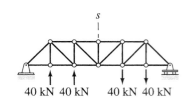

s

40 kN 40 kN   40 kN 40 kN

(e) Antisymmetric Loading Component

FIG. 10.12

Antisymmetric Loading Component The antisymmetric component of the loading is obtained by subtracting the symmetric loading component (Fig. 10.12(d)) from the total loading (Fig. 10.12(a)) and is shown in Fig. 10.12(e).

Ans.

Note that the sum of the symmetric and antisymmetric components is equal to the given loading.

Example 10.5

A beam is subjected to the loading shown in Fig. 10.13(a). Determine the symmetric and antisymmetric components of the loading with respect to the axis of symmetry of the beam.

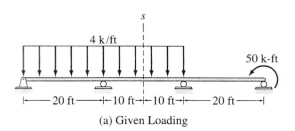

s

4 k/ft

50 k-ft

├─── 20 ft ───┼─10 ft─┼─10 ft─┼─── 20 ft ───┤

(a) Given Loading

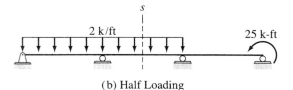

s

2 k/ft

25 k-ft

(b) Half Loading

FIG. 10.13

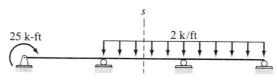

(c) Reflection of Half Loading

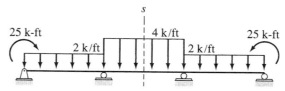

(d) Symmetric Loading Component

(e) Antisymmetric Loading Component

FIG. 10.13 (contd.)

Solution

Symmetric Loading Component The axis of symmetry (*s* axis) of the beam and the half loading are shown in Fig. 10.13(b), and the reflection of the half loading about the *s* axis is drawn in Fig. 10.13(c). The symmetric component of the given loading is determined by adding the half loading (Fig. 10.13(b)) to its reflection (Fig. 10.13(c)), as shown in Fig. 10.13(d). **Ans.**

Antisymmetric Loading Component The antisymmetric component is obtained by subtracting the symmetric component (Fig. 10.13(d)) from the total loading (Fig. 10.13(a)) and is shown in Fig. 10.13(e). **Ans.**

Note that the sum of the symmetric and antisymmetric components is equal to the given loading.

Example 10.6

A four-span continuous beam is subjected to the loading shown in Fig. 10.14(a). Determine the symmetric and antisymmetric components of the loading with respect to the axis of symmetry of the beam.

Solution

Symmetric Loading Component The half loading and its reflection are shown in Fig. 10.14(b) and (c), respectively. The symmetric component of the given loading is obtained by adding the half loading to its reflection, as shown in Fig. 10.14(d). **Ans.**

continued

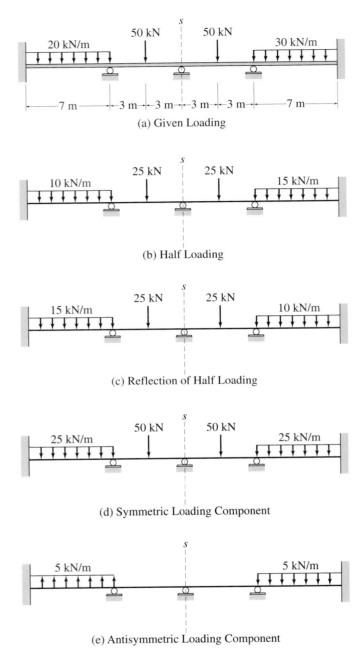

(a) Given Loading

(b) Half Loading

(c) Reflection of Half Loading

(d) Symmetric Loading Component

(e) Antisymmetric Loading Component

FIG. **10.14**

Antisymmetric Loading Component By subtracting the symmetric component from the total loading (Fig. 10.14(a)), we determine the antisymmetric component as shown in Fig. 10.14(e).   Ans.

Example 10.7

A gable frame is subjected to the loading shown in Fig. 10.15(a). Determine the symmetric and antisymmetric components of the loading with respect to the axis of symmetry of the frame.

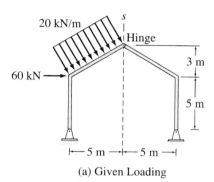

(a) Given Loading

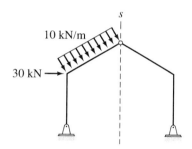

(b) Half Loading

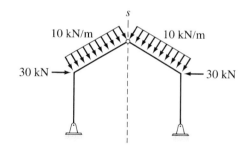

(d) Symmetric Loading Component

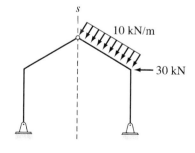

(c) Reflection of Half Loading

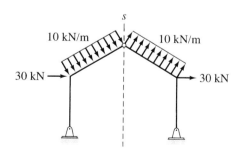

(e) Antisymmetric Loading Component

FIG. **10.15**

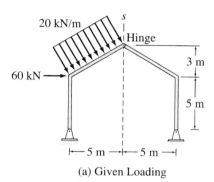 20 kN/m, Hinge, 60 kN, 3 m, 5 m, 5 m, 5 m

continued

Solution

Symmetric Loading Component The half loading and its reflection are shown in Fig. 10.15(b) and (c), respectively. The symmetric component of the given loading is determined by adding the half loading to its reflection, as shown in Fig. 10.15(d). **Ans.**

Antisymmetric Loading Component By subtracting the symmetric component from the total loading (Fig. 10.15(a)), we obtain the antisymmetric component as shown in Fig. 10.15(e). **Ans.**

Example 10.8

A two-story frame is subjected to the loading shown in Fig. 10.16(a). Determine the symmetric and antisymmetric components of the loading with respect to the axis of symmetry of the frame.

Solution

Half Loading and Its Reflection See Fig. 10.16(b) and (c), respectively.

Symmetric Loading Component See Fig. 10.16(d). **Ans.**

Antisymmetric Loading Component See Fig. 10.16(e). **Ans.**

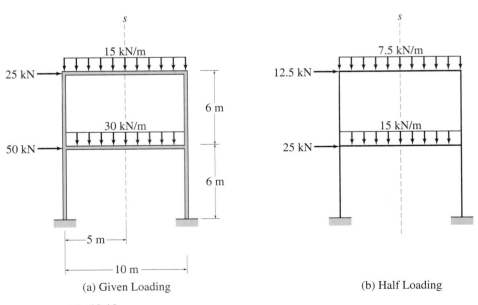

(a) Given Loading   (b) Half Loading

FIG. 10.16

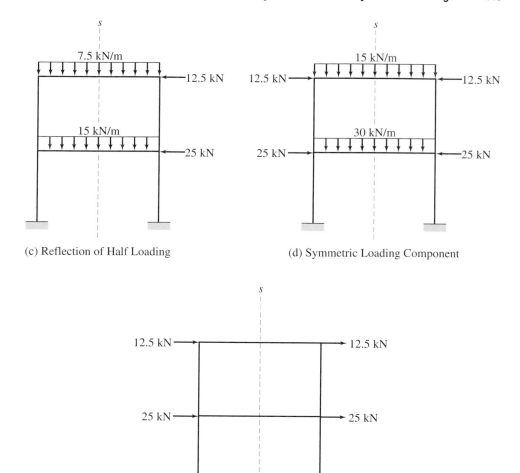

(c) Reflection of Half Loading

(d) Symmetric Loading Component

(e) Antisymmetric Loading Component

FIG. **10.16** (contd.)

10.3 BEHAVIOR OF SYMMETRIC STRUCTURES UNDER SYMMETRIC AND ANTISYMMETRIC LOADINGS

In the preceding section, we discussed how a general unsymmetric loading can be decomposed into symmetric and antisymmetric components. In this section, we examine the response characteristics of symmetric structures under symmetric and antisymmetric loading conditions. The insight gained into the behavior of symmetric structures will enable us to develop, in the following section, a general procedure that can considerably expedite the analysis of such structures.

Symmetric Structures Subjected to Symmetric Loadings

When a symmetric structure is subjected to a loading that is symmetric with respect to the structure's axis of symmetry, the response of the structure is also symmetric, with the points of the structure at the axis of symmetry neither rotating (unless there is a hinge at such a point) nor deflecting perpendicular to the axis of symmetry.

Thus, to determine the response (i.e., member forces and deformations) of the entire structure, we need to analyze only half the structure, on either side of the axis of symmetry, with symmetric boundary conditions (i.e., slopes must be either symmetric or zero, and deflections perpendicular to the axis of symmetry must be zero) at the axis. The response of the remaining half of the structure can then be obtained by reflection.

Consider, for example, a symmetric frame subjected to a loading that is symmetric with respect to the frame's axis of symmetry (s axis), as shown in Fig. 10.17(a). The deflected shape (elastic curve) of the frame is also shown in the figure. It can be seen that, like the loading, the deflected shape is symmetric with respect to the axis of symmetry of the frame. Note that the slope and the horizontal deflection are zero at point D, where the axis of symmetry intersects the frame, whereas the vertical deflection at D is not zero. The response of the entire frame can be determined by analyzing only half the frame, on either side of the axis of symmetry. The left half of the frame cut by the axis of symmetry

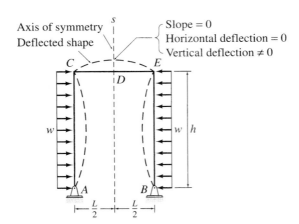

(a) Symmetric Frame Subjected
to Symmetric Loading

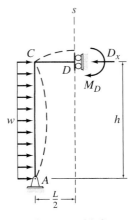

(b) Half Frame with Symmetric
Boundary Conditions

FIG. 10.17

is shown in Fig. 10.17(b). Note that the symmetric boundary conditions are imposed on this substructure by supporting it at the end D by a collar type of support (denoted by the symbol ⊢⊟ in Fig. 10.17(b)), which prevents the rotation and the horizontal deflection at the axis of symmetry but cannot prevent the vertical deflection along the axis. Once the response of the left half of the frame has been determined by analysis, the response of the right half can be obtained from that of the left half by reflection.

Consider another symmetric frame subjected to symmetric loading, as shown in Fig. 10.18(a). The left half of the frame with symmetric boundary conditions is shown in Fig. 10.18(b). As this figure indicates, the rotation and horizontal deflection at joint E have been restrained. The hinged joint B is already restrained from moving in the horizontal direction by the hinged support. Note that on the half of the frame selected for analysis (Fig. 10.18(b)), the magnitude of the concentrated load P, which acts along the axis of symmetry, has been reduced by half.

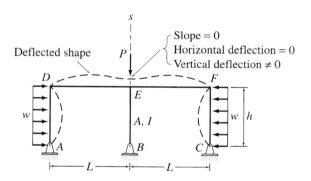

(a) Symmetric Frame Subjected
to Symmetric Loading

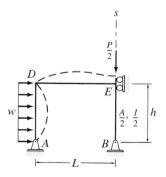

(b) Half Frame with Symmetric
Boundary Conditions

FIG. 10.18

Similarly, the cross-sectional area (A) and the moment of inertia (I) of member BE, which is located along the axis of symmetry, have been halved. Although it is usually considered convenient to reduce by half both properties A and I of the members along the axis of symmetry, we must realize that the values of the moments of inertia (I) of these members are not relevant in the analysis, because the members located along the axis of symmetry will undergo only axial deformations without bending. Once the response of the left half of the frame (Fig. 10.18(b)) has been determined by analysis, the response of the right half is obtained by reflection.

Symmetric Structures Subjected to Antisymmetric Loadings

When a symmetric structure is subjected to a loading that is antisymmetric with respect to the structure's axis of symmetry, the response of the structure is also antisymmetric, with the points of the structure at the axis of symmetry not deflecting in the direction of the axis of symmetry.

Thus to determine the response of the entire structure, we need to analyze only half the structure, on either side of the axis of symmetry, with antisymmetric boundary conditions (i.e., deflections in the direction of the axis of symmetry must be zero) at the axis. The response of the remaining half is given by the negative of the reflection of the response of the half structure that is analyzed.

Consider a symmetric frame subjected to a loading that is antisymmetric with respect to the frame's axis of symmetry (s axis), as shown in Fig. 10.19(a). It can be seen that, like the loading, the deflected shape of the frame is antisymmetric with respect to the frame's axis of

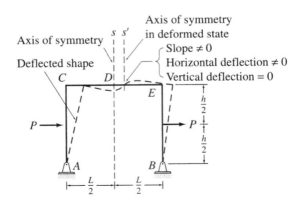

(a) Symmetric Structure Subjected to Antisymmetric Loading

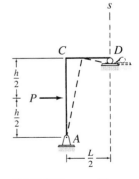

(b) Half Frame with Antisymmetric Boundary Conditions

FIG. 10.19

symmetry. Note that the vertical deflection is zero at point D, where the axis of symmetry intersects the frame, whereas the horizontal deflection and slope at D are not zero. The response of the entire frame can be determined by analyzing only half the frame, on either side of the axis of symmetry. The left half of the frame cut by the axis of symmetry is shown in Fig. 10.19(b). Note that the antisymmetric boundary conditions are imposed on this substructure by supporting it at end D by a roller support, which prevents the vertical deflection at the axis of symmetry but cannot prevent the horizontal deflection and rotation at D. Once the response of the left half of the frame has been determined by analysis, the response of the right half is given by the negative of the reflection of the response of the left half.

If a structure contains a member along the axis of symmetry, the properties of the member, I and A, should be reduced by half on the half structure selected for analysis. Note that the members along the axis of symmetry cannot undergo any axial deformations, but they can bend. Thus the axial forces in the members of trusses located along the axis of symmetry will be zero, and such members may be removed from the half structure to simplify its analysis. The magnitudes of any loads and couples acting on the structure at the axis of symmetry should be halved, on the half of the structure to be analyzed.

Symmetric Structures Subjected to General Loadings

As shown in Section 10.2, any general unsymmetric loading acting on a symmetric structure can be decomposed into symmetric and antisymmetric components with respect to the axis of symmetry of the structure. The responses of the structure due to the symmetric and antisymmetric loading components are then determined by analyzing a half of the structure, with symmetric and antisymmetric boundary conditions, respectively, as discussed in the preceding paragraphs. The symmetric and antisymmetric responses thus determined are then superimposed to obtain the total response of the structure due to the given unsymmetric loading.

10.4 PROCEDURE FOR ANALYSIS OF SYMMETRIC STRUCTURES

The following step-by-step procedure can be used to take advantage of structural symmetry in the analysis of structures.

1.   Check the given structure for symmetry, as discussed in Section 10.1. If the structure is found to be symmetric, then proceed to step 2. Otherwise, end the analysis at this stage.

2. Select a substructure (half the structure) on either side of the axis of symmetry for analysis. The cross-sectional areas and moments of inertia of the members of the substructure, which are located along the axis of symmetry, should be reduced by half, whereas full values of these properties should be used for all other members.

3. Decompose the given loading into symmetric and antisymmetric components with respect to the axis of symmetry of the structure by using the procedure described in Section 10.2.

4. Determine the response of the structure due to the symmetric loading component as follows:

   a. At each joint and end of the substructure, which is located at the axis of symmetry, apply restraints to prevent rotation and deflection perpendicular to the axis of symmetry. If there is a hinge at such a joint or end, then only the deflection, but not rotation, should be restrained at that joint or end.

   b. Apply the symmetric component of loading on the substructure with the magnitudes of the concentrated loads at the axis of symmetry reduced by half.

   c. Analyze the substructure to determine its response.

   d. Obtain the symmetric response of the complete structure by reflecting the response of the substructure to the other side of the axis of symmetry.

5. Determine the response of the structure due to the antisymmetric loading component as follows:

   a. At each joint and end of the substructure located at the axis of symmetry, apply a restraint to prevent deflection in the direction of the axis of symmetry. In the case of trusses, the axial forces in members located along the axis of symmetry will be zero. Remove such members from the substructure.

   b. Apply the antisymmetric component of loading on the substructure with the magnitudes of the loads and couples, applied at the axis of symmetry, reduced by half.

   c. Analyze the substructure to determine its response.

   d. Obtain the antisymmetric response of the complete structure by reflecting the negative of the response of the substructure to the other side of the axis of symmetry.

6. Determine the total response of the structure due to the given loading by superimposing the symmetric and antisymmetric responses obtained in steps 4 and 5, respectively.

The foregoing procedure can be applied to statically determinate as well as indeterminate symmetric structures. It will become obvious in subsequent chapters that the utilization of structural symmetry considerably reduces the computational effort required in the analysis of statically indeterminate structures.

Example 10.9

Determine the force in each member of the Warren truss shown in Fig. 10.20(a).

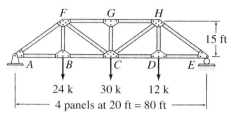

(a) Given Truss and Loading

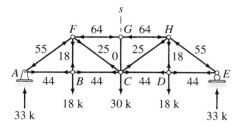

(e) Member Forces Due to Symmetric Loading Component

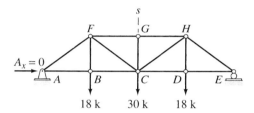

(b) Symmetric Loading Component

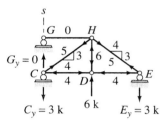

(f) Substructure with Antisymmetric Boundary Conditions

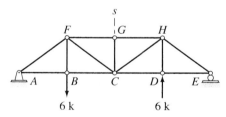

(c) Antisymmetric Loading Component

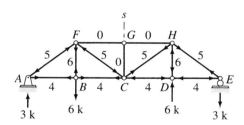

(g) Member Forces Due to Antisymmetric Loading Component

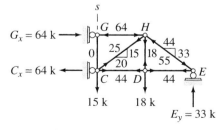

(d) Substructure with Symmetric Boundary Conditions

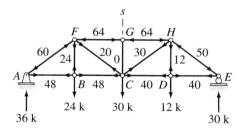

(h) Member Forces Due to Total Loading

FIG. 10.20

continued

Solution

This truss was analyzed in Example 4.4 without taking advantage of its symmetry.

Symmetry This truss is symmetric with respect to the vertical s axis passing through member CG, as shown in Fig. 10.20(b). The truss is subjected to vertical loads only, so the horizontal reaction at support A is zero ($A_x = 0$). The half of the truss to the right of the axis of symmetry, $CEHG$, will be used for analysis.

Symmetric and Antisymmetric Components of Loading The symmetric and antisymmetric components of the given loading with respect to the axis of symmetry of the truss are determined by using the procedure described in Section 10.2. These loading components are shown in Fig. 10.20(b) and (c). Note that the sum of the two components is equal to the total loading given in Fig. 10.20(a).

Member Forces Due to the Symmetric Loading Component The substructure (right half of the truss) with symmetric boundary conditions is shown in Fig. 10.20(d). Note that the joints C and G, which are located at the axis of symmetry, are supported by rollers that prevent their movements in the horizontal direction (perpendicular to the s axis). The symmetric component of loading (Fig. 10.20(b)) is applied to the substructure, with the magnitude of the 30-k concentrated load acting along the axis of symmetry reduced by half, as shown in Fig. 10.20(d). The reactions of the substructure are obtained by applying the equilibrium equations:

$$+\uparrow \sum F_y = 0 \qquad\qquad -15 - 18 + E_y = 0 \qquad E_y = 33 \text{ k} \uparrow$$

$$+\circlearrowleft \sum M_C = 0 \qquad -G_x(15) - 18(20) + 33(40) = 0 \qquad G_x = 64 \text{ k} \rightarrow$$

$$+\rightarrow \sum F_x = 0 \qquad\qquad -C_x + 64 = 0 \qquad C_x = 64 \text{ k} \leftarrow$$

The axial forces in the members of the substructure are determined by applying the method of joints. These member forces are also shown in Fig. 10.20(d).

The member axial forces in the left half of the truss can now be obtained by rotating the member forces in the right half (Fig. 10.20(d)) through 180° about the s axis, as shown in Fig. 10.20(e).

Member Forces Due to the Antisymmetric Loading Component The substructure with antisymmetric boundary conditions is shown in Fig. 10.20(f). Note that joints C and G, located at the axis of symmetry, are supported by rollers to prevent their deflections in the vertical direction. Also, member CG, which is located along the axis of symmetry, is removed from the substructure, as shown in the figure. (The force in member CG will be zero under antisymmetric loading.) The antisymmetric component of loading (Fig. 10.20(c)) is applied to the substructure, and its reactions and member axial forces are computed by applying the equilibrium equations and the method of joints (see Fig. 10.20(f)).

The member axial forces in the left half of the truss are then obtained by reflecting the negatives (i.e., the tensile forces are changed to compressive forces and vice versa) of the member forces in the right half to the left side of the axis of symmetry, as shown in Fig. 10.20(g).

Total Member Forces Finally, the total axial forces in members of the truss are obtained by superimposing the forces due to the symmetric and anti-symmetric components of the loading, as given in Fig. 10.20(e) and (g), respectively. These member forces are shown in Fig. 10.20(h).   **Ans.**

Example 10.10

Determine the member end forces of the frame shown in Fig. 10.21(a).

Solution

Symmetry The frame is symmetric with respect to the vertical *s* axis passing through the hinge at *D*, as shown in Fig. 10.21(b). The left half of the frame, *ACD*, will be used for analysis.

Symmetric and Antisymmetric Components of Loading See Fig. 10.21(b) and (c).

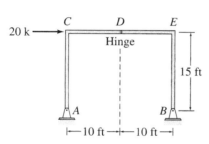

(a) Given Frame and Loading

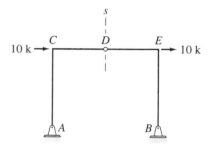

(c) Antisymmetric Loading Component

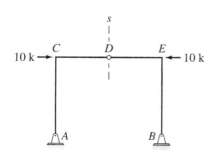

(b) Symmetric Loading Component

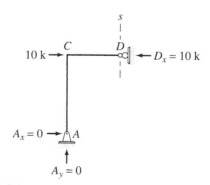

(d) Substructure with Symmetric Boundary Conditions

FIG. 10.21

continued

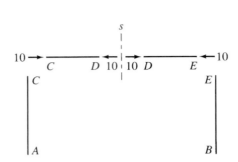

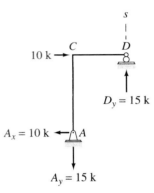

(e) Member Forces Due to Symmetric Loading Component

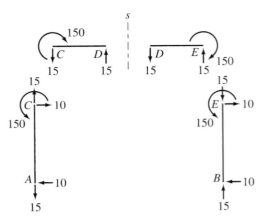

(g) Member Forces Due to Antisymmetric Loading Component

(f) Substructure with Antisymmetric Boundary Conditions

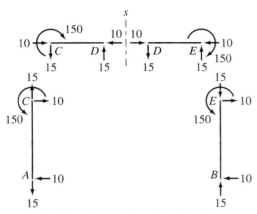

(h) Member Forces Due to Total Loading

FIG. 10.21 (contd.)

Member Forces Due to the Symmetric Loading Component The substructure with symmetric boundary conditions is shown in Fig. 10.21(d). The reactions and the member end forces of the substructure, as determined from equilibrium considerations, are shown in Fig. 10.21(d) and to the left of the s axis in Fig. 10.21(e), respectively. The member end forces to the right of the s axis are then obtained by reflection (see Fig. 10.21(e)).

Member Forces Due to the Antisymmetric Loading Component The substructure with antisymmetric boundary conditions is shown in Fig. 10.21(f). The member forces are determined by analyzing the substructure and by reflecting the negatives of the computed forces and moments about the axis of symmetry (see Fig. 10.21(g)).

Total Member Forces The total member end forces, obtained by superimposing the member forces due to the symmetric and antisymmetric components of the loading, are shown in Fig. 10.21(h).   **Ans.**

Example 10.11

Determine the substructures for the analysis of the symmetric and antisymmetric responses of the statically indeterminate beam shown in Fig. 10.22(a).

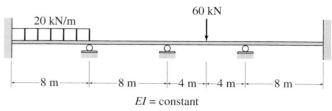

EI = constant

(a) Given Beam and Loading

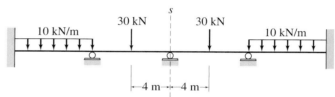

(b) Symmetric Loading Component

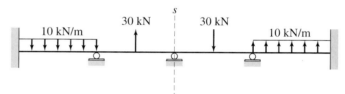

(c) Antisymmetric Loading Component

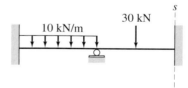

(d) Substructure for Analysis of Symmetric Response

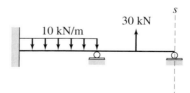

FIG. 10.22

(e) Substructure for Analysis of Antisymmetric Response

continued

Solution

Symmetry The beam is symmetric with respect to the vertical s axis shown in Fig. 10.22(b). The left half of the beam is selected for analysis.

Symmetric and Antisymmetric Components of Loading See Fig. 10.22(b) and (c).

Substructures The substructures for the analysis of the symmetric and anti-symmetric responses are shown in Fig. 10.22(d) and (e), respectively.   **Ans.**

Example 10.12

Determine the substructures for the analysis of the symmetric and antisymmetric responses of the statically indeterminate frame shown in Fig. 10.23(a).

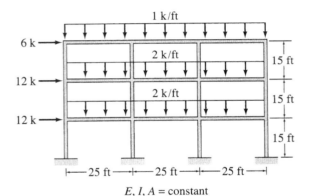

$E, I, A =$ constant

(a) Given Structure and Loading

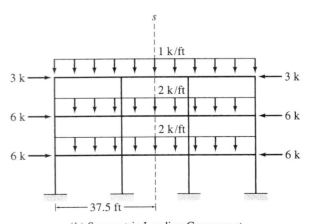

(b) Symmetric Loading Component

FIG. 10.23

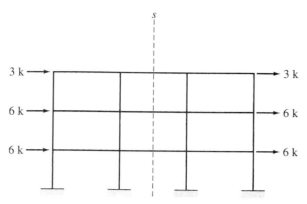

(c) Antisymmetric Loading Component

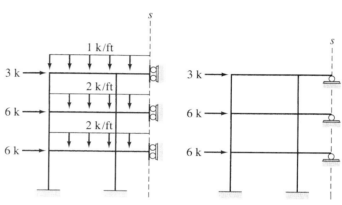

(d) Substructure for Analysis of Symmetric Response

(e) Substructure for Analysis of Antisymmetric Response

FIG. 10.23 (contd.)

Solution

Symmetry The frame is symmetric with respect to the vertical s axis shown in Fig. 10.23(b). The left half of the frame is selected for analysis.

Symmetric and Antisymmetric Components of Loading See Fig. 10.23(b) and (c).

Substructures The substructures for the analysis of the symmetric and antisymmetric responses are shown in Fig. 10.23(d) and (e), respectively.   **Ans.**

SUMMARY

In this chapter, we have learned that a plane structure is considered to be symmetric with respect to an axis in its plane if the reflection of the structure about the axis is identical in geometry, supports, and material properties to the structure itself.

A loading is considered to be symmetric with respect to an axis in its plane if the reflection of the loading about the axis is identical to the loading itself. A loading is considered to be antisymmetric with respect to an axis in its plane if the negative of the reflection of the loading about the axis is identical to the loading itself. Any general unsymmetrical loading can be decomposed into symmetric and antisymmetric components with respect to an axis.

When a symmetric structure is subjected to a loading that is symmetric with respect to the structure's axis of symmetry, the response of the structure is also symmetric. Thus we can obtain the response of the entire structure by analyzing a half of the structure, on either side of the axis of symmetry, with symmetric boundary conditions; and by reflecting the computed response about the axis of symmetry.

When a symmetric structure is subjected to a loading that is antisymmetric with respect to the structure's axis of symmetry, the response of the structure is also antisymmetric. Thus, the response of the entire structure can be obtained by analyzing a half of the structure, on either side of the axis of symmetry, with antisymmetric boundary conditions; and by reflecting the negative of the computed response about the axis of symmetry.

The response of a symmetric structure due to a general unsymmetric loading can be obtained by determining the responses of the structure due to the symmetric and antisymmetric components of the unsymmetric loading, and by superimposing the two responses.

PROBLEMS

Sections 10.1 and 10.2

10.1 through 10.15 Determine the symmetric and antisymmetric components of the loadings shown in Figs. P10.1–P10.15 with respect to the axis of symmetry of the structure.

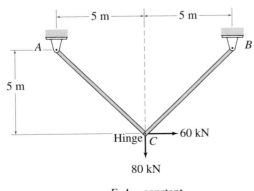

$E, A = $ constant

FIG. **P10.1, P10.16**

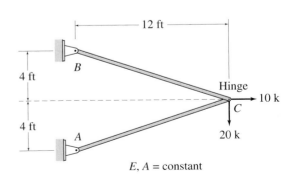

$E, A = $ constant

FIG. **P10.2 and P10.17**

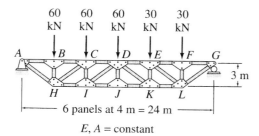

60 kN 60 kN 60 kN 30 kN 30 kN

6 panels at 4 m = 24 m

E, A = constant

FIG. **P10.3, P10.18**

Sections 10.3 and 10.4

10.16 through 10.20 Determine the force in each member of the trusses shown in Figs. P10.1–P10.5 by utilizing structural symmetry.

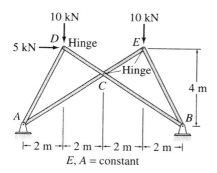

10 kN   10 kN

5 kN →

D Hinge   E

Hinge

C

4 m

A   B

⊢ 2 m ⊣ 2 m ⊣ 2 m ⊣ 2 m ⊣

E, A = constant

FIG. **P10.4, P10.19**

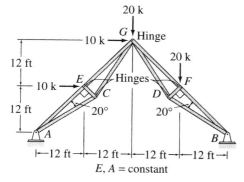

20 k

10 k →

G Hinge

20 k

12 ft

10 k → E   Hinges   F

12 ft

C   D

20°   20°

A   B

⊢12 ft⊣12 ft⊣12 ft⊣12 ft⊣

E, A = constant

FIG. **P10.5, P10.20**

10.21 through 10.23 Determine the member end forces of the frames shown in Figs. P10.6–P10.8 by utilizing structural symmetry.

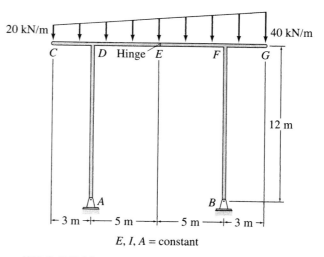

20 kN/m   40 kN/m

C   D Hinge E   F   G

12 m

A   B

⊢ 3 m ⊣ 5 m ⊢ 5 m ⊣ 3 m ⊣

E, I, A = constant

FIG. **P10.6, P10.21**

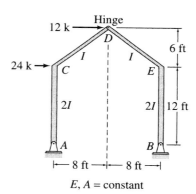

Hinge

12 k →

D

6 ft

24 k →

C   I   I   E

2I   2I   12 ft

A   B

⊢ 8 ft ⊣ 8 ft ⊣

E, A = constant

FIG. **P10.7, P10.22**

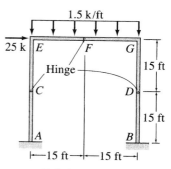

1.5 k/ft

25 k E   F   G

Hinge

15 ft

C   D

15 ft

A   B

⊢15 ft⊣15 ft⊣

E, I, A = constant

FIG. **P10.8, P10.23**

10.24 through 10.30 Determine the substructures for the analysis of the symmetric and antisymmetric responses of the structures shown in Figs. P10.9–P10.15.

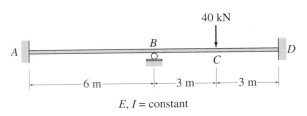

40 kN

E, I = constant

FIG. **P10.9 and P10.24**

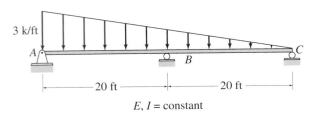

3 k/ft

20 ft

20 ft

E, I = constant

FIG. **P10.10 and P10.25**

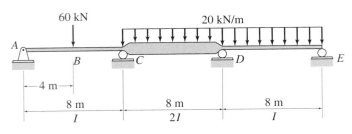

60 kN   20 kN/m

4 m

8 m   8 m   8 m

I $2I$ I

E = constant

FIG. **P10.11 and P10.26**

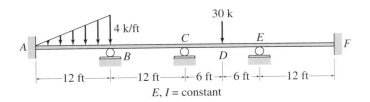

30 k

4 k/ft

12 ft   12 ft   6 ft   6 ft   12 ft

E, I = constant

FIG. **P10.12 and P10.27**

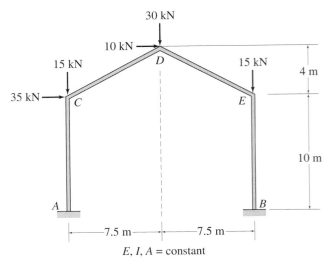

30 kN

10 kN

15 kN D 15 kN   4 m

35 kN C E

A B 10 m

7.5 m   7.5 m

E, I, A = constant

FIG. **P10.13 and P10.28**

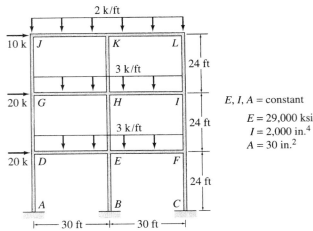

2 k/ft

10 k J K L 24 ft

3 k/ft

20 k G H I 24 ft E, I, A = constant

3 k/ft E = 29,000 ksi

20 k D E F I = 2,000 in.^{4}

A B C 24 ft A = 30 in.^{2}

30 ft   30 ft

FIG. **P10.14, P10.29**

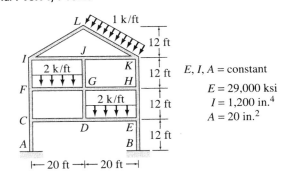

L 1 k/ft

12 ft

I J

2 k/ft K 12 ft E, I, A = constant

F G H E = 29,000 ksi

2 k/ft   12 ft I = 1,200 in.^{4}

C D E A = 20 in.^{2}

A B 12 ft

20 ft   20 ft

FIG. **P10.15, P10.30**

Part Three

Analysis of Statically
Indeterminate Structures

11

Introduction to Statically Indeterminate Structures

11.1 Advantages and Disadvantages of Indeterminate Structures
11.2 Analysis of Indeterminate Structures
Summary

Sydney Harbour, Australia
Photo courtesy of Digital Vision

In Part Two of this text, we considered the analysis of statically determinate structures. In this part (Chapters 11 through 18), we focus our attention on the analysis of statically indeterminate structures.

As discussed previously, the support reactions and internal forces of statically determinate structures can be determined from the equations of equilibrium (including equations of condition, if any). However, since indeterminate structures have more support reactions and/or members than required for static stability, the equilibrium equations alone are not sufficient for determining the reactions and internal forces of such structures, and must be supplemented by additional relationships based on the geometry of deformation of structures.

These additional relationships, which are termed the *compatibility conditions*, ensure that the continuity of the displacements is maintained throughout the structure and that the structure's various parts fit together. For example, at a rigid joint the deflections and rotations of all the members meeting at the joint must be the same. Thus the analysis of an indeterminate structure involves, in addition to the dimensions and arrangement of members of the structure, its cross-sectional and material properties (such as cross-sectional areas, moments of inertia, moduli of elasticity, etc.), which in turn, depend on the internal forces of the structure. The design of an indeterminate structure is, therefore, carried out in an iterative manner, whereby the (relative) sizes of the structural members are initially assumed and used to analyze the structure, and the internal forces thus obtained are used to revise the member sizes; if the revised member sizes are not close to those initially assumed, then the structure is reanalyzed using the latest member sizes. The iteration continues until the member sizes based on the results of an analysis are close to those assumed for that analysis.

Despite the foregoing difficulty in designing indeterminate structures, a great majority of structures being built today are statically indeterminate; for example, most modern reinforced concrete buildings are statically indeterminate. In this chapter, we discuss some of the important advantages and disadvantages of indeterminate structures as compared to determinate structures and introduce the fundamental concepts of the analysis of indeterminate structures.

11.1 ADVANTAGES AND DISADVANTAGES OF INDETERMINATE STRUCTURES

The advantages of statically indeterminate structures over determinate structures include the following.

1. Smaller Stresses The maximum stresses in statically indeterminate structures are generally lower than those in comparable determinate structures. Consider, for example, the statically determinate and indeterminate beams shown in Fig. 11.1(a) and (b), respectively. The bending moment diagrams for the beams due to a uniformly distributed load, w, are also shown in the figure. (The procedures for analyzing indeterminate beams are considered in subsequent chapters.) It can be seen from the figure that the maximum bending moment—and consequently the maximum bending stress—in the indeterminate beam is significantly lower than in the determinate beam.

2. Greater Stiffnesses Statically indeterminate structures generally have higher stiffnesses (i.e., smaller deformations), than those of comparable determinate structures. From Fig. 11.1, we observe that the maximum deflection of the indeterminate beam is only one-fifth that of the determinate beam.

3. Redundancies Statically indeterminate structures, if properly designed, have the capacity for redistributing loads when certain structural portions become overstressed or collapse in cases of overloads due to earthquakes, tornadoes, impact (e.g., gas explosions or vehicle impacts), and other such events. Indeterminate structures have more members and/or support reactions than required for static stability, so if a part (or member or support) of such a structure fails, the entire structure will

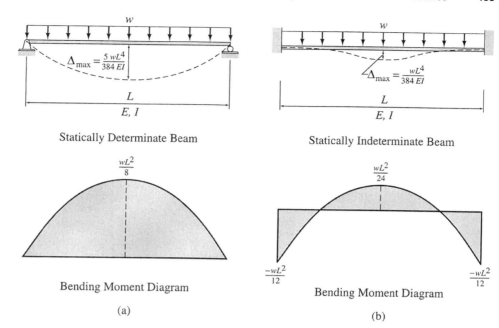

Statically Determinate Beam   Statically Indeterminate Beam

Bending Moment Diagram   Bending Moment Diagram

(a)   (b)

FIG. 11.1

not necessarily collapse, and the loads will be redistributed to the adjacent portions of the structure. Consider, for example, the statically determinate and indeterminate beams shown in Fig. 11.2(a) and (b), respectively. Suppose that the beams are supporting bridges over a waterway and that the middle pier, B, is destroyed when a barge accidentally

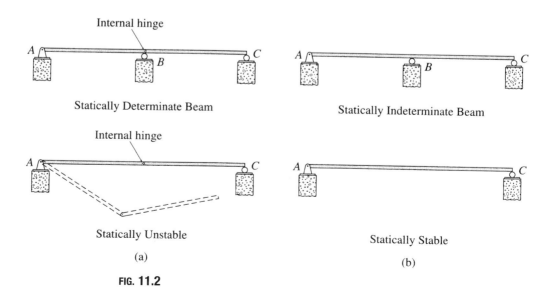

Statically Determinate Beam   Statically Indeterminate Beam

Statically Unstable   Statically Stable

(a)   (b)

FIG. 11.2

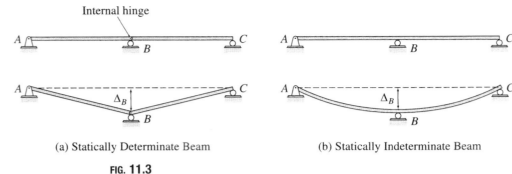

(a) Statically Determinate Beam   (b) Statically Indeterminate Beam

FIG. 11.3

rams into it. Because the statically determinate beam is supported by just the sufficient number of reactions required for static stability, the removal of support B will cause the entire structure to collapse, as shown in Fig. 11.2(a). However, the indeterminate beam (Fig. 11.2(b)) has one extra reaction in the vertical direction; therefore, the structure will not necessarily collapse and may remain stable, even after the support B has failed. Assuming that the beam has been designed to support dead loads only in case of such an accident, the bridge will be closed to traffic until pier B is repaired and then will be reopened.

The main disadvantages of statically indeterminate structures, over determinate structures, are the following.

1. Stresses Due to Support Settlements Support settlements do not cause any stresses in determinate structures; they may, however, induce significant stresses in indeterminate structures, which should be taken into account when designing indeterminate structures. Consider the determinate and indeterminate beams shown in Fig. 11.3. It can be seen from Fig. 11.3(a) that when the support B of the determinate beam undergoes a small settlement Δ_B, the portions AB and BC of the beam, which are connected together by an internal hinge at B, move as rigid bodies without bending—that is, they remain straight. Thus, no stresses develop in the determinate beam. However, when the continuous indeterminate beam of Fig. 11.3(b) is subjected to a similar support settlement, it bends, as shown in the figure; therefore, bending moments develop in the beam.

2. Stresses Due to Temperature Changes and Fabrication Errors Like support settlements, these effects do not cause stresses in determinate structures but may induce significant stresses in indeterminate ones. Consider the determinate and indeterminate beams shown in Fig. 11.4. It can be seen from Fig. 11.4(a) that when the determinate beam is subjected to a uniform temperature increase ΔT, it simply elongates, with the axial deformation given by $\delta = \alpha(\Delta T)L$ (Eq. 7.24). No stresses develop in the determinate beam, since it is free to elongate. However, when the indeterminate beam of Fig. 11.4(b), which is restrained

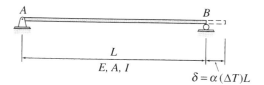

(a) Statically Determinate Beam

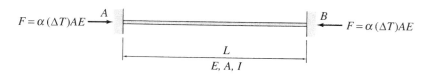

(b) Statically Indeterminate Beam

FIG. 11.4

from deforming axially by the fixed supports, is subjected to a similar temperature change, ΔT, a compressive axial force, $F = \delta(AE/L) = \alpha(\Delta T)AE$, develops in the beam, as shown in the figure. The effects of fabrication errors are similar to those of temperature changes on determinate and indeterminate structures.

11.2 ANALYSIS OF INDETERMINATE STRUCTURES

Fundamental Relationships

Regardless of whether a structure is statically determinate or indeterminate, its complete analysis requires the use of three types of relationships:

- Equilibrium equations
- Compatibility conditions
- Member force-deformation relations

The equilibrium equations relate the forces acting on the structure (or its parts), ensuring that the entire structure as well as its parts remain in equilibrium; the compatibility conditions relate the displacements of the structure so that its various parts fit together; and the member force-deformation relations, which involve the material and cross-sectional properties ($E, I,$ and A) of the members, provide the necessary link between the forces and displacements of the structure.

In the analysis of statically determinate structures, the equations of equilibrium are first used to obtain the reactions and the internal forces of the structure; then the member force-deformation relations and the compatibility conditions are employed to determine the structure's displacements. For example, consider the statically determinate truss

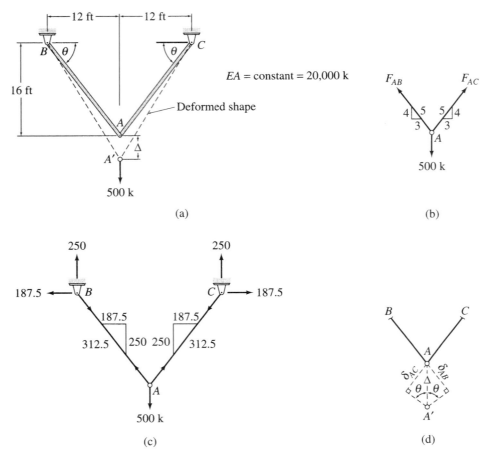

FIG. **11.5**

shown in Fig. 11.5(a). The axial forces in the truss members can be determined by considering the equilibrium of joint A (see Fig. 11.5(b)):

$$+ \rightarrow \sum F_x = 0 \qquad -0.6F_{AB} + 0.6F_{AC} = 0 \qquad F_{AB} = F_{AC}$$

$$+ \uparrow \sum F_y = 0 \qquad 2(0.8F_{AB}) - 500 = 0 \qquad F_{AB} = F_{AC} = 312.5 \text{ k (T)}$$

$$(11.1)$$

Similarly, the reactions at the supports B and C can be obtained by considering the equilibrium of joints B and C, respectively (Fig. 11.5(c)). To determine the displacement Δ of joint A of the truss, we first employ the member force-deformation relationship, $\delta = F(L/AE)$, to compute the member axial deformations:

$$\delta_{AB} = \delta_{AC} = 312.5\left(\frac{20}{20,000}\right) = 0.313 \text{ ft} \qquad (11.2)$$

Then these member axial deformations are related to the joint displacement Δ by using the compatibility condition (see Fig. 11.5(d)):

$$\delta_{AB} = \delta_{AC} = \Delta \sin \theta = 0.8\Delta \tag{11.3}$$

in which Δ is assumed to be small. Note that Eq. (11.3) states the compatibility requirement that the vertical displacements of the ends A of members AB and AC must be equal to the vertical displacement, Δ, of joint A. By substituting Eq. (11.2) into Eq. (11.3), we find the displacement of joint A to be

$$\Delta = \frac{0.313}{0.8} = 0.391 \text{ ft} = 4.69 \text{ in.} \tag{11.4}$$

The displacement Δ could also have been computed by employing the *virtual work method* formulated in Chapter 7, which automatically satisfies the member force-deformation relations and the necessary compatibility conditions.

Indeterminate Structures

In the analysis of statically indeterminate structures, the equilibrium equations alone are not sufficient for determining the reactions and internal forces. Therefore, it becomes necessary to solve the equilibrium equations in conjunction with the compatibility conditions of the structure to determine its response. Because the equilibrium equations contain the unknown forces, whereas the compatibility conditions involve displacements as the unknowns, the member force-deformation relations are utilized to express either the unknown forces in terms of the unknown displacements or vice versa. The resulting system of equations containing only one type of unknowns is then solved for the unknown forces or displacements, which are then substituted into the fundamental relationships to determine the remaining response characteristics of the structure.

Consider, for example, the indeterminate truss shown in Fig. 11.6(a). The truss is obtained by adding a vertical member AD to the determinate truss of Fig. 11.5(a), considered previously. The free-body diagram of joint A of the truss is shown in Fig. 11.6(b). The equations of equilibrium for this joint are given by

$$+ \rightarrow \textstyle\sum F_x = 0 \qquad\qquad F_{AB} = F_{AC} \tag{11.5}$$

$$+ \uparrow \textstyle\sum F_y = 0 \qquad 1.6F_{AB} + F_{AD} = 500 \tag{11.6}$$

Note that the two equilibrium equations are not sufficient for determining the three unknown member axial forces. The compatibility conditions are based on the requirement that the vertical displacements of the ends A of the three members connected to joint A must be equal to the

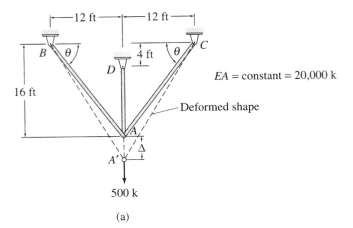

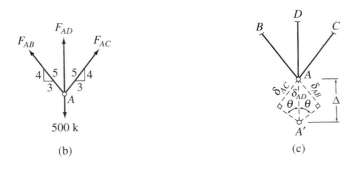

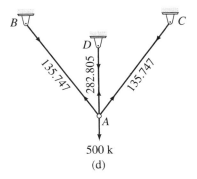

FIG. 11.6

vertical displacement Δ of joint A. The displacement diagram of joint A is shown in Fig. 11.6(c). Assuming the displacement Δ to be small, we write the compatibility conditions as

$$\delta_{AB} = \delta_{AC} = \Delta \sin \theta = 0.8\Delta \qquad (11.7)$$

$$\delta_{AD} = \Delta \qquad (11.8)$$

By substituting Eq. (11.8) into Eq. (11.7), we obtain the desired relationship between the member axial deformations:

$$\delta_{AB} = \delta_{AC} = 0.8\delta_{AD} \qquad (11.9)$$

which indicates that the axial deformations of the inclined members AB and AC are equal to 0.8 times the axial deformation of the vertical member AD. To express Eq. (11.9) in terms of member axial forces, we utilize the member force-deformation relations:

$$\delta_{AB} = F_{AB}\left(\frac{L_{AB}}{EA}\right) = F_{AB}\left(\frac{20}{20,000}\right) = 0.001F_{AB} \qquad (11.10)$$

$$\delta_{AC} = F_{AC}\left(\frac{L_{AC}}{EA}\right) = F_{AC}\left(\frac{20}{20,000}\right) = 0.001F_{AC} \qquad (11.11)$$

$$\delta_{AD} = F_{AD}\left(\frac{L_{AD}}{EA}\right) = F_{AD}\left(\frac{12}{20,000}\right) = 0.0006F_{AD} \qquad (11.12)$$

Substitution of Eqs. (11.10) through (11.12) into Eq. (11.9) yields

$$0.001F_{AB} = 0.001F_{AC} = 0.8(0.0006F_{AD})$$

or

$$F_{AB} = F_{AC} = 0.48F_{AD} \qquad (11.13)$$

Now, we can determine the axial forces in the three members of the truss by solving Eq. (11.13) simultaneously with the two equilibrium equations (Eqs. (11.5) and (11.6)). Thus (Fig. 11.6(d)),

$$F_{AB} = F_{AC} = 135.747 \text{ k (T)} \quad \text{and} \quad F_{AD} = 282.805 \text{ k (T)}$$

The member axial deformations can now be computed by substituting these values of member axial forces into the member force-deformation relations (Eqs. (11.10) through (11.12)) to obtain

$$\delta_{AB} = \delta_{AC} = 0.136 \text{ ft} = 1.629 \text{ in.} \quad \text{and} \quad \delta_{AD} = 0.17 \text{ ft} = 2.036 \text{ in.}$$

Finally, by substituting the values of member axial deformations into the compatibility conditions (Eqs. (11.7) and (11.8)), we determine the displacement of joint A as

$$\Delta = 0.17 \text{ ft} = 2.036 \text{ in.}$$

Methods of Analysis

Since the mid-1800s, many methods have been developed for analyzing statically indeterminate structures. These methods can be broadly classified into two categories, namely, the *force (flexibility) methods* and the *displacement (stiffness) methods*, depending on the type of unknowns (forces or displacements, respectively), involved in the solution of the governing equations. The force methods, which are presented in Chapters 13 and 14, are generally convenient for analyzing small structures with a few redundants (i.e., fewer excess members and/or reactions than required for static stability). These methods are also used to derive the member force-deformation relations needed to develop the displacement methods. The displacement methods are considered in Chapters 16 through 18. These methods are more systematic, can be easily implemented on computers, and are, therefore, preferred for the analysis of large and highly redundant structures.

SUMMARY

In this chapter we have learned that the advantages of statically indeterminate structures over determinate structures include smaller maximum stresses, greater stiffnesses, and redundancies. Support settlements, temperature changes, and fabrication errors may induce significant stresses in indeterminate structures, which should be taken into account when designing such structures.

The analysis of structures involves the use of three fundamental relationships: equilibrium equations, compatibility conditions, and member force-deformation relations. In the analysis of indeterminate structures, the equilibrium equations must be supplemented by the compatibility conditions based on the geometry of the deformation of the structure. The link between the equilibrium equations and the compatibility conditions is established by means of the member force-deformation relations of the structure.

The methods for the analysis of indeterminate structures can be classified into two categories, namely, the *force (flexibility) methods* and the *displacement (stiffness) methods*.

St. Louis Gateway Arch and Old Courthouse
Photo courtesy of Jefferson National Expansion Memorial, National Park Service

12

Approximate Analysis of Rectangular Building Frames

12.1 Assumptions for Approximate Analysis
12.2 Analysis for Vertical Loads
12.3 Analysis for Lateral Loads—Portal Method
12.4 Analysis for Lateral Loads—Cantilever Method
 Summary
 Problems

The analysis of statically indeterminate structures using the force and displacement methods introduced in the preceding chapter can be considered as *exact* in the sense that the compatibility and equilibrium conditions of the structure are exactly satisfied in such an analysis. However, the results of such an exact analysis represent the actual structural response only to the extent that the analytical model of the structure represents the actual structure. Experimental results have demonstrated that the response of most common types of structures under service loads can be reliably predicted by the force and displacement methods, provided an accurate analytical model of the structure is used in the analysis.

Exact analysis of indeterminate structures involves computation of deflections and solution of simultaneous equations, so it can be quite time consuming. Moreover, such an analysis depends on the relative sizes (cross-sectional areas and/or moments of inertia) of the members of the structure. Because of these difficulties associated with the exact analysis, the preliminary designs of indeterminate structures are often based on the results of *approximate analysis*, in which the internal forces are estimated by making certain assumptions about the deformations and/or the distribution of forces between the members of structures, thereby avoiding the necessity of computing deflections.

Approximate analysis proves to be quite convenient to use in the planning phase of projects, when several alternative designs of the structure are usually evaluated for relative economy. The results of approximate analysis can also be used to estimate the sizes of various structural members needed to initiate the exact analysis. The preliminary designs of members are then revised iteratively, using the results of successive exact analyses, to arrive at their final designs. Furthermore, approximate analysis is sometimes used to roughly check the results of exact analysis, which due to its complexity can be prone to errors. Finally, in recent years, there has been an increased tendency toward renovating and retrofitting older structures. Many such structures constructed prior to 1960, including many high-rise buildings, were designed solely on the basis of approximate analysis, so a knowledge and understanding of approximate methods used by the original designers is usually helpful in a renovation undertaking.

Unlike the exact methods, which are general in the sense that they can be applied to various types of structures subjected to various loading conditions, a specific method is usually required for the approximate analysis of a particular type of structure for a particular loading. For example, a different approximate method must be employed for the analysis of a rectangular frame under vertical (gravity) loads than for the analysis of the same frame subjected to lateral loads. Numerous methods have been developed for approximate analysis of indeterminate structures. Some of the more common approximate methods pertaining to rectangular frames are presented in this chapter. These methods can be expected to yield results within 20% of the exact solutions.

The objectives of this chapter are to consider the approximate analysis of rectangular building frames as well as to gain an understanding of the techniques used in the approximate analysis of structures in general. We present a general discussion of the simplifying assumptions necessary for approximate analysis and then consider the approximate analysis of rectangular frames under vertical (gravity) loads. Finally, we present the two common methods used for the approximate analysis of rectangular frames subjected to lateral loads.

12.1 ASSUMPTIONS FOR APPROXIMATE ANALYSIS

As discussed in Chapters 3 through 5, statically indeterminate structures have more support reactions and/or members than required for static stability; therefore, all the reactions and internal forces (including any moments) of such structures cannot be determined from the equations of equilibrium. The excess reactions and internal forces of an indeterminate structure are referred to as *redundants*, and the number of redundants (i.e., the difference between the total number of unknowns and the number of equilibrium equations) is termed the *degree of indetermi-*

nacy of the structure. Thus, in order to determine the reactions and internal forces of an indeterminate structure, the equilibrium equations must be supplemented by additional equations, whose number must equal the degree of indeterminacy of the structure. In an approximate analysis, these additional equations are established by using engineering judgment to make simplifying assumptions about the response of the structure. The total number of assumptions must be equal to the degree of indeterminacy of the structure, with each assumption providing an independent relationship between the unknown reactions and/or internal forces. The equations based on the simplifying assumptions are then solved in conjunction with the equilibrium equations of the structure to determine the approximate values of its reactions and internal forces.

Two types of assumptions are commonly employed in approximate analysis.

Assumptions about the Location of Points of Inflection

In the first approach, a qualitative deflected shape of the indeterminate structure is sketched and used to assume the location of the points of inflection—that is, the points where the curvature of the elastic curve changes signs, or becomes zero. Since the bending moments must be zero at the points of inflection, internal hinges are inserted in the indeterminate structure at the assumed locations of inflection points to obtain a simplified determinate structure. Each of the internal hinges provides one equation of condition, so the number of inflection points assumed should be equal to the degree of indeterminacy of the structure. Moreover, the inflection points should be selected such that the resulting determinate structure must be statically and geometrically stable. The simplified determinate structure thus obtained is then analyzed to determine the approximate values of the reactions and internal forces of the original indeterminate structure.

Consider, for example, a portal frame subjected to a lateral load P, as shown in Fig. 12.1(a). As the frame is supported by four reaction components and since there are only three equilibrium equations, it is statically indeterminate to the first degree. Therefore, we need to make one simplifying assumption about the response of the frame. By examining the deflected shape of the frame sketched in Fig. 12.1(a), we observe that an inflection point exists near the middle of the girder CD. Although the exact location of the inflection point depends on the (yet unknown) properties of the two columns of the frame and can be determined only from an exact analysis, for the purpose of approximate analysis we can *assume* that the inflection point is located at the midpoint of the girder CD. Since the bending moment at an inflection point must be zero, we insert an internal hinge at the midpoint E of girder CD to obtain the determinate frame shown in Fig. 12.1(b). The four reactions of the frame can now be determined by applying the three

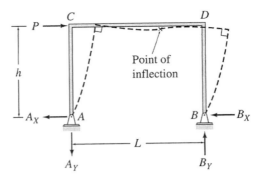

(a) Indeterminate Frame

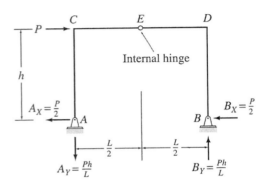

(b) Simplified Determinate Frame

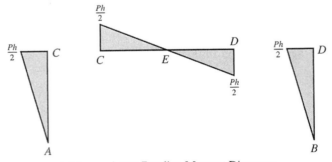

(c) Approximate Bending Moment Diagrams

FIG. **12.1**

equilibrium equations, $\sum F_X = 0$, $\sum F_Y = 0$, and $\sum M = 0$, and one equation of condition, $\sum M_E^{AE} = 0$ or $\sum M_E^{BE} = 0$, to the determinate frame (Fig. 12.1(b)):

$$+ \zeta \sum M_B = 0 \qquad\qquad A_Y(L) - Ph = 0 \qquad A_Y = \frac{Ph}{L} \downarrow$$

$$+ \uparrow \sum F_Y = 0 \qquad\qquad -\frac{Ph}{L} + B_Y = 0 \qquad B_Y = \frac{Ph}{L} \uparrow$$

$$+ \zeta \sum M_E^{BE} = 0 \qquad \frac{Ph}{L}\left(\frac{L}{2}\right) - B_X(h) = 0 \qquad B_X = \frac{P}{2} \leftarrow$$

$$+ \rightarrow \sum F_X = 0 \qquad P - A_X - \frac{P}{2} = 0 \qquad A_X = \frac{P}{2} \leftarrow$$

By using these approximate reactions, the approximate shear, bending moment, and axial force diagrams for the frame can be constructed by considering the equilibrium of its members and joints. The bending moment diagrams for the members of the frame are shown in Fig. 12.1(c).

Assumptions about Distribution of Forces among Members and/or Reactions

Approximate analysis of indeterminate structures is sometimes performed by making assumptions about the distribution of forces among the members and/or reactions of the structures. The number of such assumptions required for the analysis of a structure is equal to the degree of indeterminacy of the structure, with each assumption providing an independent equation relating the unknown member forces and/or reactions. The equations based on these assumptions are then solved simultaneously with the equilibrium equations of the structure to determine its approximate reactions and internal forces. For example, the portal frame of Fig. 12.1(a) can alternatively be analyzed by assuming that the horizontal reactions A_X and B_X are equal; that is, $A_X = B_X$. By solving this equation simultaneously with the three equilibrium equations of the frame, we obtain the same reactions as previously determined by assuming an inflection point at the midpoint of the girder CD of the frame.

The two types of assumptions described in this section can either be used individually or they can be combined with each other and/or with other types of assumptions based on the engineering judgment of the structural response to develop methods for approximate analysis of various types of structures. In the rest of this chapter, we focus our attention on the approximate analysis of rectangular building frames.

12.2 ANALYSIS FOR VERTICAL LOADS

Recall from Section 5.5 that the degree of indeterminacy of a rectangular building frame with fixed supports is equal to *three times the number of girders* in the frame provided that the frame does not contain any internal hinges or rollers. Thus, in an approximate analysis of such a rigid frame, the total number of assumptions required is equal to three times the number of girders in the frame.

A commonly used procedure for approximate analysis of rectangular building frames subjected to vertical (gravity) loads involves making three assumptions about the behavior of each girder of the frame. Consider a frame subjected to uniformly distributed loads w, as shown in Fig. 12.2(a). The free-body diagram of a typical girder DE of the frame is shown in Fig. 12.2(b). From the deflected shape of the girder sketched in the figure, we observe that two inflection points exist near both ends of the girder. These inflection points develop because the columns and the adjacent girder connected to the ends of girder DE offer partial restraint or resistance against rotation by exerting negative moments M_{DE} and M_{ED} at the girder ends D and E, respectively. Although the exact location of the inflection points depends on the relative stiffnesses of the frame members and can be determined only from an exact analysis, we can establish the regions along the girder in which these points are located by examining the two extreme conditions of rotational restraint at the girder ends shown in Fig. 12.2(c) and (d). If the girder ends were free to rotate, as in the case of a simply supported girder (Fig. 12.2(c)), the zero bending moments—and thus the inflection points—would occur at the ends. On the other extreme, if the girder ends were completely fixed against rotation, we can show by the exact analysis presented in subsequent chapters that the inflection points would occur at a distance of $0.211L$ from each end of the girder, as illustrated in Fig. 12.2(d). Therefore, when the girder ends are only partially restrained against rotation (Fig. 12.2(b)), the inflection points must occur somewhere within a distance of $0.211L$ from each end. For the purpose of approximate analysis, it is common practice to assume that the inflection points are located about halfway between the two extremes—that is, at a distance of $0.1L$ from each end of the girder. Estimating the location of two inflection points involves making two assumptions about the behavior of the girder. The third assumption is based on the experience gained from the exact analyses of rectangular frames subjected to vertical loads only, which indicates that the axial forces in girders of such frames are usually very small. Thus, in an approximate analysis, it is reasonable to assume that the girder axial forces are zero.

To summarize the foregoing discussion, in the approximate analysis of a rectangular frame subjected to vertical loads the following assumptions are made for each girder of the frame:

1. The inflection points are located at one-tenth of the span from each end of the girder.
2. The girder axial force is zero.

The effect of these simplifying assumptions is that the middle eight-tenths of the span $(0.8L)$ of each girder can be considered to be simply supported on the two end portions of the girder, each of which is of the length equal to one-tenth of the girder span $(0.1L)$, as shown in Fig. 12.2(e). Note that the girders are now statically determinate, and their end forces and moments can be determined from statics, as shown in the

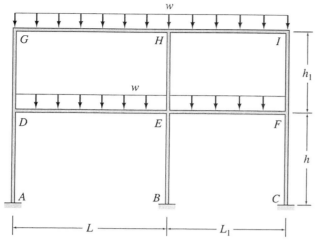

(a) Building Frame

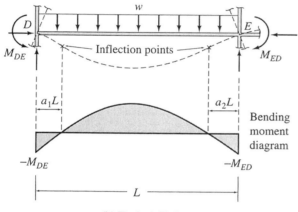

(b) Typical Girder

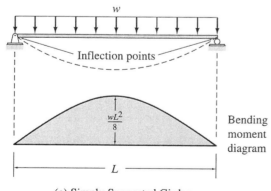

(c) Simply Supported Girder

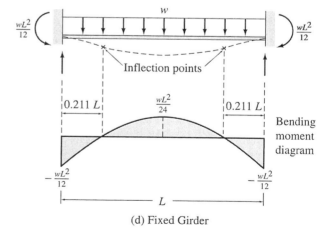

(d) Fixed Girder

FIG. 12.2

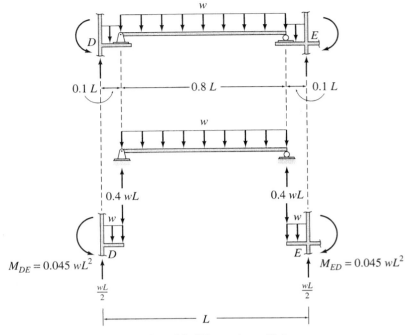

(e) Simplified Determinate Girder

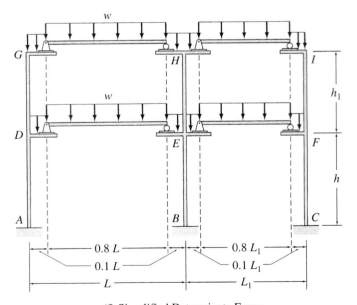

FIG. **12.2** (contd.)   (f) Simplified Determinate Frame

figure. It should be realized that by making three assumptions about the behavior of each girder of the frame, we have made a total number of assumptions equal to the degree of indeterminacy of the frame, thereby rendering the entire frame statically determinate, as shown in Fig. 12.2(f). Once the girder end forces have been computed, the end forces of the columns and the support reactions can be determined from equilibrium considerations.

Example 12.1

Draw the approximate shear and bending moment diagrams for the girders of the frame shown in Fig. 12.3(a).

Solution

As the span lengths and loads for the four girders of the frame are the same (Fig. 12.3(a)), the approximate shear and bending moment diagrams for the girders will also be the same. By applying the assumptions discussed in this section to any of the girders of the frame, we obtain the statically determinate girder shown in Fig. 12.3(b). Note that the middle portion of the girder, which has a length of $0.8L = 0.8(30) = 24$ ft, is simply supported on the two end portions, each of length $0.1L = 0.1(30) = 3$ ft.

By considering the equilibrium of the simply supported middle portion of the girder, we obtain the vertical reactions at the ends of this portion to be $1.5(24/2) = 18$ k. These forces are then applied in opposite directions (Newton's law of action and reaction) to the two end portions, as shown in the figure. The vertical forces (shears) and moments at the ends of the girder can now be determined by considering the equilibrium of the end portions. By applying the equations of equilibrium to the left end portion, we write

$$+\uparrow\sum F_Y = 0 \qquad\qquad S_L - 1.5(3) - 18 = 0 \qquad S_L = 22.5 \text{ k} \uparrow$$

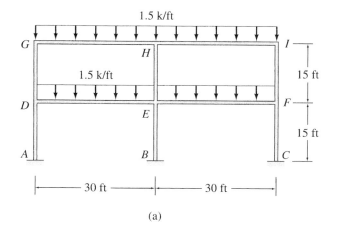

1.5 k/ft

30 ft

30 ft

(a)

FIG. **12.3**

continued

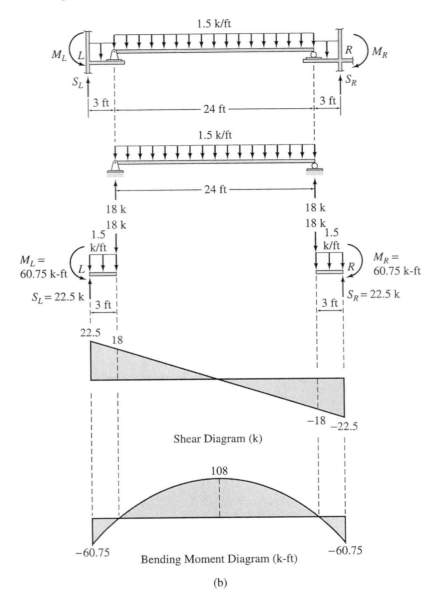

FIG. **12.3** (contd.)

(b)

$$+\zeta\sum M_L = 0 \qquad M_L - 1.5(3)\left(\frac{3}{2}\right) - 18(3) = 0 \qquad M_L = 60.75 \text{ k-ft } \zeta$$

Similarly, by applying the equilibrium equations to the right end portion, we obtain

$$S_R = 22.5 \text{ k} \uparrow \quad \text{and} \quad M_R = 60.75 \text{ k-ft } \zeta$$

By using these approximate values of the girder end forces and moments, we construct the shear and bending moment diagrams for the girder, as shown in Fig. 12.3(b).

Ans.

12.3 ANALYSIS FOR LATERAL LOADS—PORTAL METHOD

The behavior of rectangular building frames is different under lateral (horizontal) loads than under vertical loads, so different assumptions must be used in the approximate analysis for lateral loads than were used in the case of vertical loads considered previously. Two methods are commonly used for approximate analysis of rectangular frames subjected to lateral loads. These are (1) *the portal method* and (2) *the cantilever method*. The portal method is described in this section, whereas the cantilever method is considered in the following section.

The portal method was initially developed by A. Smith in 1915 and is generally considered to be appropriate for the approximate analysis of relatively low building frames. Before we consider the analysis of multistory, multibay frames by using the portal method, let us examine the behavior of a portal frame with fixed supports under a lateral load, as shown in Fig. 12.4(a). The degree of indeterminacy of this frame is three; therefore, three assumptions must be made for its approximate

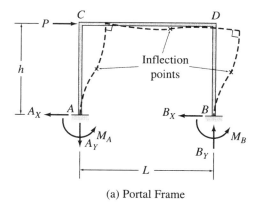

(a) Portal Frame

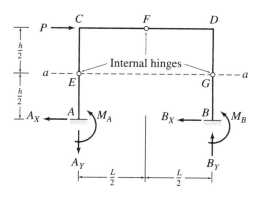

FIG. **12.4**

(b) Simplified Determinate Frame

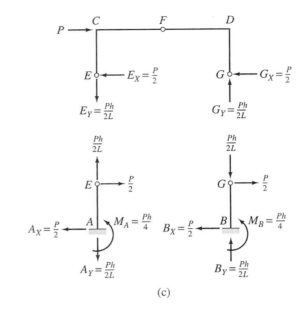

(c)

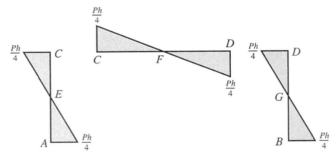

FIG. 12.4 (contd.)   (d) Approximate Bending Moment Diagrams

analysis. From the deflected shape of the frame sketched in Fig. 12.4(a), we observe that an inflection point exists near the middle of each member of the frame. Thus, in approximate analysis, it is reasonable to assume that the inflection points are located at the midpoints of the frame members. Since the bending moments at the inflection points must be zero, internal hinges are inserted at the midpoints of the three frame members to obtain the statically determinate frame shown in Fig. 12.4(b). To determine the six reactions, we pass a horizontal section aa through the hinges E and G, as shown in Fig. 12.4(b), and apply the equations of equilibrium (and condition, if any) to the three portions of the frame. Applying the three equilibrium equations and one equation of condition to the portion $ECDG$ (Fig. 12.4(c)), we compute the forces at the internal hinges E and G to be

$$+\circlearrowleft \sum M_G = 0 \qquad E_Y(L) - P\left(\frac{h}{2}\right) = 0 \qquad E_Y = \frac{Ph}{2L} \downarrow$$

$$+\uparrow \sum F_Y = 0 \qquad -\frac{Ph}{2L} + G_Y = 0 \qquad G_Y = \frac{Ph}{2L} \uparrow$$

$$+\circlearrowleft \sum M_F^{EF} = 0 \qquad \frac{Ph}{2L}\left(\frac{L}{2}\right) - E_X\left(\frac{h}{2}\right) = 0 \qquad E_X = \frac{P}{2} \leftarrow$$

$$+\rightarrow \sum F_X = 0 \qquad P - \frac{P}{2} - G_X = 0 \qquad G_X = \frac{P}{2} \leftarrow$$

The reactions at supports A and B can now be determined by considering the equilibrium of portions AE and BG, respectively. For portion AE (Fig. 12.4(c)):

$$+\rightarrow \sum F_X = 0 \qquad\qquad A_X = \frac{P}{2} \leftarrow$$

$$+\uparrow \sum F_Y = 0 \qquad\qquad A_Y = \frac{Ph}{2L} \downarrow$$

$$+\circlearrowleft \sum M_A = 0 \qquad -\frac{P}{2}\left(\frac{h}{2}\right) + M_A = 0 \qquad M_A = \frac{Ph}{4} \circlearrowright$$

Similarly, for portion BG (Fig. 12.4(c)):

$$+\rightarrow \sum F_X = 0 \qquad\qquad B_X = \frac{P}{2} \leftarrow$$

$$+\uparrow \sum F_Y = 0 \qquad\qquad B_Y = \frac{Ph}{2L} \uparrow$$

$$+\circlearrowleft \sum M_B = 0 \qquad -\frac{P}{2}\left(\frac{h}{2}\right) + M_B = 0 \qquad M_B = \frac{Ph}{4} \circlearrowright$$

Note that the horizontal reactions at the supports A and B are equal (i.e., $A_X = B_X$), indicating that the shears in the two columns of the frame must also be equal to each other. The bending moment diagrams for the members of the portal frame are shown in Fig. 12.4(d).

To develop the portal method for approximate analysis of frames, consider the two-story, three-bay building frame shown in Fig. 12.5(a). The frame contains six girders, so its degree of indeterminacy is $3(6) = 18$. From the deflected shape of the frame sketched in Fig. 12.5(a), we observe that the deflection behavior of this frame is similar to that of the portal frame considered previously (Fig. 12.4(a)), in the sense that an inflection point exists near the middle of each member of the frame. In the portal method, it is assumed that these inflection points are located at the midpoints of the members, and, therefore, an internal hinge is inserted at the middle of each of the frame members to obtain a simplified frame, as shown in Fig. 12.5(b). Note that this simplified frame is not

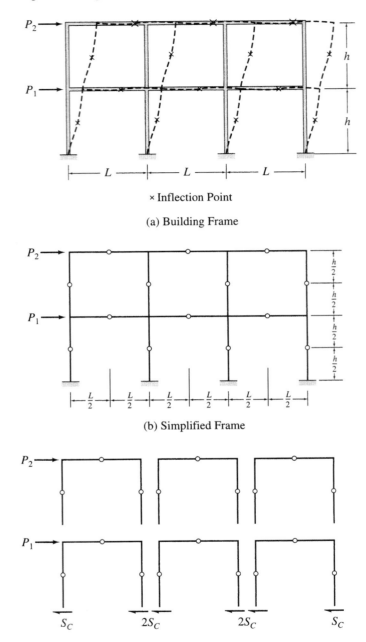

× Inflection Point

(a) Building Frame

(b) Simplified Frame

(c) Equivalent Series of Portal Frames

FIG. 12.5

statically determinate because it is obtained by inserting only 14 internal hinges (i.e., one hinge in each of the 14 members) into the original frame, which is indeterminate to the 18th degree. Thus, the degree of indeterminacy of the simplified frame of Fig. 12.5(b) is $18 - 14 = 4$;

therefore, four additional assumptions must be made before an approximate analysis involving only statics can be carried out. In the portal method, it is further assumed that the frame is composed of a series of portal frames, as shown in Fig. 12.5(c), with each interior column of the original multibay frame representing two portal legs. We showed previously (Fig. 12.4) that when a portal frame with internal hinges at the midpoints of its members is subjected to a lateral load, equal shears develop in the two legs of the portal. Since an interior column of the original multibay frame represents two portal legs, whereas an exterior column represents only one leg, we can reasonably assume that the shear in an interior column of a story of the multibay frame is twice as much as the shear in an exterior column of that story (Fig. 12.5(c)). The foregoing assumption regarding shear distribution between columns yields one more equation for each story of the frame with multiple bays than necessary for approximate analysis. For example, for each story of the frame of Fig. 12.5, this assumption can be used to express shears in any three of the columns in terms of that in the fourth. Thus, for the entire frame, this assumption provides a total of six equations—that is, two equations more than necessary for approximate analysis. However, as the extra equations are consistent with the rest, they do not cause any computational difficulty in the analysis.

From the foregoing discussion, we gather that the assumptions made in the portal method are as follows:

1. An inflection point is located at the middle of each member of the frame.
2. On each story of the frame, interior columns carry twice as much shear as exterior columns.

Procedure for Analysis

The following step-by-step procedure can be used for the approximate analysis of building frames by the portal method.

1. Draw a sketch of the simplified frame obtained by inserting an internal hinge at the midpoint of each member of the given frame.
2. Determine column shears. For each story of the frame:
    a. Pass a horizontal section through all the columns of the story, cutting the frame into two portions.
    b. Assuming that the shears in interior columns are twice as much as in exterior columns, determine the column shears by applying the equation of horizontal equilibrium ($\sum F_X = 0$) to the free body of the upper portion of the frame.
3. Draw free-body diagrams of all the members and joints of the frame, showing the external loads and the column end shears computed in the previous step.
4. Determine column moments. Determine moments at the ends of each column by applying the equations of condition that the

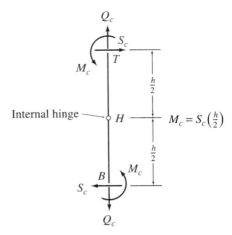

(a) Column End Forces and Moments

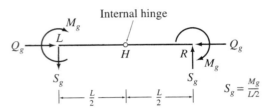

(b) Girder End Forces and Moments

FIG. 12.6

bending moment is zero at the column midheight, where an inflection point (internal hinge) has been assumed. As shown in Fig. 12.6(a), by applying the equations of condition, $\sum M_H^{BH} = 0$ and $\sum M_H^{TH} = 0$, to the free body of a column of height h, we find that the moments at the two ends of the column are equal in magnitude and have the same sense (i.e., either both end moments are clockwise or both are counterclockwise). The magnitude of the column end moments (M_C) is equal to the magnitude of the column shears (S_C) times half the column height; that is,

$$M_C = S_C\left(\frac{h}{2}\right) \tag{12.1}$$

Determine end moments for all the columns of the frame.

5. Determine girder axial forces, moments, and shears. Proceeding from the top story of the frame to the bottom, compute axial forces, moments, and shears at the ends of the girders of each successive story by starting at the far left joint of the story and working across to the right, as follows:

a. Apply the equilibrium equations, $\sum F_X = 0$ and $\sum M = 0$, to the free body of the joint under consideration to compute the axial force and moment, respectively, at the left (adjoining) end of the girder on the right side of the joint.

b. Considering the free body of the girder, determine the shear at the girder's left end by dividing the girder moment by half the girder length (see Fig. 12.6(b)); that is,

$$S_g = \frac{M_g}{(L/2)} \tag{12.2}$$

Equation (12.2) is based on the condition that the bending moment at the girder midpoint is zero.

c. By applying the equilibrium equations $\sum F_X = 0$, $\sum F_Y = 0$, and $\sum M = 0$ to the free body of the girder, determine the axial force, shear, and moment, respectively, at the right end. As shown in Fig. 12.6(b), the axial forces and shears at the ends of the girder must be equal but opposite, whereas the two end moments must be equal to each other in both magnitude and direction.

d. Select the joint to the right of the girder considered previously, and repeat steps 5(a) through 5(c) until the axial forces, moments, and shears in all the girders of the story have been determined. The equilibrium equations $\sum F_X = 0$ and $\sum M = 0$ for the right end joint have not been utilized so far, so these equations can be used to check the calculations.

e. Starting at the far left joint of the story below the one considered previously, repeat steps 5(a) through 5(d) until the axial forces, moments, and shears in all of the girders of the frame have been determined.

6. Determine column axial forces. Starting at the top story, apply the equilibrium equation $\sum F_Y = 0$ successively to the free body of each joint to determine the axial forces in the columns of the story. Repeat the procedure for each successive story, working from top to bottom, until the axial forces in all the columns of the frame have been determined.

7. Realizing that the forces and moments at the lower ends of the bottom-story columns represent the support reactions, use the three equilibrium equations of the entire frame to check the calculations. If the analysis has been performed correctly, then these equilibrium equations must be satisfied.

In steps 5 and 6 of the foregoing procedure, if we wish to compute member forces and moments by proceeding from the right end of the story toward the left, then the term *left* should be replaced by *right* and vice versa.

Example 12.2

Determine the approximate axial forces, shears, and moments for all the members of the frame shown in Fig. 12.7(a) by using the portal method.

Solution

Simplified Frame The simplified frame for approximate analysis is obtained by inserting internal hinges at the midpoints of all the members of the given frame, as shown in Fig. 12.7(b).

Column Shears To compute shears in the columns of the frame, we pass an imaginary section aa through the columns just above the support level, as shown in Fig. 12.7(b). The free-body diagram of the portion of the frame above section aa is shown in Fig. 12.7(c). Note that the shear in the interior column BE has been assumed to be twice as much as in the exterior columns AD and CF. By applying the equilibrium equation $\sum F_X = 0$, we obtain (see Fig. 12.7(c))

$$+ \rightarrow \sum F_X = 0 \qquad 60 - S - 2S - S = 0 \qquad S = 15 \text{ kN}$$

Thus, the shear forces at the lower ends of the columns are

$$S_{AD} = S_{CF} = S = 15 \text{ kN} \leftarrow \qquad S_{BE} = 2S = 30 \text{ kN} \leftarrow$$

Shear forces at the upper ends of the columns are obtained by applying the equilibrium equation $\sum F_X = 0$ to the free body of each column. For example, from the free-body diagram of column AD shown in Fig. 12.7(d), we observe that in order to satisfy $\sum F_X = 0$, the shear force at the upper end, S_{DA}, must act to the right with a magnitude of 15 kN to balance the shear force at the lower end, $S_{AD} = 15$ kN to the left. Thus, $S_{DA} = 15$ kN $\rightarrow$. Shear forces at the upper ends of the remaining columns are determined in a similar manner and are shown in Fig. 12.7(e), which depicts the free-body diagrams of all the members and joints of the frame.

Column Moments With the column shears now known, the column end moments can be computed by multiplying the column shears by half of the column heights. For example, since column AD (see Fig. 12.7(d)) is 8 m high and has end shears of 15 kN, its end moments are

$$M_{AD} = M_{DA} = 15\left(\frac{8}{2}\right) = 60 \text{ kN} \cdot \text{m} \text{ ↺}$$

Note that the end moments, M_{AD} and M_{DA}, are both counterclockwise—that is, opposite to the clockwise moments of the 15 kN end shears about the internal hinge at the column midheight. The end moments of the remaining columns of the frame are computed in a similar manner and are shown in Fig. 12.7(e).

Girder Axial Forces, Moments, and Shears We begin the calculation of girder end actions at the upper left joint D. The column shear S_{DA} and moment M_{DA} computed previously are applied to the free-body diagram of joint D in opposite directions according to Newton's third law, as shown in Fig. 12.7(d). By applying the equilibrium equation $\sum F_X = 0$, we obtain the girder axial force $Q_{DE} = 45$ kN $\leftarrow$ on joint D. Note that Q_{DE} must act in the opposite direction—that is, to the right—at end D of girder DE. From the free-body diagram of joint D (Fig. 12.7(d)), we can also see that in order to satisfy the moment equi-

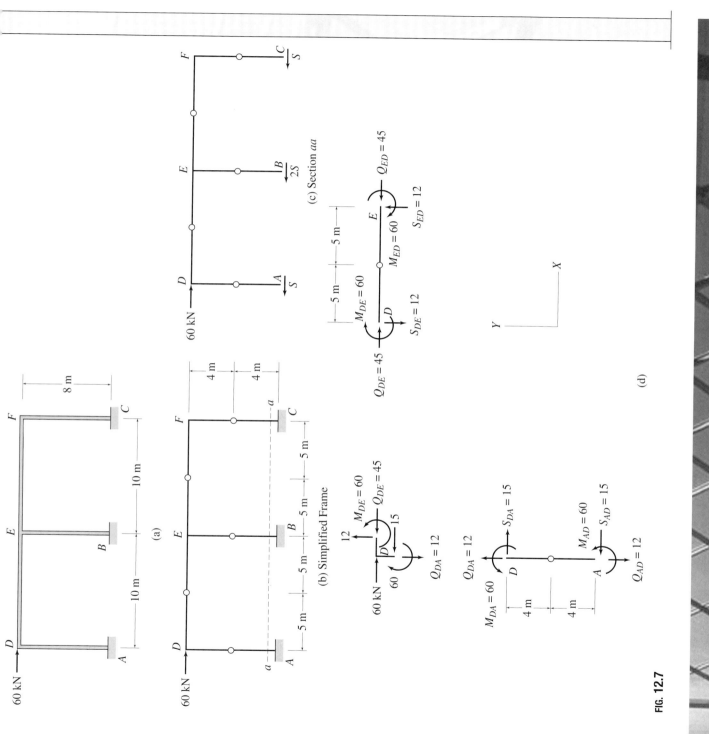

FIG. 12.7

continued

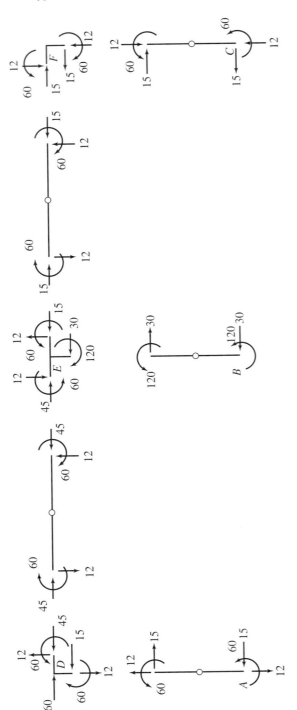

(e) Member End Forces and Moments

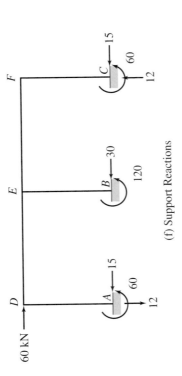

(f) Support Reactions

FIG. 12.7 (contd.)

librium equation, ($\sum M = 0$), the girder end moment M_{DE} must be equal and opposite to the 60 kN · m column end moment. Thus, $M_{DE} = 60$ kN · m, with a counterclockwise direction on joint D but a clockwise direction at the end D of girder DE.

To evaluate the girder shear S_{DE}, we consider the moment equilibrium of the left half of girder DE. From the free-body diagram of girder DE in Fig. 12.7(d), we can see that the shear force S_{DE} must act downward with a magnitude of $M_{DE}/(L/2)$ so that it can develop a counterclockwise moment of magnitude, M_{DE}, about the internal hinge to balance the clockwise end moment, M_{DE}. Thus,

$$S_{DE} = \frac{M_{DE}}{(L/2)} = \frac{60}{(10/2)} = 12 \text{ kN} \downarrow$$

The axial force, shear, and moment at the right end E can now be determined by applying the three equilibrium equations to the free body of girder DE (Fig. 12.7(d)):

$$+ \rightarrow \sum F_X = 0 \qquad\qquad 45 - Q_{ED} = 0 \qquad Q_{ED} = 45 \text{ kN} \leftarrow$$

$$+ \uparrow \sum F_Y = 0 \qquad\qquad -12 + S_{ED} = 0 \qquad S_{ED} = 12 \text{ kN} \uparrow$$

$$+ \zeta \sum M_D = 0 \qquad -60 - M_{ED} + 12(10) = 0 \qquad M_{ED} = 60 \text{ kN} \cdot \text{m} \,\rangle$$

Note that the girder end moments, M_{DE} and M_{ED}, are equal in magnitude and have the same direction.

Next, we calculate the end actions for girder EF. We first apply the equilibrium equations $\sum F_X = 0$ and $\sum M = 0$ to the free body of joint E (Fig. 12.7(e)) to obtain the axial force $Q_{EF} = 15$ kN $\rightarrow$ and the moment $M_{EF} = 60$ kN · m $\rangle$ at the left end E of the girder. We then obtain the shear $S_{EF} = 12$ kN $\downarrow$ by dividing the moment M_{EF} by half of the girder length, and we apply the three equilibrium equations to the free body of the girder to obtain $Q_{FE} = 15$ kN $\leftarrow$, $S_{FE} = 12$ kN $\uparrow$, and $M_{FE} = 60$ kN · m $\rangle$ at the right end F of the girder (see Fig. 12.7(e)).

Since all the moments and horizontal forces acting at the upper right joint F are now known, we can check the calculations that have been performed thus far by applying the two equilibrium equations $\sum F_X = 0$ and $\sum M = 0$ to the free body of this joint. From the free-body diagram of joint F shown in Fig. 12.7(e), it is obvious that these equilibrium equations are indeed satisfied.

Column Axial Forces   We begin the calculation of column axial forces at the upper left joint D. From the free-body diagram of this joint shown in Fig. 12.7(d), we observe that the axial force in column AD must be equal and opposite to the shear in girder DE. Thus, the axial force at the upper end D of column AD is $Q_{DA} = 12$ kN $\uparrow$. By applying $\sum F_Y = 0$ to the free body of column AD, we obtain the axial force at the lower end A of the column to be $Q_{AD} = 12$ kN $\downarrow$. Thus, the column AD is subjected to an axial tensile force of 12 kN. Axial forces for the remaining columns BE and CF are calculated similarly by considering the equilibrium of joints E and F, respectively. The axial forces thus obtained are shown in Fig. 12.7(e). **Ans.**

continued

Reactions The forces and moments at the lower ends of the columns AD, BE, and CF, represent the reactions at the fixed supports A, B, and C, respectively, as shown in Fig. 12.7(f). **Ans.**

Checking Computations To check our computations, we apply the three equilibrium equations to the free body of the entire frame (Fig. 12.7(f)):

$$+ \rightarrow \sum F_X = 0 \qquad\qquad 60 - 15 - 30 - 15 = 0 \qquad \text{Checks}$$

$$+ \uparrow \sum F_Y = 0 \qquad\qquad -12 + 12 = 0 \qquad \text{Checks}$$

$$+ \circlearrowleft \sum M_C = 0 \qquad -60(8) + 12(20) + 60 + 120 + 60 = 0 \qquad \text{Checks}$$

Example 12.3

Determine the approximate axial forces, shears, and moments for all the members of the frame shown in Fig. 12.8(a) by using the portal method.

Solution

Simplified Frame The simplified frame is obtained by inserting internal hinges at the midpoints of all the members of the given frame, as shown in Fig. 12.8(b).

Column Shears To compute shears in the columns of the second story of the frame, we pass an imaginary section aa through the columns DG, EH, and FI just above the floor level, as shown in Fig. 12.8(b). The free-body diagram of the portion of the frame above section aa is shown in Fig. 12.8(c). Note that the shear in the interior column EH has been assumed to be twice as much as in the exterior columns DG and FI. By applying the equilibrium equation $\sum F_X = 0$, we obtain (Fig. 12.8(c))

$$+ \rightarrow \sum F_X = 0 \qquad 10 - S_2 - 2S_2 - S_2 = 0 \qquad S_2 = 2.5 \text{ k}$$

Thus, the shear forces at the lower ends of the second-story columns are

$$S_{DG} = S_{FI} = S_2 = 2.5 \text{ k} \leftarrow \qquad S_{EH} = 2S_2 = 5 \text{ k} \leftarrow$$

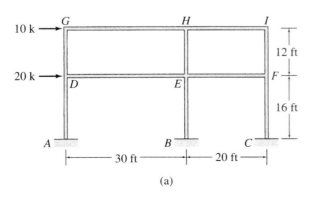

(a)

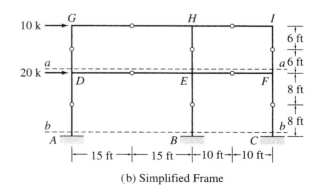

(b) Simplified Frame

FIG. 12.8

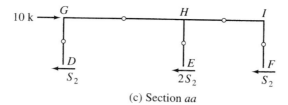

(c) Section *aa*

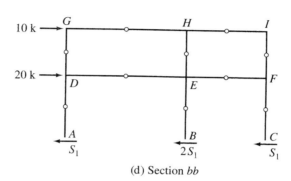

(d) Section *bb*

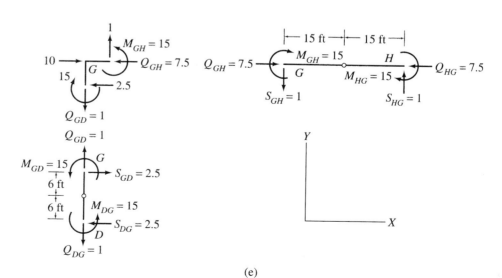

(e)

FIG. 12.8 (contd.)

continued

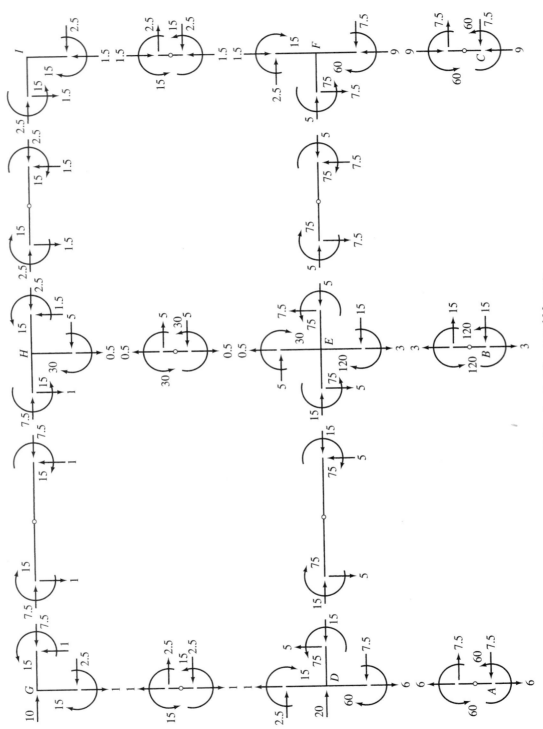

(f) Member End Forces and Moments

FIG. 12.8 (contd.)

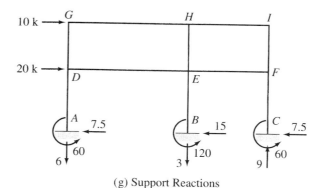

FIG. **12.8** (contd.)

(g) Support Reactions

Similarly, by employing section bb (Fig. 12.8(b)), we determine shear forces at the lower ends of the first-story columns AD, BE, and CF to be (see Fig. 12.8(d)):

$$S_{AD} = S_{CF} = S_1 = 7.5 \text{ k} \leftarrow \qquad S_{BE} = 2S_1 = 15 \text{ k} \leftarrow$$

Shear forces at the upper ends of columns are determined by applying the equilibrium equation $\sum F_X = 0$ to the free body of each column. For example, from the free-body diagram of column DG shown in Fig. 12.8(e), we can see that in order to satisfy $\sum F_X = 0$, the shear force at the upper end, S_{GD}, must act to the right with a magnitude of 2.5 k. Thus $S_{GD} = 2.5 \text{ k} \rightarrow$. Shear forces at the upper ends of the remaining columns are obtained in a similar manner and are shown in Fig. 12.8(f), which depicts the free-body diagrams of all the members and joints of the frame.

Column Moments Knowing column shears, we can now compute the column end moments by multiplying the column shears by half of the column heights. For example, since column DG (see Fig. 12.8(e)) is 12 ft high and has end shears of 2.5 k, its end moments are

$$M_{DG} = M_{GD} = 2.5\left(\frac{12}{2}\right) = 15 \text{ k-ft} \circlearrowright$$

Note that the end moments, M_{DG} and M_{GD}, are both counterclockwise—that is, opposite to the clockwise moments of the 2.5-k end shears about the internal hinge at the column midheight. The end moments of the remaining columns are computed in a similar manner and are shown in Fig. 12.8(f).

Girder Axial Forces, Moments, and Shears We begin the computation of girder end actions at the upper left joint G. The column shear S_{GD} and moment M_{GD} computed previously are applied to the free-body diagram of joint G in opposite directions in accordance with Newton's third law, as shown in Fig. 12.8(e). By summing forces in the horizontal direction, we obtain the girder axial force $Q_{GH} = 7.5 \text{ k} \leftarrow$ on joint G. Note that Q_{GH} must act in the opposite direction—that is, to the right—at the end G of girder GH. From the free-body

continued

diagram of joint G (Fig. 12.8(e)), we can also see that in order to satisfy the moment equilibrium ($\sum M = 0$), the girder end moment M_{GH} must be equal and opposite to the 15-k-ft column end moment. Thus $M_{GH} = 15$ k-ft, with a counterclockwise direction on joint G but a clockwise direction at the end G of girder GH.

To determine the girder shear S_{GH}, we consider the moment equilibrium of the left half of girder GH. From the free-body diagram of girder GH (Fig. 12.8(e)), we can see that the shear force S_{GH} must act downward with a magnitude of $M_{GH}/(L/2)$ so that it can develop a counterclockwise moment of magnitude M_{GH} about the internal hinge to balance the clockwise end moment M_{GH}. Thus

$$S_{GH} = \frac{M_{GH}}{(L/2)} = \frac{15}{(30/2)} = 1 \text{ k} \downarrow$$

The axial force, shear, and moment at the right end H can now be computed by applying the three equilibrium equations to the free body of girder GH (Fig. 12.8(e)). Applying $\sum F_X = 0$, we obtain $Q_{HG} = 7.5$ k $\leftarrow$. From $\sum F_Y = 0$, we obtain $S_{HG} = 1$ k $\uparrow$, and to compute M_{HG}, we apply the equilibrium equation:

$$+\zeta \sum M_G = 0 \qquad -15 - M_{HG} + 1(30) = 0 \qquad M_{HG} = 15 \text{ k-ft} \downarrow$$

Note that the girder end moments, M_{GH} and M_{HG}, are equal in magnitude and have the same direction.

Next, the end actions for girder HI are computed. The equilibrium equations $\sum F_X = 0$ and $\sum M = 0$ are first applied to the free body of joint H (Fig. 12.8(f)) to obtain the axial force $Q_{HI} = 2.5$ k $\rightarrow$ and the moment $M_{HI} = 15$ k-ft $\downarrow$ at the left end H of the girder. The shear $S_{HI} = 1.5$ k $\downarrow$ is then obtained by dividing the moment M_{HI} by half the girder length, and the three equilibrium equations are applied to the free body of the girder to obtain $Q_{IH} = 2.5$ k $\leftarrow$, $S_{IH} = 1.5$ k $\uparrow$, and $M_{IH} = 15$ k-ft $\downarrow$ at the right end I of the girder (see Fig. 12.8(f)).

All the moments and horizontal forces acting at the upper right joint I are now known, so we can check the calculations performed thus far by applying $\sum F_X = 0$ and $\sum M = 0$ to the free body of this joint. From the free-body diagram of joint I shown in Fig. 12.8(f), it is obvious that these equilibrium equations are indeed satisfied.

The end actions for the first-story girders DE and EF are computed in a similar manner, by starting at the left joint D and working across to the right. The girder end actions thus obtained are shown in Fig. 12.8(f).

Column Axial Forces We begin the computation of column axial forces at the upper left joint G. From the free-body diagram of joint G shown in Fig. 12.8(e), we observe that the axial force in column DG must be equal and opposite to the shear in girder GH. Thus the axial force at the upper end G of column DG is $Q_{GD} = 1$ k $\uparrow$. By applying $\sum F_Y = 0$ to the free body of column DG, we obtain the axial force at the lower end of the column to be $Q_{DG} = 1$ k $\downarrow$. Thus, the column DG is subjected to an axial tensile force of 1 k. Axial forces for the remaining second-story columns, EH and FI, are determined similarly by considering the equilibrium of joints H and I, respectively; thereafter, the axial forces for the first-story columns, AD, BE, and CF, are computed from the equilibrium consideration of joints D, E, and F, respectively. The axial forces thus obtained are shown in Fig. 12.8(f).

Ans.

Reactions The forces and moments at the lower ends of the first-story columns *AD*, *BE*, and *CF*, represent the reactions at the fixed supports *A*, *B*, and *C*, respectively, as shown in Fig. 12.8(g). **Ans.**

Checking Computations To check our computations, we apply the three equilibrium equations to the free body of the entire frame (Fig. 12.8(g)):

$$+ \rightarrow \sum F_X = 0 \qquad\qquad 10 + 20 - 7.5 - 15 - 7.5 = 0 \qquad \text{Checks}$$

$$+ \uparrow \sum F_Y = 0 \qquad\qquad -6 - 3 + 9 = 0 \qquad \text{Checks}$$

$$+ \zeta \sum M_C = 0$$
$$-10(28) - 20(16) + 60 + 6(50) + 120 + 3(20) + 60 = 0 \qquad \text{Checks}$$

12.4 ANALYSIS FOR LATERAL LOADS—CANTILEVER METHOD

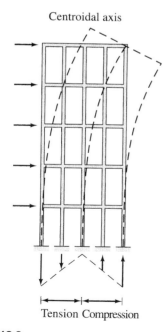

Centroidal axis

Tension Compression

FIG. 12.9

The cantilever method was initially developed by A. C. Wilson in 1908 and is generally considered to be appropriate for the approximate analysis of relatively tall building frames. The cantilever method is based on the assumption that under lateral loads, the building frames behave like cantilever beams, as shown in Fig. 12.9. Recall (from *mechanics of materials*) that the axial stress on a cross section of a cantilever beam subjected to lateral loads varies linearly with the distance from the centroidal axis (neutral surface), so that the longitudinal fibers of the beam on the concave side of the neutral surface are in compression, whereas those on the convex side undergo tension. In the cantilever method, the distribution of axial stress among the columns of a frame at the column midheights is assumed to be analogous to the axial stress distribution among the longitudinal fibers of a cantilever beam. In other words, it is assumed that the axial stress at the midheight of each column is linearly proportional to the distance of the column from the centroid of the areas of all the columns on that story. If we further assume that the cross-sectional areas of all the columns on each story of the frame are equal, then the axial force in each column will also be linearly proportional to the distance of the column from the centroid of all the columns on that story. When the lateral loads are acting on the frame toward the right, as shown in Fig. 12.9, then the columns to the right of the centroidal axis will be in compression, whereas those on the left side will be in tension and vice versa.

In addition to the foregoing assumption, the cantilever method makes the same assumption regarding the location of inflection points as used in the portal method. Thus the assumptions made in the cantilever method can be stated as follows:

1. An inflection point is located at the middle of each member of the frame.

2. On each story of the frame, the axial forces in columns are linearly proportional to their distances from the centroid of the cross-sectional areas of all the columns on that story.

Procedure for Analysis

The following step-by-step procedure can be used for the approximate analysis of building frames by the cantilever method.

1. Draw a sketch of the simplified frame obtained by inserting an internal hinge at the midpoint of each member of the given frame.
2. Determine column axial forces. For each story of the frame:
   a. Pass a horizontal section through the internal hinges at the column midheights, cutting the frame into two portions.
   b. Draw a free-body diagram of the portion of the frame above the section. Because the section passes through the columns at the internal hinges, only internal shears and axial forces (but no internal moments) act on the free body at the points where the columns have been cut.
   c. Determine the location of the centroid of all the columns on the story under consideration.
   d. Assuming that the axial forces in the columns are proportional to their distances from the centroid, determine the column axial forces by applying the moment equilibrium equation, $\sum M = 0$, to the free body of the frame above the section. To eliminate the unknown column shears from the equilibrium equation, the moments should be summed about one of the internal hinges at the column midheights through which the section has been passed.
3. Draw free-body diagrams of all the members and joints of the frame showing the external loads and the column axial forces computed in the previous step.
4. Determine girder shears and moments. For each story of the frame, the shears and moments at the ends of girders are computed by starting at the far left joint and working across to the right (or vice versa), as follows:
   a. Apply the equilibrium equation $\sum F_Y = 0$ to the free body of the joint under consideration to compute the shear at the left end of the girder that is on the right side of the joint.
   b. Considering the free body of the girder, determine the moment at the girder's left end by multiplying the girder shear by half the girder length; that is,

$$M_g = S_g \left(\frac{L}{2} \right)$$

(12.3)

Equation (12.3) is based on the condition that the bending moment at the girder midpoint is zero.

c. By applying the equilibrium equations $\sum F_Y = 0$ and $\sum M = 0$ to the free body of the girder, determine the shear and moment, respectively, at the right end.

d. Select the joint to the right of the girder considered previously, and repeat steps 4(a) through 4(c) until the shears and moments in all the girders of the story have been determined. Because the equilibrium equation $\sum F_Y = 0$ for the right end joint has not been utilized so far, it can be used to check the calculations.

5. Determine column moments and shears. Starting at the top story, apply the equilibrium equation $\sum M = 0$ to the free body of each joint of the story to determine the moment at the upper end of the column below the joint. Next, for each column of the story, calculate the shear at the upper end of the column by dividing the column moment by half the column height; that is,

$$S_C = \frac{M_C}{(h/2)} \tag{12.4}$$

Determine the shear and moment at the lower end of the column by applying the equilibrium equations $\sum F_X = 0$ and $\sum M = 0$, respectively, to the free body of the column. Repeat the procedure for each successive story, working from top to bottom, until the moments and shears in all the columns of the frame have been determined.

6. Determine girder axial forces. For each story of the frame, determine the girder axial forces by starting at the far left joint and applying the equilibrium equation $\sum F_X = 0$ successively to the free body of each joint of the story.

7. Realizing that the forces and moments at the lower ends of the bottom-story columns represent the support reactions, use the three equilibrium equations of the entire frame to check the calculations. If the analysis has been performed correctly, then these equilibrium equations must be satisfied.

Example 12.4

Determine the approximate axial forces, shears, and moments for all the members of the frame shown in Fig. 12.10(a) by using the cantilever method.

Solution

This frame was analyzed by the portal method in Example 12.3.

continued

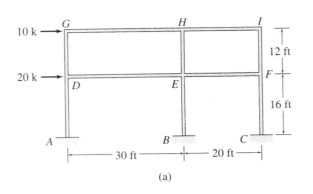

(a)

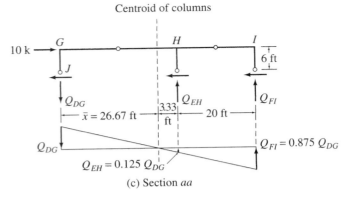

(c) Section *aa*

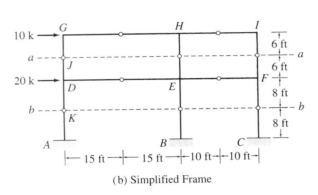

(b) Simplified Frame

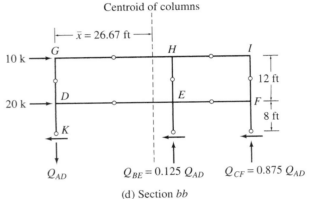

(d) Section *bb*

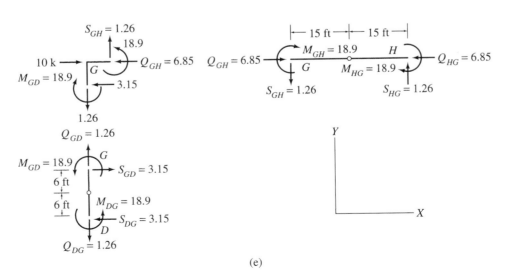

(e)

FIG. **12.10**

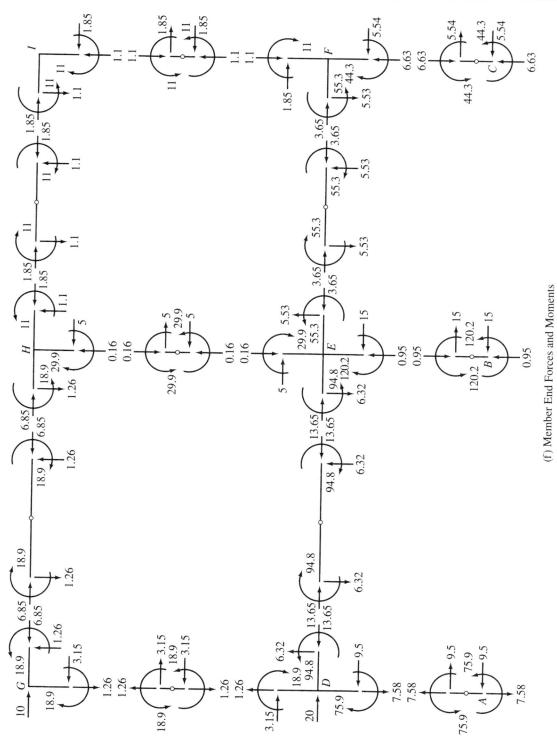

(f) Member End Forces and Moments

FIG. 12.10 (contd.)

continued

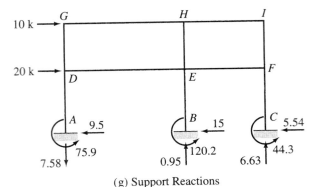

FIG. 12.10 (contd.)

(g) Support Reactions

Simplified Frame The simplified frame, obtained by inserting internal hinges at midpoints of all the members of the given frame, is shown in Fig. 12.10(b).

Column Axial Forces To compute axial forces in the columns of the second story of the frame, we pass an imaginary section *aa* through the internal hinges at the midheights of columns DG, EH, and FI, as shown in Fig. 12.10(b). The free-body diagram of the portion of the frame above this section is shown in Fig. 12.10(c). Because the section cuts the columns at the internal hinges, only internal shears and axial forces (but no internal moments) act on the free body at the points where the columns have been cut. Assuming that the cross-sectional areas of the columns are equal, we determine the location of the centroid of the three columns from the left column DG by using the relationship

$$\bar{x} = \frac{\sum Ax}{\sum A} = \frac{A(0) + A(30) + A(50)}{3A} = 26.67 \text{ ft}$$

The lateral loads are acting on the frame to the right, so the axial force in column DG, which is to the left of the centroid, must be tensile, whereas the axial forces in the columns EH and FI, located to the right of the centroid, must be compressive as shown in Fig. 12.10(c). Also, since the axial forces in the columns are assumed to be linearly proportional to their distances from the centroid, the relationships between them can be established by means of the similar triangles shown in Fig. 12.10(c); that is,

$$Q_{EH} = \frac{3.33}{26.67} Q_{DG} = 0.125 Q_{DG} \tag{1}$$

$$Q_{FI} = \frac{23.33}{26.67} Q_{DG} = 0.875 Q_{DG} \tag{2}$$

By summing moments about the left internal hinge J, we write

$$+\zeta \sum M_J = 0 \qquad -10(6) + Q_{EH}(30) + Q_{FI}(50) = 0$$

Substituting Eqs. (1) and (2) into the preceding equation and solving for Q_{DG}, we obtain

$$-60 + (0.125 Q_{DG})(30) + (0.875 Q_{DG})(50) = 0$$

$$Q_{DG} = 1.26 \text{ k}$$

Therefore, from Eqs. (1) and (2),

$$Q_{EH} = 0.125(1.26) = 0.16 \text{ k}$$

$$Q_{FI} = 0.875(1.26) = 1.1 \text{ k}$$

The axial forces in the first-story columns can be determined in a similar manner by employing section bb shown in Fig. 12.10(b). The free-body diagram of the portion of the frame above this section is shown in Fig. 12.10(d). The arrangement of columns for both stories of the frame is the same, so the location of the centroid—as well as the relationships between the axial forces—of the columns for the two stories are also the same. Thus

$$Q_{BE} = 0.125Q_{AD} \tag{3}$$

$$Q_{CF} = 0.875Q_{AD} \tag{4}$$

By summing moments about the internal hinge K, we write

$$+\circlearrowleft \sum M_K = 0 \qquad -10(20) - 20(8) + Q_{BE}(30) + Q_{CF}(50) = 0$$

Substituting Eqs. (3) and (4), we obtain

$$-360 + (0.125Q_{AD})(30) + (0.875Q_{AD})(50) = 0$$

$$Q_{AD} = 7.58 \text{ k}$$

Therefore,

$$Q_{BE} = 0.95 \text{ k}$$

$$Q_{CF} = 6.63 \text{ k}$$

The column axial forces are shown in Fig. 12.10(f), which depicts the free-body diagrams of all the members and joints of the frame.

Girder Shears and Moments Knowing column axial forces, the girder shears can now be computed by considering equilibrium in the vertical direction of the joints. Starting at the upper left joint G, we apply the equilibrium equation $\sum F_Y = 0$ to the free body of this joint (see Fig. 12.10(e)) to obtain the shear $S_{GH} = 1.26$ k $\downarrow$ at the left end of girder GH. The moment at the left end is then determined by multiplying the shear by half the girder length; that is,

$$M_{GH} = 1.26(15) = 18.9 \text{ k-ft} \circlearrowright$$

The shear and moment at the right end, H, can now be computed by applying the equilibrium equations $\sum F_Y = 0$ and $\sum M = 0$, respectively, to the free body of girder GH (Fig. 12.10(e)). By applying these equations, we obtain $S_{HG} = 1.26$ k $\uparrow$ and $M_{HG} = 18.9$ k-ft $\circlearrowright$. Note that the girder end moments, M_{GH} and M_{HG}, have the same magnitude and direction.

Next, the end shears and moments for girder HI are computed by considering the equilibrium of joints H and girder HI (see Fig. 12.10(f)), and the equilibrium equation $\sum F_Y = 0$ is applied to the free body of the right joint I to check the calculations performed thus far.

The shears and moments for the first-story girders DE and EF are computed in a similar manner by starting at the left joint D and working across to the right. The girder shears and moments thus obtained are shown in Fig. 12.10(f).

continued

Column Moments and Shears With the girder moments now known, the column moments can be determined by considering moment equilibrium of joints. Beginning at the second story and applying $\sum M = 0$ to the free body of joint G (Fig. 12.10(e)), we obtain the moment at the upper end of column DG to be $M_{GD} = 18.9$ k-ft $\circlearrowright$. The shear at the upper end of column DG is then computed by dividing M_{GD} by half the column height; that is,

$$S_{GD} = \frac{18.9}{6} = 3.15 \text{ k} \rightarrow$$

Note that S_{GD} must act to the right, so that it can develop a clockwise moment to balance the counterclockwise end moment M_{GD}. The shear and moment at the lower end D are then determined by applying the equilibrium equations $\sum F_X = 0$ and $\sum M = 0$ to the free body of column DG (see Fig. 12.10(e)). Next, the end moments and shears for columns EH and FI are computed in a similar manner; thereafter, the procedure is repeated to determine the moments and shears for the first-story columns, AD, BE, and CF (see Fig. 12.10(f)).

Girder Axial Forces We begin the computation of girder axial forces at the upper left joint G. Applying $\sum F_X = 0$ to the free-body diagram of joint G shown in Fig. 12.10(e), we find the axial force in girder GH to be 6.85 k compression. The axial force for girder HI is determined similarly by considering the equilibrium of joint H, after which the equilibrium equation $\sum F_X = 0$ is applied to the free body of the right joint I to check the calculations. The axial forces for the first-story girders DE and EF are then computed from the equilibrium consideration of joints D and E, in order. The axial forces thus obtained are shown in Fig. 12.10(f).   Ans.

Reactions The forces and moments at the lower ends of the first-story columns AD, BE, and CF represent the reactions at the fixed supports A, B, and C, respectively, as shown in Fig. 12.10(g).   Ans.

Checking Computations To check our computations, we apply the three equilibrium equations to the free body of the entire frame (Fig. 12.10(g)):

$+ \rightarrow \sum F_X = 0$	$10 + 20 - 9.5 - 15 - 5.54 = -0.04 \approx 0$	Checks
$+ \uparrow \sum F_Y = 0$	$-7.58 + 0.95 + 6.63 = 0$	Checks
$+ \circlearrowleft \sum M_C = 0$		

$$-10(28) - 20(16) + 75.9 + 7.58(50) + 120.2 - 0.95(20) + 44.3 = 0.4 \approx 0$$

Checks

SUMMARY

In this chapter, we have learned that in the approximate analysis of statically indeterminate structures, two types of simplifying assumptions are commonly employed: (1) assumptions about the location of inflection points and (2) assumptions about the distribution of forces among members and/or reactions. The total number of assumptions required is equal to the degree of indeterminacy of the structure.

The approximate analysis of rectangular frames subjected to vertical loads is based on the following assumptions for each girder of the frame: (1) the inflection points are located at one-tenth of the span from each end of the girder and (2) the girder axial force is zero.

Two methods commonly used for the approximate analysis of rectangular frames subjected to lateral loads are the portal method and the cantilever method.

The portal method involves making the assumptions that an inflection point is located at the middle of each member and that, on each story, interior columns carry twice as much shear as exterior columns.

In the cantilever method, the following assumptions are made about the behavior of the frame: that an inflection point is located at the middle of each member and that, on each story, the axial forces in the columns are linearly proportional to their distances from the centroid of the cross-sectional areas of all the columns on that story.

PROBLEMS

Section 12.2

12.1 through 12.5 Draw the approximate shear and bending moment diagrams for the girders of the frames shown in Figs. P12.1 through P12.5.

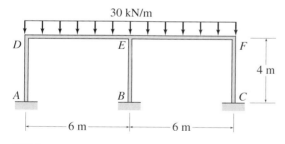

FIG. **P12.1**

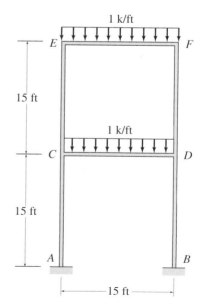

FIG. **P12.2**

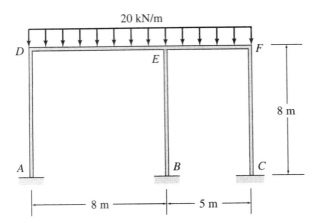

FIG. **P12.3**

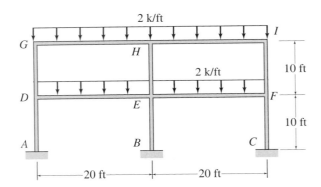

FIG. **P12.4**

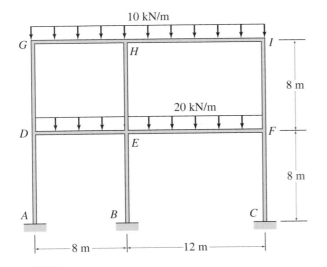

FIG. **P12.5**

Section 12.3

12.6 through 12.13 Determine the approximate axial forces, shears, and moments for all the members of the frames shown in Figs. P12.6 through P12.13 by using the portal method.

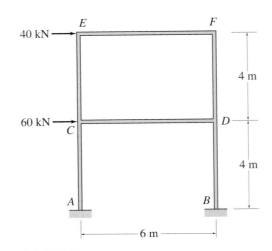

FIG. **P12.6, P12.14**

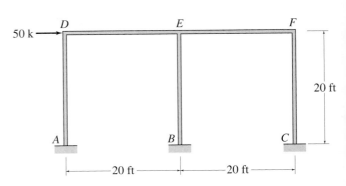

FIG. **P12.7, P12.15**

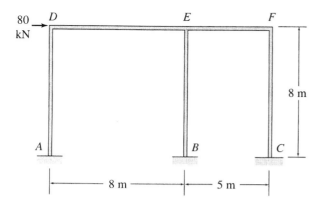

FIG. P12.8, P12.16

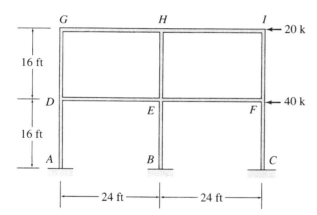

FIG. P12.9, P12.17

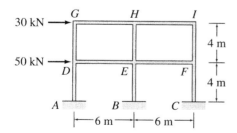

FIG. P12.10, P12.18

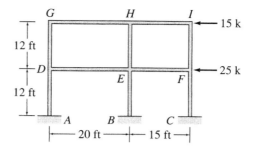

FIG. P12.11, P12.19

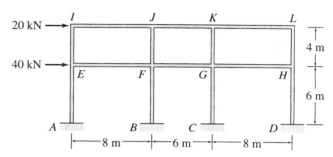

FIG. P12.12, P12.20

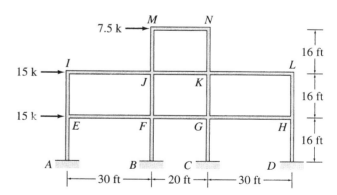

FIG. P12.13, P12.21

Section 12.4

12.14 through 12.21 Determine the approximate axial forces, shears, and moments for all the members of the frames shown in Figs. P12.6 through P12.13 by using the cantilever method.

A Cantilever Bridge
Photo courtesy of Photodisc Blue

13

Method of Consistent Deformations—Force Method

In this chapter, we study a general formulation of the force (flexibility) method called the *method of consistent deformations* for the analysis of statically indeterminate structures. The method, which was introduced by James C. Maxwell in 1864, essentially involves removing enough restraints from the indeterminate structure to render it statically determinate. This determinate structure, which must be statically stable, is referred to as the *primary structure*. The excess restraints removed from the given indeterminate structure to convert it into the determinate primary structure are called *redundant restraints*, and the reactions or internal forces associated with these restraints are termed *redundants*. The redundants are then applied as unknown loads on the primary structure, and their values are determined by solving the compatibility equations based on the condition that the deformations of the primary structure due to the combined effect of the redundants and the given external loading must be the same as the deformations of the original indeterminate structure.

Since the independent variables or unknowns in the method of consistent deformations are the redundant forces (and/or moments), which must be determined before the other response characteristics (e.g., displacements) can be evaluated, the method is classified as a *force method*.

In this chapter, we first develop the analysis of beams, frames, and trusses with a single degree of indeterminacy by using the method of consistent deformations. We then extend this method to structures with multiple degrees of indeterminacy. Finally, we consider the analysis for the effects of support settlements, temperature changes, and fabrication errors.

13.1 STRUCTURES WITH A SINGLE DEGREE OF INDETERMINACY

To illustrate the basic concept of the method of consistent deformations, consider the propped cantilever beam subjected to a concentrated load P, as shown in Fig. 13.1(a). Since the beam is supported by four support reactions (A_x, A_y, M_A, and C_y), the three equations of equilibrium ($\sum F_x = 0$, $\sum F_y = 0$, and $\sum M = 0$) are not sufficient for determining all the reactions. Therefore, the beam is statically indeterminate. The degree of indeterminacy of the beam is equal to the number of unknown reactions minus the number of equilibrium equations—that is, $4 - 3 = 1$—which indicates that the beam has one more, or *redundant*, reaction than necessary for static stability. Thus, if we can determine one of the four reactions by using a compatibility equation based on the geometry of the deformation of the beam, then the remaining three reactions can be obtained from the three equations of equilibrium.

To establish the compatibility equation, we select one of the reactions of the beam to be the redundant. Suppose that we select the vertical reaction C_y exerted by the roller support C to be the redundant. From Fig. 13.1(a), we can see that if the roller support C is removed from the beam, it will become determinate while still remaining statically stable, because the fixed support A alone can prevent it from translating and/or rotating as a rigid body. Thus, the roller support C is not necessary for the static stability of the beam, and its reaction C_y can be designated as the redundant. Note however, that the presence of support C imposes the compatibility condition on the deflected shape of the beam that the deflection at C must be zero (Fig. 13.1(a)); that is,

$$\Delta_C = 0 \tag{13.1}$$

To determine the redundant C_y by using this compatibility condition, we remove the roller support C from the indeterminate beam to convert it into the determinate cantilever beam shown in Fig. 13.1(b). This determinate beam is referred to as the *primary beam*. The redundant C_y is then applied as an unknown load on the primary beam, along with the given external load $P = 32$ k, as shown in Fig. 13.1(b). The redundant C_y can be determined by using the reasoning that if the value of the unknown load C_y acting on the primary beam (Fig. 13.1(b)) is to be the same as that of the reaction C_y exerted on the indeterminate beam by the roller support C (Fig. 13.1(a)), then the deflection at the free end C

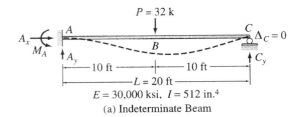

$E = 30,000$ ksi, $I = 512$ in.^{4}

(a) Indeterminate Beam

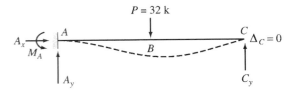

(b) Primary Beam Subjected to External Loading and Redundant C_y

$=$

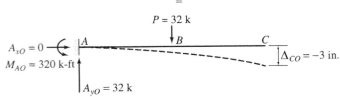

(c) Primary Beam Subjected to External Loading

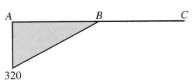

320
Bending Moment Diagram for Primary
Beam Due to External Loading (k-ft)

$+$

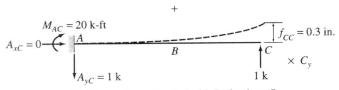

(d) Primary Beam Loaded with Redundant C_y

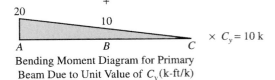

Bending Moment Diagram for Primary
Beam Due to Unit Value of C_y (k-ft/k)

$=$

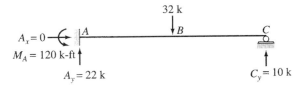

(e) Support Reactions for Indeterminate Beam

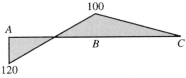

(f) Bending Moment Diagram
for Indeterminate Beam (k-ft)

FIG. **13.1**

of the primary beam due to the combined effect of the external load P and the redundant C_y must be the same as the deflection of the indeterminate beam at support C. Because the deflection Δ_C at support C of the indeterminate beam is zero (Eq. 13.1), the deflection at end C of the primary beam due to the combined effect of the external load P and the redundant C_y must also be zero. The total deflection Δ_C at end C of the primary beam due to the combined effect of P and C_y can be conveniently expressed by superimposing (algebraically adding) the deflections due to the external load P and the redundant C_y acting individually on the beam; that is,

$$\Delta_C = \Delta_{CO} + \Delta_{CC} \tag{13.2}$$

in which Δ_{CO} and Δ_{CC} represent, respectively, the deflections at the end C of the primary beam due to the external load P and the redundant C_y, each acting alone on the beam. Note that two subscripts are used to denote the deflections Δ_{CO} and Δ_{CC} of the primary beam. The first subscript, C, indicates the location of these deflections; the second subscript, O, is used to indicate that Δ_{CO} is caused by the given external loading, whereas the second subscript, C, of Δ_{CC} implies that it is due to the redundant C_y. Both of these deflections are considered to be positive if they occur in the direction of the redundant C_y, which is assumed to be upward, as shown in Fig. 13.1(b).

Since the redundant C_y is unknown, it is convenient to determine Δ_{CC} by first evaluating the deflection at C due to a unit value of the redundant C_y, as shown in Fig. 13.1(d), and then multiplying the deflection thus obtained by the unknown magnitude of the redundant. Thus,

$$\Delta_{CC} = f_{CC} C_y \tag{13.3}$$

in which f_{CC} denotes the deflection at point C of the primary beam due to the unit value of the redundant C_y. It may be recalled from Section 7.8 that f_{CC}, which has units of deflection per unit force, is referred to as a *flexibility coefficient*. By substituting Eqs. (13.1) and (13.3) into Eq. (13.2), we obtain the compatibility equation

$$\Delta_C = \Delta_{CO} + f_{CC} C_y = 0 \tag{13.4}$$

which can be solved to express the redundant C_y in terms of the deflections Δ_{CO} and f_{CC} of the primary beam:

$$C_y = -\frac{\Delta_{CO}}{f_{CC}} \tag{13.5}$$

Equations (13.4) and (13.5) can also be established intuitively by regarding the redundant C_y as the force necessary to correct the deflected shape of the primary structure so that it matches the deflected shape of the original indeterminate structure. When support C is imagined to be removed from the indeterminate beam of Fig. 13.1(a), the external load P causes a downward deflection of Δ_{CO} at end C, as shown in Fig. 13.1(c). Since the deflection at C in the original indeterminate beam is zero, the redundant force C_y must be of sufficient magnitude to push the end C back into its original position by producing an upward deflection of Δ_{CO} at end C of the primary beam. To evaluate the effect of C_y on the beam, we compute the flexibility coefficient f_{CC}, which is the deflection at C due to a unit value of the redundant (Fig. 13.1(d)). Since superposition is valid, deflection is directly proportional to load; that is, if a unit load causes a deflection of f_{CC}, then a load ten times as much will cause a deflection of $10f_{CC}$. Thus, the upward redundant of magnitude C_y causes an upward deflection of $C_y f_{CC}$ at end C of the primary beam. Since the upward deflection $(C_y f_{CC})$ caused by the redundant C_y must be equal to the downward deflection (Δ_{CO}) due to the external load P, we write

$$C_y f_{CC} = -\Delta_{CO} \qquad (13.6)$$

in which both deflections, f_{CC} and Δ_{CO}, are assumed to be positive upward. Note that Eq. (13.6) is equivalent to Eqs. (13.4) and (13.5) derived previously.

Since the primary beam is statically determinate, the deflections Δ_{CO} and f_{CC} can be computed by either using the methods previously described in Chapters 6 and 7 or by using the beam-deflection formulas given inside the front cover of the book. By using the beam-deflection formulas, we determine the deflection at end C of the primary beam due to the external load $P(= 32 \text{ k})$ to be

$$\Delta_{CO} = -\frac{5PL^3}{48EI} = -\frac{5(32)(20)^3}{48(30,000)(512)/144} = -0.25 \text{ ft} = -3 \text{ in.}$$

(see Fig. 13.1(c)) in which a negative sign has been assigned to the magnitude of Δ_{CO} to indicate that the deflection occurs in the downward direction—that is, in the direction opposite to that of the redundant C_y. Similarly, the flexibility coefficient f_{CC} is evaluated as

$$f_{CC} = \frac{L^3}{3EI} = \frac{(20)^3}{3(30,000)(512)/144} = 0.025 \text{ ft/k} = 0.3 \text{ in./k}$$

(see Fig. 13.1(d)). By substituting the expressions or the numerical values of Δ_{CO} and f_{CC} into Eq. (13.5), we determine the redundant C_y to be

$$C_y = -\left(-\frac{5PL^3}{48EI}\right)\left(\frac{3EI}{L^3}\right) = \frac{5}{16}P = 10 \text{ k} \uparrow$$

The positive answer for C_y indicates that our initial assumption about the upward direction of C_y was correct.

With the reaction C_y known, the three remaining reactions can now be determined by applying the three equilibrium equations to the free body of the indeterminate beam (Fig. 13.1(e)):

$$+ \rightarrow \sum F_x = 0 \qquad\qquad A_x = 0$$

$$+ \uparrow \sum F_y = 0 \qquad\qquad A_y - 32 + 10 = 0 \qquad A_y = 22 \text{ k} \uparrow$$

$$+ \circlearrowleft \sum M_A = 0 \qquad M_A - 32(10) + 10(20) = 0 \qquad M_A = 120 \text{ k-ft} \circlearrowleft$$

After the redundant C_y has been computed, the reactions and all other response characteristics of the beam can also be determined by employing superposition relationships similar in form to the deflection superposition relationship expressed in Eq. (13.4). Thus, the reactions can alternatively be determined by using the superposition relationships (see Fig. 13.1(a), (c), and (d)):

$$+ \rightarrow A_x = A_{xO} + A_{xC}(C_y) = 0$$

$$+ \uparrow A_y = A_{yO} + A_{yC}(C_y) = 32 - 1(10) = 22 \text{ k} \uparrow$$

$$+ \circlearrowleft M_A = M_{AO} + M_{AC}(C_y) = 320 - 20(10) = 120 \text{ k-ft} \circlearrowleft$$

Note that the second subscript O is used to denote reactions due to the external loading only (Fig. 13.1(c)), whereas the second subscript C denotes reactions due to a unit value of the redundant C_y (Fig. 13.1(d)).

Similarly, the bending moment diagram for the indeterminate beam can be obtained by superimposing the bending moment diagram of the primary beam due to external loading only, on the bending moment diagram of the primary beam due to a unit value of redundant C_y multiplied by the value of C_y. The bending moment diagram for the indeterminate beam thus constructed is shown in Fig. 13.1(f).

Moment as the Redundant

In the foregoing analysis of the propped cantilever of Fig. 13.1(a), we arbitrarily selected the vertical reaction at roller support C to be the redundant. *When analyzing a structure by the method of consistent deformations, we can choose any support reaction or internal force (or moment) as the redundant, provided that the removal of the corresponding restraint from the given indeterminate structure results in a primary structure that is statically determinate and stable.*

Considering again the propped cantilever beam of Fig. 13.1(a), which is redrawn in Fig. 13.2(a), we can see that the removal of the restraint corresponding to the horizontal reaction A_x will render the beam statically unstable. Therefore, A_x cannot be used as the redundant. However, either of the two other reactions at support A can be used as the redundant.

Let us consider the analysis of the beam by using the reaction moment M_A as the redundant. The actual sense of M_A is not known and is arbitrarily assumed to be counterclockwise, as shown in Fig. 13.2(a). To obtain the primary beam, we remove the restraint against rotation at end A by replacing the fixed support by a hinged support, as shown in

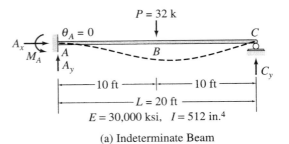

$$E = 30,000 \text{ ksi}, \quad I = 512 \text{ in.}^4$$

(a) Indeterminate Beam

=

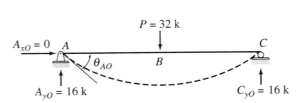

(b) Primary Beam Subjected to External Loading

+

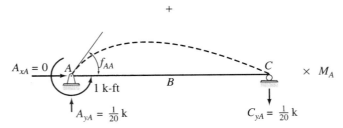

(c) Primary Beam Loaded with Redundant M_A

FIG. 13.2

Fig. 13.2(b). Note that the simply supported beam thus obtained is statically determinate and stable. The redundant M_A is now treated as an unknown load on the primary beam, and its magnitude can be determined
from the compatibility condition that the slope at A due to the combined effect of the external load P and the redundant M_A must be zero.

The primary beam is subjected separately to the external load
$P = 32$ k and a unit value of the unknown redundant M_A, as shown in
Fig. 13.2(b) and (c), respectively. As shown in these figures, θ_{AO} represents the slope at end A due to the external load P, whereas, f_{AA} denotes
the flexibility coefficient—that is, the slope at A due to a unit value of
the redundant M_A. Thus the slope at A due to M_A equals $\theta_{AA} = f_{AA}M_A$.
Because the algebraic sum of the slopes at end A due to the external
load P and the redundant M_A must be zero, we can express the compatibility equation as

$$\theta_{AO} + f_{AA}M_A = 0 \tag{13.7}$$

The slopes θ_{AO} and f_{AA} can be easily computed by using the beam-deflection formulas inside the front cover of the book. Thus

$$\theta_{AO} = -\frac{PL^2}{16EI} = -\frac{32(20)^2}{16(30,000)(512)/144} = -0.0075 \text{ rad}$$

$$f_{AA} = \frac{L}{3EI} = \frac{20}{3(30,000)(512)/144} = 0.0000625 \text{ rad/k-ft}$$

Note that a negative sign has been assigned to the magnitude of θ_{AO}, because this rotation occurs in the clockwise direction—that is, opposite to the counterclockwise direction assumed for the redundant M_A (Fig. 13.2(a)). By substituting the numerical values of θ_{AO} and f_{AA} into the compatibility equation (Eq. 13.7), we write

$$-0.0075 + (0.0000625)M_A = 0$$

from which

$$M_A = \frac{0.0075}{0.0000625} = 120 \text{ k-ft } \circlearrowleft$$

The positive answer implies that the counterclockwise sense initially assumed for M_A was correct. Note that the value of the reaction moment $M_A = 120$ k-ft $\circlearrowleft$ computed here is identical to that obtained previously by using the vertical reaction C_y as the redundant (Fig. 13.1). Once the redundant M_A is known, the remaining reactions as well as the other response characteristics of the beam can be determined either through equilibrium considerations or by superposition, as discussed previously.

Procedure for Analysis

Based on the foregoing discussion, we can develop the following step-by-step procedure for the analysis of externally indeterminate structures with a single degree of indeterminacy.

1. Determine the degree of indeterminacy of the given structure. If the degree of indeterminacy is greater than 1, and/or if the structure is internally indeterminate, then end the analysis at this stage. The analysis of internally indeterminate structures and structures with multiple degrees of indeterminacy is considered in subsequent sections.

2. Select one of the support reactions as the redundant. The choice of redundant is merely a matter of convenience, and any reaction can be selected as the redundant, provided that the removal of the corresponding restraint from the given indeterminate structure results in a primary structure that is statically determinate and stable. The sense of the redundant is not known and can be arbitrarily assumed. The actual sense of the redundant will be known after its magnitude has been determined by solving the compatibility equation. A

positive magnitude for the redundant will imply that the sense initially assumed was correct, whereas a negative value of the magnitude will indicate that the actual sense is opposite to the one assumed initially.

3. Remove the restraint corresponding to the redundant from the given indeterminate structure to obtain the primary determinate structure.

4. a. Draw a diagram of the primary structure with only the external loading applied to it. Sketch a deflected shape of the structure, and show the deflection (or slope) at the point of application and in the direction of the redundant by an appropriate symbol.

    b. Next, draw a diagram of the primary structure with only the unit value of the redundant applied to it. The unit force (or moment) must be applied in the positive direction of the redundant. Sketch a deflected shape of the structure, and show by an appropriate symbol the flexibility coefficient representing the deflection (or slope) at the point of application and in the direction of the redundant. To indicate that the load as well as the response of the structure is to be multiplied by the redundant, show the redundant preceded by a multiplication sign ($\times$) next to the diagram of the structure. The deflection (or slope) at the location of the redundant due to the unknown redundant equals the flexibility coefficient multiplied by the unknown magnitude of the redundant.

5. Write the compatibility equation by setting the algebraic sum of the deflections (or slopes) of the primary structure at the location of the redundant due to the external loading and the redundant equal to the given displacement (or rotation) of the redundant support of the actual indeterminate structure. Since we assume here that supports are unyielding, the algebraic sum of the deflections due to the external loading and the redundant can be simply set equal to zero to obtain the compatibility equation. (The case of support movements is considered in a subsequent section.)

6. Compute the deflections of the primary structure at the location of the redundant due to the external loading and due to the unit value of the redundant. A deflection is considered to be positive if it has the same sense as that assumed for the redundant. The deflections can be determined by using any of the methods discussed in Chapters 6 and 7. For beams with constant flexural rigidity EI, it is usually convenient to determine these quantities by using the deflection formulas given inside the front cover of the book, whereas the deflections of trusses and frames can be conveniently computed by using the method of virtual work.

7. Substitute the values of deflections (or slopes) computed in step 6 into the compatibility equation, and solve for the unknown redundant.

8. Determine the remaining support reactions of the indeterminate structure either by applying the three equilibrium equations to the free body of the indeterminate structure or by superposition of the reactions of the primary structure due to the external loading and due to the redundant.

9. Once the reactions have been evaluated, the other response characteristics (e.g., shear and bending diagram and/or member forces) of the indeterminate structure can be determined either through equilibrium considerations or by superposition of the responses of the primary structure due to the external loading and due to the redundant.

Example 13.1

Determine the reactions and draw the shear and bending moment diagrams for the beam shown in Fig. 13.3(a) by the method of consistent deformations.

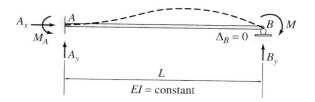

(a) Indeterminate Beam

$=$

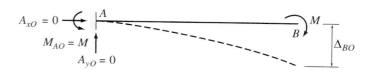

(b) Primary Beam Subjected to External Moment M

$+$

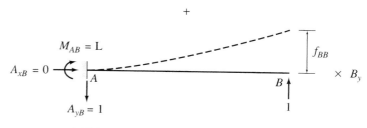

(c) Primary Beam Loaded with Redundant B_y

FIG. 13.3

continued

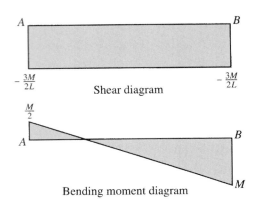

(d) Support Reactions for Indeterminate Beam

Shear diagram

Bending moment diagram

(e) Shear and Bending Moment Diagrams
for Indeterminate Beam

FIG. 13.3 (contd.)

Solution

Degree of Indeterminacy The beam is supported by four reactions, A_x, A_y, M_A, and B_y (Fig. 13.3(a)); that is, $r = 4$. Since there are only three equilibrium equations, the degree of indeterminacy of the beam is equal to $r - 3 = 1$.

Primary Beam The vertical reaction B_y at the roller support B is selected to be the redundant. The sense of B_y is assumed to be upward, as shown in Fig. 13.3(a). The primary beam obtained by removing the roller support B from the given indeterminate beam is shown in Fig. 13.3(b). Note that the primary cantilever beam is statically determinate and stable. Next, the primary beam is subjected separately to the external moment M and a unit value of the unknown redundant B_y, as shown in Fig. 13.3(b) and (c), respectively. As shown in the figure, Δ_{BO} denotes the deflection at B due to the external moment M, whereas f_{BB} denotes the flexibility coefficient representing the deflection at B due to the unit value of the redundant B_y. Thus, the deflection at B due to the unknown redundant B_y equals $f_{BB}B_y$.

Compatibility Equation The deflection at support B of the actual indeterminate beam is zero, so the algebraic sum of the deflections of the primary beam at B due to the external moment M and the redundant B_y must also be zero. Thus, the compatibility equation can be written as

$$\Delta_{BO} + f_{BB}B_y = 0 \qquad (1)$$

Deflections of Primary Beam By using the beam-deflection formulas, we obtain the deflections Δ_{BO} and f_{BB} to be

$$\Delta_{BO} = -\frac{ML^2}{2EI} \quad \text{and} \quad f_{BB} = \frac{L^3}{3EI}$$

in which the negative sign for Δ_{BO} indicates that this deflection occurs in the downward direction—that is, opposite to the upward direction assumed for the redundant B_y.

Magnitude of the Redundant By substituting the expressions for Δ_{BO} and f_{BB} into the compatibility equation (Eq. (1)), we determine the redundant B_y as

$$-\frac{ML^2}{2EI} + \left(\frac{L^3}{3EI}\right)B_y = 0 \qquad B_y = \frac{3M}{2L} \uparrow \qquad \text{Ans.}$$

The positive answer for B_y indicates that our initial assumption about the upward direction of B_y was correct.

Reactions The remaining reactions of the indeterminate beam can now be determined by superposition of the reactions of the primary beam due to the external moment M and the redundant B_y, shown in Fig. 13.3(b) and (c), respectively:

$$+ \rightarrow A_x = 0 \qquad\qquad A_x = 0 \qquad \text{Ans.}$$

$$+ \uparrow A_y = -1\left(\frac{3M}{2L}\right) = -\frac{3M}{2L} \qquad A_y = \frac{3M}{2L} \downarrow \qquad \text{Ans.}$$

$$+ \circlearrowleft M_A = M - L\left(\frac{3M}{2L}\right) = -\frac{M}{2} \qquad M_A = \frac{M}{2} \circlearrowright \qquad \text{Ans.}$$

The reactions are shown in Fig. 13.3(d).

Shear and Bending Moment Diagrams By using the reactions, the shear and bending moment diagrams for the indeterminate beam are constructed. These diagrams are shown in Fig. 13.3(e). **Ans.**

Example 13.2

Determine the reactions and draw the shear and bending moment diagrams for the beam shown in Fig. 13.4(a) by the method of consistent deformations. Select the reaction moment at the fixed support to be the redundant.

Solution

Degree of Indeterminacy The beam is supported by four reactions (Fig. 13.4(a)), so its degree of indeterminacy is equal to $4 - 3 = 1$.

Primary Beam The reaction moment M_A at the fixed support A is selected to be the redundant. The sense of M_A is assumed to be counterclockwise, as shown in Fig. 13.4(a). To obtain the primary beam, we remove the restraint against rotation at end A by replacing the fixed support by a hinged support, as shown in Fig. 13.4(b). The primary simply supported beam is then subjected

continued

separately to the external loading and a unit value of the unknown redundant M_A, as shown in Fig. 13.4(b) and (c), respectively. As shown in these figures, θ_{AO} represents the slope at A due to the external loading, whereas f_{AA} denotes the flexibility coefficient representing the slope at A due to the unit value of the redundant M_A.

Compatibility Equation By setting the algebraic sum of the slopes of the primary beam at A due to the external loading and the redundant M_A equal to the slope at the fixed support A of the actual indeterminate beam, which is zero, we write the compatibility equation:

$$\theta_{AO} + f_{AA}M_A = 0 \tag{1}$$

Slopes of Primary Beam From the beam-deflection formulas,

$$\theta_{AO} = -\frac{1{,}800 \text{ k-ft}^2}{EI} \quad \text{and} \quad f_{AA} = \frac{10 \text{ k-ft}^2/\text{k-ft}}{EI}$$

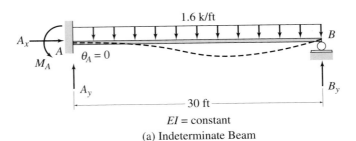

1.6 k/ft

$\theta_A = 0$

30 ft

EI = constant

(a) Indeterminate Beam

$=$

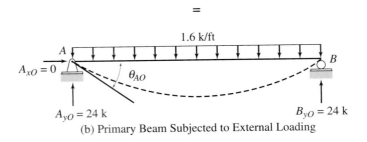

1.6 k/ft

$A_{xO} = 0$

θ_{AO}

$A_{yO} = 24$ k $B_{yO} = 24$ k

(b) Primary Beam Subjected to External Loading

$+$

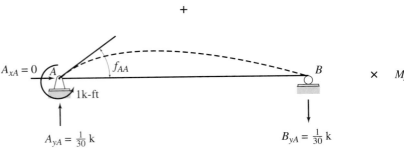

$A_{xA} = 0$ f_{AA}

1k-ft

$A_{yA} = \frac{1}{30}$ k $B_{yA} = \frac{1}{30}$ k

(c) Primary Beam Loaded with Redundant M_A

$\times$ M_A

FIG. 13.4

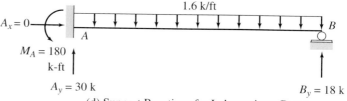

(d) Support Reactions for Indeterminate Beam

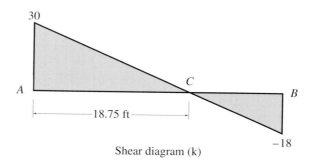

Shear diagram (k)

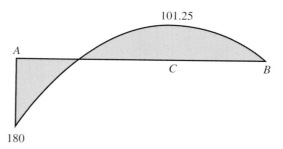

Bending moment diagram (k-ft)

(e) Shear and Bending Moment Diagrams
for Indeterminate Beam

FIG. 13.4 (contd.)

Magnitude of the Redundant By substituting the values of θ_{AO} and f_{AA} into the compatibility equation (Eq. (1)), we obtain

$$-\frac{1{,}800}{EI} + \left(\frac{10}{EI}\right)M_A = 0 \qquad M_A = 180 \text{ k-ft } \circlearrowright \qquad \text{Ans.}$$

Reactions To determine the remaining reactions of the indeterminate beam, we apply the equilibrium equations (Fig. 13.4(d)):

$$+ \rightarrow \sum F_x = 0 \qquad\qquad\qquad\qquad\qquad\qquad\quad A_x = 0 \qquad \text{Ans.}$$

$$+ \circlearrowleft M_B = 0 \qquad 180 - A_y(30) + 1.6(30)(15) = 0 \qquad A_y = 30 \text{ k} \uparrow \quad \text{Ans.}$$

$$+ \uparrow \sum F_y = 0 \qquad 30 - 1.6(30) + B_y = 0 \qquad\qquad B_y = 18 \text{ k} \uparrow \quad \text{Ans.}$$

Shear and Bending Moment Diagrams See Fig. 13.4(e). \qquad\qquad\qquad\qquad\qquad **Ans.**

Example 13.3

Determine the reactions and draw the shear and bending moment diagrams for the two-span continuous beam shown in Fig. 13.5(a) using the method of consistent deformations.

Solution

Degree of Indeterminacy The beam is supported by four reactions, so its degree of indeterminacy is equal to $4 - 3 = 1$.

Primary Beam The vertical reaction B_y at the roller support B is selected to be the redundant, and the primary beam is obtained by removing the roller support B from the given indeterminate beam, as shown in Fig. 13.5(b). Next, the primary beam is subjected separately to the external loading and a unit value of the unknown redundant B_y, as shown in Fig. 13.5(b) and (c), respectively. As shown in these figures, Δ_{BO} denotes the deflection at B due to the external loading, whereas f_{BB} denotes the flexibility coefficient representing the deflection at B due to the unit value of the redundant B_y.

Compatibility Equation Because the deflection at support B of the actual indeterminate beam is zero, the algebraic sum of the deflections of the primary beam at B due to the external loading and the redundant B_y must also be zero. Thus, the compatibility equation can be written as

$$\Delta_{BO} + f_{BB}B_y = 0 \tag{1}$$

Deflections of Primary Beam The flexural rigidity EI of the primary beam is not constant (since the moment of inertia of the right half of the beam, BD, is twice the moment of inertia of the left half, AB), so we cannot use the formulas given inside the front cover of the book for computing deflections. Therefore, we will use the conjugate-beam method, discussed in Chapter 6, for determining the deflections of the primary beam.

To determine the deflection Δ_{BO} due to the external loading, we draw the conjugate beams for the 15-kN/m uniformly distributed load and the 60-kN concentrated load, as shown in Fig. 13.5(d) and (e), respectively. Recalling that the deflection at a point on a real beam is equal to the bending moment at that point in the corresponding conjugate beam, we determine the deflection Δ_{BO} due to the combined effect of the distributed and concentrated loads as

$$EI\Delta_{BO} = \left[-4{,}218.75(10) + \left(\frac{2}{3}\right)(10)(750)\left(\frac{30}{8}\right) \right]$$

$$+ \left[-718.75(10) + \left(\frac{1}{2}\right)(10)(150)\left(\frac{10}{3}\right) \right]$$

$$\Delta_{BO} = -\frac{28{,}125 \text{ kN} \cdot \text{m}^3}{EI}$$

in which the negative sign indicates that the deflection occurs in the downward direction. Note that although the numerical values of E and I are given, it is usually convenient to carry out the analysis in terms of EI. The flexibility coefficient f_{BB} can be computed similarly by using the conjugate beam shown in Fig. 13.5(f). Thus

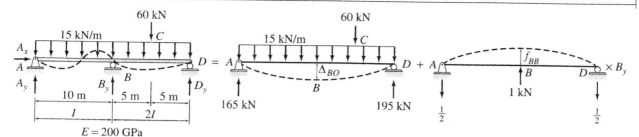

15 kN/m 60 kN ↓C

A_x

A ———— D

A_y B_y D_y

B

10 m   5 m   5 m

I   2I

$E = 200$ GPa
$I = 700(10^6)$mm^4

(a) Indeterminate Beam

60 kN ↓C

15 kN/m

A ———— D

Δ_{BO}

B

165 kN   195 kN

(b) Primary Beam Subjected
to External Loading

60 kN

f_{BB}

A ———— $\times B_y$

B D

1 kN

$\frac{1}{2}$ $\frac{1}{2}$

(c) Primary Beam Loaded
with Redundant B_y

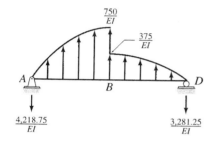

$\frac{750}{EI}$

$\frac{375}{EI}$

A   B   D

$\frac{4,218.75}{EI}$ $\frac{3,281.25}{EI}$

(d) Conjugate Beam for Uniform Load

$\frac{150}{EI}$ $\frac{75}{EI}$

A   B   C   D

$\frac{112.5}{EI}$

$\frac{718.75}{EI}$ $\frac{781.25}{EI}$

(e) Conjugate Beam for
Concentrated Load

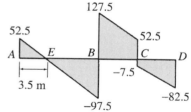

A   B   D

$\frac{20.833}{EI}$ $\frac{2.5}{EI}$ $\frac{16.667}{EI}$

$\frac{5}{EI}$

(f) Conjugate Beam for Unit
Value of Redundant B_y

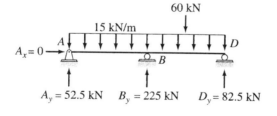

60 kN

15 kN/m

$A_x = 0$ A ———— D

B

$A_y = 52.5$ kN $B_y = 225$ kN $D_y = 82.5$ kN

(g) Support Reactions for Indeterminate Beam

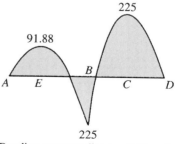

127.5

52.5   52.5

A   E   B   C   D

3.5 m   −7.5

−97.5   −82.5

Shear diagram (kN)

225

91.88

A   E   B   C   D

225

Bending moment diagram (kN · m)

(h) Shear and Bending Moment Diagrams
for Indeterminate Beam

FIG. 13.5

continued

$$EIf_{BB} = 20.833(10) - \left(\frac{1}{2}\right)(10)(5)\left(\frac{10}{3}\right) = 125 \text{ kN} \cdot \text{m}^3/\text{kN}$$

$$f_{BB} = \frac{125 \text{ kN} \cdot \text{m}^3/\text{kN}}{EI}$$

Magnitude of the Redundant By substituting the values of Δ_{BO} and f_{BB} into the compatibility equation (Eq. (1)), we obtain

$$-\frac{28,125}{EI} + \left(\frac{125}{EI}\right)B_y = 0 \qquad B_y = 225 \text{ kN} \uparrow \qquad \text{Ans.}$$

Reactions To determine the remaining reactions of the indeterminate beam, we apply the equilibrium equations (Fig. 13.5(g)):

$$+ \rightarrow \sum F_x = 0 \qquad A_x = 0 \qquad\qquad \text{Ans.}$$

$$+ \circlearrowleft \sum M_D = 0 \qquad -A_y(20) - 225(10) + 15(20)(10) + 60(5) = 0$$

$$A_y = 52.5 \text{ kN} \uparrow \qquad\qquad \text{Ans.}$$

$$+ \uparrow \sum F_y = 0 \qquad 52.5 + 225 - 15(20) - 60 + D_y = 0$$

$$D_y = 82.5 \text{ kN} \uparrow \qquad\qquad \text{Ans.}$$

Shear and Bending Moment Diagrams See Fig. 13.5(h).   Ans.

Example 13.4

Determine the reactions and the force in each member of the truss shown in Fig. 13.6(a) using the method of consistent deformations.

Solution

Degree of Indeterminacy The truss is indeterminate to the first degree.

Primary Truss The horizontal reaction D_x at the hinged support D is selected to be the redundant. The direction of D_x is arbitrarily assumed to the right, as shown in Fig. 13.6(a). The primary truss is obtained by removing the restraint against horizontal displacement at joint D by replacing the hinged support by a roller support, as shown in Fig. 13.6(b). Next, the primary truss is subjected separately to the external loading and a unit value of the unknown redundant D_x, as shown in Fig. 13.6(b) and (c), respectively.

Compatibility Equation If Δ_{DO} denotes the horizontal deflection at joint D of the primary truss due to external loading and if f_{DD} denotes the flexibility coefficient representing the horizontal deflection at D due to the unit value of the redundant D_x, then the compatibility equation can be written as

$$\Delta_{DO} + f_{DD}D_x = 0 \qquad\qquad (1)$$

Deflections of Primary Truss The deflections Δ_{DO} and f_{DD} can be evaluated by using the virtual work method. Recall from Chapter 7 that the virtual work expression for truss deflections is given by (Eq. (7.23))

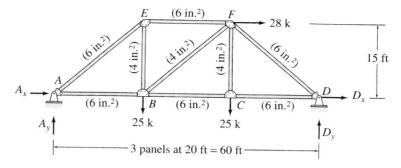

$E = 29{,}000$ ksi

(a) Indeterminate Truss

$=$

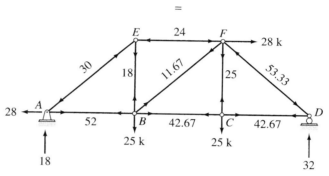

(b) Primary Truss Subjected to External Loads—F_O Forces

$+$

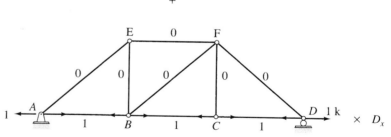

(c) Primary Truss Subjected to Unit Value of Redundant D_x — u_D Forces

FIG. **13.6**

$$\Delta = \sum \frac{FF_v L}{AE} \qquad (2)$$

in which F symbolically represents the axial forces in truss members due to the real loading that causes the deflection Δ, and F_v represents the axial forces in the truss members due to a virtual unit load acting at the joint and in the direction of the desired deflection Δ.

continued

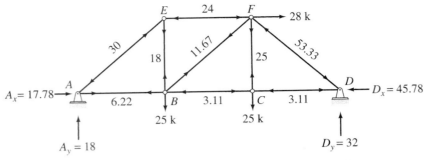

(d) Support Reactions and Member Forces for Indeterminate Truss

For computing the deflection Δ_{DO} of the primary truss, the real system consists of the given external loading, as shown in Fig. 13.6(b). The member axial forces due to this loading are symbolically denoted as F_O forces, and their numerical values, obtained by the method of joints, are shown in Fig. 13.6(b). The virtual system for Δ_{DO} consists of a unit load applied at the location and in the direction of the redundant D_x, which is the same as the system shown in Fig. 13.6(c) (without the multiplier D_x). The member axial forces due to the unit value of the redundant D_x are symbolically denoted as u_D forces, and their numerical values, obtained by the method of joints, are shown in Fig. 13.6(c). Thus, the virtual work expression for Δ_{DO} can be written as

$$\Delta_{DO} = \sum \frac{F_O u_D L}{AE} \tag{3}$$

The F_O and u_D member forces are then tabulated, and Eq. (3) is applied to determine Δ_{DO}, as shown in Table 13.1. Thus

$$\Delta_{DO} = \frac{5{,}493.6 \text{ k/in.}}{E}$$

The positive magnitude of Δ_{DO} indicates that the deflection occurs to the right—that is, in the same direction as that assumed for the redundant D_x.

For computing the flexibility coefficient f_{DD}, both the real and the virtual systems consist of a unit value of the redundant D_x applied to the primary truss, as shown in Fig. 13.6(c) (without the multiplier D_x). Thus, the virtual work expression for f_{DD} becomes

$$f_{DD} = \sum \frac{u_D^2 L}{AE} \tag{4}$$

Equation (4) is applied to determine f_{DD}, as shown in Table 13.1. Thus,

$$f_{DD} = \frac{120(1/\text{in.})}{E}$$

Magnitude of Redundant By substituting the values of Δ_{DO} and f_{DD} into the compatibility equation (Eq. (1)), we determine the redundant D_x to be

$$\frac{5{,}493.6}{E} + \left(\frac{120}{E}\right)D_x = 0$$

$$D_x = -45.78 \text{ k}$$

TABLE 13.1

Member	L (in.)	A (in.^{2})	F_O (k)	u_D (k/k)	$\dfrac{F_O u_D L}{A}$ (k/in.)	$\dfrac{u_D^2 L}{A}$ (1/in.)	$F = F_O + u_D D_x$ (k)
AB	240	6	52	1	2,080	40	6.22
BC	240	6	42.67	1	1,706.8	40	−3.11
CD	240	6	42.67	1	1,706.8	40	−3.11
EF	240	6	−24	0	0	0	−24
BE	180	4	18	0	0	0	18
CF	180	4	25	0	0	0	25
AE	300	6	−30	0	0	0	−30
BF	300	4	11.67	0	0	0	11.67
DF	300	6	−53.33	0	0	0	−53.33
				Σ	5,493.6	120	

$$\Delta_{DO} = \frac{1}{E}\sum \frac{F_O u_D L}{A} = \frac{5{,}493.6 \text{ k/in.}}{E} \qquad f_{DD} = \frac{1}{E}\sum \frac{u_D^2 L}{A} = \frac{120(1/\text{in.})}{E}$$

$$D_x = -\frac{\Delta_{DO}}{f_{DD}} = -45.78 \text{ k}$$

The negative answer for D_x indicates that our initial assumption about D_x acting to the right was incorrect and that D_x actually acts to the left.

$$D_x = 45.78 \text{ k} \leftarrow \qquad \text{Ans.}$$

Reactions The remaining reactions of the indeterminate truss can now be determined by superposition of the reactions of the primary truss due to the external loads (Fig. 13.6(b)) and due to the redundant D_x (Fig. 13.6(c)).

$$A_x = -28 - 1(-45.78) = 17.78 \text{ k} \rightarrow \qquad \text{Ans.}$$

$$A_y = 18 \text{ k} \uparrow \qquad \text{Ans.}$$

$$D_y = 32 \text{ k} \uparrow \qquad \text{Ans.}$$

The reactions are shown in Fig. 13.6(d).

Member Axial Forces The axial forces in the members of the indeterminate truss can be determined by superposition of the member forces of the primary truss due to the external loads and due to the redundant D_x; that is,

$$F = F_O + u_D D_x \qquad (5)$$

The computation of final member forces can be conveniently carried out in a tabular form, as shown in Table 13.1. For each member, the final force F is computed by algebraically adding the entry in the fourth column (F_O), to the corresponding entry in the fifth column (u_D) multiplied by the magnitude of the redundant $D_x = -45.78$ k. The value of the final force thus computed is then recorded in the eighth column, as shown in Table 13.1. The member forces thus obtained are also shown in Fig. 13.6(d).

$\qquad$ Ans.

Example 13.5

Determine the reactions and draw the shear and bending moment diagrams for the frame shown in Fig. 13.7(a) by the method of consistent deformations.

Solution

Degree of Indeterminacy The frame is indeterminate to the first degree.

Primary Frame The horizontal reaction A_X at the hinged support A is selected to be the redundant. The primary frame is obtained by removing the restraint against horizontal displacement at joint A, which is done by replacing the hinged support by a roller support, as shown in Fig. 13.7(b). Next, the primary frame is subjected separately to the external loading and a unit value of the unknown redundant A_X, as shown in Fig. 13.7(b) and (c), respectively.

Compatibility Equation From Fig. 13.7(a), (b), and (c), we observe that

$$\Delta_{AO} + f_{AA}A_X = 0 \tag{1}$$

Deflections of Primary Frame The deflections Δ_{AO} and f_{AA} of the primary frame will be evaluated by using the virtual work method discussed in Chapter 7. The virtual work expression for Δ_{AO}, which represents the horizontal deflection at joint A of the primary frame due to external loading, can be written as

$$\Delta_{AO} = \sum \int \frac{M_O m_A}{EI} dx \tag{2}$$

in which M_O denotes the bending moments due to the (real) external loading (Fig. 13.7(b)) and m_A denotes the bending moments due to a (virtual) unit load at the location and in the direction of the redundant (Fig. 13.7(c)). The x coordinates used for determining the bending moment equations for members AB and BC of the primary frame are shown in Fig. 13.7(b) and (c), and the equations for M_O and m_A are tabulated in Table 13.2. By applying Eq. (2), we obtain

$$\Delta_{AO} = \frac{1}{EI} \int_0^{30} \left(45x - \frac{3}{2}x^2 \right) \left(-20 + \frac{2}{3}x \right) dx = -\frac{67{,}500 \text{ k-ft}^3}{EI}$$

For computing the flexibility coefficient f_{AA}, both the real and virtual systems consist of a unit value of the redundant A_X applied to the primary frame, as shown in Fig. 13.7(c) (without the multiplier A_X). Thus, the virtual work expression for f_{AA} becomes

$$f_{AA} = \sum \int \frac{m_A^2}{EI} dx \tag{3}$$

By substituting the equations for m_A from Table 13.2, we obtain

$$f_{AA} = \frac{1}{EI} \left[\int_0^{20} (-x)^2 dx + \int_0^{30} \left(-20 + \frac{2}{3}x \right)^2 dx \right] = \frac{6{,}666.66 \text{ ft}^3}{EI}$$

Magnitude of the Redundant By substituting the values of Δ_{AO} and f_{AA} into the compatibility equation (Eq. (1)), we determine the redundant A_X to be

$$-\frac{67{,}500}{EI} + \left(\frac{6{,}666.66}{EI} \right) A_X = 0$$

$$A_X = 10.13 \text{ k} \rightarrow \qquad \text{Ans.}$$

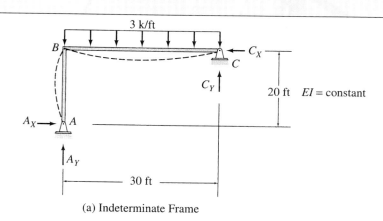

(a) Indeterminate Frame

=

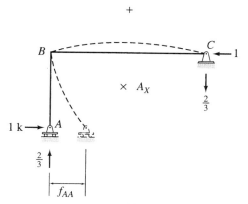

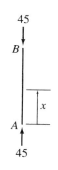

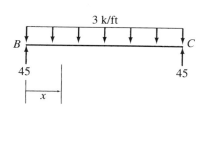

(b) Primary Frame Subjected to External Loading — M_O Moments

+

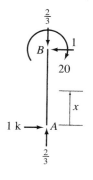

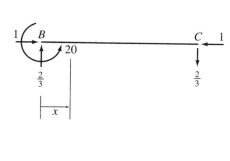

(c) Primary Frame Subjected to Unit Value of Redundant A_X — m_A Moments

FIG. 13.7

continued

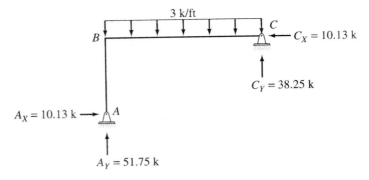

Reactions

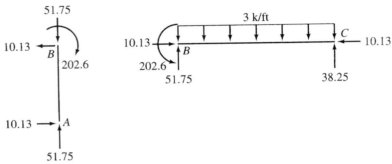

Member end forces

(d) Support Reactions and Member End Forces for Indeterminate Frame

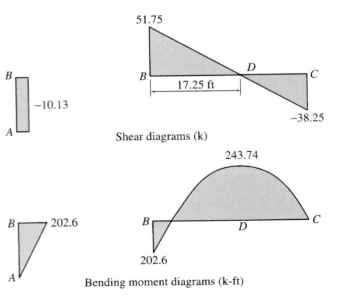

FIG. 13.7 (contd.)

(e) Shear and Bending Moment Diagrams for Indeterminate Frame

TABLE 13.2

| Member | x coordinate | | M_O (k-ft) | m_A (k-ft/k) |
	Origin	Limits (ft)		
AB	A	0–20	0	$-1x$
BC	B	0–30	$45x - \dfrac{3}{2}x^2$	$-20 + \dfrac{2}{3}x$

Reactions   The remaining reactions and member end forces of the indeterminate frame can now be determined from equilibrium. The reactions and member end forces thus obtained are shown in Fig. 13.7(d).   **Ans.**

Shear and Bending Moment Diagrams   See Fig. 13.7(e).   **Ans.**

13.2 INTERNAL FORCES AND MOMENTS AS REDUNDANTS

Thus far, we have analyzed externally indeterminate structures with a single degree of indeterminacy by selecting a support reaction as the redundant. The analysis of such structures can also be carried out by choosing an internal force or moment as the redundant, provided that the removal of the corresponding internal restraint from the indeterminate structure results in a primary structure that is statically determinate and stable.

Consider the two-span continuous beam shown in Fig. 13.8(a). The beam is indeterminate to the first degree. As discussed in the preceding section, this beam can be analyzed by treating one of the vertical reactions as the redundant. However, it is usually more convenient to analyze continuous beams (especially those with unequal spans) by selecting internal bending moments as redundants. Let us consider the analysis of the beam of Fig. 13.8(a) by using the bending moment, M_B, at the interior support B as the redundant. From Fig. 13.8(a), we can see that the slope of the elastic curve of the indeterminate beam is continuous at B. In other words, there is no change of slope of the tangents to the elastic curve at just to the left of B and at just to the right of B; that is, the angle between the tangents is zero. When the restraint corresponding to the redundant bending moment M_B is removed by inserting an internal hinge at B, as shown in Fig. 13.8(b), a discontinuity develops in the slope of the elastic curve at B, in the sense that the tangent to the elastic curve at just to the left of B rotates relative to the tangent at just to the right of B. The change of slope (or the angle) between the two tangents due to the external loads is denoted by $\theta_{BO\ rel.}$ and can be expressed as

$$\theta_{BO\ rel.} = \theta_{BL} + \theta_{BR} \tag{13.8}$$

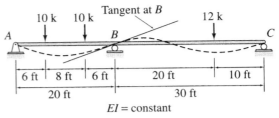

(a) Indeterminate Beam

=

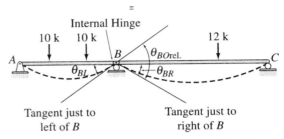

(b) Primary Beam Subjected to External Loading

+

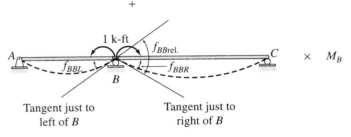

(c) Primary Beam Loaded with Redundant M_B

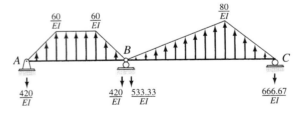

(d) Conjugate Beams for External Loading

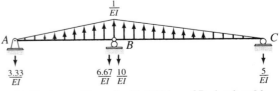

(e) Conjugate Beam for Unit Value of Redundant M_B

FIG. 13.8

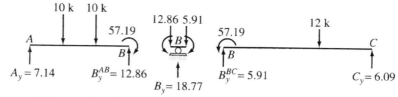

(f) Member End Forces and Support Reactions for Indeterminate Beam

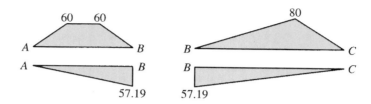

(g) Bending Moment Diagrams for Members AB and BC (k-ft)

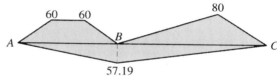

(h) Bending Moment Diagram for Continuous Beam (k-ft)

FIG. 13.8 (contd.)

(see Fig. 13.8(b)) in which θ_{BL} and θ_{BR} denote the slopes at the ends B of the left and right spans of the beam, respectively, due to the given external loading.

Since the redundant bending moment M_B provides continuity of slope of the elastic curve at B in the actual indeterminate beam, it must be of sufficient magnitude to remove the discontinuity $\theta_{BO\ rel.}$ from the primary beam by bringing the tangents back together. To evaluate the effect of M_B on the primary beam, we determine the flexibility coefficient $f_{BB\ rel.}$ representing the change of slope (or the angle) between the tangents to the elastic curve at just to the left of B and at just to the right of B due to a unit value of M_B, as shown in Fig. 13.8(c). An internal bending moment is defined by a *pair* of equal but opposite couples. Thus, two opposite couples of unit magnitude must be applied to the primary beam to determine the flexibility coefficient, as shown in Fig. 13.8(c). Note that the redundant M_B is considered to be positive in ac-cordance with the *beam convention*—that is, when it causes compression in the upper fibers and tension in the lower fibers of the beam. From Fig. 13.8(c), we can see that the flexibility coefficient can be expressed as

$$f_{BB\ rel.} = f_{BBL} + f_{BBR} \tag{13.9}$$

in which f_{BBL} and f_{BBR} denote the slopes at the ends B of the left and the right spans of the beam, respectively, due to the unit value of the redundant M_B.

The compatibility equation is based on the requirement that the slope of the elastic curve of the actual indeterminate beam is continuous at B; that is, there is no change of slope from just to the left of B to just to the right of B. Therefore, the algebraic sum of the angles between the tangents at just to the left and at just to the right of B due to the external loading and the redundant M_B must be zero. Thus,

$$\theta_{BO \text{ rel.}} + f_{BB \text{ rel.}} M_B = 0 \qquad (13.10)$$

which can be solved for the redundant bending moment M_B after the changes of slopes $\theta_{BO \text{ rel.}}$ and $f_{BB \text{ rel.}}$ have been evaluated.

Since each of the spans of the primary beam can be treated as a simply supported beam, the slopes at the ends B of the left and the right spans can be easily computed by using the conjugate-beam method. The conjugate beams for the external loading are shown in Fig. 13.8(d). Recalling that the slope at a point on a real beam is equal to the shear at that point on the corresponding conjugate beam, we determine the slopes θ_{BL} and θ_{BR} at ends B of the left and the right spans, respectively, as

$$\theta_{BL} = \frac{420 \text{ k-ft}^2}{EI} \quad \text{and} \quad \theta_{BR} = \frac{533.33 \text{ k-ft}^2}{EI}$$

Thus, from Eq. (13.8), we obtain

$$\theta_{BO \text{ rel.}} = \theta_{BL} + \theta_{BR} = \frac{420 + 533.33}{EI} = \frac{953.33 \text{ k-ft}^2}{EI}$$

The flexibility coefficient $f_{BB \text{ rel.}}$ can be computed similarly by using the conjugate beam for a unit value of the redundant M_B shown in Fig. 13.8(e). Thus

$$f_{BBL} = \frac{6.67 \text{ k-ft}^2/\text{k-ft}}{EI} \quad \text{and} \quad f_{BBR} = \frac{10 \text{ k-ft}^2/\text{k-ft}}{EI}$$

From Eq. (13.9), we obtain

$$f_{BB \text{ rel.}} = f_{BBL} + f_{BBR} = \frac{6.67 + 10}{EI} = \frac{16.67 \text{ k-ft}^2/\text{k-ft}}{EI}$$

By substituting the values of $\theta_{BO \text{ rel.}}$ and $f_{BB \text{ rel.}}$ into the compatibility equation (Eq. (13.10)), we determine the magnitude of the redundant M_B as

$$\frac{953.33}{EI} + \left(\frac{16.67}{EI}\right) M_B = 0$$

or

$$M_B = -57.19 \text{ k-ft}$$

With the redundant M_B known, the forces at the ends of the members as well as the support reactions can be determined by considering the equilibrium of the free bodies of the members AB and BC and joint B, as shown in Fig. 13.8(f). Note that the negative bending moment M_B is applied at the ends B of members AB and BC so that it causes tension in the upper fibers and compression in the lower fibers of the members.

When moments at the ends of the members of a continuous beam are known, it is usually convenient to construct its bending moment diagram in two parts; one for the external loading and another for the member end moments. This procedure is commonly referred to as constructing the *bending moment diagram by simple-beam parts*, because each member of the continuous beam is treated as a simply supported beam, to which the external loads and the end moments are applied separately and the corresponding bending moment diagrams are drawn. Such diagrams for the members AB and BC of the continuous beam under consideration are shown in Fig. 13.8(g). The member bending moment diagrams can be drawn together, as shown in Fig. 13.8(h), to obtain the bending moment diagram for the entire continuous beam.

Internally Indeterminate Structures

As the foregoing discussion indicates, structures with a single degree of indeterminacy that are externally indeterminate can be analyzed by selecting either a reaction or an internal force or moment as the redundant. However, if a structure is internally indeterminate but externally determinate, then only an internal force or moment can be used as the redundant, because the removal of an external reaction from such a structure will yield a statically unstable primary structure.

Consider, for example, the truss shown in Fig. 13.9(a). The truss consists of six members connected together by four joints and is supported by three reaction components. Thus, as discussed in Section 4.4, the degree of indeterminacy of the truss is equal to $(m + r) - 2j = (6 + 3) - 2(4) = 1$. Because the three reactions can be determined from the three equations of equilibrium of the entire truss, the truss is internally indeterminate to the first degree; that is, it contains one extra member than required for internal stability.

To analyze the truss, we must select the axial force in one of its members to be the redundant. Suppose that we select the force F_{AD} in the diagonal member AD to be the redundant. The restraint corresponding to F_{AD} is then removed from the truss by cutting member AD to obtain the primary truss shown in Fig. 13.9(b). Note that since member AD can no longer sustain a force, the primary truss is statically determinate. When the primary truss is subjected to the external load P, it deforms and a gap Δ_{ADO} opens up between the ends of the two portions of member AD, as shown in Fig. 13.9(b). Since no such gap exists

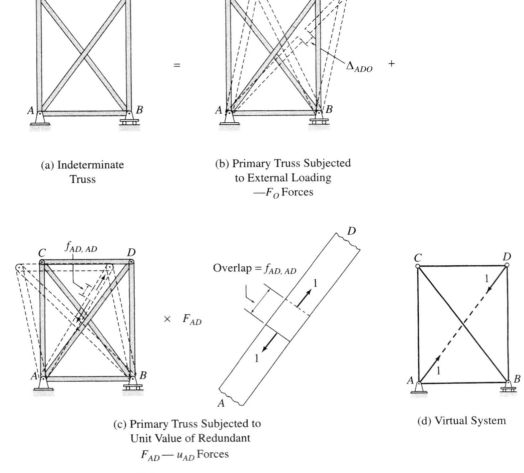

(a) Indeterminate
Truss

(b) Primary Truss Subjected
to External Loading
$-F_O$ Forces

(c) Primary Truss Subjected to
Unit Value of Redundant
$F_{AD}-u_{AD}$ Forces

(d) Virtual System

FIG. 13.9

in the actual indeterminate truss, we conclude that the redundant force F_{AD} must be of sufficient magnitude to bring the ends of the two portions of member AD back together to close the gap. To evaluate the effect of F_{AD} in closing the gap, we subject the primary truss to a unit value of F_{AD} by applying equal and opposite unit axial loads to the two portions of member AD, as shown in Fig. 13.9(c). Note that the actual sense of the redundant F_{AD} is not yet known and is arbitrarily assumed to be tensile, with the unit axial forces tending to elongate the portions of member AD, as shown in the figure. The unit value of F_{AD} deforms the primary truss and causes the ends of the two portions of member AD to overlap by an amount $f_{AD,AD}$, as shown in Fig. 13.9(c). Thus, the overlap in member AD due to the axial force of magnitude F_{AD} equals $f_{AD,AD}F_{AD}$.

Since neither a gap nor an overlap exists in member AD in the actual indeterminate truss, we can express the compatibility equation as

$$\Delta_{ADO} + f_{AD,AD}F_{AD} = 0 \tag{13.11}$$

which can be solved for the redundant axial force F_{AD} after the magnitudes of Δ_{ADO} and $f_{AD,AD}$ have been determined.

Note that Δ_{ADO} and $f_{AD,AD}$ are actually relative displacements between the joints A and D of the primary truss. These displacements can be conveniently computed using the virtual work method by employing a virtual system consisting of two unit loads applied with opposite senses in the direction of member AD at joints A and D, as shown in Fig. 13.9(d). A comparison of Fig. 13.9(c) and (d) indicates that the axial forces in the members of the primary truss due to virtual unit loads (Fig. 13.9(d)) will be the same as the u_{AD} forces due to the unit axial force in member AD (Fig. 13.9(c)). Thus the truss with a unit axial force in member AD can be used as the virtual system for computing the relative displacements. If the member axial forces due to the external load P are symbolically denoted as F_O forces (Fig. 13.9(b)), then the virtual work expression for Δ_{ADO} can be written as

$$\Delta_{ADO} = \sum \frac{F_O u_{AD}L}{AE} \tag{13.12}$$

For computing the flexibility coefficient $f_{AD,AD}$, both the real and the virtual systems consist of a unit axial force in member AD, as shown in Fig. 13.9(c). Thus, the virtual work expression for $f_{AD,AD}$ is given by

$$f_{AD,AD} = \sum \frac{u_{AD}^2 L}{AE} \tag{13.13}$$

in which the force in the redundant member AD must be included in the summation to take into account the deformation of this member.

Once the relative displacements Δ_{ADO} and $f_{AD,AD}$ have been evaluated, their values are substituted into the compatibility equation (Eq. (13.11)), which is then solved for the redundant F_{AD}. With the redundant F_{AD} known, the axial forces in the members of the indeterminate truss can be determined by superposition of the member forces of the primary truss due to the external load P and due to the redundant F_{AD}; that is,

$$F = F_O + u_{AD}F_{AD} \tag{13.14}$$

Example 13.6

Determine the reactions and draw the bending moment diagram for the two-span continuous beam shown in Fig. 13.10(a) by the method of consistent deformations. Select the bending moment at the interior support B to be the redundant.

Solution

This beam was analyzed in Example 13.3 by selecting the vertical reaction at support B as the redundant.

Primary Beam The primary beam is obtained by removing the restraint corresponding to the redundant bending moment M_B by inserting an internal hinge at B in the given indeterminate beam, as shown in Fig. 13.10(b). Next, the primary beam is subjected separately to the external loading and a unit value of the redundant M_B, as shown in Fig. 13.10(b) and (c), respectively.

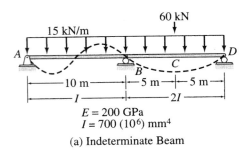

$$E = 200 \text{ GPa}$$
$$I = 700 \ (10^6) \text{ mm}^4$$

(a) Indeterminate Beam

=

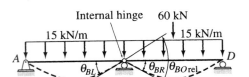

(b) Primary Beam Subjected to External Loading

+

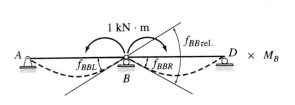

(c) Primary Beam Loaded with Redundant M_B

FIG. 13.10

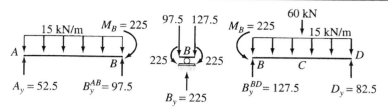

(d) Member End Forces and Support Reactions for Indeterminate Beam

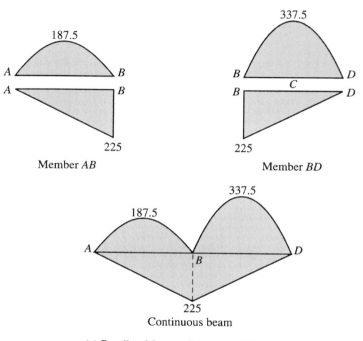

Member *AB*   Member *BD*

Continuous beam

FIG. **13.10** (contd.)   (e) Bending Moment Diagrams (kN · m)

Compatibility Equation See Fig. 13.10(b) and (c):

$$\theta_{BO \text{ rel.}} + f_{BB \text{ rel.}} M_B = 0 \tag{1}$$

Slopes of Primary Beam Each of the spans of the primary beam can be treated as a simply supported beam of constant flexural rigidity *EI*, so we can use the beam-deflection formulas given inside the front cover of the book for evaluating the changes of slopes $\theta_{BO \text{ rel.}}$ and $f_{BB \text{ rel.}}$. From Fig. 13.10(b), we can see that

$$\theta_{BO \text{ rel.}} = \theta_{BL} + \theta_{BR}$$

continued

in which θ_{BL} and θ_{BR} are the slopes at the ends B of the left and the right spans of the primary beam, respectively, due to the external loading. By using the deflection formulas, we obtain

$$\theta_{BL} = \frac{15(10)^3}{24EI} = \frac{625 \text{ kN} \cdot \text{m}^2}{EI}$$

$$\theta_{BR} = \frac{15(10)^3}{24E(2I)} + \frac{60(10)^2}{16E(2I)} = \frac{500 \text{ kN} \cdot \text{m}^2}{EI}$$

Thus,

$$\theta_{BO \text{ rel.}} = \frac{625}{EI} + \frac{500}{EI} = \frac{1,125 \text{ kN} \cdot \text{m}^2}{EI}$$

The flexibility coefficient $f_{BB \text{ rel.}}$ can be computed in a similar manner. From Fig. 13.10(c), we can see that

$$f_{BB \text{ rel.}} = f_{BBL} + f_{BBR}$$

in which

$$f_{BBL} = \frac{10}{3EI} = \frac{3.33 \text{ m}}{EI} \quad \text{and} \quad f_{BBR} = \frac{10}{3E(2I)} = \frac{1.67 \text{ m}}{EI}$$

Thus,

$$f_{BB \text{ rel.}} = \frac{3.33}{EI} + \frac{1.67}{EI} = \frac{5 \text{ m}}{EI}$$

Magnitude of the Redundant By substituting the values of $\theta_{BO \text{ rel.}}$ and $f_{BB \text{ rel.}}$ into the compatibility equation (Eq. (1)), we obtain

$$\frac{1,125}{EI} + \left(\frac{5}{EI}\right)M_B = 0$$

$$M_B = -225 \text{ kN} \cdot \text{m} \qquad \qquad \text{Ans.}$$

Reactions The forces at the ends of the members AB and BD of the continuous beam can now be determined by applying the equations of equilibrium to the free bodies of the members shown in Fig. 13.10(d). By considering the equilibrium of member AB, we obtain

$$A_y = \left(\frac{1}{2}\right)(15)(10) - \left(\frac{225}{10}\right) = 52.5 \text{ kN} \uparrow \qquad \text{Ans.}$$

$$B_y^{AB} = \left(\frac{1}{2}\right)(15)(10) + \left(\frac{225}{10}\right) = 97.5 \text{ kN} \uparrow$$

Similarly, for member BD,

$$B_y^{BD} = \left(\frac{1}{2}\right)(15)(10) + \left(\frac{60}{2}\right) + \left(\frac{225}{10}\right) = 127.5 \text{ kN} \uparrow$$

$$D_y = \left(\frac{1}{2}\right)(15)(10) + \left(\frac{60}{2}\right) - \left(\frac{225}{10}\right) = 82.5 \text{ kN} \uparrow \qquad \text{Ans.}$$

By considering the equilibrium of joint B in the vertical direction, we obtain

$$B_y = B_y^{AB} + B_y^{BD} = 97.5 + 127.5 = 225 \text{ kN} \uparrow \qquad \text{Ans.}$$

Bending Moment Diagram The bending moment diagram for the continuous beam, constructed by simple-beam parts, is shown in Fig. 13.10(e). The two parts of the diagram due to the external loading and the member end moments may be superimposed, if so desired, to obtain the resultant bending moment diagram shown in Example 13.3.   Ans.

Example 13.7

Determine the reactions and the force in each member of the truss shown in Fig. 13.11(a) by the method of consistent deformations.

Solution

Degree of Indeterminacy The truss consists of ten members connected by six joints and is supported by three reaction components. Thus the degree of indeterminacy of the truss is equal to $(m + r) - 2j = (10 + 3) - 2(6) = 1$. The three reactions can be determined from the three equations of external equilibrium, so the truss is internally indeterminate to the first degree.

Primary Truss The axial force F_{CE} in the diagonal member CE is selected to be the redundant. The sense of F_{CE} is arbitrarily assumed to be tensile. The primary truss obtained by removing member CE is shown in Fig. 13.11(b). Next, the primary truss is subjected separately to the external loading and a unit tensile force in the redundant member CE, as shown in Fig. 13.11(b) and (c), respectively.

Compatibility Equation The compatibility equation can be expressed as

$$\Delta_{CEO} + f_{CE, CE} F_{CE} = 0 \tag{1}$$

in which Δ_{CEO} denotes the relative displacement between joints C and E of the primary truss due to the external loads, and the flexibility coefficient $f_{CE, CE}$ denotes the relative displacement between the same joints due to a unit value of the redundant F_{CE}.

Deflections of Primary Truss The virtual work expression for Δ_{CEO} can be written as

$$\Delta_{CEO} = \sum \frac{F_O u_{CE} L}{AE} \tag{2}$$

in which F_O and u_{CE} represent, respectively, the member forces due to the external loads and the unit tensile force in member CE. The numerical values of these forces are computed by the method of joints (Fig. 13.11(b) and (c)) and are tabulated in Table 13.3. Equation (2) is then applied as shown in Table 13.3, to obtain

$$\Delta_{CEO} = -\frac{1,116 \text{ k-ft}}{AE}$$

continued

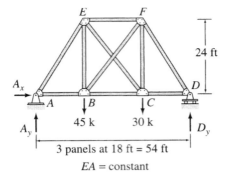

24 ft

A_x

A B C D

45 k   30 k

A_y D_y

3 panels at 18 ft = 54 ft

EA = constant

(a) Indeterminate Truss

=

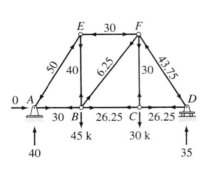

E 30 F

50   40   6.25   30   43.75

0 A D

30 B 26.25 C 26.25

45 k   30 k

40   35

(b) Primary Truss Subjected to
External Loads—F_O Forces

+

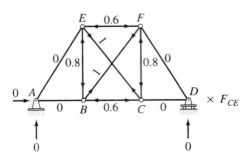

E 0.6 F

1

0   0.8   0.8   0

1

0 A D $\times F_{CE}$

0   B 0.6 C 0

0   0

(c) Primary Truss Subjected to Unit Tensile
Force in Member CE—u_{CE} Forces

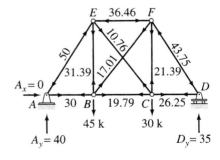

E 36.46 F

10.76

50   43.75

31.39   17.01   21.39

$A_x = 0$ D

A 30 B 19.79 C 26.25

45 k   30 k

$A_y = 40$ $D_y = 35$

(d) Support Reactions and Member
Forces for Indeterminate Truss

FIG. 13.11

TABLE 13.3

Member	L (ft)	F_O (k)	u_{CE} (k/k)	$F_O u_{CE} L$ (k-ft)	$u_{CE}^2 L$ (ft)	$F = F_O + u_{CE} F_{CE}$ (k)
AB	18	30	0	0	0	30
BC	18	26.25	−0.6	−283.5	6.48	19.79
CD	18	26.25	0	0	0	26.25
EF	18	−30	−0.6	324	6.48	−36.46
BE	24	40	−0.8	−768	15.36	31.39
CF	24	30	−0.8	−576	15.36	21.39
AE	30	−50	0	0	0	−50
BF	30	6.25	1	187.5	30	17.01
CE	30	0	1	0	30	10.76
DF	30	−43.75	0	0	0	−43.75
			Σ	−1,116	103.68	

$$\Delta_{CEO} = \frac{1}{AE} \sum F_0 u_{CE} L = -\frac{1,116 \text{ k-ft}}{AE}$$

$$f_{CE,CE} = \frac{1}{AE} \sum u_{CE}^2 L = \frac{103.68 \text{ ft}}{AE}$$

$$F_{CE} = -\frac{\Delta_{CEO}}{f_{CE,CE}} = 10.76 \text{ k (T)}$$

Next, the flexibility coefficient $f_{CE,CE}$ is computed by using the virtual work expression (see Table 13.3):

$$f_{CE,CE} = \sum \frac{u_{CE}^2 L}{AE} = \frac{103.68 \text{ ft}}{AE}$$

Magnitude of the Redundant By substituting the values of Δ_{CEO} and $f_{CE,CE}$ into the compatibility equation (Eq. (1)), we determine the redundant F_{CE} to be

$$-\frac{1,116}{AE} + \left(\frac{103.68}{AE}\right) F_{CE} = 0$$

$$F_{CE} = 10.76 \text{ k (T)} \qquad \text{Ans.}$$

Reactions See Fig. 13.11(d). Note that the reactions due to the redundant F_{CE} are zero, as shown in Fig. 13.11(c). Ans.

Member Axial Forces The forces in the remaining members of the indeterminate truss can now be determined by using the superposition relationship:

$$F = F_O + u_{CE} F_{CE}$$

The member forces thus obtained are shown in Table 13.3 and Fig. 13.11(d).
Ans.

13.3 STRUCTURES WITH MULTIPLE DEGREES OF INDETERMINACY

The method of consistent deformations developed in the preceding sections for analyzing structures with a single degree of indeterminacy can easily be extended to the analysis of structures with multiple degrees of indeterminacy. Consider, for example, the four-span continuous beam subjected to a uniformly distributed load w, as shown in Fig. 13.12(a). The beam is supported by six support reactions; thus its degree of indeterminacy is equal to $6 - 3 = 3$. To analyze the beam, we must select three support reactions as redundants. Suppose that we select the vertical reactions $B_y, C_y,$ and D_y at the interior supports $B, C,$ and D, respectively, to be the redundants. The roller supports at $B, C,$ and D are then removed from the given indeterminate beam to obtain the statically determinate and stable primary beam, as shown in Fig. 13.12(b). The three redundants are now treated as unknown loads on the primary beam, and their magnitudes can be determined from the compatibility conditions that the deflections of the primary beam at the locations $B, C,$ and D of the redundants due to the combined effect of the known external load w and the unknown redundants $B_y, C_y,$ and D_y must be equal to zero. This is because the deflections of the given indeterminate beam at the roller supports $B, C,$ and D are zero.

To establish the compatibility equations, we subject the primary beam separately to the external load w (Fig. 13.12(b)) and a unit value of each of the redundants $B_y, C_y,$ and D_y (Fig. 13.12(c), (d), and (e), respectively). As shown in Fig. 13.12(b), the deflections of the primary beam at points $B, C,$ and D due to the external load w are denoted by $\Delta_{BO}, \Delta_{CO},$ and Δ_{DO}, respectively. Note that the first subscript of a deflection Δ indicates the location of the deflection, whereas the second subscript, O, is used to indicate that the deflection is due to the external loading. The flexibility coefficients representing the deflections of the primary beam due to unit values of the redundants are also defined by using double subscripts, as shown in Fig. 13.12(c) through (e). The first subscript of a flexibility coefficient denotes the location of the deflection, and the second subscript indicates the location of the unit load causing the deflection. For example, the flexibility coefficient f_{CB} denotes the deflection at point C of the primary beam due to a unit load at point B (Fig. 13.12(c)), whereas f_{BC} denotes the deflection at B due to a unit load at C (Fig. 13.12(d)), and so on. Alternatively, a flexibility coefficient f_{ij} may also be interpreted as the deflection corresponding to a redundant i due to a unit value of a redundant j; for example, f_{CB} denotes the deflection corresponding to the redundant C_y due to a unit value of the redundant B_y (Fig. 13.12(c)), f_{BC} denotes the deflection corresponding to the redundant B_y due to a unit value of the redundant C_y, and so on. A deflection or flexibility coefficient at the location of a redundant is considered to be positive if it has the same sense as that assumed for the redundant.

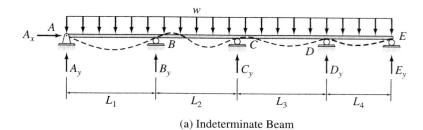

(a) Indeterminate Beam

=

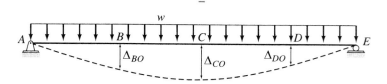

(b) Primary Beam Subjected to External Loading

+

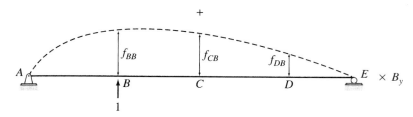

(c) Primary Beam Loaded with Redundant B_y

+

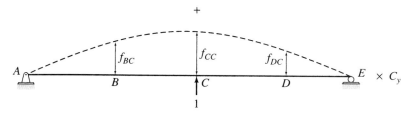

(d) Primary Beam Loaded with Redundant C_y

+

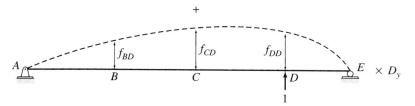

FIG. 13.12

(e) Primary Beam Loaded with Redundant D_y

Focusing our attention at point B of the primary beam, we see that the deflection at this point due to the external load is Δ_{BO} (Fig. 13.12(b)), the deflection due to B_y is $f_{BB}B_y$ (Fig. 13.12(c)), the deflection due to C_y is $f_{BC}C_y$ (Fig. 13.12(d)), and the deflection due to D_y is $f_{BD}D_y$ (Fig. 13.12(e)). Thus, the total deflection at B due to the combined effect of the external load and all of the redundants is $\Delta_{BO} + f_{BB}B_y + f_{BC}C_y + f_{BD}D_y$. Since the deflection of the actual indeterminate beam (Fig. 13.12(a)) at support B is zero, we set the algebraic sum of the deflections of the primary beam at B equal to zero to obtain the compatibility equation, $\Delta_{BO} + f_{BB}B_y + f_{BC}C_y + f_{BD}D_y = 0$. Next, we focus our attention at point C of the primary beam; by algebraically adding the deflections at C due to the external load and the redundants and by setting the sum equal to zero, we obtain the second compatibility equation, $\Delta_{CO} + f_{CB}B_y + f_{CC}C_y + f_{CD}D_y = 0$. Similarly, by setting equal to zero the algebraic sum of the deflections of the primary beam at D due to the external load and the redundants, we obtain the third compatibility equation, $\Delta_{DO} + f_{DB}B_y + f_{DC}C_y + f_{DD}D_y = 0$. The three compatibility equations thus obtained are

$$\Delta_{BO} + f_{BB}B_y + f_{BC}C_y + f_{BD}D_y = 0 \tag{13.15}$$

$$\Delta_{CO} + f_{CB}B_y + f_{CC}C_y + f_{CD}D_y = 0 \tag{13.16}$$

$$\Delta_{DO} + f_{DB}B_y + f_{DC}C_y + f_{DD}D_y = 0 \tag{13.17}$$

Since the number of compatibility equations is equal to the number of unknown redundants, these equations can be solved for the redundants. As Eqs. (13.15) through (13.17) indicate, the compatibility equations of structures with multiple degrees of indeterminacy are, in general, *coupled*, in the sense that each equation may contain more than one unknown redundant. The coupling occurs because the deflection at the location of a redundant may be caused not just by that particular redundant (and the external load), but also by some, or all, of the remaining redundants. Because of such coupling, the compatibility equations must be solved simultaneously to determine the unknown redundants.

The primary beam is statically determinate, so its deflections due to the external loading as well as the flexibility coefficients can be evaluated by using the methods discussed previously in this text. The total number of deflections (including flexibility coefficients) involved in a system of compatibility equations depends on the degree of indeterminacy of the structure. From Eqs. (13.15) through (13.17), we can see that for the beam under consideration, which is indeterminate to the third degree, the compatibility equations contain a total of 12 deflections (i.e., 3 deflections due to the external loading plus 9 flexibility coefficients). However, according to *Maxwell's law of reciprocal deflections* (Section 7.8), $f_{CB} = f_{BC}$, $f_{DB} = f_{BD}$, and $f_{DC} = f_{CD}$. Thus, three of the flexibility coefficients can be obtained by the application of Maxwell's law, thereby reducing the number of deflections to be computed to 9. Using similar reasoning, it can be shown that the total number of deflections needed for the analysis of a structure with the degree of indeterminacy of i equals $(i + i^2)$, of which $(3i + i^2)/2$ deflections must be computed, whereas the remaining can be obtained by the application of Maxwell's law of reciprocal deflections.

Once the redundants have been determined by solving the compatibility equations, the other response characteristics of the structure can be evaluated either by equilibrium or by superposition.

Procedure for Analysis

Based on the foregoing discussion, we can develop the following step-by-step procedure for the analysis of structures by the method of consistent deformations:

1. Determine the degree of indeterminacy of the structure.
2. Select redundant forces and/or moments. The total number of redundants must be equal to the degree of indeterminacy of the structure. Also, the redundants must be chosen so that the removal of the corresponding restraints from the given indeterminate structure results in a primary structure that is statically determinate and stable. The senses of the redundants are not known and can be arbitrarily assumed. A positive answer for a redundant will imply that the sense initially assumed for the redundant was correct.
3. Remove the restraints corresponding to the redundants from the given indeterminate structure to obtain the primary (determinate) structure.
4. a. Draw a diagram of the primary structure with only the external loading applied to it. Sketch a deflected shape of the structure, and show the deflection (or slope) at the point of application and in the direction of each redundant by an appropriate symbol.
   b. Next, for each redundant, draw a diagram of the primary structure with only the unit value of the redundant applied to it. The unit force (or moment) must be applied in the positive direction of the redundant. Sketch a deflected shape of the structure, and show by appropriate symbols the flexibility coefficients at the locations of all the redundants. To indicate that the load as well as the structural response is to be multiplied by the redundant under consideration, show the redundant preceded by a multiplication sign ($\times$) next to the diagram of the structure. The deflection (or slope) at the location of any redundant due to the redundant under consideration equals the flexibility coefficient at that location multiplied by the unknown magnitude of the redundant.

5. Write a compatibility equation for the location of each redundant by setting the algebraic sum of the deflections (or slopes) of the primary structure due to the external loading and each of the redundants equal to the known displacement (or rotation) at the corresponding location on the actual indeterminate structure. The total number of compatibility equations thus obtained must be equal to the number of redundants.

6. Compute the deflections (and the flexibility coefficients) involved in the compatibility equations by using the methods discussed previously in this text and by the application of Maxwell's law of reciprocal deflections. A deflection (or flexibility coefficient) at the location of a redundant is considered to be positive if it has the same sense as that assumed for the redundant.

7. Substitute the values of deflections computed in step 6 into the compatibility equations, and solve them for the unknown redundants.

8. Once the redundants have been determined, the other response characteristics (e.g., reactions, shear and bending moment diagrams, and/or member forces) of the indeterminate structure can be evaluated either through equilibrium considerations or by superposition of the responses of the primary structure due to the external loading and due to each of the redundants.

Example 13.8

Determine the reactions and draw the shear and bending moment diagrams for the three-span continuous beam shown in Fig. 13.13(a) using the method of consistent deformations.

Solution

Degree of Indeterminacy $i = 2$.

Primary Beam The vertical reactions B_y and C_y at the interior supports B and C, respectively, are selected as the redundants. The roller supports at B and C are then removed to obtain the primary beam shown in Fig. 13.13(b). Next, the primary beam is subjected separately to the 2-k/ft external load and the unit values of the redundants B_y and C_y, as shown in Fig. 13.13(b), (c), and (d), respectively.

Compatibility Equations Since the deflections of the actual indeterminate beam at supports B and C are zero, we set equal to zero the algebraic sum of the deflections at points B and C, respectively, of the primary beam due to the 2-k/ft external load and each of the redundants to obtain the compatibility equations:

$$\Delta_{BO} + f_{BB}B_y + f_{BC}C_y = 0 \qquad (1)$$

$$\Delta_{CO} + f_{CB}B_y + f_{CC}C_y = 0 \qquad (2)$$

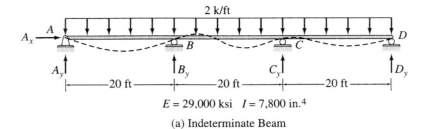

E = 29,000 ksi I = 7,800 in.^{4}

(a) Indeterminate Beam

=

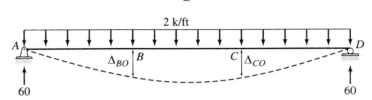

(b) Primary Beam Subjected to External Load

+

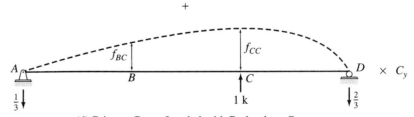

(c) Primary Beam Loaded with Redundant B_y

+

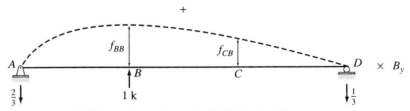

(d) Primary Beam Loaded with Redundant C_y

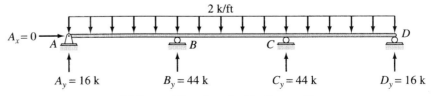

(e) Support Reactions for Continuous Beam

FIG. 13.13

continued

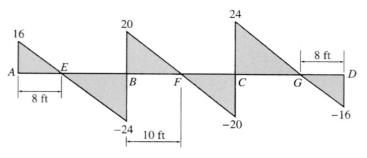

Shear diagram (k)

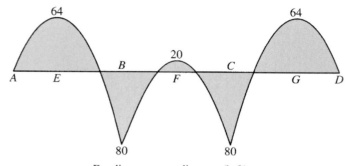

Bending moment diagram (k-ft)

FIG. 13.13 (contd.)

(f) Shear and Bending Moment Diagrams for Continuous Beam

Deflections of the Primary Beam By using the beam-deflection formulas, we obtain

$$\Delta_{BO} = \Delta_{CO} = -\frac{293{,}333.333 \text{ k-ft}^3}{EI}$$

$$f_{BB} = f_{CC} = \frac{3{,}555.556 \text{ ft}^3}{EI}$$

$$f_{CB} = \frac{3{,}111.111 \text{ ft}^3}{EI}$$

By applying Maxwell's law,

$$f_{BC} = \frac{3{,}111.111 \text{ ft}^3}{EI}$$

Magnitudes of the Redundants By substituting the values of the deflections and flexibility coefficients of the primary beam just computed into the compatibility equations (Eqs. (1) and (2)), we obtain

$$-293{,}333.333 + 3{,}555.556 B_y + 3{,}111.111 C_y = 0$$

$$-293{,}333.333 + 3{,}111.111 B_y + 3{,}555.556 C_y = 0$$

or

$$3{,}555.556 B_y + 3{,}111.111 C_y = 293{,}333.333 \tag{1a}$$

$$3{,}111.111 B_y + 3{,}555.556 C_y = 293{,}333.333 \tag{2a}$$

Solving Eqs. (1a) and (2a) simultaneously for B_y and C_y, we obtain

$$B_y = C_y = 44 \text{ k} \uparrow \qquad \text{Ans.}$$

Reactions The remaining reactions can now be determined by applying the three equations of equilibrium to the free body of the continuous beam as follows (Fig. 13.13(e)):

$$+ \rightarrow \sum F_x = 0 \qquad A_x = 0 \qquad \text{Ans.}$$

$$+ \circlearrowleft \sum M_D = 0 \qquad -A_y(60) + 2(60)(30) - 44(40 + 20) = 0$$

$$A_y = 16 \text{ k} \uparrow \quad \text{Ans.}$$

$$+ \uparrow \sum F_y = 0 \qquad 16 - 2(60) + 44 + 44 + D_y = 0$$

$$D_y = 16 \text{ k} \uparrow \qquad \text{Ans.}$$

Shear and Bending Moment Diagrams The shear and bending moment diagrams of the beam are shown in Fig. 13.13(f).   Ans.

The shapes of the shear and bending moment diagrams for continuous beams, in general, are similar to those for the three-span continuous beam shown in Fig. 13.13(f). As shown in this figure, negative bending moments generally develop at the interior supports of continuous beams, whereas the bending moment diagram is usually positive over the middle portions of the spans. The bending moment at a hinged support at an end of the beam must be zero, and it is generally negative at a fixed end support. Also, the shape of the bending moment diagram is parabolic for the spans subjected to uniformly distributed loads, and it consists of linear segments for spans subjected to concentrated loads. The actual values of the bending moments, of course, depend on the magnitude of the loading as well as on the lengths and flexural rigidities of the spans of the continuous beam.

Example 13.9

Determine the reactions and draw the shear and bending moment diagrams for the beam shown in Fig. 13.14(a) by the method of consistent deformations.

Solution

Degree of Indeterminacy $i = 2$.

Primary Beam The vertical reactions C_y and E_y at the roller supports C and E, respectively, are selected as the redundants. These supports are then removed to obtain the cantilever primary beam shown in Fig. 13.14(b). Next, the primary beam is subjected separately to the external loading and the unit

continued

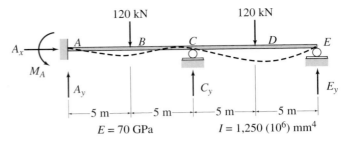

(a) Indeterminate Beam

=

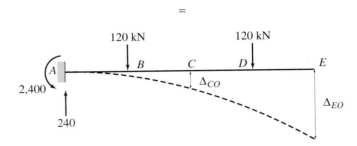

(b) Primary Beam Subjected to External Loading

+

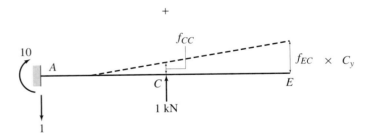

(c) Primary Beam Loaded with Redundant C_y

+

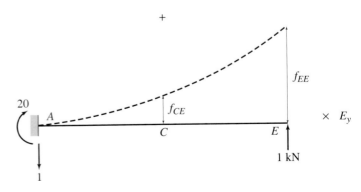

(d) Primary Beam Loaded with Redundant E_y

FIG. 13.14

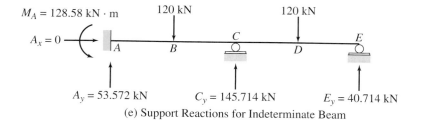

(e) Support Reactions for Indeterminate Beam

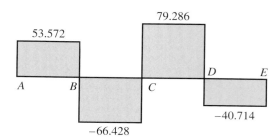

Shear diagram (kN)

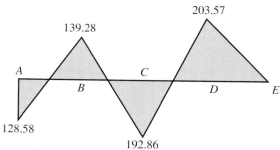

Bending moment diagram (kN · m)

(f) Shear and Bending Moment Diagrams for
Indeterminate Beam

FIG. 13.14 (contd.)

values of the redundants C_y and E_y, as shown in Fig. 13.14(b), (c), and (d), respectively.

Compatibility Equations See Fig. 13.14(a) through (d).

$$\Delta_{CO} + f_{CC}C_y + f_{CE}E_y = 0 \quad (1)$$

$$\Delta_{EO} + f_{EC}C_y + f_{EE}E_y = 0 \quad (2)$$

Deflections of Primary Beam By using the deflection formulas, we obtain

$$\Delta_{CO} = -\frac{82{,}500 \text{ kN} \cdot \text{m}^3}{EI} \qquad \Delta_{EO} = -\frac{230{,}000 \text{ kN} \cdot \text{m}^3}{EI}$$

continued

$$f_{CC} = \frac{333.333 \text{ m}^3}{EI} \qquad f_{EC} = \frac{833.333 \text{ m}^3}{EI}$$

$$f_{EE} = \frac{2,666.667 \text{ m}^3}{EI}$$

By applying Maxwell's law,

$$f_{CE} = \frac{833.333 \text{ m}^3}{EI}$$

Magnitudes of the Redundants By substituting the deflections of the primary beam into the compatibility equations, we obtain

$$-82,500 + 333.333C_y + 833.333E_y = 0$$

$$-230,000 + 833.333C_y + 2,666.667E_y = 0$$

or

$$333.333C_y + 833.333E_y = 82,500 \qquad \text{(1a)}$$

$$833.333C_y + 2,666.667E_y = 230,000 \qquad \text{(2a)}$$

Solving Eqs. (1a) and (2a) simultaneously for C_y and E_y, we obtain

$$C_y = 145.714 \text{ kN} \uparrow \qquad E_y = 40.714 \text{ kN} \uparrow \qquad \text{Ans.}$$

Reactions The remaining reactions can now be determined by applying the three equations of equilibrium to the free body of the indeterminate beam (Fig. 13.14(e)):

$$+ \rightarrow \sum F_x = 0 \qquad A_x = 0 \qquad\qquad \text{Ans.}$$

$$+\uparrow \sum F_y = 0 \qquad A_y - 120 + 145.714 - 120 + 40.714 = 0$$

$$A_y = 53.572 \text{ kN} \uparrow \qquad\qquad \text{Ans.}$$

$$+ \zeta \sum M_A = 0 \qquad M_A - 120(5) + 145.714(10) - 120(15) + 40.714(20) = 0$$

$$M_A = 128.58 \text{ kN} \cdot \text{m} \; \zeta \qquad\qquad \text{Ans.}$$

Shear and Bending Moment Diagrams See Fig. 13.14(f).   Ans.

Example 13.10

Determine the moments at the supports of the fixed beam shown in Fig. 13.15(a) by the method of consistent deformations. Also, draw the bending moment diagram for the beam.

Solution

Degree of Indeterminacy The beam is supported by six support reactions; thus, its degree of indeterminacy is $i = 6 - 3 = 3$. However, since the beam is subjected only to vertical loading, the horizontal reactions A_x and C_x must be

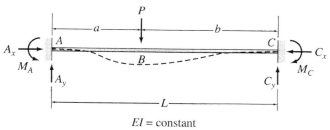

EI = constant

(a) Indeterminate Beam

=

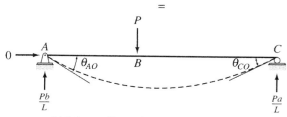

(b) Primary Beam Subjected to External Load

+

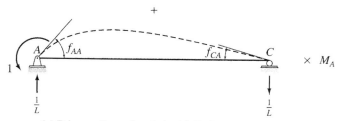

(c) Primary Beam Loaded with Redundant M_A

+

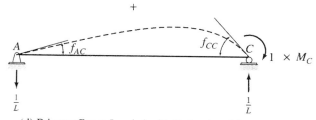

(d) Primary Beam Loaded with Redundant M_C

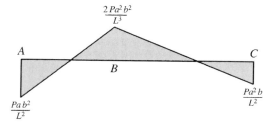

(e) Bending Moment Diagram for Fixed Beam

FIG. **13.15**

continued

zero. Therefore, to analyze this beam, we need to select only two of the remaining four reactions as the redundants.

Primary Beam The moments M_A and M_C at the fixed supports A and C, respectively, are selected as the redundants. The restraints against rotation at ends A and C of the fixed beam are then removed to obtain the simply supported primary beam shown in Fig. 13.15(b). Next, the primary beam is subjected separately to the external load P and the unit values of redundants M_A and M_C, as shown in Fig. 13.15(b), (c), and (d), respectively.

Compatibility Equations Noting that the slopes of the actual indeterminate beam at the fixed supports A and C are zero, we write the compatibility equations:

$$\theta_{AO} + f_{AA}M_A + f_{AC}M_C = 0 \tag{1}$$

$$\theta_{CO} + f_{CA}M_A + f_{CC}M_C = 0 \tag{2}$$

Slopes of the Primary Beam The slopes at ends A and C of the primary beam due to the external load P and due to the unit value of each of the redundants obtained by using either the deflection formulas or the conjugate-beam method are

$$\theta_{AO} = -\frac{Pb(L^2 - b^2)}{6EIL}$$

$$\theta_{CO} = -\frac{Pa(L^2 - a^2)}{6EIL}$$

$$f_{AA} = f_{CC} = \frac{L}{3EI}$$

$$f_{CA} = \frac{L}{6EI}$$

By applying Maxwell's law,

$$f_{AC} = \frac{L}{6EI}$$

Magnitudes of the Redundants By substituting the expressions for slopes into the compatibility equations (Eqs. (1) and (2)), we obtain

$$-\frac{Pb(L^2 - b^2)}{6EIL} + \left(\frac{L}{3EI}\right)M_A + \left(\frac{L}{6EI}\right)M_C = 0 \tag{1a}$$

$$-\frac{Pa(L^2 - a^2)}{6EIL} + \left(\frac{L}{6EI}\right)M_A + \left(\frac{L}{3EI}\right)M_C = 0 \tag{2a}$$

which can be simplified as

$$2M_A + M_C = \frac{Pb(L^2 - b^2)}{L^2} \tag{1b}$$

$$M_A + 2M_C = \frac{Pa(L^2 - a^2)}{L^2} \tag{2b}$$

To solve Eqs. (1b) and (2b) for M_A and M_C, we multiply Eq. (1b) by 2 and subtract it from Eq. (2b):

$$M_A = -\frac{P}{3L^2}[a(L^2 - a^2) - 2b(L^2 - b^2)]$$

$$= -\frac{P}{3L^2}[a(L - a)(L + a) - 2b(L - b)(L + b)]$$

$$= -\frac{Pab}{3L^2}[(L + a) - 2(L + b)]$$

$$= \frac{Pab^2}{L^2}$$

$$M_A = \frac{Pab^2}{L^2} \; \circlearrowleft \qquad\qquad \text{Ans.}$$

By substituting the expression for M_A into Eq. (1b) or Eq. (2b) and solving for M_C, we obtain the following.

$$M_C = \frac{Pa^2b}{L^2} \; \circlearrowright \qquad\qquad \text{Ans.}$$

Bending Moment Diagram   The vertical reactions A_y and C_y can now be determined by superposition of the reactions of the primary beam due to the external load P and due to each of the redundants (Fig. 13.15(b) through (d)). Thus

$$A_y = \frac{Pb}{L} + \frac{1}{L}(M_A - M_C) = \frac{Pb^2}{L^3}(3a + b)$$

$$C_y = \frac{Pa}{L} - \frac{1}{L}(M_A - M_C) = \frac{Pa^2}{L^3}(a + 3b)$$

The bending moment diagram of the beam is shown in Fig. 13.15(e).   Ans.

The moments at the ends of beams whose ends are fixed against rotation are usually referred to as *fixed-end moments*. Such moments play an important role in the analysis of structures by the displacement method, to be considered in subsequent chapters. As illustrated here, the expressions for fixed-end moments due to various loading conditions can be conveniently derived by using the method of consistent deformations. The fixed-end-moment expressions for some common types of loading conditions are given inside the back cover of the book for convenient reference.

Example 13.11

Determine the reactions and draw the shear and bending moment diagrams for the four-span continuous beam shown in Fig. 13.16(a) using the method of consistent deformations.

continued

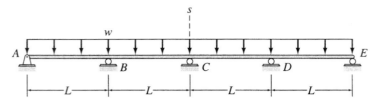

EI = constant

(a) Indeterminate Beam

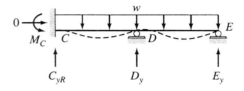

(b) Substructure for Analysis

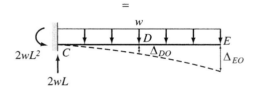

(c) Primary Beam Subjected to External Load

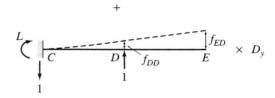

(d) Primary Beam Loaded with Redundant D_y

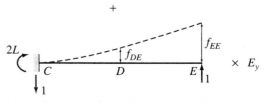

(e) Primary Beam Loaded with Redundant E_y

FIG. 13.16

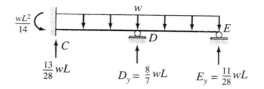

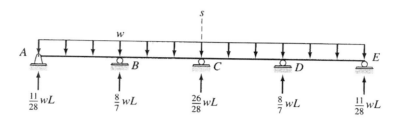

(f) Reactions for Substructure

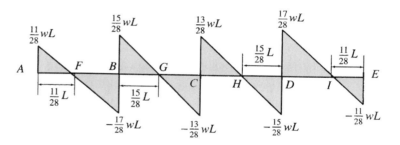

(g) Support Reactions for Continuous Beam

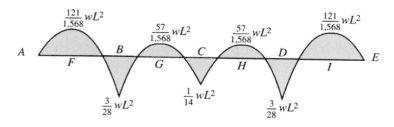

Shear diagram

Bending moment diagram

(h) Shear and Bending Moment Diagrams for Continuous Beam

FIG. 13.16 (contd.)

continued

Solution

Symmetry As the beam and the loading are symmetric with respect to the vertical *s* axis passing through roller support *C* (Fig. 13.16(a)), we will analyze only the right half of the beam with symmetric boundary conditions, as shown in Fig. 13.16(b). The response of the left half of the beam will then be obtained by reflecting the response of the right half to the other side of the axis of symmetry.

Degree of Indeterminacy The degree of indeterminacy of the substructure (Fig. 13.16(b)) is 2. Note that, since the degree of indeterminacy of the complete continuous beam (Fig. 13.16(a)) is three, the utilization of structural symmetry will reduce the computational effort required in the analysis.

Primary Beam The vertical reactions D_y and E_y at the roller supports *D* and *E*, respectively, of the substructure are selected as the redundants. The roller supports at *D* and *E* are then removed to obtain the cantilever primary beam shown in Fig. 13.16(c).

Compatibility Equations See Fig. 13.16(b) through (e).

$$\Delta_{DO} + f_{DD}D_y + f_{DE}E_y = 0 \tag{1}$$

$$\Delta_{EO} + f_{ED}D_y + f_{EE}E_y = 0 \tag{2}$$

Deflections of the Primary Beam By using the deflection formulas, we obtain

$$\Delta_{DO} = -\frac{17wL^4}{24EI} \qquad \Delta_{EO} = -\frac{2wL^4}{EI}$$

$$f_{DD} = \frac{L^3}{3EI} \qquad f_{ED} = \frac{5L^3}{6EI}$$

$$f_{EE} = \frac{8L^3}{3EI}$$

By applying Maxwell's law,

$$f_{DE} = \frac{5L^3}{6EI}$$

Magnitudes of the Redundants By substituting the deflections of the primary beam into the compatibility equations, we obtain

$$-\frac{17wL^4}{24EI} + \left(\frac{L^3}{3EI}\right)D_y + \left(\frac{5L^3}{6EI}\right)E_y = 0 \tag{1a}$$

$$-\frac{2wL^4}{EI} + \left(\frac{5L^3}{6EI}\right)D_y + \left(\frac{8L^3}{3EI}\right)E_y = 0 \tag{2a}$$

which can be simplified to

$$8D_y + 20E_y = 17wL \tag{1b}$$

$$5D_y + 16E_y = 12wL \tag{2b}$$

Solving Eqs. (1b) and (2b) simultaneously for D_y and E_y, we obtain

$$D_y = \frac{8}{7}wL \uparrow \qquad E_y = \frac{11}{28}wL \uparrow \qquad \text{Ans.}$$

Reactions The remaining reactions of the substructure, obtained by applying the equations of equilibrium, are shown in Fig. 13.16(f). The reactions to the left of the *s* axis are then obtained by reflection, as shown in Fig. 13.16(g).

Ans.

Shear and Bending Moment Diagrams By using the reactions of the continuous beam, its shear and bending moment diagrams are constructed. These diagrams are shown in Fig. 13.16(h).

Ans.

Example 13.12

Determine the reactions and the force in each member of the truss shown in Fig. 13.17(a) by the method of consistent deformations.

Solution

Degree of Indeterminacy $i = (m + r) - 2j = (14 + 4) - 2(8) = 2$.

Primary Truss The vertical reaction D_y at the roller support D and the axial force F_{BG} in the diagonal member BG are selected as the redundants. The roller support D and member BG are then removed from the given indeterminate truss to obtain the primary truss shown in Fig. 13.17(b). The primary truss is subjected separately to the external loading (Fig. 13.17(b)), a unit value of the redundant D_y (Fig. 13.17(c)), and a unit tensile force in the redundant member BG (Fig. 13.17(d)).

Compatibility Equations The compatibility equations can be expressed as

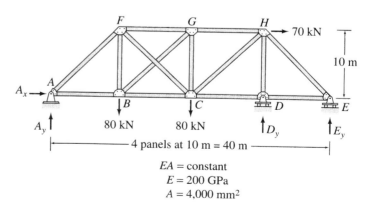

$EA = $ constant
$E = 200$ GPa
$A = 4{,}000$ mm²

FIG. 13.17

(a) Indeterminate Truss

continued

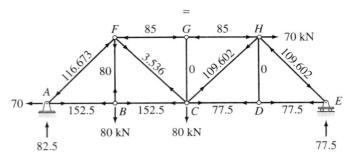

(b) Primary Truss Subjected to External Loads — F_O Forces

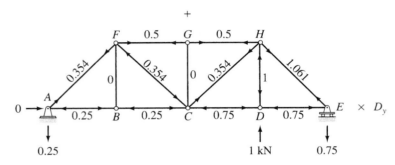

(c) Primary Truss Subjected to Unit Value of Redundant D_y— u_D Forces

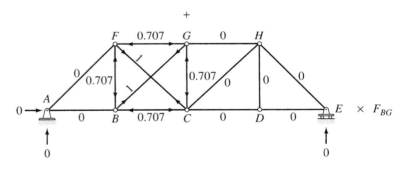

(d) Primary Truss Subjected to Unit Tensile Force in Member BG —u_{BG} Forces

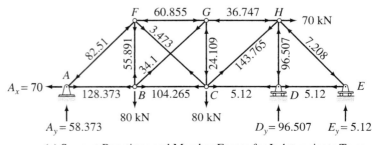

FIG. 13.17 (contd.)

(e) Support Reactions and Member Forces for Indeterminate Truss

$$\Delta_{DO} + f_{DD}D_y + f_{D,BG}F_{BG} = 0 \qquad (1)$$

$$\Delta_{BGO} + f_{BG,D}D_y + f_{BG,BG}F_{BG} = 0 \qquad (2)$$

in which Δ_{DO} = vertical deflection at joint D of the primary truss due to the external loading; Δ_{BGO} = relative displacement between joints B and G due to the external loading; f_{DD} = vertical deflection at joint D due to a unit load at joint D; $f_{BG,D}$ = relative displacement between joints B and G due to a unit load at joint D; $f_{BG,BG}$ = relative displacement between joints B and G due to a unit tensile force in member BG; and $f_{D,BG}$ = vertical deflection at joint D due to a unit tensile force in member BG.

Deflections of Primary Truss   The virtual work expressions for the preceding deflections are

$$\Delta_{DO} = \sum \frac{F_O u_D L}{AE} \qquad\qquad \Delta_{BGO} = \sum \frac{F_O u_{BG} L}{AE}$$

$$f_{DD} = \sum \frac{u_D^2 L}{AE} \qquad\qquad f_{BG,BG} = \sum \frac{u_{BG}^2 L}{AE}$$

$$f_{BG,D} = f_{D,BG} = \sum \frac{u_D u_{BG} L}{AE}$$

in which F_O, u_D, and u_{BG} represent the member forces due to the external loading, a unit load at joint D, and a unit tensile force in member BG, respectively. The numerical values of the member forces, as computed by the method of joints (Fig. 13.17(b) through (d)), are tabulated in Table 13.4. Note that since the axial rigidity EA is the same for all the members, only the numerators of the virtual work expressions are evaluated in Table 13.4. Thus

$$\Delta_{DO} = -\frac{4{,}472.642 \text{ kN} \cdot \text{m}}{AE} \qquad \Delta_{BGO} = -\frac{992.819 \text{ kN} \cdot \text{m}}{AE}$$

$$f_{DD} = \frac{48.736 \text{ m}}{AE} \qquad\qquad f_{BG,BG} = \frac{48.284 \text{ m}}{AE}$$

$$f_{BG,D} = f_{D,BG} = -\frac{6.773 \text{ m}}{AE}$$

Magnitudes of the Redundants   By substituting these deflections and flexibility coefficients into the compatibility equations (Eqs. (1) and (2)), we write

$$-4{,}472.642 + 48.736D_y - 6.773F_{BG} = 0 \qquad (1a)$$

$$-992.819 - 6.773D_y + 48.284F_{BG} = 0 \qquad (2a)$$

Solving Eqs. (1a) and (2a) simultaneously for D_y and F_{BG}, we obtain

$$D_y = 96.507 \text{ kN} \uparrow \qquad F_{BG} = 34.1 \text{ kN (T)} \qquad\qquad \text{Ans.}$$

Reactions   The remaining reactions of the indeterminate truss can now be determined by superposition of reactions of the primary truss due to the external loading and due to each of the redundants. The reactions thus obtained are shown in Fig. 13.17(e). 　　　　　　　　　　　　　　　　　　　　　　　　　　　**Ans.**

continued

TABLE 13.4

Member	L (m)	F_O (kN)	u_D (kN/kN)	u_{BG} (kN/kN)	$F_O u_D L$ (kN·m)	$F_O u_{BG} L$ (kN·m)	$u_D^2 L$ (m)	$u_{BG}^2 L$ (m)	$u_D u_{BG} L$ (m)	$F = F_O + u_D D_y$ $+ u_{BG} F_{BG}$ (kN)
AB	10	152.5	−0.25	0	−381.25	0	0.625	0	0	128.373
BC	10	152.5	−0.25	−0.707	−381.25	−1,078.175	0.625	5	1.768	104.265
CD	10	77.5	−0.75	0	−581.25	0	5.625	0	0	5.12
DE	10	77.5	−0.75	0	−581.25	0	5.625	0	0	5.12
FG	10	−85	0.5	−0.707	−425	600.95	2.5	5	−3.535	−60.855
GH	10	−85	0.5	0	−425	0	2.5	0	0	−36.747
BF	10	80	0	−0.707	0	−565.60	0	5	0	55.891
CG	10	0	0	−0.707	0	0	0	5	0	−24.109
DH	10	0	−1	0	0	0	10	0	0	−96.507
AF	14.142	−116.673	0.354	0	−584.096	0	1.772	0	0	−82.51
BG	14.142	0	0	1	0	0	0	14.142	0	34.1
CF	14.142	3.536	−0.354	1	−17.702	50.006	1.772	14.142	−5.006	3.473
CH	14.142	109.602	0.354	0	548.697	0	1.772	0	0	143.765
EH	14.142	−109.602	1.061	0	−1,644.541	0	15.92	0	0	−7.208
				Σ	−4,472.642	−992.819	48.736	48.284	−6.773	

Member Axial Forces The forces in the remaining members of the indeterminate truss can be determined by using the superposition relationship:

$$F = F_O + u_D D_y + u_{BG} F_{BG}$$

The member forces thus obtained are shown in Table 13.4 and Fig. 13.17(e).

Ans.

Example 13.13

Determine the reactions and draw the shear and bending moment diagrams for the frame shown in Fig. 13.18(a) by the method of consistent deformations.

Solution

Degree of Indeterminacy $i = 2$.

Primary Frame The reactions D_X and D_Y at the hinged support D are selected as the redundants. The hinged support D is then removed to obtain the primary frame shown in Fig. 13.18(b). Next, the primary frame is subjected separately to the external loading and the unit values of the redundants D_X and D_Y, as shown in Fig. 13.18(b), (c), and (d), respectively.

Compatibility Equations Noting that the horizontal and vertical deflections of the actual indeterminate frame at the hinged support D are zero, we write the compatibility equations:

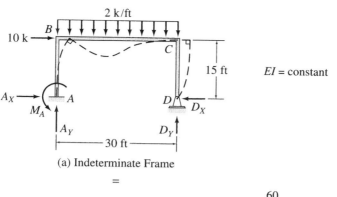

(a) Indeterminate Frame

=

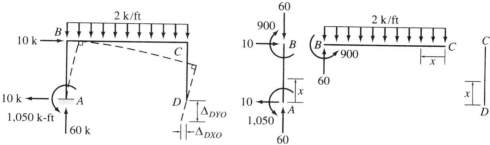

(b) Primary Frame Subjected to External Loading — M_O Moments

+

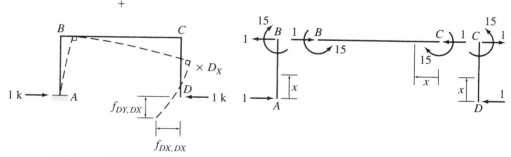

(c) Primary Frame Subjected to Unit Value of Redundant D_X — m_{DX} Moments

+

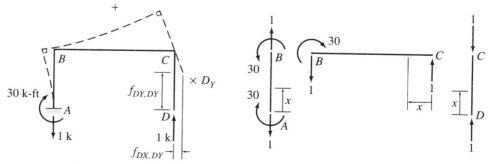

(d) Primary Frame Subjected to Unit Value of Redundant D_Y — m_{DY} Moments

FIG. 13.18

continued

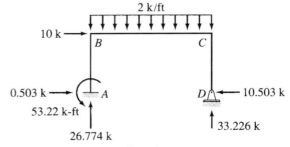

Reactions

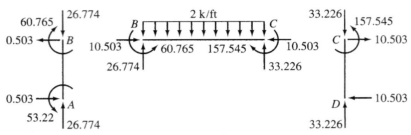

Member end forces

(e) Support Reactions and Member End Forces for Indeterminate Frame

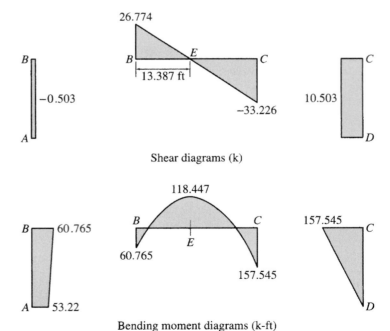

Shear diagrams (k)

Bending moment diagrams (k-ft)

FIG. 13.18 (contd.)   (f) Shear and Bending Moment Diagrams for Indeterminate Frame

TABLE 13.5

Member	Origin	x coordinate Limits (ft)	M_O (k-ft)	m_{DX} (k-ft/k)	m_{DY} (k-ft/k)
AB	A	0–15	$-1{,}050 + 10x$	$-x$	30
CB	C	0–30	$-x^2$	-15	x
DC	D	0–15	0	x	0

$$\Delta_{DXO} + f_{DX,DX}D_X + f_{DX,DY}D_Y = 0 \qquad (1)$$

$$\Delta_{DYO} + f_{DY,DX}D_X + f_{DY,DY}D_Y = 0 \qquad (2)$$

Deflections of Primary Frame The equations for bending moments for the members of the frame due to the external loading and unit values of the redundants are tabulated in Table 13.5. By applying the virtual work method, we obtain

$$\Delta_{DXO} = \sum \int \frac{M_O m_{DX}}{EI}\, dx = \frac{241{,}875 \text{ k-ft}^3}{EI}$$

$$\Delta_{DYO} = \sum \int \frac{M_O m_{DY}}{EI}\, dx = -\frac{641{,}250 \text{ k-ft}^3}{EI}$$

$$f_{DX,DX} = \sum \int \frac{m_{DX}^2}{EI}\, dx = \frac{9{,}000 \text{ ft}^3}{EI}$$

$$f_{DY,DY} = \sum \int \frac{m_{DY}^2}{EI}\, dx = \frac{22{,}500 \text{ ft}^3}{EI}$$

$$f_{DX,DY} = f_{DY,DX} = \sum \int \frac{m_{DX} m_{DY}}{EI}\, dx = -\frac{10{,}125 \text{ ft}^3}{EI}$$

Magnitudes of the Redundants By substituting these deflections and flexibility coefficients into the compatibility equations, we write

$$241{,}875 + 9{,}000 D_X - 10{,}125 D_Y = 0 \qquad (1\text{a})$$

$$-641{,}250 - 10{,}125 D_X + 22{,}500 D_Y = 0 \qquad (2\text{a})$$

Solving Eqs. (1a) and (2a) simultaneously for D_X and D_Y, we obtain

$$D_X = 10.503 \text{ k} \leftarrow \qquad D_Y = 33.226 \text{ k} \uparrow \qquad \text{Ans.}$$

Reactions The remaining reactions and the member end forces of the indeterminate frame can now be determined by applying the equations of equilibrium. The reactions and member and forces thus obtained are shown in Fig. 13.18(e). **Ans.**

Shear and Bending Moment Diagrams See Fig. 13.18(f). **Ans.**

13.4 SUPPORT SETTLEMENTS, TEMPERATURE CHANGES, AND FABRICATION ERRORS

Support Settlements

Thus far, we have considered the analysis of structures with unyielding supports. As discussed in Chapter 11, support movements due to weak foundations and the like may induce significant stresses in externally indeterminate structures and must be considered in their designs. Support settlements, however, do not have any effect on the stress conditions of structures that are internally indeterminate but externally determinate. This lack of effect is due to the fact that the settlements cause such structures to displace and/or rotate as rigid bodies without changing their shapes. The method of consistent deformations, as developed in the preceding sections, can be easily modified to include the effect of support settlements in the analysis.

Consider, for example, a two-span continuous beam subjected to a uniformly distributed load w, as shown in Fig. 13.19(a). Suppose that the supports B and C of the beam undergo small settlements Δ_B and Δ_C, respectively, as shown in the figure. To analyze the beam, we consider the vertical reactions B_y and C_y to be the redundants. The supports B and C are removed from the indeterminate beam to obtain the primary beam, which is then subjected separately to the external load w and the unit values of the redundants B_y and C_y, as shown in Fig. 13.19(b), (c), and (d), respectively. By realizing that the deflections of the actual indeterminate beam at supports B and C are equal to the settlements Δ_B and Δ_C, respectively, we obtain the compatibility equations

$$\Delta_{BO} + f_{BB}B_y + f_{BC}C_y = \Delta_B \qquad (13.18)$$

$$\Delta_{CO} + f_{CB}B_y + f_{CC}C_y = \Delta_C \qquad (13.19)$$

which can be solved for the redundants B_y and C_y. Note that the right-hand sides of the compatibility equations (Eqs. (13.18) and (13.19)) are no longer equal to zero, as in the case of unyielding supports considered in the previous sections, but are equal to the prescribed values of settlements at supports B and C, respectively. Once the redundants have been determined by solving the compatibility equations, the other response characteristics of the beam can be evaluated either by equilibrium or by superposition.

Although support settlements are usually specified with respect to the undeformed position of the indeterminate structure, the magnitudes of such displacements to be used in the compatibility equations must be measured from the chord connecting the deformed positions of the supports of the primary structure to the deformed positions of the redundant supports. Any such support displacement is considered to be positive if it has the same sense as that assumed for the redundant. In

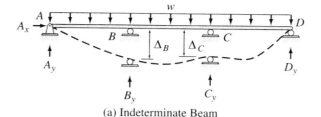

(a) Indeterminate Beam

=

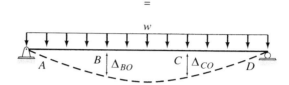

(b) Primary Beam Subjected to External Loading

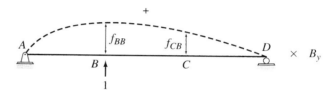

(c) Primary Beam Loaded with Redundant B_y

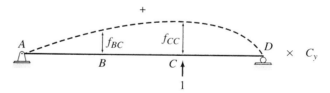

(d) Primary Beam Loaded with Redundant C_y

FIG. 13.19

the case of the beam of Fig. 13.19(a), since the end supports A and D do not undergo any settlement, the chord AD of the primary beam coincides with the undeformed position of the indeterminate beam; therefore, the settlements of supports B and C relative to the chord of the primary beam are equal to the prescribed settlements Δ_B and Δ_C, respectively.

Now, suppose that all of the supports of a beam undergo settlement as shown in Fig. 13.20. If we consider the reactions B_y and C_y to be the redundants, then the displacements Δ_{BR} and Δ_{CR} of supports B and C, respectively, relative to the chord of the primary beam should be used in

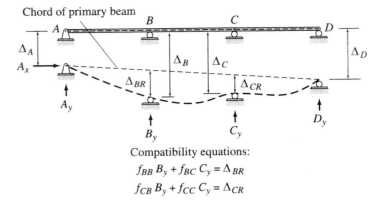

FIG. 13.20

Chord of primary beam

Compatibility equations:

$$f_{BB}\, B_y + f_{BC}\, C_y = \Delta_{BR}$$
$$f_{CB}\, B_y + f_{CC}\, C_y = \Delta_{CR}$$

the compatibility equations instead of the specified displacements Δ_B and Δ_C. This is because only the displacements relative to the chord cause stresses in the beam. In other words, if the supports of the beam would have settled either by equal amounts or by amounts so that the deformed positions of all of the supports would lie on a straight line, then the beam would remain straight without bending, and no stresses would develop in the beam.

Example 13.14

Determine the reactions and draw the shear and bending moment diagrams for the three-span continuous beam shown in Fig. 13.21(a) due to the uniformly distributed load and due to the support settlements of 5/8 in. at B, $1\frac{1}{2}$ in. at C, and 3/4 in. at D. Use the method of consistent deformations.

Solution

This beam was previously analyzed in Example 13.8 for the 2-k/ft uniformly distributed loading by selecting the vertical reactions at the interior supports B and C as the redundants. We will use the same primary beam as used previously.

Relative Settlements The specified support settlements are depicted in Fig. 13.21(b) using an exaggerated scale. It can be seen from this figure that the settlements of supports B and C relative to the chord of the primary beam (which is the line connecting the displaced positions of supports A and D) are

$$\Delta_{BR} = -0.375 \text{ in.} \quad \text{and} \quad \Delta_{CR} = -1.0 \text{ in.}$$

in which the negative signs for the magnitudes of Δ_{BR} and Δ_{CR} indicate that these settlements occur in the downward direction—that is, opposite to the upward direction assumed for the redundants B_y and C_y.

Compatibility Equations The compatibility equations for the beam remain the same as in Example 13.8, except that the right-hand sides of the equations must now be set equal to the settlements Δ_{BR} and Δ_{CR}. Thus

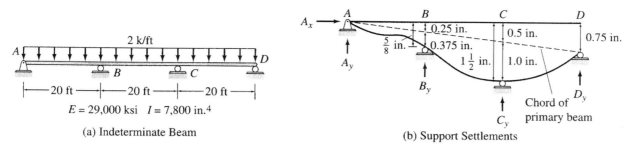

2 k/ft

A ↓↓↓↓↓↓↓↓↓↓↓↓↓↓↓↓↓↓ D

B   C

|— 20 ft —|— 20 ft —|— 20 ft —|

E = 29,000 ksi   I = 7,800 in.⁴

(a) Indeterminate Beam

(b) Support Settlements

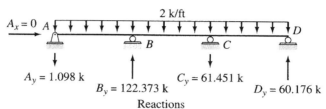

$A_x = 0$ A ↓↓↓↓↓↓↓↓↓↓↓↓↓↓↓↓↓↓ D

2 k/ft

B   C

$A_y = 1.098$ k $B_y = 122.373$ k $C_y = 61.451$ k $D_y = 60.176$ k

Reactions

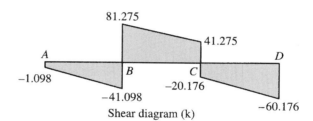

81.275

41.275

A   D

−1.098   B   C

−41.098   −20.176

Shear diagram (k)   −60.176

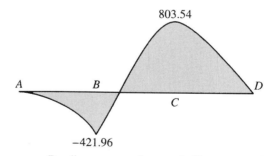

803.54

A   B   D

C

−421.96

Bending moment diagram (k-ft)

(c) Support Reactions and Shear and Bending
Moment Diagrams for Continuous Beam

FIG. 13.21

$$\Delta_{BO} + f_{BB}B_y + f_{BC}C_y = \Delta_{BR} \tag{1}$$

$$\Delta_{CO} + f_{CB}B_y + f_{CC}C_y = \Delta_{CR} \tag{2}$$

Deflections of Primary Beam In Example 13.8, the deflections and the
flexibility coefficients of the beam were expressed in terms of EI. Since the

continued

right-hand sides of the compatibility equations were zero, the EI terms simply canceled out of the computations. In the present example, however, because of the presence of support settlements on the right-hand sides of the compatibility equations, the EI terms cannot be canceled out; therefore, the actual numerical values of deflections and flexibility coefficients must be computed.

$$\Delta_{BO} = \Delta_{CO} = -\frac{293{,}333.333 \text{ k-ft}^3}{EI} = -\frac{293{,}333.333(12)^3}{(29{,}000)(7{,}800)} = -2.241 \text{ in.}$$

$$f_{BB} = f_{CC} = \frac{3{,}555.556 \text{ ft}^3}{EI} = \frac{3{,}555.556(12)^3}{(29{,}000)(7{,}800)} = 0.0272 \text{ in./k}$$

$$f_{CB} = f_{BC} = \frac{3{,}111.111 \text{ ft}^3}{EI} = \frac{3{,}111.111(12)^3}{(29{,}000)(7{,}800)} = 0.0238 \text{ in./k}$$

Magnitudes of the Redundants By substituting the numerical values into the compatibility equations, we write

$$-2.241 + 0.0272B_y + 0.0238C_y = -0.375 \tag{1a}$$

$$-2.241 + 0.0238B_y + 0.0272C_y = -1 \tag{2a}$$

By solving Eqs. (1a) and (2a) simultaneously for B_y and C_y, we obtain

$$B_y = 122.373 \text{ k} \uparrow \quad \text{and} \quad C_y = -61.451 \text{ k} = 61.451 \text{ k} \downarrow \qquad \text{Ans.}$$

Reactions and Shear and Bending Moment Diagrams The remaining reactions of the continuous beam can now be determined by equilibrium. The reactions and the shear and bending moment diagrams of the beam are shown in Fig. 13.21(c). A comparison of these results with those of Example 13.8 (without settlement) indicates that even small support settlements may have a significant effect on the reactions and the shear and bending moment diagrams of indeterminate structures.   Ans.

Example 13.15

Determine the reactions and draw the shear and bending moment diagrams for the beam shown in Fig. 13.22(a) due to the loading shown and due to the support settlements of 40 mm at C and 25 mm at E. Use the method of consistent deformations.

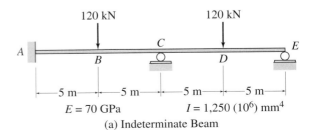

$E = 70$ GPa $I = 1{,}250 \ (10^6) \text{ mm}^4$

(a) Indeterminate Beam

FIG. **13.22**

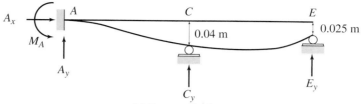

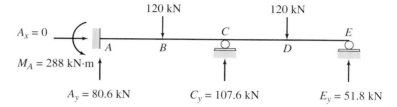

(b) Support Settlements

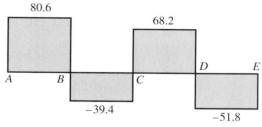

Reactions

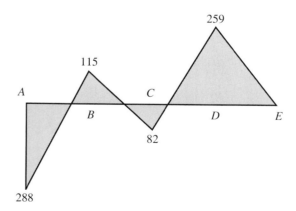

Shear diagram (kN)

Bending moment diagram (kN · m)

(c) Support Reactions and Shear and Bending Moment
Diagrams for Indeterminate Beam

FIG. 13.22 (contd.)

continued

Solution

This beam was previously analyzed in Example 13.9 for the external loading by selecting the vertical reactions at the roller supports C and E as the redundants. We will use the same primary beam as used previously.

Support Settlements The specified support settlements are depicted in Fig. 13.22(b), from which it can be seen that the chord AE of the primary beam coincides with the undeformed position of the indeterminate beam; therefore, the settlements of supports C and E relative to the chord of the primary beam are equal to the prescribed settlements, that is

$$\Delta_{CR} = \Delta_C = -0.04 \text{ m} \quad \text{and} \quad \Delta_{ER} = \Delta_E = -0.025 \text{ m}$$

Compatibility Equations

$$\Delta_{CO} + f_{CC}C_y + f_{CE}E_y = \Delta_{CR} \tag{1}$$

$$\Delta_{EO} + f_{EC}C_y + f_{EE}E_y = \Delta_{ER} \tag{2}$$

Deflections of Primary Beam From Example 13.9,

$$\Delta_{CO} = -\frac{82,500 \text{ kN} \cdot \text{m}^3}{EI} = -\frac{82,500}{70(10^6)(1,250)(10^{-6})} = -0.943 \text{ m}$$

$$\Delta_{EO} = -\frac{230,000 \text{ kN} \cdot \text{m}^3}{EI} = -\frac{230,000}{70(10^6)(1,250)(10^{-6})} = -2.629 \text{ m}$$

$$f_{CC} = \frac{333.333 \text{ m}^3}{EI} = \frac{333.333}{70(10^6)(1,250)(10^{-6})} = 0.00381 \text{ m/kN}$$

$$f_{EC} = f_{CE} = \frac{833.333 \text{ m}^3}{EI} = \frac{833.333}{70(10^6)(1,250)(10^{-6})} = 0.00952 \text{ m/kN}$$

$$f_{EE} = \frac{2,666.667 \text{ m}^3}{EI} = \frac{2,666.667}{70(10^6)(1,250)(10^{-6})} = 0.0305 \text{ m/kN}$$

Magnitudes of the Redundants By substituting the numerical values into the compatibility equations, we write

$$-0.943 + 0.00381C_y + 0.00952E_y = -0.04 \tag{1a}$$

$$-2.629 + 0.00952C_y + 0.0305E_y = -0.025 \tag{2a}$$

Solving Eqs. (1a) and (2a) simultaneously for C_y and E_y, we obtain

$$C_y = 107.6 \text{ kN} \uparrow \quad \text{and} \quad E_y = 51.8 \text{ kN} \uparrow \qquad \text{Ans.}$$

Reactions and Shear and Bending Moment Diagrams The remaining reactions of the indeterminate beam can now be determined by equilibrium. The reactions and the shear and bending moment diagrams are shown in Fig. 13.22(c).

Ans.

Temperature Changes and Fabrication Errors

Unlike support settlements, which affect only externally indeterminate structures, temperature changes and fabrication errors may affect the stress conditions of externally and/or internally indeterminate structures. The procedure for the analysis of structures subjected to temperature changes and/or fabrication errors is the same as used previously for the case of external loads. The only difference is that the primary structure is now subjected to the prescribed temperature changes and/or fabrication errors (instead of external loads) to evaluate its deflection at the locations of redundants due to these effects. The redundants are then determined by applying the usual compatibility conditions that the deflections of the primary structure at the locations of the redundants due to the combined effect of temperature changes and/or fabrication errors and the redundants must equal the known deflections at the corresponding locations on the actual indeterminate structure. The procedure is illustrated by the following example.

Example 13.16

Determine the reactions and the force in each member of the truss shown in Fig. 13.23(a) due to a temperature increase of 45°C in member AB and a temperature drop of 20°C in member CD. Use the method of consistent deformations.

Solution

Degree of Indeterminacy $i = (m + r) - 2j = (6 + 3) - 2(4) = 1$. The truss is internally indeterminate to the first degree.

Primary Truss The axial force F_{AD} in the diagonal member AD is selected to be the redundant. The primary truss obtained by removing member AD is shown in Fig. 13.23(b). Next, the primary truss is subjected separately to the prescribed temperature changes and a 1-kN tensile force in the redundant member AD, as shown in Fig. 13.23(b) and (c), respectively.

Compatibility Equation The compatibility equation can be expressed as

$$\Delta_{ADO} + f_{AD,AD}F_{AD} = 0 \qquad (1)$$

in which Δ_{ADO} denotes the relative displacement between joints A and D of the primary truss due to temperature changes and the flexibility coefficient $f_{AD,AD}$ denotes the relative displacement between the same joints due to a unit value of the redundant F_{AD}.

Deflections of Primary Truss As discussed in Section 7.3, the virtual work expression for Δ_{ADO} can be written as

$$\Delta_{ADO} = \sum \alpha(\Delta T)Lu_{AD}$$

in which the product $\alpha(\Delta T)L$ equals the axial deformation of a member of the primary truss due to a change in temperature ΔT, and u_{AD} represents the axial force in the same member due to a 1-kN tensile force in member AD. The

continued

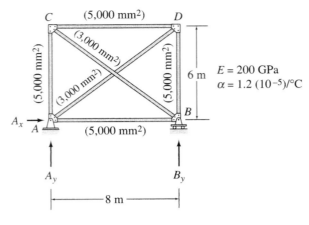

$E = 200$ GPa
$\alpha = 1.2\ (10^{-5})/°C$

(a) Indeterminate Truss

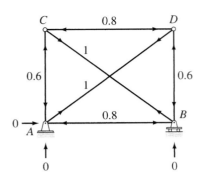

(c) Primary Truss Subjected to Unit
Tensile Force in Member AD — u_{AD}
Forces (kN)

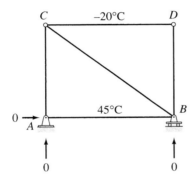

(b) Primary Truss Subjected
to Temperature Changes

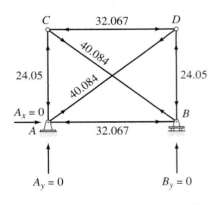

(d) Member Forces for Indeterminate Truss

FIG. 13.23

numerical values of these quantities are tabulated in Table 13.6, from which Δ_{ADO} is determined to be

$$\Delta_{ADO} = -1.92 \text{ mm}$$

Next, the flexibility coefficient $f_{AD, AD}$ is computed by using the virtual work expression (see Table 13.6)

$$f_{AD, AD} = \sum \frac{u_{AD}^2 L}{AE} = 0.0479 \text{ mm}$$

Magnitude of the Redundant By substituting the values of Δ_{ADO} and $f_{AD, AD}$ into the compatibility equation (Eq. (1)), we obtain

$$-1.92 + (0.0479)F_{AD} = 0$$

$$F_{AD} = 40.084 \text{ kN (T)} \qquad \text{Ans.}$$

TABLE 13.6

Member	L (m)	A (m²)	ΔT (°C)	u_{AD} (kN/kN)	$(\Delta T)Lu_{AD}$ (°C·m)	u_{AD}^2L/A (1/m)	$F = u_{AD}F_{AD}$ (kN)
AB	8	0.005	45	−0.8	−288	1,024	−32.067
CD	8	0.005	−20	−0.8	128	1,024	−32.067
AC	6	0.005	0	−0.6	0	432	−24.05
BD	6	0.005	0	−0.6	0	432	−24.05
AD	10	0.003	0	1.0	0	3,333.333	40.084
BC	10	0.003	0	1.0	0	3,333.333	40.084
				Σ	−160	9,578.667	

$$\Delta_{ADO} = \alpha\sum(\Delta T)Lu_{AD} = 1.2(10^{-5})(-160) = -0.00192 \text{ m} = -1.92 \text{ mm}$$

$$f_{AD,AD} = \frac{1}{E}\sum\frac{u_{AD}^2L}{A} = \frac{9,578.667}{200(10^6)} = 47.893(10^{-6}) \text{ m/kN} = 0.0479 \text{ mm/kN}$$

$$F_{AD} = -\frac{\Delta_{ADO}}{f_{AD,AD}} = 40.084 \text{ kN (T)}$$

Reactions Since the truss is statically determinate externally, its reactions due to the temperature changes are zero. **Ans.**

Member Axial Forces The forces in the members of the primary truss due to the temperature changes are zero, so the forces in the members of the indeterminate truss can be expressed as

$$F = u_{AD}F_{AD}$$

The member forces thus obtained are shown in Table 13.6 and Fig. 13.23(d). **Ans.**

SUMMARY

In this chapter we have discussed a general formulation of the force (flexibility) method of analysis of statically indeterminate structures, called the method of consistent deformations. The method involves removing enough restraints from the indeterminate structure to render it statically determinate. The determinate structure is called the primary structure, and the reactions or internal forces associated with the excess restraints removed from the indeterminate structure are termed redundants. The redundants are now treated as unknown loads applied to the primary structure, and their magnitudes are determined by solving the compatibility equations based on the condition that the deflections of the primary structure at the locations (and in the directions) of the redundants, due to the combined effect of the prescribed external loading and the unknown redundants, must be equal to the known deflections at the corresponding locations on the original indeterminate structure. Once the redundants have been determined, the other response characteristics of the indeterminate structure can be evaluated either through equilibrium considerations or by superposition of the responses of the primary structure due to the external loading and due to each of the redundants.

PROBLEMS

Section 13.1

13.1 through 13.4 Determine the reactions and draw the shear and bending moment diagrams for the beams shown in Figs. P13.1–P13.4 using the method of consistent deformations. Select the reaction at the roller support to be the redundant.

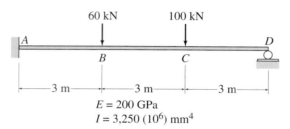

$E = 200$ GPa
$I = 3,250 \ (10^6)$ mm^4

FIG. P13.1, P13.5, P13.49

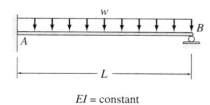

$EI = $ constant

FIG. P13.2, P13.6

13.5 through 13.8 Determine the reactions and draw the shear and bending moment diagrams for the beams shown in Figs. P13.1–P13.4 by using the method of consistent deformations. Select the reaction moment at the fixed support to be the redundant.

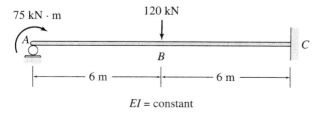

$EI = $ constant

FIG. P13.3, P13.7

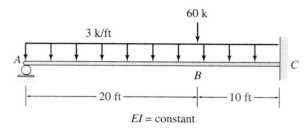

$EI = $ constant

FIG. P13.4, P13.8

13.9 through 13.12 Determine the reactions and draw the shear and bending moment diagrams for the beams shown in Figs. P13.9–P13.12 using the method of consistent deformations. Select the reaction at the interior support to be the redundant.

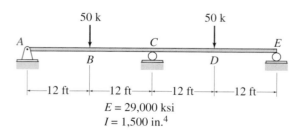

$E = 29,000$ ksi
$I = 1,500$ in.^{4}

FIG. P13.9, P13.30, P13.50

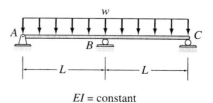

$EI = $ constant

FIG. P13.10, P13.31

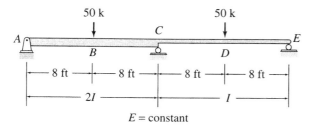

FIG. **P13.11, P13.32**

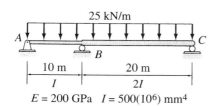

FIG. **P13.12, P13.33, P13.51**

13.13 through 13.25 Determine the reactions and draw the shear and bending moment diagrams for the structures shown in Figs. P13.13–P13.25 using the method of consistent deformations.

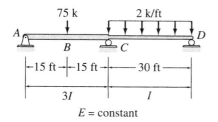

FIG. **P13.13**

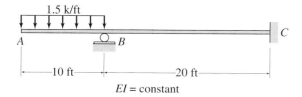

FIG. **P13.14**

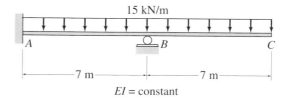

FIG. **P13.15**

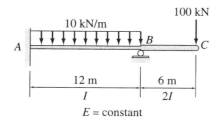

FIG. **P13.16**

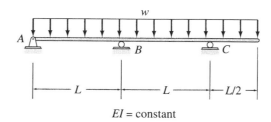

FIG. **P13.17**

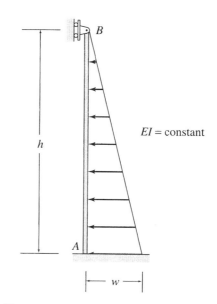

FIG. **P13.18**

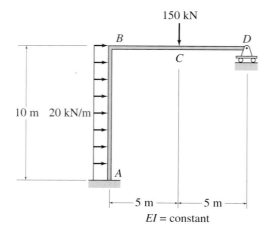

FIG. **P13.19**

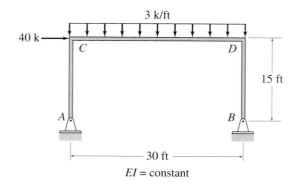

FIG. **P13.22**

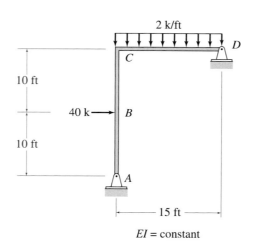

FIG. **P13.20**

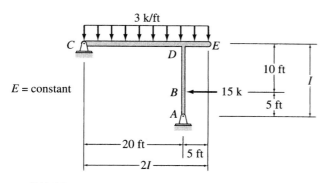

FIG. **P13.21**

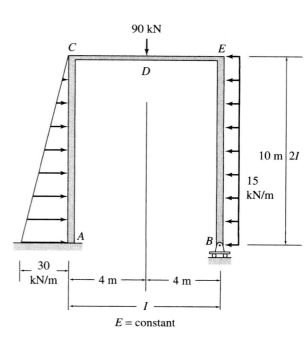

FIG. **P13.23**

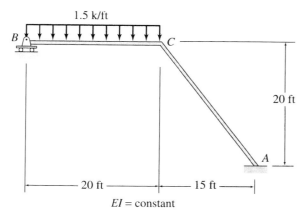

FIG. **P13.24**

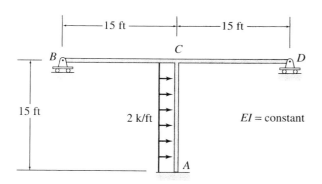

FIG. **P13.25**

13.26 through 13.29 Determine the reactions and the force in each member of the trusses shown in Figs. P13.26–P13.29 using the method of consistent deformations.

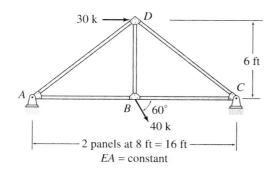

FIG. **P13.26**

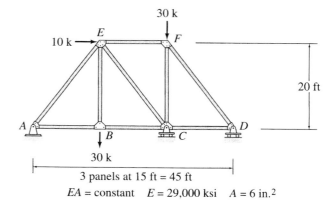

FIG. **P13.27, P13.52**

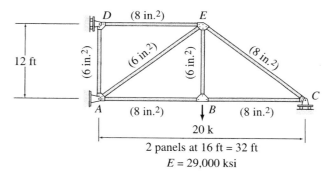

FIG. **P13.28**

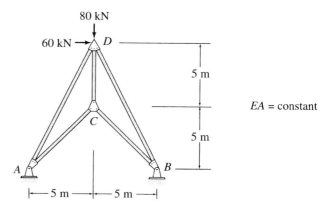

FIG. **P13.29**

Section 13.2

13.30 through 13.33 Solve Problems 13.9 through 13.12 by selecting the bending moment at the interior support to be the redundant. See Figs. P13.9–P13.12.

13.34 through 13.36 Determine the reactions and the force in each member of the trusses shown in Figs. P13.34–P13.36 using the method of consistent deformations.

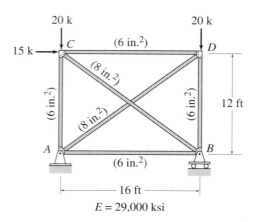

$E = 29,000$ ksi

FIG. **P13.34**

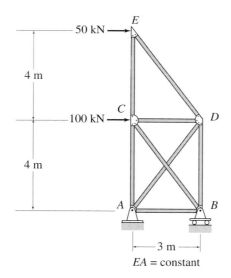

EA = constant

FIG. **P13.35**

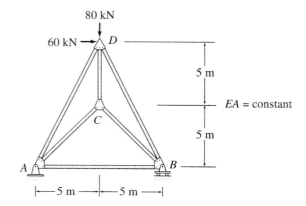

FIG. **P13.36**

Section 13.3

13.37 through 13.45 Determine the reactions and draw the shear and bending moment diagrams for the structures shown in Figs. P13.37–P13.45 using the method of consistent deformations.

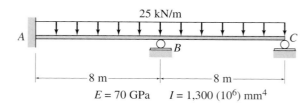

$E = 70$ GPa $I = 1,300 \ (10^6)$ mm^4

FIG. **P13.37, P13.53**

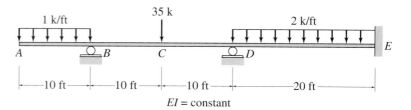

EI = constant

FIG. **P13.38**

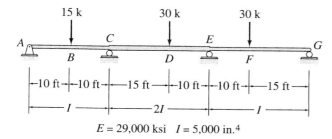

FIG. **P13.39, P13.54**

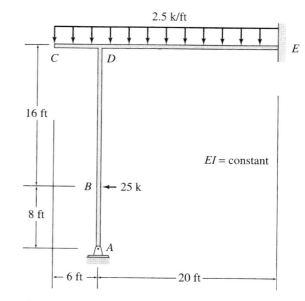

FIG. **P13.43**

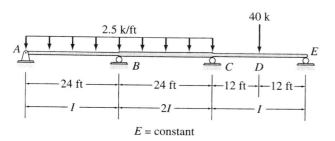

FIG. **P13.40**

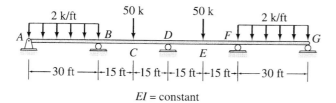

FIG. **P13.41**

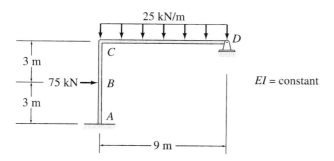

FIG. **P13.42**

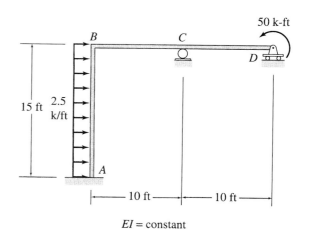

FIG. **P13.44**

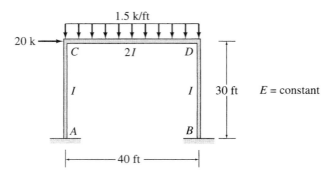

FIG. **P13.45**

13.46 and 13.47 Determine the reactions and the force in each member of the trusses shown in Figs. P13.46 and P13.47 using the method of consistent deformations.

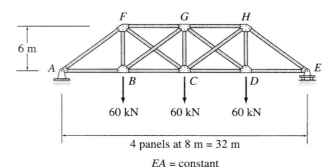

FIG. **P13.46**

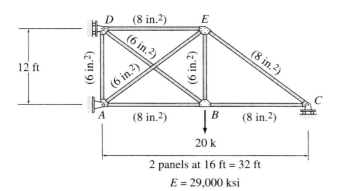

FIG. **P13.47**

Section 13.4

13.48 Determine the reactions for the beam shown in Fig. P13.48 due to a small settlement Δ at the roller support C.

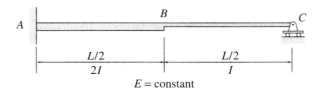

FIG. **P13.48**

13.49 Solve Problem 13.1 for the loading shown and a settlement of 30 mm at support D. See Fig. P13.1.

13.50 Solve Problem 13.9 for the loading shown in Fig. P13.9 and a settlement of $1\frac{1}{4}$ in. at support C.

13.51 Solve Problem 13.12 for the loading shown in Fig. P13.12 and the support settlements of 10 mm at A, 65 mm at B, and 40 mm at C.

13.52 Solve Problem 13.27 for the loading shown in Fig. P13.27 and the support settlements of 1 in. at A, 3 in. at C, and $1\frac{3}{4}$ in. at D.

13.53 Solve Problem 13.37 for the loading shown in Fig. P13.37 and the support settlements of 50 mm at B and 25 mm at C.

13.54 Solve Problem 13.39 for the loading shown in Fig. P13.39 and the support settlements of $\frac{1}{2}$ in. at A, 4 in. at C, 3 in. at E, and $2\frac{1}{2}$ in. at G.

13.55 Determine the reactions and the force in each member of the truss shown in Fig. P13.55 due to a temperature drop of 25°C in members AB, BC, and CD and a temperature increase of 60°C in member EF. Use the method of consistent deformations.

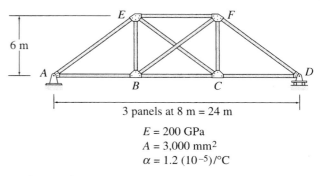

FIG. **P13.55, P13.56**

13.56 Determine the reactions and the force in each member of the truss shown in Fig. P13.55 if member *EF* is 30 mm too short. Use the method of consistent deformations.

13.57 Determine the reactions and the force in each member of the truss shown in Fig. P13.57 due to a temperature increase of 70°F in member *AB*. Use the method of consistent deformations.

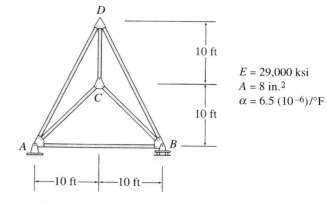

$E = 29{,}000$ ksi
$A = 8$ in.^{2}
$\alpha = 6.5\ (10^{-6})/°F$

FIG. **P13.57**

Three-Moment Equation and the Method of Least Work

14.1 Derivation of Three-Moment Equation
14.2 Application of Three-Moment Equation
14.3 Method of Least Work
Summary
Problems

A Continuous Beam Bridge
Photo courtesy of the Illinois Department of Transportation

In this chapter, we consider two alternate formulations of the force (flexibility) method of analysis of statically indeterminate structures: (1) the *three-moment equation* and (2) the *method of least work*.

The three-moment equation, which was initially presented by Clapeyron in 1857, provides a convenient tool for analyzing continuous beams. The three-moment equation represents, in a general form, the compatibility condition that the slope of the elastic curve be continuous at an interior support of the continuous beam. Since the equation involves three moments—the bending moments at the support under consideration and at the two adjacent supports—it commonly is referred to as the *three-moment equation*. When using this method, the bending moments at the interior (and any fixed) supports of the continuous beam are treated as the redundants. The three-moment equation is then applied at the location of each redundant to obtain a set of compatibility equations which can be solved for the unknown redundant moments.

Another formulation of the force method, called the method of least work, is also discussed in this chapter. This method, which is based on Castigliano's second theorem, essentially is similar to the method of consistent deformations, except that the compatibility equations in the method of least work are established by minimizing the structure's strain energy expressed in terms of the unknown redundants instead of by deflection superposition, as in the method of consistent deformations.

We begin this chapter with the derivation of the three-moment equation for beams with prismatic spans and subjected to external loads and support settlements. Next, we present a procedure for the application of this equation for the analysis of continuous beams. Finally, we consider the method of least work.

14.1 DERIVATION OF THREE-MOMENT EQUATION

Consider an arbitrary continuous beam subjected to external loads and support settlements as shown in Fig. 14.1(a). As discussed in the previous chapter, this beam can be analyzed by the method of consistent deformations by treating the bending moments at the interior supports to be the redundants. From Fig. 14.1(a), we can see that the slope of the elastic curve of the indeterminate beam is continuous at the interior supports. When the restraints corresponding to the redundant bending moments are removed by inserting internal hinges at the interior support points, the primary structure thus obtained consists of a series of simply supported beams. As shown in Figs. 14.1(b) and (c), respectively, when this primary structure is subjected to the known external loading and support settlements, discontinuities develop in the slope of the elastic curve at the locations of the interior supports. Since the redundant bending moments provide continuity of the slope of the elastic curve, these unknown moments are applied as loads on the primary structure as shown in Fig. 14.1(d), and their magnitudes are determined by solving the compatibility equations based on the condition that, at each interior support of the primary structure, the slope of the elastic curve, due to the combined effect of the external loading, support settlements, and unknown redundants, must be continuous.

The three-moment equation uses the foregoing compatibility condition of slope continuity at an interior support to provide a general relationship between the unknown bending moments at the support where compatibility is being considered and at the adjacent supports to the left and to the right, in terms of the loads on the intermediate spans and any settlements of the three supports.

To derive the three-moment equation, we focus our attention on the compatibility equation at an interior support c of the continuous beam, with prismatic spans and a constant modulus of elasticity, shown in Fig. 14.1(a). As indicated in this figure, the adjacent supports to the left and to the right of c are identified as ℓ and r, respectively; the subscripts ℓ and r are used to refer to the loads and properties of the left span, ℓc, and the right span, cr, respectively; and the settlements of supports ℓ, c, and r are denoted by Δ_ℓ, Δ_c, and Δ_r, respectively. The support settlements

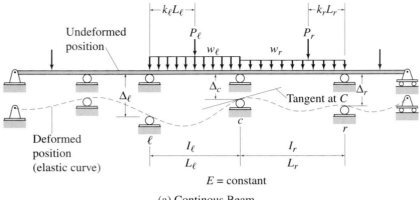

(a) Continous Beam

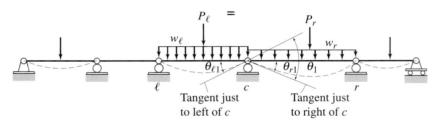

(b) Primary Structure Subjected to External Loading

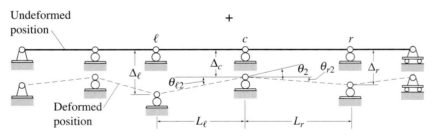

(c) Primary Structure Subjected to Support Settlements

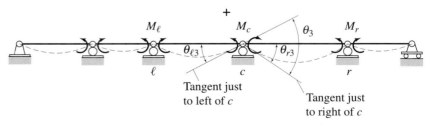

(d) Primary Structure Loaded with Redundant Bending Moments

FIG. 14.1

are considered positive when in the downward direction, as shown in the figure.

From Fig. 14.1(a), we can see that the slope of the elastic curve of the indeterminate beam is continuous at c. In other words, there is no change of slope of the tangents to the elastic curve at just to the left of c and just to the right of c; that is, the angle between the tangents is zero. However, when the primary structure, obtained by inserting internal hinges at the interior support points, is subjected to external loads, as shown in Fig. 14.1(b), a discontinuity develops in the slope of the elastic curve at c, in the sense that the tangent to the elastic curve at just to the left of c rotates relative to the tangent at just to the right of c. The change of slope (or the angle) between the two tangents due to external loads is denoted by θ_1 and can be expressed as (see Fig. 14.1(b))

$$\theta_1 = \theta_{\ell 1} + \theta_{r1} \tag{14.1}$$

in which $\theta_{\ell 1}$ and θ_{r1} denote, respectively, the slopes at the ends c of the spans to the left and to the right of the support c, due to external loads. Similarly, the slope discontinuity at c in the primary structure, due to support settlements (Fig. 14.1(c)), can be written as

$$\theta_2 = \theta_{\ell 2} + \theta_{r2} \tag{14.2}$$

in which $\theta_{\ell 2}$ and θ_{r2} represent, respectively, the slopes of the spans to the left and to the right of c, due to support settlements. Finally, when the primary structure is loaded with the redundant support bending moments, as shown in Fig. 14.1(d), the slope discontinuity at c can be expressed as

$$\theta_3 = \theta_{\ell 3} + \theta_{r3} \tag{14.3}$$

in which $\theta_{\ell 3}$ and θ_{r3} denote, respectively, the slopes at end c of the spans to the left and to the right of the support c, due to unknown redundant moments.

The compatibility equation is based on the requirement that the slope of the elastic curve of the actual indeterminate beam is continuous at c; that is, there is no change of slope from just to the left of c to just to the right of c. Therefore, the algebraic sum of the angles between the tangents at just to the left and at just to the right of c due to the external loading, support settlements and the redundant bending moments must be zero. Thus,

$$\theta_1 + \theta_2 + \theta_3 = 0 \tag{14.4}$$

By substituting Eqs. (14.1) through (14.3) into Eq. (14.4), we obtain

$$(\theta_{\ell 1} + \theta_{r1}) + (\theta_{\ell 2} + \theta_{r2}) + (\theta_{\ell 3} + \theta_{r3}) = 0 \tag{14.5}$$

Since each span of the primary structure can be treated as a simply supported beam, the slopes at the ends c of the left and the right spans, due to the external loads (Fig. 14.1(b)), can be conveniently determined either by the conjugate-beam method or by using the beam-deflection formulas given inside the front cover of the book. By using the deflection formulas, we obtain

$$\theta_{\ell 1} = \sum \frac{P_\ell L_\ell^2 k_\ell (1 - k_\ell^2)}{6EI_\ell} + \frac{w_\ell L_\ell^3}{24EI_\ell} \tag{14.6a}$$

$$\theta_{r1} = \sum \frac{P_r L_r^2 k_r (1 - k_r^2)}{6EI_r} + \frac{w_r L_r^3}{24EI_r} \tag{14.6b}$$

in which the summation signs have been added to the first terms on the right sides of these equations, so that multiple concentrated loads can be applied to each span (instead of a single concentrated load as shown in Figs. 14.1(a) and (b) for simplicity). As continuous beams usually are loaded with uniformly distributed loads over entire spans and concentrated loads, the effects of only these two types of loadings generally are considered in the three-moment equation. However, the effects of other types of loads can be included simply by adding the expressions of slopes due to these loads to the right sides of Eqs. (14.6a) and (14.6b).

The slopes $\theta_{\ell 2}$ and θ_{r2}, of the left and the right spans, respectively, due to support settlements, can be obtained directly from the deformed positions of the spans depicted in Fig. 14.1(c). Since the settlements are assumed to be small, the slopes can be expressed as

$$\theta_{\ell 2} = \frac{\Delta_\ell - \Delta_c}{L_\ell} \qquad \theta_{r2} = \frac{\Delta_r - \Delta_c}{L_r} \tag{14.7}$$

The slopes at ends c of the left and the right spans, due to redundant support bending moments, (Fig. 14.1(d)), can be determined conveniently by using the beam-deflection formulas. Thus,

$$\theta_{\ell 3} = \frac{M_\ell L_\ell}{6EI_\ell} + \frac{M_c L_\ell}{3EI_\ell} \tag{14.8a}$$

$$\theta_{r3} = \frac{M_c L_r}{3EI_r} + \frac{M_r L_r}{6EI_r} \tag{14.8b}$$

in which M_ℓ, M_c and M_r denote the bending moments at supports ℓ, c and r, respectively. As shown in Fig. 14.1(d), these redundant bending moments are considered to be positive in accordance with the *beam convention*—that is, when causing compression in the upper fibers and tension in the lower fibers of the beam.

By substituting Eqs. (14.6) through (14.8) into Eq. (14.5), we write the compatibility equation as

$$\sum \frac{P_\ell L_\ell^2 k_\ell (1 - k_\ell^2)}{6EI_\ell} + \frac{w_\ell L_\ell^3}{24EI_\ell} + \sum \frac{P_r L_r^2 k_r (1 - k_r^2)}{6EI_r} + \frac{w_r L_r^3}{24EI_r} + \frac{\Delta_l - \Delta_c}{L_\ell}$$

$$+ \frac{\Delta_r - \Delta_c}{L_r} + \frac{M_\ell L_\ell}{6EI_\ell} + \frac{M_c L_\ell}{3EI_\ell} + \frac{M_c L_r}{3EI_r} + \frac{M_r L_r}{6EI_r} = 0$$

By simplifying the foregoing equation and rearranging it to separate the terms containing redundant moments from those involving loads and support settlements, we obtain the general form of the *three-moment equation*:

$$
\frac{M_\ell L_\ell}{I_\ell} + 2M_c\left(\frac{L_\ell}{I_\ell} + \frac{L_r}{I_r}\right) + \frac{M_r L_r}{I_r}
$$

$$
= -\sum \frac{P_\ell L_\ell^2 k_\ell}{I_\ell}(1 - k_\ell^2) - \sum \frac{P_r L_r^2 k_r}{I_r}(1 - k_r^2) - \frac{w_\ell L_\ell^3}{4I_\ell} - \frac{w_r L_r^3}{4I_r}
$$

$$
- 6E\left(\frac{\Delta_\ell - \Delta_c}{L_\ell} + \frac{\Delta_r - \Delta_c}{L_r}\right)
$$

$$(14.9)$$

in which M_c = bending moment at support c where the compatibility is being considered; M_ℓ, M_r = bending moments at the adjacent supports to the left and to the right of c, respectively; E = modulus of elasticity; L_ℓ, L_r = lengths of the spans to the left and to the right of c, respectively; I_ℓ, I_r = moments of inertia of the spans to the left and to the right of c, respectively; P_ℓ, P_r = concentrated loads acting on the left and the right spans, respectively; k_ℓ (or k_r) = ratio of the distance of P_ℓ (or P_r) from the left (or right) support to the span length; w_ℓ, w_r = uniformly distributed loads applied to the left and the right spans, respectively; Δ_c = settlement of the support c under consideration; and Δ_ℓ, Δ_r = settlements of the adjacent supports to the left and to the right of c, respectively. As noted before, the support bending moments are considered to be positive in accordance with the *beam convention*—that is, when causing compression in the upper fibers and tension in the lower fibers of the beam. Furthermore, the external loads and support settlements are considered positive when in the downward direction, as shown in Fig. 14.1(a).

If the moments of inertia of two adjacent spans of a continuous beam are equal (i.e., $I_\ell = I_r = I$), then the three-moment equation simplifies to

$$
M_\ell L_\ell + 2M_c(L_\ell + L_r) + M_r L_r
$$

$$
= -\sum P_\ell L_\ell^2 k_\ell(1 - k_\ell^2) - \sum P_r L_r^2 k_r(1 - k_r^2) - \frac{1}{4}(w_\ell L_\ell^3 + w_r L_r^3)
$$

$$
- 6EI\left(\frac{\Delta_\ell - \Delta_c}{L_\ell} + \frac{\Delta_r - \Delta_c}{L_r}\right)
$$

$$(14.10)$$

If both the moments of inertia and the lengths of two adjacent spans are equal (i.e., $I_\ell = I_r = I$ and $L_\ell = L_r = L$), then the three-moment equation becomes

$$M_\ell + 4M_c + M_r$$

$$= -\sum P_\ell L k_\ell (1 - k_\ell^2) - \sum P_r L k_r (1 - k_r^2) \quad (14.11)$$

$$- \frac{L^2}{4}(w_\ell + w_r) - \frac{6EI}{L^2}(\Delta_\ell - 2\Delta_c + \Delta_r)$$

The foregoing three-moment equations are applicable to any three consecutive supports, ℓ, c and r, of a continuous beam, provided that there are no discontinuities, such as internal hinges, in the beam between the left support ℓ and the right support r.

14.2 APPLICATION OF THREE-MOMENT EQUATION

The following step-by-step procedure can be used for analyzing continuous beams by the three-moment equation.

1. Select the unknown bending moments at all interior supports of the beam as the redundants.
2. By treating each interior support successively as the intermediate support c, write a three-moment equation. When writing these equations, it should be realized that bending moments at the simple end supports are known. For such a support with a cantilever overhang, the bending moment equals that due to the external loads acting on the cantilever portion about the end support. The total number of three-moment equations thus obtained must be equal to the number of redundant support bending moments, which must be the only unknowns in these equations.
3. Solve the system of three-moment equations for the unknown support bending moments.
4. Compute the span end shears. For each span of the beam, (a) draw a free-body diagram showing the external loads and end moments and (b) apply the equations of equilibrium to calculate the shear forces at the ends of the span.
5. Determine support reactions by considering the equilibrium of the support joints of the beam.
6. If so desired, draw shear and bending moment diagrams of the beam by using the *beam sign convention*.

Fixed Supports

The three-moment equations, as given by Eqs. (14.9) through (14.11), were derived to satisfy the compatibility condition of slope continuity at the interior supports of continuous beams. These equations can, how-

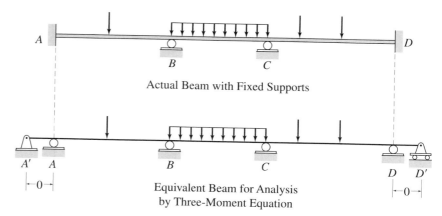

FIG. **14.2**

Actual Beam with Fixed Supports

Equivalent Beam for Analysis
by Three-Moment Equation

ever, be used to satisfy the compatibility condition of zero slope at the
fixed end supports of beams. This can be achieved by replacing the fixed
support by an imaginary interior roller support with an adjoining end
span of zero length simply supported at its outer end, as shown in Fig.
14.2. The reaction moment at the actual fixed support is now treated as
the redundant bending moment at the imaginary interior support, and
the three-moment equation when applied to this imaginary support
satisfies the compatibility condition of zero slope of the elastic curve at
the actual fixed support. When analyzing a beam for support settlements, both imaginary supports—that is, the interior roller support and
the outer simple end support—are considered to undergo the same settlement as the actual fixed support.

Example 14.1

Determine the reactions and draw the shear and bending moment diagrams for
the beam shown in Fig. 14.3(a) by using the three-moment equation.

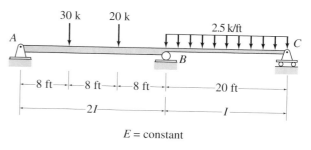

$E = $ constant

(a) Indeterminate Beam

FIG. **14.3**

continued

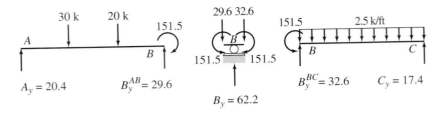

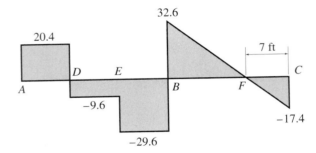

(b) Span End Moments and Shears

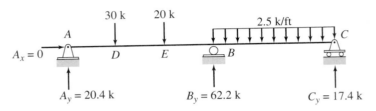

(c) Support Reactions

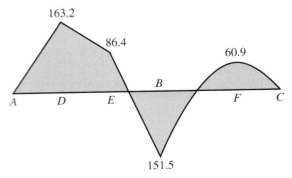

Shear diagram (k)

Bending moment diagram (k-ft)

(d) Shear and Bending Moment Diagrams

FIG. **14.3** (contd.)

Solution

Redundant The beam has one degree of indeterminacy. The bending moment M_B, at the interior support B, is the redundant.

Three-Moment Equation at Joint B Considering the supports, A, B, and C as ℓ, c, and r, respectively, and substituting $L_\ell = 24$ ft, $L_r = 20$ ft, $I_\ell = 2I$, $I_r = I$, $P_{\ell 1} = 30$ k, $k_{\ell 1} = 1/3$, $P_{\ell 2} = 20$ k, $k_{\ell 2} = 2/3$, $w_r = 2.5$ k/ft, and $P_r = w_\ell = \Delta_\ell = \Delta_c = \Delta_r = 0$, into Eq. (14.9), we obtain

$$\frac{M_A(24)}{2I} + 2M_B\left(\frac{24}{2I} + \frac{20}{I}\right) + \frac{M_C(20)}{I} = -\frac{30(24)^2(1/3)}{2I}[1 - (1/3)^2]$$

$$-\frac{20(24)^2(2/3)}{2I}[1 - (2/3)^2] - \frac{2.5(20)^3}{4I}$$

Since A and C are simple end supports, we have by inspection

$$M_A = M_C = 0$$

Thus, the three-moment equation becomes

$$64M_B = -9{,}693.33$$

from which we obtain the redundant bending moment to be

$$M_B = -151.5 \text{ k-ft} \qquad \text{Ans.}$$

Span End Shears and Reactions The shears at the ends of the spans AB and BC of the continuous beam can now be determined by applying the equations of equilibrium to the free bodies of the spans shown in Fig. 14.3(b). Note that the negative bending moment M_B is applied at the ends B of spans AB and BC so that it causes tension in the upper fibers and compression in the lower fibers of the beam. By considering the equilibrium of span AB, we obtain

$$+\circlearrowleft \sum M_B = 0 \qquad -A_y(24) + 30(16) + 20(8) - 151.5 = 0$$

$$A_y = 20.4 \text{ k} \uparrow \quad \text{Ans.}$$

$$+\uparrow \sum F_y = 0 \qquad 20.4 - 30 - 20 + B_y^{AB} = 0$$

$$B_y^{AB} = 29.6 \text{ k} \uparrow$$

Similarly, for span BC,

$$+\circlearrowleft \sum M_C = 0 \qquad -B_y^{BC}(20) + 151.5 + 2.5(20)(10) = 0$$

$$B_y^{BC} = 32.6 \text{ k} \uparrow$$

$$+\uparrow \sum F_y = 0 \qquad 32.6 - 2.5(20) + C_y = 0$$

$$C_y = 17.4 \text{ k} \uparrow \qquad \text{Ans.}$$

By considering the equilibrium of joint B in the vertical direction, we obtain

$$B_y = B_y^{AB} + B_y^{BC} = 29.6 + 32.6 = 62.2 \text{ k} \uparrow \qquad \text{Ans.}$$

The reactions are shown in Fig. 14.3(c).

Shear and Bending Moment Diagrams See Fig. 14.3(d). $\qquad$ Ans.

Example 14.2

Determine the reactions for the continuous beam shown in Fig. 14.4(a) due to the uniformly distributed load and due to the support settlements of 10 mm at A, 50 mm at B, 20 mm at C, and 40 mm at D. Use the three-moment equation.

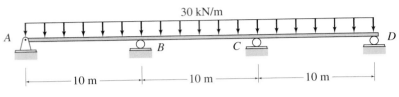

$EI = $ constant

$E = 200$ GPa $I = 700 \, (10^6) \, \text{mm}^4$

(a) Indeterminate Beam

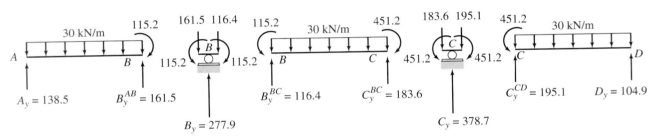

(b) Span End Moments and Shears

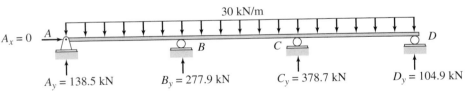

(c) Support Reactions

FIG. **14.4**

Solution

Redundants The bending moments M_B and M_C, at the interior supports B and C, respectively, are the redundants.

Three-Moment Equation at Joint B By considering the supports A, B, and C as ℓ, c, and r, respectively, and substituting $L = 10$ m, $E = 200$ GPa $= 200(10^6) \, \text{kN/m}^2$, $I = 700(10^6) \, \text{mm}^4 = 700(10^{-6}) \, \text{m}^4$, $w_\ell = w_r = 30 \, \text{kN/m}$, $\Delta_\ell = \Delta_A = 10 \, \text{mm} = 0.01 \, \text{m}$, $\Delta_c = \Delta_B = 50 \, \text{mm} = 0.05 \, \text{m}$, $\Delta_r = \Delta_C = 20 \, \text{mm} = 0.02$ m and $P_\ell = P_r = 0$, into Eq. (14.11), we write

$$M_A + 4M_B + M_C = -\frac{(10)^2}{4}(30+30) - \frac{6(200)(700)}{(10)^2}[0.01 - 2(0.05) + 0.02]$$

Since A is a simple end support, $M_A = 0$. The foregoing equation thus simplifies to

$$4M_B + M_C = -912 \tag{1}$$

Three-Moment Equation at Joint C Similarly, by considering the supports B, C, and D as ℓ, c, and r, respectively, and by substituting the appropriate numerical values in Eq. (14.11), we obtain

$$M_B + 4M_C + M_D = -\frac{(10)^2}{4}(30+30) - \frac{6(200)(700)}{(10)^2}[0.05 - 2(0.02) + 0.04]$$

Since D is a simple end support, $M_D = 0$. Thus, the foregoing equation becomes

$$M_B + 4M_C = -1,920 \tag{2}$$

Support Bending Moments Solving Eqs. (1) and (2) simultaneously for M_B and M_C, we obtain

$$M_B = -115.2 \text{ kN} \cdot \text{m} \qquad\qquad \textbf{Ans.}$$

$$M_C = -451.2 \text{ kN} \cdot \text{m} \qquad\qquad \textbf{Ans.}$$

Span End Shears and Reactions With the redundants M_B and M_C known, the span end shears and the support reactions can be determined by considering the equilibrium of the free bodies of the spans AB, BC, and CD, and joints B and C, as shown in Fig. 14.4(b). The reactions are shown in Fig. 14.4(c). **Ans.**

Example 14.3

Determine the reactions for the continuous beam shown in Fig. 14.5(a) by the three-moment equation.

Solution

Since support A of the beam is fixed, we replace it with an imaginary interior roller support with an adjoining end span of zero length, as shown in Fig. 14.5(b).

Redundants From Fig. 14.5(b), we can see that the bending moments M_A and M_B at the supports A and B, respectively, are the redundants.

Three-Moment Equation at Joint A By using Eq. (14.10) for supports A', A, and B, we obtain

$$2M_A(0+20) + M_B(20) = -45(20)^2(1/2)[1-(1/2)^2]$$

or

$$2M_A + M_B = -337.5 \tag{1}$$

continued

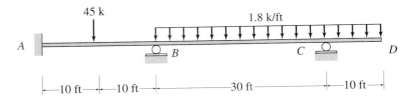

EI = constant

(a) Indeterminate Beam

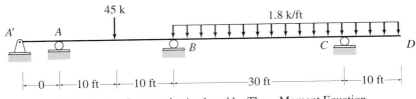

(b) Equivalent Beam to be Analyzed by Three-Moment Equation

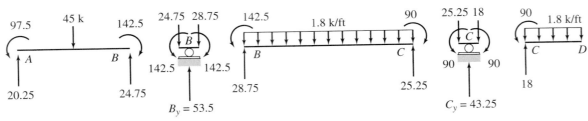

(c) Span End Moments and Shears

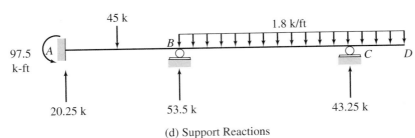

(d) Support Reactions

FIG. 14.5

Three-Moment Equation at Joint B Similarly, applying Eq. (14.10) for supports A, B, and C, we write

$$M_A(20) + 2M_B(20 + 30) + M_C(30)$$
$$= -45(20)^2(1/2)[1 - (1/2)^2] - (1/4)(1.8)(30)^3$$

The bending moment at end C of the cantilever overhang CD is computed as

$$M_C = -1.8(10)(5) = -90 \text{ k-ft} \qquad \text{Ans.}$$

By substituting $M_C = -90$ k-ft into the foregoing three-moment equation and simplifying, we obtain

$$M_A + 5M_B = -810 \qquad (2)$$

Support Bending Moments Solving Eqs. (1) and (2), we obtain

$$M_A = -97.5 \text{ k-ft} \qquad \text{Ans.}$$

$$M_B = -142.5 \text{ k-ft} \qquad \text{Ans.}$$

Span End Shears and Reactions See Figs. 14.5(c) and (d). Ans.

14.3 METHOD OF LEAST WORK

In this section, we consider another formulation of the force method called the *method of least work*. In this method, the compatibility equations are established by using *Castigliano's second theorem* instead of by deflection superposition, as in the method of consistent deformations considered in the previous chapter. With this exception, the two methods are similar and require essentially the same amount of computational effort. The method of least work usually proves to be more convenient for analyzing composite structures that contain both axial force members and flexural members (e.g., beams supported by cables). However, the method is not as general as the method of consistent deformations in the sense that, in its original form (as presented here), the method of least work cannot be used for analyzing the effects of support settlements, temperature changes, and fabrication errors.

To develop the method of least work, let us consider a statically indeterminate beam with unyielding supports subjected to an external loading w, as shown in Fig. 14.6. Suppose that we select the vertical reaction B_y at the interior support B to be the redundant. By treating the redundant as an unknown load applied to the beam along with the prescribed loading w, an expression for the strain energy can be written in terms of the known load w and the unknown redundant B_y as

$$U = f(w, B_y) \qquad (14.12)$$

Equation (14.12) indicates symbolically that the strain energy for the beam is expressed as a function of the known external load w and the unknown redundant B_y.

According to Castigliano's second theorem (Section 7.7), the partial derivative of the strain energy with respect to a force equals the deflection of the point of application of the force along its line of action. Since

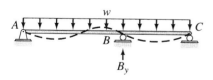

FIG. 14.6

the deflection at the point of application of the redundant B_y is zero, by applying Castigliano's second theorem, we can write

$$\frac{\partial U}{\partial B_y} = 0 \tag{14.13}$$

It should be realized that Eq. (14.13) represents the compatibility equation in the direction of redundant B_y, and it can be solved for the redundant.

As Eq. (14.13) indicates, the first partial derivative of the strain energy with respect to the redundant must be equal to zero. This implies that for the value of the redundant that satisfies the equations of equilibrium and compatibility, the strain energy of the structure is a minimum or maximum. Since for a linearly elastic structure there is no maximum value of strain energy, because it can be increased indefinitely by increasing the value of the redundant, we conclude that for the true value of the redundant the strain energy must be a minimum. This conclusion is known as the *principle of least work:*

The magnitudes of the redundants of a statically indeterminate structure must be such that the strain energy stored in the structure is a minimum (i.e., the internal work done is the least).

The method of least work, as described here, can be easily extended to the analysis of structures with multiple degrees of indeterminacy. If a structure is indeterminate to the nth degree, then n redundants are selected, and the strain energy for the structure is expressed in terms of the known external loading and the n unknown redundants as

$$U = f(w, R_1, R_2, \dots, R_n) \tag{14.14}$$

in which w represents all the known loads and $R_1, R_2, \dots, R_n$ denote the n redundants. Next, the principle of least work is applied separately for each redundant by partially differentiating the strain energy expression (Eq. (14.14)) with respect to each of the redundants and by setting each partial derivative equal to zero; that is,

$$\frac{\partial U}{\partial R_1} = 0$$

$$\frac{\partial U}{\partial R_2} = 0 \tag{14.15}$$

$$\vdots$$

$$\frac{\partial U}{\partial R_n} = 0$$

which represents a system of n simultaneous equations in terms of n redundants and can be solved for the redundants.

The procedure for the analysis of indeterminate structures by the method of least work is illustrated by the following examples.

Example 14.4

Determine the reactions for the beam shown in Fig. 14.7 by the method of least work.

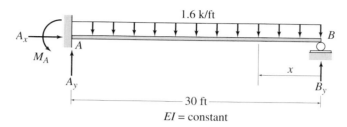

FIG. 14.7

Solution

This beam was analyzed in Example 13.2 by the method of consistent deformations.

The beam is supported by four reactions, so its degree of indeterminacy is equal to 1. The vertical reaction B_y, at the roller support B, is selected as the redundant. We will evaluate the magnitude of the redundant by minimizing the strain energy of the beam with respect to B_y.

As discussed in Section 7.6, the strain energy of a beam subjected only to bending can be expressed as

$$U = \int_0^L \frac{M^2}{2EI}\, dx \tag{1}$$

According to the principle of least work, the partial derivative of strain energy with respect to B_y must be zero; that is,

$$\frac{\partial U}{\partial B_y} = \int_0^L \left(\frac{\partial M}{\partial B_y}\right) \frac{M}{EI}\, dx = 0 \tag{2}$$

Using the x coordinate shown in Fig. 14.7, we write the equation for bending moment, M, in terms of B_y, as

$$M = B_y(x) - \frac{1.6x^2}{2}$$

Next, we partially differentiate the expression for M with respect to B_y, to obtain

$$\frac{\partial M}{\partial B_y} = x$$

By substituting the expressions for M and $\partial M / \partial B_y$ into Eq. (2), we write

$$\frac{1}{EI}\left[\int_0^{30} x(B_y x - 0.8x^2)\, dx\right] = 0$$

continued

By integrating, we obtain

$$9{,}000B_y - 162{,}000 = 0$$

from which

$$B_y = 18 \text{ k} \uparrow \qquad \text{Ans.}$$

To determine the remaining reactions of the indeterminate beam, we apply the equilibrium equations (Fig. 14.7):

$$+ \rightarrow \Sigma F_x = 0 \qquad\qquad\qquad\qquad\qquad\qquad\quad A_x = 0 \qquad \text{Ans.}$$

$$+ \uparrow \Sigma F_y = 0 \qquad A_y - 1.6(30) + 18 = 0 \qquad\qquad A_y = 30 \text{ k} \uparrow \qquad \text{Ans.}$$

$$+ \circlearrowleft \Sigma M_A = 0 \qquad M_A - 1.6(30)(15) + 18(30) = 0 \qquad M_A = 180 \text{ k-ft} \circlearrowleft$$

$$\text{Ans.}$$

Example 14.5

Determine the reactions for the two-span continuous beam shown in Fig. 14.8 by the method of least work.

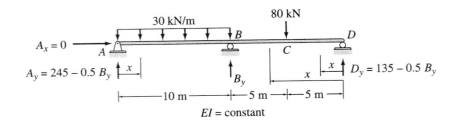

FIG. 14.8

$$EI = \text{constant}$$

Solution

The beam is supported by four reactions, A_x, A_y, B_y, and D_y. Since there are only three equilibrium equations, the degree of indeterminacy of the beam is equal to 1. Let us select the reaction B_y to be the redundant. The magnitude of the redundant will be determined by minimizing the strain energy of the beam with respect to B_y.

The strain energy of a beam subjected only to bending is expressed as

$$U = \int_0^L \frac{M^2}{2EI}\,dx \qquad (1)$$

According to the principle of least work,

$$\frac{\partial U}{\partial B_y} = \int_0^L \left(\frac{\partial M}{\partial B_y}\right)\frac{M}{EI}\,dx = 0 \qquad (2)$$

Before we can obtain the equations for bending moments, M, we must express the reactions at the supports A and D of the beam in terms of the redundant B_y. Applying the three equilibrium equations, we write

$$+ \rightarrow \Sigma F_x = 0 \qquad A_x = 0 \qquad\qquad\qquad \text{Ans.}$$

TABLE 14.1

Segment	x coordinate Origin	x coordinate Limits (m)	M	$\partial M/\partial B_y$
AB	A	0–10	$(245 - 0.5B_y)x - 15x^2$	$-0.5x$
DC	D	0–5	$(135 - 0.5B_y)x$	$-0.5x$
CB	D	5–10	$(135 - 0.5B_y)x - 80(x - 5)$	$-0.5x$

$$+\circlearrowleft \sum M_D = 0$$

$$-A_y(20) + 30(10)(15) - B_y(10) + 80(5) = 0$$

$$A_y = 245 - 0.5B_y \qquad (3)$$

$$+\uparrow \sum F_y = 0$$

$$(245 - 0.5B_y) - 30(10) + B_y - 80 + D_y = 0$$

$$D_y = 135 - 0.5B_y \qquad (4)$$

To determine the equations for bending moments, M, the beam is divided into three segments, AB, BC, and CD. The x coordinates used for determining the equations are shown in Fig. 14.8, and the bending moment equations, in terms of B_y, are tabulated in Table 14.1. Next, the derivatives of the bending moments with respect to B_y are evaluated. These derivatives are listed in the last column of Table 14.1.

By substituting the expressions for M and $\partial M/\partial B_y$ into Eq. (2), we write

$$\frac{1}{EI}\left[\int_0^{10} (-0.5x)(245x - 0.5B_y x - 15x^2)\, dx\right.$$

$$+ \int_0^5 (-0.5x)(135x - 0.5B_y x)\, dx$$

$$\left. + \int_5^{10} (-0.5x)(55x - 0.5B_y x + 400)\, dx\right] = 0$$

By integrating, we obtain

$$-40{,}416.667 + 166.667B_y = 0$$

from which

$$B_y = 242.5 \text{ kN} \uparrow \qquad \text{Ans.}$$

By substituting the value of B_y into Eqs. (3) and (4), respectively, we determine the vertical reactions at supports A and D.

$$A_y = 123.75 \text{ kN} \uparrow \qquad \text{Ans.}$$

$$D_y = 13.75 \text{ kN} \uparrow \qquad \text{Ans.}$$

Example 14.6

Determine the force in each member of the truss shown in Fig. 14.9(a) by the method of least work.

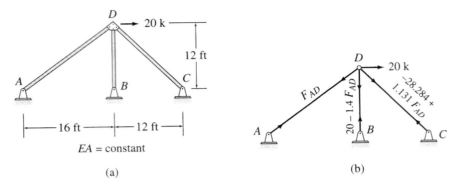

EA = constant

(a)   (b)

FIG. 14.9

Solution

The truss contains one more member than necessary for internal stability; therefore, its degree of indeterminacy is equal to 1. Let us select the force F_{AD} in member AD to be the redundant. We will determine the magnitude of F_{AD} by minimizing the strain energy of the truss with respect to F_{AD}.

As discussed in Section 7.6, the strain energy of a truss can be expressed as

$$U = \sum \frac{F^2 L}{2AE} \tag{1}$$

According to the principle of least work, the partial derivative of strain energy with respect to F_{AD} must be zero; that is,

$$\frac{\partial U}{\partial F_{AD}} = \sum \left(\frac{\partial F}{\partial F_{AD}} \right) \frac{FL}{AE} = 0 \tag{2}$$

TABLE 14.2

Member	L (ft)	F	$\dfrac{\partial F}{\partial F_{AD}}$	$\left(\dfrac{\partial F}{\partial F_{AD}} \right) FL$	F (k)
AD	20	F_{AD}	1	$20F_{AD}$	13.474
BD	12	$20 - 1.4F_{AD}$	-1.4	$-336 + 23.52F_{AD}$	1.136
CD	16.971	-28.284 $+ 1.131F_{AD}$	1.131	-542.889 $+ 21.709F_{AD}$	-13.045
			$\sum$	$-878.889 + 65.229F_{AD}$	

$$\frac{1}{AE} \sum \left(\frac{\partial F}{\partial F_{AD}} \right) FL = 0$$

$$-878.889 + 65.229F_{AD} = 0$$

$$F_{AD} = 13.474 \text{ k (T)}$$

The axial forces in members BD and CD are expressed in terms of the redundant F_{AD} by considering the equilibrium of joint D (Fig. 14.9(b)). These member forces F, along with their partial derivatives with respect to F_{AD}, are tabulated in Table 14.2. To apply Eq. (2), the terms $(\partial F/\partial F_{AD})\,FL$ are computed for the individual members and are added as shown in Table 14.2. Note that since EA is constant, it is not included in the summation. Equation (2) is then solved, as shown in Table 14.2, to determine the magnitude of the redundant.

$$F_{AD} = 13.474 \text{ k (T)} \qquad \text{Ans.}$$

Finally, the forces in members BD and CD are evaluated by substituting the value of F_{AD} into the expressions for the member forces given in the third column of Table 14.2.

$$F_{BD} = 1.136 \text{ k (T)} \qquad \text{Ans.}$$

$$F_{CD} = 13.045 \text{ k (C)} \qquad \text{Ans.}$$

Example 14.7

A beam is supported by a fixed support A and a cable BD, as shown in Fig. 14.10(a). Determine the tension in the cable by the method of least work.

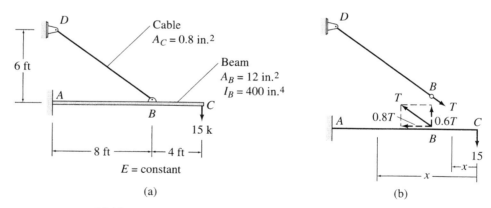

FIG. 14.10

Solution

We will analyze the structure by considering the tension T in cable BD to be redundant. The magnitude of the redundant will be determined by minimizing the strain energy of the structure with respect to T.

Because the structure contains both axially loaded and flexural members, its total strain energy is expressed as the sum of the strain energy due to axial forces and the strain energy due to bending; that is,

continued

TABLE 14.3

Segment	Origin	Limits (ft)	M	F	$\dfrac{\partial M}{\partial T}$	$\dfrac{\partial F}{\partial T}$
		x coordinate				
CB	C	0–4	$-15x$	0	0	0
BA	C	4–12	$-15x + 0.6T(x-4)$	$-0.8T$	$0.6(x-4)$	-0.8
BD	—	—	0	T	0	1

$$U = \sum \frac{F^2 L}{2AE} + \sum \int \frac{M^2}{2EI}\,dx \tag{1}$$

According to the principle of least work,

$$\frac{\partial U}{\partial T} = \sum \left(\frac{\partial F}{\partial T}\right)\frac{FL}{AE} + \sum \int \left(\frac{\partial M}{\partial T}\right)\frac{M}{EI}\,dx = 0 \tag{2}$$

The expressions for the bending moments M and the axial forces F in terms of the redundant T and their derivatives with respect to T are tabulated in Table 14.3. By substituting these expressions and derivatives into Eq. (2), we write

$$\frac{1}{E}\left[\frac{(-0.8)(-0.8T)(8)(12)^2}{12} + \frac{1(T)(10)(12)^2}{0.8}\right.$$

$$\left. + \frac{(12)^4}{400}\int_4^{12} 0.6(x-4)(-15x+0.6Tx-2.4T)\,dx\right] = 0$$

$$T = 27.612 \text{ k} \quad \textbf{Ans.}$$

SUMMARY

In this chapter, we have studied two formulations of the force (flexibility) method of analysis of statically indeterminate structures, namely, the three-moment equation and the method of least work.

The three-moment equation represents, in a general form, the compatibility condition that the slope of the elastic curve be continuous at an interior support of the continuous beam. This method, which can be used for analyzing continuous beams subjected to external loads and support settlements, involves treating the bending moments at the interior (and any fixed) supports of the beam as the redundants. The three-moment equation is then applied at the location of each redundant to obtain a set of compatibility equations which can then be solved for the redundant bending moments.

The principle of least work states that *the magnitudes of the redundants of an indeterminate structure must be such that the strain energy stored in the structure is a minimum.* To analyze an indeterminate structure by the method of least work, the strain energy of the structure is first expressed in terms of the redundants. Then the partial derivatives of the strain energy with respect to each of the redundants are determined and set equal to zero to obtain a system of simultaneous equations that can be solved for the redundants. The method of least work cannot be used for analyzing the effects of support settlements, temperature changes, and fabrication errors.

PROBLEMS

Section 14.2

14.1 through 14.8 Determine the reactions and draw the shear and bending moment diagrams for the beams shown in Figs. P14.1 through P14.8 using the three-moment equation.

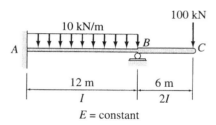

FIG. **P14.4, P14.11**

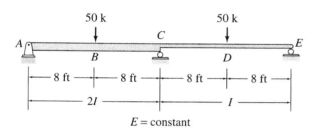

FIG. **P14.1**

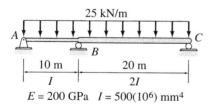

FIG. **P14.2, P14.9**

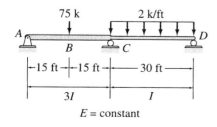

FIG. **P14.3**

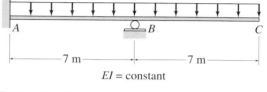

FIG. **P14.5, P14.12**

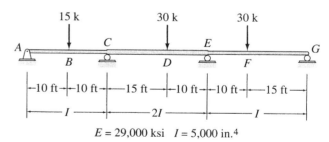

FIG. **P14.6, P14.10**

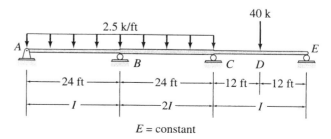

FIG. **P14.7**

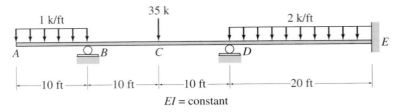

FIG. **P14.8**

14.9 Solve Problem 14.2 for the loading shown in Fig. P14.2 and the support settlements of 10 mm at A, 65 mm at B, and 40 mm at C.

14.10 Solve Problem 14.6 for the loading shown in Fig. P14.6 and the support settlements of $\frac{1}{2}$ in. at A, 4 in. at C, 3 in. at E, and 2-1/2 in. at G.

Section 14.3

14.11 Solve Problem 14.4 by the method of least work. See Fig. P14.4.

14.12 Solve Problem 14.5 by the method of least work. See Fig. P14.5.

14.13 Determine the reactions and the force in each member of the truss shown in Fig. P14.13 using the method of least work.

14.14 A beam is supported by a fixed support A and a cable BC, as shown in Fig. P14.14. Determine the tension in the cable by the method of least work.

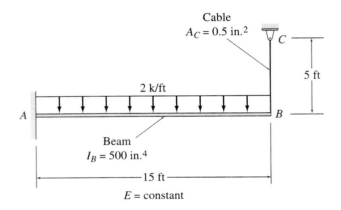

FIG. **P14.14**

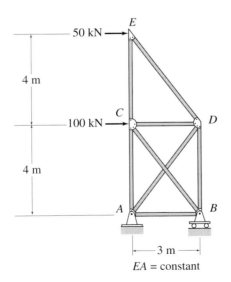

FIG. **P14.13**

15

Influence Lines for Statically Indeterminate Structures

15.1 Influence Lines for Beams and Trusses
15.2 Qualitative Influence Lines by Müller-Breslau's Principle
Summary
Problems

The Golden Gate Bridge, San Francisco
Photo courtesy of Digital Vision

In this chapter, we discuss the procedures for constructing influence lines for statically indeterminate structures. It may be recalled from Chapter 8 that *an influence line is a graph of a response function of a structure as a function of the position of a downward unit load moving across the structure.*

The basic procedure for constructing influence lines for indeterminate structures is the same as that for determinate structures considered in Chapter 8. The procedure essentially involves computing the values of the response function of interest for various positions of a unit load on the structure and plotting the response function values as ordinates against the position of the unit load as abscissa to obtain the influence line. Since the influence lines for forces and moments of determinate structures consist of straight-line segments, such influence lines were constructed in Chapter 8 by evaluating the ordinates for only a few positions of the unit load and by connecting them with straight lines. The influence lines for indeterminate structures, however, are generally curved lines. (For indeterminate girders with floor systems and trusses and for other indeterminate structures to which moving loads are transmitted via framing systems, the influence lines usually consist of chords of curved lines.) Thus the construction of influence lines for indeterminate structures requires computation of many more ordinates than necessary in the case of determinate structures.

Although any of the methods of analysis of indeterminate structures presented in Part Three can be used for computing the ordinates of influence lines, we will use the method of consistent deformations, discussed in Chapter 13, for such purposes. Once the influence lines for indeterminate structures have been constructed, they can be used in the same manner as those for determinate structures discussed in Chapter 9. In this chapter, the procedure for constructing influence lines for statically indeterminate beams and trusses is developed, and the application of Müller-Breslau's principle for constructing qualitative influence lines for indeterminate beams and frames is discussed.

15.1 INFLUENCE LINES FOR BEAMS AND TRUSSES

Consider the continuous beam shown in Fig. 15.1(a). Suppose that we wish to draw the influence line for the vertical reaction at the interior support B of the beam. The beam is subjected to a downward-moving concentrated load of unit magnitude, the position of which is defined by the coordinate x measured from the left end A of the beam, as shown in the figure.

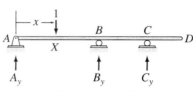

(a) Indeterminate Beam

To develop the influence line for the reaction B_y, we need to determine the expression for B_y in terms of the variable position x of the unit load. Noting that the beam is statically indeterminate to the first degree, we select the reaction B_y to be the redundant. The roller support at B is then removed from the actual indeterminate beam to obtain the statically determinate primary beam shown in Fig. 15.1(b). Next, the primary beam is subjected, separately, to the unit load positioned at an arbitrary point X at a distance x from the left end, and the redundant B_y, as shown in Fig. 15.1(b) and (c), respectively. The expression for B_y can now be determined by using the compatibility condition that the deflection of the primary beam at B due to the combined effect of the external unit load and the unknown redundant B_y must be equal to zero. Thus

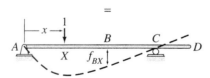

(b) Primary Beam Subjected to Unit Load

$$f_{BX} + f_{BB}B_y = 0$$

from which

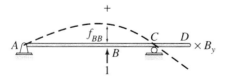

(c) Primary Beam Loaded with Redundant B_y

$$B_y = -\frac{f_{BX}}{f_{BB}} \tag{15.1}$$

in which the flexibility coefficient f_{BX} denotes the deflection of the primary beam at B due to the unit load at X (Fig. 15.1(b)), whereas the flexibility coefficient f_{BB} denotes the deflection at B due to the unit value of the redundant B_y (Fig. 15.1(c)).

We can use Eq. (15.1) for constructing the influence line for B_y by placing the unit load successively at a number of positions X along the beam, evaluating f_{BX} for each position of the unit load, and plotting the values of the ratio $-f_{BX}/f_{BB}$. However, a more efficient procedure can

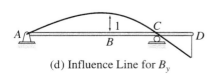

(d) Influence Line for B_y

FIG. **15.1**

be devised by applying *Maxwell's law of reciprocal deflections* (Section 7.8), according to which the deflection at B due to a unit load at X must be equal to the deflection at X due to a unit load B; that is, $f_{BX} = f_{XB}$. Thus, Eq. (15.1) can be rewritten as

$$B_y = -\frac{f_{XB}}{f_{BB}} \qquad (15.2)$$

which represents the equation of the influence line for B_y. Note that the deflections f_{XB} and f_{BB} are considered to be positive when in the upward direction (i.e., in the positive direction of the redundant B_y) in accordance with the sign convention adopted for the method of consistent deformations in Chapter 13.

Equation (15.2) is more convenient to apply than Eq. (15.1) in constructing the influence line, because according to Eq. (15.2), the unit load needs to be placed on the primary beam only at B, and the deflections f_{XB} at a number of points X along the beam are to be computed. The influence line can then be constructed by plotting the values of the ratio $-f_{XB}/f_{BB}$ as ordinates against the distance x, which represents the position of point X, as abscissa.

The equation of an influence line, when expressed in the form of Eq. (15.2), shows the validity of *Müller-Breslau's principle* for statically indeterminate structures. It can be seen from Eq. (15.2) for the influence line for B_y that since f_{BB} is a constant, the ordinate of the influence line at any point X is proportional to the deflection f_{XB} of the primary beam at that point due to the unit load at B. Furthermore, this equation indicates that the influence line for B_y can be obtained by multiplying the deflected shape of the primary beam due to the unit load at B by the scaling factor $-1/f_{BB}$. Note that this scaling yields a deflected shape, with a unit displacement at B, as shown in Fig. 15.1(d). The foregoing observation shows the validity of Müller-Breslau's principle for indeterminate structures. Recall from Section 8.2 that, according to this principle, the influence line for B_y can be obtained by removing the support B from the original beam and by giving the released beam a unit displacement in the direction of B_y. Also, note from Fig. 15.1(d) that, unlike the case of statically determinate structures considered in Chapter 8, the removal of support B from the indeterminate beam does not render it statically unstable; therefore, the influence line for its reaction B_y is a curved line. Once the influence line for the redundant B_y has been determined, the influence lines for the remaining reactions and the shears and bending moments of the beam can be obtained through equilibrium considerations.

Influence Lines for Structures with Multiple Degrees of Indeterminacy

The procedure for constructing the influence lines for structures with multiple degrees of indeterminacy is similar to that for structures with a single degree of indeterminacy. Consider, for example, the three-span

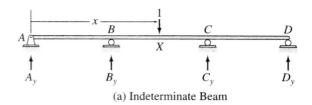

(a) Indeterminate Beam

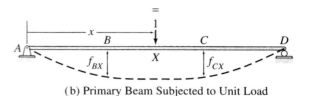

(b) Primary Beam Subjected to Unit Load

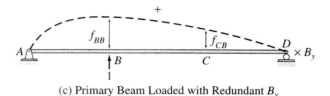

(c) Primary Beam Loaded with Redundant B_y

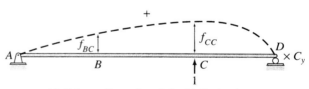

(d) Primary Beam Loaded with Redundant C_y

FIG. 15.2

continuous beam shown in Fig. 15.2(a). Because the beam is statically indeterminate to the second degree, we select the reactions B_y and C_y to be the redundants. To determine the influence lines for the redundants, we place a unit load successively at a number of positions X along the beam; and for each position of the unit load, the ordinates of the influence lines for B_y and C_y are evaluated by applying the compatibility equations (see Fig. 15.2(a) through (d))

$$f_{BX} + f_{BB}B_y + f_{BC}C_y = 0 \qquad (15.3)$$

$$f_{CX} + f_{CB}B_y + f_{CC}C_y = 0 \qquad (15.4)$$

Once the influence lines for the redundants have been obtained, the influence lines for the remaining reactions and the shears and bending moments of the beam can be determined by statics.

As discussed previously, the analysis can be considerably expedited by the application of *Maxwell's law of reciprocal deflections*, according to which $f_{BX} = f_{XB}$ and $f_{CX} = f_{XC}$. Thus, the unit load needs to be placed successively only at points B and C, and the deflections f_{XB} and f_{XC} at a number of points X along the beam are computed instead of computing the deflections f_{BX} and f_{CX} at points B and C, respectively, for each of a number of positions of the unit load.

Procedure for Analysis

The procedure for constructing influence lines for statically indeterminate structures by the method of consistent deformations can be summarized as follows:

1. Determine the degree of indeterminacy of the structure and select redundants.
2. Select a number of points along the length of the structure at which the numerical values of the ordinates of the influence lines will be evaluated.
3. To construct the influence lines for the redundants, place a unit load successively at each of the points selected in step 2; and for each position of the unit load, apply the method of consistent deformations to compute the values of the redundants. Plot the values of the redundants thus obtained as ordinates against the position of the unit load as abscissa, to construct the influence lines for the redundants. (Evaluation of the deflections involved in the compatibility equations can be considerably expedited by the application of *Maxwell's law of reciprocal deflections*, as illustrated by Examples 15.1 through 15.3.)
4. Once the influence lines for the redundants have been determined, the influence lines for the other force and/or moment response functions of the structure can be obtained through equilibrium considerations.

Example 15.1

Draw the influence lines for the reaction at support B and the bending moment at point C of the beam shown in Fig. 15.3(a).

Solution

The beam has one degree of indeterminacy. We select the vertical reaction B_y at the roller support B to be the redundant. The ordinates of the influence lines will be computed at 3-m intervals at points A through E, as shown in Fig. 15.3(a).

Influence Line for Redundant B_y   The value of the redundant B_y for an arbitrary position X of the unit load can be determined by solving the compatibility equation (see Fig. 15.3(b) and (c))

$$f_{BX} + \bar{f}_{BB}B_y = 0$$

continued

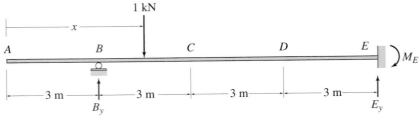

1 kN

A   B   C   D   E   ) M_E

|← 3 m →|← 3 m →|← 3 m →|← 3 m →|

B_y E_y

EI = constant

(a) Indeterminate Beam

=

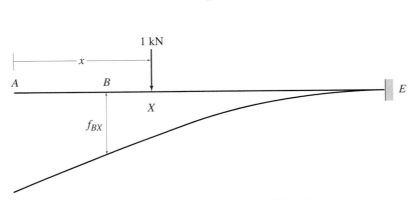

1 kN

A   B

f_{BX}

X

E

(b) Primary Beam Subjected to Unit Load

+

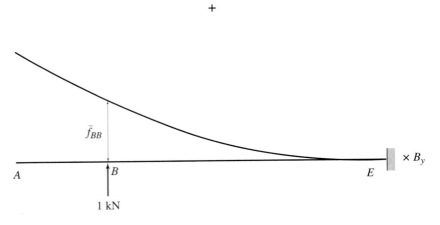

$\bar{f}_{BB}$

A   B   E   × B_y

1 kN

(c) Primary Beam Loaded with Redundant B_y

FIG. **15.3**

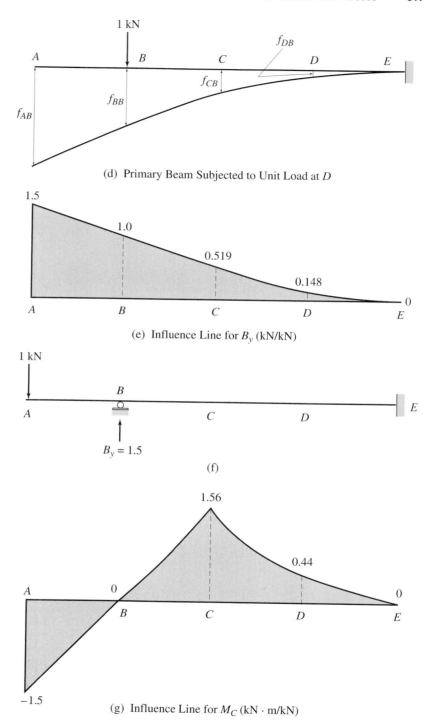

(d) Primary Beam Subjected to Unit Load at D

(e) Influence Line for B_y (kN/kN)

(f)

(g) Influence Line for M_C (kN · m/kN)

FIG. 15.3 (contd.)

continued

from which

$$B_y = -\frac{f_{BX}}{\bar{f}_{BB}} \tag{1}$$

Since by Maxwell's law of reciprocal deflections, $f_{BX} = f_{XB}$, we place the unit load at B on the primary beam (Fig. 15.3(d)) and compute the deflections at points A through E by using the beam-deflection formulas given inside the front cover of the book. Thus,

$$f_{BA} = f_{AB} = -\frac{364.5 \text{ kN} \cdot \text{m}^3/\text{kN}}{EI}$$

$$f_{BB} = -\frac{243 \text{ kN} \cdot \text{m}^3/\text{kN}}{EI}$$

$$f_{BC} = f_{CB} = -\frac{126 \text{ kN} \cdot \text{m}^3/\text{kN}}{EI}$$

$$f_{BD} = f_{DB} = -\frac{36 \text{ kN} \cdot \text{m}^3/\text{kN}}{EI}$$

$$f_{BE} = f_{EB} = 0$$

in which the negative signs indicate that these deflections are in the downward direction. Note that the flexibility coefficient $\bar{f}_{BB}$ in Eq. (1) denotes the upward (positive) deflection of the primary beam at B due to the unit value of the redundant B_y (Fig. 15.3(c)), whereas the deflection f_{BB} represents the downward (negative) deflection at B due to the external unit load at B (Fig. 15.3(d)). Thus,

$$\bar{f}_{BB} = -f_{BB} = +\frac{243 \text{ kN} \cdot \text{m}^3/\text{kN}}{EI}$$

The ordinates of the influence line for B_y can now be evaluated by applying Eq. (1) successively for each position of the unit load. For example, when the unit load is located at A, the value of B_y is obtained as

$$B_y = -\frac{f_{BA}}{\bar{f}_{BB}} = \frac{364.5}{243} = 1.5 \text{ kN/kN}$$

The remaining ordinates of the influence line for B_y are calculated in a similar manner. These ordinates are tabulated in Table 15.1, and the influence line for B_y is shown in Fig. 15.3(e). **Ans.**

Influence Line for M_C With the influence line for B_y known, the ordinates of the influence line for the bending moment at C can now be evaluated by placing the unit load successively at points A through E on the indeterminate beam and by using the corresponding values of B_y computed previously. For example, as depicted in Fig. 15.3(f), when the unit load is located at point A, the value of the reaction at B is $B_y = 1.5$ kN/kN. By considering the equilibrium of the free body of the portion of the beam to the left of C, we obtain

$$M_C = -1(6) + 1.5(3) = -1.5 \text{ kN} \cdot \text{m/kN}$$

The values of the remaining ordinates of the influence line are calculated in a similar manner. These ordinates are listed in Table 15.1, and the influence line for M_C is shown in Fig. 15.3(g). **Ans.**

TABLE 15.1

Unit Load at	Influence Line Ordinates	
	B_y (kN/kN)	M_C (kN · m/kN)
A	1.5	−1.5
B	1.0	0
C	0.519	1.56
D	0.148	0.44
E	0	0

Example 15.2

Draw the influence lines for the vertical reactions at the supports and the shear and bending moment at point C of the two-span continuous beam shown in Fig. 15.4(a).

Solution

The beam is indeterminate to the first degree. We select the vertical reaction D_y at the interior support D as the redundant. The influence line ordinates will be evaluated at 10-ft intervals at points A through F shown in Fig. 15.4(a).

Influence Line for Redundant D_y The value of the redundant D_y for an arbitrary position X of the unit load can be determined by solving the compatibility equation (see Fig. 15.4(b) and (c))

$$f_{DX} + \bar{f}_{DD}D_y = 0$$

from which

$$D_y = -\frac{f_{DX}}{\bar{f}_{DD}} \tag{1}$$

Since $f_{DX} = f_{XD}$ in accordance with Maxwell's law, we place the unit load at D on the primary beam (Fig. 15.4(d)) and compute the deflections at points A through F by using the conjugate-beam method. The conjugate beam is shown in Fig. 15.4(e), from which we obtain the following:

$$f_{DA} = f_{AD} = 0$$

$$f_{DB} = f_{BD} = -\frac{1}{EI}\left[86(10) - \left(\frac{1}{2}\right)(10)(2)\left(\frac{10}{3}\right)\right] = -\frac{826.667 \text{ k-ft}^3/\text{k}}{EI}$$

$$f_{DC} = f_{CD} = -\frac{1}{EI}\left[86(20) - \left(\frac{1}{2}\right)(20)(4)\left(\frac{20}{3}\right)\right] = -\frac{1{,}453.333 \text{ k-ft}^3/\text{k}}{EI}$$

$$f_{DD} = -\frac{1}{EI}\left[86(30) - \left(\frac{1}{2}\right)(30)(6)(10)\right] = -\frac{1{,}680 \text{ k-ft}^3/\text{k}}{EI}$$

$$f_{DE} = f_{ED} = -\frac{1}{EI}\left[124(10) - \left(\frac{1}{2}\right)(10)(6)\left(\frac{10}{3}\right)\right] = -\frac{1{,}140 \text{ k-ft}^3/\text{k}}{EI}$$

$$f_{DF} = f_{FD} = 0$$

in which the negative signs indicate that these deflections occur in the downward direction. Note that the flexibility coefficient $\bar{f}_{DD}$ in Eq. (1) denotes the upward (positive) deflection of the primary beam at D due to the unit value of the redundant D_y (Fig. 15.4(c)), whereas the deflection f_{DD} represents the downward (negative) deflection at D due to the external unit load at D (Fig. 15.4(d)). Thus

$$\bar{f}_{DD} = -f_{DD} = +\frac{1{,}680 \text{ k-ft}^3/\text{k}}{EI}$$

continued

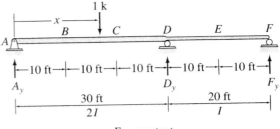

(a) Indeterminate Beam

=

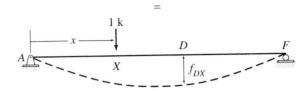

(b) Primary Beam Subjected to Unit Load

+

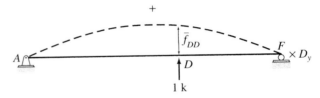

(c) Primary Beam Loaded with Redundant D_y

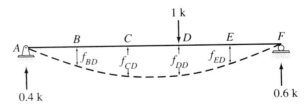

(d) Primary Beam Subjected to Unit Load at D

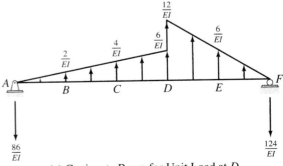

(e) Conjugate Beam for Unit Load at D

FIG. 15.4

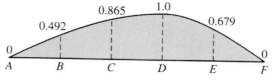

(f) Influence Line for D_y (k/k)

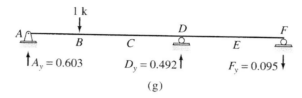

(g)

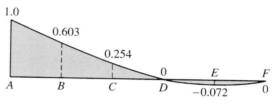

(h) Influence Line for A_y (k/k)

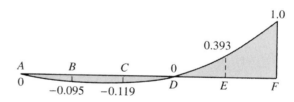

(i) Influence Line for F_y (k/k)

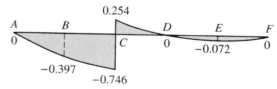

(j) Influence Line for S_C (k/k)

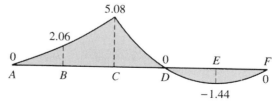

FIG. **15.4** (contd.)

(k) Influence Line for M_C (k-ft/k)

continued

TABLE 15.2

Unit Load at	Influence Line Ordinates				
	D_y (k/k)	A_y (k/k)	F_y (k/k)	S_C (k/k)	M_C (k-ft/k)
A	0	1.0	0	0	0
B	0.492	0.603	−0.095	−0.397	2.06
C	0.865	0.254	−0.119	−0.746 (left)	5.08
				0.254 (right)	
D	1.0	0	0	0	0
E	0.679	−0.072	0.393	−0.072	−1.44
F	0	0	1.0	0	0

The ordinates of the influence line for D_y can now be computed by applying Eq. (1) successively for each position of the unit load. For example, when the unit load is located at B, the value of D_y is given by

$$D_y = -\frac{f_{DB}}{\bar{f}_{DD}} = \frac{826.667}{1,680} = 0.492 \text{ k/k}$$

The remaining ordinates of the influence line for D_y are computed in a similar manner. These ordinates are tabulated in Table 15.2, and the influence line for D_y is shown in Fig. 15.4(f).   Ans.

Influence Lines for A_y and F_y With the influence line for D_y known, the influence lines for the remaining reactions can now be determined by applying the equations of equilibrium. For example, for the position of the unit load at point B as shown in Fig. 15.4(g), the value of the reaction D_y has been found to be 0.492 k/k. By applying the equilibrium equations, we determine the values of the reactions A_y and F_y to be

$$+\zeta \sum M_F = 0 \qquad -A_y(50) + 1(40) - 0.492(20) = 0$$

$$A_y = 0.603 \text{ k/k} \uparrow$$

$$+\uparrow \sum F_y = 0 \qquad 0.603 - 1 + 0.492 + F_y = 0$$

$$F_y = -0.095 \text{ k/k} = 0.095 \text{ k/k} \downarrow$$

The values of the remaining influence line ordinates are computed in a similar manner. These ordinates are listed in Table 15.2, and the influence lines for A_y and F_y are shown in Fig. 15.4(h) and (i), respectively.   Ans.

Influence Lines for S_C and M_C The ordinates of the influence lines for the shear and bending moment at C can now be evaluated by placing the unit load successively at points A through F on the indeterminate beam and by using the corresponding values of the reactions computed previously. For example, as shown in Fig. 15.4(g), when the unit load is located at point B, the values of the reactions are $A_y = 0.603$ k/k; $D_y = 0.492$ k/k; and $F_y = -0.095$ k/k. By considering the equilibrium of the free body of the portion of the beam to the left of C, we obtain

$$S_C = 0.603 - 1 = -0.397 \text{ k/k}$$

$$M_C = 0.603(20) - 1(10) = 2.06 \text{ k-ft/k}$$

The values of the remaining ordinates of the influence lines are computed in a similar manner. These ordinates are listed in Table 15.2, and the influence lines for the shear and bending moment at C are shown in Fig. 15.4(j) and (k), respectively.

Ans.

Example 15.3

Draw the influence lines for the reactions at supports for the beam shown in Fig. 15.5(a).

Solution

The beam is indeterminate to the second degree. We select the vertical reactions D_y and G_y at the roller supports D and G, respectively, to be the redundants. The influence line ordinates will be evaluated at 5-m intervals at points A through G shown in Fig. 15.5(a).

Influence Lines for Redundants D_y and G_y The values of the redundants D_y and G_y for an arbitrary position X of the unit load can be determined by solving the compatibility equations (see Fig. 15.5(b) through (d)):

$$f_{DX} + \bar{f}_{DD}D_y + \bar{f}_{DG}G_y = 0 \tag{1}$$

$$f_{GX} + \bar{f}_{GD}D_y + \bar{f}_{GG}G_y = 0 \tag{2}$$

Since by Maxwell's law, $f_{DX} = f_{XD}$, we place the unit load at D on the primary beam (Fig. 15.5(e)) and compute the deflections at points A through G by using the beam-deflection formulas given inside the front cover of the book. Thus,

$$f_{DA} = f_{AD} = 0$$

$$f_{DB} = f_{BD} = -\frac{166.667 \text{ kN} \cdot \text{m}^3/\text{kN}}{EI}$$

$$f_{DC} = f_{CD} = -\frac{583.333 \text{ kN} \cdot \text{m}^3/\text{kN}}{EI}$$

$$f_{DD} = -\frac{1{,}125 \text{ kN} \cdot \text{m}^3/\text{kN}}{EI}$$

$$f_{DE} = f_{ED} = -\frac{1{,}687.5 \text{ kN} \cdot \text{m}^3/\text{kN}}{EI}$$

$$f_{DF} = f_{FD} = -\frac{2{,}250 \text{ kN} \cdot \text{m}^3/\text{kN}}{EI}$$

$$f_{DG} = f_{GD} = -\frac{2{,}812.5 \text{ kN} \cdot \text{m}^3/\text{kN}}{EI}$$

continued

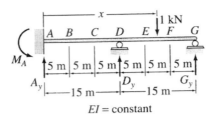

(a) Indeterminate Beam

$=$

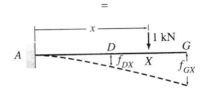

(b) Primary Beam Subjected to Unit Load

$+$

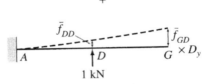

(c) Primary Beam Subjected to
Redundant D_y

$+$

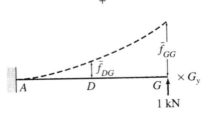

(d) Primary Beam Subjected to
Redundant G_y

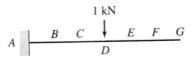

(e) Primary Beam Subjected to
Unit Load at D

(f) Primary Beam Subjected to
Unit Load at G

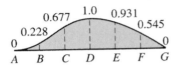

(g) Influence Line for D_y (kN/kN)

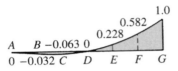

(h) Influence Line for G_y (kN/kN)

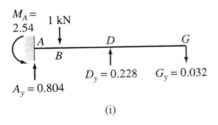

(i)

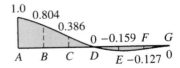

(j) Influence Line for A_y (kN/kN)

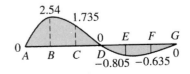

(k) Influence Line for M_A (kN · m/kN)

FIG. **15.5**

Similarly, the deflections $f_{GX} = f_{XG}$ are computed by placing the unit load at G (Fig. 15.5(f)):

$$f_{GA} = f_{AG} = 0$$

$$f_{GB} = f_{BG} = -\frac{354.167 \text{ kN} \cdot \text{m}^3/\text{kN}}{EI}$$

$$f_{GC} = f_{CG} = -\frac{1{,}333.333 \text{ kN} \cdot \text{m}^3/\text{kN}}{EI}$$

$$f_{GE} = f_{EG} = -\frac{4{,}666.667 \text{ kN} \cdot \text{m}^3/\text{kN}}{EI}$$

$$f_{GF} = f_{FG} = -\frac{6{,}770.833 \text{ kN} \cdot \text{m}^3/\text{kN}}{EI}$$

$$f_{GG} = -\frac{9{,}000 \text{ kN} \cdot \text{m}^3/\text{kN}}{EI}$$

In these equations the negative signs indicate that these deflections are in the downward direction.

The upward deflections due to the unit values of the redundants (Fig. 15.5(c) and (d)) are given by

$$\bar{f}_{DD} = +\frac{1{,}125 \text{ kN} \cdot \text{m}^3/\text{kN}}{EI}$$

$$\bar{f}_{DG} = \bar{f}_{GD} = +\frac{2{,}812.5 \text{ kN} \cdot \text{m}^3/\text{kN}}{EI}$$

$$\bar{f}_{GG} = +\frac{9{,}000 \text{ kN} \cdot \text{m}^3/\text{kN}}{EI}$$

By substituting the numerical values of these flexibility coefficients into the compatibility equations (Eqs. (1) and (2)) and solving for D_y and G_y, we obtain

$$D_y = \frac{EI}{1{,}968.75}(-8f_{DX} + 2.5f_{GX}) \tag{3}$$

$$G_y = \frac{EI}{1{,}968.75}(2.5f_{DX} - f_{GX}) \tag{4}$$

The values of the redundants D_y and G_y for each position of the unit load can now be determined by substituting the corresponding values of the deflections f_{DX} and f_{GX} into Eqs. (3) and (4). For example, the ordinates of the influence lines for D_y and G_y for the position of the unit load at B can be computed by substituting $f_{DX} = f_{DB} = -166.667/EI$ and $f_{GX} = f_{GB} = -354.167/EI$ into Eqs. (3) and (4):

$$D_y = \frac{EI}{1{,}968.75}\left[-8\left(-\frac{166.667}{EI}\right) + 2.5\left(-\frac{354.167}{EI}\right)\right] = 0.228 \text{ kN/kN} \uparrow$$

$$G_y = \frac{EI}{1{,}968.75}\left[2.5\left(-\frac{166.667}{EI}\right) + \frac{354.167}{EI}\right] = -0.032 \text{ kN/kN}$$

$$= 0.032 \text{ kN/kN} \downarrow$$

continued

TABLE 15.3

	Influence Line Ordinates			
Unit Load at	D_y (kN/kN)	G_y (kN/kN)	A_y (kN/kN)	M_A (kN · m/kN)
A	0	0	1.0	0
B	0.228	−0.032	0.804	2.540
C	0.677	−0.063	0.386	1.735
D	1.0	0	0	0
E	0.931	0.228	−0.159	−0.805
F	0.545	0.582	−0.127	−0.635
G	0	1.0	0	0

The remaining ordinates of the influence lines for the redundants are computed in a similar manner. These ordinates are tabulated in Table 15.3, and the influence lines for D_y and G_y are shown in Fig. 15.5(g) and (h), respectively.   **Ans.**

Influence Lines for A_y and M_A The ordinates of the influence lines for the remaining reactions can now be determined by placing the unit load successively at points A through G on the indeterminate beam and by applying the equations of equilibrium. For example, for the position of the unit load at B (Fig. 15.5(i)), the values of the reactions D_y and G_y have been found to be 0.228 kN/kN and −0.032 kN/kN, respectively. By considering the equilibrium of the beam, we determine the values of the reactions A_y and M_A to be as follows:

$$+\uparrow \textstyle\sum F_y = 0 \qquad A_y - 1 + 0.228 - 0.032 = 0$$

$$A_y = 0.804 \text{ kN/kN} \uparrow$$

$$+\circlearrowleft \textstyle\sum M_A = 0 \qquad M_A - 1(5) + 0.228(15) - 0.032(30) = 0$$

$$M_A = 2.54 \text{ kN} \cdot \text{m/kN} \circlearrowright$$

The values of the remaining influence line ordinates are computed in a similar manner. These ordinates are listed in Table 15.3, and the influence lines for A_y and M_A are shown in Fig. 15.5(j) and (k), respectively.   **Ans.**

Example 15.4

Draw the influence lines for the forces in members BC, BE, and CE of the truss shown in Fig. 15.6(a). Live loads are transmitted to the top chord of the truss.

Solution

The truss is internally indeterminate to the first degree. We select the axial force F_{CE} in the diagonal member CE to be the redundant.

Influence Line for Redundant F_{CE} To determine the influence line for F_{CE}, we place a unit load successively at joints B and C of the truss, and for each position of the unit load, we apply the method of consistent deformations to compute the value of F_{CE}. The primary truss, obtained by removing member CE, is subjected separately to the unit load at B and C, as shown in Fig. 15.6(b)

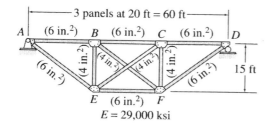

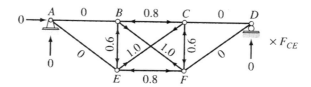

(a) Indeterminate Truss

(d) Primary Truss Subjected to Unit Tensile Force in Member CE — u_{CE} Forces

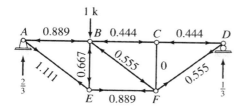

(b) Primary Truss Subjected to Unit Load at B — u_B Forces

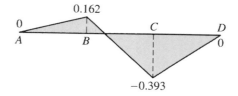

(e) Influence Line for F_{CE} (k/k)

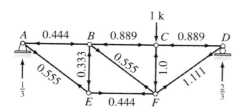

(c) Primary Truss Subjected to Unit Load at C — u_C Forces

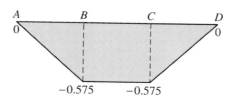

(f) Influence Line for F_{BC} (k/k)

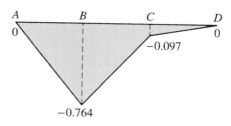

(g) Influence Line for F_{BE} (k/k)

FIG. 15.6

and (c), respectively, and a unit tensile force in the redundant member CE, as shown in Fig. 15.6(d).

When the unit load is located at B, the compatibility equation can be expressed as

$$f_{CE,B} + f_{CE,CE}F_{CE} = 0$$

in which $f_{CE,B}$ denotes the relative displacement between joints C and E of the primary truss due to the unit load at B and $f_{CE,CE}$ denotes the relative

continued

TABLE 15.4

Member	L (in.)	A (in.^{2})	u_B (k/k)	u_C (k/k)	u_{CE} (k/k)	$\dfrac{u_B u_{CE} L}{A}$	$\dfrac{u_C u_{CE} L}{A}$	$\dfrac{u_{CE}^2 L}{A}$
AB	240	6	−0.889	−0.444	0	0	0	0
BC	240	6	−0.444	−0.889	−0.8	14.208	28.448	25.6
CD	240	6	−0.444	−0.889	0	0	0	0
EF	240	6	0.889	0.444	−0.8	−28.448	−14.208	25.6
BE	180	4	−0.667	−0.333	−0.6	18.009	8.991	16.2
CF	180	4	0	−1.0	−0.6	0	27.0	16.2
AE	300	6	1.111	0.555	0	0	0	0
BF	300	4	−0.555	0.555	1.0	−41.625	41.625	75.0
CE	300	4	0	0	1.0	0	0	75.0
DF	300	6	0.555	1.111	0	0	0	0
					Σ	−37.856	91.856	233.6

displacements between the same joints due to a unit value of the redundant F_{CE}. Applying the virtual work method (see Fig. 15.6(b) and (d) and Table 15.4), we obtain

$$f_{CE,B} = \frac{1}{E}\sum \frac{u_B u_{CE} L}{A} = -\frac{37.856}{E}$$

$$f_{CE,CE} = \frac{1}{E}\sum \frac{u_{CE}^2 L}{A} = \frac{233.6}{E}$$

By substituting these numerical values into the compatibility equation, we determine the ordinate of the influence line for F_{CE} at B to be

$$F_{CE} = 0.162 \text{ k/k (T)}$$

Similarly, when the unit load is located at C, the compatibility equation is given by

$$f_{CE,C} + f_{CE,CE} F_{CE} = 0$$

(see Fig. 15.6(c) and (d) and Table 15.4) in which

$$f_{CE,C} = \frac{1}{E}\sum \frac{u_C u_{CE} L}{A} = \frac{91.856}{E}$$

By substituting the numerical values of $f_{CE,C}$ and $f_{CE,CE}$ into the compatibility equation, we determine the ordinate of the influence line for F_{CE} at C to be

$$F_{CE} = -0.393 \text{ k/k} = 0.393 \text{ k/k (C)}$$

The influence line for F_{CE} is shown in Fig. 15.6(e).   Ans.

Influence Lines for F_{BC} and F_{BE} The ordinate at B of the influence line for force in any member of the truss can be determined by the superposition relationship (see Fig. 15.6(b) and (d) and Table 15.4)

$$F = u_B + u_{CE} F_{CE}$$

in which F_{CE} denotes the ordinate at B of the influence line for the redundant F_{CE}. Thus the ordinates at B of the influence lines for F_{BC} and F_{BE} are

$$F_{BC} = -0.444 + (-0.8)(0.162) = -0.575 \text{ k/k} = 0.575 \text{ k/k (C)}$$

$$F_{BE} = -0.667 + (-0.6)(0.162) = -0.764 \text{ k/k} = 0.764 \text{ k/k (C)}$$

Similarly, the ordinates of the influence lines for F_{BC} and F_{BE} at C can be determined by using the superposition relationship (see Fig. 15.6(c) and (d) and Table 15.4)

$$F = u_C + u_{CE}F_{CE}$$

in which F_{CE} now denotes the ordinate at C of the influence line for the redundant F_{CE}. Thus

$$F_{BC} = -0.889 + (-0.8)(-0.393) = -0.575 \text{ k/k} = 0.575 \text{ k/k (C)}$$

$$F_{BE} = -0.333 + (-0.6)(-0.393) = -0.097 \text{ k/k} = 0.097 \text{ k/k (C)}$$

The influence lines for F_{BC} and F_{BE} are shown in Fig. 15.6(f) and (g), respectively.   Ans.

15.2 QUALITATIVE INFLUENCE LINES BY MÜLLER-BRESLAU'S PRINCIPLE

In many practical applications, such as when designing continuous beams or building frames subjected to uniformly distributed live loads, it is usually sufficient to draw only the qualitative influence lines to decide where to place the live loads to maximize the response functions of interest. As in the case of statically determinate structures (Section 8.2), *Müller-Breslau's principle* provides a convenient means of establishing qualitative influence lines for indeterminate structures.

Recall from Section 8.2 that Müller-Breslau's principle can be stated as follows:

The influence line for a force (or moment) response function is given by the deflected shape of the released structure obtained by removing the restraint corresponding to the response function from the original structure and by giving the released structure a unit displacement (or rotation) at the location and in the direction of the response function, so that only the response function and the unit load perform external work.

The procedure for constructing qualitative influence lines for indeterminate structures is the same as that for determinate structures discussed in Section 8.2. The procedure essentially involves: (1) removing from the given structure the restraint corresponding to the response function of interest to obtain the released structure; (2) applying a small displacement (or rotation) to the released structure at the location and in the positive direction of the response function; and (3) drawing a deflected shape of the released structure consistent with its support and continuity conditions. The influence lines for indeterminate structures are generally curved lines.

Once a qualitative influence line for a structural response function has been constructed, it can be used to decide where to place the live loads to maximize the value of the response function. As discussed in Section 9.2, the value of a response function due to a uniformly distributed live load is maximum positive (or negative) when the load is placed over those portions of the structure where the ordinates of the response function influence line are positive (or negative). Because the influence-line ordinates tend to diminish rapidly with distance from the point of application of the response function, live loads placed more than three span lengths away from the location of the response function generally have a negligible effect on the value of the response function. With the live-load pattern known, an indeterminate analysis of the structure can be performed to determine the maximum value of the response function.

Example 15.5

Draw qualitative influence lines for the vertical reactions at supports A and B, the bending moment at point B, and the shear and bending moment at point C of the four-span continuous beam shown in Fig. 15.7(a). Also, show the arrangements of a uniformly distributed downward live load w_ℓ to cause the maximum positive reactions at supports A and B, the maximum negative bending moment at B, the maximum negative shear at C, and the maximum positive bending moment at C.

Solution

Influence Line for A_y To determine the qualitative influence line for the vertical reaction A_y at support A, we remove the vertical restraint at A from the actual beam and give the released beam a small displacement in the positive

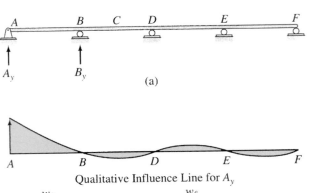

(a)

Qualitative Influence Line for A_y

Arrangement of Live Load for Maximum Positive A_y

(b)

FIG. 15.7

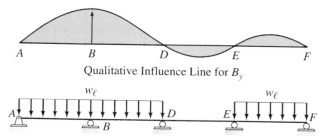

Qualitative Influence Line for B_y

Arrangement of Live Load for Maximum Positive B_y

(c)

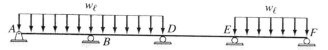

Qualitative Influence Line for M_B

Arrangement of Live Load for Maximum Negative M_B

(d)

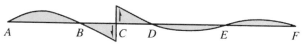

Qualitative Influence Line for S_C

Arrangement of Live Load for Maximum Negative S_C

(e)

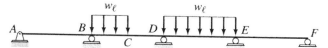

Qualitative Influence Line for M_C

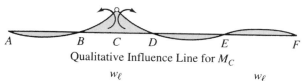

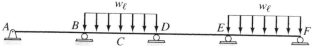

Arrangement of Live Load for Maximum Positive M_C

FIG. **15.7** (contd.)

(f)

continued

direction of A_y. The deflected shape of the released beam thus obtained (Fig. 15.7(b)) represents the general shape of the influence line (i.e., the qualitative influence line) for A_y. Note that the deflected shape is consistent with the support conditions of the released beam; that is, points B, D, E, and F of the released beam, which are attached to roller supports, do not displace.   **Ans.**

To maximize the positive value of A_y, the live load w_ℓ is placed over spans AB and DE of the beam, where the ordinates of the influence line for A_y are positive, as shown in Fig. 15.7(b).   **Ans.**

Influence Line for B_y The qualitative influence line for B_y and the live-load arrangement for the maximum positive value of B_y are determined in a similar manner and are shown in Fig. 15.7(c).   **Ans.**

Influence Line for M_B To determine the qualitative influence line for the bending moment at B, we insert a hinge at B in the actual beam and give the released beam a small rotation in the positive direction of M_B by rotating the portion to the left of B counterclockwise and the portion to the right of B clockwise, as shown in Fig. 15.7(d). The deflected shape of the released beam thus obtained represents the qualitative influence line for M_B.   **Ans.**

To cause the maximum negative bending moment at B, we place the live load w_ℓ over spans AB, BD, and EF of the beam, where the ordinates of the influence line for M_B are negative, as shown in Fig. 15.7(d).   **Ans.**

Influence Line for S_C The qualitative influence line for S_C is determined by cutting the actual beam at C and by giving the released beam a small relative displacement in the positive direction of S_C by moving end C of the left portion of the beam downward and end C of the right portion upward, as shown in Fig. 15.7(e).   **Ans.**

To obtain the maximum negative shear at C, the live load is placed over span DE and the portion BC of the span BD of the beam, where the ordinates of the influence line for S_c are negative, as shown in Fig. 15.7(e).   **Ans.**

Influence Line for M_C The qualitative influence line for the bending moment at C and the live-load arrangement for the maximum positive value of M_C are shown in Fig. 15.7(f).   **Ans.**

Example 15.6

Draw qualitative influence lines for the bending moment and shear at point A of the building frame shown in Fig. 15.8(a). Also, show the arrangements of a uniformly distributed downward live load w_ℓ that will cause the maximum positive bending moment and the maximum negative shear at A.

Solution

Influence Line for M_A The qualitative influence line for the bending moment at A is shown in Fig. 15.8(b). Note that since the members of the frame are connected together by rigid joints, the original angles between the members intersecting at a joint must be maintained in the deflected shape of the frame. To obtain the maximum positive bending moment at A, the live load w_ℓ is placed over those spans of the frame where the ordinates of the influence line for M_A

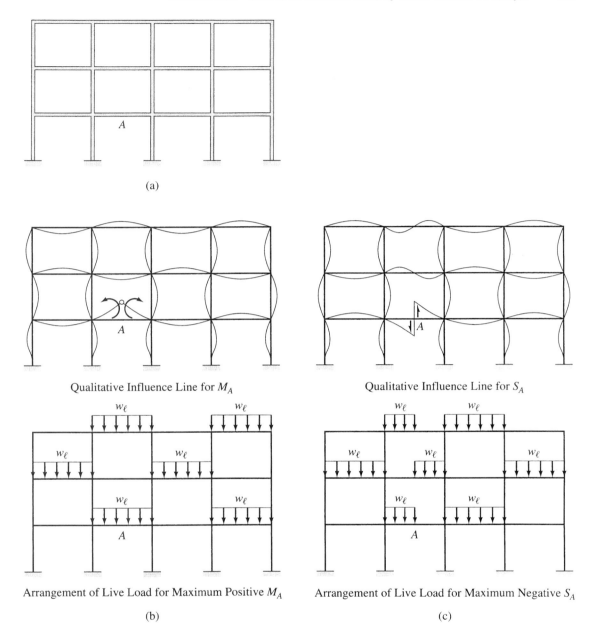

(a)

Qualitative Influence Line for M_A

Qualitative Influence Line for S_A

Arrangement of Live Load for Maximum Positive M_A

Arrangement of Live Load for Maximum Negative S_A

(b)   (c)

FIG. **15.8**

are positive, as shown in Fig. 15.8(b). This type of live-load pattern is sometimes referred to as a *checker-board load pattern*.   Ans.

Influence Line for S_A The qualitative influence line for the shear at A and the live-load arrangement for the maximum negative value of S_A are shown in Fig. 15.8(c).   Ans.

SUMMARY

In this chapter we have discussed influence lines for statically indeterminate structures. The procedure for constructing such influence lines by the method of consistent deformations essentially involves (1) constructing the influence lines for the redundants by placing a unit load successively at a number of points along the length of the structure and, for each position of the unit load, computing the values of the redundants by applying the method of consistent deformations, and (2) using the influence lines for the redundants and, by applying the equations of equilibrium, determining the influence lines for other response functions of the structure.

Evaluation of the deflections involved in the application of the method of consistent deformations can be considerably expedited by using Maxwell's law of reciprocal deflections. The procedure for constructing qualitative influence lines for indeterminate structures by Müller-Breslau's principle is presented in Section 15.2.

PROBLEMS

Section 15.1

15.1 Draw the influence lines for the reactions at the supports and the shear and bending moment at point B of the beam shown in Fig. P15.1. Determine the influence line ordinates at 3-m intervals. Select the reaction at support C to be the redundant.

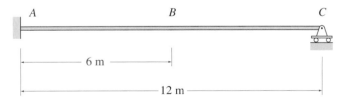

FIG. P15.1, P15.2

15.2 Determine the influence lines for the reactions at the supports for the beam of Problem 15.1 by selecting the moment at support A to be the redundant. See Fig. P15.1.

15.3 Draw the influence lines for the reaction at support C and the shear and bending moment at point B of the beam

shown in Fig. P15.3. Determine the influence line ordinates at 5-ft intervals.

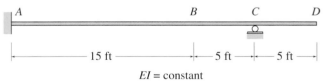

FIG. P15.3

15.4 Draw the influence lines for the reactions at the supports and the shear and bending moment at point C of the beam shown in Fig. P15.4. Determine the influence line ordinates at 10-ft intervals.

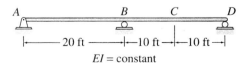

FIG. P15.4

15.5 Draw the influence lines for the reactions at the supports and the shear and bending moment at point C of the beam shown in Fig. P15.5. Determine the influence line ordinates at 4-m intervals.

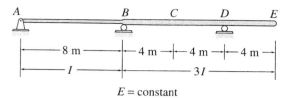

$E = $ constant

FIG. **P15.5**

15.6 Draw the influence lines for the reactions at the supports and the forces in members BC, CE, and EF of the truss shown in Fig. P15.6. Live loads are transmitted to the bottom chord of the truss.

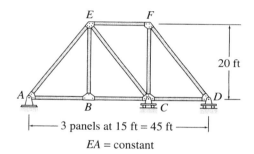

$EA = $ constant

FIG. **P15.6**

15.7 Draw the influence lines for the forces in members BC and CD of the truss shown in Fig. P15.7. Live loads are transmitted to the top chord of the truss.

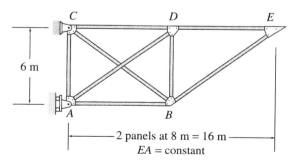

FIG. **P15.7**

15.8 Draw the influence lines for the forces in members BC, BF, and CF of the truss shown in Fig. P15.8. Live loads are transmitted to the bottom chord of the truss.

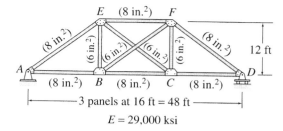

$E = 29{,}000$ ksi

FIG. **P15.8**

15.9 Draw the influence lines for the reactions at supports B and D and the shear and bending moment at point C of the beam shown in Fig. P15.9. Determine the influence line ordinates at 5-ft intervals.

$EI = $ constant

FIG. **P15.9**

15.10 Draw the influence lines for the reactions at the supports for the beam shown in Fig. P15.10. Determine the influence line ordinates at 3-m intervals.

$E = $ constant

FIG. **P15.10**

15.11 Draw the influence lines for the reaction at support C and the forces in members BC, CE, and EF of the truss shown in Fig. P15.11. Live loads are transmitted to the bottom chord of the truss.

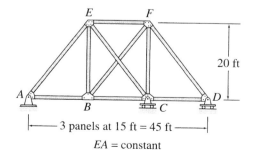

$EA = $ constant

FIG. **P15.11**

15.12 Draw the influence lines for the forces in members BG, CD, and DG of the truss shown in Fig. P15.12. Live loads are transmitted to the bottom chord of the truss.

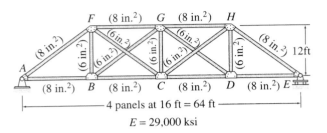

FIG. **P15.12**

Section 15.2

15.13 through 15.15 Draw qualitative influence lines for the vertical reactions at supports A and B, the bending moment at point B, and the shear and bending moment at point C of the beams shown in Figs. P15.13–P15.15. Also, show the arrangements of a uniformly distributed downward live load w_ℓ to cause the maximum upward reactions at supports A and B, the maximum negative bending moment at B, the maximum negative shear at C, and the maximum positive bending moment at C.

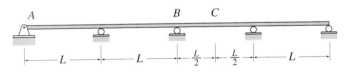

FIG. **P15.13**

FIG. **P15.14**

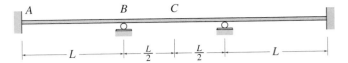

FIG. **P15.15**

15.16 Draw qualitative influence lines for the bending moment and shear at point A of the building frame shown in Fig. P15.16. Also, show the arrangements of a uniformly distributed downward live load w_ℓ to cause the maximum positive bending moment at A, and the maximum negative shear at A.

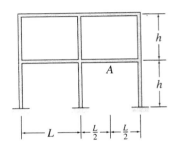

FIG. **P15.16**

15.17 For the building frame shown in Fig. P15.17, determine the arrangements of a uniformly distributed downward live load w_ℓ that will cause the maximum negative bending moment at point A and the maximum positive bending moment at point B.

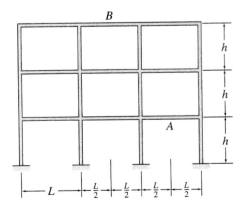

FIG. **P15.17**

16

Slope-Deflection Method

16.1 Slope-Deflection Equations
16.2 Basic Concept of the Slope-Deflection Method
16.3 Analysis of Continuous Beams
16.4 Analysis of Frames without Sidesway
16.5 Analysis of Frames with Sidesway
Summary
Problems

388 Market Street Building, San Francisco
Photo courtesy of the American Institute of Steel Construction, Inc.

In Chapters 13 and 14, we considered various formulations of the force (flexibility) method of analysis of statically indeterminate structures. Recall that in the force method, the unknown redundant *forces* are determined first by solving the structure's compatibility equations; then the other response characteristics of the structure are evaluated by equilibrium equations or superposition. An alternative approach that can be used for analyzing indeterminate structures is termed the *displacement* (*stiffness*) *method*. Unlike the force method, in the displacement method the unknown *displacements* are determined first by solving the structure's equilibrium equations; then the other response characteristics are evaluated through compatibility considerations and member force-deformation relations.

In this chapter, we consider a classical formulation of the displacement method, called the *slope-deflection method*. An alternative classical formulation, the *moment-distribution method*, is presented in the next chapter, followed by an introduction to the modern *matrix stiffness method* in Chapter 18.

The slope-deflection method for the analysis of indeterminate beams and frames was introduced by George A. Maney in 1915. The method takes into account only the bending deformations of structures. Although the slope-deflection method is itself considered to be a useful tool for analyzing indeterminate beams and frames, an understanding of the fundamentals of this method provides a valuable introduction to the matrix stiffness method, which forms the basis of most computer software currently used for structural analysis.

We first derive the fundamental relationships necessary for the application of the slope-deflection method and then develop the basic concept of the slope-deflection method. We consider the application of the method to the analysis of continuous beams and present the analysis of the frames in which joint translations are prevented. Finally, we consider the analysis of frames with joint translations.

16.1 SLOPE-DEFLECTION EQUATIONS

When a continuous beam or a frame is subjected to external loads, internal moments generally develop at the ends of its individual members. *The slope-deflection equations relate the moments at the ends of a member to the rotations and displacements of its ends and the external loads applied to the member.*

To derive the slope-deflection equations, let us focus our attention on an arbitrary member AB of the continuous beam shown in Fig. 16.1(a). When the beam is subjected to external loads and support settlements, member AB deforms, as shown in the figure, and internal moments are induced at its ends. The free-body diagram and the elastic curve for member AB are shown using an exaggerated scale in Fig. 16.1(b). As indicated in this figure, double-subscript notation is used for member end moments, with the first subscript identifying the member end at which the moment acts and the second subscript indicating the other end of the member. Thus, M_{AB} denotes the moment at end A of member AB, whereas M_{BA} represents the moment at end B of member AB. Also, as shown in Fig. 16.1(b), θ_A and θ_B denote, respectively, the rotations of ends A and B of the member with respect to the undeformed (horizontal) position of the member; Δ denotes the relative translation between the two ends of the member in the direction perpendicular to the undeformed axis of the member; and the angle ψ denotes the rotation of the member's chord (i.e., the straight line connecting the deformed positions of the member ends) due to the relative translation Δ. Since the deformations are assumed to be small, the chord rotation can be expressed as

$$\psi = \frac{\Delta}{L} \qquad (16.1)$$

The sign convention used in this chapter is as follows:

> *The member end moments, end rotations, and chord rotation are positive when counterclockwise.*

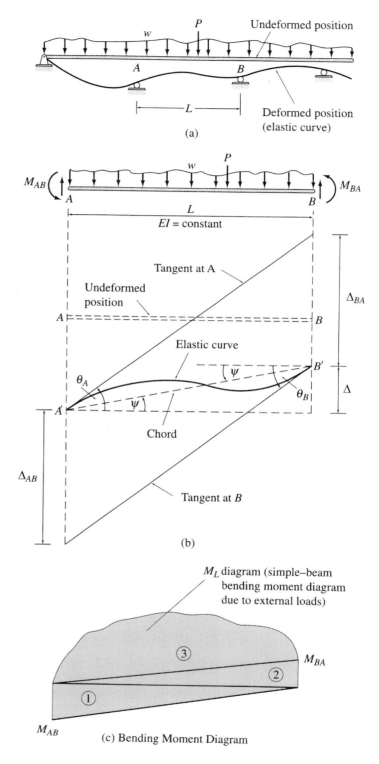

(a)

(b)

(c) Bending Moment Diagram

FIG. 16.1

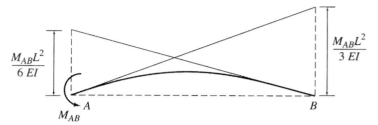

Tangential Deviations Due to M_{AB}

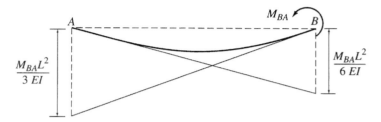

Tangential Deviations Due to M_{BA}

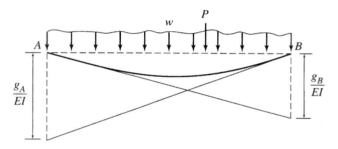

Tangential Deviations Due to External Loading

(d)

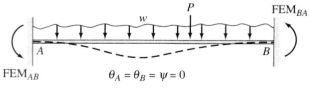

(e) Fixed-End Moments

FIG. 16.1 (contd.)

Note that all the moments and rotations are shown in the positive sense in Fig. 16.1(b).

The slope-deflection equations can be derived by relating the member end moments to the end rotations and chord rotation by applying the second moment-area theorem (Section 6.4). From Fig. 16.1(b), we can see that

$$\theta_A = \frac{\Delta_{BA} + \Delta}{L} \qquad \theta_B = \frac{\Delta_{AB} + \Delta}{L} \tag{16.2}$$

By substituting $\Delta/L = \psi$ into the preceding equations, we write

$$\theta_A - \psi = \frac{\Delta_{BA}}{L} \qquad \theta_B - \psi = \frac{\Delta_{AB}}{L} \tag{16.3}$$

in which, as shown in Fig. 16.1(b), Δ_{BA} is the tangential deviation of end B from the tangent to the elastic curve at end A and Δ_{AB} is the tangential deviation of end A from the tangent to the elastic curve at end B. According to the second moment-area theorem, the expressions for the tangential deviations Δ_{BA} and Δ_{AB} can be obtained by summing the moments about the ends B and A, respectively, of the area under the M/EI diagram between the two ends.

The bending moment diagram for the member is constructed in parts by applying M_{AB}, M_{BA}, and the external loading separately on the member with simply supported ends. The three *simple-beam bending moment diagrams* thus obtained are shown in Fig. 16.1(c). Assuming that the member is prismatic—that is, EI is constant along the length of the member—we sum the moments of the area under the M/EI diagram about the ends B and A, respectively, to determine the tangential deviations:

$$\Delta_{BA} = \frac{1}{EI}\left[\left(\frac{M_{AB}L}{2}\right)\left(\frac{2L}{3}\right) - \left(\frac{M_{BA}L}{2}\right)\left(\frac{L}{3}\right) - g_B\right]$$

or

$$\Delta_{BA} = \frac{M_{AB}L^2}{3EI} - \frac{M_{BA}L^2}{6EI} - \frac{g_B}{EI} \tag{16.4a}$$

and

$$\Delta_{AB} = \frac{1}{EI}\left[-\left(\frac{M_{AB}L}{2}\right)\left(\frac{L}{3}\right) + \left(\frac{M_{BA}L}{2}\right)\left(\frac{2L}{3}\right) + g_A\right]$$

or

$$\Delta_{AB} = -\frac{M_{AB}L^2}{6EI} + \frac{M_{BA}L^2}{3EI} + \frac{g_A}{EI} \tag{16.4b}$$

in which g_B and g_A are the moments about the ends B and A, respectively, of the area under the simple-beam bending moment diagram due to external loading (M_L diagram in Fig. 16.1(c)). The three terms in Eqs. (16.4a) and (16.4b) represent the tangential deviations due to M_{AB}, M_{BA}, and the external loading, acting separately on the member (Fig. 16.1(d)), with a negative term indicating that the corresponding tangential deviation is in the direction opposite to that shown on the elastic curve of the member in Fig. 16.1(b).

By substituting the expressions for Δ_{BA} and Δ_{AB} (Eqs. (16.4)) into Eq. (16.3), we write

$$\theta_A - \psi = \frac{M_{AB}L}{3EI} - \frac{M_{BA}L}{6EI} - \frac{g_B}{EIL} \tag{16.5a}$$

$$\theta_B - \psi = -\frac{M_{AB}L}{6EI} + \frac{M_{BA}L}{3EI} + \frac{g_A}{EIL} \tag{16.5b}$$

To express the member end moments in terms of the end rotations, the chord rotation, and the external loading, we solve Eqs. (16.5a) and (16.5b) simultaneously for M_{AB} and M_{BA}. Rewriting Eq. (16.5a) as

$$\frac{M_{BA}L}{3EI} = \frac{2M_{AB}L}{3EI} - \frac{2g_B}{EIL} - 2(\theta_A - \psi)$$

By substituting this equation into Eq. (16.5b) and solving the resulting equation for M_{AB}, we obtain

$$M_{AB} = \frac{2EI}{L}(2\theta_A + \theta_B - 3\psi) + \frac{2}{L^2}(2g_B - g_A) \tag{16.6a}$$

and by substituting Eq. (16.6a) into either Eq. (16.5a) or Eq. (16.5b), we obtain the expression for M_{BA}:

$$M_{BA} = \frac{2EI}{L}(\theta_A + 2\theta_B - 3\psi) + \frac{2}{L^2}(g_B - 2g_A) \tag{16.6b}$$

As Eqs. (16.6) indicate, the moments that develop at the ends of a member depend on the rotations and translations of the member's ends as well as on the external loading applied between the ends.

Now, suppose that the member under consideration, instead of being a part of a larger structure, was an isolated beam with both its ends completely fixed against rotations and translations, as shown in Fig. 16.1(e). The moments that would develop at the ends of such a *fixed beam* are referred to as *fixed-end moments*, and their expressions can be obtained from Eqs. (16.6) by setting $\theta_A = \theta_B = \psi = 0$; that is,

$$\text{FEM}_{AB} = \frac{2}{L^2}(2g_B - g_A) \tag{16.7a}$$

$$\text{FEM}_{BA} = \frac{2}{L^2}(g_B - 2g_A) \tag{16.7b}$$

in which FEM_{AB} and FEM_{BA} denote the fixed-end moments due to external loading at the ends A and B, respectively, of the fixed beam AB (see Fig. 16.1(e)).

By comparing Eqs. (16.6) and (16.7), we find that the second terms on the right sides of Eqs. (16.6) are equal to the fixed-end moments that would develop if the ends of the member were fixed against rotations and translations. Thus, by substituting Eqs. (16.7) into Eqs. (16.6), we obtain

$$M_{AB} = \frac{2EI}{L}(2\theta_A + \theta_B - 3\psi) + \text{FEM}_{AB} \qquad (16.8\text{a})$$

$$M_{BA} = \frac{2EI}{L}(\theta_A + 2\theta_B - 3\psi) + \text{FEM}_{BA} \qquad (16.8\text{b})$$

Equations (16.8), which express the moments at the ends of a member in terms of its end rotations and translations for a specified external loading, are called the *slope-deflection equations*. These equations are valid only for prismatic members composed of linearly elastic material and subjected to small deformations. Also, although the equations take into account the bending deformations of members, the deformations due to axial forces and shears are neglected.

From Eqs. (16.8), we observe that the two slope-deflection equations have the same form and that either one of the equations can be obtained from the other simply by switching the subscripts A and B. Thus it is usually convenient to express these equations by the following single slope-deflection equation:

$$M_{nf} = \frac{2EI}{L}(2\theta_n + \theta_f - 3\psi) + \text{FEM}_{nf} \qquad (16.9)$$

in which the subscript n refers to the *near* end of the member where the moment M_{nf} acts and the subscript f identifies the *far* (other) end of the member.

Fixed-End Moments

The expressions for fixed-end moments due to any loading condition can be derived by using the method of consistent deformations, as discussed in Chapter 13 (see Example 13.10). However, it is usually more convenient to determine the fixed-end moment expressions by applying Eqs. (16.7), which require only the computation of the moments of the area under the simple-beam bending moment diagram about the ends of the beam.

To illustrate the application of Eqs. (16.7), consider a fixed beam subjected to a concentrated load P, as shown in Fig. 16.2(a). The fixed-end moments of this beam were previously determined in Example 13.10 by the method of consistent deformations. To apply Eqs. (16.7), we replace the fixed ends of the beam by simple supports and construct the simple-beam bending moment diagram, as shown in Fig. 16.2(b). The

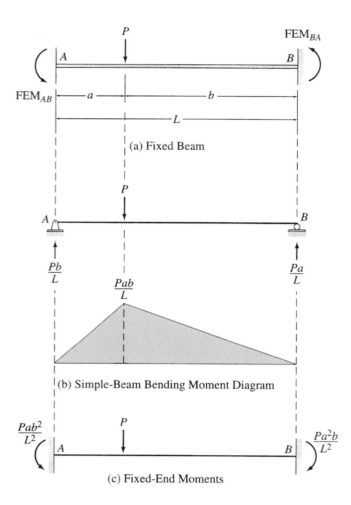

FIG. 16.2

(a) Fixed Beam

(b) Simple-Beam Bending Moment Diagram

(c) Fixed-End Moments

moments of the area under the simple-beam bending moment diagram about the ends A and B are given by

$$g_A = \frac{1}{2}a\left(\frac{Pab}{L}\right)\left(\frac{2a}{3}\right) + \frac{1}{2}b\left(\frac{Pab}{L}\right)\left(a + \frac{b}{3}\right)$$

$$g_B = \frac{1}{2}a\left(\frac{Pab}{L}\right)\left(\frac{a}{3} + b\right) + \frac{1}{2}b\left(\frac{Pab}{L}\right)\left(\frac{2b}{3}\right)$$

By substituting $L = a + b$ into these equations and simplifying, we obtain

$$g_A = \frac{Pab}{6}(2a + b) \qquad g_B = \frac{Pab}{6}(a + 2b)$$

By substituting the expressions for g_A and g_B into Eqs. (16.7), we determine the fixed-end moments to be

$$\text{FEM}_{AB} = \frac{2}{L^2}\left[\frac{2Pab}{6}(a+2b) - \frac{Pab}{6}(2a+b)\right] = \frac{Pab^2}{L^2} \circlearrowleft$$

$$\text{FEM}_{BA} = \frac{2}{L^2}\left[\frac{Pab}{6}(a+2b) - \frac{2Pab}{6}(2a+b)\right] = -\frac{Pa^2b}{L^2}$$

Recall that Eqs. (16.7) are based on the sign convention that the counterclockwise end moments are positive. Thus the negative answer for FEM_{BA} indicates that its correct sense is clockwise; that is,

$$\text{FEM}_{BA} = \frac{Pa^2b}{L^2} \circlearrowright$$

as shown in Fig. 16.2(c).

The fixed-end moment expressions for some common types of loading conditions are given inside the back cover of the book for convenient reference.

Members with One End Hinged

The slope-deflection equations derived previously (Eqs. (16.8) or Eq. (16.9)) are based on the condition that the member is rigidly connected to joints at both ends, so that the member end rotations θ_A and θ_B are equal to the rotations of the adjacent joints. When one of the member's ends is connected to the adjacent joint by a hinged connection, the moment at the hinged end must be zero. The slope-deflection equations can be easily modified to reflect this condition. With reference to Fig. 16.1(b), if the end B of member AB is hinged, then the moment at B must be zero. By substituting $M_{BA} = 0$ into Eqs. (16.8), we write

$$M_{AB} = \frac{2EI}{L}(2\theta_A + \theta_B - 3\psi) + \text{FEM}_{AB} \tag{16.10a}$$

$$M_{BA} = 0 = \frac{2EI}{L}(\theta_A + 2\theta_B - 3\psi) + \text{FEM}_{BA} \tag{16.10b}$$

Solving Eq. (16.10b) for θ_B, we obtain

$$\theta_B = -\frac{\theta_A}{2} + \frac{3}{2}\psi - \frac{L}{4EI}(\text{FEM}_{BA}) \tag{16.11}$$

To eliminate θ_B from the slope-deflection equations, we substitute Eq. (16.11) into Eq. (16.10a), thus obtaining the *modified slope-deflection equations* for member AB with a hinge at end B:

$$M_{AB} = \frac{3EI}{L}(\theta_A - \psi) + \left(\text{FEM}_{AB} - \frac{\text{FEM}_{BA}}{2}\right) \tag{16.12a}$$

$$M_{BA} = 0 \tag{16.12b}$$

Similarly, it can be shown that for a member AB with a hinge at end A, the rotation of the hinged end is given by

$$\theta_A = -\frac{\theta_B}{2} + \frac{3}{2}\psi - \frac{L}{4EI}(\text{FEM}_{AB}) \qquad (16.13)$$

and the modified slope-deflection equations can be expressed as

$$M_{BA} = \frac{3EI}{L}(\theta_B - \psi) + \left(\text{FEM}_{BA} - \frac{\text{FEM}_{AB}}{2}\right) \qquad (16.14a)$$

$$M_{AB} = 0 \qquad (16.14b)$$

Because the modified slope-deflection equations given by Eqs. (16.12) and (16.14) are similar in form, they can be conveniently summarized as

$$M_{rh} = \frac{3EI}{L}(\theta_r - \psi) + \left(\text{FEM}_{rh} - \frac{\text{FEM}_{hr}}{2}\right) \qquad (16.15a)$$

$$M_{hr} = 0 \qquad (16.15b)$$

in which the subscript r refers to the *rigidly connected* end of the member where the moment M_{rh} acts and the subscript h identifies the *hinged* end of the member. The rotation of the hinged end can now be written as

$$\theta_h = -\frac{\theta_r}{2} + \frac{3}{2}\psi - \frac{L}{4EI}(\text{FEM}_{hr}) \qquad (16.16)$$

16.2 BASIC CONCEPT OF THE SLOPE-DEFLECTION METHOD

To illustrate the basic concept of the slope-deflection method, consider the three-span continuous beam shown in Fig. 16.3(a). Although the

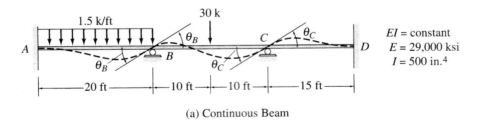

(a) Continuous Beam

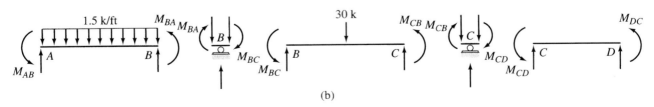

(b)

FIG. 16.3

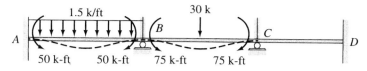

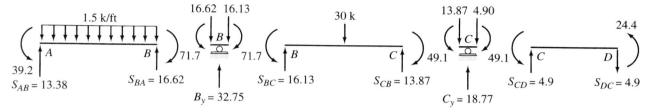

50 k-ft   50 k-ft   75 k-ft   75 k-ft

(c) Fixed-End Moments

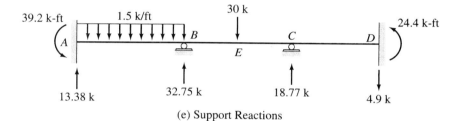

16.62 16.13

1.5 k/ft

39.2

$S_{AB} = 13.38$

$S_{BA} = 16.62$

$B_y = 32.75$

$S_{BC} = 16.13$

$S_{CB} = 13.87$

13.87 4.90

$C_y = 18.77$

$S_{CD} = 4.9$

$S_{DC} = 4.9$

24.4

(d) Member End Moments and Shears

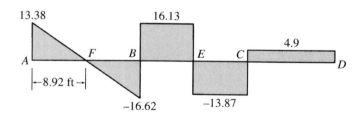

39.2 k-ft   1.5 k/ft   30 k   24.4 k-ft

13.38 k   32.75 k   18.77 k   4.9 k

(e) Support Reactions

13.38   16.13   4.9

$\leftarrow$8.92 ft$\rightarrow$

−16.62   −13.87

(f) Shear Diagram (k)

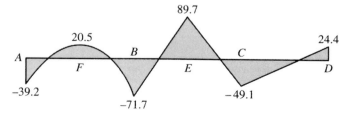

89.7

20.5   24.4

−39.2   −49.1

−71.7

(g) Bending Moment Diagram (k-ft)

FIG. 16.3 (contd.)

structure actually consists of a single continuous beam between the fixed supports A and D, for the purpose of analysis it is considered to be composed of three members, AB, BC, and CD, rigidly connected at joints A, B, C, and D located at the supports of the structure. Note that the continuous beam has been divided into members and joints, so that the unknown external reactions act only at the joints.

Degrees of Freedom

With the joint locations now established, we identify the unknown independent displacements (translations and rotations) of the joints of the structure. These unknown joint displacements are referred to as the *degrees of freedom* of the structure. From the qualitative deflected shape of the continuous beam shown in Fig. 16.3(a), we can see that none of its joints can translate. Furthermore, the fixed joints A and D cannot rotate, whereas joints B and C are free to rotate. Thus the continuous beam has two degrees of freedom, θ_B and θ_C, which represent the unknown rotations of joints B and C, respectively.

The number of degrees of freedom is sometimes called the *degree of kinematic indeterminacy* of the structure. Since the beam of Fig. 16.3(a) has two degrees of freedom, it is considered to be kinematically indeterminate to the second degree. A structure without any degrees of freedom is termed *kinematically determinate*. In other words, if the displacements of all the joints of a structure are either zero or known, the structure is considered to be kinematically determinate.

Equations of Equilibrium

The unknown joint rotations are determined by solving the equations of equilibrium of the joints that are free to rotate. The free-body diagrams of the members and joints B and C of the continuous beam are shown in Fig. 16.3(b). In addition to the external loads, each member is subjected to an internal moment at each of its ends. Since the correct senses of the member end moments are not yet known, it is assumed that the moments at the ends of all the members are positive (counterclockwise) in accordance with the slope-deflection sign convention adopted in the preceding section. Note that the free-body diagrams of the joints show the member end moments acting in an opposite (clockwise) direction, in accordance with Newton's law of action and reaction.

Because the entire structure is in equilibrium, each of its members and joints must also be in equilibrium. By applying the moment equilibrium equations $\sum M_B = 0$ and $\sum M_C = 0$, respectively, to the free bodies of joints B and C, we obtain the equilibrium equations

$$M_{BA} + M_{BC} = 0 \tag{16.17a}$$

$$M_{CB} + M_{CD} = 0 \tag{16.17b}$$

Slope-Deflection Equations

The foregoing equilibrium equations (Eqs. (16.17)) can be expressed in terms of the unknown joint rotations, θ_B and θ_C, by using slope-deflection equations that relate member end moments to the unknown joint rotations. However, before we can write the slope-deflection equations, we need to compute the fixed-end moments due to the external loads acting on the members of the continuous beam.

To calculate the fixed-end moments, we apply imaginary clamps at joints B and C to prevent them from rotating, as shown in Fig. 16.3(c). The fixed-end moments that develop at the ends of the members of this fully restrained or kinematically determinate structure can easily be evaluated either by applying Eqs. (16.7) or by using the fixed-end moment expressions given inside the back cover of the book. By using the fixed-end moment expressions, we calculate the fixed-end moments as follows:

For member AB:

$$\text{FEM}_{AB} = \frac{wL^2}{12} = \frac{1.5(20)^2}{12} = 50 \text{ k-ft } \circlearrowleft \qquad \text{or} \qquad +50 \text{ k-ft}$$

$$\text{FEM}_{BA} = 50 \text{ k-ft } \circlearrowright \qquad \text{or} \qquad -50 \text{ k-ft}$$

For member BC:

$$\text{FEM}_{BC} = \frac{PL}{8} = \frac{30(20)}{8} = 75 \text{ k-ft } \circlearrowleft \qquad \text{or} \qquad +75 \text{ k-ft}$$

$$\text{FEM}_{CB} = 75 \text{ k-ft } \circlearrowright \qquad \text{or} \qquad -75 \text{ k-ft}$$

Note that, in accordance with the slope-deflection sign convention, the counterclockwise fixed-end moments are considered to be positive. Since no external loads act on member CD, its fixed-end moments are zero; that is,

$$\text{FEM}_{CD} = \text{FEM}_{DC} = 0$$

The fixed-end moments are shown on the diagram of the restrained structure in Fig. 16.3(c).

The slope-deflection equations for the three members of the continuous beam can now be written by using Eq. (16.9). Since none of the supports of the continuous beam translates, the chord rotations of the three members are zero (i.e., $\psi_{AB} = \psi_{BC} = \psi_{CD} = 0$). Also, since supports A and D are fixed, the rotations $\theta_A = \theta_D = 0$. By applying Eq. (16.9) for member AB, with A as the near end and B as the far end, we obtain the slope-deflection equation

$$M_{AB} = \frac{2EI}{20}(0 + \theta_B - 0) + 50 = 0.1EI\theta_B + 50 \qquad (16.18a)$$

Next, by considering B as the near end and A as the far end, we write

$$M_{BA} = \frac{2EI}{20}(2\theta_B + 0 - 0) - 50 = 0.2EI\theta_B - 50 \qquad (16.18b)$$

Similarly, by applying Eq. (16.9) for member BC, we obtain

$$M_{BC} = \frac{2EI}{20}(2\theta_B + \theta_C) + 75 = 0.2EI\theta_B + 0.1EI\theta_C + 75 \qquad (16.18c)$$

$$M_{CB} = \frac{2EI}{20}(2\theta_C + \theta_B) - 75 = 0.2EI\theta_C + 0.1EI\theta_B - 75 \qquad (16.18d)$$

and for member CD,

$$M_{CD} = \frac{2EI}{15}(2\theta_C) = 0.267EI\theta_C \qquad (16.18e)$$

$$M_{DC} = \frac{2EI}{15}(\theta_C) = 0.133EI\theta_C \qquad (16.18f)$$

These slope-deflection equations automatically satisfy the compatibility conditions of the structure. Since the member ends are rigidly connected to the adjacent joints, the rotations of member ends are equal to the rotations of the adjacent joints. Thus, the θ terms in the slope-deflection equations (Eqs. (16.18)) represent the rotations of the member ends as well as those of the joints.

Joint Rotations

To determine the unknown joint rotations θ_B and θ_C, we substitute the slope-deflection equations (Eqs. (16.18)) into the joint equilibrium equations (Eqs. (16.17)) and solve the resulting system of equations simultaneously for θ_B and θ_C. Thus by substituting Eqs. (16.18b) and (16.18c) into Eq. (16.17a), we obtain

$$(0.2EI\theta_B - 50) + (0.2EI\theta_B + 0.1EI\theta_C + 75) = 0$$

or

$$0.4EI\theta_B + 0.1EI\theta_C = -25 \qquad (16.19a)$$

and by substituting Eqs. (16.18d) and (16.18e) into Eq. (16.17b), we get

$$(0.2EI\theta_C + 0.1EI\theta_B - 75) + 0.267EI\theta_C = 0$$

or

$$0.1EI\theta_B + 0.467EI\theta_C = 75 \qquad (16.19b)$$

Solving Eqs. (16.19a) and (16.19b) simultaneously for $EI\theta_B$ and $EI\theta_C$, we obtain

$$EI\theta_B = -108.46 \text{ k-ft}^2$$

$$EI\theta_C = 183.82 \text{ k-ft}^2$$

By substituting the numerical values of $E = 29{,}000$ ksi $= 29{,}000(12)^2$ ksf and $I = 500$ in.$^4 = (500/12^4)$ ft^4, we determine the rotations of joints B and C to be

$$\theta_B = -0.0011 \text{ rad} \quad \text{or} \quad 0.0011 \text{ rad} \circlearrowright$$

$$\theta_C = 0.0018 \text{ rad} \circlearrowleft$$

Member End Moments

The moments at the ends of the three members of the continuous beam can now be determined by substituting the numerical values of $EI\theta_B$ and $EI\theta_C$ into the slope-deflection equations (Eqs. (16.18)). Thus

$$M_{AB} = 0.1(-108.46) + 50 = 39.2 \text{ k-ft} \circlearrowleft$$

$$M_{BA} = 0.2(-108.46) - 50 = -71.7 \text{ k-ft} \quad \text{or} \quad 71.7 \text{ k-ft} \circlearrowright$$

$$M_{BC} = 0.2(-108.46) + 0.1(183.82) + 75 = 71.7 \text{ k-ft} \circlearrowleft$$

$$M_{CB} = 0.2(183.82) + 0.1(-108.46) - 75$$
$$= -49.1 \text{ k-ft} \quad \text{or} \quad 49.1 \text{ k-ft} \circlearrowright$$

$$M_{CD} = 0.267(183.82) = 49.1 \text{ k-ft} \circlearrowleft$$

$$M_{DC} = 0.133(183.82) = 24.4 \text{ k-ft} \circlearrowleft$$

Note that a positive answer for an end moment indicates that its sense is counterclockwise, whereas a negative answer for an end moment implies a clockwise sense.

To check that the solution of simultaneous equations (Eqs. (16.19)) has been carried out correctly, the numerical values of member end moments should be substituted into the joint equilibrium equations (Eqs. (16.17)). If the solution is correct, then the equilibrium equations should be satisfied.

$$M_{BA} + M_{BC} = -71.7 + 71.7 = 0 \qquad \text{Checks}$$

$$M_{CB} + M_{CD} = -49.1 + 49.1 = 0 \qquad \text{Checks}$$

Member End Shears

The member end moments just computed are shown on the free-body diagrams of the members and joints in Fig. 16.3(d). The shear forces at the ends of members can now be determined by applying the equations of equilibrium to the free bodies of the members. Thus, for member AB,

$$+\circlearrowleft \sum M_B = 0 \qquad 39.2 - S_{AB}(20) + 1.5(20)(10) - 71.7 = 0$$
$$S_{AB} = 13.38 \text{ k} \uparrow$$

$$+\uparrow \sum F_y = 0 \qquad 13.38 - 1.5(20) + S_{BA} = 0$$
$$S_{BA} = 16.62 \text{ k} \uparrow$$

Similarly, for member BC,

$$+\circlearrowleft \sum M_C = 0 \qquad 71.7 - S_{BC}(20) + 30(10) - 49.1 = 0$$

$$S_{BC} = 16.13 \text{ k} \uparrow$$

$$+\uparrow \sum F_y = 0 \qquad 16.13 - 30 + S_{CB} = 0$$

$$S_{CB} = 13.87 \text{ k} \uparrow$$

and for member CD,

$$+\circlearrowleft \sum M_D = 0 \qquad 49.1 - S_{CD}(15) + 24.4 = 0 \qquad S_{CD} = 4.9 \text{ k} \uparrow$$

$$+\uparrow \sum F_y = 0 \qquad \qquad S_{DC} = 4.9 \text{ k} \downarrow$$

The foregoing member end shears can, alternatively, be evaluated by superposition of end shears due to the external load and each of the end moments acting separately on the member. For example, the shear at end A of member AB is given by

$$S_{AB} = \frac{1.5(20)}{2} + \frac{39.2}{20} - \frac{71.7}{20} = 13.38 \text{ k} \uparrow$$

in which the first term equals the shear due to the 1.5-k/ft uniformly distributed load, whereas the second and third terms are the shears due to the 39.2-k-ft and 71.7-k-ft moments, respectively, at the ends A and B of the member.

Support Reactions

From the free-body diagram of joint B in Fig. 16.3(d), we can see that the vertical reaction at the roller support B is equal to the sum of the shears at ends B of members AB and BC; that is,

$$B_y = S_{BA} + S_{BC} = 16.62 + 16.13 = 32.75 \text{ k} \uparrow$$

Similarly, the vertical reaction at the roller support C equals the sum of the shears at ends C of members BC and CD. Thus

$$C_y = S_{CB} + S_{CD} = 13.87 + 4.9 = 18.77 \text{ k} \uparrow$$

The reactions at the fixed support A are equal to the shear and moment at the end A of member AB; that is,

$$A_y = S_{AB} = 13.38 \text{ k} \uparrow$$

$$M_A = M_{AB} = 39.2 \text{ k-ft} \circlearrowright$$

Similarly, the reactions at the fixed support D equal the shear and moment at end D of member CD. Thus

$$D_y = S_{DC} = 4.9 \text{ k} \downarrow$$

$$M_D = M_{DC} = 24.4 \text{ k-ft} \circlearrowright$$

The support reactions are shown in Fig. 16.3(e).

Equilibrium Check

To check our computations of member end shears and support reactions, we apply the equations of equilibrium to the free body of the
entire structure. Thus (Fig. 16.3(e)),

$$+\uparrow \sum F_y = 0$$

$$13.38 - 1.5(20) + 32.75 - 30 + 18.77 - 4.9 = 0 \qquad \text{Checks}$$

$$+\curvearrowleft \sum M_D = 0$$

$$39.2 - 13.38(55) + 1.5(20)(45) - 32.75(35) + 30(25)$$

$$- 18.77(15) + 24.4 = -0.1 \approx 0 \qquad \text{Checks}$$

This equilibrium check, as well as the check performed previously on
the solution of simultaneous equations, does not detect any errors involved in the slope-deflection equations. Therefore, the slope-deflection
equations should be developed very carefully and should always be
checked before proceeding with the rest of the analysis.

Shear and Bending Moment Diagrams

With the support reactions known, the shear and bending moment diagrams can now be constructed in the usual manner by using the *beam
sign convention* described in Section 5.1. The shear and bending moment diagrams thus obtained for the continuous beam are shown in Fig.
16.3(f) and (g), respectively.

16.3 ANALYSIS OF CONTINUOUS BEAMS

Based on the discussion presented in the preceding section, the procedure
for the analysis of continuous beams by the slope-deflection method can
be summarized as follows:

1. Identify the degrees of freedom of the structure. For continuous
   beams, the degrees of freedom consist of the unknown rotations of
   the joints.
2. Compute fixed-end moments. For each member of the structure,
   evaluate the fixed-end moments due to the external loads by using
   the expressions given inside the back cover of the book. The counterclockwise fixed-end moments are considered to be positive.
3. In the case of support settlements, determine the rotations of the
   chords of members adjacent to the supports that settle by dividing
   the relative translation between the two ends of the member by the
   member length ($\psi = \Delta/L$). The chord rotations are measured from

the undeformed (horizontal) positions of members, with counterclockwise rotations considered as positive.

4. Write slope-deflection equations. For each member, apply Eq. (16.9) to write two slope-deflection equations relating member end moments to the unknown rotations of the adjacent joints.

5. Write equilibrium equations. For each joint that is free to rotate, write a moment equilibrium equation, $\sum M = 0$, in terms of the moments at the member ends connected to the joint. The total number of such equilibrium equations must be equal to the number of degrees of freedom of the structure.

6. Determine the unknown joint rotations. Substitute the slope-deflection equations into the equilibrium equations, and solve the resulting system of equations for the unknown joint rotations.

7. Calculate member end moments by substituting the numerical values of joint rotations determined in step 6 into the slope-deflection equations. A positive answer for an end moment indicates that its sense is counterclockwise, whereas a negative answer for an end moment implies a clockwise sense.

8. To check whether or not the solution of simultaneous equations was carried out correctly in step 6, substitute the numerical values of member end moments into the joint equilibrium equations developed in step 5. If the solution is correct, then the equilibrium equations should be satisfied.

9. Compute member end shears. For each member, (a) draw a free-body diagram showing the external loads and end moments and (b) apply the equations of equilibrium to calculate the shear forces at the ends of the member.

10. Determine support reactions by considering the equilibrium of the joints of the structure.

11. To check the calculations of member end shears and support reactions, apply the equations of equilibrium to the free body of the entire structure. If the calculations have been carried out correctly, then the equilibrium equations should be satisfied.

12. Draw shear and bending moment diagrams of the structure by using the *beam sign convention*.

Beams with Simple Supports at Their Ends

Although the foregoing procedure can be used to analyze continuous beams that are simply supported at one or both ends, the analysis of such structures can be considerably expedited by using the modified slope-deflection equations (Eqs. (16.15)) for spans adjacent to the simple end supports, thereby eliminating the rotations of simple supports from the analysis (see Example 16.3). However, this simplified approach can be used only for those simple end supports at which no external moment is applied. This is because the modified slope-deflection equations for a member with one end hinged (Eqs. (16.15)) are based on the condition that the moment at the hinged end is zero.

Structures with Cantilever Overhangs

Consider a continuous beam with a cantilever overhang, as shown in Fig. 16.4(a). Since the cantilever portion CD of the beam is statically determinate in the sense that the shear and moment at its end C can be obtained by applying the equations of equilibrium (Fig. 16.4(b)), it is not necessary to include this portion in the analysis. Thus, for the purpose of analysis, the cantilever portion CD can be removed from the structure, provided that the moment and the force exerted by the cantilever on the remaining structure are included in the analysis. The indeterminate part AC of the structure, which needs to be analyzed, is shown in Fig. 16.4(c).

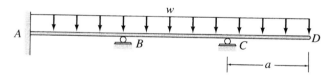

(a) Actual Beam

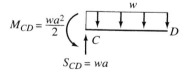

(b) Statically Determinate Cantilever Portion

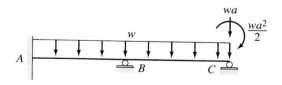

FIG. **16.4**

(c) Statically Indeterminate Part to Be Analyzed

Example 16.1

Determine the reactions and draw the shear and bending moment diagrams for the two-span continuous beam shown in Fig. 16.5(a) by the slope-deflection method.

Solution

Degrees of Freedom From Fig. 16.5(a), we can see that only joint B of the beam is free to rotate. Thus, the structure has only one degree of freedom, which is the unknown joint rotation, θ_B.

Fixed-End Moments By using the fixed-end moment expressions given inside the back cover of the book, we evaluate the fixed-end moments due to the external loads for each member:

$$\text{FEM}_{AB} = \frac{Pab^2}{L^2} = \frac{18(10)(15)^2}{(25)^2} = 64.8 \text{ k-ft} \circlearrowleft \quad \text{or} \quad +64.8 \text{ k-ft}$$

$$\text{FEM}_{BA} = \frac{Pa^2b}{L^2} = \frac{18(10)^2(15)}{(25)^2} = 43.2 \text{ k-ft} \circlearrowright \quad \text{or} \quad -43.2 \text{ k-ft}$$

$$\text{FEM}_{BC} = \frac{wL^2}{12} = \frac{2(30)^2}{12} = 150 \text{ k-ft} \circlearrowleft \quad \text{or} \quad +150 \text{ k-ft}$$

$$\text{FEM}_{CB} = 150 \text{ k-ft} \circlearrowright \quad \text{or} \quad -150 \text{ k-ft}$$

Note that in accordance with the slope-deflection sign convention, the counterclockwise fixed-end moments are considered as positive, whereas the clockwise fixed-end moments are considered to be negative.

Chord Rotations Since no support settlements occur, the chord rotations of both members are zero; that is, $\psi_{AB} = \psi_{BC} = 0$.

Slope-Deflection Equations To relate the member end moments to the unknown joint rotation, θ_B, we write the slope-deflection equations for the two members of the structure by applying Eq. (16.9). Note that since the supports A and C are fixed, the rotations $\theta_A = \theta_C = 0$. Thus the slope-deflection equations for member AB can be expressed as

$$M_{AB} = \frac{2EI}{25}(\theta_B) + 64.8 = 0.08EI\theta_B + 64.8 \tag{1}$$

$$M_{BA} = \frac{2EI}{25}(2\theta_B) - 43.2 = 0.16EI\theta_B - 43.2 \tag{2}$$

Similarly, by applying Eq. (16.9) for member BC, we obtain the slope-deflection equations

$$M_{BC} = \frac{2EI}{30}(2\theta_B) + 150 = 0.133EI\theta_B + 150 \tag{3}$$

$$M_{CB} = \frac{2EI}{30}(\theta_B) - 150 = 0.0667EI\theta_B - 150 \tag{4}$$

Equilibrium Equation The free-body diagram of joint B is shown in Fig. 16.5(b). Note that the member end moments, which are assumed to be in a counterclockwise direction on the ends of the members, must be applied in the

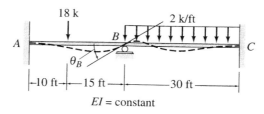

(a) Continuous Beam

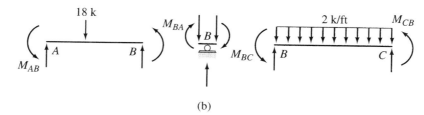

(b)

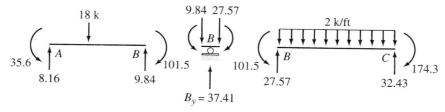

$B_y = 37.41$

(c) Member End Moments and Shears

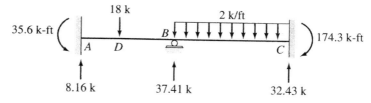

(d) Support Reactions

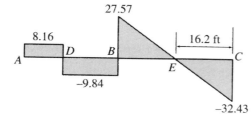

(e) Shear Diagram (k)

(f) Bending Moment Diagram (k-ft)

FIG. **16.5**

continued

(opposite) clockwise direction on the free body of the joint, in accordance with Newton's third law. By applying the moment equilibrium equation $\sum M_B = 0$ to the free body of joint B, we obtain the equilibrium equation

$$M_{BA} + M_{BC} = 0 \qquad (5)$$

Joint Rotation To determine the unknown joint rotation, θ_B, we substitute the slope-deflection equations (Eqs. (2) and (3)) into the equilibrium equation (Eq. (5)) to obtain

$$(0.16EI\theta_B - 43.2) + (0.133EI\theta_B + 150) = 0$$

or

$$0.293EI\theta_B = -106.8$$

from which

$$EI\theta_B = -364.5 \text{ k-ft}^2$$

Member End Moments The member end moments can now be computed by substituting the numerical value of $EI\theta_B$ back into the slope-deflection equations (Eqs. (1) through (4)). Thus,

$$M_{AB} = 0.08(-364.5) + 64.8 = 35.6 \text{ k-ft } \circlearrowright$$

$$M_{BA} = 0.16(-364.5) - 43.2 = -101.5 \text{ k-ft} \qquad \text{or} \qquad 101.5 \text{ k-ft } \circlearrowleft$$

$$M_{BC} = 0.133(-364.5) + 150 = 101.5 \text{ k-ft } \circlearrowright$$

$$M_{CB} = 0.0667(-364.5) - 150 = -174.3 \text{ k-ft} \qquad \text{or} \qquad 174.3 \text{ k-ft } \circlearrowleft$$

Note that a positive answer for an end moment indicates that its sense is counterclockwise, whereas a negative answer for an end moment implies a clockwise sense. Since the end moments M_{BA} and M_{BC} are equal in magnitude but opposite in sense, the equilibrium equation, $M_{BA} + M_{BC} = 0$, is indeed satisfied.

Member End Shears The member end shears, obtained by considering the equilibrium of each member, are shown in Fig. 16.5(c).

Support Reactions The reactions at the fixed supports A and C are equal to the forces and moments at the ends of the members connected to these joints. To determine the reaction at the roller support B, we consider the equilibrium of the free body of joint B in the vertical direction (see Fig. 16.5(c)), to obtain

$$B_y = S_{BA} + S_{BC} = 9.84 + 27.57 = 37.41 \text{ k } \uparrow$$

The support reactions are shown in Fig. 16.5(d).   **Ans.**

Equilibrium Check To check our calculations of member end shears and support reactions, we apply the equations of equilibrium to the free body of the entire structure. Thus (see Fig. 16.5(d)),

$$+\uparrow \sum F_y = 0$$

$$8.16 - 18 + 37.41 - 2(30) + 32.43 = 0 \qquad \text{Checks}$$

$$+\circlearrowleft \sum M_C = 0$$

$$35.6 - 8.16(55) + 18(45) - 37.41(30) + 2(30)(15) - 174.3 = 0.2 \approx 0 \qquad \text{Checks}$$

Shear and Bending Moment Diagrams The shear and bending moment diagrams can now be constructed by using the *beam sign convention* described in Section 5.1. These diagrams are shown in Fig. 16.5(e) and (f).   **Ans.**

Example 16.2

Determine the reactions and draw the shear and bending moment diagrams for the three-span continuous beam shown in Fig. 16.6(a) by the slope-deflection method.

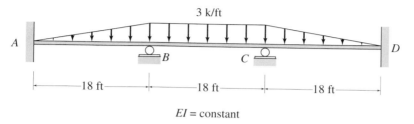

EI = constant

(a) Continous Beam

(b) Free-Body Diagrams of Joints B and C

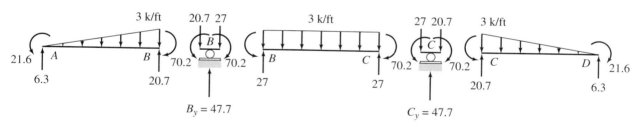

$B_y = 47.7$ $C_y = 47.7$

(c) Member End Moments and Shears

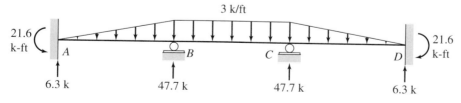

(d) Support Reactions

FIG. 16.6

continued

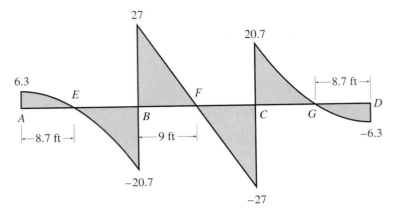

(e) Shear Diagram (k)

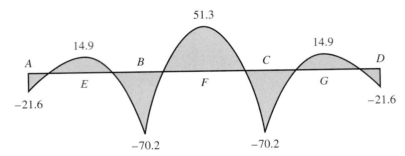

(f) Bending Moment Diagram (k-ft)

FIG. 16.6 (contd.)

Solution

Degrees of Freedom θ_B and θ_C.

Fixed-End Moments

$$\text{FEM}_{AB} = \frac{3(18)^2}{30} = 32.4 \text{ k-ft} \ \circlearrowright \qquad \text{or} \qquad +32.4 \text{ k-ft}$$

$$\text{FEM}_{BA} = \frac{3(18)^2}{20} = 48.6 \text{ k-ft} \ \circlearrowleft \qquad \text{or} \qquad -48.6 \text{ k-ft}$$

$$\text{FEM}_{BC} = \frac{3(18)^2}{12} = 81 \text{ k-ft} \ \circlearrowright \qquad \text{or} \qquad +81 \text{ k-ft}$$

$$\text{FEM}_{CB} = 81 \text{ k-ft} \ \circlearrowleft \qquad \text{or} \qquad -81 \text{ k-ft}$$

$$\text{FEM}_{CD} = \frac{3(18)^2}{20} = 48.6 \text{ k-ft} \ \circlearrowright \qquad \text{or} \qquad +48.6 \text{ k-ft}$$

$$\text{FEM}_{DC} = \frac{3(18)^2}{30} = 32.4 \text{ k-ft} \ \circlearrowleft \qquad \text{or} \qquad -32.4 \text{ k-ft}$$

Slope-Deflection Equations Using Eq. (16.9) for members *AB*, *BC*, and *CD*, we write

$$M_{AB} = \frac{2EI}{18}(\theta_B) + 32.4 = 0.111EI\theta_B + 32.4 \qquad (1)$$

$$M_{BA} = \frac{2EI}{18}(2\theta_B) - 48.6 = 0.222EI\theta_B - 48.6 \qquad (2)$$

$$M_{BC} = \frac{2EI}{18}(2\theta_B + \theta_C) + 81 = 0.222EI\theta_B + 0.111EI\theta_C + 81 \qquad (3)$$

$$M_{CB} = \frac{2EI}{18}(\theta_B + 2\theta_C) - 81 = 0.111EI\theta_B + 0.222EI\theta_C - 81 \qquad (4)$$

$$M_{CD} = \frac{2EI}{18}(2\theta_C) + 48.6 = 0.222EI\theta_C + 48.6 \qquad (5)$$

$$M_{DC} = \frac{2EI}{18}(\theta_C) - 32.4 = 0.111EI\theta_C - 32.4 \qquad (6)$$

Equilibrium Equations See Fig. 16.6(b).

$$M_{BA} + M_{BC} = 0 \qquad (7)$$

$$M_{CB} + M_{CD} = 0 \qquad (8)$$

Joint Rotations By substituting the slope-deflection equations (Eqs. (1) through (6)) into the equilibrium equations (Eqs. (7) and (8)), we obtain

$$0.444EI\theta_B + 0.111EI\theta_C = -32.4 \qquad (9)$$

$$0.111EI\theta_B + 0.444EI\theta_C = 32.4 \qquad (10)$$

By solving Eqs. (9) and (10) simultaneously, we determine the values of $EI\theta_B$ and $EI\theta_C$ to be

$$EI\theta_B = -97.3 \text{ k-ft}^2$$

$$EI\theta_C = 97.3 \text{ k-ft}^2$$

Member End Moments To compute the member end moments, we substitute the numerical values of $EI\theta_B$ and $EI\theta_C$ back into the slope-deflection equations (Eqs. (1) through (6)) to obtain

$$M_{AB} = 0.111(-97.3) + 32.4 = 21.6 \text{ k-ft} \circlearrowleft \qquad \text{Ans.}$$

$$M_{BA} = 0.222(-97.3) - 48.6 = -70.2 \text{ k-ft} \qquad \text{or} \qquad 70.2 \text{ k-ft} \circlearrowright \quad \text{Ans.}$$

$$M_{BC} = 0.222(-97.3) + 0.111(97.3) + 81 = 70.2 \text{ k-ft} \circlearrowleft \qquad \text{Ans.}$$

$$M_{CB} = 0.111(-97.3) + 0.222(97.3) - 81$$

$$= -70.2 \text{ k-ft} \qquad \text{or} \qquad 70.2 \text{ k-ft} \circlearrowright \qquad \text{Ans.}$$

$$M_{CD} = 0.222(97.3) + 48.6 = 70.2 \text{ k-ft} \circlearrowleft \qquad \text{Ans.}$$

$$M_{DC} = 0.111(97.3) - 32.4 = -21.6 \text{ k-ft} \qquad \text{or} \qquad 21.6 \text{ k-ft} \circlearrowright \quad \text{Ans.}$$

Note that the numerical values of M_{BA}, M_{BC}, M_{CB}, and M_{CD} do satisfy the equilibrium equations (Eqs. (7) and (8)).

Member End Shears and Support Reactions See Fig. 16.6(c) and (d).   **Ans.**

Equilibrium Check The equilibrium equations check.

Shear and Bending Moment Diagrams See Fig. 16.6(e) and (f).   **Ans.**

Example 16.3

Determine the member end moments and reactions for the continuous beam shown in Fig. 16.7(a) by the slope-deflection method.

Solution

This beam was previously analyzed in Example 13.6 by the method of consistent deformations.

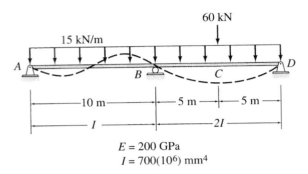

(a) Continuous Beam

(b) Free-Body Diagrams of Joints

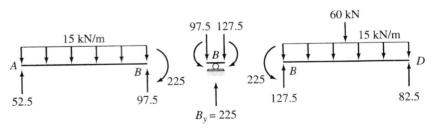

(c) Member End Moments and Shears

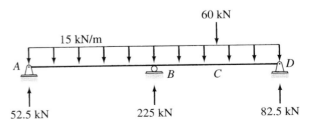

(d) Support Reactions

FIG. **16.7**

From Fig. 16.7(a), we can see that all three joints of the beam are free to rotate. Thus the beam can be considered to have three degrees of freedom, θ_A, θ_B, and θ_D, and it can be analyzed by using the usual slope-deflection equations (Eq. (16.9)) for members rigidly connected at both ends. However, this approach is quite time consuming, since it requires solving three simultaneous equations to determine the three unknown joint rotations.

Since the end supports A and D of the beam are simple supports at which no external moment is applied, the moments at the end A of member AB and at the end D of member BD must be zero. (This can easily be verified by considering moment equilibrium of the free bodies of joints A and D shown in Fig. 16.7(b).) Thus the end A of member AB and the end D of member BD can be considered to be hinged ends, and the modified slope-deflection equations (Eqs. (16.15)) can be used for these members. Furthermore, since the modified slope-deflection equations do not contain the rotations of the hinged ends, by using these equations the rotations θ_A and θ_D of the simple supports can be eliminated from the analysis, which will then involve only one unknown joint rotation, θ_B. It should be noted that once θ_B has been evaluated, the values of the rotations θ_A and θ_D, if desired, can be computed by using Eq. (16.16). In the following, we use this simplified approach to analyze the continuous beam.

Degrees of Freedom θ_B.

Fixed-End Moments

$$\text{FEM}_{AB} = \frac{15(10)^2}{12} = 125 \text{ kN} \cdot \text{m} \; \circlearrowright \quad \text{or} \quad +125 \text{ kN} \cdot \text{m}$$

$$\text{FEM}_{BA} = 125 \text{ kN} \cdot \text{m} \; \circlearrowleft \quad \text{or} \quad -125 \text{ kN} \cdot \text{m}$$

$$\text{FEM}_{BD} = \frac{60(10)}{8} + \frac{15(10)^2}{12} = 200 \text{ kN} \cdot \text{m} \; \circlearrowright \quad \text{or} \quad +200 \text{ kN} \cdot \text{m}$$

$$\text{FEM}_{DB} = 200 \text{ kN} \cdot \text{m} \; \circlearrowleft \quad \text{or} \quad -200 \text{ kN} \cdot \text{m}$$

Slope-Deflection Equations Since both members of the beam have one end hinged, we use Eqs. (16.15) to obtain the slope-deflection equations for both members. Thus

$$M_{AB} = 0 \qquad\qquad\qquad\qquad\qquad\qquad\qquad\qquad \text{Ans.}$$

$$M_{BA} = \frac{3EI}{10}(\theta_B) + \left(-125 - \frac{125}{2}\right) = 0.3EI\theta_B - 187.5 \qquad (1)$$

$$M_{BD} = \frac{3E(2I)}{10}(\theta_B) + \left(200 + \frac{200}{2}\right) = 0.6EI\theta_B + 300 \qquad (2)$$

$$M_{DB} = 0 \qquad\qquad\qquad\qquad\qquad\qquad\qquad\qquad \text{Ans.}$$

Equilibrium Equation By considering the moment equilibrium of the free body of joint B (Fig. 16.7(b)), we obtain the equilibrium equation

$$M_{BA} + M_{BD} = 0 \qquad (3)$$

continued

Joint Rotation To determine the unknown joint rotation θ_B, we substitute the slope-deflection equations (Eqs. (1) and (2)) into the equilibrium equation (Eq. (3)) to obtain

$$(0.3EI\theta_B - 187.5) + (0.6EI\theta_B + 300) = 0$$

or

$$0.9EI\theta_B = -112.5$$

from which

$$EI\theta_B = -125 \text{ kN} \cdot \text{m}^2$$

Member End Moments The member end moments can now be computed by substituting the numerical value of $EI\theta_B$ into the slope-deflection equations (Eqs. (1) and (2)). Thus

$$M_{BA} = 0.3(-125) - 187.5 = -225 \text{ kN} \cdot \text{m} \qquad \text{or} \qquad 225 \text{ kN} \cdot \text{m} \curvearrowright \text{ Ans.}$$

$$M_{BD} = 0.6(-125) + 300 = 225 \text{ kN} \cdot \text{m} \curvearrowleft \qquad\qquad\qquad\qquad \text{Ans.}$$

Member End Shears and Support Reactions See Fig. 16.7(c) and (d).

Equilibrium Check See Fig. 16.7(d).

$$+\uparrow \sum F_y = 0 \qquad 52.5 - 15(20) + 225 - 60 + 82.5 = 0 \qquad \text{Checks}$$

$$+\curvearrowleft \sum M_D = 0$$

$$-52.5(20) + 15(20)(10) - 225(10) + 60(5) = 0 \qquad \text{Checks}$$

Example 16.4

Determine the member end moments and reactions for the continuous beam shown in Fig. 16.8(a) by the slope-deflection method.

Solution

Since the moment and shear at end C of the cantilever member CD of the beam can be computed directly by applying the equations of equilibrium (see Fig. 16.8(b)), it is not necessary to include this member in the analysis. Thus, only the indeterminate part AC of the beam, shown in Fig. 16.8(c), needs to be analyzed. Note that, as shown in this figure, the 120-kN · m moment and the 30-kN force exerted at joint C by the cantilever CD must be included in the analysis.

Degrees of Freedom From Fig. 16.8(c), we can see that joints B and C are free to rotate. Thus, the structure to be analyzed has two degrees of freedom, which are the unknown joint rotations θ_B and θ_C.

Fixed-End Moments

$$\text{FEM}_{AB} = \text{FEM}_{BA} = 0$$

$$\text{FEM}_{BC} = \frac{10(9)^2}{12} = 67.5 \text{ kN} \cdot \text{m} \curvearrowleft \qquad \text{or} \qquad +67.5 \text{ kN} \cdot \text{m}$$

$$\text{FEM}_{CB} = 67.5 \text{ kN} \cdot \text{m} \curvearrowright \qquad \text{or} \qquad -67.5 \text{ kN} \cdot \text{m}$$

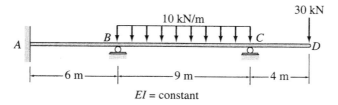

(a) Continuous Beam

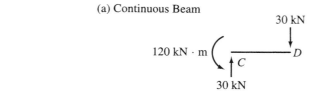

(b) Statically Determinate
Cantilever Portion

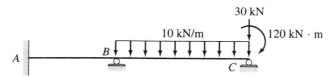

(c) Statically Indeterminate Part to be Analyzed

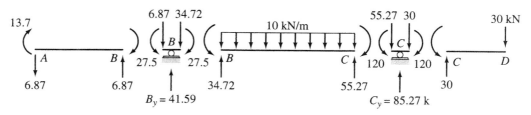

(d) Free-Body Diagrams of Joints B and C

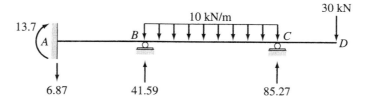

(e) Member End Moments and Shears

(f) Support Reactions

FIG. 16.8

continued

Slope-Deflection Equations By applying Eq. (16.9) to members AB and BC, we write the slope-deflection equations:

$$M_{AB} = \frac{2EI}{6}(\theta_B) = 0.333EI\theta_B \tag{1}$$

$$M_{BA} = \frac{2EI}{6}(2\theta_B) = 0.667EI\theta_B \tag{2}$$

$$M_{BC} = \frac{2EI}{9}(2\theta_B + \theta_C) + 67.5 = 0.444EI\theta_B + 0.222EI\theta_C + 67.5 \tag{3}$$

$$M_{CB} = \frac{2EI}{9}(2\theta_C + \theta_B) - 67.5 = 0.222EI\theta_B + 0.444EI\theta_C - 67.5 \tag{4}$$

Equilibrium Equations By considering the moment equilibrium of the free bodies of joints B and C (Fig. 16.8(d)), we obtain the equilibrium equations

$$M_{BA} + M_{BC} = 0 \tag{5}$$

$$M_{CB} + 120 = 0 \tag{6}$$

Joint Rotations Substitution of the slope-deflection equations (Eqs. (2) through (4)) into the equilibrium equations (Eqs. (5) and (6)) yields

$$1.111EI\theta_B + 0.222EI\theta_C = -67.5 \tag{7}$$

$$0.222EI\theta_B + 0.444EI\theta_C = -52.5 \tag{8}$$

By solving Eqs. (7) and (8) simultaneously, we determine the values of $EI\theta_B$ and $EI\theta_C$ to be

$$EI\theta_B = -41.25 \text{ kN} \cdot \text{m}^2$$

$$EI\theta_C = -97.62 \text{ kN} \cdot \text{m}^2$$

Member End Moments The member end moments can now be computed by substituting the numerical values of $EI\theta_B$ and $EI\theta_C$ into the slope-deflection equations (Eqs. (1) through (4)):

$M_{AB} = 0.333(-41.25) = -13.7 \text{ kN} \cdot \text{m}$ or $13.7 \text{ kN} \cdot \text{m} \;\curvearrowright$ **Ans.**

$M_{BA} = 0.667(-41.25) = -27.5 \text{ kN} \cdot \text{m}$ or $27.5 \text{ kN} \cdot \text{m} \;\curvearrowright$ **Ans.**

$M_{BC} = 0.444(-41.25) + 0.222(-97.62) + 67.5$

       $= 27.5 \text{ kN} \cdot \text{m} \;\curvearrowleft$ **Ans.**

$M_{CB} = 0.222(-41.25) + 0.444(-97.62) - 67.5$

       $= -120 \text{ kN} \cdot \text{m}$ or $120 \text{ kN} \cdot \text{m} \;\curvearrowright$ **Ans.**

Note that the numerical values of M_{BA}, M_{BC}, and M_{CB} do satisfy the equilibrium equations (Eqs. (5) and (6)).

Member End Shears and Support Reactions See Fig. 16.8(e) and (f).   **Ans.**

Equilibrium Check The equilibrium equations check.

Example 16.5

Determine the reactions and draw the shear and bending moment diagrams for the continuous beam shown in Fig. 16.9(a) due to a settlement of 20 mm at support B. Use the slope-deflection method.

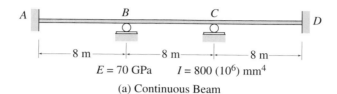

$E = 70$ GPa $I = 800 \ (10^6)$ mm^4

(a) Continuous Beam

(b) Chord Rotations Due to Support Settlement

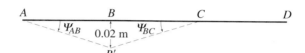

(c) Free-Body Diagrams of Joints B and C

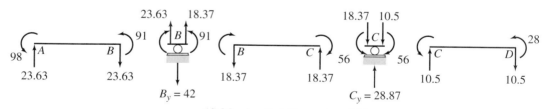

(d) Member End Moments and Shears

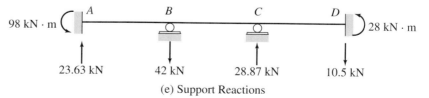

(e) Support Reactions

FIG. 16.9

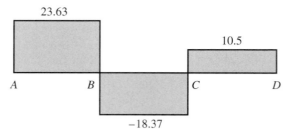

(f) Shear Diagram (kN)

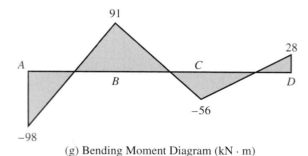

(g) Bending Moment Diagram (kN · m)

FIG. **16.9** (contd.)

Solution

Degrees of Freedom θ_B and θ_C.

Fixed-End Moments Since no external loads act on the beam, the fixed-end moments are zero.

Chord Rotations The specified support settlement is depicted in Fig. 16.9(b), using an exaggerated scale. The inclined dashed lines in this figure indicate the chords (not the elastic curves) of the members in the deformed positions. Because the length of member AB is 8 m, the rotation of its chord is

$$\psi_{AB} = -\frac{0.02}{8} = -0.0025$$

in which the negative sign has been assigned to the value of ψ_{AB} to indicate that its direction is clockwise, as shown in Fig. 16.9(b). Similarly, the chord rotation for member BC is

$$\psi_{BC} = \frac{0.02}{8} = 0.0025$$

From Fig. 16.9(b), we can see that

$$\psi_{CD} = 0$$

Slope-Deflection Equations Applying Eq. (16.9) to members AB, BC, and CD, we write

$$M_{AB} = \frac{2EI}{8}(\theta_B + 0.0075) \tag{1}$$

$$M_{BA} = \frac{2EI}{8}(2\theta_B + 0.0075) \tag{2}$$

$$M_{BC} = \frac{2EI}{8}(2\theta_B + \theta_C - 0.0075) \tag{3}$$

$$M_{CB} = \frac{2EI}{8}(\theta_B + 2\theta_C - 0.0075) \tag{4}$$

$$M_{CD} = \frac{2EI}{8}(2\theta_C) \tag{5}$$

$$M_{DC} = \frac{2EI}{8}(\theta_C) \tag{6}$$

Equilibrium Equations See Fig. 16.9(c).

$$M_{BA} + M_{BC} = 0 \tag{7}$$

$$M_{CB} + M_{CD} = 0 \tag{8}$$

Joint Rotations Substitution of the slope-deflection equations (Eqs. (1) through (6)) into the equilibrium equations (Eqs. (7) and (8)) yields

$$4\theta_B + \theta_C = 0 \tag{9}$$

$$\theta_B + 4\theta_C = 0.0075 \tag{10}$$

By solving Eqs. (9) and (10) simultaneously, we determine

$$\theta_B = -0.0005 \text{ rad}$$

$$\theta_C = 0.002 \text{ rad}$$

Member End Moments To compute the member end moments, we substitute the numerical values of θ_B, θ_C, and $EI = (70)(800) = 56{,}000$ kN·m^2 into the right sides of the slope-deflection equations (Eqs. (1) through (6)) to obtain

$M_{AB} = 98$ kN·m ↻   Ans.

$M_{BA} = 91$ kN·m ↻   Ans.

$M_{BC} = -91$ kN·m   or 91 kN·m ↺   Ans.

$M_{CB} = -56$ kN·m   or 56 kN·m ↺   Ans.

$M_{CD} = 56$ kN·m ↻   Ans.

$M_{DC} = 28$ kN·m ↻   Ans.

Member End Shears and Support Reactions See Fig. 16.9(d) and (e). Ans.

Equilibrium Check See Fig. 16.9(e).

$$+\uparrow \sum F_y = 0 \qquad 23.63 - 42 + 28.87 - 10.5 = 0 \qquad \text{Checks}$$

$$+\circlearrowleft M_A = 0$$

$$98 - 42(8) + 28.87(16) - 10.5(24) + 28 = -0.08 \approx 0 \qquad \text{Checks}$$

Shear and Bending Moment Diagrams See Fig. 16.9(f) and (g). Ans.

Example 16.6

Determine the member end moments and reactions for the three-span continuous beam shown in Fig. 16.10(a) due to the uniformly distributed load and due to the support settlements of $\frac{5}{8}$ in. at B, $1\frac{1}{2}$ in. at C, and $\frac{3}{4}$ in. at D. Use the slope-deflection method.

Solution

Degrees of Freedom Although all four joints of the beam are free to rotate, we will eliminate the rotations of the simple supports at the ends A and D from the analysis by using the modified slope-deflection equations for members AB and CD, respectively. Thus, the analysis will involve only two unknown joint rotations, θ_B and θ_C.

Fixed-End Moments

$$\text{FEM}_{AB} = \text{FEM}_{BC} = \text{FEM}_{CD} = \frac{2(20)^2}{12} = 66.7 \text{ k-ft } \circlearrowleft \quad \text{or} \quad +66.7 \text{ k-ft}$$

$$\text{FEM}_{BA} = \text{FEM}_{CB} = \text{FEM}_{DC} = 66.7 \text{ k-ft } \circlearrowright \quad \text{or} \quad -66.7 \text{ k-ft}$$

Chord Rotations The specified support settlements are depicted in Fig. 16.10(b) using an exaggerated scale. The inclined dashed lines in this figure indicate the chords (not the elastic curves) of the members in the deformed positions. It can be seen from this figure that since support A does not settle but support B settles by $\frac{5}{8}$ in., the relative settlement between the two ends of member AB is $\frac{5}{8}$ in. $= 0.0521$ ft. Because the length of member AB is 20 ft, the rotation of the chord of member AB is

$$\psi_{AB} = -\frac{0.0521}{20} = -0.0026$$

in which the negative sign has been assigned to the value of ψ_{AB} to indicate that its direction is clockwise, as shown in Fig. 16.10(b). The chord rotation for member BC can be computed in a similar manner by using the settlement of supports B and C. From Fig. 16.10(b), we observe that the relative settlement between the ends of member BC is $1\frac{1}{2}$ in. $- \frac{5}{8}$ in. $= 0.875$ in. $= 0.0729$ ft, and so

$$\psi_{BC} = -\frac{0.0729}{20} = -0.00365$$

Similarly, the chord rotation for member CD is

$$\psi_{CD} = \frac{1.5 - 0.75}{(12)(20)} = 0.00313$$

Slope-Deflection Equations

$$M_{AB} = 0 \qquad\qquad\qquad\qquad\qquad\qquad\qquad\qquad\qquad \textbf{Ans.}$$

$$M_{BA} = \frac{3EI}{20}(\theta_B + 0.0026) - 100 = 0.15EI\theta_B + 0.00039EI - 100 \qquad (1)$$

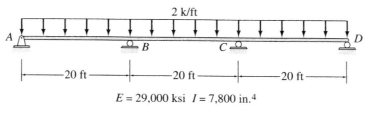

$E = 29{,}000$ ksi $I = 7{,}800$ in.⁴

(a) Continuous Beam

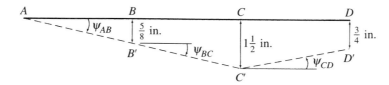

(b) Chord Rotations Due to Support Settlements

M_{BA} (B) M_{BC} M_{CB} (C) M_{CD}

(c) Free-Body Diagrams of Joints B and C

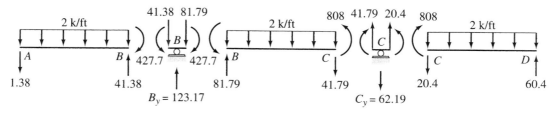

$B_y = 123.17$ $C_y = 62.19$

(d) Member End Moments and Shears

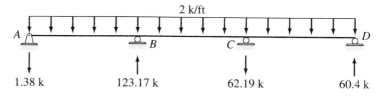

(e) Support Reactions

FIG. **16.10**

continued

$$M_{BC} = \frac{2EI}{20}[2\theta_B + \theta_C - 3(-0.00365)] + 66.7$$

$$= 0.2EI\theta_B + 0.1EI\theta_C + 0.0011EI + 66.7 \qquad (2)$$

$$M_{CB} = \frac{2EI}{20}[2\theta_C + \theta_B - 3(-0.00365)] - 66.7$$

$$= 0.1EI\theta_B + 0.2EI\theta_C + 0.0011EI - 66.7 \qquad (3)$$

$$M_{CD} = \frac{3EI}{20}(\theta_C - 0.00313) + 100 = 0.15EI\theta_C - 0.00047EI + 100 \qquad (4)$$

$$M_{DC} = 0 \qquad \text{Ans.}$$

Equilibrium Equations See Fig. 16.10(c).

$$M_{BA} + M_{BC} = 0 \qquad (5)$$

$$M_{CB} + M_{CD} = 0 \qquad (6)$$

Joint Rotations By substituting the slope-deflection equations (Eqs. (1) through (4)) into the equilibrium equations (Eqs. (5) and (6)), we obtain

$$0.35EI\theta_B + 0.1EI\theta_C = -0.00149EI + 33.3$$

$$0.1EI\theta_B + 0.35EI\theta_C = -0.00063EI - 33.3$$

Substituting $EI = (29,000)(7,800)/(12)^2$ k-ft^2 into the right sides of the above equations yields

$$0.35EI\theta_B + 0.1EI\theta_C = -2,307.24 \qquad (7)$$

$$0.1EI\theta_B + 0.35EI\theta_C = -1,022.93 \qquad (8)$$

By solving Eqs. (7) and (8) simultaneously, we determine the values of $EI\theta_B$ and $EI\theta_C$ to be

$$EI\theta_B = -6,268.81 \text{ k-ft}^2$$

$$EI\theta_C = -1,131.57 \text{ k-ft}^2$$

Member End Moments To compute the member end moments, we substitute the numerical values of $EI\theta_B$ and $EI\theta_C$ back into the slope-deflection equations (Eqs. (1) through (4)) to obtain

$$M_{BA} = -427.7 \text{ k-ft} \qquad \text{or} \qquad 427.7 \text{ k-ft} \circlearrowright \qquad \text{Ans.}$$

$$M_{BC} = 427.7 \text{ k-ft} \circlearrowleft \qquad \text{Ans.}$$

$$M_{CB} = 808 \text{ k-ft} \circlearrowleft \qquad \text{Ans.}$$

$$M_{CD} = -808 \text{ k-ft} \qquad \text{or} \qquad 808 \text{ k-ft} \circlearrowright \qquad \text{Ans.}$$

Member End Shears and Support Reactions See Fig. 16.10(d) and (e). **Ans.**

Equilibrium Check The equilibrium equations check.

We previously analyzed the continuous beam considered here in Example 13.14 by the method of consistent deformations. Theoretically, the slope-deflection method and the method of consistent deformations should yield identical results for a given structure. The small differences between the results determined here and those obtained in Example 13.14 are due to the round-off errors.

Example 16.7

Determine the reactions and draw the shear and bending moment diagrams for the four-span continuous beam shown in Fig. 16.11(a).

Solution

Because the beam and the loading are symmetric with respect to the vertical s axis passing through roller support C (Fig. 16.11(a)), the response of the complete beam can be determined by analyzing only the left half, AC, of the beam, with symmetric boundary conditions as shown in Fig. 16.11(b). Furthermore, from Fig. 16.11(b), we can see that the one-half of the beam with symmetric boundary conditions is also symmetric with respect to the s' axis passing through roller support B. Therefore, we need to analyze only one-fourth of the beam—that is, the portion AB—with symmetric boundary conditions, as shown in Fig. 16.11(c).

Since the substructure to be analyzed consists simply of the fixed beam AB (Fig. 16.11(c)), its end moments can be obtained directly from the fixed-end moment expressions given inside the back cover of the book. Thus

$$M_{AB} = \text{FEM}_{AB} = \frac{wL^2}{12} \; \circlearrowright$$

$$M_{BA} = \text{FEM}_{BA} = \frac{wL^2}{12} \; \circlearrowleft$$

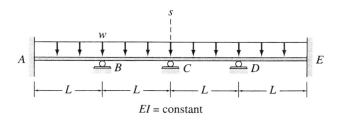

(a) Continuous Beam

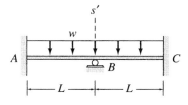

(b) One-Half Beam with Symmetric Boundary Conditions

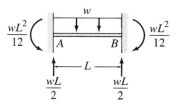

(c) One-Fourth Beam with Symmetric Boundary Conditions

FIG. 16.11

continued

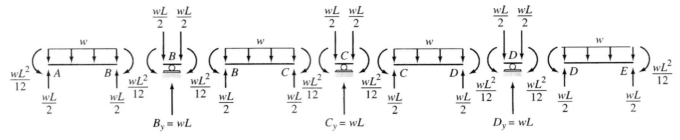

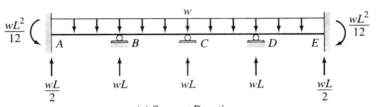

(d) Member End Moments and Shears

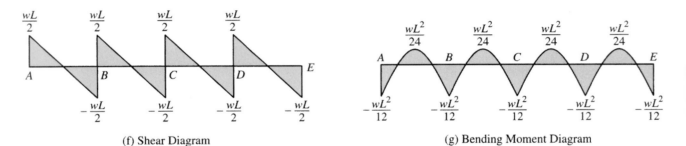

(e) Support Reactions

(f) Shear Diagram

(g) Bending Moment Diagram

FIG. 16.11 (contd.)

The shears at the ends of member AB are determined by considering the equilibrium of the member.

The shears and moments at the ends of member BC can now be obtained by reflecting the corresponding responses of member AB to the right of the s' axis, and the member end moments and shears on the right half of the beam can be determined by reflecting the corresponding responses on the left half to the other side of the s axis. The member end moments and shears thus obtained are shown in Fig. 16.11(d), and the support reactions are given in Fig. 16.11(e).

The shear and bending moment diagrams for the beam are shown in Fig. 16.11(f) and (g), respectively.   Ans.

As this example shows, the utilization of structural symmetry can considerably reduce the computational effort required in the analysis. The beam considered in this example (Fig. 16.11(a)) has three degrees of freedom, $\theta_B, \theta_C,$ and θ_D. However, by taking advantage of the structure's symmetry, we were able to eliminate all the degrees of freedom from the analysis.

16.4 ANALYSIS OF FRAMES WITHOUT SIDESWAY

The slope-deflection method can also be used for the analysis of frames. Since the axial deformations of the members of frames composed of common engineering materials are generally much smaller than the bending deformations, the axial deformations of members are neglected in the analysis, and the members are assumed to be *inextensible* (i.e., they cannot undergo any axial elongation or shortening).

Consider the frame shown in Fig. 16.12(a). A qualitative deflected shape of the frame for an arbitrary load P is also shown. From the figure, we can see that the fixed joints A and B can neither rotate nor translate, whereas joint C, which is located at the hinged support, can rotate, but it cannot translate. As for joint D, while it is free to rotate, its translation in any direction is prevented by members AD and CD, which are assumed to be inextensible. Similarly, joint E is free to rotate, but since members BE and DE cannot deform axially and since joints B and D do not translate, joint E also cannot translate. Thus none of the joints of the frame can translate.

Now suppose that we remove member CD from the frame of Fig. 16.12(a) to obtain the frame shown in Fig. 16.12(b). Since the axial deformations of columns AD and BE are neglected, joints D and E cannot translate in the vertical direction. However, there are no restraints to prevent these joints from rotating, and displacing in the horizontal direction, as shown in Fig. 16.12(b). Note that since the girder DE is assumed to be inextensible, the horizontal displacements of joints D and E must be the same.

The lateral displacements of building frames, like that of the frame of Fig. 16.12(b), are commonly referred to as *sidesways* and the frames whose joints undergo translations are termed *frames with sidesway*, whereas the frames without joint translations are called *frames without sidesway*. In applying the slope-deflection method, it is usually convenient to distinguish between the frames without sidesway (i.e., without unknown joint translations), and those with sidesway. For an arbitrary plane frame subjected to a general coplanar loading, the number of independent joint translations—which are commonly referred to as the *sidesway degrees of freedom, ss*—can be expressed as

$$ss = 2j - [2(f + h) + r + m] \qquad (16.20)$$

in which j = number of joints; f = number of fixed supports; h = number of hinged supports; r = number of roller supports; and m = number of (inextensible) members. The foregoing expression is based on the reasoning that two translations (e.g., in the horizontal and vertical directions) are needed to specify the deformed position of each free joint of a plane frame; and that each fixed and hinged support prevents both translations, each roller support prevents translation in one direction (of the joint attached to it), and each inextensible member connecting two

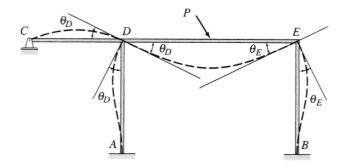

(a) Frame without Sidesway

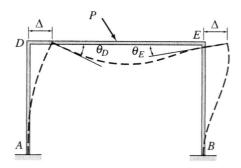

(b) Frame with Sidesway

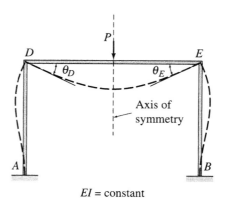

EI = constant

(c) Symmetric Frame Subjected to Symmetric
Loading — No Sidesway

FIG. 16.12

joints prevents one joint translation in its axial direction. The number of
independent joint translations, *ss*, is then obtained by subtracting from
the total number of possible translations of *j* free joints the number of
translations restrained by the supports and members of the frame. We
can verify our conclusions about the frames of Figs. 16.12(a) and (b)

by applying Eq. (16.20). Since the frame of Fig. 16.12(a) consists of five joints ($j = 5$), four members ($m = 4$), two fixed supports ($f = 2$), and one hinged support ($h = 1$), the application of Eq. (16.20) yields $ss = 2(5) - [2(2 + 1) + 4] = 0$, which indicates that this frame can be considered as without sidesway. As for the frame of Fig. 16.12(b), since it has $j = 4$, $m = 3$, and $f = 2$, the number of its sidesway degrees of freedom is given by $ss = 2(4) - [2(2) + 3] = 1$, which indicates that the frame can undergo one independent joint translation. Note that this independent joint translation is identified as the horizontal displacement Δ of joints D and E in Fig. 16.12(b).

It is important to realize that a frame may contain joints that are free to translate, but it may still be considered for analytical purposes as one without sidesway under a particular loading condition if no joint translations occur when the frame is subjected to that loading condition. An example of such a frame is shown in Fig. 16.12(c). Although joints D and E of the symmetric frame are free to translate horizontally, they will not translate when the frame is subjected to a loading that is symmetric with respect to the structure's axis of symmetry. Thus this frame, when subjected to a symmetric loading, can be analyzed as a frame without sidesway. In the following, we discuss the application of the slope-deflection method to the analysis of frames without sidesway. The analysis of frames with sidesway is considered in the next section.

The procedure for the analysis of frames without sidesway is almost identical to that for the analysis of continuous beams presented in the preceding section. This similarity occurs because, like the continuous beams, the degrees of freedom of frames without sidesway consist of only the unknown joint rotations, with the joint translations being either zero or known (as in the case of support settlements). However, unlike the continuous beams, more than two members may be connected to a joint of a frame, and the equilibrium equation for such a joint would involve more than two member end moments. The analysis of frames without sidesway is illustrated by the following examples.

Example 16.8

Determine the member end moments and reactions for the frame shown in Fig. 16.13(a) by the slope-deflection method.

Solution

Degrees of Freedom The joints C, D, and E of the frame are free to rotate. However, we will eliminate the rotation of the simple support at end E by using the modified slope-deflection equations for member DE. Thus the analysis will involve only two unknown joint rotations, θ_C and θ_D.

continued

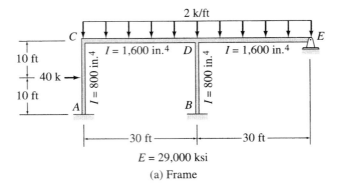

2 k/ft

C   E

I = 1,600 in.⁴   D   I = 1,600 in.⁴

10 ft

I = 800 in.⁴   40 k →   I = 800 in.⁴

10 ft

A   B

|——— 30 ft ———|——— 30 ft ———|

E = 29,000 ksi

(a) Frame

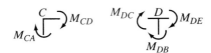

(b) Free-Body Diagrams of Joints C and D

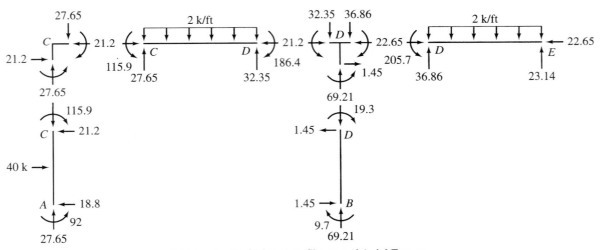

(c) Member End Moments, Shears and Axial Forces

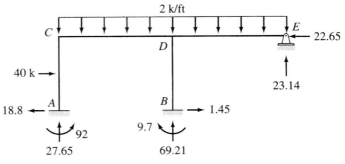

(d) Support Reactions

FIG. 16.13

Fixed-End Moments By using the fixed-end moment expressions given inside the back cover of the book, we obtain

$$\text{FEM}_{AC} = \frac{40(20)}{8} = 100 \text{ k-ft} \circlearrowright \quad \text{or} \quad +100 \text{ k-ft}$$

$$\text{FEM}_{CA} = 100 \text{ k-ft} \circlearrowleft \quad \text{or} \quad -100 \text{ k-ft}$$

$$\text{FEM}_{BD} = \text{FEM}_{DB} = 0$$

$$\text{FEM}_{CD} = \text{FEM}_{DE} = \frac{2(30)^2}{12} = 150 \text{ k-ft} \circlearrowright \quad \text{or} \quad +150 \text{ k-ft}$$

$$\text{FEM}_{DC} = \text{FEM}_{ED} = 150 \text{ k-ft} \circlearrowleft \quad \text{or} \quad -150 \text{ k-ft}$$

Slope-Deflection Equations As indicated in Fig. 16.13(a), the moments of inertia of the columns and the girders of the frame are 800 in.[4] and 1,600 in.[4], respectively. Using $I = I_{\text{column}} = 800$ in.[4] as the reference moment of inertia, we express I_{girder} in terms of I as

$$I_{\text{girder}} = 1{,}600 = 2(800) = 2I$$

Next, we write the slope-deflection equations by applying Eq. (16.9) to members $AC, BD,$ and CD, and Eqs. (16.15) to member DE. Thus

$$M_{AC} = \frac{2EI}{20}(\theta_C) + 100 = 0.1EI\theta_C + 100 \tag{1}$$

$$M_{CA} = \frac{2EI}{20}(2\theta_C) - 100 = 0.2EI\theta_C - 100 \tag{2}$$

$$M_{BD} = \frac{2EI}{20}(\theta_D) = 0.1EI\theta_D \tag{3}$$

$$M_{DB} = \frac{2EI}{20}(2\theta_D) = 0.2EI\theta_D \tag{4}$$

$$M_{CD} = \frac{2E(2I)}{30}(2\theta_C + \theta_D) + 150 = 0.267EI\theta_C + 0.133EI\theta_D + 150 \tag{5}$$

$$M_{DC} = \frac{2E(2I)}{30}(2\theta_D + \theta_C) - 150 = 0.133EI\theta_C + 0.267EI\theta_D - 150 \tag{6}$$

$$M_{DE} = \frac{3E(2I)}{30}(\theta_D) + \left(150 + \frac{150}{2}\right) = 0.2EI\theta_D + 225 \tag{7}$$

$$M_{ED} = 0 \hspace{5cm} \textbf{Ans.}$$

Equilibrium Equations By applying the moment equilibrium equation, $\sum M = 0$, to the free bodies of joints C and D (Fig. 16.13(b)), we obtain the equilibrium equations

$$M_{CA} + M_{CD} = 0 \tag{8}$$

$$M_{DB} + M_{DC} + M_{DE} = 0 \tag{9}$$

continued

Joint Rotations Substitution of the slope-deflection equations into the equilibrium equations yields

$$0.467EI\theta_C + 0.133EI\theta_D = -50 \tag{10}$$

$$0.133EI\theta_C + 0.667EI\theta_D = -75 \tag{11}$$

By solving Eqs. (10) and (11) simultaneously, we determine the values of $EI\theta_C$ and $EI\theta_D$ to be

$$EI\theta_C = -79.545 \text{ k-ft}^2$$

$$EI\theta_D = -96.591 \text{ k-ft}^2$$

Member End Moments The member end moments can now be computed by substituting the numerical values of $EI\theta_C$ and $EI\theta_D$ into the slope-deflection equations (Eqs. (1) through (7)).

$$M_{AC} = 92 \text{ k-ft} \circlearrowleft \qquad \text{Ans.}$$

$$M_{CA} = -115.9 \text{ k-ft} \qquad \text{or} \qquad 115.9 \text{ k-ft} \circlearrowright \qquad \text{Ans.}$$

$$M_{BD} = -9.7 \text{ k-ft} \qquad \text{or} \qquad 9.7 \text{ k-ft} \circlearrowright \qquad \text{Ans.}$$

$$M_{DB} = -19.3 \text{ k-ft} \qquad \text{or} \qquad 19.3 \text{ k-ft} \circlearrowright \qquad \text{Ans.}$$

$$M_{CD} = 115.9 \text{ k-ft} \circlearrowleft \qquad \text{Ans.}$$

$$M_{DC} = -186.4 \text{ k-ft} \qquad \text{or} \qquad 186.4 \text{ k-ft} \circlearrowright \qquad \text{Ans.}$$

$$M_{DE} = 205.7 \text{ k-ft} \circlearrowleft \qquad \text{Ans.}$$

To check that the solution of the simultaneous equations (Eqs. (10) and (11)) has been carried out correctly, we substitute the numerical values of member end moments back into the equilibrium equations (Eqs. (8) and (9)) to obtain

$$M_{CA} + M_{CD} = -115.9 + 115.9 = 0 \qquad \text{Checks}$$

$$M_{DB} + M_{DC} + M_{DE} = -19.3 - 186.4 + 205.7 = 0 \qquad \text{Checks}$$

Member End Shears The member end shears, obtained by considering the equilibrium of each member, are shown in Fig. 16.13(c).

Member Axial Forces With end shears known, member axial forces can now be evaluated by considering the equilibrium of joints C and D in order. The axial forces thus obtained are shown in Fig. 16.13(c).

Support Reactions See Fig. 16.13(d). Ans.

Equilibrium Check The equilibrium equations check.

Example 16.9

Determine the member end moments and reactions for the frame of Example 16.8 due to a settlement of $\frac{3}{4}$ in. at support B. Use the slope-deflection method.

Solution

The frame is shown in Fig. 16.14(a).

Degrees of Freedom θ_C and θ_D are the degrees of freedom.

Chord Rotations Since the axial deformation of member BD is neglected, the $\frac{3}{4}$-in. settlement of support B causes the joint D to displace downward by the same amount, as shown in Fig. 16.14(b). The inclined dashed lines in this figure represent the chords (not the elastic curves) of members CD and DE in the deformed positions. The rotation of the chord of member CD is

$$\psi_{CD} = -\frac{\frac{3}{4}}{(12)(30)} = -0.00208$$

in which the negative sign has been assigned to the value of ψ_{CD} to indicate that its sense is clockwise. Similarly, for member DE,

$$\psi_{DE} = 0.00208$$

Slope-Deflection Equations

$$M_{AC} = 0.1EI\theta_C \tag{1}$$

$$M_{CA} = 0.2EI\theta_C \tag{2}$$

$$M_{BD} = 0.1EI\theta_D \tag{3}$$

$$M_{DB} = 0.2EI\theta_D \tag{4}$$

$$M_{CD} = \frac{2E(2I)}{30}[2\theta_C + \theta_D - 3(-0.00208)]$$

$$= 0.267EI\theta_C + 0.133EI\theta_D + 0.000832EI \tag{5}$$

$$M_{DC} = \frac{2E(2I)}{30}[2\theta_D + \theta_C - 3(-0.00208)]$$

$$= 0.133EI\theta_C + 0.267EI\theta_D + 0.000832EI \tag{6}$$

$$M_{DE} = \frac{3E(2I)}{30}(\theta_D - 0.00208) = 0.2EI\theta_D - 0.000416EI \tag{7}$$

$$M_{ED} = 0 \qquad\qquad \textbf{Ans.}$$

Equilibrium Equations See Fig. 16.14(c).

$$M_{CA} + M_{CD} = 0 \tag{8}$$

$$M_{DB} + M_{DC} + M_{DE} = 0 \tag{9}$$

continued

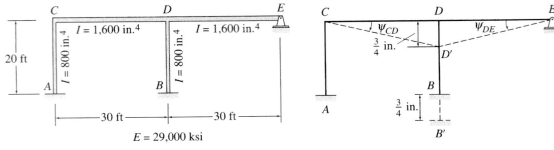

(a) Frame

(b) Chord Rotations Due to Support Settlements

$E = 29,000$ ksi

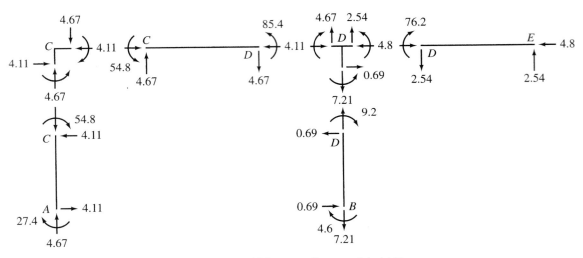

(c) Free-Body Diagrams of Joints C and D

(d) Member End Moments, Shears and Axial Forces

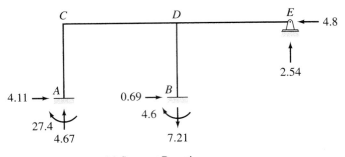

(e) Support Reactions

FIG. 16.14

Joint Rotations By substituting the slope-deflection equations into the equilibrium equations, we obtain

$$0.467EI\theta_C + 0.133EI\theta_D = -0.000832EI$$

$$0.133EI\theta_C + 0.667EI\theta_D = -0.000416EI$$

Substitution of $EI = (29{,}000)(800)/(12)^2$ k-ft^2 into the right sides of the preceding equations yields

$$0.467EI\theta_C + 0.133EI\theta_D = -134 \tag{10}$$

$$0.133EI\theta_C + 0.667EI\theta_D = -67 \tag{11}$$

Solving Eqs. (10) and (11) simultaneously, we obtain

$$EI\theta_C = -273.883 \text{ k-ft}^2$$

$$EI\theta_D = -45.838 \text{ k-ft}^2$$

Member End Moments By substituting the numerical values of $EI\theta_C$ and $EI\theta_D$ into the slope-deflection equations, we obtain

$$M_{AC} = -27.4 \text{ k-ft} \quad \text{or} \quad 27.4 \text{ k-ft} \; \circlearrowright \qquad \text{Ans.}$$

$$M_{CA} = -54.8 \text{ k-ft} \quad \text{or} \quad 54.8 \text{ k-ft} \; \circlearrowright \qquad \text{Ans.}$$

$$M_{BD} = -4.6 \text{ k-ft} \quad \text{or} \quad 4.6 \text{ k-ft} \; \circlearrowright \qquad \text{Ans.}$$

$$M_{DB} = -9.2 \text{ k-ft} \quad \text{or} \quad 9.2 \text{ k-ft} \; \circlearrowright \qquad \text{Ans.}$$

$$M_{CD} = 54.8 \text{ k-ft} \; \circlearrowleft \qquad \text{Ans.}$$

$$M_{DC} = 85.4 \text{ k-ft} \; \circlearrowleft \qquad \text{Ans.}$$

$$M_{DE} = -76.2 \text{ k-ft} \quad \text{or} \quad 76.2 \text{ k-ft} \; \circlearrowright \qquad \text{Ans.}$$

Back substitution of the numerical values of member end moments into the equilibrium equations (Eqs. (8) and (9)) yields

$$M_{CA} + M_{CD} = -54.8 + 54.8 = 0 \qquad \text{Checks}$$

$$M_{DB} + M_{DC} + M_{DE} = -9.2 + 85.4 - 76.2 = 0 \qquad \text{Checks}$$

Member End Shears and Axial Forces See Fig. 16.14(d).

Support Reactions See Fig. 16.14(e).

Equilibrium Check The equilibrium equations check.

16.5 ANALYSIS OF FRAMES WITH SIDESWAY

A frame, in general, will undergo sidesway if its joints are not restrained against translation, unless it is a symmetric frame subjected to symmetric loading. To develop the analysis of frames with sidesway, consider the rectangular frame shown in Fig. 16.15(a). A qualitative deflected shape of the frame for an arbitrary loading is also shown in the figure

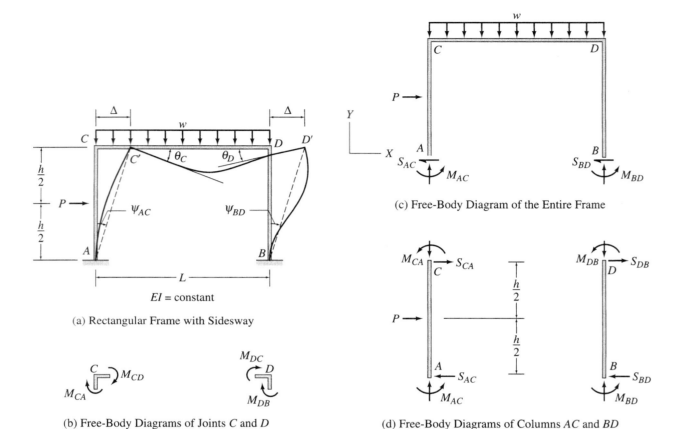

(a) Rectangular Frame with Sidesway

EI = constant

(b) Free-Body Diagrams of Joints C and D

(c) Free-Body Diagram of the Entire Frame

(d) Free-Body Diagrams of Columns AC and BD

FIG. **16.15**

using an exaggerated scale. While the fixed joints A and B of the frame are completely restrained against rotation as well as translation, the joints C and D are free to rotate and translate. However, since the columns AC and BD are assumed to be inextensible and the deformations of the frame are assumed to be small, the joints C and D can translate only in the horizontal direction—that is, in the direction perpendicular to the columns AC and BD, respectively. Furthermore, since the girder CD is also assumed to be inextensible, the horizontal displacements of joints C and D must be the same. Thus the frame has three unknown joint displacements or degrees of freedom, the rotations θ_C and θ_D of joints C and D, respectively, and the horizontal displacement Δ of both joints C and D.

As shown in Fig. 16.15(a), the displacement Δ of the joints C and D causes the chords of the columns AC and BD to rotate, and these chord rotations can be expressed in terms of the unknown displacement Δ as

$$\psi_{AC} = \psi_{BD} = -\frac{\Delta}{h} \tag{16.21}$$

in which the negative sign indicates that the chord rotations are clockwise. Since joints C and D cannot displace vertically, the chord rotation of the girder CD is zero; that is, $\psi_{CD} = 0$.

To relate the member end moments to the unknown joint displacements, θ_C, θ_D, and Δ, we write the slope-deflection equations for the three members of the frame. Thus by applying Eq. (16.9), we obtain

$$M_{AC} = \frac{2EI}{h}\left(\theta_C + \frac{3\Delta}{h}\right) + \text{FEM}_{AC} \tag{16.22a}$$

$$M_{CA} = \frac{2EI}{h}\left(2\theta_C + \frac{3\Delta}{h}\right) + \text{FEM}_{CA} \tag{16.22b}$$

$$M_{BD} = \frac{2EI}{h}\left(\theta_D + \frac{3\Delta}{h}\right) \tag{16.22c}$$

$$M_{DB} = \frac{2EI}{h}\left(2\theta_D + \frac{3\Delta}{h}\right) \tag{16.22d}$$

$$M_{CD} = \frac{2EI}{L}(2\theta_C + \theta_D) + \text{FEM}_{CD} \tag{16.22e}$$

$$M_{DC} = \frac{2EI}{L}(2\theta_D + \theta_C) + \text{FEM}_{DC} \tag{16.22f}$$

Note that the foregoing slope-deflection equations contain three unknowns, θ_C, θ_D, and Δ, which must be determined by solving three independent equations of equilibrium before the values of the member end moments can be computed. Two of the three equilibrium equations necessary for the solution of the unknown joint displacements are obtained by considering the moment equilibrium of joints C and D (Fig. 16.15(b)):

$$M_{CA} + M_{CD} = 0 \tag{16.23a}$$

$$M_{DB} + M_{DC} = 0 \tag{16.23b}$$

The third equilibrium equation, commonly termed the *shear equation*, is based on the condition that the sum of all the horizontal forces acting on the free body of the entire frame must be zero. The free-body diagram of the frame, obtained by passing an imaginary section just above the support level, is shown in Fig. 16.15(c). By applying the equilibrium equation $\sum F_X = 0$, we write

$$P - S_{AC} - S_{BD} = 0 \tag{16.23c}$$

in which S_{AC} and S_{BD} are the shears at the lower ends of the columns AC and BD, respectively, as shown in Fig. 16.15(c). To express the third equilibrium equation (Eq. (16.23c)) in terms of column end moments, we consider the equilibrium of the free bodies of the columns AC and BD shown in Fig. 16.15(d). By summing moments about the top of each column, we obtain the following:

$$+\circlearrowleft \sum M_C^{AC} = 0 \qquad M_{AC} - S_{AC}(h) + P\left(\frac{h}{2}\right) + M_{CA} = 0$$

$$S_{AC} = \frac{M_{AC} + M_{CA}}{h} + \frac{P}{2} \qquad\qquad (16.24\text{a})$$

$$+\circlearrowleft \sum M_D^{BD} = 0 \qquad M_{BD} + M_{DB} - S_{BD}(h) = 0$$

$$S_{BD} = \frac{M_{BD} + M_{DB}}{h} \qquad\qquad (16.24\text{b})$$

By substituting Eqs. (16.24a) and (16.24b) into Eq. (16.23c), we obtain the third equilibrium equation in terms of member end moments:

$$P - \left(\frac{M_{AC} + M_{CA}}{h} + \frac{P}{2}\right) - \left(\frac{M_{BD} + M_{DB}}{h}\right) = 0$$

which reduces to

$$M_{AC} + M_{CA} + M_{BD} + M_{DB} - \frac{Ph}{2} = 0 \qquad\qquad (16.25)$$

With the three equilibrium equations (Eqs. (16.23a), (16.23b), and (16.25)) now established, we can proceed with the rest of the analysis in the usual manner. By substituting the slope-deflection equations (Eqs. (16.22)) into the equilibrium equations, we obtain the system of equations that can be solved for the unknown joint displacements θ_C, θ_D, and Δ. The joint displacements thus obtained can then be back substituted into the slope-deflection equations to determine the member end moments, from which the end shears and axial forces of members and the support reactions can be computed, as discussed previously.

Frames with Inclined Legs

The analysis of frames with inclined legs is similar to that of the rectangular frames considered previously, except that when frames with inclined legs are subjected to sidesway, their horizontal members also undergo chord rotations, which must be included in the analysis. Recall from our previous discussion that the chord rotations of the horizontal members of rectangular frames, subjected to sidesway, are zero.

Consider the frame with inclined legs shown in Fig. 16.16(a). In order to analyze this frame by the slope-deflection method, we must relate the chord rotations of its three members to each other or to an independent joint translation. To that end, we subject the joint C of the

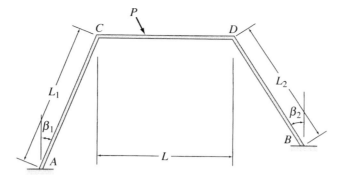

(a) Frame with Inclined Legs

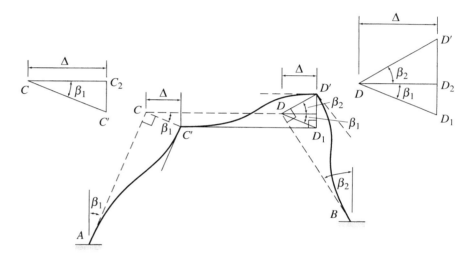

(b) Deflected Shape of the Frame Due to Sidesway

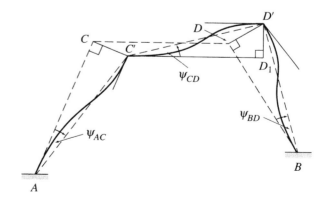

(c) Chord Rotations Due to Sidesway

FIG. 16.16

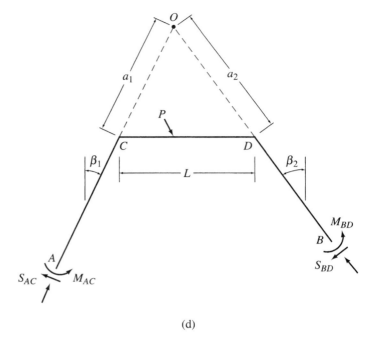

(d)

FIG. 16.16 (contd.)

frame to an arbitrary horizontal displacement Δ and draw a qualitative deflected shape of the frame, which is consistent with its support conditions as well as with our assumption that the members of the frame are inextensible. To draw the deflected shape, which is shown in Fig. 16.16(b), we first imagine that the members BD and CD are disconnected at joint D. Since member AC is assumed to be inextensible, joint C can move only in an arc about point A. Furthermore, since the translation of joint C is assumed to be small, we can consider the arc to be a straight line perpendicular to member AC.

Thus, in order to move joint C horizontally by a distance Δ, we must displace it in a direction perpendicular to member AC by a distance CC' (Fig. 16.16(b)), so that the horizontal component of CC' equals Δ. Note that although joint C is free to rotate, its rotation is ignored at this stage of the analysis, and the elastic curve AC' of member AC is drawn with the tangent at C' parallel to the undeformed direction of the member. The member CD remains horizontal and translates as a rigid body into the position $C'D_1$ with the displacement DD_1 equal to CC', as shown in the figure. Since the horizontal member CD is assumed to be inextensible and the translation of joint D is assumed to be small, the end D of this member can be moved from its deformed position D_1 only in the vertical direction. Similarly, since member BD is also assumed to be inextensible, its end D can be moved only in the direction perpendicular to the member. Therefore, to obtain the deformed position of joint D, we move the end D of member CD from its deformed position D_1 in the vertical direction and the end D of member BD in the direction perpendicular to BD, until the two ends meet at point D', where they are reconnected to obtain the displaced position D' of joint D. By assuming that joint D does not rotate, we draw the elastic curves $C'D'$ and BD', respectively, of members CD and BD, to complete the deflected shape of the entire frame.

The chord rotation of a member can be obtained by dividing the relative displacement between the two ends of the member in the direction perpendicular to the member, by the member's length. Thus we can see from Fig. 16.16(b) that the chord rotations of the three members of the frame are given by

$$\psi_{AC} = -\frac{CC'}{L_1} \qquad \psi_{BD} = -\frac{DD'}{L_2} \qquad \psi_{CD} = \frac{D_1 D'}{L} \qquad (16.26)$$

in which the chord rotations of members AC and BD are considered to be negative because they are clockwise (Fig. 16.16(c)). The three chord rotations can be expressed in terms of the joint displacement Δ by considering the displacement diagrams of joints C and D, shown in Fig. 16.16(b). Since CC' is perpendicular to AC, which is inclined at an angle β_1 with the vertical, CC' must make the same angle β_1 with the horizontal. Thus, from the displacement diagram of joint C (triangle $CC'C_2$), we can see that

$$CC' = \frac{\Delta}{\cos \beta_1} \qquad (16.27)$$

Next, let us consider the displacement diagram of joint D (triangle DD_1D'). It has been shown previously that DD_1 is equal in magnitude and parallel to CC'. Therefore,

$$DD_2 = DD_1 \cos \beta_1 = \Delta$$

Since DD' is perpendicular to member BD, it makes an angle β_2 with the horizontal. Thus, from the displacement diagram of joint D,

$$DD' = \frac{DD_2}{\cos \beta_2} = \frac{\Delta}{\cos \beta_2} \qquad (16.28)$$

and

$$D_1 D' = DD_1 \sin \beta_1 + DD' \sin \beta_2 = \frac{\Delta}{\cos \beta_1} \sin \beta_1 + \frac{\Delta}{\cos \beta_2} \sin \beta_2$$

or

$$D_1 D' = \Delta(\tan \beta_1 + \tan \beta_2) \qquad (16.29)$$

By substituting Eqs. (16.27) through (16.29) into Eq. (16.26), we obtain the chord rotations of the three members in terms of Δ:

$$\psi_{AC} = -\frac{\Delta}{L_1 \cos \beta_1} \tag{16.30a}$$

$$\psi_{BD} = -\frac{\Delta}{L_2 \cos \beta_2} \tag{16.30b}$$

$$\psi_{CD} = \frac{\Delta}{L}(\tan \beta_1 + \tan \beta_2) \tag{16.30c}$$

The foregoing expressions of chord rotations can be used to write the slope-deflection equations, thereby relating member end moments to the three unknown joint displacements, θ_C, θ_D, and Δ. As in the case of the rectangular frames considered previously, the three equilibrium equations necessary for the solution of the unknown joint displacements can be established by summing the moments acting on joints C and D and by summing the horizontal forces acting on the entire frame. However, for frames with inclined legs, it is usually more convenient to establish the third equilibrium equation by summing the moments of all the forces and couples acting on the entire frame about a moment center O, which is located at the intersection of the longitudinal axes of the two inclined members, as shown in Fig. 16.16(d). The location of the moment center O can be determined by using the conditions (see Fig. 16.16(d))

$$a_1 \cos \beta_1 = a_2 \cos \beta_2 \tag{16.31a}$$

$$a_1 \sin \beta_1 + a_2 \sin \beta_2 = L \tag{16.31b}$$

By solving Eqs. (16.31a) and (16.31b) simultaneously for a_1 and a_2, we obtain

$$a_1 = \frac{L}{\cos \beta_1 (\tan \beta_1 + \tan \beta_2)} \tag{16.32a}$$

$$a_2 = \frac{L}{\cos \beta_2 (\tan \beta_1 + \tan \beta_2)} \tag{16.32b}$$

Once the equilibrium equations have been established, the analysis can be completed in the usual manner, as discussed previously.

Multistory Frames

The foregoing method can be extended to the analysis of multistory frames subjected to sidesway, as illustrated by Example 16.12. However, because of the considerable amount of computational effort involved, the analysis of such structures today is performed on computers using the matrix formulation of the displacement method presented in Chapter 18.

Example 16.10

Determine the member end moments and reactions for the frame shown in Fig. 16.17(a) by the slope-deflection method.

Solution

Degrees of Freedom The degrees of freedom are θ_C, θ_D, and Δ (see Fig. 16.17(b)).

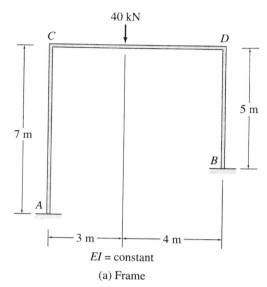

(a) Frame

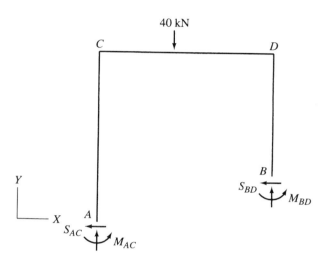

(c) Free-Body Diagram of the Entire Frame

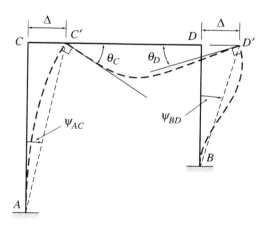

(b) Qualitative Deflected Shape of the Frame

(d) Free-Body Diagrams of Columns AC and BD

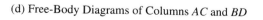

FIG. 16.17

continued

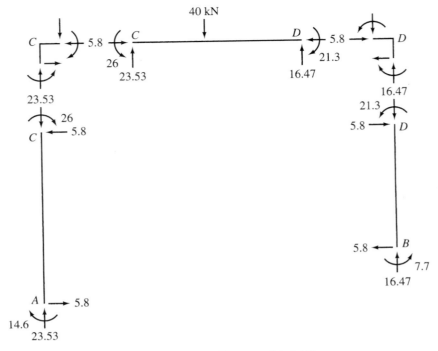

(e) Member End Moments, Shears, and Axial Forces

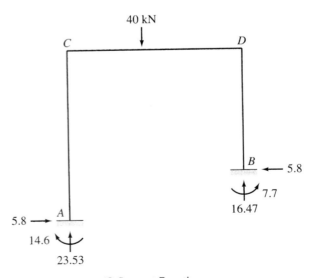

(f) Support Reactions

FIG. 16.17 (contd.)

Fixed-End Moments By using the fixed-end moment expressions given inside the back cover of the book, we obtain

$$\text{FEM}_{CD} = \frac{40(3)(4)^2}{(7)^2} = 39.2 \text{ kN} \cdot \text{m} \circlearrowright \quad \text{or} \quad +39.2 \text{ kN} \cdot \text{m}$$

$$\text{FEM}_{DC} = \frac{40(3)^2(4)}{(7)^2} = 29.4 \text{ kN} \cdot \text{m} \circlearrowleft \quad \text{or} \quad -29.4 \text{ kN} \cdot \text{m}$$

$$\text{FEM}_{AC} = \text{FEM}_{CA} = \text{FEM}_{BD} = \text{FEM}_{DB} = 0$$

Chord Rotations From Fig. 16.17(b), we can see that

$$\psi_{AC} = -\frac{\Delta}{7} \qquad \psi_{BD} = -\frac{\Delta}{5} \qquad \psi_{CD} = 0$$

Slope-Deflection Equations

$$M_{AC} = \frac{2EI}{7}\left[\theta_C - 3\left(-\frac{\Delta}{7}\right)\right] = 0.286EI\theta_C + 0.122EI\Delta \tag{1}$$

$$M_{CA} = \frac{2EI}{7}\left[2\theta_C - 3\left(-\frac{\Delta}{7}\right)\right] = 0.571EI\theta_C + 0.122EI\Delta \tag{2}$$

$$M_{BD} = \frac{2EI}{5}\left[\theta_D - 3\left(-\frac{\Delta}{5}\right)\right] = 0.4EI\theta_D + 0.24EI\Delta \tag{3}$$

$$M_{DB} = \frac{2EI}{5}\left[2\theta_D - 3\left(-\frac{\Delta}{5}\right)\right] = 0.8EI\theta_D + 0.24EI\Delta \tag{4}$$

$$M_{CD} = \frac{2EI}{7}(2\theta_C + \theta_D) + 39.2 = 0.571EI\theta_C + 0.286EI\theta_D + 39.2 \tag{5}$$

$$M_{DC} = \frac{2EI}{7}(\theta_C + 2\theta_D) - 29.4 = 0.286EI\theta_C + 0.571EI\theta_D - 29.4 \tag{6}$$

Equilibrium Equations By considering the moment equilibrium of joints C and D, we obtain the equilibrium equations

$$M_{CA} + M_{CD} = 0 \tag{7}$$

$$M_{DB} + M_{DC} = 0 \tag{8}$$

To establish the third equilibrium equation, we apply the force equilibrium equation $\sum F_X = 0$ to the free body of the entire frame (Fig. 16.17(c)), to obtain

$$S_{AC} + S_{BD} = 0$$

in which S_{AC} and S_{BD} represent the shears at the lower ends of columns AC and BD, respectively, as shown in Fig. 16.17(c). To express the column end shears in terms of column end moments, we draw the free-body diagrams of the two columns (Fig. 16.17(d)) and sum the moments about the top of each column:

$$S_{AC} = \frac{M_{AC} + M_{CA}}{7} \qquad \text{and} \qquad S_{BD} = \frac{M_{BD} + M_{DB}}{5}$$

continued

By substituting these equations into the third equilibrium equation, we obtain

$$\frac{M_{AC} + M_{CA}}{7} + \frac{M_{BD} + M_{DB}}{5} = 0$$

which can be rewritten as

$$5(M_{AC} + M_{CA}) + 7(M_{BD} + M_{DB}) = 0 \qquad (9)$$

Joint Displacements To determine the unknown joint displacements θ_C, θ_D, and Δ, we substitute the slope-deflection equations (Eqs. (1) through (6)) into the equilibrium equations (Eqs. (7) through (9)) to obtain

$$1.142EI\theta_C + 0.286EI\theta_D + 0.122EI\Delta = -39.2 \qquad (10)$$

$$0.286EI\theta_C + 1.371EI\theta_D + 0.24EI\Delta = 29.4 \qquad (11)$$

$$4.285EI\theta_C + 8.4EI\theta_D + 4.58EI\Delta = 0 \qquad (12)$$

Solving Eqs. (10) through (12) simultaneously yields

$$EI\theta_C = -40.211 \text{ kN} \cdot \text{m}^2$$

$$EI\theta_D = 34.24 \text{ kN} \cdot \text{m}^2$$

$$EI\Delta = -25.177 \text{ kN} \cdot \text{m}^3$$

Member End Moments By substituting the numerical values of $EI\theta_C, EI\theta_D$, and $EI\Delta$ into the slope-deflection equations (Eqs. (1) through (6)), we obtain

$$M_{AC} = -14.6 \text{ kN} \cdot \text{m} \qquad \text{or} \qquad 14.6 \text{ kN} \cdot \text{m} \;\circlearrowright \qquad \text{Ans.}$$

$$M_{CA} = -26 \text{ kN} \cdot \text{m} \qquad \text{or} \qquad 26 \text{ kN} \cdot \text{m} \;\circlearrowright \qquad \text{Ans.}$$

$$M_{BD} = 7.7 \text{ kN} \cdot \text{m} \;\circlearrowleft \qquad \text{Ans.}$$

$$M_{DB} = 21.3 \text{ kN} \cdot \text{m} \;\circlearrowleft \qquad \text{Ans.}$$

$$M_{CD} = 26 \text{ kN} \cdot \text{m} \;\circlearrowleft \qquad \text{Ans.}$$

$$M_{DC} = -21.3 \text{ kN} \cdot \text{m} \qquad \text{or} \qquad 21.3 \text{ kN} \cdot \text{m} \;\circlearrowright \qquad \text{Ans.}$$

To check that the solution of the simultaneous equations (Eqs. (10) through (12)) has been carried out correctly, we substitute the numerical values of member end moments back into the equilibrium equations (Eqs. (7) through (9)):

$$M_{CA} + M_{CD} = -26 + 26 = 0 \qquad \text{Checks}$$

$$M_{DB} + M_{DC} = 21.3 - 21.3 = 0 \qquad \text{Checks}$$

$$5(M_{AC} + M_{CA}) + 7(M_{BD} + M_{DB}) = 5(-14.6 - 26) + 7(7.7 + 21.3) = 0 \qquad \text{Checks}$$

Member End Shears The member end shears, obtained by considering the equilibrium of each member, are shown in Fig. 16.17(e).

Member Axial Forces With end shears known, member axial forces can now be evaluated by considering the equilibrium of joints C and D. The axial forces thus obtained are shown in Fig. 16.17(e).

Support Reactions See Fig. 16.17(f). Ans.

Equilibrium Check The equilibrium equations check.

Example 16.11

Determine the member end moments and reactions for the frame shown in Fig. 16.18(a) by the slope-deflection method.

Solution

Degrees of Freedom Degrees of freedom are θ_C, θ_D, and Δ.

Fixed-End Moments Since no external loads are applied to the members, the fixed-end moments are zero.

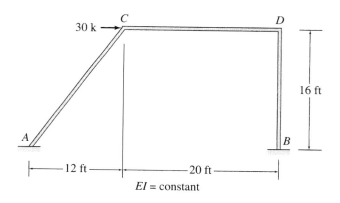

(a) Frame

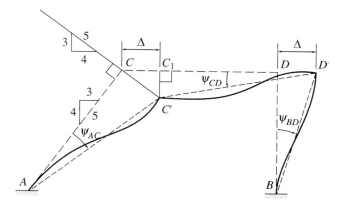

(b) Chord Rotations Due to Sidesway

FIG. **16.18**

continued

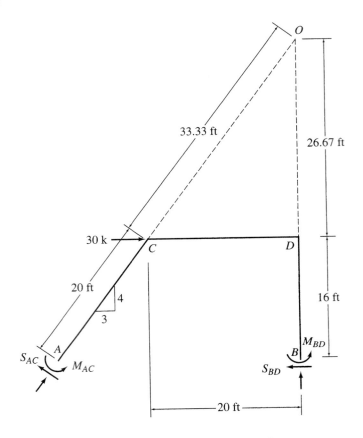

(c) Free-Body Diagram of the Entire Frame

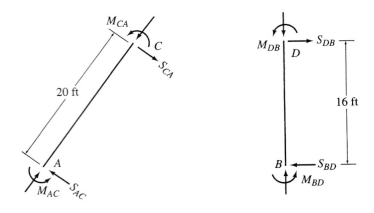

FIG. **16.18** (contd.)

(d) Free-Body Diagrams of Columns *AC* and *BD*

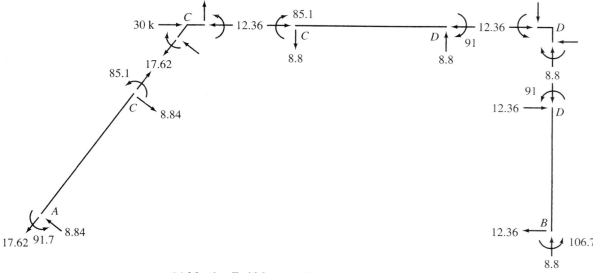

(e) Member End Moments, Shears, and Axial Forces

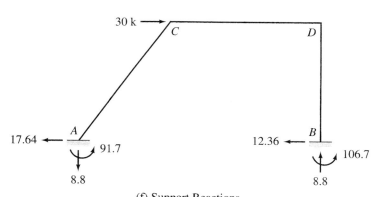

(f) Support Reactions

FIG. 16.18 (contd.)

Chord Rotations From Fig. 16.18(b), we can see that

$$\psi_{AC} = -\frac{CC'}{20} = -\frac{\left(\frac{5}{4}\right)\Delta}{20} = -0.0625\Delta$$

$$\psi_{BD} = -\frac{DD'}{16} = -\frac{\Delta}{16} = -0.0625\Delta$$

$$\psi_{CD} = \frac{C'C_1}{20} = \frac{\left(\frac{3}{4}\right)\Delta}{20} = 0.0375\Delta$$

continued

Slope-Deflection Equations

$$M_{AC} = \frac{2EI}{20}[\theta_C - 3(-0.0625\Delta)] = 0.1EI\theta_C + 0.0188EI\Delta \tag{1}$$

$$M_{CA} = \frac{2EI}{20}[2\theta_C - 3(-0.0625\Delta)] = 0.2EI\theta_C + 0.0188EI\Delta \tag{2}$$

$$M_{BD} = \frac{2EI}{16}[\theta_D - 3(-0.0625\Delta)] = 0.125EI\theta_D + 0.0234EI\Delta \tag{3}$$

$$M_{DB} = \frac{2EI}{16}[2\theta_D - 3(-0.0625\Delta)] = 0.25EI\theta_D + 0.0234EI\Delta \tag{4}$$

$$M_{CD} = \frac{2EI}{20}[2\theta_C + \theta_D - 3(0.0375\Delta)] = 0.2EI\theta_C + 0.1EI\theta_D - 0.0113EI\Delta \tag{5}$$

$$M_{DC} = \frac{2EI}{20}[2\theta_D + \theta_C - 3(0.0375\Delta)] = 0.2EI\theta_D + 0.1EI\theta_C - 0.0113EI\Delta \tag{6}$$

Equilibrium Equations By considering the moment equilibrium of joints C and D, we obtain the equilibrium equations

$$M_{CA} + M_{CD} = 0 \tag{7}$$

$$M_{DB} + M_{DC} = 0 \tag{8}$$

The third equilibrium equation is established by summing the moments of all the forces and couples acting on the free body of the entire frame about point O, which is located at the intersection of the longitudinal axes of the two columns, as shown in Fig. 16.18(c). Thus

$$+\circlearrowleft \sum M_O = 0 \qquad M_{AC} - S_{AC}(53.33) + M_{BD} - S_{BD}(42.67) + 30(26.67) = 0$$

in which the shears at the lower ends of the columns can be expressed in terms of column end moments as (see Fig. 16.18(d))

$$S_{AC} = \frac{M_{AC} + M_{CA}}{20} \qquad \text{and} \qquad S_{BD} = \frac{M_{BD} + M_{DB}}{16}$$

By substituting these expressions into the third equilibrium equation, we obtain

$$1.67M_{AC} + 2.67M_{CA} + 1.67M_{BD} + 2.67M_{DB} = 800 \tag{9}$$

Joint Displacements Substitution of the slope-deflection equations (Eqs. (1) through (6)) into the equilibrium equations (Eqs. (7) through (9)) yields

$$0.4EI\theta_C + 0.1EI\theta_D + 0.0075EI\Delta = 0 \tag{10}$$

$$0.1EI\theta_C + 0.45EI\theta_D + 0.0121EI\Delta = 0 \tag{11}$$

$$0.71EI\theta_C + 0.877EI\theta_D + 0.183EI\Delta = 800 \tag{12}$$

By solving Eqs. (10) through (12) simultaneously, we determine

$$EI\theta_C = -66.648 \text{ k-ft}^2$$

$$EI\theta_D = -125.912 \text{ k-ft}^2$$

$$EI\Delta = 5{,}233.6 \text{ k-ft}^3$$

Member End Moments By substituting the numerical values of $EI\theta_C, EI\theta_D$, and $EI\Delta$ into the slope-deflection equations (Eqs. (1) through (6)), we obtain

$$M_{AC} = 91.7 \text{ k-ft} \curvearrowright \qquad \text{Ans.}$$

$$M_{CA} = 85.1 \text{ k-ft} \curvearrowright \qquad \text{Ans.}$$

$$M_{BD} = 106.7 \text{ k-ft} \curvearrowright \qquad \text{Ans.}$$

$$M_{DB} = 91 \text{ k-ft} \curvearrowright \qquad \text{Ans.}$$

$$M_{CD} = -85.1 \text{ k-ft} \qquad \text{or} \qquad 85.1 \text{ k-ft} \curvearrowleft \qquad \text{Ans.}$$

$$M_{DC} = -91 \text{ k-ft} \qquad \text{or} \qquad 91 \text{ k-ft} \curvearrowleft \qquad \text{Ans.}$$

Back substitution of the numerical values of member end moments into the equilibrium equations yields

$$M_{CA} + M_{CD} = 85.1 - 85.1 = 0 \qquad \text{Checks}$$

$$M_{DB} + M_{DC} = 91 - 91 = 0 \qquad \text{Checks}$$

$$1.67M_{AC} + 2.67M_{CA} + 1.67M_{BD} + 2.67M_{DB} = 1.67(91.7) + 2.67(85.1)$$

$$+ 1.67(106.7) + 2.67(91)$$

$$= 801.5 \approx 800 \qquad \text{Checks}$$

Member End Shears and Axial Forces See Fig. 16.18(e).

Support Reactions See Fig. 16.18(f). Ans.

Equilibrium Check The equilibrium equations check.

Example 16.12

Determine the member end moments, the support reactions, and the horizontal deflection of joint F of the two-story frame shown in Fig. 16.19(a) by the slope-deflection method.

Solution

Degrees of Freedom From Fig. 16.19(a), we can see that the joints C, D, E, and F of the frame are free to rotate, and translate in the horizontal direction. As shown in Fig. 16.19(b), the horizontal displacement of the first-story joints C and D is designated as Δ_1, whereas the horizontal displacement of the second-story joints E and F is expressed as $\Delta_1 + \Delta_2$, with Δ_2 representing the displacement of the second-story joints relative to the first-story joints. Thus, the frame has six degrees of freedom—that is, $\theta_C, \theta_D, \theta_E, \theta_F, \Delta_1$, and Δ_2.

Fixed-End Moments The nonzero fixed-end moments are

$$\text{FEM}_{CD} = \text{FEM}_{EF} = 200 \text{ k-ft}$$

$$\text{FEM}_{DC} = \text{FEM}_{FE} = -200 \text{ k-ft}$$

continued

Chord Rotations See Fig. 16.19(b).

$$\psi_{AC} = \psi_{BD} = -\frac{\Delta_1}{20}$$

$$\psi_{CE} = \psi_{DF} = -\frac{\Delta_2}{20}$$

$$\psi_{CD} = \psi_{EF} = 0$$

Slope-Deflection Equations Using $I_{\text{column}} = I$ and $I_{\text{girder}} = 2I$, we write

$$M_{AC} = 0.1EI\theta_C + 0.015EI\Delta_1 \tag{1}$$

$$M_{CA} = 0.2EI\theta_C + 0.015EI\Delta_1 \tag{2}$$

$$M_{BD} = 0.1EI\theta_D + 0.015EI\Delta_1 \tag{3}$$

$$M_{DB} = 0.2EI\theta_D + 0.015EI\Delta_1 \tag{4}$$

$$M_{CE} = 0.2EI\theta_C + 0.1EI\theta_E + 0.015EI\Delta_2 \tag{5}$$

$$M_{EC} = 0.2EI\theta_E + 0.1EI\theta_C + 0.015EI\Delta_2 \tag{6}$$

$$M_{DF} = 0.2EI\theta_D + 0.1EI\theta_F + 0.015EI\Delta_2 \tag{7}$$

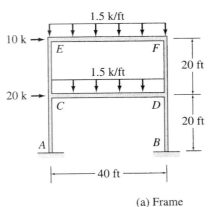

$E = 29{,}000$ ksi
$I_{\text{column}} = 1{,}000$ in.^{4}
$I_{\text{girder}} = 2{,}000$ in.^{4}

(a) Frame

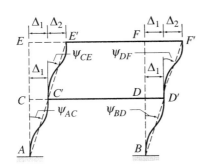

(b) Chord Rotations Due to Sidesway

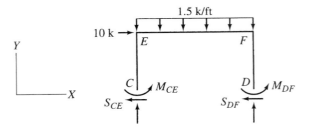

(c) Free-Body Diagram of the Top Story

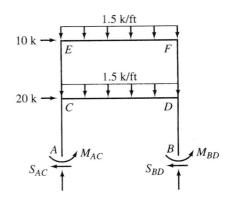

(d) Free-Body Diagram of the Entire Frame

FIG. 16.19

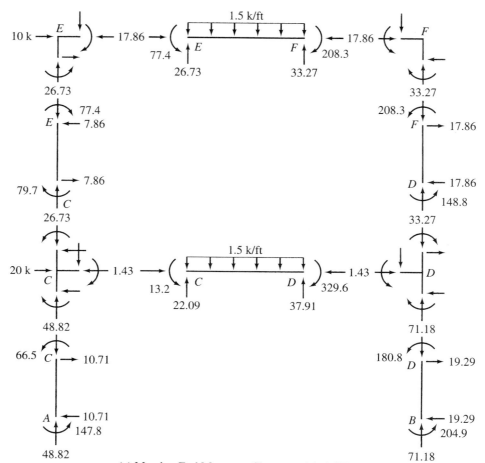

(e) Member End Moments, Shears, and Axial Forces

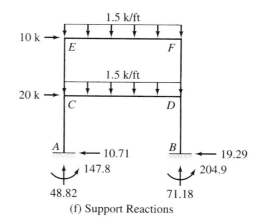

(f) Support Reactions

FIG. **16.19** (contd.)

continued

$$M_{FD} = 0.2EI\theta_F + 0.1EI\theta_D + 0.015EI\Delta_2 \tag{8}$$

$$M_{CD} = 0.2EI\theta_C + 0.1EI\theta_D + 200 \tag{9}$$

$$M_{DC} = 0.2EI\theta_D + 0.1EI\theta_C - 200 \tag{10}$$

$$M_{EF} = 0.2EI\theta_E + 0.1EI\theta_F + 200 \tag{11}$$

$$M_{FE} = 0.2EI\theta_F + 0.1EI\theta_E - 200 \tag{12}$$

Equilibrium Equations By considering the moment equilibrium of joints C, D, E, and F, we obtain

$$M_{CA} + M_{CD} + M_{CE} = 0 \tag{13}$$

$$M_{DB} + M_{DC} + M_{DF} = 0 \tag{14}$$

$$M_{EC} + M_{EF} = 0 \tag{15}$$

$$M_{FD} + M_{FE} = 0 \tag{16}$$

To establish the remaining two equilibrium equations, we successively pass a horizontal section just above the lower ends of the columns of each story of the frame and apply the equation of horizontal equilibrium ($\sum F_X = 0$) to the free body of the portion of the frame above the section. The free-body diagrams thus obtained are shown in Fig. 16.19(c) and (d). By applying the equilibrium equation $\sum F_X = 0$ to the top story of the frame (Fig. 16.19(c)), we obtain

$$S_{CE} + S_{DF} = 10$$

Similarly, by applying $\sum F_X = 0$ to the entire frame (Fig. 16.19(d)), we write

$$S_{AC} + S_{BD} = 30$$

By expressing column end shears in terms of column end moments as

$$S_{AC} = \frac{M_{AC} + M_{CA}}{20} \qquad S_{BD} = \frac{M_{BD} + M_{DB}}{20}$$

$$S_{CE} = \frac{M_{CE} + M_{EC}}{20} \qquad S_{DF} = \frac{M_{DF} + M_{FD}}{20}$$

and by substituting these expressions into the force equilibrium equations, we obtain

$$M_{CE} + M_{EC} + M_{DF} + M_{FD} = 200 \tag{17}$$

$$M_{AC} + M_{CA} + M_{BD} + M_{DB} = 600 \tag{18}$$

Joint Displacements Substitution of the slope-deflection equations (Eqs. (1) through (12)) into the equilibrium equations (Eqs. (13) through (18)) yields

$$0.6EI\theta_C + 0.1EI\theta_D + 0.1EI\theta_E + 0.015EI\Delta_1 + 0.015EI\Delta_2 = -200 \tag{19}$$

$$0.1EI\theta_C + 0.6EI\theta_D + 0.1EI\theta_F + 0.015EI\Delta_1 + 0.015EI\Delta_2 = 200 \tag{20}$$

$$0.1EI\theta_C + 0.4EI\theta_E + 0.1EI\theta_F + 0.015EI\Delta_2 = -200 \tag{21}$$

$$0.1EI\theta_D + 0.1EI\theta_E + 0.4EI\theta_F + 0.015EI\Delta_2 = 200 \tag{22}$$

$$0.3EI\theta_C + 0.3EI\theta_D + 0.3EI\theta_E + 0.3EI\theta_F + 0.06EI\Delta_2 = 200 \tag{23}$$

$$0.1EI\theta_C + 0.1EI\theta_D + 0.02EI\Delta_1 = 200 \tag{24}$$

By solving Eqs. (19) through (24) by the Gauss-Jordan elimination method (Appendix B), we determine

$$EI\theta_C = -812.988 \text{ k-ft}^2$$

$$EI\theta_D = -241.556 \text{ k-ft}^2$$

$$EI\theta_E = -789.612 \text{ k-ft}^2$$

$$EI\theta_F = 353.248 \text{ k-ft}^2$$

$$EI\Delta_1 = 15{,}272.728 \text{ k-ft}^3 \qquad \text{or} \qquad \Delta_1 = 0.0758 \text{ ft} = 0.91 \text{ in.} \rightarrow$$

$$EI\Delta_2 = 10{,}787.878 \text{ k-ft}^3 \qquad \text{or} \qquad \Delta_2 = 0.0536 \text{ ft} = 0.643 \text{ in.} \rightarrow$$

Thus, the horizontal deflection of joint F of the frame is as follows:

$$\Delta_F = \Delta_1 + \Delta_2 = 0.91 + 0.643 = 1.553 \text{ in.} \rightarrow \qquad \textbf{Ans.}$$

Member End Moments By substituting the numerical values of the joint displacements into the slope-deflection equations (Eqs. (1) through (12)), we obtain

$$M_{AC} = 147.8 \text{ k-ft } \circlearrowright \qquad\qquad\qquad \textbf{Ans.}$$

$$M_{CA} = 66.5 \text{ k-ft } \circlearrowright \qquad\qquad\qquad \textbf{Ans.}$$

$$M_{BD} = 204.9 \text{ k-ft } \circlearrowright \qquad\qquad\qquad \textbf{Ans.}$$

$$M_{DB} = 180.8 \text{ k-ft } \circlearrowright \qquad\qquad\qquad \textbf{Ans.}$$

$$M_{CE} = -79.7 \text{ k-ft} \qquad \text{or} \qquad 79.7 \text{ k-ft } \circlearrowleft \qquad \textbf{Ans.}$$

$$M_{EC} = -77.4 \text{ k-ft} \qquad \text{or} \qquad 77.4 \text{ k-ft } \circlearrowleft \qquad \textbf{Ans.}$$

$$M_{DF} = 148.8 \text{ k-ft } \circlearrowright \qquad\qquad\qquad \textbf{Ans.}$$

$$M_{FD} = 208.3 \text{ k-ft } \circlearrowright \qquad\qquad\qquad \textbf{Ans.}$$

$$M_{CD} = 13.2 \text{ k-ft } \circlearrowright \qquad\qquad\qquad \textbf{Ans.}$$

$$M_{DC} = -329.6 \text{ k-ft} \qquad \text{or} \qquad 329.6 \text{ k-ft } \circlearrowleft \qquad \textbf{Ans.}$$

$$M_{EF} = 77.4 \text{ k-ft } \circlearrowright \qquad\qquad\qquad \textbf{Ans.}$$

$$M_{FE} = -208.3 \text{ k-ft} \qquad \text{or} \qquad 208.3 \text{ k-ft } \circlearrowleft \qquad \textbf{Ans.}$$

Back substitution of the numerical values of member end moments into the equilibrium equations yields

$$M_{CA} + M_{CD} + M_{CE} = 66.5 + 13.2 - 79.7 = 0 \qquad \text{Checks}$$

$$M_{DB} + M_{DC} + M_{DF} = 180.8 - 329.6 + 148.8 = 0 \qquad \text{Checks}$$

$$M_{EC} + M_{EF} = -77.4 + 77.4 = 0 \qquad \text{Checks}$$

$$M_{FD} + M_{FE} = 208.3 - 208.3 = 0 \qquad \text{Checks}$$

$$M_{CE} + M_{EC} + M_{DF} + M_{FD} = -79.7 - 77.4 + 148.8 + 208.3 = 200 \qquad \text{Checks}$$

$$M_{AC} + M_{CA} + M_{BD} + M_{DB} = 147.8 + 66.5 + 204.9 + 180.8 = 600 \qquad \text{Checks}$$

Member End Shears and Axial Forces See Fig. 16.19(e).

Support Reactions See Fig. 16.19(f). \qquad\qquad\qquad\qquad\qquad\qquad **Ans.**

Equilibrium Check The equilibrium equations check.

SUMMARY

In this chapter, we have studied a classical formulation of the displacement (stiffness) method, called the slope-deflection method, for the analysis of beams and frames. The method is based on the slope-deflection equation:

$$M_{nf} = \frac{2EI}{L}(2\theta_n + \theta_f - 3\psi) + \text{FEM}_{nf} \qquad (16.9)$$

which relates the moments at the ends of a member to the rotations and displacements of its ends and the external loads applied to the member.

The procedure for analysis essentially involves (1) identifying the unknown joint displacements (degrees of freedom) of the structure; (2) for each member, writing slope-deflection equations relating member end moments to the unknown joint displacements; (3) establishing the equations of equilibrium of the structure in terms of member end moments; (4) substituting the slope-deflection equations into the equilibrium equations and solving the resulting system of equations to determine the unknown joint displacements; and (5) computing member end moments by substituting the values of joint displacements back into the slope-deflection equations. Once member end moments have been evaluated, member end shears and axial forces, and support reactions, can be determined through equilibrium considerations.

PROBLEMS

Section 16.3

16.1 through 16.5 Determine the reactions and draw the shear and bending moment diagrams for the beams shown in Figs. P16.1–P16.5 by using the slope-deflection method.

16.6 Solve Problem 16.2 for the loading shown in Fig. P16.2 and a settlement of $\frac{1}{2}$ in. at support B.

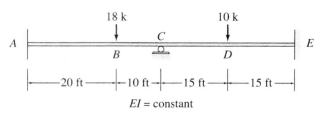

EI = constant

FIG. P16.1

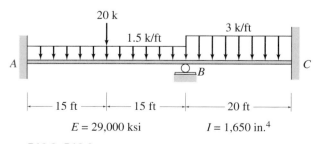

E = 29,000 ksi I = 1,650 in.^{4}

FIG. P16.2, P16.6

16.7 Solve Problem 16.4 for the loading shown in Fig. P16.4 and the support settlements of 50 mm at *B* and 25 mm at *C*.

16.8 through 16.14 Determine the reactions and draw the shear and bending moment diagrams for the beams shown in Figs. P16.8–P16.14 by using the slope-deflection method.

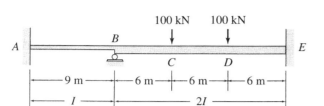

FIG. **P16.3**

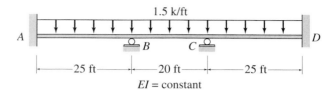

FIG. **P16.8**

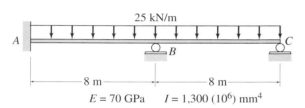

FIG. **P16.4, P16.7**

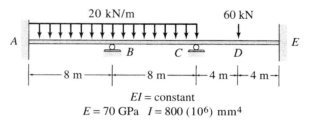

FIG. **P16.9, P16.15**

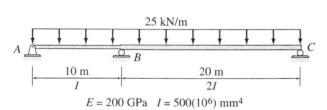

FIG. **P16.5**

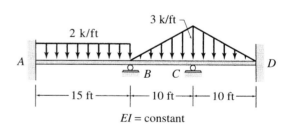

FIG. **P16.10**

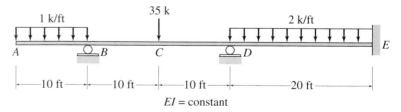

FIG. **P16.11**

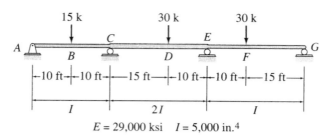

$E = 29,000$ ksi $I = 5,000$ in.^{4}

FIG. **P16.12, P16.16**

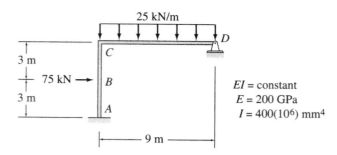

FIG. **P16.17, P16.21**

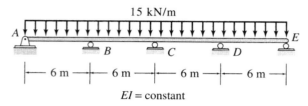

EI = constant

FIG. **P16.13**

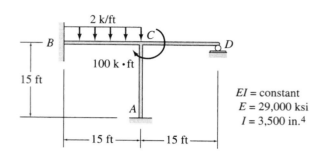

FIG. **P16.18, P16.22**

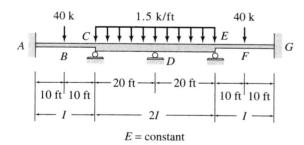

E = constant

FIG. **P16.14**

16.15 Solve Problem 16.9 for the loading shown in Fig. P16.9 and a settlement of 25 mm at support C.

16.16 Solve Problem 16.12 for the loading shown in Fig. P16.12 and support settlements of $\frac{1}{2}$ in. at A; 4 in. at C; 3 in. at E; and $2\frac{1}{2}$ in. at G.

Section 16.4

16.17 through 16.20 Determine the member end moments and reactions for the frames shown in Figs. P16.17–P16.20 by using the slope-deflection method.

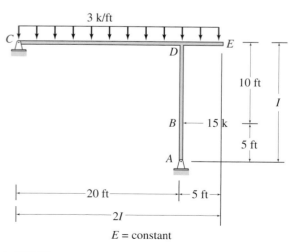

E = constant

FIG. **P16.19**

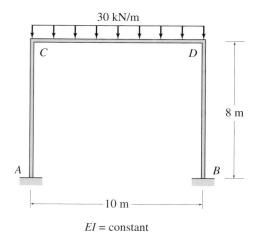

$EI = $ constant

FIG. P16.20

16.21 Solve Problem 16.17 for the loading shown in Fig. P16.17 and a settlement of 50 mm at support D.

16.22 Solve Problem 16.18 for the loading shown in Fig. P16.18 and a settlement of $\frac{1}{4}$ in. at support A.

16.23 Determine the member end moments and reactions for the frame in Fig. P16.23 for the loading shown and the support settlements of 1 in. at A and $1\frac{1}{2}$ in. at D. Use the slope-deflection method.

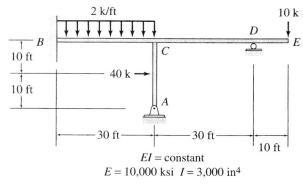

$EI = $ constant
$E = 10,000$ ksi $I = 3,000$ in⁴

FIG. P16.23

Section 16.5

16.24 through 16.31 Determine the member end moments and reactions for the frames shown in Figs. P16.24–P16.31 by using the slope-deflection method.

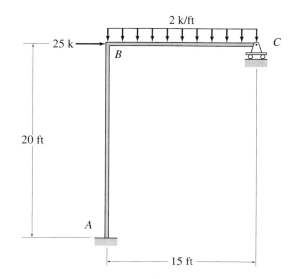

$EI = $ constant

FIG. P16.24

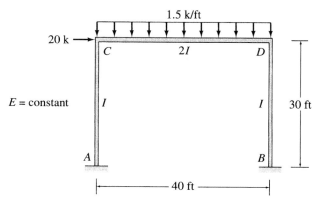

FIG. P16.25

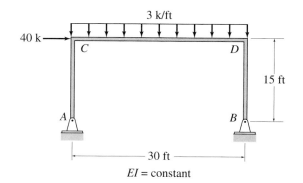

$EI = $ constant

FIG. P16.26

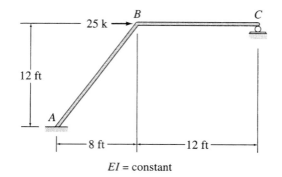

FIG. **P16.27**

25 k →

B

C

12 ft

A

8 ft

12 ft

EI = constant

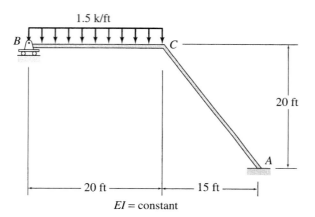

1.5 k/ft

B

C

20 ft

A

20 ft

15 ft

EI = constant

FIG. **P16.28**

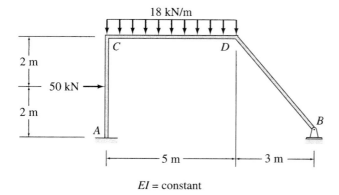

18 kN/m

C

D

2 m

50 kN →

2 m

A

B

5 m

3 m

EI = constant

FIG. **P16.29**

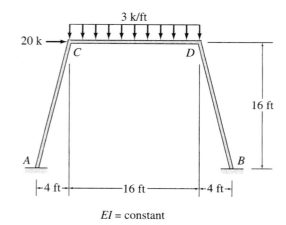

3 k/ft

20 k →

C

D

16 ft

A

B

4 ft

16 ft

4 ft

EI = constant

FIG. **P16.30**

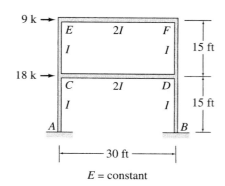

9 k →

E $2I$ F

I I 15 ft

18 k →

C $2I$ D

I I 15 ft

A   B

30 ft

E = constant

FIG. **P16.31**

The Empire State Building,
New York
Photo courtesy of Bethlehem Steel Corporation

17

Moment-Distribution Method

17.1 Definitions and Terminology
17.2 Basic Concept of the Moment-Distribution Method
17.3 Analysis of Continuous Beams
17.4 Analysis of Frames without Sidesway
17.5 Analysis of Frames with Sidesway
Summary
Problems

In this chapter, we consider another classical formulation of the displacement method, the *moment-distribution method.* Like the slope-deflection method, the moment-distribution method can be used only for the analysis of continuous beams and frames, taking into account their bending deformations only. This method, which was initially developed by Hardy Cross in 1924, was the most widely used method for analysis of structures from 1930, when it was first published, through the 1960s. Since the early 1970s, with the increasing availability of computers, the use of the moment-distribution method has declined in favor of the computer-oriented matrix methods of structural analysis. Nonetheless, the moment-distribution method is still preferred by many engineers for analyzing smaller structures, since it provides a better insight into the behavior of structures. Furthermore, this method may also be used for preliminary designs as well as for checking the results of computerized analyses.

The main reason for the popularity of the moment-distribution method in the precomputer era was due to the fact that it does not involve the solution of as many simultaneous equations as required by the other classical methods. In the analysis of continuous beams and frames without sidesway, the moment-distribution method completely avoids the solution of simultaneous equations, whereas in the case of frames with sidesway, the number of simultaneous equations involved usually equals the number of independent joint translations.

The moment-distribution method is classified as a displacement method, and from a theoretical viewpoint, it is very similar to the slope-deflection method considered in the preceding chapter. However, unlike the slope-deflection method in which all the structure's equilibrium equations are satisfied simultaneously, in the moment-distribution method the moment equilibrium equations of the joints are solved iteratively by successively considering the moment equilibrium at one joint at a time, while the remaining joints of the structure are assumed to be restrained against displacement.

We first derive the fundamental relations necessary for the application of the moment-distribution method and then develop the basic concept of the method. We next consider the application of the method to the analysis of continuous beams and frames without sidesway and, finally, discuss the analysis of frames with sidesway.

17.1   DEFINITIONS AND TERMINOLOGY

Before we can develop the moment-distribution method, it is necessary to adopt a sign convention and define the various terms used in the analysis.

Sign Convention

In applying the moment-distribution method, we will adopt the same sign convention as used previously for the slope-deflection method:

> Counterclockwise member end moments are considered positive.

Since a counterclockwise moment at an end of a member must act in a clockwise direction on the adjacent joint, the foregoing sign convention implies that *clockwise moments on joints are considered positive.*

Member Stiffness

Consider a prismatic beam AB, which is hinged at end A and fixed at end B, as shown in Fig. 17.1(a). If we apply a moment M at the end A, the beam rotates by an angle θ at the hinged end A and develops a moment M_{BA} at the fixed end B, as shown in the figure. The relationship between the applied moment M and the rotation θ can be established by

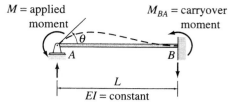

(a) Beam with Far End Fixed

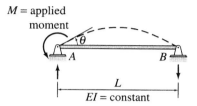

(b) Beam with Far End Hinged

FIG. **17.1**

using the slope-deflection equation derived in Section 16.1. By substituting $M_{nf} = M$, $\theta_n = \theta$, and $\theta_f = \psi = \text{FEM}_{nf} = 0$ into the slope-deflection equation (Eq. (16.9)), we obtain

$$M = \left(\frac{4EI}{L}\right)\theta \tag{17.1}$$

The *bending stiffness*, $\bar{K}$, *of a member is defined as the moment that must be applied at an end of the member to cause a unit rotation of that end.* Thus, by setting $\theta = 1$ rad in Eq. (17.1), we obtain the expression for the bending stiffness of the beam of Fig. 17.1(a) to be

$$\bar{K} = \frac{4EI}{L} \tag{17.2}$$

When the modulus of elasticity for all the members of a structure is the same (i.e., $E = $ constant), it is usually convenient to work with the *relative bending stiffnesses* of members in the analysis. *The relative bending stiffness, K, of a member is obtained by dividing its bending stiffness, $\bar{K}$, by 4E.* Thus, the relative bending stiffness of the beam of Fig. 17.1(a) is given by

$$K = \frac{\bar{K}}{4E} = \frac{I}{L} \tag{17.3}$$

Now, suppose that the far end B of the beam of Fig. 17.1(a) is hinged, as shown in Fig. 17.1(b). The relationship between the applied moment M and the rotation θ of the end A of the beam can now be determined by using the modified slope-deflection equation (Eqs.

(16.15)) derived in Section 16.1. By substituting $M_{rh} = M$, $\theta_r = \theta$, and $\psi = FEM_{rh} = FEM_{hr} = 0$ into Eq. 16.15(a), we obtain

$$M = \left(\frac{3EI}{L}\right)\theta \tag{17.4}$$

By setting $\theta = 1$ rad, we obtain the expression for the bending stiffness of the beam of Fig. 17.1(b) to be

$$\bar{K} = \frac{3EI}{L} \tag{17.5}$$

A comparison of Eqs. (17.2) and (17.5) indicates that the stiffness of the beam is reduced by 25 percent when the fixed support at B is replaced by a hinged support. The relative bending stiffness of the beam can now be obtained by dividing its bending stiffness by $4E$:

$$K = \frac{3}{4}\left(\frac{I}{L}\right) \tag{17.6}$$

From Eqs. (17.1) and (17.4), we can see that the relationship between the applied end moment M and the rotation θ of the corresponding end of a member can be summarized as follows:

$$M = \begin{cases} \left(\dfrac{4EI}{L}\right)\theta & \text{if far end of member is fixed} \\[2ex] \left(\dfrac{3EI}{L}\right)\theta & \text{if far end of member is hinged} \end{cases} \tag{17.7}$$

Similarly, based on Eqs. (17.2) and (17.5), the bending stiffness of a member is given by

$$\bar{K} = \begin{cases} \dfrac{4EI}{L} & \text{if far end of member is fixed} \\[2ex] \dfrac{3EI}{L} & \text{if far end of member is hinged} \end{cases} \tag{17.8}$$

and the relative bending stiffness of a member can be expressed as (see Eqs. (17.3) and (17.6))

$$K = \begin{cases} \dfrac{I}{L} & \text{if far end of member is fixed} \\[2ex] \dfrac{3}{4}\left(\dfrac{I}{L}\right) & \text{if far end of member is hinged} \end{cases} \tag{17.9}$$

Carryover Moment

Let us consider again the hinged-fixed beam of Fig. 17.1(a). When a moment M is applied at the hinged end A of the beam, a moment M_{BA} develops at the fixed end B, as shown in the figure. The moment M_{BA} is termed the *carryover moment*. To establish the relationship between the applied moment M and the carryover moment M_{BA}, we write the slope-deflection equation for M_{BA} by substituting $M_{nf} = M_{BA}$, $\theta_f = \theta$, and $\theta_n = \psi = \text{FEM}_{nf} = 0$ into Eq. (16.9):

$$M_{BA} = \left(\frac{2EI}{L}\right)\theta \tag{17.10}$$

By substituting $\theta = ML/(4EI)$ from Eq. (17.1) into Eq. (17.10), we obtain

$$M_{BA} = \frac{M}{2} \tag{17.11}$$

As Eq. (17.11) indicates, when a moment of magnitude M is applied at the hinged end of a beam, one-half of the applied moment is *carried over* to the far end, provided that the far end is fixed. Note that the direction of the carryover moment, M_{BA}, is the same as that of the applied moment, M.

When the far end of the beam is hinged, as shown in Fig. 17.1(b), the carryover moment M_{BA} is zero. Thus, we can express the carryover moment as

$$M_{BA} = \begin{cases} \dfrac{M}{2} & \text{if far end of member is fixed} \\[2mm] 0 & \text{if far end of member is hinged} \end{cases} \tag{17.12}$$

The ratio of the carryover moment to the applied moment (M_{BA}/M) is called the *carryover factor* of the member. It represents the fraction of the applied moment M that is *carried over* to the far end of the member. By dividing Eq. (17.12) by M, we can express the carryover factor (COF) as

$$\text{COF} = \begin{cases} \dfrac{1}{2} & \text{if far end of member is fixed} \\[2mm] 0 & \text{if far end of member is hinged} \end{cases} \tag{17.13}$$

Derivation of Member Stiffness and Carryover Moment by the Moment-Area Method

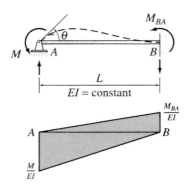

(a) Beam with Far End Fixed

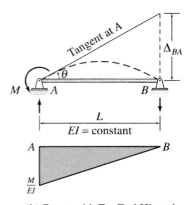

(b) Beam with Far End Hinged

FIG. 17.2

The foregoing expressions of member bending stiffness and carryover moment can, alternatively, be derived by applying the moment-area method discussed in Chapter 6.

The hinged-fixed beam of Fig. 17.1(a) is redrawn in Fig. 17.2(a), which also shows the M/EI diagram of the beam. Because the right end B of the beam is fixed, the tangent to the elastic curve at B is horizontal, and it passes through the left end A. Therefore, the tangential deviation of end A from the tangent at end B is equal to zero (i.e., $\Delta_{AB} = 0$). Since according to the second moment-area theorem, this tangential deviation is equal to the moment of the M/EI diagram between A and B about A, we can write

$$\Delta_{AB} = \frac{1}{2}\left(\frac{M}{EI}\right)L\left(\frac{L}{3}\right) - \frac{1}{2}\left(\frac{M_{BA}}{EI}\right)L\left(\frac{2L}{3}\right) = 0$$

from which

$$M_{BA} = \frac{M}{2}$$

Note that the preceding expression for carryover moment is identical to Eq. (17.11), which was derived previously by using the slope-deflection equations.

With the tangent at B horizontal, the angle θ at A equals the change in slope θ_{BA} between A and B. Since, according to the first moment-area theorem, θ_{BA} is equal to the area of the M/EI diagram between A and B, we write

$$\theta = \frac{1}{2}\left(\frac{M}{EI}\right)L - \frac{1}{2}\left(\frac{M_{BA}}{EI}\right)L$$

By substituting $M_{BA} = M/2$, we obtain

$$\theta = \left(\frac{L}{4EI}\right)M$$

from which

$$M = \left(\frac{4EI}{L}\right)\theta$$

which is the same as Eq. (17.1), derived previously.

The elastic curve and the M/EI diagram for the beam, when its far end B is hinged, are shown in Fig. 17.2(b). From the elastic curve we can see that

$$\theta = \frac{\Delta_{BA}}{L}$$

in which, according to the second moment-area theorem,

$$\Delta_{BA} = \text{moment of } M/EI \text{ diagram between } A \text{ and } B \text{ about } B$$

$$= \frac{1}{2}\left(\frac{M}{EI}\right)L\left(\frac{2L}{3}\right) = \left(\frac{L^2}{3EI}\right)M$$

Therefore,

$$\theta = \frac{\Delta_{BA}}{L} = \left(\frac{L}{3EI}\right)M$$

from which

$$M = \left(\frac{3EI}{L}\right)\theta$$

which is identical to Eq. (17.4), derived previously by using the slope-deflection equations.

Distribution Factors

When analyzing a structure by the moment-distribution method, an important question that arises is how to distribute a moment applied at a joint among the various members connected to that joint. Consider the three-member frame shown in Fig. 17.3(a), and suppose that a moment M is applied to the joint B, causing it to rotate by an angle θ, as shown in the figure. To determine what fraction of the applied moment M is resisted by each of the three members connected to the joint, we draw free-body diagrams of joint B and of the three members AB, BC, and BD, as shown in Fig. 17.3(b). By considering the moment equilibrium of the free body of joint B (i.e., $\sum M_B = 0$), we write

$$M + M_{BA} + M_{BC} + M_{BD} = 0$$

or

$$M = -(M_{BA} + M_{BC} + M_{BD}) \tag{17.14}$$

Since members AB, BC, and BD are rigidly connected to joint B, the rotations of the ends B of these members are the same as that of the joint. The moments at the ends B of the members can be expressed in terms of the joint rotation θ by applying Eq. (17.7). Noting that the far ends A and C, respectively, of members AB and BC are fixed, whereas the far end D of member BD is hinged, we apply Eqs. (17.7) through (17.9) to each member to obtain

$$M_{BA} = \left(\frac{4EI_1}{L_1}\right)\theta = \bar{K}_{BA}\theta = 4EK_{BA}\theta \tag{17.15}$$

$$M_{BC} = \left(\frac{4EI_2}{L_2}\right)\theta = \bar{K}_{BC}\theta = 4EK_{BC}\theta \tag{17.16}$$

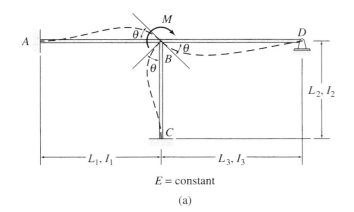

$$E = \text{constant}$$

(a)

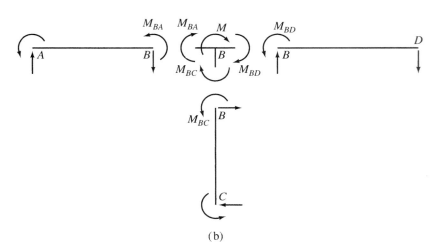

FIG. 17.3   (b)

$$M_{BD} = \left(\frac{3EI_3}{L_3}\right)\theta = \bar{K}_{BD}\theta = 4EK_{BD}\theta \qquad (17.17)$$

Substitution of Eqs. (17.15) through (17.17) into the equilibrium equation (Eq. (17.14)) yields

$$M = -\left(\frac{4EI_1}{L_1} + \frac{4EI_2}{L_2} + \frac{3EI_3}{L_3}\right)\theta$$

$$= -(\bar{K}_{BA} + \bar{K}_{BC} + \bar{K}_{BD})\theta = -\left(\sum \bar{K}_B\right)\theta \qquad (17.18)$$

in which $\sum \bar{K}_B$ represents the sum of the bending stiffnesses of all the members connected to joint B.

The rotational stiffness of a joint is defined as the moment required to cause a unit rotation of the joint. From Eq. (17.18), we can see that the rotational stiffness of a joint is equal to the sum of the bending stiffnesses of all the members rigidly connected to the joint. The negative sign in Eq. (17.18) appears because of the sign convention we have adopted, according to which the member end moments are considered positive when in the counterclockwise direction, whereas the moments acting on the joints are considered positive when they act in the clockwise direction.

To express member end moments in terms of the applied moment M, we first rewrite Eq. (17.18) in terms of the relative bending stiffnesses of members as

$$M = -4E(K_{BA} + K_{BC} + K_{BD})\theta = -4E(\textstyle\sum K_B)\theta$$

from which

$$\theta = -\frac{M}{4E\sum K_B} \tag{17.19}$$

By substituting Eq. (17.19) into Eqs. (17.15) through (17.17), we obtain

$$M_{BA} = -\left(\frac{K_{BA}}{\sum K_B}\right)M \tag{17.20}$$

$$M_{BC} = -\left(\frac{K_{BC}}{\sum K_B}\right)M \tag{17.21}$$

$$M_{BD} = -\left(\frac{K_{BD}}{\sum K_B}\right)M \tag{17.22}$$

From Eqs. (17.20) through (17.22), we can see that the applied moment M is distributed to the three members in proportion to their relative bending stiffnesses. The ratio $K/\sum K_B$ for a member is termed the *distribution factor* of that member for end B, and it represents the fraction of the applied moment M that is distributed to end B of the member. Thus Eqs. (17.20) through (17.22) can be expressed as

$$M_{BA} = -\text{DF}_{BA}M \tag{17.23}$$

$$M_{BC} = -\text{DF}_{BC}M \tag{17.24}$$

$$M_{BD} = -\text{DF}_{BD}M \tag{17.25}$$

in which $\text{DF}_{BA} = K_{BA}/\sum K_B$, $\text{DF}_{BC} = K_{BC}/\sum K_B$, and $\text{DF}_{BD} = K_{BD}/\sum K_B$ are the distribution factors for ends B of members AB, BC, and BD, respectively.

For example, if joint B of the frame of Fig. 17.3(a) is subjected to a clockwise moment of 150 k-ft (i.e., $M = 150$ k-ft) and if $L_1 = L_2 = 20$ ft, $L_3 = 30$ ft, and $I_1 = I_2 = I_3 = I$, so that

$$K_{BA} = K_{BC} = \frac{I}{20} = 0.05I$$

$$K_{BD} = \frac{3}{4}\left(\frac{I}{30}\right) = 0.025I$$

then the distribution factors for the ends B of members AB, BC, and BD are given by

$$DF_{BA} = \frac{K_{BA}}{K_{BA} + K_{BC} + K_{BD}} = \frac{0.05I}{(0.05 + 0.05 + 0.025)I} = 0.4$$

$$DF_{BC} = \frac{K_{BC}}{K_{BA} + K_{BC} + K_{BD}} = \frac{0.05I}{0.125I} = 0.4$$

$$DF_{BD} = \frac{K_{BD}}{K_{BA} + K_{BC} + K_{BD}} = \frac{0.05I}{0.125I} = 0.2$$

These distribution factors indicate that 40 percent of the 150-k-ft moment applied to joint B is exerted at end B of member AB, 40 percent at end B of member BC, and the remaining 20 percent at end B of member BD. Thus, the moments at ends B of the three members are

$$M_{BA} = -DF_{BA}M = -0.4(150) = -60 \text{ k-ft} \qquad \text{or} \qquad 60 \text{ k-ft} \, \circlearrowleft$$

$$M_{BC} = -DF_{BC}M = -0.4(150) = -60 \text{ k-ft} \qquad \text{or} \qquad 60 \text{ k-ft} \, \circlearrowleft$$

$$M_{BD} = -DF_{BD}M = -0.2(150) = -30 \text{ k-ft} \qquad \text{or} \qquad 30 \text{ k-ft} \, \circlearrowleft$$

Based on the foregoing discussion, we can state that, in general, the distribution factor (DF) for an end of a member that is rigidly connected to the adjacent joint equals the ratio of the relative bending stiffness of the member to the sum of the relative bending stiffnesses of all the members framing into the joint; that is,

$$DF = \frac{K}{\sum K} \tag{17.26}$$

Furthermore, the moment distributed to (or resisted by) a rigidly connected end of a member equals the distribution factor for that end times the negative of the moment applied to the adjacent joint.

Fixed-End Moments

The fixed-end moment expressions for some common types of loading conditions as well as for relative displacements of member ends are given inside the back cover of the book for convenient reference. In the moment-distribution method, the effects of joint translations due to support settlements and sidesway are also taken into account by means of fixed-end moments.

Consider the fixed beam of Fig. 17.4(a). As shown in this figure, a small settlement Δ of the left end A of the beam with respect to the right end B causes the beam's chord to rotate counterclockwise by an angle $\psi = \Delta/L$. By writing the slope-deflection equations (Eq. (16.9)) for the two end moments with $\psi = \Delta/L$ and by setting θ_A, θ_B, and fixed-

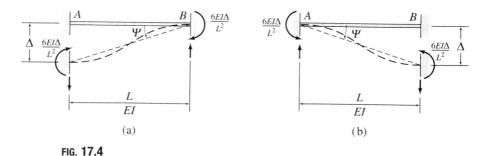

FIG. 17.4

end moments FEM_{AB} and FEM_{BA} due to external loading, equal to zero, we obtain

$$\text{FEM}_{AB} = \text{FEM}_{BA} = -\frac{6EI\Delta}{L^2}$$

in which FEM_{AB} and FEM_{BA} now denote the fixed-end moments due to the relative translation Δ between the two ends of the beam. Note that the magnitudes as well as the directions of the two fixed-end moments are the same. It can be seen from Fig. 17.4(a) that when a relative displacement causes a chord rotation in the counterclockwise direction, then the two fixed-end moments act in the clockwise (negative) direction to maintain zero slopes at the two ends of the beam. Conversely, if the chord rotation due to a relative displacement is clockwise, as shown in Fig. 17.4(b), then both fixed-end moments act in the counterclockwise (positive) direction to prevent the ends of the beam from rotating.

17.2 BASIC CONCEPT OF THE MOMENT-DISTRIBUTION METHOD

The moment-distribution method is an iterative procedure, in which it is initially assumed that all the joints of the structure that are free to rotate are temporarily restrained against rotation by imaginary clamps applied to them. External loads and joint translations (if any) are applied to this hypothetical fixed structure, and fixed-end moments at the ends of its members are computed. These fixed-end moments generally are not in equilibrium at those joints of the structure that are actually free to rotate. The conditions of equilibrium at such joints are then satisfied iteratively by releasing one joint at a time, with the remaining joints assumed to remain clamped. A joint at which the moments are not in balance is selected, and its unbalanced moment is evaluated. The joint is then released by removing the clamp, thereby allowing it to rotate under the unbalanced moment until the equilibrium state is reached. The rotation of the joint induces moments at the ends of the members connected to it. Such member end moments are referred to as *distributed moments*, and their values are determined by multiplying the negative of the unbalanced joint moment by the distribution factors for the member ends connected to the joint. The bending of these members due to the distributed moments causes carryover moments to develop at the far ends of the members, which can easily be evaluated by using the member carryover factors. The joint, which is now in equilibrium, is reclamped in its rotated position. Next, another joint with an unbalanced moment is selected and is released, balanced, and reclamped in the same manner. The procedure is repeated until the unbalanced moments at all the joints of the structure are negligibly small. The final member end moments are obtained by algebraically summing the fixed-end moment and all the distributed and carryover moments at each member end. This iterative process of determining member end moments by successively distributing the unbalanced moment at each joint is called the *moment-distribution process*.

With member end moments known, member end shears, member axial forces, and support reactions can be determined through equilibrium considerations, as discussed in Chapter 16.

To illustrate the moment-distribution method, consider the three-span continuous beam shown in Fig. 17.5(a). This structure was previously analyzed in Section 16.2 by the slope-deflection method. It is usually convenient to carry out the moment-distribution analysis in a tabular form, as shown in Fig. 17.5(a). Note that the table, which is sometimes referred to as a *moment-distribution table*, consists of six columns, one for each member end of the structure. All the computations for a particular member end moment are recorded in the column for that member end.

Distribution Factors

The first step in the analysis is to calculate the distribution factors at those joints of the structure that are free to rotate.

As discussed in Section 17.1 (Eq. (17.26)), the distribution factor for an end of a member is equal to the relative bending stiffness of the member divided by the sum of the relative bending stiffnesses of all the members connected to the joint. From Fig. 17.5(a), we can see that only joints B and C of the continuous beam are free to rotate. The distribution factors at joint B are

$$\mathrm{DF}_{BA} = \frac{K_{BA}}{K_{BA} + K_{BC}} = \frac{I/20}{2I/20} = 0.5$$

$$\mathrm{DF}_{BC} = \frac{K_{BC}}{K_{BA} + K_{BC}} = \frac{I/20}{2I/20} = 0.5$$

Similarly, at joint C,

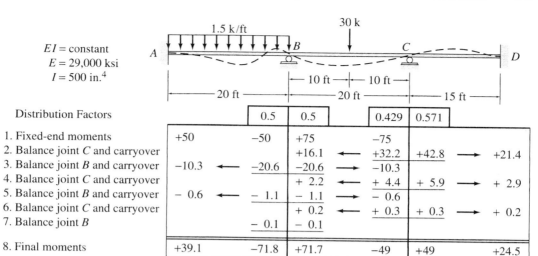

EI = constant
E = 29,000 ksi
I = 500 in.^{4}

Distribution Factors

		0.5	0.5		0.429	0.571	
1. Fixed-end moments	+50	−50	+75	−75			
2. Balance joint C and carryover			+16.1 ←	+32.2	+42.8 →	+21.4	
3. Balance joint B and carryover	−10.3 ←	−20.6	−20.6 →	−10.3			
4. Balance joint C and carryover			+ 2.2 ←	+ 4.4	+ 5.9 →	+ 2.9	
5. Balance joint B and carryover	− 0.6 ←	− 1.1	− 1.1 →	− 0.6			
6. Balance joint C and carryover			+ 0.2 ←	+ 0.3	+ 0.3 →	+ 0.2	
7. Balance joint B			− 0.1	− 0.1			
8. Final moments	+39.1	−71.8	+71.7	−49	+49	+24.5	

(a) Continuous Beam and Moment-Distribution Table

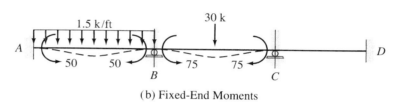

(b) Fixed-End Moments

(c) Unbalanced Moment at Joint C

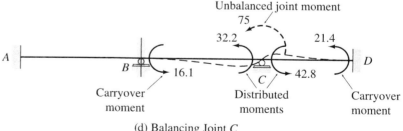

(d) Balancing Joint C

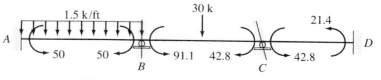

(e) Member End Moments with Joint C Balanced

FIG. 17.5

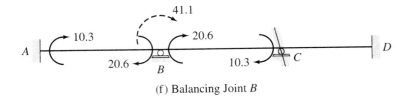

(f) Balancing Joint B

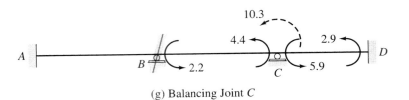

(g) Balancing Joint C

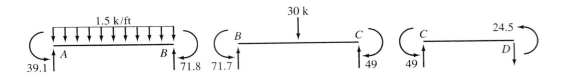

(h) Final Member End Moments (k-ft)

Member Ends	AB	BA	BC	CB	CD	DC
Distribution Factors		0.5	0.5	0.429	0.571	
1. Fixed-end moments	+50	−50	+75	−75		
2. Balance joints		−12.5	−12.5	+32.2	+42.8	
3. Carryover	− 6.3		+16.1	− 6.3		+21.4
4. Balance joints		− 8.1	− 8.1	+ 2.7	+ 3.6	
5. Carryover	− 4.1		+ 1.4	− 4.1		+ 1.8
6. Balance joints		− 0.7	− 0.7	+ 1.8	+ 2.3	
7. Carryover	− 0.4		+ 0.9	− 0.4		+ 1.2
8. Balance joints		− 0.5	− 0.5	+ 0.2	+ 0.2	
9. Carryover	− 0.3		+ 0.1	− 0.3		+ 0.1
10. Balance joints		− 0.05	− 0.05	+ 0.1	+ 0.2	
11. Final moments	+38.9	−71.8	+71.7	−49.1	+49.1	+24.5

(i)

FIG. 17.5 (contd.)

$$DF_{CB} = \frac{K_{CB}}{K_{CB} + K_{CD}} = \frac{I/20}{(I/20) + (I/15)} = 0.429$$

$$DF_{CD} = \frac{K_{CD}}{K_{CB} + K_{CD}} = \frac{I/15}{(I/20) + (I/15)} = 0.571$$

Note that the sum of the distribution factors at each joint must always equal 1. The distribution factors are recorded in boxes directly beneath the corresponding member ends on top of the moment-distribution table, as shown in Fig. 17.5(a).

Fixed-End Moments

Next, by assuming that joints B and C are restrained against rotation by imaginary clamps applied to them (Fig. 17.5(b)), we calculate the fixed-end moments that develop at the ends of each member. By using the fixed-end moment expressions given inside the back cover of the book, we obtain

$$FEM_{AB} = \frac{1.5(20)^2}{12} = 50 \text{ k-ft} \circlearrowleft \qquad \text{or} \qquad +50 \text{ k-ft}$$

$$FEM_{BA} = 50 \text{ k-ft} \circlearrowright \qquad \text{or} \qquad -50 \text{ k-ft}$$

$$FEM_{BC} = \frac{30(20)}{8} = 75 \text{ k-ft} \circlearrowleft \qquad \text{or} \qquad +75 \text{ k-ft}$$

$$FEM_{CB} = 75 \text{ k-ft} \circlearrowright \qquad \text{or} \qquad -75 \text{ k-ft}$$

$$FEM_{CD} = FEM_{DC} = 0$$

Note that in accordance with the moment-distribution sign convention, the counterclockwise fixed-end moments are considered to be positive. The fixed-end moments are recorded on the first line of the moment-distribution table, as shown in Fig. 17.5(a).

Balancing Joint *C*

Since joints B and C are actually not clamped, we release them, one at a time. We can release either joint B or joint C; let us begin at joint C. From Fig. 17.5(b), we can see that there is a -75-k-ft (clockwise) fixed-end moment at end C of member BC, whereas no moment exists at end C of member CD. As long as joint C is restrained against rotation by the clamp, the -75-k-ft unbalanced moment is absorbed by the clamp. However, when the imaginary clamp is removed to release the joint, the -75-k-ft unbalanced moment acts at the joint, as shown in Fig. 17.5(c), causing it to rotate in the counterclockwise direction until it is in equilibrium (Fig. 17.5(d)). The rotation of joint C causes the distributed moments, DM_{CB} and DM_{CD}, to develop at ends C of members BC and CD, which can be evaluated by multiplying the negative of the unbalanced moment (i.e., $+75$-k-ft) by the distribution factors DF_{CB} and DF_{CD}, respectively. Thus

$$DM_{CB} = 0.429(+75) = +32.2 \text{ k-ft}$$

$$DM_{CD} = 0.571(+75) = +42.8 \text{ k-ft}$$

These distributed moments are recorded on line 2 of the moment-distribution table (Fig. 17.5(a)), and a line is drawn beneath them to indicate that joint C is now balanced. Note that the sum of the three moments above the line at joint C is equal to zero (i.e., $-75 + 32.2 + 42.8 = 0$).

The distributed moment at end C of member BC induces a carry-over moment at the far end B (Fig. 17.5(d)), which can be determined by multiplying the distributed moment by the carryover factor of the member. Since joint B remains clamped, the carryover factor of member BC is $\frac{1}{2}$ (Eq. (17.13)). Thus, the carryover moment at the end B of member BC is

$$COM_{BC} = COF_{CB}(DM_{CB}) = \frac{1}{2}(+32.2) = +16.1 \text{ k-ft}$$

Similarly, the carryover moment at the end D of member CD is computed as

$$COM_{DC} = COF_{CD}(DM_{CD}) = \frac{1}{2}(+42.8) = +21.4 \text{ k-ft}$$

These carryover moments are recorded on the same line of the moment-distribution table as the distributed moments, with a horizontal arrow from each distributed moment to its carryover moment, as shown in Fig. 17.5(a).

The total member end moments at this point in the analysis are depicted in Fig. 17.5(e). It can be seen from this figure that joint C is now in equilibrium, because it is subjected to two equal, but opposite, moments. Joint B, however, is not in equilibrium, and it needs to be balanced. Before we release joint B, an imaginary clamp is applied to joint C in its rotated position, as shown in Fig. 17.5(e).

Balancing Joint *B*

Joint B is now released. The unbalanced moment at this joint is obtained by summing all the moments acting at the ends B of members AB and BC, which are rigidly connected to joint B. From the moment-distribution table (lines 1 and 2), we can see that there is a -50-k-ft fixed-end moment at end B of member AB, whereas the end B of member BC is subjected to a $+75$-k-ft fixed-end moment and a $+16.1$-k-ft carryover moment. Thus the unbalanced moment at joint B is

$$UM_B = -50 + 75 + 16.1 = +41.1 \text{ k-ft}$$

This unbalanced moment causes joint B to rotate, as shown in Fig. 17.5(f), and induces distributed moments at ends B of members AB and BC. As discussed previously, the distributed moments are evaluated by multiplying the negative of the unbalanced moment by the distribution factors:

$$DM_{BA} = 0.5(-41.1) = -20.6 \text{ k-ft}$$

$$DM_{BC} = 0.5(-41.1) = -20.6 \text{ k-ft}$$

These distributed moments are recorded on line 3 of the moment-distribution table, and a line is drawn beneath them to indicate that joint B is now balanced. One-half of the distributed moments are then carried over to the far ends A and C of members AB and BC, respectively, as indicated by horizontal arrows on line 3 of the table. Joint B is then reclamped in its rotated position.

Balancing Joint *C*

With joint B now balanced, we can see from the moment-distribution table (line 3) that, due to the carryover effect, there is a -10.3-k-ft unbalanced moment at joint C. Recall that the moments above the horizontal line at joint C were balanced previously. Thus we release joint C again and distribute the unbalanced moment to ends C of members BC and CD as (Fig. 17.5(g))

$$DM_{CB} = 0.429(+10.3) = +4.4 \text{ k-ft}$$

$$DM_{CD} = 0.571(+10.3) = +5.9 \text{ k-ft}$$

These distributed moments are recorded on line 4 of the moment-distribution table, and one-half of these moments are carried over to the ends B and D of members BC and CD, respectively, as indicated on the table. Joint C is then reclamped.

Balancing Joint *B*

The $+2.2$-k-ft unbalanced moment at joint B (line 4 of the moment-distribution table) is balanced in a similar manner. The distributed and the carryover moments thus computed are shown on line 5 of the table. Joint B is then reclamped.

It can be seen from line 5 of the moment-distribution table that the unbalanced moment at joint C has now been reduced to only -0.6 k-ft. Another balancing of joint C produces an even smaller unbalanced moment of $+0.2$ k-ft at joint B, as shown on line 6 of the moment-distribution table. Since the distributed moments induced by this unbalanced moment are negligibly small, we end the moment-distribution process. The final member end moments are obtained by algebraically summing the entries in each column of the moment-distribution table. The final moments thus obtained are recorded on line 8 of the table and are shown on the free-body diagrams of the members in Fig. 17.5(h). Note that the final moments satisfy the equations of moment equilibrium at joints B and C.

With the member end moments known, member end shears and support reactions can now be determined by considering the equilibrium of the free bodies of the members and joints of the continuous beam, as discussed in Section 16.2. The shear and bending moment diagrams can then be constructed in the usual manner by using the *beam sign convention* (see Fig. 16.3).

Practical Application of the Moment-Distribution Process

In the foregoing discussion, we determined the member end moments by successively balancing one joint of the structure at a time. Although this approach provides a clearer insight into the basic concept of the moment-distribution process, from a practical viewpoint, it is usually more convenient to use an alternative approach in which all the joints of the structure that are free to rotate are balanced simultaneously in the same step. All the carryover moments that are induced at the far ends of the members are then computed simultaneously in the following step, and the process of balancing the joints and carrying over moments is repeated until the unbalanced moments at the joints are negligibly small.

To illustrate this alternative approach, consider again the three-span continuous beam of Fig. 17.5(a). The moment-distribution table used for carrying out the computations is shown in Fig. 17.5(i). The previously computed distribution factors and fixed-end moments are recorded on the top and the first line, respectively, of the table, as shown in the figure. The moment-distribution process is started by balancing joints B and C. From line 1 of the moment-distribution table (Fig. 17.5(i)), we can see that the unbalanced moment at joint B is

$$\text{UM}_B = -50 + 75 = +25 \text{ k-ft}$$

As discussed previously, the balancing of joint B induces distributed moments at ends B of the members AB and BC, which can be evaluated by multiplying the negative of the unbalanced moment by the distribution factors. Thus,

$$\text{DM}_{BA} = 0.5(-25) = -12.5 \text{ k-ft}$$
$$\text{DM}_{BC} = 0.5(-25) = -12.5 \text{ k-ft}$$

Joint C is then balanced in a similar manner. From line 1 of the moment-distribution table, we can see that the unbalanced moment at joint C is

$$\text{UM}_C = -75 \text{ k-ft}$$

Thus, the balancing of joint C induces the following distributed moments at ends C of members BC and CD, respectively:

$$DM_{CB} = 0.429(+75) = +32.2 \text{ k-ft}$$

$$DM_{CD} = 0.571(+75) = +42.8 \text{ k-ft}$$

The four distributed moments are recorded on line 2 of the moment-distribution table, and a line is drawn beneath them, across the entire width of the table, to indicate that all the joints are now balanced.

In the next step of the analysis, the carryover moments that develop at the far ends of the members are computed by multiplying the distributed moments by the carryover factors:

$$COM_{AB} = \frac{1}{2}(DM_{BA}) = \frac{1}{2}(-12.5) = -6.3 \text{ k-ft}$$

$$COM_{CB} = \frac{1}{2}(DM_{BC}) = \frac{1}{2}(-12.5) = -6.3 \text{ k-ft}$$

$$COM_{BC} = \frac{1}{2}(DM_{CB}) = \frac{1}{2}(+32.2) = +16.1 \text{ k-ft}$$

$$COM_{DC} = \frac{1}{2}(DM_{CD}) = \frac{1}{2}(+42.8) = +21.4 \text{ k-ft}$$

These carryover moments are recorded on the next line (line 3) of the moment-distribution table, with an inclined arrow pointing from each distributed moment to its carryover moment, as shown in Fig. 17.5(i). We can see from line 3 of the moment-distribution table that, due to the carryover effect, there are now +16.1-k-ft and −6.3-k-ft unbalanced moments at joints B and C, respectively. Thus these joints are balanced again, and the distributed moments thus obtained are recorded on line 4 of the moment-distribution table. One-half of the distributed moments are then carried over to the far ends of the members (line 5), and the process is continued until the unbalanced moments are negligibly small. The final member end moments, obtained by algebraically summing the entries in each column of the moment-distribution table, are recorded on line 11 of the table (Fig. 17.5(i)). Note that these final moments are in agreement with those determined previously in Fig. 17.5(a) and in Section 16.2 by the slope-deflection method. The small differences between the results obtained by different approaches are due to the round-off errors.

17.3 ANALYSIS OF CONTINUOUS BEAMS

Based on the discussion presented in the preceding section, the procedure for the analysis of continuous beams by the moment-distribution method can be summarized as follows:

1. Calculate distribution factors. At each joint that is free to rotate, calculate the distribution factor for each of the members rigidly connected to the joint. The distribution factor for a member end is computed by dividing the relative bending stiffness (I/L) of the member by the sum of the relative bending stiffnesses of all the members rigidly connected to the joint. The sum of all the distribution factors at a joint must equal 1.

2. Compute fixed-end moments. Assuming that all the free joints are clamped against rotation, evaluate, for each member, the fixed-end moments due to the external loads and support settlements (if any) by using the fixed-end moment expressions given inside the back cover of the book. The counterclockwise fixed-end moments are considered to be positive.

3. Balance the moments at all the joints that are free to rotate by applying the moment-distribution process as follows:

   a. At each joint, evaluate the unbalanced moment and distribute the unbalanced moment to the members connected to the joint. The distributed moment at each member end rigidly connected to the joint is obtained by multiplying the negative of the unbalanced moment by the distribution factor for the member end.

   b. Carry over one-half of each distributed moment to the opposite (far) end of the member.

   c. Repeat steps 3(a) and 3(b) until either all the free joints are balanced or the unbalanced moments at these joints are negligibly small.

4. Determine the final member end moments by algebraically summing the fixed-end moment and all the distributed and carryover moments at each member end. If the moment distribution has been carried out correctly, then the final moments must satisfy the equations of moment equilibrium at all the joints of the structure that are free to rotate.

5. Compute member end shears by considering the equilibrium of the members of the structure.

6. Determine support reactions by considering the equilibrium of the joints of the structure.

7. Draw shear and bending moment diagrams by using the *beam sign convention*.

Beams with Simple Supports at the Ends

Although the foregoing procedure can be used to analyze continuous beams that are simply supported at one or both ends, the analysis of such structures can be considerably simplified by using the reduced relative bending stiffnesses, $K = 3I(4L)$, for spans adjacent to the simple

end supports, in accordance with Eq. (17.9). When using reduced stiffnesses, the joints at the simple end supports are balanced only once
during the moment-distribution process, after which they are left un-
clamped so that no moments can be carried over to them as the interior
joints of the structure are balanced (see Example 17.3).

Structures with Cantilever Overhangs

Consider a continuous beam with a cantilever overhang, as shown in
Fig. 17.6(a). Since the cantilever portion CD does not contribute to the
rotational stiffness of joint C, the distribution factor for its end C is zero
($\text{DF}_{CD} = 0$). Thus, joint C can be treated as a simple end support in the
analysis. The moment at end C of the cantilever, however, does affect
the unbalanced moment at joint C and must be included along with the
other fixed-end moments in the analysis (Fig. 17.6(b)). Note that the
cantilever portion CD is statically determinate; therefore, the moment at
its end C can be easily evaluated by applying the equation of moment
equilibrium (Fig. 17.6(c)).

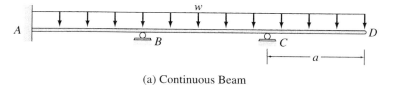

(a) Continuous Beam

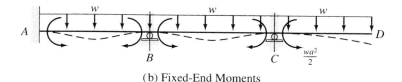

(b) Fixed-End Moments

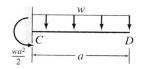

(c) Statically Determinate
Cantilever Portion

FIG. 17.6

Example 17.1

Determine the member end moments for the two-span continuous beam shown in Fig. 17.7(a) by using the moment-distribution method.

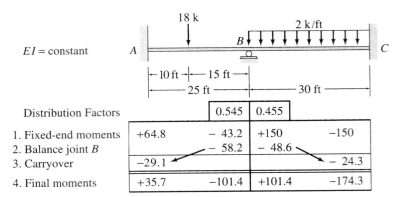

Distribution Factors		0.545	0.455	
1. Fixed-end moments	+64.8	− 43.2	+150	−150
2. Balance joint B		− 58.2	− 48.6	
3. Carryover	−29.1			− 24.3
4. Final moments	+35.7	−101.4	+101.4	−174.3

(a) Continuous Beam and Moment-Distribution Table

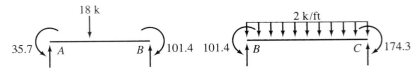

(b) Final Member End Moments (k-ft)

FIG. 17.7

Solution

This beam was previously analyzed in Example 16.1 by the slope-deflection method.

Distribution Factors Only joint B is free to rotate. The distribution factors at this joint are

$$DF_{BA} = \frac{K_{BA}}{K_{BA} + K_{BC}} = \frac{I/25}{(I/25) + (I/30)} = 0.545$$

$$DF_{BC} = \frac{K_{BC}}{K_{BA} + K_{BC}} = \frac{I/30}{(I/25) + (I/30)} = 0.455$$

Note that the sum of the distribution factors at joint B is equal to 1; that is,

$$DF_{BA} + DF_{BC} = 0.545 + 0.455 = 1 \qquad \text{Checks}$$

The distribution factors are recorded in boxes beneath the corresponding member ends on top of the moment-distribution table, as shown in Fig. 17.7(a).

Fixed-End Moments Assuming that joint B is clamped against rotation, we calculate the fixed-end moments due to the external loads by using the fixed-end moment expressions given inside the back cover of the book:

$$FEM_{AB} = \frac{18(10)(15)^2}{(25)^2} = 64.8 \text{ k-ft } \circlearrowleft \qquad \text{or} \qquad +64.8 \text{ k-ft}$$

$$\text{FEM}_{BA} = \frac{18(10)^2(15)}{(25)^2} = 43.2 \text{ k-ft} \circlearrowright \quad \text{or} \quad -43.2 \text{ k-ft}$$

$$\text{FEM}_{BC} = \frac{2(30)^2}{12} = 150 \text{ k-ft} \circlearrowleft \quad \text{or} \quad +150 \text{ k-ft}$$

$$\text{FEM}_{CB} = 150 \text{ k-ft} \circlearrowright \quad \text{or} \quad -150 \text{ k-ft}$$

These fixed-end moments are recorded on the first line of the moment-distribution table, as shown in Fig. 17.7(a).

Moment Distribution Since joint B is actually not clamped, we release the joint and determine the unbalanced moment acting on it by summing the moments at ends B of members AB and BC:

$$\text{UM}_B = -43.2 + 150 = +106.8 \text{ k-ft}$$

This unbalanced moment at joint B induces distributed moments at the ends B of members AB and BC, which can be determined by multiplying the negative of the unbalanced moment by the distribution factors:

$$\text{DM}_{BA} = \text{DF}_{BA}(-\text{UM}_B) = 0.545(-106.8) = -58.2 \text{ k-ft}$$

$$\text{DM}_{BC} = \text{DF}_{BC}(-\text{UM}_B) = 0.455(-106.8) = -48.6 \text{ k-ft}$$

These distributed moments are recorded on line 2 of the moment-distribution table, and a line is drawn beneath them to indicate that joint B is now balanced. The carryover moments at the far ends A and C of members AB and BC, respectively, are then computed as

$$\text{COM}_{AB} = \frac{1}{2}(\text{DM}_{BA}) = \frac{1}{2}(-58.2) = -29.1 \text{ k-ft}$$

$$\text{COM}_{CB} = \frac{1}{2}(\text{DM}_{BC}) = \frac{1}{2}(-48.6) = -24.3 \text{ k-ft}$$

The carryover moments are recorded on the next line (line 3) of the moment-distribution table, with an inclined arrow pointing from each distributed moment to its carryover moment, as shown in Fig. 17.7(a).

Joint B is the only joint of the structure that is free to rotate, and because it has been balanced, we end the moment-distribution process.

Final Moments The final member end moments are obtained by algebraically summing all the moments in each column of the moment-distribution table. The final moments thus obtained are recorded on the last line of the table in Fig. 17.7(a). Note that these final moments satisfy the equation of moment equilibrium at joint B. A positive answer for an end moment indicates that its sense is counterclockwise, whereas a negative answer for an end moment implies a clockwise sense. The final member end moments are depicted in Fig. 17.7(b).

Ans.

The member end shears and support reactions can now be determined by considering the equilibrium of the members and joints of the continuous beam, as discussed in Example 16.1. The shear and bending moment diagrams of the beam were also constructed in Example 16.1.

Example 17.2

Determine the member end moments for the three-span continuous beam shown in Fig. 17.8(a) by the moment-distribution method.

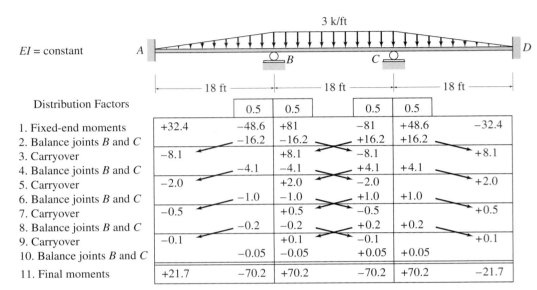

Distribution Factors		0.5	0.5		0.5	0.5	
1. Fixed-end moments	+32.4	−48.6	+81	−81	+48.6	−32.4
2. Balance joints B and C		−16.2	−16.2	+16.2	+16.2	
3. Carryover	−8.1		+8.1	−8.1		+8.1
4. Balance joints B and C		−4.1	−4.1	+4.1	+4.1	
5. Carryover	−2.0		+2.0	−2.0		+2.0
6. Balance joints B and C		−1.0	−1.0	+1.0	+1.0	
7. Carryover	−0.5		+0.5	−0.5		+0.5
8. Balance joints B and C		−0.2	−0.2	+0.2	+0.2	
9. Carryover	−0.1		+0.1	−0.1		+0.1
10. Balance joints B and C		−0.05	−0.05	+0.05	+0.05	
11. Final moments	+21.7	−70.2	+70.2	−70.2	+70.2	−21.7

(a) Continuous-Beam and Moment-Distribution Table

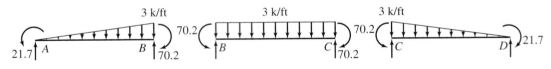

(b) Final Member End Moments (k-ft)

FIG. 17.8

Solution

This beam was analyzed previously in Example 16.2 by using the slope-deflection method.

Distribution Factors From Fig. 17.8(a), we can see that joints B and C of the beam are free to rotate. The distribution factors at joint B are

$$DF_{BA} = \frac{K_{BA}}{K_{BA} + K_{BC}} = \frac{I/18}{(I/18) + (I/18)} = 0.5$$

$$DF_{BC} = \frac{K_{BC}}{K_{BA} + K_{BC}} = \frac{I/18}{(I/18) + (I/18)} = 0.5$$

Similarly, at joint C,

$$DF_{CB} = \frac{K_{CB}}{K_{CB} + K_{CD}} = \frac{I/18}{(I/18) + (I/18)} = 0.5$$

$$DF_{CD} = \frac{K_{CD}}{K_{CB} + K_{CD}} = \frac{I/18}{(I/18) + (I/18)} = 0.5$$

Fixed-End Moments

$$FEM_{AB} = +\frac{3(18)^2}{30} = +32.4 \text{ k-ft}$$

$$FEM_{BA} = -\frac{3(18)^2}{20} = -48.6 \text{ k-ft}$$

$$FEM_{BC} = +\frac{3(18)^2}{12} = +81 \text{ k-ft}$$

$$FEM_{CB} = -81 \text{ k-ft}$$

$$FEM_{CD} = +\frac{3(18)^2}{20} = +48.6 \text{ k-ft}$$

$$FEM_{DC} = -\frac{3(18)^2}{30} = -32.4 \text{ k-ft}$$

Moment Distribution After recording the distribution factors and the fixed-end moments in the moment-distribution table shown in Fig. 17.8(a), we begin the moment-distribution process by balancing joints B and C. The unbalanced moment at joint B is equal to $-48.6 + 81 = +32.4$ k-ft. Thus, the distributed moments at the ends B of members AB and BC are

$$DM_{BA} = DF_{BA}(-UM_B) = 0.5(-32.4) = -16.2 \text{ k-ft}$$

$$DM_{BC} = DF_{BC}(-UM_B) = 0.5(-32.4) = -16.2 \text{ k-ft}$$

Similarly, noting that the unbalanced moment at joint C equals $-81 + 48.6 = -32.4$ k-ft, we determine the distributed moments at the ends C of members BC and CD to be

$$DM_{CB} = DF_{CB}(-UM_C) = 0.5(+32.4) = +16.2 \text{ k-ft}$$

$$DM_{CD} = DF_{CD}(-UM_C) = 0.5(+32.4) = +16.2 \text{ k-ft}$$

One-half of these distributed moments are then carried over to the far ends of the members, as shown on the third line of the moment-distribution table in Fig. 17.8(a). This process is repeated, as shown in the figure, until the unbalanced moments are negligibly small.

Final Moments The final member end moments, obtained by summing the moments in each column of the moment-distribution table, are recorded on the last line of the table in Fig. 17.8(a). These moments are depicted in Fig. 17.8(b).

Ans.

The member end shears, support reactions, and shear and bending moment diagrams of the beam were determined in Example 16.2.

Example 17.3

Determine the reactions and draw the shear and bending moment diagrams for the two-span continuous beam shown in Fig. 17.9(a) by using the moment-distribution method.

Solution

Distribution Factors From Fig. 17.9(a), we can see that joints *B* and *C* of the continuous beam are free to rotate. The distribution factors at joint *B* are

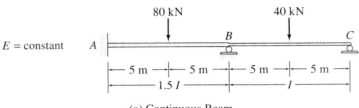

(a) Continuous Beam

Distribution Factors		0.6	0.4		1.0
1. Fixed-end moments	+100	−100	+50		−50
2. Balance joints *B* and *C*		+ 30	+20		+50
3. Carryover	+ 15		+25		+10
4. Balance joints *B* and *C*		− 15	−10		−10
5. Carryover	− 7.5		− 5		− 5
6. Balance joints *B* and *C*		+ 3	+ 2		+ 5
7. Carryover	+ 1.5		+ 2.5		+ 1
8. Balance joints *B* and *C*		− 1.5	− 1		− 1
9. Carryover	− 0.8		− 0.5		− 0.5
10. Balance joints *B* and *C*		+ 0.3	+ 0.2		+ 0.5
11. Carryover	+ 0.2		+ 0.3		+ 0.1
12. Balance joints *B* and *C*		− 0.2	− 0.1		− 0.1
13. Final moments	+108.4	− 83.4	+83.4		0

(b) Moment-Distribution Table: $K_{BC} = \frac{I}{10}$

Distribution Factors		$\frac{2}{3}$	$\frac{1}{3}$		1
1. Fixed-end moments	+100	−100	+50		−50
2. Balance joints *B* and *C*		+ 33.3	+16.7		+50
3. Carryover	+ 16.7		+25		
4. Balance joint *B*		− 16.7	− 8.3		
5. Carryover	− 8.3				
6. Final moments	+108.4	− 83.4	+83.4		0

(c) Moment-Distribution Table: $K_{BC} = \frac{3}{4}\left(\frac{I}{10}\right)$

FIG. 17.9

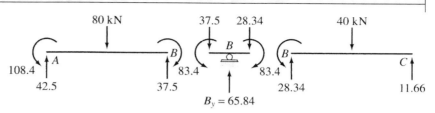

(d) Member End Moments and Shears

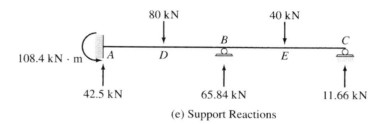

(e) Support Reactions

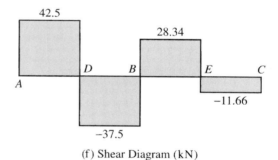

(f) Shear Diagram (kN)

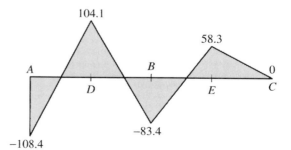

FIG. **17.9** (contd.)

(g) Bending Moment Diagram (kN · m)

continued

$$DF_{BA} = \frac{K_{BA}}{K_{BA} + K_{BC}} = \frac{1.5I/10}{(1.5I/10) + (I/10)} = 0.6$$

$$DF_{BC} = \frac{K_{BC}}{K_{BA} + K_{BC}} = \frac{I/10}{(1.5I/10) + (I/10)} = 0.4$$

Similarly, at joint C,

$$DF_{CB} = \frac{K_{CB}}{K_{CB}} = \frac{0.1I}{0.1I} = 1$$

Fixed-End Moments

$$FEM_{AB} = +\frac{80(10)}{8} = +100 \text{ kN} \cdot \text{m}$$

$$FEM_{BA} = -100 \text{ kN} \cdot \text{m}$$

$$FEM_{BC} = +\frac{40(10)}{8} = +50 \text{ kN} \cdot \text{m}$$

$$FEM_{CB} = -50 \text{ kN} \cdot \text{m}$$

Moment Distribution After recording the distribution factors and the fixed-end moments in the moment-distribution table shown in Fig. 17.9(b), we begin the moment-distribution process by balancing joints B and C. The unbalanced moment at joint B is equal to $-100 + 50 = -50$ kN · m. Thus the distributed moments at the ends B of members AB and BC are

$$DM_{BA} = DF_{BA}(-UM_B) = 0.6(+50) = +30 \text{ kN} \cdot \text{m}$$

$$DM_{BC} = DF_{BC}(-UM_B) = 0.4(+50) = +20 \text{ kN} \cdot \text{m}$$

Similarly, noting that the unbalanced moment at joint C is -50 kN · m, we determine the distributed moment at end C of member BC to be

$$DM_{CB} = DF_{CB}(-UM_C) = 1(+50) = +50 \text{ kN} \cdot \text{m}$$

One-half of these distributed moments are then carried over to the far ends of the members, as shown on the third line of the moment-distribution table in Fig. 17.9(b). This process is repeated, as shown in the figure, until the unbalanced moments are negligibly small.

Final Moments The final member end moments, obtained by summing the moments in each column of the moment-distribution table, are recorded on the last line of the table in Fig. 17.9(b).   **Ans.**

Alternative Method Because the end support C of the continuous beam is a simple support, the analysis can be simplified by using the reduced relative bending stiffness for member BC, which is adjacent to the simple support C:

$$K_{BC} = \frac{3}{4}\left(\frac{I}{10}\right)$$

Note that the relative bending stiffness of member AB remains the same as before. The distribution factors at joint B are now given by

$$DF_{BA} = \frac{K_{BA}}{K_{BA} + K_{BC}} = \frac{1.5I/10}{(1.5I/10) + (3I/40)} = \frac{2}{3}$$

$$DF_{BC} = \frac{K_{BC}}{K_{BA} + K_{BC}} = \frac{3I/40}{(1.5I/10) + (3I/40)} = \frac{1}{3}$$

At joint C, $DF_{CB} = K_{CB}/K_{CB} = 1$. These distribution factors, and the fixed-end moments that remain the same as before, are recorded in the moment-distribution table, as shown in Fig. 17.9(c).

Since we are using the reduced relative bending stiffness for member BC, joint C needs to be balanced only once in the moment-distribution process. Thus joints B and C are balanced and the distributed moments are computed in the usual manner, as indicated on the second line of the moment-distribution table (Fig. 17.9(c)). However, as shown on the third line of the table in Fig. 17.9(c), no moment is carried over to end C of member BC. Joint B is balanced once more, and the moment is carried over to the end A of member AB (lines 4 and 5). Because both joints B and C are now balanced, we can end the moment-distribution process and determine the final moments by summing the moments in each column of the moment-distribution table.   Ans.

Member End Shears The member end shears, obtained by considering the equilibrium of each member, are shown in Fig. 17.9(d).   Ans.

Support Reactions See Fig. 17.9(e).   Ans.

Shear and Bending Moment Diagrams See Fig. 17.9(f) and (g).   Ans.

Example 17.4

Determine the member end moments for the continuous beam shown in Fig. 17.10(a) by using the moment-distribution method.

Solution

This beam was previously analyzed in Example 16.4 by the slope-deflection method.

Distribution Factors Since the cantilever portion CD does not contribute to the rotational stiffness of joint C, we can treat joint C as a simple end support and use the reduced relative bending stiffness of member BC in the analysis:

$$K_{BA} = \frac{I}{6} \quad \text{and} \quad K_{BC} = \frac{3}{4}\left(\frac{I}{9}\right) = \frac{I}{12}$$

At joint B,

$$DF_{BA} = \frac{I/6}{(I/6) + (I/12)} = \frac{2}{3}$$

$$DF_{BC} = \frac{I/12}{(I/6) + (I/12)} = \frac{1}{3}$$

continued

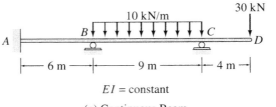

EI = constant

(a) Continuous Beam

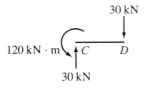

(b) Statically Determinate
Cantilever Portion

AB		BA	BC		CB	CD
		$\frac{2}{3}$	$\frac{1}{3}$		1	
			+67.5		− 67.5	+120
	−45		−22.5		− 52.5	
−22.5			−26.3			
	+17.5		+ 8.8			
+ 8.8						
−13.7		−27.5	+27.5		−120	+120

(c) Moment-Distribution Table

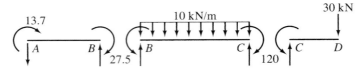

(d) Final Member End Moments (kN · m)

FIG. 17.10

At joint C,

$$DF_{CB} = 1$$

Fixed-End Moments Using the fixed-end moment expressions and Fig. 17.10(b), we obtain

$$FEM_{AB} = FEM_{BA} = 0$$

$$FEM_{BC} = +67.5 \text{ kN} \cdot \text{m} \qquad FEM_{CB} = -67.5 \text{ kN} \cdot \text{m}$$

$$FEM_{CD} = +30(4) = +120 \text{ kN} \cdot \text{m}$$

Moment Distribution The moment distribution is carried out as shown on the moment-distribution table in Fig. 17.10(c).

Final Moments See the moment-distribution table and Fig. 17.10(d). **Ans.**

Example 17.5

Determine the member end moments for the continuous beam shown in Fig. 17.11(a) due to a settlement of 20 mm at support *B*. Use the moment-distribution method.

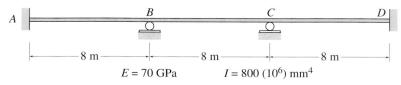

(a) Continuous Beam

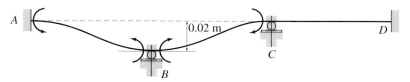

(b) Fixed-End Moments Due to Support Settlement

AB		BA	BC		CB	CD		DC
		0.5	0.5		0.5	0.5		
+105		+105	−105		−105			
					+52.5	+52.5		+26.3
			+26.3					
−6.6		−13.1	−13.1		−6.6			
					+3.3	+3.3		+1.6
			+1.6					
−0.4		−0.8	−0.8		−0.4			
					+0.2	+0.2		+0.1
			+0.1					
		−0.05	−0.05					
+98		+91.1	−91		−56	+56		+28

(c) Moment-Distribution Table

(d) Final Member End Moments (kN · m)

FIG. 17.11

continued

Solution

This beam was analyzed previously in Example 16.5 by using the slope-deflection method.

Distribution Factors At joint B,

$$DF_{BA} = \frac{I/8}{(I/8) + (I/8)} = 0.5$$

$$DF_{BC} = \frac{I/8}{(I/8) + (I/8)} = 0.5$$

At joint C,

$$DF_{CB} = \frac{I/8}{(I/8) + (I/8)} = 0.5$$

$$DF_{CD} = \frac{I/8}{(I/8) + (I/8)} = 0.5$$

Fixed-End Moments A qualitative deflected shape of the continuous beam with all joints clamped against rotation and subjected to the specified support settlement is depicted in Fig. 17.11(b) using an exaggerated scale. It can be seen from this figure that the relative settlements for the three members are $\Delta_{AB} = \Delta_{BC} = 0.02$ m, and $\Delta_{CD} = 0$.

By using the fixed-end moment expressions, we determine the fixed-end moments due to the support settlement to be

$$FEM_{AB} = FEM_{BA} = +\frac{6EI\Delta}{L^2} = +\frac{6(70)(800)(0.02)}{(8)^2} = +105 \text{ kN} \cdot \text{m}$$

$$FEM_{BC} = FEM_{CB} = -\frac{6EI\Delta}{L^2} = -\frac{6(70)(800)(0.02)}{(8)^2} = -105 \text{ kN} \cdot \text{m}$$

$$FEM_{CD} = FEM_{DC} = 0$$

Moment Distribution The moment distribution is carried out in the usual manner, as shown on the moment-distribution table in Fig. 17.11(c).

Final Moments See the moment-distribution table and Fig. 17.11(d). **Ans.**

Example 17.6

Determine the member end moments for the three-span continuous beam shown in Fig. 17.12(a) due to the uniformly distributed load and due to the support settlements of $\frac{5}{8}$ in. at B, $1\frac{1}{2}$ in. at C, and $\frac{3}{4}$ in. at D. Use the moment-distribution method.

Solution

This beam was previously analyzed in Example 16.6 by the slope-deflection method.

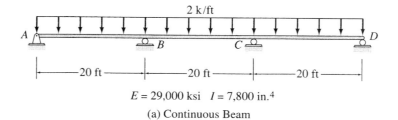

$E = 29{,}000$ ksi $I = 7{,}800$ in.^{4}

(a) Continuous Beam

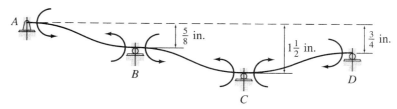

(b) Fixed-End Moments Due to Support Settlements

AB	BA	BC		CB	CD		DC
1	0.429	0.571		0.571	0.429		1
+1293.9	+1160.5	+1784.8		+1651.4	−1406		−1539.4
−1293.9	−1263.5	−1681.8		− 140.1	− 105.3		+1539.4
	− 647	− 70.1		− 840.9	+ 769.7		
	+ 307.6	+ 409.5		+ 40.7	+ 30.5		
		+ 20.4		+ 204.8			
	− 8.8	− 11.6		− 116.9	− 87.9		
		− 58.5		− 5.8			
	+ 25.1	+ 33.4		+ 3.3	+ 2.5		
		+ 1.7		+ 16.7			
	− 0.7	− 1.0		− 9.5	− 7.2		
		− 4.8		− 0.5			
	+ 2.1	+ 2.7		+ 0.3	+ 0.2		
		+ 0.2		+ 1.4			
	− 0.1	− 0.1		− 0.8	− 0.6		
		− 0.4					
	+ 0.2	+ 0.2					
0	− 424.6	+ 424.6		+ 804.1	− 804.1		0

(c) Moment-Distribution Table

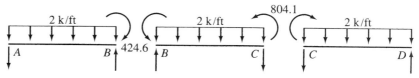

(d) Final Member End Moments (k-ft)

FIG. 17.12

continued

Distribution Factors At joint A,

$$DF_{AB} = 1$$

At joint B,

$$DF_{BA} = \frac{3I/80}{(3I/80) + (I/20)} = 0.429$$

$$DF_{BC} = \frac{I/20}{(3I/80) + (I/20)} = 0.571$$

At joint C,

$$DF_{CB} = \frac{I/20}{(3I/80) + (I/20)} = 0.571$$

$$DF_{CD} = \frac{3I/80}{(3I/80) + (I/20)} = 0.429$$

At joint D,

$$DF_{DC} = 1$$

Fixed-End Moments A qualitative deflected shape of the continuous beam with all joints clamped against rotation and subjected to the specified support settlements is depicted in Fig. 17.12(b) using an exaggerated scale. It can be seen from this figure that the relative settlements for the three members are $\Delta_{AB} = \frac{5}{8}$ in., $\Delta_{BC} = 1\frac{1}{2} - \frac{5}{8} = \frac{7}{8}$ in., and $\Delta_{CD} = 1\frac{1}{2} - \frac{3}{4} = \frac{3}{4}$ in. By using the fixed-end-moment expressions, we determine the fixed-end moments due to the support settlements to be

$$FEM_{AB} = FEM_{BA} = +\frac{6EI\Delta}{L^2} = +\frac{6(29{,}000)(7{,}800)\left(\frac{5}{8}\right)}{(20)^2(12)^3}$$

$$= +1{,}227.2 \text{ k-ft}$$

$$FEM_{BC} = FEM_{CB} = +\frac{6(29{,}000)(7{,}800)\left(\frac{7}{8}\right)}{(20)^2(12)^3} = +1{,}718.1 \text{ k-ft}$$

$$FEM_{CD} = FEM_{DC} = -\frac{6(29{,}000)(7{,}800)\left(\frac{3}{4}\right)}{(20)^2(12)^3} = -1{,}472.7 \text{ k-ft}$$

The fixed-end moments due to the 2-k/ft external load are

$$FEM_{AB} = FEM_{BC} = FEM_{CD} = +\frac{2(20)^2}{12} = +66.7 \text{ k-ft}$$

$$FEM_{BA} = FEM_{CB} = FEM_{DC} = -66.7 \text{ k-ft}$$

Thus, the total fixed-end moments due to the combined effect of the external load and the support settlements are

$$\text{FEM}_{AB} = +1{,}293.9 \text{ k-ft} \qquad \text{FEM}_{BA} = +1{,}160.5 \text{ k-ft}$$

$$\text{FEM}_{BC} = +1{,}784.8 \text{ k-ft} \qquad \text{FEM}_{CB} = +1{,}651.4 \text{ k-ft}$$

$$\text{FEM}_{CD} = -1{,}406 \text{ k-ft} \qquad \text{FEM}_{DC} = -1{,}539.4 \text{ k-ft}$$

Moment Distribution The moment distribution is carried out in the usual manner, as shown on the moment-distribution table in Fig. 17.12(c). Note that the joints A and D at the simple end supports are balanced only once and that no moments are carried over to these joints.

Final Moments See the moment-distribution table and Fig. 17.12(d). Ans.

17.4 ANALYSIS OF FRAMES WITHOUT SIDESWAY

The procedure for the analysis of frames without sidesway is similar to that for the analysis of continuous beams presented in the preceding section. However, unlike the continuous beams, more than two members may be connected to a joint of a frame. In such cases, care must be taken to record the computations in such a manner that mistakes are avoided. Whereas some engineers like to record the moment-distribution computations directly on a sketch of the frame, others prefer to use a tabular format for such purposes. We will use a tabular form for calculations, as illustrated by the following example.

Example 17.7

Determine the member end moments for the frame shown in Fig. 17.13(a) by using the moment-distribution method.

Solution

This frame was analyzed in Example 16.8 by the slope-deflection method.

Distribution Factors At joint C,

$$\text{DF}_{CA} = \frac{\left(\dfrac{800}{20}\right)}{\left(\dfrac{800}{20}\right) + \left(\dfrac{1600}{30}\right)} = 0.429 \qquad \text{DF}_{CD} = \frac{\left(\dfrac{1600}{30}\right)}{\left(\dfrac{800}{20}\right) + \left(\dfrac{1600}{30}\right)} = 0.571$$

$$\text{DF}_{CA} + \text{DF}_{CD} = 0.429 + 0.571 = 1 \qquad\qquad \text{Checks}$$

continued

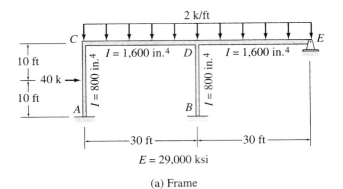

(a) Frame

Member Ends	AC		CA	CD		DC	DB	DE		ED		BD
Distribution Factors			0.429	0.571		0.4	0.3	0.3		1		
1. Fixed-end moments	+100		−100	+150		−150		+150		−150		
2. Balance joints			− 21.4	− 28.6						+150		
3. Carryover	− 10.7					− 14.3		+ 75				
4. Balance joints						− 24.3	−18.2	− 18.2				
5. Carryover				− 12.2								−9.1
6. Balance joints			+ 5.2	+ 7								
7. Carryover	+ 2.6					+ 3.5						
8. Balance joints						− 1.4	− 1.1	− 1.1				
9. Carryover				− 0.7								−0.6
10. Balance joints			+ 0.3	+ 0.4								
11. Carryover	+ 0.2					+ 0.2						
12. Balance joints						− 0.1	− 0.1	− 0.1				
13. Final moments	+ 92.1		−115.9	+115.9		−186.4	−19.4	+205.6		0		−9.7

Carryover

(b) Moment-Distribution Table

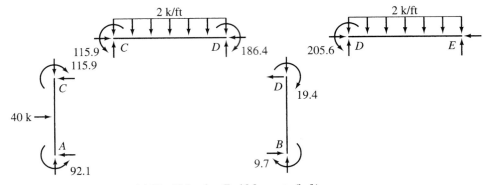

(c) Final Member End Moments (k-ft)

FIG. 17.13

At joint D,

$$DF_{DB} = \frac{\left(\dfrac{800}{20}\right)}{\left(\dfrac{800}{20}\right)+\left(\dfrac{1600}{30}\right)+\left(\dfrac{3}{4}\right)\left(\dfrac{1600}{30}\right)} = 0.3$$

$$DF_{DC} = \frac{\left(\dfrac{1600}{30}\right)}{\left(\dfrac{800}{20}\right)+\left(\dfrac{1600}{30}\right)+\left(\dfrac{3}{4}\right)\left(\dfrac{1600}{30}\right)} = 0.4$$

$$DF_{DE} = \frac{\left(\dfrac{3}{4}\right)\left(\dfrac{1600}{30}\right)}{\left(\dfrac{800}{20}\right)+\left(\dfrac{1600}{30}\right)+\left(\dfrac{3}{4}\right)\left(\dfrac{1600}{30}\right)} = 0.3$$

$$DF_{DB} + DF_{DE} + DF_{DC} = 2(0.3) + 0.4 = 1 \qquad \text{Checks}$$

At joint E,

$$DF_{ED} = 1$$

Fixed-End Moments By using the fixed-end moment expressions, we obtain

$FEM_{AC} = +100$ k-ft $\qquad\qquad FEM_{CA} = -100$ k-ft

$FEM_{BD} = FEM_{DB} = 0$

$FEM_{CD} = FEM_{DE} = +150$ k-ft $\qquad FEM_{DC} = FEM_{ED} = -150$ k-ft

Moment Distribution The moment-distribution process is carried out in tabular form, as shown in Fig. 17.13(b). The table, which is similar in form to those used previously for the analysis of continuous beams, contains one column for each member end of the structure. Note that the columns for all member ends, which are connected to the same joint, are grouped together, so that any unbalanced moment at the joint can be conveniently distributed among the members connected to it. Also, when the columns for two ends of a member cannot be located adjacent to each other, then an overhead arrow connecting the columns for the member ends may serve as a reminder to carry over moments from one end of the member to the other. In Fig. 17.13(b), such an arrow is used between the columns for the ends of member BD. This arrow indicates that a distributed moment at end D of member BD induces a carryover moment at the far end B. Note, however, that no moment can be carried over from end B to end D of member BD, because joint B, which is at a fixed support, will not be released during the moment-distribution process.

The moment distribution is carried out in the same manner as discussed previously for continuous beams. Note that any unbalanced moment at joint D must be distributed to the ends D of the three members connected to it in accordance with their distribution factors.

Final Moments The final member end moments are obtained by summing all the moments in each column of the moment-distribution table. Note that the final moments, which are recorded on the last line of the moment-distribution table and are depicted in Fig. 17.13(c), satisfy the equations of moment equilibrium at joints C and D of the frame. Ans.

17.5 ANALYSIS OF FRAMES WITH SIDESWAY

Thus far, we have considered the analysis of structures in which the translations of the joints were either zero or known (as in the case of support settlements). In this section, we apply the moment-distribution method to analyze frames whose joints may undergo both rotations and translations that have not been prescribed. As discussed in Section 16.4, such frames are commonly referred to as frames with sidesway.

Consider, for example, the rectangular frame shown in Fig. 17.14(a). A qualitative deflected shape of the frame for an arbitrary loading is also shown in the figure using an exaggerated scale. While the fixed joints A and B of the frame are completely restrained against rotation as well as translation, the joints C and D are free to rotate and translate. However, since the members of the frame are assumed to be inextensible and the deformations are assumed to be small, the joints C and D displace by the same amount, Δ, in the horizontal direction only, as shown in the figure.

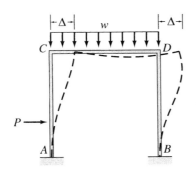

(a) Actual Frame —
M Moments

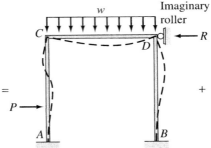

(b) Frame with Sidesway Prevented —
M_O Moments

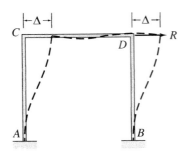

(c) Frame Subjected to R —
M_R Moments

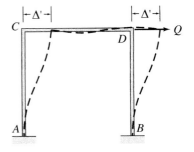

(d) Frame Subjected to an
Arbitrary Translation Δ' —
M_Q Moments

FIG. **17.14**

The moment-distribution analysis of such a frame, with sidesway, is carried out in two parts. In the first part, the sidesway of the frame is prevented by adding an imaginary roller to the structure, as shown in Fig. 17.14(b). External loads are then applied to this frame, and member end moments are computed by applying the moment-distribution process in the usual manner. With the member end moments known, the restraining force (reaction) R that develops at the imaginary support is evaluated by applying the equations of equilibrium.

In the second part of the analysis, the frame is subjected to the force R, which is applied in the opposite direction, as shown in Fig. 17.14(c). The moments that develop at the member ends are determined and superimposed on the moments computed in the first part (Fig. 17.14(b)) to obtain the member end moments in the actual frame (Fig. 17.14(a)). If M, M_O, and M_R denote, respectively, the member end moments in the actual frame, the frame with sidesway prevented, and the frame subjected to R, then we can write (see Fig. 17.14(a), (b), and (c))

$$M = M_O + M_R \qquad (17.27)$$

An important question that arises in the second part of the analysis is how to determine the member end moments M_R that develop when the frame undergoes sidesway under the action of R (Fig. 17.14(c)). Since the moment-distribution method cannot be used directly to compute the moments due to the known lateral load R, we employ an indirect approach in which the frame is subjected to an arbitrary known joint translation Δ' caused by an unknown load Q acting at the location and in the direction of R, as shown in Fig. 17.14(d). From the known joint translation, Δ', we determine the relative translation between the ends of each member, and we calculate the member fixed-end moments in the same manner as done previously in the case of support settlements. The fixed-end moments thus obtained are distributed by the moment-distribution process to determine the member end moments M_Q caused by the yet-unknown load Q. Once the member end moments M_Q have been determined, the magnitude of Q can be evaluated by the application of equilibrium equations.

With the load Q and the corresponding moments M_Q known, the desired moments M_R due to the lateral load R can now be determined easily by multiplying M_Q by the ratio R/Q; that is,

$$M_R = \left(\frac{R}{Q}\right) M_Q \qquad (17.28)$$

By substituting Eq. (17.28) into Eq. (17.27), we can express the member end moments in the actual frame (Fig. 17.14(a)) as

$$M = M_O + \left(\frac{R}{Q}\right) M_Q \qquad (17.29)$$

This method of analysis is illustrated by the following examples.

Example 17.8

Determine the member end moments for the frame shown in Fig. 17.15(a) by using the moment-distribution method.

Solution

This frame was analyzed in Example 16.10 by the slope-deflection method.

Distribution Factors At joint C,

$$DF_{CA} = DF_{CD} = \frac{I/7}{2(I/7)} = 0.5$$

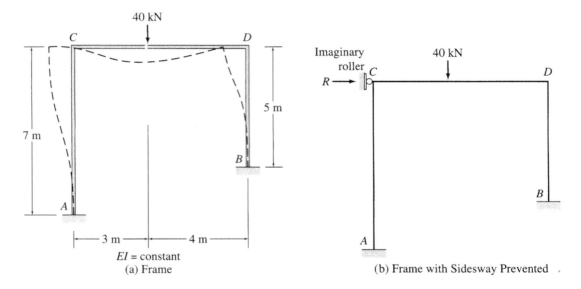

(a) Frame

(b) Frame with Sidesway Prevented

AC		CA	CD		DC	DB		BD
		0.5	0.5		0.417	0.583		
			+39.2		−29.4			
		−19.6	−19.6		+12.3	+17.1		
− 9.8			+ 6.2		− 9.8			+ 8.6
		− 3.1	− 3.1		+ 4.1	+ 5.7		
− 1.6			+ 2.1		− 1.6			+ 2.9
		− 1.1	− 1.1		+ 0.7	+ 0.9		
− 0.6			+ 0.4		− 0.6			+ 0.5
		− 0.2	− 0.2		+ 0.3	+ 0.3		
−12		−24	+23.9		−24	+24		+12

(c) Member End Moments for Frame with Sidesway Prevented —
M_O Moments

FIG. 17.15

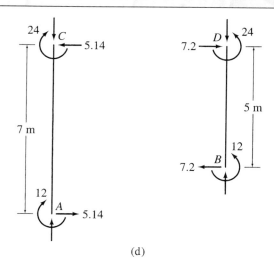

(d)

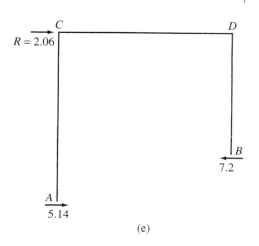

(e)

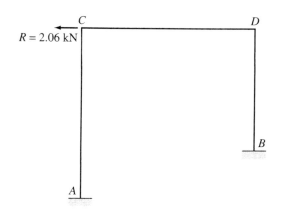

(f) Frame Subjected to $R = 2.06$ kN —
M_R Moments

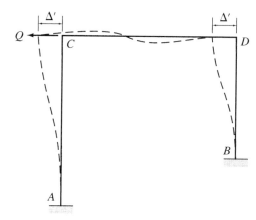

(g) Frame Subjected to an Arbitrary Translation
$\Delta' - M_Q$ Moments

FIG. 17.15 (contd.)

continued

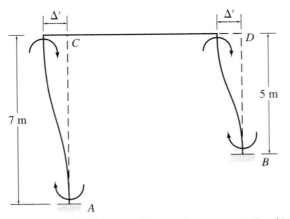

(h) Fixed-End Moments Due to Known Translation Δ′

AC		CA	CD		DC	DB		BD
		0.5	0.5		0.417	0.583		
−50		−50				−98		−98
		+25	+25	+40.9	+57.1			
+12.5			+20.5	+12.5				+28.6
		−10.3	−10.3	− 5.2	− 7.3			
− 5.2			− 2.6	− 5.2				− 3.7
		+ 1.3	+ 1.3	+ 2.2	+ 3			
+ 0.7			+ 1.1	+ 0.7				+ 1.5
		− 0.6	− 0.6	− 0.3	− 0.4			
− 0.3			− 0.2	− 0.3				− 0.2
		+ 0.1	+ 0.1	+ 0.1	+ 0.2			
−42.3		−34.5	+34.3		+45.4	−45.4		−71.8

(i) Member End Moments Due to Known Translation Δ′ — M_Q Moments

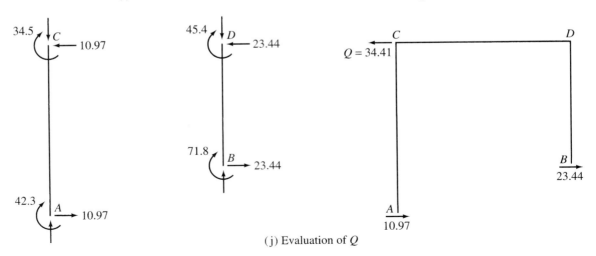

(j) Evaluation of Q

FIG. **17.15** (contd.)

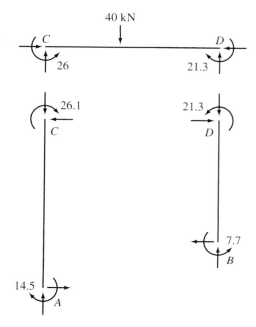

FIG. **17.15** (contd.)

(k) Actual Member End Moments (kN · m)

At joint D,

$$\text{DF}_{DC} = \frac{I/7}{(I/7)+(I/5)} = 0.417$$

$$\text{DF}_{DB} = \frac{I/5}{(I/7)+(I/5)} = 0.583$$

$$\text{DF}_{DC} + \text{DF}_{DB} = 0.417 + 0.583 = 1 \qquad \text{Checks}$$

Part I: Sidesway Prevented In the first part of the analysis, the sidesway of the frame is prevented by adding an imaginary roller at joint C, as shown in Fig. 17.15(b). Assuming that joints C and D of this frame are clamped against rotation, we calculate the fixed-end moments due to the external load to be

$$\text{FEM}_{CD} = +39.2 \text{ kN} \cdot \text{m} \qquad \text{FEM}_{DC} = -29.4 \text{ kN} \cdot \text{m}$$

$$\text{FEM}_{AC} = \text{FEM}_{CA} = \text{FEM}_{BD} = \text{FEM}_{DB} = 0$$

The moment-distribution of these fixed-end moments is then performed, as shown on the moment-distribution table in Fig. 17.15(c), to determine the member end moments M_O in the frame with sidesway prevented.

To evaluate the restraining force R that develops at the imaginary roller support, we first calculate the shears at the lower ends of the columns AC and

continued

BD by considering the moment equilibrium of the free bodies of the columns shown in Fig. 17.15(d). Next, by considering the equilibrium of the horizontal forces acting on the entire frame (Fig. 17.15(e)), we determine the restraining force *R* to be

$$+ \rightarrow \sum F_X = 0 \qquad R + 5.14 - 7.2 = 0$$

$$R = 2.06 \text{ kN} \rightarrow$$

Note that the restraining force acts to the right, indicating that if the roller would not have been in place, the frame would have swayed to the left.

Part II: Sidesway Permitted Since the actual frame is not supported by a roller at joint *C*, we neutralize the effect of the restraining force by applying a lateral load *R* = 2.06 kN in the opposite direction (i.e., to the left) to the frame, as shown in Fig. 17.15(f). As discussed previously, since the moment-distribution method cannot be used directly to compute member end moments M_R due to the lateral load *R* = 2.06 kN, we use an indirect approach in which the frame is subjected to an arbitrary known joint translation Δ′ caused by an unknown load *Q* acting at the location and in the direction of *R*, as shown in Fig. 17.15(g). Assuming that the joints *C* and *D* of the frame are clamped against rotation, as shown in Fig. 17.15(h), the fixed-end moments due to the translation Δ′ are given by

$$\text{FEM}_{AC} = \text{FEM}_{CA} = -\frac{6EI\Delta'}{(7)^2} = -\frac{6EI\Delta'}{49}$$

$$\text{FEM}_{BD} = \text{FEM}_{DB} = -\frac{6EI\Delta'}{(5)^2} = -\frac{6EI\Delta'}{25}$$

$$\text{FEM}_{CD} = \text{FEM}_{DC} = 0$$

in which negative signs have been assigned to the fixed-end moments for the columns, because these moments must act in the clockwise direction, as shown in Fig. 17.15(h).

Instead of arbitrarily assuming a numerical value for Δ′ to compute the fixed-end moments, it is usually more convenient to assume a numerical value for one of the fixed-end moments, evaluate Δ′ from the expression of that fixed-end moment, and use the value of Δ′ thus obtained to compute the remaining fixed-end moments. Thus, we arbitrarily assume the fixed-end moment FEM$_{AC}$ to be −50 kN · m; that is,

$$\text{FEM}_{AC} = \text{FEM}_{CA} = -\frac{6EI\Delta'}{49} = -50 \text{ kN} \cdot \text{m}$$

By solving for Δ′, we obtain

$$\Delta' = \frac{408.33}{EI}$$

By substituting this value of Δ' into the expressions for FEM_{BD} and FEM_{DB}, we determine the consistent values of these moments to be

$$\text{FEM}_{BD} = \text{FEM}_{DB} = -\frac{6(408.33)}{25} = -98 \text{ kN} \cdot \text{m}$$

The foregoing fixed-end moments are then distributed by the usual moment-distribution process, as shown in Fig. 17.15(i), to determine the member end moments M_Q caused by the yet-unknown load Q.

To evaluate the magnitude of Q that corresponds to these member end moments, we first calculate shears at the lower ends of the columns by considering their moment equilibrium (Fig. 17.15(j)) and then apply the equation of equilibrium in the horizontal direction to the entire frame:

$$+ \rightarrow \sum F_X = 0$$

$$-Q + 10.97 + 23.44 = 0$$

$$Q = 34.41 \text{ kN} \leftarrow$$

which indicates that the moments M_Q computed in Fig. 17.15(i) are caused by a lateral load $Q = 34.41$ kN. Since the moments are linearly proportional to the magnitude of the load, the desired moments M_R due to the lateral load $R = 2.06$ kN must be equal to the moments M_Q (Fig. 17.15(i)) multiplied by the ratio $R/Q = 2.06/34.41$.

Actual Member End Moments The actual member end moments, M, can now be determined by algebraically summing the member end moments M_O computed in Fig. 17.15(c) and 2.06/34.41 times the member end moments M_Q computed in Fig. 17.15(i). Thus

$$M_{AC} = -12 + \left(\frac{2.06}{34.41}\right)(-42.3) = -14.5 \text{ kN} \cdot \text{m} \qquad \text{Ans.}$$

$$M_{CA} = -24 + \left(\frac{2.06}{34.41}\right)(-34.5) = -26.1 \text{ kN} \cdot \text{m} \qquad \text{Ans.}$$

$$M_{CD} = 23.9 + \left(\frac{2.06}{34.41}\right)(34.3) = 26 \text{ kN} \cdot \text{m} \qquad \text{Ans.}$$

$$M_{DC} = -24 + \left(\frac{2.06}{34.41}\right)(45.4) = -21.3 \text{ kN} \cdot \text{m} \qquad \text{Ans.}$$

$$M_{DB} = 24 + \left(\frac{2.06}{34.41}\right)(-45.4) = 21.3 \text{ kN} \cdot \text{m} \qquad \text{Ans.}$$

$$M_{BD} = 12 + \left(\frac{2.06}{34.41}\right)(-71.8) = 7.7 \text{ kN} \cdot \text{m} \qquad \text{Ans.}$$

These moments are depicted in Fig. 17.15(k).

Example 17.9

Determine the reactions for the nonprismatic beam shown in Fig. 17.16(a) by using the moment-distribution method.

Solution

Since the stiffness and carryover relationships derived in Section 17.1 as well as the expressions of fixed-end moments given inside the back cover of the book are valid only for prismatic members, we will analyze the given nonprismatic beam as if it were composed of two prismatic members, AB and BC, rigidly connected at joint B. Note that joint B is free to rotate as well as translate in the vertical direction, as shown in Fig. 17.16(a).

Distribution Factors The distribution factors at joint B are

$$DF_{BA} = \frac{I/30}{(I/30) + (2I/18)} = 0.231$$

$$DF_{BC} = \frac{2I/18}{(I/30) + (2I/18)} = 0.769$$

Part I: Joint Translation Prevented In this part of the analysis, the translation of joint B is prevented by an imaginary roller, as shown in Fig. 17.16(b). The fixed-end moments due to the external load are

$$FEM_{AB} = +150 \text{ k-ft} \qquad FEM_{BA} = -150 \text{ k-ft}$$

$$FEM_{BC} = +54 \text{ k-ft} \qquad FEM_{CB} = -54 \text{ k-ft}$$

The moment distribution of these fixed-end moments is performed, as shown in Fig. 17.16(b), to determine the member end moments M_O. The restraining force R at the imaginary roller support is then evaluated by considering the equilibrium of members AB and BC and of joint B as shown in Fig. 17.16(c). The restraining force is found to be

$$R = 53.04 \text{ k} \uparrow$$

Part II: Joint Translation Permitted Since the actual beam is not supported by a roller at joint B, we neutralize its restraining effect by applying a downward load $R = 53.04$ k to the beam, as shown in Fig. 17.16(d). To determine the member end moments M_R due to R, we subject the beam to an arbitrary known translation Δ', as shown in Fig. 17.16(e). The fixed-end moments due to Δ' are given by (see Fig. 17.16(f))

$$FEM_{AB} = FEM_{BA} = \frac{6EI\Delta'}{(30)^2} = \frac{EI\Delta'}{150}$$

$$FEM_{BC} = FEM_{CB} = -\frac{6E(2I)\Delta'}{(18)^2} = -\frac{EI\Delta'}{27}$$

If we arbitrarily assume that

$$FEM_{BC} = FEM_{CB} = -\frac{EI\Delta'}{27} = -100 \text{ k-ft}$$

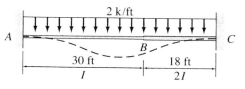

(a) Beam

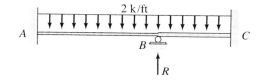

(b) Beam with Joint Translation Prevented —
M_O Moments

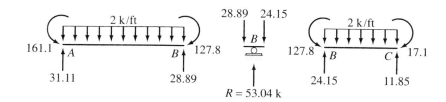

(c) Evaluation of Restraining Force R

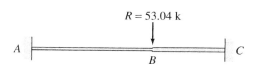

(d) Beam Subjected to $R = 53.04$ k —
M_R Moments

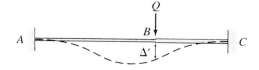

(e) Beam Subjected to an Arbitrary
Translation Δ' — M_Q Moments

FIG. **17.16**

continued

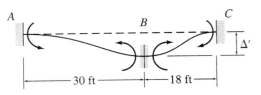

(f) Fixed-End Moments Due to
Known Translation Δ'

	0.231	0.769	
+18	+18	−100	−100
	+18.9	+ 63.1	
+ 9.5			+ 31.5
+27.5	+36.9	− 36.9	− 68.5

(g) Member End Moments Due to Known
Translation Δ' — M_Q Moments

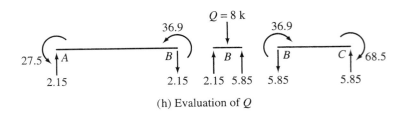

(h) Evaluation of Q

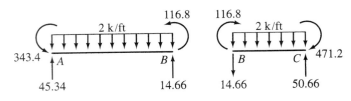

(i) Actual Member End Moments (k-ft)

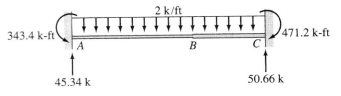

(j) Support Reactions

FIG. 17.16 (contd.)

then

$$EI\Delta' = 2{,}700$$

and, therefore,

$$\text{FEM}_{AB} = \text{FEM}_{BA} = \frac{2{,}700}{150} = 18 \text{ k-ft}$$

These fixed-end moments are distributed by the moment-distribution process, as shown in Fig. 17.16(g), to determine the member end moments M_Q. The load Q at the location and in the direction of R that corresponds to these moments can now be evaluated by considering equilibrium of members AB and BC and of joint B, as shown in Fig. 17.16(h). The magnitude of Q is found to be

$$Q = 8 \text{ k} \downarrow$$

Thus, the desired moments M_R due to the vertical load $R = 53.04$ k (Fig. 17.16(d)) must be equal to the moments M_Q (Fig. 17.16(g)) multiplied by the ratio $R/Q = 53.04/8 = 6.63$.

Actual Member End Moments The actual member end moments, M, can now be determined by algebraically summing the member end moments M_O computed in Fig. 17.16(b) and 6.63 times the member end moments M_Q computed in Fig. 17.16(g).

$$M_{AB} = 161.1 + 6.63(27.5) = 343.4 \text{ k-ft} \qquad \text{Ans.}$$
$$M_{BA} = -127.8 + 6.63(36.9) = 116.8 \text{ k-ft} \qquad \text{Ans.}$$
$$M_{BC} = 127.8 + 6.63(-36.9) = -116.8 \text{ k-ft} \qquad \text{Ans.}$$
$$M_{CB} = -17.1 + 6.63(-68.5) = -471.2 \text{ k-ft} \qquad \text{Ans.}$$

The member end shears obtained by applying equations of equilibrium are shown in Fig. 17.16(i).

Support Reactions See Fig. 17.16(j).   **Ans.**

Equilibrium Check The equilibrium equations check.

Example 17.10

Determine the member end moments and reactions for the frame shown in Fig. 17.17(a) by using the moment-distribution method.

Solution

Distribution Factors At joint C,

$$\text{DF}_{CA} = \text{DF}_{CD} = \frac{I/20}{2(I/20)} = 0.5$$

continued

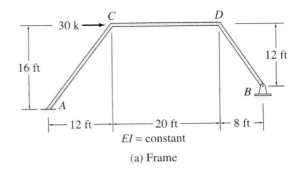

(a) Frame

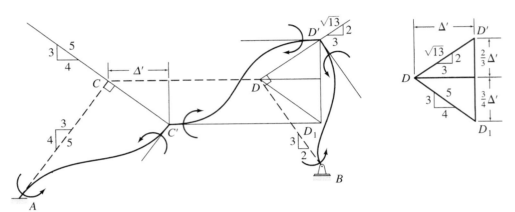

(b) Fixed-End Moments Due to an Arbitrary Translation Δ'

AC		CA	CD		DC	DB		BD
		0.5	0.5		0.49	0.51		1
+54.1		+54.1	−61.3		−61.3	+100		+100
		+ 3.6	+ 3.6		−19	− 19.7		−100
+ 1.8			− 9.5		+ 1.8	− 50		
		+ 4.8	+ 4.8		+23.6	+ 24.6		
+ 2.4			+11.8		+ 2.4			
		− 5.9	− 5.9		− 1.2	− 1.2		
− 3			− 0.6		− 3			
		+ 0.3	+ 0.3		+ 1.5	+ 1.5		
+ 0.2			+ 0.8		+ 0.2			
		− 0.4	− 0.4		− 0.1	− 0.1		
− 0.2					− 0.2			
					+ 0.1	+ 0.1		
+55.3		+56.5	−56.4		−55.2	+ 55.2		0

(c) Member End Moments Due to Known Translation Δ' — M_Q Moments

FIG. 17.17

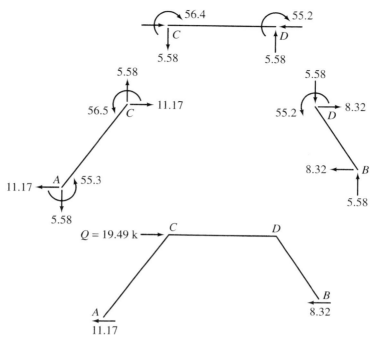

(d) Evaluation of Q

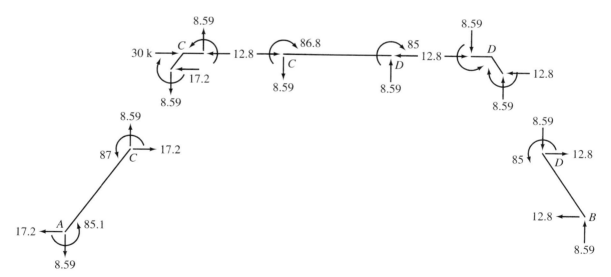

(e) Member End Moments and Forces

FIG. **17.17** (contd.)

continued

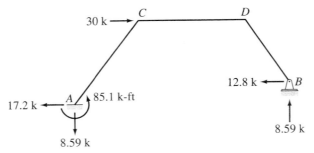

30 k →

C   D

12.8 k ← B

17.2 k ← A ↑ 85.1 k-ft

8.59 k

8.59 k

FIG. 17.17 (contd.)   (f) Support Reactions

At joint D,

$$DF_{DC} = \frac{I/20}{(I/20) + (3/4)(I/14.42)} = 0.49$$

$$DF_{DB} = \frac{(3/4)(I/14.42)}{(I/20) + (3/4)(I/14.42)} = 0.51$$

Member End Moments Due to an Arbitrary Sidesway Δ' Since no external loads are applied to the members of the frame, the member end moments M_O in the frame restrained against sidesway will be zero. To determine the member end moments M due to the 30-k lateral load, we subject the frame to an arbitrary known horizontal translation Δ' at joint C. Figure 17.17(b) shows a qualitative deflected shape of the frame with all joints clamped against rotation and subjected to the horizontal displacement Δ' at joint C. The procedure for constructing such deflected shapes was discussed in Section 16.5. Note that, since the frame members are assumed to be inextensible and deformations are assumed to be small, an end of a member can translate only in the direction perpendicular to the member. From this figure, we can see that the relative translation Δ_{AC} between the ends of member AC in the direction perpendicular to the member can be expressed in terms of the joint translation Δ' as

$$\Delta_{AC} = CC' = \frac{5}{4}\Delta' = 1.25\Delta'$$

Similarly, the relative translations for members CD and BD are given by

$$\Delta_{CD} = D_1D' = \frac{2}{3}\Delta' + \frac{3}{4}\Delta' = 1.417\Delta'$$

$$\Delta_{BD} = DD' = \frac{\sqrt{13}}{3}\Delta' = 1.202\Delta'$$

The fixed-end moments due to the relative translations are

$$FEM_{AC} = FEM_{CA} = \frac{6EI(1.25\Delta')}{(20)^2}$$

$$FEM_{CD} = FEM_{DC} = -\frac{6EI(1.417\Delta')}{(20)^2}$$

$$FEM_{BD} = FEM_{DB} = \frac{6EI(1.202\Delta')}{(14.42)^2}$$

in which, as shown in Fig. 17.17(b), the fixed-end moments for members AC and BD are counterclockwise (positive), whereas those for member CD are clockwise (negative). If we arbitrarily assume that

$$\text{FEM}_{BD} = \text{FEM}_{DB} = \frac{6EI(1.202\Delta')}{(14.42)^2} = 100 \text{ k-ft}$$

then

$$EI\Delta' = 2{,}883.2$$

and, therefore,

$$\text{FEM}_{AC} = \text{FEM}_{CA} = 54.1 \text{ k-ft}$$

$$\text{FEM}_{CD} = \text{FEM}_{DC} = -61.3 \text{ k-ft}$$

These fixed-end moments are distributed by the moment-distribution process, as shown in Fig. 17.17(c), to determine the member end moments M_Q.

To determine the magnitude of the load Q that corresponds to the member end moments computed in Fig. 17.17(c), we first calculate the shears at the ends of the girder CD by considering the moment equilibrium of the free body of the girder shown in Fig. 17.17(d). The girder shears (5.58 k) thus obtained are then applied to the free bodies of the inclined members AC and BD, as shown in the figure. Next, we apply the equations of moment equilibrium to members AC and BD to calculate the horizontal forces at the lower ends of these members. The magnitude of Q can now be determined by considering the equilibrium of horizontal forces acting on the entire frame as (see Fig. 17.17(d))

$$+ \rightarrow \sum F_x = 0$$

$$Q - 11.17 - 8.32 = 0$$

$$Q = 19.49 \text{ k} \rightarrow$$

Actual Member End Moments   The actual member end moments, M, due to the 30-k lateral load can now be evaluated by multiplying the moments M_Q computed in Fig. 17.17(c) by the ratio $30/Q = 30/19.49$:

$$M_{AC} = \frac{30}{19.49}(55.3) = 85.1 \text{ k-ft} \qquad \text{Ans.}$$

$$M_{CA} = \frac{30}{19.49}(56.5) = 87 \text{ k-ft} \qquad \text{Ans.}$$

$$M_{CD} = \frac{30}{19.49}(-56.4) = -86.8 \text{ k-ft} \qquad \text{Ans.}$$

$$M_{DC} = \frac{30}{19.49}(-55.2) = -85 \text{ k-ft} \qquad \text{Ans.}$$

$$M_{DB} = \frac{30}{19.49}(55.2) = 85 \text{ k-ft} \qquad \text{Ans.}$$

$$M_{BD} = 0 \qquad \text{Ans.}$$

Member End Forces   See Fig. 17.17(e).

Support Reactions   See Fig. 17.17(f).   Ans.

Equilibrium Check   The equilibrium equations check.

Analysis of Multistory Frames

The foregoing procedure can be extended to the analysis of structures with multiple degrees of freedom of sidesway. Consider the two-story rectangular frame shown in Fig. 17.18(a). The moment-distribution analysis of this frame is carried out in three parts. In the first part, the sidesway of both floors of the frame is prevented by adding imaginary rollers at the floor levels, as shown in Fig. 17.18(b). Member end moments M_O that develop in this frame due to the external loads are computed by the moment-distribution process, and the restraining forces R_1 and R_2 at the imaginary supports are evaluated by applying the equations of equilibrium. In the second part of the analysis, the lower floor of the frame is allowed to displace by a known amount Δ_1' while the sidesway of the upper floor is prevented, as shown in Fig. 17.18(c). The fixed-end moments caused by this displacement are computed and distributed to obtain the member end moments M_{Q1}. With the member end

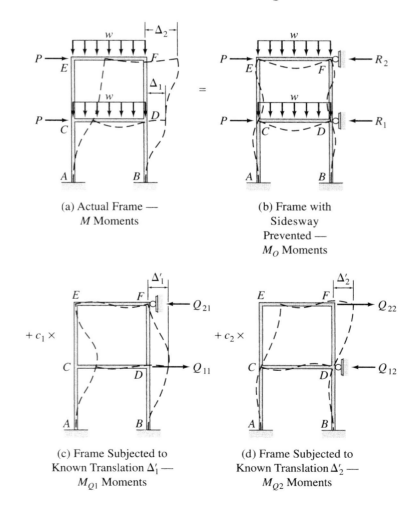

(a) Actual Frame —
M Moments

(b) Frame with
Sidesway
Prevented —
M_O Moments

(c) Frame Subjected to
Known Translation Δ_1' —
M_{Q1} Moments

(d) Frame Subjected to
Known Translation Δ_2' —
M_{Q2} Moments

FIG. 17.18

moments known, the forces Q_{11} and Q_{21} at the locations of the roller supports are determined from the equilibrium equations. Similarly, in the third part of the analysis, the upper floor of the frame is allowed to displace by a known amount Δ_2', as shown in Fig. 17.18(d), and the corresponding member end moments M_{Q2}, and the forces Q_{12} and Q_{22}, are evaluated. The member end moments M in the actual frame (Fig. 17.18(a)) are determined by superposition of the moments computed in the three parts as

$$M = M_O + c_1 M_{Q1} + c_2 M_{Q2} \qquad (17.30)$$

in which c_1 and c_2 are the constants whose values are obtained by solving the equations of superposition of horizontal forces at the locations of the imaginary supports. By superimposing the horizontal forces shown in Fig. 17.18(a) through (d) at joints D and F, respectively, we obtain

$$-R_1 + c_1 Q_{11} - c_2 Q_{12} = 0$$
$$-R_2 - c_1 Q_{21} + c_2 Q_{22} = 0$$

By solving these equations simultaneously, we obtain the values of the constants c_1 and c_2, which are then used in Eq. (17.30) to determine the desired member end moments, M.

As the foregoing discussion indicates, the analysis of multistory frames by the moment-distribution method can be quite tedious and time consuming. Therefore, the analysis of such structures is performed today on computers using the matrix formulation of the displacement method presented in Chapter 18.

SUMMARY

In this chapter we have studied a classical formulation of the displacement (stiffness) method, called the moment-distribution method, for the analysis of beams and frames.

The procedure for the analysis of continuous beams and frames without sidesway essentially involves computing fixed-end moments due to the external loads by assuming that all the free joints of the structure are temporarily restrained against rotation and balancing the moments at free joints by the moment-distribution process. In the moment-distribution process, at each free joint of the structure, the unbalanced moment is evaluated and distributed to the member ends connected to it. Carryover moments induced at the far ends of the members are then computed, and the process of balancing the joints and carrying over moments is repeated until the unbalanced moments are negligibly small. The final member end moments are obtained by algebraically summing the fixed-end moment and all the distributed and carryover moments at each member end.

The analysis of frames with a single degree of freedom of sidesway is carried out in two parts. In the first part, the sidesway is prevented by the addition of an imaginary roller to the structure. Member end moments that develop in this restrained frame, due to the external loads, are computed by the moment-distribution process; and the restraining force R at the imaginary roller is evaluated by the application of the equations of equilibrium. In the second part of the analysis, to calculate the member moments due to the force R applied in the opposite direction, the structure is allowed to displace by an arbitrarily assumed *known* amount; and the member moments and the corresponding force Q at the location of R are evaluated as before. The actual member end moments are determined by algebraically summing the moments computed in the first part and R/Q times the moments computed in the second part.

Once member end moments are known, member end shears, member axial forces, and support reactions can be evaluated through equilibrium considerations.

PROBLEMS

Section 17.3

17.1 through 17.5 Determine the reactions and draw the shear and bending moment diagrams for the beams shown in Figs. P17.1–P17.5 by using the moment-distribution method.

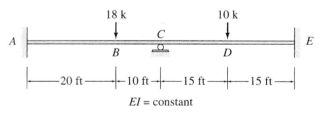

EI = constant

FIG. P17.1

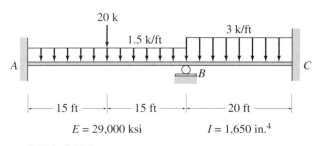

E = 29,000 ksi I = 1,650 in.^{4}

FIG. P17.2, P17.6

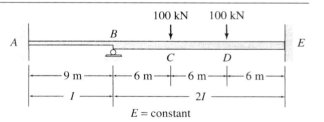

E = constant

FIG. P17.3

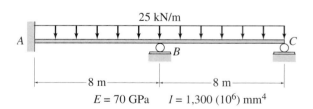

E = 70 GPa I = 1,300 (10^6) mm^4

FIG. P17.4, P17.7

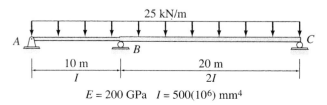

E = 200 GPa I = 500(10^6) mm^4

FIG. P17.5

17.6 Solve Problem 17.2 for the loading shown in Fig. P17.2 and a settlement of $\frac{1}{2}$ in. at support B.

17.7 Solve Problem 17.4 for the loading shown in Fig. P17.4 and the support settlements of 50 mm at B and 25 mm at C.

17.8 through 17.14 Determine the reactions and draw the shear and bending moment diagrams for the beams shown in Figs. P17.8–P17.14 by using the moment-distribution method.

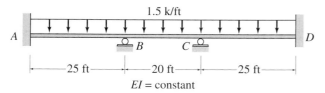

FIG. **P17.8**

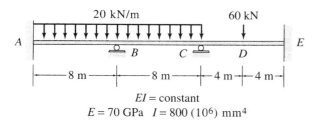

FIG. **P17.9, P17.15**

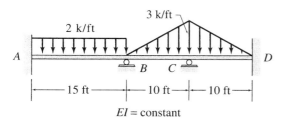

FIG. **P17.10**

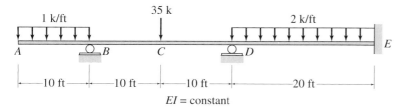

FIG. **P17.11**

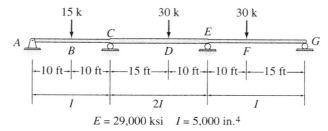

$E = 29{,}000$ ksi $I = 5{,}000$ in.^{4}

FIG. **P17.12, P17.16**

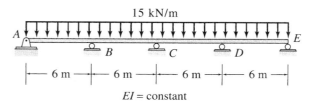

FIG. **P17.13**

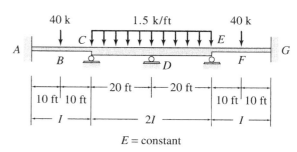

FIG. **P17.14**

17.15 Solve Problem 17.9 for the loading shown in Fig. P17.9 and a settlement of 25 mm at support C.

17.16 Solve Problem 17.12 for the loading shown in Fig. P17.12 and the support settlements of $\frac{1}{2}$ in. at A, 4 in. at C, 3 in. at E, and $2\frac{1}{2}$ in. at G.

Section 17.4

17.17 through 17.20 Determine the member end moments and reactions for the frames shown in Figs. P17.17–P17.20 by using the moment-distribution method.

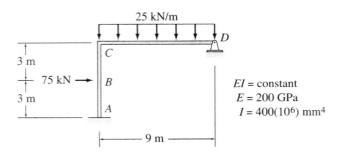

FIG. **P17.17, P17.21**

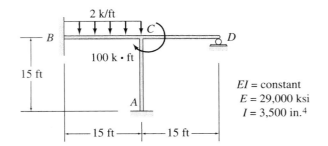

FIG. **P17.18, P17.22**

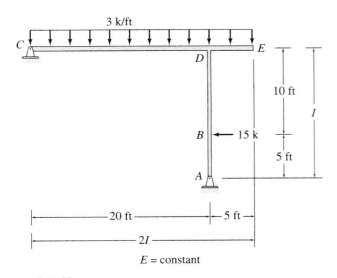

FIG. **P17.19**

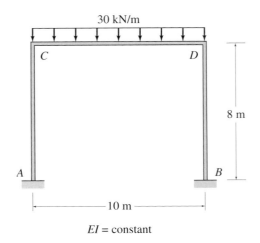

FIG. **P17.20**

17.21 Solve Problem 17.17 for the loading shown in Fig. P17.17 and a settlement of 50 mm at support D.

17.22 Solve Problem 17.18 for the loading shown in Fig. P17.18 and a settlement of $\frac{1}{4}$ in. at support A.

17.23 Determine the member end moments and reactions for the frame of Fig. P17.23 for the loading shown in the figure and the support settlements of 1 in. at A and $1\frac{1}{2}$ in. at D. Use the moment-distribution method.

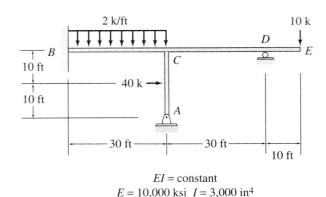

FIG. **P17.23**

Section 17.5

17.24 through 17.31 Determine the member end moments and reactions for the frames shown in Figs. P17.24–P17.31 by using the moment-distribution method.

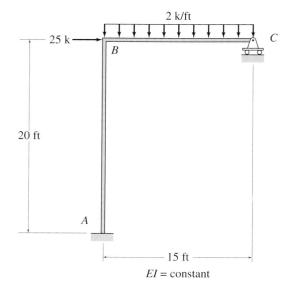

FIG. P17.24

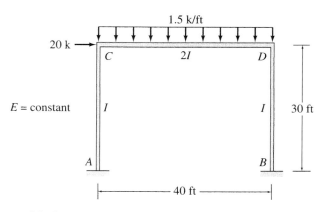

FIG. P17.25

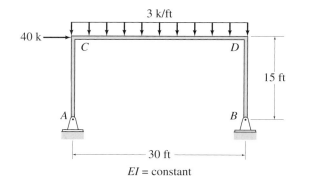

FIG. P17.26

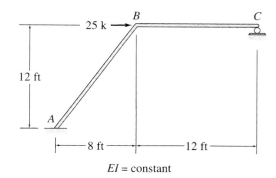

FIG. P17.27

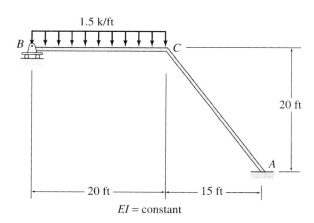

FIG. P17.28

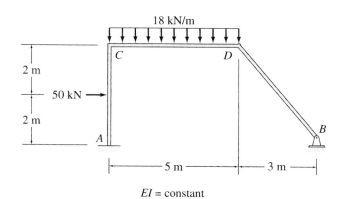

FIG. P17.29

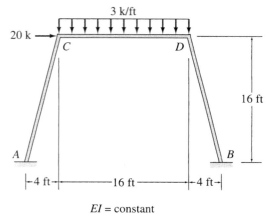

FIG. P17.30

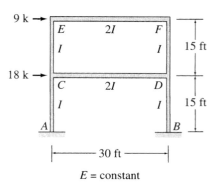

FIG. P17.31

18

Introduction to Matrix Structural Analysis

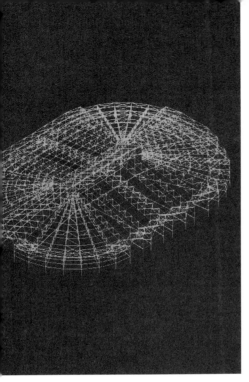

Analytical Model of the Superstructure of the Anaheim Arena, California
Photo Courtesy of Thornton-Tomasetti Engineers

In this text we have focused our attention on the classical methods of structural analysis. Although a study of classical methods is essential for developing an understanding of structural behavior and the principles of structural analysis, the analysis of large structures by using these hand-calculation methods can be quite time consuming. With the availability of inexpensive, yet powerful, microcomputers, the analysis of structures in most design offices is routinely performed today on computers using software based on matrix methods of structural analysis.

The objective of this chapter is to introduce the reader to the exciting and still-growing field of matrix structural analysis. However, only the basic concepts of matrix analysis are presented herein. For a more detailed study, the reader should refer to one of the many textbooks devoted entirely to the subject of matrix structural analysis.

Matrix methods do not involve any new fundamental principles; but the fundamental relationships of equilibrium, compatibility, and member force-displacement relations are now expressed in the form of matrix equations, so that the numerical computations can be efficiently performed on a computer. Therefore, familiarity with the basic operations of matrix algebra is a prerequisite to understanding matrix structural analysis. A review of the concepts of matrix algebra necessary for formulating the matrix methods of structural analysis is presented in Appendix B for the convenience of the reader.

Although both the flexibility (force) and the stiffness (displacement) methods can be expressed in matrix form, the stiffness method is more systematic and can be more easily implemented on computers. Thus, most of the commercially available computer programs for structural analysis are based on the stiffness method. In this chapter, we will consider only the matrix stiffness (displacement) method of structural analysis. This method can be used to analyze statically determinate as well as indeterminate structures.

We begin by discussing the process of preparing an analytical model of the structure to be analyzed. We also define global and local coordinate systems and explain the concept of degrees of freedom. Next we derive member force-displacement relations in local coordinates. We consider the transformation of member end forces and end displacements from local to global coordinates and vice versa, and develop the member stiffness relations in global coordinates. We formulate the stiffness relations for the entire structure by combining the member stiffness relations and, finally, develop a step-by-step procedure for the analysis of trusses, continuous beams, and frames by the matrix stiffness method.

18.1 ANALYTICAL MODEL

In the matrix stiffness method of analysis, the structure is considered to be an assemblage of straight members connected at their ends to joints. *A member is defined as a part of the structure for which the member force-displacement relations to be used in the analysis are valid.* In other words, given the displacements of the ends of a member, one should be able to determine the forces and moments at its ends by using the force-displacement relations. Such relations for prismatic members will be derived in the following section. *A joint is defined as a structural part of infinitesimal size to which the member ends are connected.* The members and joints of structures are also referred to as *elements* and *nodes*, respectively.

Before proceeding with the analysis, an analytical model of the structure must be prepared. The model is represented by a line diagram of the structure, on which all the joints and members are identified by numbers. Consider, for example, the frame shown in Fig. 18.1(a). The analytical model of the frame is shown in Fig. 18.1(b), in which the joint numbers are enclosed within circles to distinguish them from the member numbers, which are enclosed within rectangles. As shown in this figure, the frame is considered to be composed of four members and five joints for the purpose of analysis. Note that, since the member force-displacement relations to be used in the analysis are valid for prismatic members only, the vertical column of the frame has been subdivided into two members, each with constant cross-sectional properties (I and A) along its length.

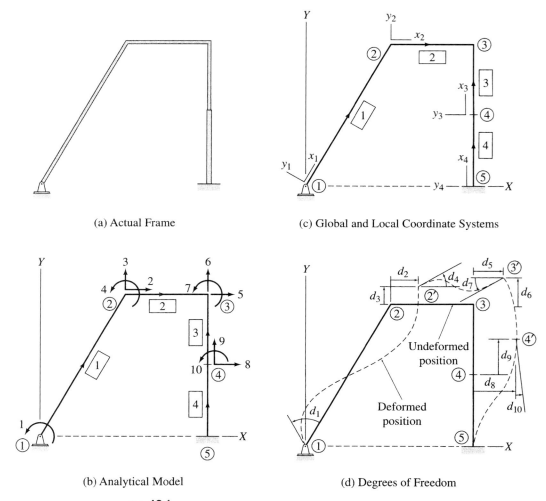

(a) Actual Frame

(c) Global and Local Coordinate Systems

(b) Analytical Model

(d) Degrees of Freedom

FIG. 18.1

Global and Local Coordinate Systems

In the stiffness method, the overall geometry and behavior of the structure are described with reference to a Cartesian or rectangular *global (or structural) coordinate system*. The global coordinate system used in this chapter is a right-handed XYZ coordinate system, with the plane structure lying in the XY plane, as shown in Fig. 18.1(b).

Since it is usually convenient to derive the basic force-displacement relations in terms of the forces and displacements in the directions along and perpendicular to members, a *local (or member) coordinate system* is defined for each member of the structure. The origin of the local xyz coordinate system for a member may be arbitrarily located at one of the ends of the member, with the x axis directed along the centroidal axis of the member. The positive direction of the y axis is chosen so that the coordinate system is right-handed, with the local z axis pointing in the positive direction of the global Z axis. In Fig. 18.1(b), the positive direction of the x axis for each member is indicated by drawing an arrow along each member on the line diagram of the structure. For example, this figure indicates that the origin of the local coordinate system for member 1 is located at its end connected to joint 1, with the x_1 axis directed from joint 1 to joint 2. The joint to which the member end with the origin of the local coordinate system is connected is referred to as the *beginning joint* for the member, whereas the joint adjacent to the opposite end of the member is termed the *end joint*. For example, in Fig. 18.1(b), member 1 begins at joint 1 and ends at joint 2, whereas member 2 begins at joint 2 and ends at joint 3, and so on. Once the local x axis is defined for a member, the corresponding y axis can be established by applying the right-hand rule. The local y axes thus obtained for the members of the frame under consideration are shown in Fig. 18.1(c). Note that, for each member, if we curl the fingers of our right hand from the direction of the x axis toward the direction of the corresponding y axis, then our extended thumb points out of the plane of the page, which is the positive direction of the global Z axis.

Degrees of Freedom

The degrees of freedom of a structure are the independent joint displacements (translations and rotations) that are necessary to specify the deformed shape of the structure when subjected to an arbitrary loading. Consider again the plane frame of Fig. 18.1(a). The deformed shape of the frame, for an arbitrary loading, is depicted in Fig. 18.1(d) using an exaggerated scale. Unlike in the case of the classical methods of analysis considered previously, it is usually not necessary to neglect member axial deformations when analyzing frames by the matrix stiffness method. From Fig. 18.1(d), we can see that joint 1, which is located at the hinged support, can rotate, but it cannot translate. Thus joint 1 has only one degree of freedom, which is designated as d_1 in the figure. Since joint 2 of the frame is not attached to a support, three displacements—the translations d_2 and d_3 in the X and Y directions, respectively, and the rotation d_4 about the Z axis—are needed to completely specify its deformed position $2'$. Thus joint 2 has three degrees of freedom. Similarly, joints 3 and 4, which are also *free* joints, have three degrees of freedom each. Finally, joint 5, which is attached to the fixed support, can neither translate nor rotate; therefore, it does not have any degrees of freedom. Thus, the entire frame has a total of ten degrees of freedom. As shown in Fig. 18.1(d), the joint displacements are defined relative to the global coordinate system, with joint translations considered as positive when in the positive directions of the X and Y axes and joint rotations considered as positive when counterclockwise. Note that all the joint displacements are shown in the positive sense in Fig. 18.1(d). The joint displacements of the frame can be collectively written in matrix form as

$$\mathbf{d} = \begin{bmatrix} d_1 \\ d_2 \\ \vdots \\ d_9 \\ d_{10} \end{bmatrix}$$

in which $\mathbf{d}$ is termed the *joint displacement vector* of the structure.

When applying the stiffness method, it is not necessary to draw the deformed shape of the structure, as shown in Fig. 18.1(d), to identify its degrees of freedom. Instead, the degrees of freedom can be directly specified on the line diagram of the structure by drawing arrows at the joints, as shown in Fig. 18.1(b). As indicated in this figure, the degrees of freedom are numbered by starting at the lowest joint number and proceeding sequentially to the highest joint number. In the case of more than one degree of freedom at a joint, the translation in the X direction is numbered first, followed by the translation in the Y direction, and then the rotation.

In continuous beams subjected to lateral loads, the axial deformations of members are zero. Therefore, it is not necessary to consider the joint displacements in the direction of the beam's centroidal axis in the analysis. Thus a joint of a plane continuous beam can have up to two degrees of freedom, namely, a translation perpendicular to the beam's centroidal axis and a rotation. For example, the continuous beam of Fig. 18.2(a) has four degrees of freedom, as shown in Fig. 18.2(b).

Since the joints of trusses are assumed to be frictionless hinges, they are not subjected to moments; therefore, their rotations are zero. Thus, when analyzing plane trusses, only two degrees of freedom, namely, translations in the global X and Y directions, need to be considered for each joint. For example, the truss of Fig. 18.3(a) has three degrees of freedom, as shown in Fig. 18.3(b).

(a) Actual Continuous Beam

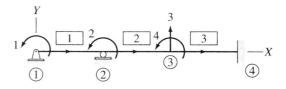

FIG. 18.2

(b) Analytical Model and Degrees of Freedom

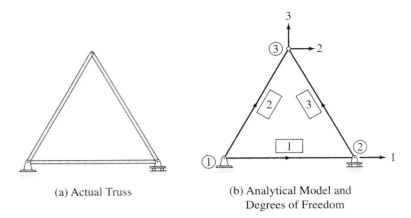

(a) Actual Truss

(b) Analytical Model and
Degrees of Freedom

FIG. **18.3**

18.2 MEMBER STIFFNESS RELATIONS IN LOCAL COORDINATES

In the matrix stiffness method of analysis, the joint displacements of the structure are determined by solving a system of simultaneous equations, which is expressed in the form

$$\bar{\mathbf{P}} = \mathbf{Sd} \tag{18.1}$$

in which $\mathbf{d}$ denotes the joint displacement vector, as discussed previously; $\bar{\mathbf{P}}$ represents the effects of external loads at the joints of the structure; and $\mathbf{S}$ is called the *structure stiffness matrix*. As will be discussed in Section 18.5, the stiffness matrix for the entire structure, $\mathbf{S}$, is obtained by assembling the stiffness matrices for the individual members of the structure. *The stiffness matrix for a member is used to express the forces at the ends of the member as functions of the displacements of the member's ends.* Note that the terms *forces* and *displacements* are used here in the general sense to include moments and rotations, respectively. In this section, we derive stiffness matrices for the members of plane frames, continuous beams, and plane trusses in the local coordinate systems of the members.

Frame Members

To establish the stiffness relationships for the members of plane frames, let us focus our attention on an arbitrary prismatic member m of the frame shown in Fig. 18.4(a). When the frame is subjected to external loads, member m deforms and internal forces are induced at its ends. The undeformed and deformed positions of the member are shown in Fig. 18.4(b). As indicated in this figure, three displacements—translations in the x and y directions and rotation about the z axis—are needed to completely specify the deformed position of each end of the

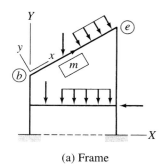

(a) Frame

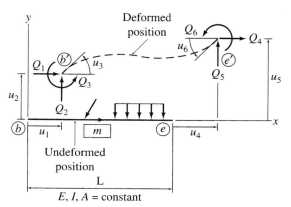

(b) Frame Member — Local Coordinates

$=$

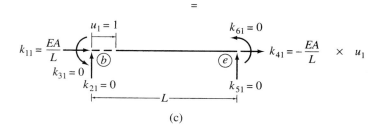

(c)

$+$

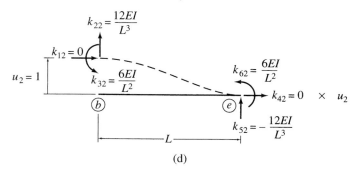

(d)

FIG. 18.4

$+$

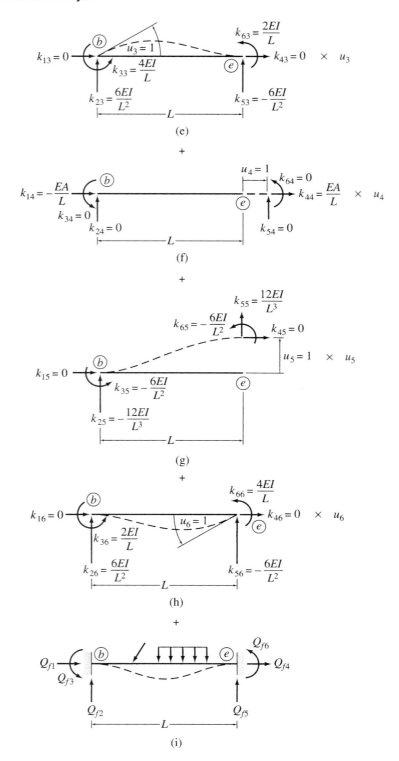

FIG. **18.4** (contd.)

member. Thus the member has a total of six end displacements or degrees of freedom. As shown in Fig. 18.4(b), the member end displacements are denoted by u_1 through u_6, and the corresponding member end forces are denoted by Q_1 through Q_6. Note that these end displacements and forces are defined relative to the local coordinate system of the member, with translations and forces considered as positive when in the positive directions of the local x and y axes, and rotations and moments considered as positive when counterclockwise. As indicated in Fig. 18.4(b), the member end displacements and forces are numbered by beginning at the member end b, where the origin of the local coordinate system is located, with the translation and force in the x direction numbered first, followed by the translation and force in the y direction, and then the rotation and moment. The displacements and forces at the opposite end e of the member are then numbered in the same sequential order.

Our objective here is to determine the relationships between the member end forces and end displacements in terms of the external loads applied to the member. Such relationships can be conveniently established by subjecting the member, separately, to each of the six end displacements and external loads, and by expressing the total member end forces as the algebraic sums of the end forces required to cause the individual end displacements and the forces caused by the external loads. Thus, from Fig. 18.4(b) through (i), we can see that

$$Q_1 = k_{11}u_1 + k_{12}u_2 + k_{13}u_3 + k_{14}u_4 + k_{15}u_5 + k_{16}u_6 + Q_{f1} \tag{18.2a}$$

$$Q_2 = k_{21}u_1 + k_{22}u_2 + k_{23}u_3 + k_{24}u_4 + k_{25}u_5 + k_{26}u_6 + Q_{f2} \tag{18.2b}$$

$$Q_3 = k_{31}u_1 + k_{32}u_2 + k_{33}u_3 + k_{34}u_4 + k_{35}u_5 + k_{36}u_6 + Q_{f3} \tag{18.2c}$$

$$Q_4 = k_{41}u_1 + k_{42}u_2 + k_{43}u_3 + k_{44}u_4 + k_{45}u_5 + k_{46}u_6 + Q_{f4} \tag{18.2d}$$

$$Q_5 = k_{51}u_1 + k_{52}u_2 + k_{53}u_3 + k_{54}u_4 + k_{55}u_5 + k_{56}u_6 + Q_{f5} \tag{18.2e}$$

$$Q_6 = k_{61}u_1 + k_{62}u_2 + k_{63}u_3 + k_{64}u_4 + k_{65}u_5 + k_{66}u_6 + Q_{f6} \tag{18.2f}$$

in which k_{ij} *represents the force at the location and in the direction of Q_i required, along with other end forces, to cause a unit value of the displacement u_j while all other end displacements are zero.* These forces per unit displacement are referred to as *stiffness coefficients*. Note that a double-subscript notation is used for stiffness coefficients, with the first subscript identifying the force and the second subscript identifying the displacement. The last terms on the right sides of Eqs. (18.2) represent the fixed-end forces due to external loads (Fig. 18.4(i)), which can be determined by using the expressions for fixed-end moments given inside the back cover of the book and by applying the equations of equilibrium.

By using the definition of matrix multiplication, Eqs. (18.2) can be expressed in matrix form as

$$\begin{bmatrix} Q_1 \\ Q_2 \\ Q_3 \\ Q_4 \\ Q_5 \\ Q_6 \end{bmatrix} = \begin{bmatrix} k_{11} & k_{12} & k_{13} & k_{14} & k_{15} & k_{16} \\ k_{21} & k_{22} & k_{23} & k_{24} & k_{25} & k_{26} \\ k_{31} & k_{32} & k_{33} & k_{34} & k_{35} & k_{36} \\ k_{41} & k_{42} & k_{43} & k_{44} & k_{45} & k_{46} \\ k_{51} & k_{52} & k_{53} & k_{54} & k_{55} & k_{56} \\ k_{61} & k_{62} & k_{63} & k_{64} & k_{65} & k_{66} \end{bmatrix} \begin{bmatrix} u_1 \\ u_2 \\ u_3 \\ u_4 \\ u_5 \\ u_6 \end{bmatrix} + \begin{bmatrix} Q_{f1} \\ Q_{f2} \\ Q_{f3} \\ Q_{f4} \\ Q_{f5} \\ Q_{f6} \end{bmatrix} \qquad (18.3)$$

or, symbolically as

$$\mathbf{Q} = \mathbf{ku} + \mathbf{Q}_f \qquad (18.4)$$

in which $\mathbf{Q}$ and $\mathbf{u}$ are the member end force and member end displacement vectors, respectively, in local coordinates; $\mathbf{k}$ is called the *member stiffness matrix in local coordinates*, and $\mathbf{Q}_f$ is the *member fixed-end force vector in local coordinates*.

The stiffness coefficients, k_{ij}, can be evaluated by subjecting the member, separately, to unit values of each of the six end displacements. The member end forces required to cause the individual unit displacements are then determined by using the principles of mechanics of materials and the slope-deflection equations (Chapter 16) and by applying the equations of equilibrium. The member end forces thus obtained represent the stiffness coefficients for the member.

Let us evaluate the stiffness coefficients corresponding to a unit value of the displacement u_1 at end b of the member, as shown in Fig. 18.4(c). Note that all other displacements of the member are zero. Recalling from *mechanics of materials* that the axial deformation u_1 of a member caused by an axial force Q_1 is given by $u_1 = Q_1 L/EA$, we determine the force k_{11} that must be applied at end b of the member (Fig. 18.4(c)) to cause a displacement $u_1 = 1$ to be

$$k_{11} = \frac{EA}{L}$$

The axial force k_{41} at the far end e of the member can now be obtained by applying the equation of equilibrium:

$$+ \rightarrow \sum F_x = 0 \qquad k_{11} + k_{41} = 0$$

$$k_{41} = -k_{11} = -\frac{EA}{L}$$

in which the negative sign indicates that this force acts in the negative x direction. Since the imposition of end displacement $u_1 = 1$ does not cause the member to bend, no moments or forces in the y direction develop at the member ends. Therefore,

$$k_{21} = k_{31} = k_{51} = k_{61} = 0$$

Similarly, the end forces required to cause an axial displacement $u_4 = 1$ at end e of the member are (Fig. 18.4(f))

$$K_{14} = -\frac{EA}{L} \qquad k_{44} = \frac{EA}{L} \qquad k_{24} = k_{34} = k_{54} = k_{64} = 0$$

The deformed shape of the beam due to a unit value of displacement u_2 while all other displacements are zero is shown in Fig. 18.4(d). The end moments required (along with end forces in the y direction) to cause this deflected shape can be determined by using the slope-deflection equations derived in Section 16.1. By substituting $M_{AB} = k_{32}$, $M_{BA} = k_{62}$, $\theta_A = \theta_B = 0$, $\psi = -1/L$, and $\text{FEM}_{AB} = \text{FEM}_{BA} = 0$ into Eqs. (16.8), we obtain

$$k_{32} = k_{62} = \frac{6EI}{L^2}$$

The end forces in the y direction can now be obtained by applying the following equilibrium equations:

$$+\zeta \sum M_e = 0 \qquad 2\left(\frac{6EI}{L^2}\right) - k_{22}(L) = 0$$

$$k_{22} = \frac{12EI}{L^3}$$

$$+\uparrow \sum F_y = 0 \qquad \frac{12EI}{L^3} + k_{52} = 0$$

$$k_{52} = -\frac{12EI}{L^3}$$

Since no axial deformations are induced in the member, the axial forces at the member ends are zero; that is,

$$k_{12} = k_{42} = 0$$

The member end forces required to cause a displacement $u_5 = 1$ (Fig. 18.4(g)) can be determined in a similar manner:

$$k_{15} = k_{45} = 0 \qquad k_{25} = -\frac{12EI}{L^3} \qquad k_{35} = k_{65} = -\frac{6EI}{L^2} \qquad k_{55} = \frac{12EI}{L^3}$$

The deformed shape of the member due to a rotation $u_3 = 1$, with $u_1 = u_2 = u_4 = u_5 = u_6 = 0$, is shown in Fig. 18.4(e). By substituting $M_{AB} = k_{33}$, $M_{BA} = k_{63}$, $\theta_A = 1$, and $\theta_B = \psi = \text{FEM}_{AB} = \text{FEM}_{BA} = 0$ into the slope-deflection equations (Eqs. (16.8)), we obtain the member end moments to be

$$k_{33} = \frac{4EI}{L} \qquad k_{63} = \frac{2EI}{L}$$

By applying the equations of equilibrium, we determine

$$k_{23} = \frac{6EI}{L^2} \qquad k_{53} = -\frac{6EI}{L^2}$$

Proceeding in the same manner, the stiffness coefficients corresponding to the unit displacement $u_6 = 1$ are found to be (Fig. 18.4(h))

$$k_{16} = k_{46} = 0 \qquad k_{26} = -k_{56} = \frac{6EI}{L^2} \qquad k_{36} = \frac{2EI}{L} \qquad k_{66} = \frac{4EI}{L}$$

Substitution of the foregoing values of the stiffness coefficients into Eq. (18.3) yields the following stiffness matrix for the members of plane frames in local coordinates:

$$\mathbf{k} = \frac{EI}{L^3} \begin{bmatrix} \dfrac{AL^2}{I} & 0 & 0 & -\dfrac{AL^2}{I} & 0 & 0 \\ 0 & 12 & 6L & 0 & -12 & 6L \\ 0 & 6L & 4L^2 & 0 & -6L & 2L^2 \\ -\dfrac{AL^2}{I} & 0 & 0 & \dfrac{AL^2}{I} & 0 & 0 \\ 0 & -12 & -6L & 0 & 12 & -6L \\ 0 & 6L & 2L^2 & 0 & -6L & 4L^2 \end{bmatrix} \qquad (18.5)$$

Note that the ith column of the member stiffness matrix consists of the end forces required to cause a unit value of the displacement u_i while all other displacements are zero. For example, the second column of $\mathbf{k}$ consists of the six end forces required to cause the displacement $u_2 = 1$, as shown in Fig. 18.4(d), and so on. From Eq. (18.5), we can see that the stiffness matrix $\mathbf{k}$ is symmetric; that is, $k_{ij} = k_{ji}$. It can be shown by using Betti's law (Section 7.8) that stiffness matrices for linearly elastic structures are always symmetric.

Continuous Beam Members

Since the axial deformations of the members of continuous beams subjected to lateral loads are zero, we do not need to consider the degrees of freedom in the direction of the member's centroidal axis in the analysis. Thus, only four degrees of freedom need to be considered for the members of plane continuous beams. The degrees of freedom and the corresponding end forces for a continuous beam member are shown in Fig. 18.5.

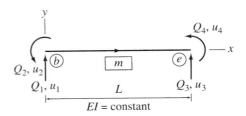

FIG. 18.5

Continuous beam member – local coordinates

The stiffness relations expressed in symbolic or condensed matrix form in Eq. (18.4) remain valid for continuous beam members. However, $\mathbf{Q}, \mathbf{u}$, and $\mathbf{Q}_f$ are now 4×1 vectors, and the member stiffness matrix in local coordinates, $\mathbf{k}$, is given by

$$\mathbf{k} = \frac{EI}{L^3} \begin{bmatrix} 12 & 6L & -12 & 6L \\ 6L & 4L^2 & -6L & 2L^2 \\ -12 & -6L & 12 & -6L \\ 6L & 2L^2 & -6L & 4L^2 \end{bmatrix} \qquad (18.6)$$

Note that the foregoing 4×4 $\mathbf{k}$ matrix is obtained by deleting the first and fourth columns and the first and fourth rows from the corresponding matrix for frame members derived previously (Eq. (18.5)).

Truss Members

A member of a truss is subjected to only axial forces, which can be determined from the displacements of the member's ends in the direction of the centroidal axis of the member. Thus only two axial degrees of freedom need to be considered for the members of plane trusses. The degrees of freedom and the corresponding end forces for a truss member are shown in Fig. 18.6.

The stiffness relationships for truss members in local coordinates are expressed as

$$\mathbf{Q} = \mathbf{ku} \qquad (18.7)$$

Note that Eq. (18.7) is obtained from Eq. (18.4) by setting $\mathbf{Q}_f = \mathbf{0}$. This is because the members of trusses are not subjected to any external loads and, therefore, the member fixed-end forces are zero. In Eq. (18.7), $\mathbf{Q}$ and $\mathbf{u}$ are 2×1 vectors consisting of the member end forces and end displacements, respectively (Fig. 18.6); and $\mathbf{k}$ is the member stiffness matrix in local coordinates, which is given by

$$\mathbf{k} = \frac{EA}{L} \begin{bmatrix} 1 & -1 \\ -1 & 1 \end{bmatrix} \qquad (18.8)$$

The foregoing stiffness matrix for truss members can either be derived directly by using the procedure discussed previously (see Fig. 18.4(c) and (f)) or it can be obtained by deleting columns 2, 3, 5, and 6 and rows 2, 3, 5, and 6 from the corresponding matrix for frame members (Eq. 18.5)).

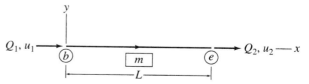

FIG. 18.6

Truss member–local coordinates

18.3 COORDINATE TRANSFORMATIONS

When members of a structure are oriented in different directions, it becomes necessary to transform the stiffness relations for each member from the member's local coordinate system to a common global coordinate system. The member stiffness relations in global coordinates thus obtained are then combined to establish the stiffness relations for the entire structure. In this section, we discuss the transformation of member end forces and end displacements from local to global coordinates, and vice versa, for the members of plane frames, continuous beams, and plane trusses. Coordinate transformation of the stiffness relationships is considered in the following section.

Frame Members

Consider an arbitrary member m of the frame shown in Fig. 18.7(a). The orientation of the member with respect to the global XY coordinate system is defined by an angle θ measured counterclockwise from the positive direction of the global X axis to the positive direction of the local x axis, as shown in the figure. The stiffness relations derived in the preceding section are valid only for member end forces $\mathbf{Q}$ and end displacements $\mathbf{u}$ described with reference to the local xy coordinate system of the member, as shown in Fig. 18.7(b).

Now, suppose that the member end forces and end displacements are specified relative to the global XY coordinate system (Fig. 18.7(c)) and we wish to determine the equivalent system of end forces and end displacements, in local xy coordinates, that has the same effect on the member. As shown in Fig. 18.7(c), the member end forces in global coordinates are denoted by F_1 through F_6, and the corresponding member end displacements are denoted by v_1 through v_6. These global member end forces and end displacements are numbered by beginning at the member end b, where the origin of the local coordinate system is located, with the force and translation in the X direction numbered first, followed by the force and translation in the Y direction and then the moment and rotation. The forces and displacements at the opposite end e of the member are then numbered in the same sequential order.

A comparison of Fig. 18.7(b) and (c) indicates that at the end b of the member, the local force Q_1 must be equal to the algebraic sum of the components of the global forces F_1 and F_2 in the direction of the local x axis. Thus

$$Q_1 = F_1 \cos\theta + F_2 \sin\theta \tag{18.9a}$$

In a similar manner, the local force Q_2 equals the algebraic sum of the components of F_1 and F_2 in the direction of the local y axis; that is,

$$Q_2 = -F_1 \sin\theta + F_2 \cos\theta \tag{18.9b}$$

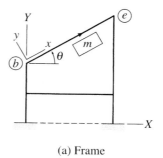

(a) Frame

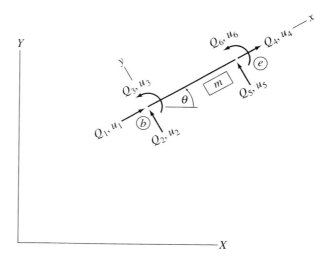

(b) Member End Forces and End Displacements in Local Coordinates

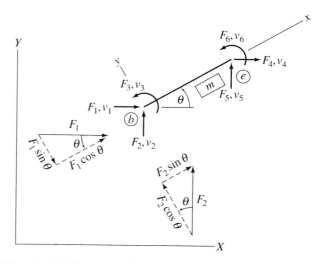

FIG. 18.7

(c) Member End Forces and End Displacements in Global Coordinates

Since the local z axis and the global Z axis are in the same direction—that is, directed out of the plane of the page—the local end moment Q_3 is equal to the global end moment F_3. Thus

$$Q_3 = F_3 \tag{18.9c}$$

By using a similar procedure at end e of the member, we express the local forces in terms of the global forces as

$$Q_4 = F_4 \cos \theta + F_5 \sin \theta \tag{18.9d}$$

$$Q_5 = -F_4 \sin \theta + F_5 \cos \theta \tag{18.9e}$$

$$Q_6 = F_6 \tag{18.9f}$$

Equations (18.9a) through (18.9f) can be written in matrix form as

$$
\begin{bmatrix} Q_1 \\ Q_2 \\ Q_3 \\ Q_4 \\ Q_5 \\ Q_6 \end{bmatrix} =
\begin{bmatrix}
\cos \theta & \sin \theta & 0 & 0 & 0 & 0 \\
-\sin \theta & \cos \theta & 0 & 0 & 0 & 0 \\
0 & 0 & 1 & 0 & 0 & 0 \\
0 & 0 & 0 & \cos \theta & \sin \theta & 0 \\
0 & 0 & 0 & -\sin \theta & \cos \theta & 0 \\
0 & 0 & 0 & 0 & 0 & 1
\end{bmatrix}
\begin{bmatrix} F_1 \\ F_2 \\ F_3 \\ F_4 \\ F_5 \\ F_6 \end{bmatrix} \tag{18.10}
$$

or symbolically as

$$\mathbf{Q} = \mathbf{TF} \tag{18.11}$$

in which

$$
\mathbf{T} =
\begin{bmatrix}
\cos \theta & \sin \theta & 0 & 0 & 0 & 0 \\
-\sin \theta & \cos \theta & 0 & 0 & 0 & 0 \\
0 & 0 & 1 & 0 & 0 & 0 \\
0 & 0 & 0 & \cos \theta & \sin \theta & 0 \\
0 & 0 & 0 & -\sin \theta & \cos \theta & 0 \\
0 & 0 & 0 & 0 & 0 & 1
\end{bmatrix} \tag{18.12}
$$

is referred to as the *transformation matrix*. The member's direction cosines, necessary for the evaluation of $\mathbf{T}$, can be easily determined by using the relationships

$$\cos \theta = \frac{X_e - X_b}{L} = \frac{X_e - X_b}{\sqrt{(X_e - X_b)^2 + (Y_e - Y_b)^2}} \tag{18.13a}$$

$$\sin \theta = \frac{Y_e - Y_b}{L} = \frac{Y_e - Y_b}{\sqrt{(X_e - X_b)^2 + (Y_e - Y_b)^2}} \tag{18.13b}$$

in which X_b and Y_b represent the global coordinates of the beginning joint b for the member; X_e and Y_e denote the global coordinates of the end joint e; and L is the length of the member.

Like end forces, the member end displacements are vectors which are defined in the same directions as the corresponding forces. Therefore, the transformation matrix **T** developed for the case of end forces (Eq. (18.12)) can also be used to transform member end displacements from global to local coordinates:

$$\mathbf{u} = \mathbf{Tv} \qquad (18.14)$$

Next, we determine the transformations of member end forces and end displacements from local to global coordinates. From Fig. 18.7(b) and (c), we observe that at end b of the member, the global force F_1 must be equal to the algebraic sum of the components of the local forces Q_1 and Q_2 in the direction of the global X axis. Thus

$$F_1 = Q_1 \cos \theta - Q_2 \sin \theta \qquad (18.15a)$$

Similarly, the global force F_2 equals the algebraic sum of the components of Q_1 and Q_2 in the direction of the global Y axis; that is,

$$F_2 = Q_1 \sin \theta + Q_2 \cos \theta \qquad (18.15b)$$

and, as discussed previously,

$$F_3 = Q_3 \qquad (18.15c)$$

Similarly, at end e of the member,

$$F_4 = Q_4 \cos \theta - Q_5 \sin \theta \qquad (18.15d)$$

$$F_5 = Q_4 \sin \theta + Q_5 \cos \theta \qquad (18.15e)$$

$$F_6 = Q_6 \qquad (18.15f)$$

Equations (18.15a) through (18.15f) can be expressed in matrix form as

$$\begin{bmatrix} F_1 \\ F_2 \\ F_3 \\ F_4 \\ F_5 \\ F_6 \end{bmatrix} = \begin{bmatrix} \cos \theta & -\sin \theta & 0 & 0 & 0 & 0 \\ \sin \theta & \cos \theta & 0 & 0 & 0 & 0 \\ 0 & 0 & 1 & 0 & 0 & 0 \\ 0 & 0 & 0 & \cos \theta & -\sin \theta & 0 \\ 0 & 0 & 0 & \sin \theta & \cos \theta & 0 \\ 0 & 0 & 0 & 0 & 0 & 1 \end{bmatrix} \begin{bmatrix} Q_1 \\ Q_2 \\ Q_3 \\ Q_4 \\ Q_5 \\ Q_6 \end{bmatrix} \qquad (18.16)$$

A comparison of Eqs. (18.10) and (18.16) indicates that the transformation matrix in Eq. (18.16), which transforms the forces from local to global coordinates, is the transpose of the transformation matrix **T** in Eq. (18.10), which transforms the forces from global to local coordinates. Thus Eq. (18.16) can be written as

$$\mathbf{F} = \mathbf{T}^T \mathbf{Q} \qquad (18.17)$$

The matrix $\mathbf{T}^T$ can also define the transformation of member end displacements from local to global coordinates; that is,

$$\mathbf{v} = \mathbf{T}^T\mathbf{u} \qquad (18.18)$$

Continuous Beam Members

When analyzing continuous beams, the member local coordinates are oriented so that the positive directions of the local x and y axes are the same as the positive directions of the global X and Y axes, respectively (Fig. 18.8). This orientation enables us to avoid coordinate transformations because the member end forces and end displacements in the global and local coordinates are the same; that is,

$$\mathbf{F} = \mathbf{Q} \qquad \mathbf{v} = \mathbf{u} \qquad (18.19)$$

Truss Members

Consider an arbitrary member m of the truss shown in Fig. 18.9(a). The end forces and end displacements for the member, in local and global coordinates, are shown in Fig. 18.9(b) and (c), respectively. Note that at each member end, two degrees of freedom and two end forces are needed in global coordinates to represent the components of the member axial displacement and axial force, respectively. Thus, in global coordinates, the truss member has a total of four degrees of freedom, v_1 through v_4, and four end forces, F_1 through F_4, as shown in Fig. 18.9(c).

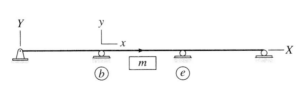

(a) Continuous Beam

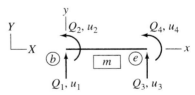

(b) Member End Forces and End Displacements in Local Coordinates

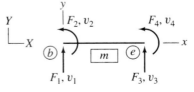

(c) Member End Forces and End Displacements in Global Coordinates

FIG. 18.8

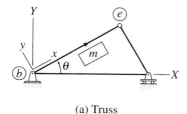

(a) Truss

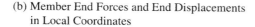

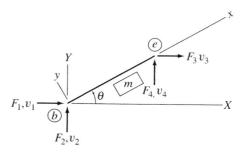

(b) Member End Forces and End Displacements
in Local Coordinates

(c) Member End Forces and End Displacements
in Global Coordinates

FIG. 18.9

The transformation matrix **T** for truss members can be established
by expressing the local end forces, **Q**, in terms of the global end forces,
F, as (Fig. 18.9(b) and (c))

$$Q_1 = F_1 \cos \theta + F_2 \sin \theta \tag{18.20a}$$

$$Q_2 = F_3 \cos \theta + F_4 \sin \theta \tag{18.20b}$$

or in matrix form as

$$\begin{bmatrix} Q_1 \\ Q_2 \end{bmatrix} = \begin{bmatrix} \cos \theta & \sin \theta & 0 & 0 \\ 0 & 0 & \cos \theta & \sin \theta \end{bmatrix} \begin{bmatrix} F_1 \\ F_2 \\ F_3 \\ F_4 \end{bmatrix} \tag{18.21}$$

from which we obtain the transformation matrix,

$$\mathbf{T} = \begin{bmatrix} \cos \theta & \sin \theta & 0 & 0 \\ 0 & 0 & \cos \theta & \sin \theta \end{bmatrix} \tag{18.22}$$

The transformation relations given in symbolic or condensed matrix
form in Eqs. (18.11), (18.14), (18.17), and (18.18) remain valid for a
truss member, with the vectors **Q**, **F**, **u**, and **v** now representing the end
forces and end displacements of the truss member, as shown in Figs.
18.9(b) and (c), and the matrix **T** representing the transformation matrix
defined in Eq. (18.22).

18.4 MEMBER STIFFNESS RELATIONS IN GLOBAL COORDINATES

By using the member stiffness relations in local coordinates (Section 18.2) and the transformation relations (Section 18.3), we can now develop the stiffness relations for members in global coordinates.

Frame Members

To establish the member stiffness relations in global coordinates, we first substitute the stiffness relations in local coordinates $\mathbf{Q} = \mathbf{ku} + \mathbf{Q}_f$ (Eq. (18.4)) into the force transformation relations $\mathbf{F} = \mathbf{T}^T\mathbf{Q}$ (Eq. (18.17)) to obtain

$$\mathbf{F} = \mathbf{T}^T\mathbf{Q} = \mathbf{T}^T(\mathbf{ku} + \mathbf{Q}_f) = \mathbf{T}^T\mathbf{ku} + \mathbf{T}^T\mathbf{Q}_f \tag{18.23}$$

Then, by substituting the displacement transformation relations $\mathbf{u} = \mathbf{Tv}$ (Eq. (18.14)) into Eq. (18.23), we determine the desired relations between the member end forces, $\mathbf{F}$, and end displacements, $\mathbf{v}$, to be

$$\mathbf{F} = \mathbf{T}^T\mathbf{kTv} + \mathbf{T}^T\mathbf{Q}_f \tag{18.24}$$

Equation (18.24) can be conveniently written as

$$\mathbf{F} = \mathbf{Kv} + \mathbf{F}_f \tag{18.25}$$

where

$$\mathbf{K} = \mathbf{T}^T\mathbf{kT} \tag{18.26}$$

$$\mathbf{F}_f = \mathbf{T}^T\mathbf{Q}_f \tag{18.27}$$

The matrix $\mathbf{K}$ is called the *member stiffness matrix in global coordinates* and $\mathbf{F}_f$ is the *member fixed-end force vector in global coordinates*.

Continuous Beam Members

As stated previously, the local coordinates of the members of continuous beams are oriented so that the positive directions of the local x and y axes are the same as the positive directions of the global X and Y axes, respectively. Thus no transformations of coordinates are needed, and the member stiffness relations in the local and global coordinates are the same.

Truss Members

The stiffness relations for truss members in global coordinates are expressed as

$$\mathbf{F} = \mathbf{Kv} \qquad (18.28)$$

Note that Eq. (18.28) is obtained from Eq. (18.25) by setting the fixed-end force vector $\mathbf{F}_f = \mathbf{0}$.

When analyzing trusses, it is usually convenient to use the explicit form of the member stiffness matrix $\mathbf{K}$. By substituting Eqs. (18.8) and (18.22) into Eq. (18.26), we write

$$\mathbf{K} = \begin{bmatrix} \cos\theta & 0 \\ \sin\theta & 0 \\ 0 & \cos\theta \\ 0 & \sin\theta \end{bmatrix} \frac{EA}{L} \begin{bmatrix} 1 & -1 \\ -1 & 1 \end{bmatrix} \begin{bmatrix} \cos\theta & \sin\theta & 0 & 0 \\ 0 & 0 & \cos\theta & \sin\theta \end{bmatrix}$$

By performing the matrix multiplications, we obtain

$$\mathbf{K} = \frac{EA}{L} \begin{bmatrix} \cos^2\theta & \cos\theta\sin\theta & -\cos^2\theta & -\cos\theta\sin\theta \\ \cos\theta\sin\theta & \sin^2\theta & -\cos\theta\sin\theta & -\sin^2\theta \\ -\cos^2\theta & -\cos\theta\sin\theta & \cos^2\theta & \cos\theta\sin\theta \\ -\cos\theta\sin\theta & -\sin^2\theta & \cos\theta\sin\theta & \sin^2\theta \end{bmatrix}$$

$$(18.29)$$

The matrix $\mathbf{K}$ of Eq. (18.29) could have been determined alternatively by subjecting an inclined truss member, separately, to unit values of each of the four global end displacements and by evaluating the end forces in global coordinates required to cause the individual unit displacements. The end forces required to cause a unit value of the displacement v_i while all other displacements are zero represent the ith column of the member global stiffness matrix $\mathbf{K}$.

18.5 STRUCTURE STIFFNESS RELATIONS

Once the member stiffness relations in global coordinates have been determined, the stiffness relations for the entire structure can be established by writing equilibrium equations for the joints of the structure and by applying the compatibility conditions that the displacements of the member ends rigidly connected to joints must be the same as the corresponding joint displacements.

To illustrate this procedure, consider the two-member frame shown in Fig. 18.10(a). The analytical model of the frame is given in Fig. 18.10(b), which indicates that the structure has three degrees of freedom, d_1, d_2, and d_3. The joint loads corresponding to these degrees of freedom are designated as P_1, P_2, and P_3, respectively. The global end forces $\mathbf{F}^{(i)}$ and end displacements $\mathbf{v}^{(i)}$ for the two members of the frame are shown in Fig. 18.10(c), in which the superscript (i) denotes the member number. Our objective is to express the joint loads $\mathbf{P}$ as functions of the joint displacements $\mathbf{d}$.

Equilibrium Equations

By applying the three equations of equilibrium, $\sum F_X = 0$, $\sum F_Y = 0$, and $\sum M = 0$, to the free body of joint 2 shown in Fig. 18.10(c), we obtain the equilibrium equations

$$P_1 = F_4^{(1)} + F_1^{(2)} \tag{18.30a}$$

$$P_2 = F_5^{(1)} + F_2^{(2)} \tag{18.30b}$$

$$P_3 = F_6^{(1)} + F_3^{(2)} \tag{18.30c}$$

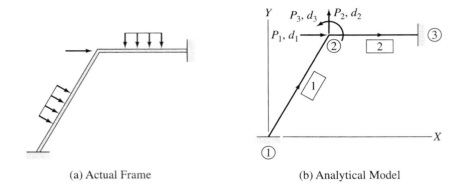

(a) Actual Frame   (b) Analytical Model

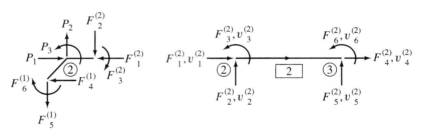

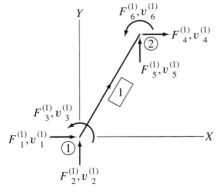

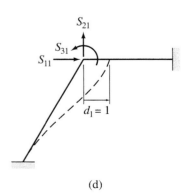

(c)   (d)

FIG. 18.10

$$\mathbf{K}_1 = \begin{matrix} & 0 & 0 & 0 & 1 & 2 & 3 \\ & \begin{bmatrix} K_{11}^{(1)} & K_{12}^{(1)} & K_{13}^{(1)} & K_{14}^{(1)} & K_{15}^{(1)} & K_{16}^{(1)} \\ K_{21}^{(1)} & K_{22}^{(1)} & K_{23}^{(1)} & K_{24}^{(1)} & K_{25}^{(1)} & K_{26}^{(1)} \\ K_{31}^{(1)} & K_{32}^{(1)} & K_{33}^{(1)} & K_{34}^{(1)} & K_{35}^{(1)} & K_{36}^{(1)} \\ K_{41}^{(1)} & K_{42}^{(1)} & K_{43}^{(1)} & K_{44}^{(1)} & K_{45}^{(1)} & K_{46}^{(1)} \\ K_{51}^{(1)} & K_{52}^{(1)} & K_{53}^{(1)} & K_{54}^{(1)} & K_{55}^{(1)} & K_{56}^{(1)} \\ K_{61}^{(1)} & K_{62}^{(1)} & K_{63}^{(1)} & K_{64}^{(1)} & K_{65}^{(1)} & K_{66}^{(1)} \end{bmatrix} & \begin{matrix} 0 \\ 0 \\ 0 \\ 1 \\ 2 \\ 3 \end{matrix} \end{matrix}$$

$$\mathbf{K}_2 = \begin{matrix} & 1 & 2 & 3 & 0 & 0 & 0 \\ & \begin{bmatrix} K_{11}^{(2)} & K_{12}^{(2)} & K_{13}^{(2)} & K_{14}^{(2)} & K_{15}^{(2)} & K_{16}^{(2)} \\ K_{21}^{(2)} & K_{22}^{(2)} & K_{23}^{(2)} & K_{24}^{(2)} & K_{25}^{(2)} & K_{26}^{(2)} \\ K_{31}^{(2)} & K_{32}^{(2)} & K_{33}^{(2)} & K_{34}^{(2)} & K_{35}^{(2)} & K_{36}^{(2)} \\ K_{41}^{(2)} & K_{42}^{(2)} & K_{43}^{(2)} & K_{44}^{(2)} & K_{45}^{(2)} & K_{46}^{(2)} \\ K_{51}^{(2)} & K_{52}^{(2)} & K_{53}^{(2)} & K_{54}^{(2)} & K_{55}^{(2)} & K_{56}^{(2)} \\ K_{61}^{(2)} & K_{62}^{(2)} & K_{63}^{(2)} & K_{64}^{(2)} & K_{65}^{(2)} & K_{66}^{(2)} \end{bmatrix} & \begin{matrix} 1 \\ 2 \\ 3 \\ 0 \\ 0 \\ 0 \end{matrix} \end{matrix}$$

$$\mathbf{S} = \begin{matrix} & 1 & 2 & 3 \\ & \begin{bmatrix} K_{44}^{(1)} + K_{11}^{(2)} & K_{45}^{(1)} + K_{12}^{(2)} & K_{46}^{(1)} + K_{13}^{(2)} \\ K_{54}^{(1)} + K_{21}^{(2)} & K_{55}^{(1)} + K_{22}^{(2)} & K_{56}^{(1)} + K_{23}^{(2)} \\ K_{64}^{(1)} + K_{31}^{(2)} & K_{65}^{(1)} + K_{32}^{(2)} & K_{66}^{(1)} + K_{33}^{(2)} \end{bmatrix} & \begin{matrix} 1 \\ 2 \\ 3 \end{matrix} \end{matrix}$$

(e)

$$\mathbf{F}_{f1} = \begin{bmatrix} F_{f1}^{(1)} \\ F_{f2}^{(1)} \\ F_{f3}^{(1)} \\ F_{f4}^{(1)} \\ F_{f5}^{(1)} \\ F_{f6}^{(1)} \end{bmatrix} \begin{matrix} 0 \\ 0 \\ 0 \\ 1 \\ 2 \\ 3 \end{matrix} \qquad \mathbf{F}_{f2} = \begin{bmatrix} F_{f1}^{(2)} \\ F_{f2}^{(2)} \\ F_{f3}^{(2)} \\ F_{f4}^{(2)} \\ F_{f5}^{(2)} \\ F_{f6}^{(2)} \end{bmatrix} \begin{matrix} 1 \\ 2 \\ 3 \\ 0 \\ 0 \\ 0 \end{matrix}$$

$$\mathbf{P}_f = \begin{bmatrix} F_{f4}^{(1)} + F_{f1}^{(2)} \\ F_{f5}^{(1)} + F_{f2}^{(2)} \\ F_{f6}^{(1)} + F_{f3}^{(2)} \end{bmatrix} \begin{matrix} 1 \\ 2 \\ 3 \end{matrix}$$

(f)

FIG. 18.10 (contd.)

Member Stiffness Relations

To express the joint loads $\mathbf{P}$ in terms of the joint displacements $\mathbf{d}$, we first relate the member end forces $\mathbf{F}^{(i)}$ to end displacements $\mathbf{v}^{(i)}$, by using the member stiffness relations in global coordinates derived in the preceding section. By writing Eq. (18.25) in expanded form for member 1, we obtain

$$
\begin{bmatrix}
F_1^{(1)} \\
F_2^{(1)} \\
F_3^{(1)} \\
F_4^{(1)} \\
F_5^{(1)} \\
F_6^{(1)}
\end{bmatrix}
=
\begin{bmatrix}
K_{11}^{(1)} & K_{12}^{(1)} & K_{13}^{(1)} & K_{14}^{(1)} & K_{15}^{(1)} & K_{16}^{(1)} \\
K_{21}^{(1)} & K_{22}^{(1)} & K_{23}^{(1)} & K_{24}^{(1)} & K_{25}^{(1)} & K_{26}^{(1)} \\
K_{31}^{(1)} & K_{32}^{(1)} & K_{33}^{(1)} & K_{34}^{(1)} & K_{35}^{(1)} & K_{36}^{(1)} \\
K_{41}^{(1)} & K_{42}^{(1)} & K_{43}^{(1)} & K_{44}^{(1)} & K_{45}^{(1)} & K_{46}^{(1)} \\
K_{51}^{(1)} & K_{52}^{(1)} & K_{53}^{(1)} & K_{54}^{(1)} & K_{55}^{(1)} & K_{56}^{(1)} \\
K_{61}^{(1)} & K_{62}^{(1)} & K_{63}^{(1)} & K_{64}^{(1)} & K_{65}^{(1)} & K_{66}^{(1)}
\end{bmatrix}
\begin{bmatrix}
v_1^{(1)} \\
v_2^{(1)} \\
v_3^{(1)} \\
v_4^{(1)} \\
v_5^{(1)} \\
v_6^{(1)}
\end{bmatrix}
+
\begin{bmatrix}
F_{f1}^{(1)} \\
F_{f2}^{(1)} \\
F_{f3}^{(1)} \\
F_{f4}^{(1)} \\
F_{f5}^{(1)} \\
F_{f6}^{(1)}
\end{bmatrix}
$$

$$(18.31)$$

from which we determine the expressions for forces at end 2 of the member to be

$$
\begin{aligned}
F_4^{(1)} = {}& K_{41}^{(1)} v_1^{(1)} + K_{42}^{(1)} v_2^{(1)} + K_{43}^{(1)} v_3^{(1)} + K_{44}^{(1)} v_4^{(1)} \\
& + K_{45}^{(1)} v_5^{(1)} + K_{46}^{(1)} v_6^{(1)} + F_{f4}^{(1)}
\end{aligned}
\tag{18.32a}
$$

$$
\begin{aligned}
F_5^{(1)} = {}& K_{51}^{(1)} v_1^{(1)} + K_{52}^{(1)} v_2^{(1)} + K_{53}^{(1)} v_3^{(1)} + K_{54}^{(1)} v_4^{(1)} \\
& + K_{55}^{(1)} v_5^{(1)} + K_{56}^{(1)} v_6^{(1)} + F_{f5}^{(1)}
\end{aligned}
\tag{18.32b}
$$

$$
\begin{aligned}
F_6^{(1)} = {}& K_{61}^{(1)} v_1^{(1)} + K_{62}^{(1)} v_2^{(1)} + K_{63}^{(1)} v_3^{(1)} + K_{64}^{(1)} v_4^{(1)} \\
& + K_{65}^{(1)} v_5^{(1)} + K_{66}^{(1)} v_6^{(1)} + F_{f6}^{(1)}
\end{aligned}
\tag{18.32c}
$$

Similarly, by writing Eq. (18.25) for member 2, we obtain

$$
\begin{bmatrix}
F_1^{(2)} \\
F_2^{(2)} \\
F_3^{(2)} \\
F_4^{(2)} \\
F_5^{(2)} \\
F_6^{(2)}
\end{bmatrix}
=
\begin{bmatrix}
K_{11}^{(2)} & K_{12}^{(2)} & K_{13}^{(2)} & K_{14}^{(2)} & K_{15}^{(2)} & K_{16}^{(2)} \\
K_{21}^{(2)} & K_{22}^{(2)} & K_{23}^{(2)} & K_{24}^{(2)} & K_{25}^{(2)} & K_{26}^{(2)} \\
K_{31}^{(2)} & K_{32}^{(2)} & K_{33}^{(2)} & K_{34}^{(2)} & K_{35}^{(2)} & K_{36}^{(2)} \\
K_{41}^{(2)} & K_{42}^{(2)} & K_{43}^{(2)} & K_{44}^{(2)} & K_{45}^{(2)} & K_{46}^{(2)} \\
K_{51}^{(2)} & K_{52}^{(2)} & K_{53}^{(2)} & K_{54}^{(2)} & K_{55}^{(2)} & K_{56}^{(2)} \\
K_{61}^{(2)} & K_{62}^{(2)} & K_{63}^{(2)} & K_{64}^{(2)} & K_{65}^{(2)} & K_{66}^{(2)}
\end{bmatrix}
\begin{bmatrix}
v_1^{(2)} \\
v_2^{(2)} \\
v_3^{(2)} \\
v_4^{(2)} \\
v_5^{(2)} \\
v_6^{(2)}
\end{bmatrix}
+
\begin{bmatrix}
F_{f1}^{(2)} \\
F_{f2}^{(2)} \\
F_{f3}^{(2)} \\
F_{f4}^{(2)} \\
F_{f5}^{(2)} \\
F_{f6}^{(2)}
\end{bmatrix}
$$

$$(18.33)$$

from which we determine the forces at end 2 of the member to be

$$
\begin{aligned}
F_1^{(2)} = {}& K_{11}^{(2)} v_1^{(2)} + K_{12}^{(2)} v_2^{(2)} + K_{13}^{(2)} v_3^{(2)} + K_{14}^{(2)} v_4^{(2)} \\
& + K_{15}^{(2)} v_5^{(2)} + K_{16}^{(2)} v_6^{(2)} + F_{f1}^{(2)}
\end{aligned}
\tag{18.34a}
$$

$$F_2^{(2)} = K_{21}^{(2)} v_1^{(2)} + K_{22}^{(2)} v_2^{(2)} + K_{23}^{(2)} v_3^{(2)} + K_{24}^{(2)} v_4^{(2)}$$
$$+ K_{25}^{(2)} v_5^{(2)} + K_{26}^{(2)} v_6^{(2)} + F_{f2}^{(2)} \tag{18.34b}$$
$$F_3^{(2)} = K_{31}^{(2)} v_1^{(2)} + K_{32}^{(2)} v_2^{(2)} + K_{33}^{(2)} v_3^{(2)} + K_{34}^{(2)} v_4^{(2)}$$
$$+ K_{35}^{(2)} v_5^{(2)} + K_{36}^{(2)} v_6^{(2)} + F_{f3}^{(2)} \tag{18.34c}$$

Compatibility Equations

By comparing Fig. 18.10(b) and (c), we observe that since the lower end 1 of member 1 is rigidly connected to the fixed joint 1, which can neither translate nor rotate, the three displacements of end 1 of the member must be zero. Similarly, since end 2 of this member is rigidly connected to joint 2, the displacements of end 2 must be the same as the displacements of joint 2. Thus, the compatibility equations for member 1 are

$$v_1^{(1)} = v_2^{(1)} = v_3^{(1)} = 0 \qquad v_4^{(1)} = d_1 \qquad v_5^{(1)} = d_2 \qquad v_6^{(1)} = d_3 \tag{18.35}$$

In a similar manner, the compatibility equations for member 2 are found to be

$$v_1^{(2)} = d_1 \qquad v_2^{(2)} = d_2 \qquad v_3^{(2)} = d_3 \qquad v_4^{(2)} = v_5^{(2)} = v_6^{(2)} = 0 \tag{18.36}$$

By substituting the compatibility equations for member 1 (Eq. (18.35)) into the member's force-displacement relations as given by Eqs. (18.32), we express the member end forces $\mathbf{F}^{(1)}$ in terms of the joint displacements $\mathbf{d}$ as

$$F_4^{(1)} = K_{44}^{(1)} d_1 + K_{45}^{(1)} d_2 + K_{46}^{(1)} d_3 + F_{f4}^{(1)} \tag{18.37a}$$
$$F_5^{(1)} = K_{54}^{(1)} d_1 + K_{55}^{(1)} d_2 + K_{56}^{(1)} d_3 + F_{f5}^{(1)} \tag{18.37b}$$
$$F_6^{(1)} = K_{64}^{(1)} d_1 + K_{65}^{(1)} d_2 + K_{66}^{(1)} d_3 + F_{f6}^{(1)} \tag{18.37c}$$

Similarly, for member 2, substitution of Eq. (18.36) into Eqs. (18.34) yields

$$F_1^{(2)} = K_{11}^{(2)} d_1 + K_{12}^{(2)} d_2 + K_{13}^{(2)} d_3 + F_{f1}^{(2)} \tag{18.38a}$$
$$F_2^{(2)} = K_{21}^{(2)} d_1 + K_{22}^{(2)} d_2 + K_{23}^{(2)} d_3 + F_{f2}^{(2)} \tag{18.38b}$$
$$F_3^{(2)} = K_{31}^{(2)} d_1 + K_{32}^{(2)} d_2 + K_{33}^{(2)} d_3 + F_{f3}^{(2)} \tag{18.38c}$$

Structure Stiffness Relations

Finally, by substituting Eqs. (18.37) and (18.38) into the joint equilibrium equations (Eqs. (18.30)), we obtain the desired relationships between the joint loads $\mathbf{P}$ and the joint displacement $\mathbf{d}$ of the frame as

$$P_1 = (K_{44}^{(1)} + K_{11}^{(2)}) d_1 + (K_{45}^{(1)} + K_{12}^{(2)}) d_2 + (K_{46}^{(1)} + K_{13}^{(2)}) d_3$$
$$+ (F_{f4}^{(1)} + F_{f1}^{(2)}) \tag{18.39a}$$

$$P_2 = (K_{54}^{(1)} + K_{21}^{(2)})d_1 + (K_{55}^{(1)} + K_{22}^{(2)})d_2 + (K_{56}^{(1)} + K_{23}^{(2)})d_3$$
$$+ (F_{f5}^{(1)} + F_{f2}^{(2)}) \tag{18.39b}$$

$$P_3 = (K_{64}^{(1)} + K_{31}^{(2)})d_1 + (K_{65}^{(1)} + K_{32}^{(2)})d_2 + (K_{66}^{(1)} + K_{33}^{(2)})d_3$$
$$+ (F_{f6}^{(1)} + F_{f3}^{(2)}) \tag{18.39c}$$

Equations (18.39) can be conveniently expressed in condensed matrix form as

$$\mathbf{P} = \mathbf{Sd} + \mathbf{P}_f \tag{18.40}$$

or

$$\mathbf{P} - \mathbf{P}_f = \mathbf{Sd} \tag{18.41}$$

in which

$$\mathbf{S} = \begin{bmatrix} K_{44}^{(1)} + K_{11}^{(2)} & K_{45}^{(1)} + K_{12}^{(2)} & K_{46}^{(1)} + K_{13}^{(2)} \\ K_{54}^{(1)} + K_{21}^{(2)} & K_{55}^{(1)} + K_{22}^{(2)} & K_{56}^{(1)} + K_{23}^{(2)} \\ K_{64}^{(1)} + K_{31}^{(2)} & K_{65}^{(1)} + K_{32}^{(2)} & K_{66}^{(1)} + K_{33}^{(2)} \end{bmatrix} \tag{18.42}$$

is called the *structure stiffness matrix* and

$$\mathbf{P}_f = \begin{bmatrix} F_{f4}^{(1)} + F_{f1}^{(2)} \\ F_{f5}^{(1)} + F_{f2}^{(2)} \\ F_{f6}^{(1)} + F_{f3}^{(2)} \end{bmatrix} \tag{18.43}$$

is termed the *structure fixed-joint force vector*. The foregoing procedure of determining the structure stiffness relations by combining the member stiffness relations is often referred to as the *direct stiffness method* [39].

The structure stiffness matrix $\mathbf{S}$ is interpreted in a manner analogous to the member stiffness matrix; that is, a *structure stiffness coefficient* S_{ij} *represents the force at the location and in the direction of* P_i *required, along with other joint forces, to cause a unit value of the displacement* d_j *while all other joint displacements are zero.* Thus the jth column of matrix $\mathbf{S}$ consists of the joint loads required to cause a unit value of the displacement d_j while all other displacements are zero. For example, the first column of $\mathbf{S}$ consists of the three joint loads required to cause the displacement $d_1 = 1$, as shown in Fig. 18.10(d), and so on.

The foregoing interpretation of the structural stiffness matrix $\mathbf{S}$ indicates that such a matrix can, alternatively, be determined by subjecting the structure, separately, to unit values of each of its joint displacements and by evaluating the joint loads required to cause the individual displacements. However, such a procedure cannot be easily implemented on computers and is seldom used in practice. Therefore, this alternative procedure is not pursued in this chapter.

Assembly of S and P$_f$ by Using Member Code Numbers

In the preceding paragraphs, we determined the structure stiffness matrix $\mathbf{S}$ (Eq. (18.42)) and the structure fixed-joint force vector $\mathbf{P}_f$ (Eq. (18.43)) by substituting the member compatibility equations into the member global stiffness relations and then substituting the resulting relationships into the joint equilibrium equations. This process of writing three types of equations and then making substitutions can be quite tedious and time consuming for large structures.

From Eq. (18.42), we observe that the stiffness of a joint in a direction equals the sum of the stiffnesses in that direction of the members meeting at the joint. This fact indicates that the structure stiffness matrix $\mathbf{S}$ can be formulated directly by adding the elements of the member stiffness matrices into their proper positions in the structure matrix, thereby avoiding the necessity of writing any equations. The technique of directly forming a structure stiffness matrix by assembling the elements of the member global stiffness matrices was introduced by S. S. Tezcan in 1963 [38] and is sometimes referred to as the *code number technique*.

To illustrate this technique, consider again the two-member frame of Fig. 18.10. The stiffness matrices in global coordinates for the members 1 and 2 of the frame are designated as $\mathbf{K}_1$ and $\mathbf{K}_2$, respectively (Fig. 18.10(e)). Our objective is to form the structure stiffness matrix $\mathbf{S}$ by assembling the elements of $\mathbf{K}_1$ and $\mathbf{K}_2$. Before we can determine the positions of the elements of a member matrix $\mathbf{K}$ in the structure matrix $\mathbf{S}$, we need to identify, for each of the member's degrees of freedom in global coordinates, the number of the corresponding structure degree of freedom. If the structure degree of freedom corresponding to a member degree of freedom is not defined (i.e., the corresponding joint displacement is zero), then a zero is used for the structure degree of freedom number. Thus by comparing the global degrees of freedom of member 1 shown in Fig. 18.10(c) with the structure degrees of freedom given in Fig. 18.10(b), we determine the structure degree of freedom numbers for the member to be $0, 0, 0, 1, 2, 3$. Note that these numbers are in the same order as the member degrees of freedom; for example, the fourth number, 1, corresponds to the fourth degree of freedom, $v_4^{(1)}$, of the member, and so on. In other words, the first three numbers identify, in order, the X translation, the Y translation, and the rotation of the beginning joint of the member, whereas the last three numbers identify the X translation, the Y translation, and the rotation, respectively, of the end joint. In a similar manner, we determine the structure degree of freedom numbers for member 2 to be $1, 2, 3, 0, 0, 0$.

The structure degree of freedom numbers for a member can be used to define the compatibility equations for the member. For example, the structure degree of freedom numbers, $0, 0, 0, 1, 2, 3$, imply the following compatibility equations for member 1:

$$v_1^{(1)} = v_2^{(1)} = v_3^{(1)} = 0 \qquad v_4^{(1)} = d_1 \qquad v_5^{(1)} = d_2 \qquad v_6^{(1)} = d_3$$

which are identical to those given in Eq. (18.35).

The positions of the elements of the member stiffness matrix $\mathbf{K}_1$ in the structure stiffness matrix $\mathbf{S}$ can now be determined by writing the member's structure degree of freedom numbers $(0, 0, 0, 1, 2, 3)$ on the right side and at the top of $\mathbf{K}_1$, as shown in Fig. 18.10(e). Note that the numbers on the right side of $\mathbf{K}_1$ represent the row numbers of the $\mathbf{S}$ matrix, whereas the numbers at the top represent the column numbers of $\mathbf{S}$. For example, the element $K_{65}^{(1)}$ of $\mathbf{K}_1$ must be located in row 3 and column 2 of $\mathbf{S}$, as shown in Fig. 18.10(e). By using this approach, the remaining elements of $\mathbf{K}_1$, except those corresponding to zero row or column number of $\mathbf{S}$, are stored in their proper positions in the structure stiffness matrix $\mathbf{S}$.

The same procedure is then repeated for member 2. When two or more member stiffness coefficients are located in the same position in $\mathbf{S}$, then the coefficients must be algebraically added. The completed structure stiffness matrix $\mathbf{S}$ is shown in Fig. 18.10(e). Note that this matrix is identical to the one obtained previously (Eq. (18.42)) by substituting the member compatibility equations and stiffness relations into the joint equilibrium equations.

The foregoing procedure of directly forming the structure stiffness matrix by assembling member stiffness coefficients can be easily implemented on computers. To save computer storage space, one member stiffness matrix is generated at a time; it is stored in the structure stiffness matrix, and the space is reused to generate the stiffness matrix for the next member, and so on.

The structure fixed-joint force vector, $\mathbf{P}_f$, can be assembled by using a procedure similar to that for forming the structure stiffness matrix. To generate the $\mathbf{P}_f$ vector for the frame under consideration, the structure degree of freedom numbers for member 1 are first written on the right side of the member's fixed-end force vector $\mathbf{F}_{f1}$, as shown in Fig. 18.10(f). Each of these numbers now represents the row number of $\mathbf{P}_f$ in which the corresponding member force is to be stored. For example, the element $F_{f5}^{(1)}$ must be located in row 2 of $\mathbf{P}_f$, as shown in the figure. In a similar manner, the remaining elements of $\mathbf{F}_{f1}$, except those corresponding to zero row number of $\mathbf{P}_f$, are stored in their proper positions in $\mathbf{P}_f$. The same procedure is then repeated for member 2. The structure fixed-joint force vector $\mathbf{P}_f$ thus obtained is shown in Fig. 18.10(f). Note that this vector is identical to that given in Eq. (18.43).

Once $\mathbf{S}$ and $\mathbf{P}_f$ have been evaluated, the structure stiffness relations (Eq. (18.41)), which now represent a system of simultaneous linear algebraic equations, can be solved for the unknown joint displacements $\mathbf{d}$. With $\mathbf{d}$ known, the end displacements for each member can be determined by applying the compatibility equations defined by its structure degree of freedom numbers; then the corresponding end forces can be computed by using the member's stiffness relations.

The procedure for generating the structure stiffness matrix $\mathbf{S}$ and fixed-joint force vector $\mathbf{P}_f$, as described here for frames, can be applied to continuous beams and trusses as well, except that in the case of trusses $\mathbf{P}_f = \mathbf{0}$.

18.6 PROCEDURE FOR ANALYSIS

Based on the discussion presented in the previous sections, we can develop the following step-by-step procedure for the analysis of structures by the matrix stiffness method.

1. Prepare an analytical model of the structure as follows:
   a. Draw a line diagram of the structure, on which each joint and member must be identified by a number.
   b. Select a global XY coordinate system, with the X and Y axes oriented in the horizontal (positive to the right) and vertical (positive upward) directions, respectively. It is usually convenient to locate the origin of this coordinate system at a lower left joint of the structure, so that the X and Y coordinates of most of the joints are positive.
   c. For each member, establish a local xy coordinate system by selecting one of the joints at its ends as the beginning joint and the other as the end joint. On the line diagram of the structure, for each member indicate the positive direction of the local x axis by drawing an arrow along the member pointing toward the end joint. For horizontal members, the coordinate transformations can be avoided by selecting the joint at the left end of the member as the beginning joint.
   d. Identify the degrees of freedom or unknown joint displacements, $\mathbf{d}$, of the structure. The degrees of freedom are specified on the structure's line diagram by drawing arrows at the joints and are numbered by starting at the lowest joint number and proceeding sequentially to the highest joint number. In the case of more than one degree of freedom at a joint, the X translation is numbered first, followed by the Y translation, and then the rotation. Recall that a joint of a plane frame can have up to three degrees of freedom (two translations and a rotation); a joint of a continuous beam can have up to two degrees of freedom (a translation perpendicular to the beam's centroidal axis and a rotation); and a joint of a plane truss can have up to two degrees of freedom (two translations). Note that joint translations are considered as positive when in the positive directions of the X and Y axes; joint rotations are considered as positive when counterclockwise.

2. Evaluate the structure stiffness matrix $\mathbf{S}$ and fixed-joint force vector $\mathbf{P}_f$. For each member of the structure, perform the following operations:

   a. For trusses, go directly to step 2(d). Otherwise, compute the member stiffness matrix in local coordinates, $\mathbf{k}$. Expressions of $\mathbf{k}$ for the members of frames and continuous beams are given in Eqs. (18.5) and (18.6), respectively.

   b. If the member is subjected to external loads, then evaluate its fixed-end force vector in local coordinates, $\mathbf{Q}_f$, by using the expressions for fixed-end moments given inside the back cover of the book and by applying the equations of equilibrium (see Examples 18.2 and 18.3).

   c. For horizontal members with the local x axis positive to the right (i.e., in the same direction as the global X axis), the member stiffness relations in the local and global coordinates are the same (i.e., $\mathbf{K} = \mathbf{k}$ and $\mathbf{F}_f = \mathbf{Q}_f$); go to step 2(e). Otherwise, compute the member's transformation matrix $\mathbf{T}$ by using Eq. (18.12).

   d. Determine the member stiffness matrix in global coordinates, $\mathbf{K} = \mathbf{T}^T \mathbf{k} \mathbf{T}$ (Eq. (18.26)), and the corresponding fixed-end force vector, $\mathbf{F}_f = \mathbf{T}^t \mathbf{Q}_f$ (Eq. (18.27)). The matrix $\mathbf{K}$ must be symmetric. For trusses, it is usually more convenient to use the explicit form of $\mathbf{K}$ given in Eq. (18.29). Also, for trusses, $\mathbf{F}_f = \mathbf{0}$.

   e. Identify the member's structure degree of freedom numbers and store the pertinent elements of $\mathbf{K}$ and $\mathbf{F}_f$ in their proper positions in the structure stiffness matrix $\mathbf{S}$ and the fixed-joint force vector $\mathbf{P}_f$, respectively, by using the procedure described in Section 18.5. The complete structure stiffness matrix $\mathbf{S}$ obtained by assembling the stiffness coefficients of all the members of the structure must be symmetric.

3. Form the joint load vector, $\mathbf{P}$.

4. Determine the unknown joint displacements. Substitute $\mathbf{P}, \mathbf{P}_f$, and $\mathbf{S}$ into the structure stiffness relations, $\mathbf{P} - \mathbf{P}_f = \mathbf{S}\mathbf{d}$ (Eq. (18.41)), and solve the resulting system of simultaneous equations for the unknown joint displacements $\mathbf{d}$.

5. Compute member end displacements and end forces. For each member, do the following:

   a. Obtain member end displacements in global coordinates, $\mathbf{v}$, from the joint displacements, $\mathbf{d}$, by using the member's structure degree of freedom numbers.

   b. Determine member end displacements in local coordinates by using the relationship $\mathbf{u} = \mathbf{T}\mathbf{v}$ (Eq. (18.14)). For horizontal members with the local x axis positive to the right, $\mathbf{u} = \mathbf{v}$.

   c. Compute member end forces in local coordinates by using the relationship $\mathbf{Q} = \mathbf{k}\mathbf{u} + \mathbf{Q}_f$ (Eq. (18.4)). For trusses, $\mathbf{Q}_f = \mathbf{0}$.

   d. Calculate member end forces in global coordinates by using the transformation relationship $\mathbf{F} = \mathbf{T}^T \mathbf{Q}$ (Eq. (18.17)). For horizontal members with the local x axis positive to the right, $\mathbf{F} = \mathbf{Q}$.

6. Determine support reactions by considering the equilibrium of the joints located at the supports of the structure.

Computer Program

A computer program for the analysis of plane framed structures using the stiffness method is provided on the CD-ROM enclosed with this book. A brief description of the program as well as information on how to use this program, including an illustrative example, are presented in Appendix C.

Example 18.1

Determine the reactions and the force in each member of the truss shown in Fig. 18.11(a) by the matrix stiffness method.

Solution

Degrees of Freedom From the analytical model of the truss shown in Fig. 18.11(b), we observe that only joint 3 is free to translate. Thus the truss has two degrees of freedom, d_1 and d_2, which are the unknown translations of joint 3 in the X and Y directions, respectively.

Structure Stiffness Matrix

Member 1 As shown in Fig. 18.11(b), joint 1 has been selected as the beginning joint and joint 3 as the end joint for member 1. By applying Eqs. (18.13), we determine

$$L = \sqrt{(X_3 - X_1)^2 + (Y_3 - Y_1)^2} = \sqrt{(15 - 0)^2 + (20 - 0)^2} = 25 \text{ ft}$$

$$\cos \theta = \frac{X_3 - X_1}{L} = \frac{15}{25} = 0.6$$

$$\sin \theta = \frac{Y_3 - Y_1}{L} = \frac{20}{25} = 0.8$$

The member stiffness matrix in global coordinates can now be evaluated by using Eq. (18.29)

$$\mathbf{K}_1 = \frac{(29{,}000)(9)}{(25)(12)} \begin{bmatrix} 0.36 & 0.48 & -0.36 & -0.48 \\ 0.48 & 0.64 & -0.48 & -0.64 \\ -0.36 & -0.48 & 0.36 & 0.48 \\ -0.48 & -0.64 & 0.48 & 0.64 \end{bmatrix}$$

continued

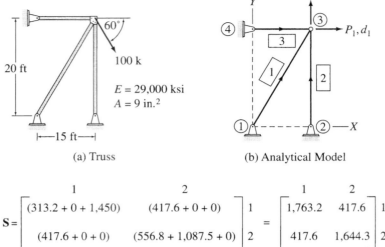

(a) Truss

(b) Analytical Model

$$S = \begin{bmatrix} (313.2 + 0 + 1,450) & (417.6 + 0 + 0) \\ (417.6 + 0 + 0) & (556.8 + 1,087.5 + 0) \end{bmatrix} \begin{matrix} 1 \\ 2 \end{matrix} = \begin{bmatrix} 1,763.2 & 417.6 \\ 417.6 & 1,644.3 \end{bmatrix} \begin{matrix} 1 \\ 2 \end{matrix}$$

(c) Structure Stiffness Matrix

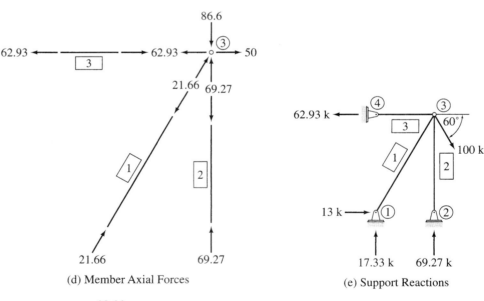

(d) Member Axial Forces

(e) Support Reactions

FIG. **18.11**

or

$$K_1 = \begin{bmatrix} 313.2 & 417.6 & -313.2 & -417.6 \\ 417.6 & 556.8 & -417.6 & -556.8 \\ -313.2 & -417.6 & 313.2 & 417.6 \\ -417.6 & -556.8 & 417.6 & 556.8 \end{bmatrix} \begin{matrix} 0 \\ 0 \\ 1 \\ 2 \end{matrix}$$ (1)

From Fig. 18.11(b), we observe that the displacements of the beginning joint 1 for the member are zero, whereas the displacements of the end joint 3 are d_1 and d_2. Thus the structure degree of freedom numbers for this member are $0, 0, 1, 2$. These numbers are written on the right side and at the top of $\mathbf{K}_1$ (see Eq. (1)) to indicate the rows and columns, respectively, of the structure stiffness matrix S, where the elements of $\mathbf{K}_1$ must be stored. Note that the elements of $\mathbf{K}_1$, which correspond to the zero structure degree of freedom number, are simply disregarded. Thus, the element in row 3 and column 3 of $\mathbf{K}_1$ is stored in row 1 and column 1 of **S**, as shown in Fig. 18.11(c). Similarly, the element in row 3 and column 4 of $\mathbf{K}_1$ is stored in row 1 and column 2 of **S**. The remaining elements of $\mathbf{K}_1$ are stored in **S** in a similar manner (Fig. 18.11(c)).

Member 2   From Fig. 18.11(b), we can see that joint 2 is the beginning joint and joint 3 is the end joint for member 2. By applying Eqs. (18.13), we obtain

$$\cos \theta = \frac{X_3 - X_2}{L} = \frac{15 - 15}{20} = 0$$

$$\sin \theta = \frac{Y_3 - Y_2}{L} = \frac{20 - 0}{20} = 1$$

Thus, by using Eq. (18.29)

$$
\mathbf{K}_2 =
\begin{array}{c}
\begin{array}{cccc} 0 & \quad 0 & 1 & \quad 2 \end{array} \\
\left[
\begin{array}{cccc}
0 & 0 & 0 & 0 \\
0 & 1{,}087.5 & 0 & -1{,}087.5 \\
0 & 0 & 0 & 0 \\
0 & -1{,}087.5 & 0 & 1{,}087.5
\end{array}
\right]
\begin{array}{c} 0 \\ 0 \\ 1 \\ 2 \end{array}
\end{array}
$$

From Fig. 18.11(b), we can see that the structure degree of freedom numbers for this member are $0, 0, 1, 2$. These numbers are used to store the pertinent elements of $\mathbf{K}_2$ in their proper positions in the structure stiffness matrix **S**, as shown in Fig. 18.11(c).

Member 3 $\cos \theta = 1$ $\sin \theta = 0$

By using Eq. (18.29),

$$
\mathbf{K}_3 =
\begin{array}{c}
\begin{array}{cccc} 0 & 0 & 1 & 2 \end{array} \\
\left[
\begin{array}{cccc}
1{,}450 & 0 & -1{,}450 & 0 \\
0 & 0 & 0 & 0 \\
-1{,}450 & 0 & 1{,}450 & 0 \\
0 & 0 & 0 & 0
\end{array}
\right]
\begin{array}{c} 0 \\ 0 \\ 1 \\ 2 \end{array}
\end{array}
$$

The structure degree of freedom numbers for this member are $0, 0, 1, 2$. By using these numbers, the elements of $\mathbf{K}_3$ are stored in **S**, as shown in Fig. 18.11(c).

Note that the structure stiffness matrix **S** (Fig. 18.11(c)), obtained by assembling the stiffness coefficients of the three members, is symmetric.

Joint Load Vector By comparing Fig. 18.11(a) and (b), we realize that

$$P_1 = 100 \cos 60^\circ = 50 \text{ k} \qquad P_2 = -100 \sin 60^\circ = -86.6 \text{ k}$$

continued

Thus the joint load vector is

$$\mathbf{P} = \begin{bmatrix} 50 \\ -86.6 \end{bmatrix} \tag{2}$$

Joint Displacements The stiffness relations for the entire truss can be expressed as (Eq. (18.41) with $\mathbf{P}_f = 0$)

$$\mathbf{P} = \mathbf{Sd} \tag{3}$$

By substituting $\mathbf{P}$ from Eq. (2) and $\mathbf{S}$ from Fig. 18.11(c), we write Eq. (3) in expanded form as

$$\begin{bmatrix} 50 \\ -86.6 \end{bmatrix} = \begin{bmatrix} 1,763.2 & 417.6 \\ 417.6 & 1,644.3 \end{bmatrix} \begin{bmatrix} d_1 \\ d_2 \end{bmatrix}$$

By solving these equations simultaneously, we determine the joint displacements to be

$$d_1 = 0.0434 \text{ in.} \qquad d_2 = -0.0637 \text{ in.}$$

or

$$\mathbf{d} = \begin{bmatrix} 0.0434 \\ -0.0637 \end{bmatrix} \text{in.}$$

Member End Displacements and End Forces
Member 1   The member end displacements in global coordinates, $\mathbf{v}$, can be obtained by simply comparing the member's global degree of freedom numbers with the structure degree of freedom numbers for the member, as follows:

$$\mathbf{v}_1 = \begin{bmatrix} v_1 \\ v_2 \\ v_3 \\ v_4 \end{bmatrix} \begin{matrix} 0 \\ 0 \\ 1 \\ 2 \end{matrix} = \begin{bmatrix} 0 \\ 0 \\ d_1 \\ d_2 \end{bmatrix} = \begin{bmatrix} 0 \\ 0 \\ 0.0434 \\ -0.0637 \end{bmatrix} \text{in.} \tag{4}$$

Note that the structure degree of freedom numbers for the member $(0, 0, 1, 2)$ are written on the right side of $\mathbf{v}$, as shown in Eq. (4). Since the structure degree of freedom numbers corresponding to v_1 and v_2 are zero, this indicates that $v_1 = v_2 = 0$. Similarly, the numbers 1 and 2 corresponding to v_3 and v_4, respectively, indicate that $v_3 = d_1$ and $v_4 = d_2$. It should be realized that these compatibility equations could have been established alternatively simply by a visual inspection of the line diagram of the structure (Fig. 18.11(b)). However, the use of the structure degree of freedom numbers enables us conveniently to program this procedure on a computer.

The member end displacements in local coordinates can now be determined by using the relationship $\mathbf{u} = \mathbf{Tv}$ (Eq. (18.14)), with $\mathbf{T}$ as defined in Eq. (18.22):

$$\mathbf{u}_1 = \begin{bmatrix} u_1 \\ u_2 \end{bmatrix} = \begin{bmatrix} 0.6 & 0.8 & 0 & 0 \\ 0 & 0 & 0.6 & 0.8 \end{bmatrix} \begin{bmatrix} 0 \\ 0 \\ 0.0434 \\ -0.0637 \end{bmatrix} = \begin{bmatrix} 0 \\ -0.0249 \end{bmatrix} \text{in.}$$

By using Eq. (18.7), we compute member end forces in local coordinates as

$$\mathbf{Q} = \mathbf{ku}$$

$$\mathbf{Q}_1 = \begin{bmatrix} Q_1 \\ Q_2 \end{bmatrix} = 870 \begin{bmatrix} 1 & -1 \\ -1 & 1 \end{bmatrix} \begin{bmatrix} 0 \\ -0.0249 \end{bmatrix} = \begin{bmatrix} 21.66 \\ -21.66 \end{bmatrix} k$$

Thus, as shown in Fig. 18.11(d), the axial force in member 1 is

$$21.66 \text{ k (C)} \qquad\qquad \text{Ans.}$$

By applying Eq. (18.17), we can determine member end forces in global coordinates as

$$\mathbf{F} = \mathbf{T}^T\mathbf{Q}$$

$$\mathbf{F}_1 = \begin{bmatrix} F_1 \\ F_2 \\ F_3 \\ F_4 \end{bmatrix} = \begin{bmatrix} 0.6 & 0 \\ 0.8 & 0 \\ 0 & 0.6 \\ 0 & 0.8 \end{bmatrix} \begin{bmatrix} 21.66 \\ -21.66 \end{bmatrix} = \begin{bmatrix} 13 \\ 17.33 \\ -13 \\ -17.33 \end{bmatrix} k$$

Member 2   The member end displacements in global coordinates are given by

$$\mathbf{v}_2 = \begin{bmatrix} v_1 \\ v_2 \\ v_3 \\ v_4 \end{bmatrix} \begin{matrix} 0 \\ 0 \\ 1 \\ 2 \end{matrix} = \begin{bmatrix} 0 \\ 0 \\ d_1 \\ d_2 \end{bmatrix} = \begin{bmatrix} 0 \\ 0 \\ 0.0434 \\ -0.0637 \end{bmatrix} \text{in.}$$

By using the relationship $\mathbf{u} = \mathbf{Tv}$, we determine the member end displacements in local coordinates to be

$$\mathbf{u}_2 = \begin{bmatrix} u_1 \\ u_2 \end{bmatrix} = \begin{bmatrix} 0 & 1 & 0 & 0 \\ 0 & 0 & 0 & 1 \end{bmatrix} \begin{bmatrix} 0 \\ 0 \\ 0.0434 \\ -0.0637 \end{bmatrix} = \begin{bmatrix} 0 \\ -0.0637 \end{bmatrix} \text{in.}$$

Next, the member end forces in local coordinates are computed by using the relationship $\mathbf{Q} = \mathbf{ku}$:

$$\mathbf{Q}_2 = \begin{bmatrix} Q_1 \\ Q_2 \end{bmatrix} = 1{,}087.5 \begin{bmatrix} 1 & -1 \\ -1 & 1 \end{bmatrix} \begin{bmatrix} 0 \\ -0.0637 \end{bmatrix} = \begin{bmatrix} 69.27 \\ -69.27 \end{bmatrix} k$$

Thus, as shown in Fig. 18.11(d), the axial force in member 2 is

$$69.27 \text{ k (C)} \qquad\qquad \text{Ans.}$$

By using the relationship $\mathbf{F} = \mathbf{T}^T\mathbf{Q}$, we calculate the member end forces in global coordinates to be

$$\mathbf{F}_2 = \begin{bmatrix} F_1 \\ F_2 \\ F_3 \\ F_4 \end{bmatrix} = \begin{bmatrix} 0 & 0 \\ 1 & 0 \\ 0 & 0 \\ 0 & 1 \end{bmatrix} \begin{bmatrix} 69.27 \\ -69.27 \end{bmatrix} = \begin{bmatrix} 0 \\ 69.27 \\ 0 \\ -69.27 \end{bmatrix} k$$

continued

Member 3

$$\mathbf{v}_3 = \begin{bmatrix} v_1 \\ v_2 \\ v_3 \\ v_4 \end{bmatrix} \begin{matrix} 0 \\ 0 \\ 1 \\ 2 \end{matrix} = \begin{bmatrix} 0 \\ 0 \\ d_1 \\ d_2 \end{bmatrix} = \begin{bmatrix} 0 \\ 0 \\ 0.0434 \\ -0.0637 \end{bmatrix} \text{in.}$$

$$\mathbf{u} = \mathbf{T}\mathbf{v}$$

$$\mathbf{u}_3 = \begin{bmatrix} u_1 \\ u_2 \end{bmatrix} = \begin{bmatrix} 1 & 0 & 0 & 0 \\ 0 & 0 & 1 & 0 \end{bmatrix} \begin{bmatrix} 0 \\ 0 \\ 0.0434 \\ -0.0637 \end{bmatrix} = \begin{bmatrix} 0 \\ 0.0434 \end{bmatrix} \text{in.}$$

$$\mathbf{Q} = \mathbf{k}\mathbf{u}$$

$$\mathbf{Q}_3 = \begin{bmatrix} Q_1 \\ Q_2 \end{bmatrix} = 1{,}450 \begin{bmatrix} 1 & -1 \\ -1 & 1 \end{bmatrix} \begin{bmatrix} 0 \\ 0.0434 \end{bmatrix} = \begin{bmatrix} -62.93 \\ 62.93 \end{bmatrix} \text{k}$$

Thus, the axial force in member 3 is (Fig. 18.11(d))

$$62.93 \text{ k (T)} \qquad \text{Ans.}$$

$$\mathbf{F} = \mathbf{T}^T \mathbf{Q}$$

$$\mathbf{F}_3 = \begin{bmatrix} F_1 \\ F_2 \\ F_3 \\ F_4 \end{bmatrix} = \begin{bmatrix} 1 & 0 \\ 0 & 0 \\ 0 & 1 \\ 0 & 0 \end{bmatrix} \begin{bmatrix} -62.93 \\ 62.93 \end{bmatrix} = \begin{bmatrix} -62.93 \\ 0 \\ 62.93 \\ 0 \end{bmatrix} \text{k}$$

Support Reactions As shown in Fig. 18.11(e), the reactions at the support joints 1, 2, and 4 are equal to the forces in global coordinates at the ends of the members connected to these joints. **Ans.**

Equilibrium Check Applying the equations of equilibrium to the free body of the entire structure (Fig. 18.11(e)), we obtain

$+ \rightarrow \sum F_X = 0 \qquad 13 - 62.93 + 100 \cos 60° = 0.07 \approx 0$ Checks

$+ \uparrow \sum F_Y = 0 \qquad 17.33 + 69.27 - 100 \sin 60° = 0$ Checks

$+ \circlearrowleft \sum M_① = 0 \qquad 69.27(15) + 62.93(20) - 100 \cos 60°(20) - 100 \sin 60°(15)$

$$= -1.39 \approx 0 \qquad \text{Checks}$$

Example 18.2

Determine the reactions and the member end forces for the three-span continuous beam shown in Fig. 18.12(a) by using the matrix stiffness method.

Solution

Degrees of Freedom From the analytical model of the beam shown in Fig. 18.12(b), we observe that the structure has two degrees of freedom, d_1 and d_2, which are the unknown rotations of joints 2 and 3, respectively. Note that the member local coordinate systems are chosen so that the positive directions of the

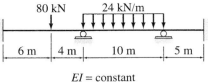

80 kN   24 kN/m

| 6 m | 4 m | 10 m | 5 m |

EI = constant

(a) Continuous Beam

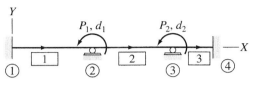

Y

P_1, d_1 P_2, d_2

— X

(b) Analytical Model

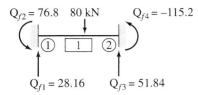

$Q_{f2} = 76.8$ 80 kN $Q_{f4} = -115.2$

① 1 ②

$Q_{f1} = 28.16$ $Q_{f3} = 51.84$

24 kN/m

$Q_{f2} = 200$ ② 2 ③ $Q_{f4} = -200$

$Q_{f1} = 120$ $Q_{f3} = 120$

(c) Member Fixed-End Forces

$$\mathbf{S} = EI \begin{bmatrix} (0.4+0.4) & 0.2 \\ 0.2 & (0.4+0.8) \end{bmatrix} \begin{matrix} 1 \\ 2 \end{matrix} = EI \begin{bmatrix} 0.8 & 0.2 \\ 0.2 & 1.2 \end{bmatrix} \begin{matrix} 1 \\ 2 \end{matrix}$$

$$\mathbf{P}_f = \begin{bmatrix} (-115.2+200) \\ -200 \end{bmatrix} \begin{matrix} 1 \\ 2 \end{matrix} = \begin{bmatrix} 84.8 \\ -200 \end{bmatrix} \begin{matrix} 1 \\ 2 \end{matrix}$$

(d) Structure Stiffness Matrix and Fixed-Joint Force Vector

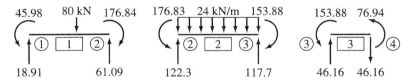

45.98   80 kN   176.84   176.83   24 kN/m   153.88   153.88   76.94

① 1 ②   ② 2 ③   ③ 3 ④

18.91   61.09   122.3   117.7   46.16   46.16

(e) Member End Forces

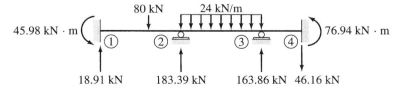

80 kN   24 kN/m

45.98 kN · m   ① ② ③ ④   76.94 kN · m

18.91 kN   183.39 kN   163.86 kN   46.16 kN

(f) Support Reactions

FIG. 18.12

continued

local and global axes are the same. Therefore, no coordinate transformations are needed; that is, the member stiffness relations in the local and global coordinates are the same.

Structure Stiffness Matrix
Member 1   By substituting $L = 10$ m into Eq. (18.6), we obtain

$$\mathbf{K}_1 = \mathbf{k}_1 = EI \begin{array}{cccc} 0 & 0 & 0 & 1 \\ \begin{bmatrix} 0.012 & 0.06 & -0.012 & 0.06 \\ 0.06 & 0.4 & -0.06 & 0.2 \\ -0.012 & -0.06 & 0.012 & -0.06 \\ 0.06 & 0.2 & -0.06 & 0.4 \end{bmatrix} & \begin{array}{c} 0 \\ 0 \\ 0 \\ 1 \end{array} \end{array}$$

By using the fixed-end moment expressions given inside the back cover of the book, we evaluate the fixed-end moments due to the 80-kN load as

$$Q_{f2} = \frac{80(6)(4)^2}{(10)^2} = 76.8 \text{ kN} \cdot \text{m}$$

$$Q_{f4} = -\frac{80(6)^2(4)}{(10)^2} = -115.2 \text{ kN} \cdot \text{m}$$

The fixed-end shears Q_{f1} and Q_{f3} can now be determined by considering the equilibrium of the free body of member 1, shown in Fig. 18.12(c):

$$+\zeta \sum M_{\circled{2}} = 0 \qquad 76.8 - Q_{f1}(10) + 80(4) - 115.2 = 0$$

$$Q_{f1} = 28.16 \text{ kN}$$

$$+\uparrow \sum F_y = 0 \qquad 28.16 - 80 + Q_{f3} = 0$$

$$Q_{f3} = 51.84 \text{ kN}$$

Thus, the fixed-end force vector for member 1 is

$$\mathbf{F}_{f1} = \mathbf{Q}_{f1} = \begin{array}{c} \begin{bmatrix} 28.16 \\ 76.8 \\ 51.84 \\ -115.2 \end{bmatrix} & \begin{array}{c} 0 \\ 0 \\ 0 \\ 1 \end{array} \end{array}$$

From Fig. 18.12(b), we observe that the structure degree of freedom numbers for this member are $0, 0, 0, 1$. By using these numbers, the pertinent elements of $\mathbf{K}_1$ and $\mathbf{F}_{f1}$ are stored in their proper positions in the structure stiffness matrix $\mathbf{S}$ and the fixed-joint force vector $\mathbf{P}_f$, respectively, as shown in Fig. 18.12(d).
Member 2   By substituting $L = 10$ m into Eq. (18.6), we obtain

$$\mathbf{K}_2 = \mathbf{k}_2 = EI \begin{array}{cccc} 0 & 1 & 0 & 2 \\ \begin{bmatrix} 0.012 & 0.06 & -0.012 & 0.06 \\ 0.06 & 0.4 & -0.06 & 0.2 \\ -0.012 & -0.06 & 0.012 & -0.06 \\ 0.06 & 0.2 & -0.06 & 0.4 \end{bmatrix} & \begin{array}{c} 0 \\ 1 \\ 0 \\ 2 \end{array} \end{array}$$

The fixed-end moments due to the 24-kN/m load are

$$Q_{f2} = -Q_{f4} = \frac{24(10)^2}{12} = 200 \text{ kN} \cdot \text{m}$$

Application of the equations of equilibrium to the free body of member 2 yields (Fig. 18.12(c))

$$Q_{f1} = Q_{f3} = 120 \text{ kN}$$

Thus,

$$\mathbf{F}_{f2} = \mathbf{Q}_{f2} = \begin{bmatrix} 120 \\ 200 \\ 120 \\ -200 \end{bmatrix} \begin{matrix} 0 \\ 1 \\ 0 \\ 2 \end{matrix}$$

By using the structure degree of freedom numbers, $0, 1, 0, 2$, for this member, we store the relevant elements of $\mathbf{K}_2$ and $\mathbf{F}_{f2}$ into $\mathbf{S}$ and $\mathbf{P}_f$, respectively, as shown in Fig. 18.12(d).

Member 3 $L = 5$ m:

$$\mathbf{K}_3 = \mathbf{k}_3 = EI \begin{matrix} & 0 & 2 & 0 & 0 & \\ & \begin{bmatrix} 0.096 & 0.24 & -0.096 & 0.24 \\ 0.24 & 0.8 & -0.24 & 0.4 \\ -0.096 & -0.24 & 0.096 & -0.24 \\ 0.24 & 0.4 & -0.24 & 0.8 \end{bmatrix} & \begin{matrix} 0 \\ 2 \\ 0 \\ 0 \end{matrix} \end{matrix}$$

The elements of $\mathbf{K}_3$ are stored in $\mathbf{S}$ using the structure degree of freedom numbers $0, 2, 0, 0$. Note that since member 3 is not subjected to any external loads,

$$\mathbf{F}_{f3} = \mathbf{Q}_{f3} = \mathbf{0}$$

Joint Load Vector Since no external moments are applied to the beam at joints 2 and 3, the joint load vector is zero; that is,

$$\mathbf{P} = \mathbf{0}$$

Joint Displacements The stiffness relations for the entire continuous beam, $\mathbf{P} - \mathbf{P}_f = \mathbf{Sd}$, are written in expanded form as

$$\begin{bmatrix} -84.8 \\ 200 \end{bmatrix} = EI \begin{bmatrix} 0.8 & 0.2 \\ 0.2 & 1.2 \end{bmatrix} \begin{bmatrix} d_1 \\ d_2 \end{bmatrix}$$

By solving these equations simultaneously, we determine the joint displacements to be

$$EId_1 = -154.09 \text{ kN} \cdot \text{m}^2 \qquad EId_2 = 192.35 \text{ kN} \cdot \text{m}^2$$

or

$$\mathbf{d} = \frac{1}{EI} \begin{bmatrix} -154.09 \\ 192.35 \end{bmatrix} \text{kN} \cdot \text{m}^2$$

Member End Displacements and End Forces

Member 1   By using the member's structure degree of freedom numbers, we obtain the member end displacements:

continued

$$\mathbf{u}_1 = \mathbf{v}_1 = \begin{bmatrix} v_1 \\ v_2 \\ v_3 \\ v_4 \end{bmatrix} \begin{matrix} 0 \\ 0 \\ 0 \\ 1 \end{matrix} = \begin{bmatrix} 0 \\ 0 \\ 0 \\ d_1 \end{bmatrix} = \frac{1}{EI} \begin{bmatrix} 0 \\ 0 \\ 0 \\ -154.09 \end{bmatrix}$$

By using the member stiffness relations $\mathbf{Q} = \mathbf{ku} + \mathbf{Q}_f$ (Eq. (18.4)), we compute member end forces as

$$\mathbf{F}_1 = \mathbf{Q}_1 = EI \begin{bmatrix} 0.012 & 0.06 & -0.012 & 0.06 \\ 0.06 & 0.4 & -0.06 & 0.2 \\ -0.012 & -0.06 & 0.012 & -0.06 \\ 0.06 & 0.2 & -0.06 & 0.4 \end{bmatrix} \frac{1}{EI} \begin{bmatrix} 0 \\ 0 \\ 0 \\ -154.09 \end{bmatrix} + \begin{bmatrix} 28.16 \\ 76.8 \\ 51.84 \\ -115.2 \end{bmatrix}$$

$$= \begin{bmatrix} 18.91 \text{ kN} \\ 45.98 \text{ kN} \cdot \text{m} \\ 61.09 \text{ kN} \\ -176.84 \text{ kN} \cdot \text{m} \end{bmatrix} \qquad \text{Ans.}$$

Member 2

$$\mathbf{u}_2 = \mathbf{v}_2 = \begin{bmatrix} v_1 \\ v_2 \\ v_3 \\ v_4 \end{bmatrix} \begin{matrix} 0 \\ 1 \\ 0 \\ 2 \end{matrix} = \begin{bmatrix} 0 \\ d_1 \\ 0 \\ d_2 \end{bmatrix} = \frac{1}{EI} \begin{bmatrix} 0 \\ -154.09 \\ 0 \\ 192.35 \end{bmatrix}$$

$$\mathbf{Q} = \mathbf{ku} + \mathbf{Q}_f$$

$$\mathbf{F}_2 = \mathbf{Q}_2 = \begin{bmatrix} 0.012 & 0.06 & -0.012 & 0.06 \\ 0.06 & 0.4 & -0.06 & 0.2 \\ -0.012 & -0.06 & 0.012 & -0.06 \\ 0.06 & 0.2 & -0.06 & 0.4 \end{bmatrix} \begin{bmatrix} 0 \\ -154.09 \\ 0 \\ 192.35 \end{bmatrix} + \begin{bmatrix} 120 \\ 200 \\ 120 \\ -200 \end{bmatrix}$$

$$= \begin{bmatrix} 122.3 \text{ kN} \\ 176.83 \text{ kN} \cdot \text{m} \\ 117.7 \text{ kN} \\ -153.88 \text{ kN} \cdot \text{m} \end{bmatrix} \qquad \text{Ans.}$$

Member 3

$$\mathbf{u}_3 = \mathbf{v}_3 = \begin{bmatrix} v_1 \\ v_2 \\ v_3 \\ v_4 \end{bmatrix} \begin{matrix} 0 \\ 2 \\ 0 \\ 0 \end{matrix} = \begin{bmatrix} 0 \\ d_2 \\ 0 \\ 0 \end{bmatrix} = \frac{1}{EI} \begin{bmatrix} 0 \\ 192.35 \\ 0 \\ 0 \end{bmatrix}$$

$$\mathbf{Q} = \mathbf{ku} + \mathbf{Q}_f$$

$$\mathbf{F}_3 = \mathbf{Q}_3 = \begin{bmatrix} 0.096 & 0.24 & -0.096 & 0.24 \\ 0.24 & 0.8 & -0.24 & 0.4 \\ -0.096 & -0.24 & 0.096 & -0.24 \\ 0.24 & 0.4 & -0.24 & 0.8 \end{bmatrix} \begin{bmatrix} 0 \\ 192.35 \\ 0 \\ 0 \end{bmatrix} = \begin{bmatrix} 46.16 \text{ kN} \\ 153.88 \text{ kN} \cdot \text{m} \\ -46.16 \text{ kN} \\ 76.94 \text{ kN} \cdot \text{m} \end{bmatrix}$$

$$\text{Ans.}$$

The end forces for the three members of the continuous beam are shown in Fig. 18.12(e).

Support Reactions Since support joint 1 is the beginning joint for member 1, equilibrium considerations require that the reactions at joint 1, $\mathbf{R}_①$, be equal to the upper half of $\mathbf{F}_1$ (i.e., the forces at end 1 of member 1).

$$\mathbf{R}_① = \begin{bmatrix} 18.91 \text{ kN} \\ 45.98 \text{ kN} \cdot \text{m} \end{bmatrix} \quad \text{Ans.}$$

in which the first element of $\mathbf{R}_①$ represents the vertical force and the second element represents the moment, as shown in Fig. 18.12(f). In a similar manner, since support joint 2 is the end joint for member 1 but the beginning joint for member 2, the reaction vector at joint 2, $\mathbf{R}_②$, must be equal to the algebraic sum of the lower half of $\mathbf{F}_1$ and the upper half of $\mathbf{F}_2$.

$$\mathbf{R}_② = \begin{bmatrix} 61.09 \\ -176.84 \end{bmatrix} + \begin{bmatrix} 122.3 \\ 176.83 \end{bmatrix} = \begin{bmatrix} 183.39 \text{ kN} \\ -0.01 \approx 0 \end{bmatrix} \quad \text{Ans.}$$

Similarly, at support joint 3, $\mathbf{R}_③$ can be determined by algebraically summing the lower half of $\mathbf{F}_2$ and the upper half of $\mathbf{F}_3$.

$$\mathbf{R}_③ = \begin{bmatrix} 117.7 \\ -153.88 \end{bmatrix} + \begin{bmatrix} 46.16 \\ 153.88 \end{bmatrix} = \begin{bmatrix} 163.86 \text{ kN} \\ 0 \end{bmatrix} \quad \text{Ans.}$$

Finally, the reaction vector at joint 4 must be equal to the lower half of $\mathbf{F}_3$:

$$\mathbf{R}_④ = \begin{bmatrix} -46.16 \text{ kN} \\ 76.94 \text{ kN} \cdot \text{m} \end{bmatrix} \quad \text{Ans.}$$

The support reactions are shown in Fig. 18.12(f). **Ans.**

Equilibrium Check Applying the equations of equilibrium to the entire structure (Fig. 18.12(f)), we obtain

$$+\uparrow \sum F_Y = 0$$
$$18.91 - 80 + 183.39 - 24(10) + 163.86 - 46.16 = 0 \quad \text{Checks}$$
$$+\zeta \sum M_④ = 0$$
$$45.98 - 18.91(25) + 80(19) - 183.39(15)$$
$$+ 24(10)(10) - 163.86(5) + 76.94 = 0.02 \approx 0 \quad \text{Checks}$$

Example 18.3

Determine the reactions and the member end forces for the frame shown in Fig. 18.13(a) by using the matrix stiffness method.

Solution

Degrees of Freedom From the analytical model of the frame shown in Fig. 18.13(b), we observe that while joints 1 and 3 of the structure can neither trans-

continued

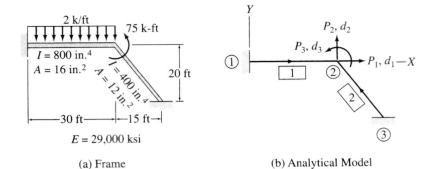

(a) Frame   (b) Analytical Model

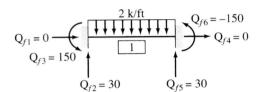

(c) Member Fixed-End Forces

$$\mathbf{S} = \begin{bmatrix} (15,466.67 + 5,050.8) & -6,651.9 & 618.67 \\ -6,651.9 & (71.6 + 8,931.07) & (-1,074.07 + 464) \\ 618.67 & (-1,074.07 + 464) & (21,481.48 + 12,888.89) \end{bmatrix} \begin{matrix} 1 \\ 2 \\ 3 \end{matrix}$$

$$= \begin{bmatrix} 20,517.47 & -6,651.9 & 618.67 \\ -6,651.9 & 9,002.67 & -610.07 \\ 618.67 & -610.07 & 34,370.37 \end{bmatrix} \begin{matrix} 1 \\ 2 \\ 3 \end{matrix}$$

$$\mathbf{P}_f = \begin{bmatrix} 0 \\ 30 \\ -150 \end{bmatrix} \begin{matrix} 1 \\ 2 \\ 3 \end{matrix}$$

FIG. **18.13**

(d) Structure Stiffness Matrix and Fixed-Joint Force Vector

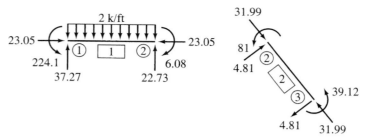

(e) Member End Forces in Local Coordinates

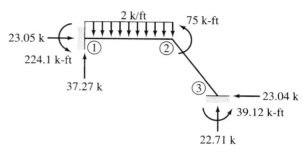

FIG. 18.13 (contd.)

(f) Support Reactions

late nor rotate, joint 2 is free to translate as well as rotate. Thus the frame has three degrees of freedom: the translations d_1 and d_2 in the X and Y directions, respectively, and the rotation d_3 of joint 2.

Structure Stiffness Matrix

Member 1   Since the local xy coordinate system for this member coincides with the global XY coordinate system, no coordinate transformations are needed; that is, the member stiffness relations in the local and global coordinates are the same. By substituting $E = 29,000(12)^2$ ksf, $I = 800/(12)^4$ ft^4, $A = 16/(12)^2$ ft^2, and $L = 30$ ft into Eq. (18.5), we obtain

$$
\mathbf{K}_1 = \mathbf{k}_1 =
\begin{array}{c}
\begin{array}{cccccc}
0 & 0 & 0 & 1 & 2 & 3
\end{array} \\
\left[
\begin{array}{cccccc}
15,466.67 & 0 & 0 & -15,466.67 & 0 & 0 \\
0 & 71.6 & 1,074.07 & 0 & -71.6 & 1,074.07 \\
0 & 1,074.07 & 21,481.48 & 0 & -1,074.07 & 10,740.74 \\
-15,466.67 & 0 & 0 & 15,466.67 & 0 & 0 \\
0 & -71.6 & -1,074.07 & 0 & 71.6 & -1,074.07 \\
0 & 1,074.07 & 10,740.74 & 0 & -1,074.07 & 21,481.48
\end{array}
\right]
\begin{array}{c}
0 \\ 0 \\ 0 \\ 1 \\ 2 \\ 3
\end{array}
\end{array}
$$

(1)

By using the fixed-end moment expressions given inside the back cover of the book, we evaluate the fixed-end moments due to the 2-k/ft load as

$$
Q_{f3} = -Q_{f6} = \frac{2(30)^2}{12} = 150 \text{ k-ft}
$$

continued

By applying equilibrium equations to the free body of the member, we obtain (Fig. 18.13(c))

$$\mathbf{Q}_{f2} = \mathbf{Q}_{f5} = 30 \text{ k}$$

Thus,

$$\mathbf{F}_{f1} = \mathbf{Q}_{f1} = \begin{bmatrix} 0 \\ 30 \\ 150 \\ \hline 0 \\ 30 \\ -150 \end{bmatrix} \begin{matrix} 0 \\ 0 \\ 0 \\ 1 \\ 2 \\ 3 \end{matrix} \tag{2}$$

By using the structure degree of freedom numbers, $0, 0, 0, 1, 2, 3$, for this member, the pertinent elements of $\mathbf{K}_1$ and $\mathbf{F}_{f1}$ are stored in their proper positions in the structure stiffness matrix $\mathbf{S}$ and the fixed-joint force vector $\mathbf{P}_f$, respectively, as shown in Fig. 18.13(d).

Member 2   By substituting $E = 29,000(12)^2$ ksf, $I = 400/(12)^4$ ft^4, $A = 12/(12)^2$ ft^2, and $L = 25$ ft into Eq. (18.5), we obtain

$$\mathbf{k}_2 = \begin{bmatrix} 13,920 & 0 & 0 & -13,920 & 0 & 0 \\ 0 & 61.87 & 773.33 & 0 & -61.87 & 773.33 \\ 0 & 773.33 & 12,888.89 & 0 & -773.33 & 6,444.44 \\ -13,920 & 0 & 0 & 13,920 & 0 & 0 \\ 0 & -61.87 & -773.33 & 0 & 61.87 & -773.33 \\ 0 & 773.33 & 6,444.44 & 0 & -773.33 & 12,888.89 \end{bmatrix} \tag{3}$$

Since member 2 is not subjected to any external loads,

$$\mathbf{Q}_{f2} = \mathbf{0} \tag{4}$$

By using the global coordinates of the beginning joint 3 and the end joint 2, we determine the direction cosines of member 2 as (Eq. (18.13))

$$\cos \theta = \frac{X_2 - X_3}{L} = \frac{30 - 45}{25} = -0.6$$

$$\sin \theta = \frac{Y_2 - Y_3}{L} = \frac{0 - (-20)}{25} = 0.8$$

Substitution of these values into Eq. (18.12) yields the following transformation matrix for the member:

$$\mathbf{T}_2 = \begin{bmatrix} -0.6 & 0.8 & 0 & 0 & 0 & 0 \\ -0.8 & -0.6 & 0 & 0 & 0 & 0 \\ 0 & 0 & 1 & 0 & 0 & 0 \\ 0 & 0 & 0 & -0.6 & 0.8 & 0 \\ 0 & 0 & 0 & -0.8 & -0.6 & 0 \\ 0 & 0 & 0 & 0 & 0 & 1 \end{bmatrix} \tag{5}$$

To determine the member stiffness matrix in global coordinates, $\mathbf{K}_2$, we substitute the matrices $\mathbf{k}_2$ and $\mathbf{T}_2$ into the relationship $\mathbf{K} = \mathbf{T}^T \mathbf{k} \mathbf{T}$ (Eq. (18.26)) and carry out the necessary matrix multiplications to obtain

$$
\mathbf{K}_2 = \begin{array}{c} \\ \end{array}
\begin{bmatrix}
5{,}050.8 & -6{,}651.9 & -618.67 & -5{,}050.8 & 6{,}651.9 & -618.67 \\
-6{,}651.9 & 8{,}931.07 & -464 & 6{,}651.9 & -8{,}931.07 & -464 \\
-618.67 & -464 & 12{,}888.89 & 618.67 & 464 & 6{,}444.44 \\
-5{,}050.8 & 6{,}651.9 & 618.67 & 5{,}050.8 & -6{,}651.9 & 618.67 \\
6{,}651.9 & -8{,}931.07 & 464 & -6{,}651.9 & 8{,}931.07 & 464 \\
-618.67 & -464 & 6{,}444.44 & 618.67 & 464 & 12{,}888.89
\end{bmatrix}
\begin{array}{c} 0 \\ 0 \\ 0 \\ 1 \\ 2 \\ 3 \end{array}
$$

with column headers $0\ 0\ 0\ 1\ 2\ 3$.

(6)

Note that $\mathbf{K}_2$ is symmetric. By using the structure degree of freedom numbers, $0,0,0,1,2,3$, for member 2, the relevant elements of $\mathbf{K}_2$ are added into their positions in the $\mathbf{S}$ matrix, as shown in Fig. 18.13(d). Note that $\mathbf{F}_{f2} = \mathbf{0}$.

Joint Load Vector By comparing Fig. 18.13(a) and (b), we write

$$
\mathbf{P} = \begin{bmatrix} 0 \\ 0 \\ 75 \end{bmatrix}
$$

Joint Displacements The stiffness relations for the entire frame, $\mathbf{P} - \mathbf{P}_f = \mathbf{Sd}$, are written in expanded form as

$$
\begin{bmatrix} 0 \\ 0 \\ 75 \end{bmatrix} - \begin{bmatrix} 0 \\ 30 \\ -150 \end{bmatrix} = \begin{bmatrix} 20{,}517.47 & -6{,}651.9 & 618.67 \\ -6{,}651.9 & 9{,}002.67 & -610.07 \\ 618.67 & -610.07 & 34{,}370.37 \end{bmatrix} \begin{bmatrix} d_1 \\ d_2 \\ d_3 \end{bmatrix}
$$

or

$$
\begin{bmatrix} 0 \\ -30 \\ 225 \end{bmatrix} = \begin{bmatrix} 20{,}517.47 & -6{,}651.9 & 618.67 \\ -6{,}651.9 & 9{,}002.67 & -610.07 \\ 618.67 & -610.07 & 34{,}370.37 \end{bmatrix} \begin{bmatrix} d_1 \\ d_2 \\ d_3 \end{bmatrix}
$$

By solving these equations simultaneously, we determine the joint displacements to be

$$
\mathbf{d} = \begin{bmatrix} -0.00149 \text{ ft} \\ -0.00399 \text{ ft} \\ 0.0065 \text{ rad} \end{bmatrix}
$$

Member End Displacements and End Forces
Member 1

$$
\mathbf{u}_1 = \mathbf{v}_1 = \begin{bmatrix} v_1 \\ v_2 \\ v_3 \\ v_4 \\ v_5 \\ v_6 \end{bmatrix} \begin{array}{c} 0 \\ 0 \\ 0 \\ 1 \\ 2 \\ 3 \end{array} = \begin{bmatrix} 0 \\ 0 \\ 0 \\ d_1 \\ d_2 \\ d_3 \end{bmatrix} = \begin{bmatrix} 0 \\ 0 \\ 0 \\ -0.00149 \text{ ft} \\ -0.00399 \text{ ft} \\ 0.0065 \text{ rad} \end{bmatrix}
$$

continued

By substituting $\mathbf{k}_1$, $\mathbf{Q}_{f1}$, and $\mathbf{u}_1$ in the member stiffness relationship $\mathbf{Q} = \mathbf{ku} + \mathbf{Q}_f$ (Eq. (18.4)), we determine the member end forces to be

$$\mathbf{F}_1 = \mathbf{Q}_1 = \begin{bmatrix} 23.05 \text{ k} \\ 37.27 \text{ k} \\ 224.1 \text{ k-ft} \\ -23.05 \text{ k} \\ 22.73 \text{ k} \\ -6.08 \text{ k-ft} \end{bmatrix} \qquad \text{Ans.}$$

Member 2

$$\mathbf{v}_2 = \begin{bmatrix} v_1 \\ v_2 \\ v_3 \\ v_4 \\ v_5 \\ v_6 \end{bmatrix} \begin{matrix} 0 \\ 0 \\ 0 \\ 1 \\ 2 \\ 3 \end{matrix} = \begin{bmatrix} 0 \\ 0 \\ 0 \\ d_1 \\ d_2 \\ d_3 \end{bmatrix} = \begin{bmatrix} 0 \\ 0 \\ 0 \\ -0.00149 \text{ ft} \\ -0.00399 \text{ ft} \\ 0.0065 \text{ rad} \end{bmatrix}$$

By substituting $\mathbf{K}_2$, $\mathbf{v}_2$, and $\mathbf{F}_{f2} = \mathbf{0}$ into the member stiffness relationship in global coordinates, $\mathbf{F} = \mathbf{Kv} + \mathbf{F}_f$ (Eq. (18.25)), we determine the member end forces in global coordinates to be

$$\mathbf{F}_2 = \begin{bmatrix} -23.04 \text{ k} \\ 22.71 \text{ k} \\ 39.12 \text{ k-ft} \\ 23.04 \text{ k} \\ -22.71 \text{ k} \\ 81 \text{ k-ft} \end{bmatrix}$$

The member end forces in local coordinates can now be evaluated by substituting $\mathbf{F}_2$ and $\mathbf{T}_2$ into the relationship $\mathbf{Q} = \mathbf{TF}$ (Eq. (18.11)).

$$\mathbf{Q}_2 = \begin{bmatrix} 31.99 \text{ k} \\ 4.81 \text{ k} \\ 39.12 \text{ k-ft} \\ -31.99 \text{ k} \\ -4.81 \text{ k} \\ 81 \text{ k-ft} \end{bmatrix} \qquad \text{Ans.}$$

The end forces in the local coordinates of the members are shown in Fig. 18.13(e). **Ans.**

Support Reactions Since support joints 1 and 3 are the beginning joints for members 1 and 2, respectively, the reaction vectors $\mathbf{R}_{\textcircled{1}}$ and $\mathbf{R}_{\textcircled{3}}$ must be equal to the upper halves of $\mathbf{F}_1$ and $\mathbf{F}_2$, respectively.

$$\mathbf{R}_{\textcircled{1}} = \begin{bmatrix} 23.05 \text{ k} \\ 37.27 \text{ k} \\ 224.1 \text{ k-ft} \end{bmatrix}, \qquad \mathbf{R}_{\textcircled{3}} = \begin{bmatrix} -23.04 \text{ k} \\ 22.71 \text{ k} \\ 39.12 \text{ k-ft} \end{bmatrix} \qquad \text{Ans.}$$

The support reactions are shown in Fig. 18.13(f). **Ans.**

Equilibrium Check Applying the equations of equilibrium to the entire frame (Fig. 18.13(f)), we obtain

$$+ \rightarrow \sum F_X = 0 \qquad 23.05 - 23.04 = 0.01 \approx 0 \qquad \text{Checks}$$

$$+ \uparrow \sum F_Y = 0 \qquad 37.27 - 2(30) + 22.71 = -0.02 \approx 0 \qquad \text{Checks}$$

$$+ \circlearrowleft \sum M_{\textcircled{1}} = 0 \qquad 224.1 - 2(30)(15) + 75 - 23.04(20) + 22.71(45) + 39.12$$

$$= -0.63 \approx 0 \qquad \text{Checks}$$

SUMMARY

In this chapter we have studied the basic concepts of the matrix stiffness method for the analysis of plane framed structures. A block diagram summarizing the various steps involved in the analysis is presented in Fig. 18.14.

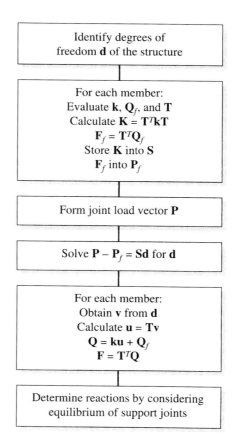

FIG. **18.14**

PROBLEMS

Section 18.6

18.1 through 18.3 Determine the reactions and the force in each member of the trusses shown in Figs. P18.1–P18.3 by using the matrix stiffness method.

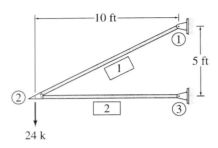

24 k

EA = constant

FIG. **P18.1**

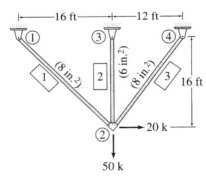

50 k

E = 10,000 ksi

FIG. **P18.2**

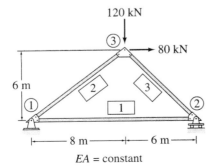

EA = constant

FIG. **P18.3**

18.4 through 18.6 Determine the reactions and the member end forces for the beams shown in Figs. P18.4–P18.6 by using the matrix stiffness method.

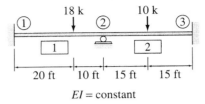

EI = constant

FIG. **P18.4**

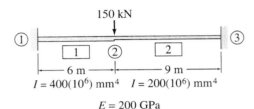

$I = 400(10^6)$ mm^4 $I = 200(10^6)$ mm^4

E = 200 GPa

FIG. **P18.5**

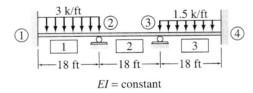

EI = constant

FIG. **P18.6**

18.7 through 18.9 Determine the reactions and the member end forces in local coordinates for the frames shown in Figs. P18.7–P18.9 by using the matrix stiffness method.

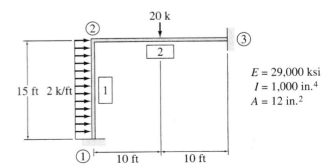

E = 29,000 ksi
I = 1,000 in.^{4}
A = 12 in.^{2}

FIG. **P18.7**

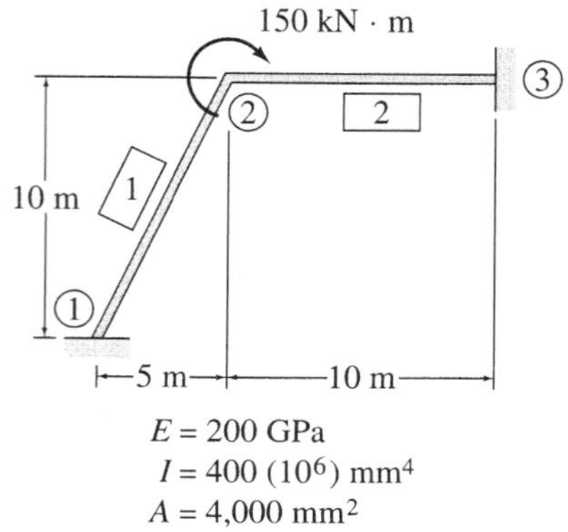
150 kN · m
10 m
5 m
10 m
1
2
3
1
2
E = 200 GPa
I = 400 (10^6) mm^4
A = 4,000 mm^2

FIG. P18.8

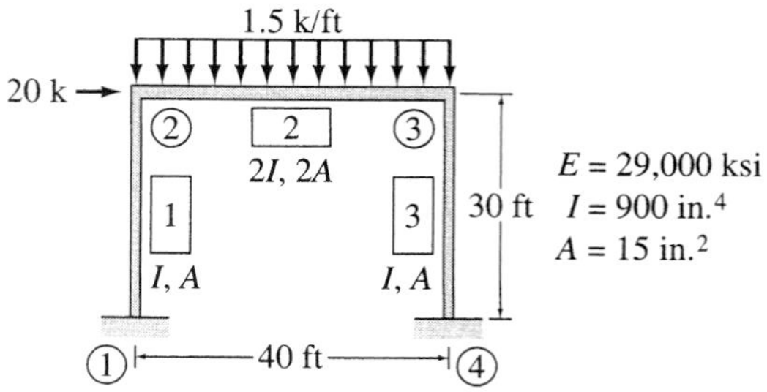
1.5 k/ft
20 k
2
3
2
2I, 2A
1
3
I, A
I, A
30 ft
40 ft
1
4
E = 29,000 ksi
I = 900 in.^4
A = 15 in.^2

FIG. P18.9

A

Areas and Centroids of Geometric Shapes

Shape	Area	Centroid
Right-angled triangle	$A = \dfrac{bh}{2}$	$\bar{x} = \dfrac{2b}{3}$
Triangle	$A = \dfrac{bh}{2}$	$\bar{x} = \dfrac{a+b}{3}$

Shape	Area	Centroid
Trapezoid	$A = \dfrac{b(h_1 + h_2)}{2}$	$\bar{x} = \dfrac{b(h_1 + 2h_2)}{3(h_1 + h_2)}$
Semi-parabola	$A = \dfrac{2bh}{3}$	$\bar{x} = \dfrac{3b}{8}$
Parabolic spandrel	$A = \dfrac{bh}{3}$	$\bar{x} = \dfrac{3b}{4}$
Parabolic segment	$A = \dfrac{2bh}{3}$	$\bar{x} = \dfrac{b}{2}$
	Note: When the segment represents a part of the bending moment diagram of a member subjected to uniformly distributed load w, then $h = wb^2/8$.	
Cubic	$A = \dfrac{3bh}{4}$	$\bar{x} = \dfrac{2b}{5}$

Shape	Area	Centroid
Cubic spandrel	$A = \dfrac{bh}{4}$	$\bar{x} = \dfrac{4b}{5}$
nth-degree curve $y = ax^n,\ n \geq 1$	$A = \dfrac{bh}{n+1}$	$\bar{x} = \dfrac{(n+1)b}{(n+2)}$

B

Review of Matrix Algebra

B.1 Definition of a Matrix
B.2 Types of Matrices
B.3 Matrix Operations
B.4 Solution of Simultaneous Equations by the Gauss-Jordan Method
Problems

In this appendix, some basic concepts of matrix algebra necessary for formulating the computerized analysis of structures are briefly reviewed. A more comprehensive and mathematically rigorous treatment of these concepts can be found in any textbook on matrix algebra, such as [11] and [28].

B.1 DEFINITION OF A MATRIX

A matrix is a rectangular array of quantities arranged in rows and columns. A matrix containing m rows and n columns can be expressed as:

$$\mathbf{A} = [A] = \begin{bmatrix} A_{11} & A_{12} & \cdots & & \cdots & A_{1n} \\ A_{21} & A_{22} & \cdots & & \cdots & A_{2n} \\ \cdots & & & \cdots & A_{ij} & \cdots \\ A_{m1} & A_{m2} & \cdots & | & \cdots & A_{mn} \end{bmatrix} \begin{matrix} \\ \\ i\text{th row} \\ \\ \end{matrix} \qquad (\text{B.1})$$
$$\qquad\qquad\qquad\qquad\qquad j\text{th column} \qquad m \times n$$

As Eq. (B.1) indicates, matrices are usually denoted either by *boldface letters* (e.g., $\mathbf{A}$) or by *italic letters* enclosed within brackets (e.g., $[A]$). The quantities that form a matrix are referred to as the *elements* of the matrix, and each element is represented by a double-subscripted letter, with the first subscript identifying the row and the second subscript identifying the column in which the element is located. Thus in Eq. (B.1), A_{12} represents the element located in the first row and the second column of the matrix $\mathbf{A}$, and A_{21} represents the element in the second row and the first column of $\mathbf{A}$. In general, an element located in the ith row and the jth column of matrix $\mathbf{A}$ is designated as A_{ij}. It is common practice to enclose the entire array of elements between brackets, as shown in Eq. (B.1).

The *size* of a matrix is measured by its *order*, which refers to the number of rows and columns of the matrix. Thus the matrix $\mathbf{A}$ in Eq. (B.1), which consists of m rows and n columns, is considered to be of order $m \times n$ (m by n). As an example, consider a matrix $\mathbf{B}$ given by

$$\mathbf{B} = \begin{bmatrix} 5 & 21 & 3 & -7 \\ 40 & -6 & 19 & 23 \\ -8 & 12 & 50 & 22 \end{bmatrix}$$

The order of this matrix is 3×4, and its elements can be symbolically represented by B_{ij}, with $i = 1$ to 3 and $j = 1$ to 4; for example, $B_{23} = 19$, $B_{31} = -8$, $B_{34} = 22$, etc.

B.2 TYPES OF MATRICES

Row Matrix

If all the elements of a matrix are arranged in a single row (i.e., $m = 1$), then the matrix is called a *row matrix*. An example of a row matrix is

$$\mathbf{C} = \begin{bmatrix} 50 & -3 & -27 & 35 \end{bmatrix}$$

Column Matrix

A matrix with only one column of elements (i.e., $n = 1$) is called a *column matrix*. For example,

$$\mathbf{D} = \{D\} = \begin{bmatrix} -10 \\ 33 \\ -6 \\ 15 \end{bmatrix}$$

Column matrices are also referred to as *vectors* and are sometimes denoted by *italic* letters enclosed within braces (e.g., $\{D\}$).

Square Matrix

A matrix with the same number of rows and columns ($m = n$) is called a *square matrix*. An example of a 3×3 square matrix is

$$\mathbf{A} = \begin{bmatrix} 5 & 21 & 3 \\ 40 & -6 & 19 \\ -8 & 12 & 50 \end{bmatrix} \qquad \text{(B.2)}$$

Main diagonal

The elements with the same subscripts—that is, $A_{11}, A_{22}, \ldots, A_{nn}$—form the *main diagonal* of the square matrix $\mathbf{A}$. These elements are referred to as the *diagonal elements*. As shown in Eq. (B.2), the main diagonal extends from the upper left corner to the lower right corner of the square matrix. The remaining elements of the matrix (i.e., A_{ij} with $i \neq j$) that are not along the main diagonal are termed the *off-diagonal elements*.

Symmetric Matrix

If the elements of a square matrix are symmetric about its main diagonal (i.e., $A_{ij} = A_{ji}$), the matrix is called a *symmetric matrix*. An example of a 4×4 symmetric matrix is

$$\mathbf{A} = \begin{bmatrix} -12 & -6 & 13 & 5 \\ -6 & 7 & -28 & 31 \\ 13 & -28 & 10 & -9 \\ 5 & 31 & -9 & -2 \end{bmatrix}$$

Diagonal Matrix

If all the off-diagonal elements of a square matrix are zero (i.e., $A_{ij} = 0$ for $i \neq j$), the matrix is referred to as a *diagonal matrix*. For example,

$$\mathbf{A} = \begin{bmatrix} 3 & 0 & 0 \\ 0 & -8 & 0 \\ 0 & 0 & 14 \end{bmatrix}$$

Unit or Identity Matrix

A diagonal matrix with all its diagonal elements equal to 1 (i.e., $I_{ii} = 1$ and $I_{ij} = 0$ for $i \neq j$) is called a *unit*, or *identity*, *matrix*. Unit matrices usually are denoted by $\mathbf{I}$ or $[I]$. An example of a 4×4 unit matrix is

$$\mathbf{I} = \begin{bmatrix} 1 & 0 & 0 & 0 \\ 0 & 1 & 0 & 0 \\ 0 & 0 & 1 & 0 \\ 0 & 0 & 0 & 1 \end{bmatrix}$$

Null Matrix

When all the elements of a matrix are zero (i.e., $O_{ij} = 0$), the matrix is called a *null matrix*. Null matrices are commonly denoted by $\mathbf{O}$ or $[O]$. For example,

$$\mathbf{O} = \begin{bmatrix} 0 & 0 & 0 & 0 \\ 0 & 0 & 0 & 0 \\ 0 & 0 & 0 & 0 \end{bmatrix}$$

B.3 MATRIX OPERATIONS

Equality

Two matrices $\mathbf{A}$ and $\mathbf{B}$ are equal if they are of the same order and if their corresponding elements are identical (i.e., $A_{ij} = B_{ij}$). Consider, for example, the matrices

$$\mathbf{A} = \begin{bmatrix} -3 & 5 & 6 \\ 4 & 7 & 9 \\ 12 & 0 & 1 \end{bmatrix} \quad \text{and} \quad \mathbf{B} = \begin{bmatrix} -3 & 5 & 6 \\ 4 & 7 & 9 \\ 12 & 0 & 1 \end{bmatrix}$$

Since both $\mathbf{A}$ and $\mathbf{B}$ are of order 3×3 and since each element of $\mathbf{A}$ is equal to the corresponding element of $\mathbf{B}$, the matrices are considered to be equal to each other; that is, $\mathbf{A} = \mathbf{B}$.

Addition and Subtraction

The addition (or subtraction) of two matrices $\mathbf{A}$ and $\mathbf{B}$, which must be of the same order, is carried out by adding (or subtracting) the corresponding elements of the two matrices. Thus if $\mathbf{A} + \mathbf{B} = \mathbf{C}$, then $C_{ij} = A_{ij} + B_{ij}$; and if $\mathbf{A} - \mathbf{B} = \mathbf{D}$, then $D_{ij} = A_{ij} - B_{ij}$. For example, if

$$\mathbf{A} = \begin{bmatrix} 2 & 5 \\ 3 & 0 \\ 8 & 1 \end{bmatrix} \quad \text{and} \quad \mathbf{B} = \begin{bmatrix} 10 & 4 \\ 6 & 7 \\ 9 & 2 \end{bmatrix}$$

then

$$\mathbf{A} + \mathbf{B} = \mathbf{C} = \begin{bmatrix} 12 & 9 \\ 9 & 7 \\ 17 & 3 \end{bmatrix}$$

and

$$\mathbf{A} - \mathbf{B} = \mathbf{D} = \begin{bmatrix} -8 & 1 \\ -3 & -7 \\ -1 & -1 \end{bmatrix}$$

Note that matrices $\mathbf{C}$ and $\mathbf{D}$ have the same order as matrices $\mathbf{A}$ and $\mathbf{B}$.

Multiplication by a Scalar

To obtain the product of a scalar and a matrix, each element of the matrix must be multiplied by the scalar. Thus, if

$$\mathbf{B} = \begin{bmatrix} 7 & 3 \\ -1 & 4 \end{bmatrix} \qquad \text{and} \qquad c = -3$$

then

$$c\mathbf{B} = \begin{bmatrix} -21 & -9 \\ 3 & -12 \end{bmatrix}$$

Multiplication of Matrices

The multiplication of two matrices can be carried out only if the number of columns of the first matrix equals the number of rows of the second matrix. Such matrices are referred to as being *conformable* for multiplication. Consider, for example, the matrices

$$\mathbf{A} = \begin{bmatrix} -1 & 5 \\ 7 & -3 \end{bmatrix} \qquad \text{and} \qquad \mathbf{B} = \begin{bmatrix} 2 & 3 & -6 \\ 4 & -8 & 9 \end{bmatrix} \qquad (B.3)$$

in which $\mathbf{A}$ is of order 2×2 and $\mathbf{B}$ is of order 2×3. Note that the product $\mathbf{AB}$ of these matrices is defined, because the first matrix, $\mathbf{A}$, of the sequence $\mathbf{AB}$ has two columns and the second matrix, $\mathbf{B}$, has two rows. However, if the sequence of the matrices is reversed, the product $\mathbf{BA}$ does not exist, because now the first matrix, $\mathbf{B}$, has three columns and the second matrix, $\mathbf{A}$, has two rows. The product $\mathbf{AB}$ is usually referred to either as $\mathbf{A}$ *postmultiplied* by $\mathbf{B}$ or as $\mathbf{B}$ *premultiplied* by $\mathbf{A}$. Conversely, the product $\mathbf{BA}$ is referred to either as $\mathbf{B}$ postmultiplied by $\mathbf{A}$ or as $\mathbf{A}$ premultiplied by $\mathbf{B}$.

When two conformable matrices are multiplied, the product matrix thus obtained will have the number of rows of the first matrix and the number of columns of the second matrix. Thus, if a matrix $\mathbf{A}$ of order $m \times n$ is postmultiplied by a matrix $\mathbf{B}$ of order $n \times s$, then the product matrix $\mathbf{C}$ will be of order $m \times s$; that is,

$$\begin{array}{ccc} \mathbf{A} & \mathbf{B} = & \mathbf{C} \\ m \times n \longleftarrow \text{equal} \longrightarrow n \times s & & m \times s \end{array}$$

$$i\text{th row} \begin{bmatrix} \cdots & \overline{(A_{i1} \to A_{in})} & \cdots \end{bmatrix} \begin{bmatrix} \begin{array}{c} B_{1j} \\ \downarrow \\ B_{nj} \end{array} \end{bmatrix} = \begin{bmatrix} C_{ij} \end{bmatrix} i\text{th row}$$

$$j\text{th column} \qquad\qquad j\text{th column} \qquad (B.4)$$

As illustrated in Eq. (B.4), any element C_{ij} of the product matrix $\mathbf{C}$ can be evaluated by multiplying each element of the ith row of $\mathbf{A}$ by the corresponding element of the jth column of B and by algebraically summing the resulting products; that is,

$$C_{ij} = A_{i1}B_{1j} + A_{i2}B_{2j} + \cdots + A_{in}B_{nj} \tag{B.5}$$

Equation (B.5) can be conveniently expressed as

$$C_{ij} = \sum_{k=1}^{n} A_{ik}B_{kj} \tag{B.6}$$

in which n represents the number of columns of the matrix $\mathbf{A}$ and the number of rows of the matrix $\mathbf{B}$. Note that Eq. (B.6) can be used to determine any element of the product matrix $\mathbf{C} = \mathbf{AB}$.

To illustrate the procedure of matrix multiplication, we compute the product $\mathbf{C} = \mathbf{AB}$ of the matrices $\mathbf{A}$ and $\mathbf{B}$ given in Eq. (B.3) as

$$\mathbf{C} = \mathbf{AB} = \underset{2 \times 2}{\begin{bmatrix} -1 & 5 \\ 7 & -3 \end{bmatrix}} \underset{2 \times 3}{\begin{bmatrix} 2 & 3 & -6 \\ 4 & -8 & 9 \end{bmatrix}} = \underset{2 \times 3}{\begin{bmatrix} 18 & -43 & 51 \\ 2 & 45 & -69 \end{bmatrix}}$$

in which the element C_{11} of the product matrix $\mathbf{C}$ is obtained by multiplying each element of the first row of $\mathbf{A}$ by the corresponding element of the first column of $\mathbf{B}$ and summing the resulting products; that is,

$$C_{11} = -1(2) + 5(4) = 18$$

Similarly, the element C_{21} is determined by multiplying the elements of the second row of $\mathbf{A}$ by the corresponding elements of the first column of $\mathbf{B}$ and adding the resulting products; that is,

$$C_{21} = 7(2) - 3(4) = 2$$

The remaining elements of $\mathbf{C}$ are determined in a similar manner:

$$C_{12} = -1(3) + 5(-8) = -43$$
$$C_{22} = 7(3) - 3(-8) = 45$$
$$C_{13} = -1(-6) + 5(9) = 51$$
$$C_{23} = 7(-6) - 3(9) = -69$$

Note that the order of the product matrix $\mathbf{C}$ is 2×3, which equals the number of rows of $\mathbf{A}$ and the number of columns of $\mathbf{B}$.

A common application of matrix multiplication is to express simultaneous equations in compact matrix form. Consider the system of simultaneous linear equations:

$$A_{11}x_1 + A_{12}x_2 + A_{13}x_3 = P_1$$
$$A_{21}x_1 + A_{22}x_2 + A_{23}x_3 = P_2 \tag{B.7}$$
$$A_{31}x_1 + A_{32}x_2 + A_{33}x_3 = P_3$$

in which x_1, x_2, and x_3 are the unknowns and A's and P's represent the coefficients and constants, respectively. By using the definition of matrix multiplication, this system of simultaneous equations can be written in matrix form as

$$\begin{bmatrix} A_{11} & A_{12} & A_{13} \\ A_{21} & A_{22} & A_{23} \\ A_{31} & A_{32} & A_{33} \end{bmatrix} \begin{bmatrix} x_1 \\ x_2 \\ x_3 \end{bmatrix} = \begin{bmatrix} P_1 \\ P_2 \\ P_3 \end{bmatrix} \tag{B.8}$$

or, symbolically, as

$$\mathbf{Ax} = \mathbf{P} \tag{B.9}$$

Even when two matrices $\mathbf{A}$ and $\mathbf{B}$ are of such orders that both products $\mathbf{AB}$ and $\mathbf{BA}$ can be determined, the two products are generally not equal; that is,

$$\mathbf{AB} \neq \mathbf{BA} \tag{B.10}$$

It is, therefore, necessary to maintain the proper sequential order of matrices when computing matrix products. Although matrix multiplication is generally not commutative, as indicated by Eq. (B.10), it is associative and distributive, provided that the sequential order in which the matrices are to be multiplied is maintained. Thus

$$\mathbf{ABC} = (\mathbf{AB})\mathbf{C} = \mathbf{A}(\mathbf{BC}) \tag{B.11}$$

and

$$\mathbf{A}(\mathbf{B} + \mathbf{C}) = \mathbf{AB} + \mathbf{AC} \tag{B.12}$$

Multiplication of any matrix $\mathbf{A}$ by a conformable null matrix $\mathbf{O}$ yields a null matrix; that is,

$$\mathbf{OA} = \mathbf{O} \qquad \text{and} \qquad \mathbf{AO} = \mathbf{O} \tag{B.13}$$

For example,

$$\begin{bmatrix} 0 & 0 \\ 0 & 0 \end{bmatrix} \begin{bmatrix} 5 & -7 \\ 9 & 2 \end{bmatrix} = \begin{bmatrix} 0 & 0 \\ 0 & 0 \end{bmatrix}$$

Multiplication of any matrix $\mathbf{A}$ by a conformable unit matrix $\mathbf{I}$ yields the same matrix $\mathbf{A}$, that is,

$$\mathbf{IA} = \mathbf{A} \qquad \text{and} \qquad \mathbf{AI} = \mathbf{A} \tag{B.14}$$

For example,

$$\begin{bmatrix} 1 & 0 \\ 0 & 1 \end{bmatrix} \begin{bmatrix} 5 & -7 \\ 9 & 2 \end{bmatrix} = \begin{bmatrix} 5 & -7 \\ 9 & 2 \end{bmatrix}$$

and

$$\begin{bmatrix} 5 & -7 \\ 9 & 2 \end{bmatrix} \begin{bmatrix} 1 & 0 \\ 0 & 1 \end{bmatrix} = \begin{bmatrix} 5 & -7 \\ 9 & 2 \end{bmatrix}$$

As Eqs. (B.13) and (B.14) indicate, the null and unit matrices serve the purposes in matrix algebra that are analogous to those of the numbers 0 and 1, respectively, in scalar algebra.

Inverse of a Square Matrix

The inverse of a square matrix $\mathbf{A}$ is defined as a matrix $\mathbf{A}^{-1}$ with elements of such magnitudes that the multiplication of the original matrix $\mathbf{A}$ by its inverse $\mathbf{A}^{-1}$ yields a unit matrix $\mathbf{I}$; that is,

$$\mathbf{A}^{-1}\mathbf{A} = \mathbf{A}\mathbf{A}^{-1} = \mathbf{I} \tag{B.15}$$

Consider, for example, the square matrix

$$\mathbf{A} = \begin{bmatrix} 1 & -2 \\ 3 & -4 \end{bmatrix}$$

The inverse of $\mathbf{A}$ is given by

$$\mathbf{A}^{-1} = \begin{bmatrix} -2 & 1 \\ -1.5 & 0.5 \end{bmatrix}$$

so that the products $\mathbf{A}^{-1}\mathbf{A}$ and $\mathbf{A}\mathbf{A}^{-1}$ satisfy Eq. (B.15):

$$\mathbf{A}^{-1}\mathbf{A} = \begin{bmatrix} -2 & 1 \\ -1.5 & 0.5 \end{bmatrix} \begin{bmatrix} 1 & -2 \\ 3 & -4 \end{bmatrix}$$

$$= \begin{bmatrix} (-2+3) & (4-4) \\ (-1.5+1.5) & (3-2) \end{bmatrix} = \begin{bmatrix} 1 & 0 \\ 0 & 1 \end{bmatrix} = \mathbf{I}$$

and

$$\mathbf{A}\mathbf{A}^{-1} = \begin{bmatrix} 1 & -2 \\ 3 & -4 \end{bmatrix} \begin{bmatrix} -2 & 1 \\ -1.5 & 0.5 \end{bmatrix} = \begin{bmatrix} (-2+3) & (1-1) \\ (-6+6) & (3-2) \end{bmatrix} = \begin{bmatrix} 1 & 0 \\ 0 & 1 \end{bmatrix} = \mathbf{I}$$

The operation of inversion is defined only for square matrices. The inverse of such a matrix is also a square matrix of the same order as the original matrix. A procedure for determining inverses of matrices is presented in the following section. The operation of matrix inversion serves the same purpose as the operation of division in scalar algebra. Consider a system of simultaneous equations expressed in the matrix form as

$$\mathbf{A}\mathbf{x} = \mathbf{P}$$

in which $\mathbf{A}$ represents the square matrix of known coefficients; $\mathbf{x}$ represents the vector of the unknowns; and $\mathbf{P}$ represents the vector of the constants. Since the operation of division is not defined in matrix algebra, we cannot solve the foregoing matrix equation for $\mathbf{x}$ by dividing $\mathbf{P}$ by $\mathbf{A}$ (i.e., $\mathbf{x} = \mathbf{P}/\mathbf{A}$). Instead, to determine the unknowns $\mathbf{x}$, we premultiply both sides of the equation by $\mathbf{A}^{-1}$ to obtain

$$\mathbf{A}^{-1}\mathbf{A}\mathbf{x} = \mathbf{A}^{-1}\mathbf{P}$$

Since $\mathbf{A}^{-1}\mathbf{A} = \mathbf{I}$ and $\mathbf{I}\mathbf{x} = \mathbf{x}$, we can write

$$\mathbf{x} = \mathbf{A}^{-1}\mathbf{P}$$

which indicates that a system of simultaneous equations can be solved by premultiplying the vector of the constants by the inverse of the coefficient matrix.

An important property of matrix inversion is that *the inverse of a symmetric matrix is always a symmetric matrix.*

Transpose of a Matrix

The *transpose* of a matrix is obtained by interchanging its corresponding rows and columns. The transposed matrix is usually identified by the superscript T placed on the symbol of the original matrix. Consider, for example, the 2×3 matrix

$$\mathbf{A} = \begin{bmatrix} 6 & -2 & 4 \\ 1 & 8 & -3 \end{bmatrix}$$

The transpose of $\mathbf{A}$ is given by

$$\mathbf{A}^T = \begin{bmatrix} 6 & 1 \\ -2 & 8 \\ 4 & -3 \end{bmatrix}$$

Note that the first column of $\mathbf{A}$ becomes the first row of $\mathbf{A}^T$. Similarly, the second and third columns of $\mathbf{A}$ become, respectively, the second and third rows of $\mathbf{A}^T$. The order of $\mathbf{A}^T$ thus obtained is 3×2.

As another example, consider the 3×3 matrix

$$\mathbf{B} = \begin{bmatrix} 9 & 7 & -5 \\ 7 & -3 & 2 \\ -5 & 2 & 6 \end{bmatrix}$$

Since the elements of $\mathbf{B}$ are symmetric about the main diagonal (i.e., $B_{ij} = B_{ji}$), interchanging the rows and the columns of this matrix produces a matrix $\mathbf{B}^T$ that is identical to the matrix $\mathbf{B}$ itself; that is,

$$\mathbf{B}^T = \mathbf{B}$$

Thus, *the transpose of a symmetric matrix yields the same matrix.*

Another useful property of matrix transposition is that *the transpose of a product of matrices equals the product of the transposes in reverse order*; that is,

$$(\mathbf{AB})^T = \mathbf{B}^T \mathbf{A}^T \tag{B.16}$$

Similarly,

$$(\mathbf{ABC})^T = \mathbf{C}^T \mathbf{B}^T \mathbf{A}^T \tag{B.17}$$

Partitioning of Matrices

Partitioning is a process by which a matrix is subdivided into a number of smaller matrices called *submatrices*. For example, a 3×4 matrix $\mathbf{A}$ is partitioned into four submatrices by drawing horizontal and vertical dashed partition lines:

$$\mathbf{A} = \begin{bmatrix} 3 & 5 & -1 & 2 \\ -2 & 4 & 7 & 9 \\ 6 & 1 & 3 & 4 \end{bmatrix} = \begin{bmatrix} \mathbf{A}_{11} & \mathbf{A}_{12} \\ \mathbf{A}_{21} & \mathbf{A}_{22} \end{bmatrix} \tag{B.18}$$

in which the submatrices are

$$\mathbf{A}_{11} = \begin{bmatrix} 3 & 5 & -1 \\ -2 & 4 & 7 \end{bmatrix} \qquad \mathbf{A}_{12} = \begin{bmatrix} 2 \\ 9 \end{bmatrix}$$

$$\mathbf{A}_{21} = \begin{bmatrix} 6 & 1 & 3 \end{bmatrix} \qquad \mathbf{A}_{22} = [4]$$

Matrix operations such as addition, subtraction, and multiplication can be preformed on partitioned matrices in the same manner as described previously by treating the submatrices as elements, provided that the matrices are partitioned in such a way that their corresponding submatrices are conformable for the particular operation. For example, suppose that we wish to postmultiply the 3×4 matrix $\mathbf{A}$ of Eq. (B.18) by a 4×2 matrix $\mathbf{B}$, which is partitioned into two submatrices as

$$\mathbf{B} = \begin{bmatrix} 1 & 8 \\ -5 & 2 \\ -3 & 6 \\ 7 & -1 \end{bmatrix} = \begin{bmatrix} \mathbf{B}_{11} \\ \mathbf{B}_{21} \end{bmatrix} \tag{B.19}$$

The product $\mathbf{AB}$ is expressed in terms of the submatrices as

$$\mathbf{AB} = \begin{bmatrix} \mathbf{A}_{11} & \mathbf{A}_{12} \\ \mathbf{A}_{21} & \mathbf{A}_{22} \end{bmatrix} \begin{bmatrix} \mathbf{B}_{11} \\ \mathbf{B}_{21} \end{bmatrix} = \begin{bmatrix} \mathbf{A}_{11}\mathbf{B}_{11} + \mathbf{A}_{12}\mathbf{B}_{21} \\ \mathbf{A}_{21}\mathbf{B}_{11} + \mathbf{A}_{22}\mathbf{B}_{21} \end{bmatrix} \tag{B.20}$$

Note that the matrices $\mathbf{A}$ and $\mathbf{B}$ have been partitioned in such a way that their corresponding submatrices are conformable for multiplication; that is, the orders of the submatrices are such that the products $\mathbf{A}_{11}\mathbf{B}_{11}, \mathbf{A}_{12}\mathbf{B}_{21}, \mathbf{A}_{21}\mathbf{B}_{11}$, and $\mathbf{A}_{22}\mathbf{B}_{21}$ are defined. As shown in Eqs. (B.18) and (B.19), this is achieved by partitioning the rows of the second matrix $\mathbf{B}$ of the product $\mathbf{AB}$ in the same way that the columns of the first matrix $\mathbf{A}$ are partitioned. The products of the submatrices are given by

$$\mathbf{A}_{11}\mathbf{B}_{11} = \begin{bmatrix} 3 & 5 & -1 \\ -2 & 4 & 7 \end{bmatrix} \begin{bmatrix} 1 & 8 \\ -5 & 2 \\ -3 & 6 \end{bmatrix} = \begin{bmatrix} -19 & 28 \\ -43 & 34 \end{bmatrix}$$

$$\mathbf{A}_{12}\mathbf{B}_{21} = \begin{bmatrix} 2 \\ 9 \end{bmatrix} \begin{bmatrix} 7 & -1 \end{bmatrix} = \begin{bmatrix} 14 & -2 \\ 63 & -9 \end{bmatrix}$$

$$\mathbf{A}_{21}\mathbf{B}_{11} = \begin{bmatrix} 6 & 1 & 3 \end{bmatrix} \begin{bmatrix} 1 & 8 \\ -5 & 2 \\ -3 & 6 \end{bmatrix} = \begin{bmatrix} -8 & 68 \end{bmatrix}$$

$$\mathbf{A}_{22}\mathbf{B}_{21} = \begin{bmatrix} 4 \end{bmatrix} \begin{bmatrix} 7 & -1 \end{bmatrix} = \begin{bmatrix} 28 & -4 \end{bmatrix}$$

Substitution into Eq. (B.20) yields

$$\mathbf{AB} = \begin{bmatrix} \begin{bmatrix} -19 & 28 \\ -43 & 34 \end{bmatrix} + \begin{bmatrix} 14 & -2 \\ 63 & -9 \end{bmatrix} \\ \begin{bmatrix} -8 & 68 \end{bmatrix} + \begin{bmatrix} 28 & -4 \end{bmatrix} \end{bmatrix} = \begin{bmatrix} -5 & 26 \\ 20 & 25 \\ 20 & 64 \end{bmatrix}$$

B.4 SOLUTION OF SIMULTANEOUS EQUATIONS BY THE GAUSS-JORDAN METHOD

The *Gauss-Jordan elimination method* is one of the most commonly used procedures for solving simultaneous linear algebraic equations. To illustrate the method, consider the following system of three simultaneous equations:

$$\begin{aligned} 2x_1 - 5x_2 + 4x_3 &= 44 \\ 3x_1 + x_2 - 8x_3 &= -35 \\ 4x_1 - 7x_2 - x_3 &= 28 \end{aligned} \tag{B.21a}$$

To solve for the unknowns $x_1, x_2,$ and x_3, we begin by dividing the first equation by the coefficient of its x_1 term:

$$\begin{aligned} x_1 - 2.5x_2 + 2x_3 &= 22 \\ 3x_1 + x_2 - 8x_3 &= -35 \\ 4x_1 - 7x_2 - x_3 &= 28 \end{aligned} \tag{B.21b}$$

Next, the unknown x_1 is eliminated from the remaining equations by successively subtracting from each remaining equation the product of the coefficient of its x_1 term and the first equation. Thus, to eliminate x_1 from the second equation, we multiply the first equation by 3 and subtract it from the second equation. Similarly, we eliminate x_1 from the third equation by multiplying the first equation by 4 and subtracting it from the third equation. The system of equations thus obtained is

$$
\begin{aligned}
x_1 - 2.5x_2 + 2x_3 &= 22 \\
8.5x_2 - 14x_3 &= -101 \\
3x_2 - 9x_3 &= -60
\end{aligned}
\tag{B.21c}
$$

With x_1 eliminated from all but the first equation, we now divide the second equation by the coefficient of its x_2 term:

$$
\begin{aligned}
x_1 - 2.5x_2 + 2x_3 &= 22 \\
x_2 - 1.647x_3 &= -11.882 \\
3x_2 - 9x_3 &= -60
\end{aligned}
\tag{B.21d}
$$

Next, we eliminate x_2 from the first and the third equations, successively, by multiplying the second equation by -2.5 and subtracting it from the first equation, and then by multiplying the second equation by 3 and subtracting it from the third equation. This yields

$$
\begin{aligned}
x_1 - 2.118x_3 &= -7.705 \\
x_2 - 1.647x_3 &= -11.882 \\
- 4.059x_3 &= -24.354
\end{aligned}
\tag{B.21e}
$$

By dividing the third equation by the coefficient of its x_3 term, we obtain

$$
\begin{aligned}
x_1 - 2.118x_3 &= -7.705 \\
x_2 - 1.647x_3 &= -11.882 \\
x_3 &= 6
\end{aligned}
\tag{B.21f}
$$

Finally, by multiplying the third equation by -2.118 and subtracting it from the first equation, and by multiplying the third equation by -1.647 and subtracting it from the second equation, we determine the solution of the given system of equations (Eq. (B.21a)) to be

$$
\begin{aligned}
x_1 &= 5 \\
x_2 &= -2 \\
x_3 &= 6
\end{aligned}
\tag{B.21g}
$$

That is, $x_1 = 5$, $x_2 = -2$, and $x_3 = 6$. To check that the solution is carried out correctly, we substitute the numerical values of $x_1, x_2,$ and x_3 back into the original equations (Eq. (B.21a)):

$$
\begin{aligned}
2(5) - 5(-2) + 4(6) &= 44 \qquad &\text{Checks} \\
3(5) - 2 - 8(6) &= -35 \qquad &\text{Checks} \\
4(5) - 7(-2) - 6 &= 28 \qquad &\text{Checks}
\end{aligned}
$$

As the foregoing example illustrates, the Gauss-Jordan method essentially involves successively eliminating each unknown from all but one of the equations of the system by performing the following operations: (1) dividing an equation by a scalar; and (2) multiplying an equation by a scalar and subtracting the resulting equation from another equation. These operations, which do not change the solution of the original system of equations, are applied repeatedly until a system with each equation containing only one unknown is obtained.

The solution of simultaneous equations is usually carried out in matrix form by operating on the rows of the coefficient matrix and the vector containing the constant terms of the equations. The foregoing operations are then referred to as *elementary row operations*. These operations are applied to both the coefficient matrix and the vector of the constants simultaneously, until the coefficient matrix is reduced to a unit matrix. The elements of the vector, which initially contained the constant terms of the original equations, now represent the solution of the original simultaneous equations. To illustrate this procedure, consider again the system of three simultaneous equations given in Eq. (B.21a). The system can be expressed in matrix form as

$$\mathbf{Ax} = \mathbf{P}$$

$$\begin{bmatrix} 2 & -5 & 4 \\ 3 & 1 & -8 \\ 4 & -7 & -1 \end{bmatrix} \begin{bmatrix} x_1 \\ x_2 \\ x_3 \end{bmatrix} = \begin{bmatrix} 44 \\ -35 \\ 28 \end{bmatrix} \qquad \text{(B.22)}$$

When applying the Gauss-Jordan method, it is usually convenient to write the coefficient matrix $\mathbf{A}$ and the vector of constants $\mathbf{P}$ as submatrices of a partitioned *augmented matrix*:

$$\left[\begin{array}{ccc|c} 2 & -5 & 4 & 44 \\ 3 & 1 & -8 & -35 \\ 4 & -7 & -1 & 28 \end{array} \right] \qquad \text{(B.23a)}$$

To determine the solution, we begin by dividing row 1 of the augmented matrix by $A_{11} = 2$:

$$\left[\begin{array}{ccc|c} 1 & -2.5 & 2 & 22 \\ 3 & 1 & -8 & -35 \\ 4 & -7 & -1 & 28 \end{array} \right] \qquad \text{(B.23b)}$$

Next, we multiply row 1 by $A_{21} = 3$ and subtract it from row 2 and then multiply row 1 by $A_{31} = 4$ and subtract it from row 3. This yields

$$\left[\begin{array}{ccc|c} 1 & -2.5 & 2 & 22 \\ 0 & 8.5 & -14 & -101 \\ 0 & 3 & -9 & -60 \end{array} \right] \qquad \text{(B.23c)}$$

Divide row 2 by $A_{22} = 8.5$, obtaining

$$\begin{bmatrix} 1 & -2.5 & 2 & \vdots & 22 \\ 0 & 1 & -1.647 & \vdots & -11.882 \\ 0 & 3 & -9 & \vdots & -60 \end{bmatrix} \tag{B.23d}$$

Multiply row 2 by $A_{12} = -2.5$ and subtract it from row 1; then multiply row 2 by $A_{32} = 3$ and subtract it from row 3. This yields

$$\begin{bmatrix} 1 & 0 & -2.118 & \vdots & -7.705 \\ 0 & 1 & -1.647 & \vdots & -11.882 \\ 0 & 0 & -4.059 & \vdots & -24.354 \end{bmatrix} \tag{B.23e}$$

Divide row 3 by $A_{33} = -4.059$:

$$\begin{bmatrix} 1 & 0 & -2.118 & \vdots & -7.705 \\ 0 & 1 & -1.647 & \vdots & -11.882 \\ 0 & 0 & 1 & \vdots & 6 \end{bmatrix} \tag{B.23f}$$

Multiply row 3 by $A_{13} = -2.118$ and subtract it from row 1; then multiply row 3 by $A_{23} = -1.647$ and subtract it from row 2. This yields

$$\begin{bmatrix} 1 & 0 & 0 & \vdots & 5 \\ 0 & 1 & 0 & \vdots & -2 \\ 0 & 0 & 1 & \vdots & 6 \end{bmatrix} \tag{B.23g}$$

Thus $x_1 = 5$, $x_2 = -2$, and $x_3 = 6$.

Matrix Inversion

The Gauss-Jordan elimination method can also be used to determine the inverses of square matrices. The procedure is similar to that described previously for solving simultaneous equations, except that in the augmented matrix, the coefficient matrix is now replaced by the matrix $\mathbf{A}$ that is to be inverted and the vector of constants $\mathbf{P}$ is replaced by a unit matrix $\mathbf{I}$ of the same order as the matrix $\mathbf{A}$. Elementary row operations are then performed on the augmented matrix to reduce the matrix $\mathbf{A}$ to a unit matrix. The matrix $\mathbf{I}$, which was initially the unit matrix, now represents the inverse of the original matrix $\mathbf{A}$.

To illustrate the foregoing procedure, let us compute the inverse of the 2×2 matrix

$$\mathbf{A} = \begin{bmatrix} 1 & -2 \\ 3 & -4 \end{bmatrix} \tag{B.24}$$

The augmented matrix is given by

$$\begin{bmatrix} 1 & -2 & \vdots & 1 & 0 \\ 3 & -4 & \vdots & 0 & 1 \end{bmatrix} \tag{B.25a}$$

By multiplying row 1 by $A_{21} = 3$ and subtracting it from row 2, we obtain

$$\begin{bmatrix} 1 & -2 & | & 1 & 0 \\ 0 & 2 & | & -3 & 1 \end{bmatrix} \tag{B.25b}$$

Next, by dividing row 2 by $A_{22} = 2$, we obtain

$$\begin{bmatrix} 1 & -2 & | & 1 & 0 \\ 0 & 1 & | & -1.5 & 0.5 \end{bmatrix} \tag{B.25c}$$

Finally, by multiplying row 2 by -2 and subtracting it from row 1, we obtain

$$\begin{bmatrix} 1 & 0 & | & -2 & 1 \\ 0 & 1 & | & -1.5 & 0.5 \end{bmatrix} \tag{B.25d}$$

Thus,

$$\mathbf{A}^{-1} = \begin{bmatrix} -2 & 1 \\ -1.5 & 0.5 \end{bmatrix}$$

The computations can be checked by using the relationship $\mathbf{A}^{-1}\mathbf{A} = \mathbf{I}$. We showed in Section B.3 that the matrix $\mathbf{A}^{-1}$, as computed here, does indeed satisfy this relationship.

PROBLEMS

Section B.3

B.1 Determine the matrix $\mathbf{C} = \mathbf{A} + 3\mathbf{B}$ if

$$\mathbf{A} = \begin{bmatrix} 12 & -8 & 15 \\ -8 & 7 & 10 \\ 15 & 10 & -5 \end{bmatrix} \quad \mathbf{B} = \begin{bmatrix} 2 & -1 & 1 \\ -1 & 4 & 6 \\ 1 & 6 & 3 \end{bmatrix}$$

B.2 Determine the matrix $\mathbf{C} = 2\mathbf{A} - \mathbf{B}$ if

$$\mathbf{A} = \begin{bmatrix} 3 & 7 \\ 8 & 4 \\ 2 & -2 \end{bmatrix} \quad \mathbf{B} = \begin{bmatrix} -1 & 6 \\ 5 & 1 \\ 3 & -4 \end{bmatrix}$$

B.3 Determine the products $\mathbf{C} = \mathbf{AB}$ and $\mathbf{D} = \mathbf{BA}$ if

$$\mathbf{A} = [-6 \quad 4 \quad -2] \quad \mathbf{B} = \begin{bmatrix} 2 \\ -1 \\ 5 \end{bmatrix}$$

B.4 Determine the products $\mathbf{C} = \mathbf{AB}$ and $\mathbf{D} = \mathbf{BA}$ if

$$\mathbf{A} = \begin{bmatrix} 2 & -5 \\ -5 & 3 \end{bmatrix} \quad \mathbf{B} = \begin{bmatrix} -3 & 4 \\ 4 & 1 \end{bmatrix}$$

B.5 Show that $(\mathbf{AB})^T = \mathbf{B}^T\mathbf{A}^T$ by using the matrices $\mathbf{A}$ and $\mathbf{B}$ given here.

$$\mathbf{A} = \begin{bmatrix} 8 & -2 & 5 \\ 1 & -4 & 3 \\ 2 & 0 & 6 \end{bmatrix} \quad \mathbf{B} = \begin{bmatrix} 1 & -5 \\ 7 & 0 \\ 0 & -3 \end{bmatrix}$$

Section B.4

B.6 Solve the following system of simultaneous equations by the Gauss-Jordan method.

$$2x_1 + 5x_2 - x_3 = 15$$
$$5x_1 - x_2 + 3x_3 = 27$$
$$-x_1 + 3x_2 + 4x_3 = 14$$

B.7 Solve the following system of simultaneous equations by the Gauss-Jordan method.

$$-12x_1 - 3x_2 + 6x_3 = 45$$
$$5x_1 + 2x_2 - 4x_3 = -9$$
$$10x_1 + x_2 - 7x_3 = -32$$

B.8 Solve the following system of simultaneous equations by the Gauss-Jordan method.

$$5x_1 - 2x_2 + 6x_3 \qquad\quad = \quad 0$$
$$-2x_1 + 4x_2 + \quad x_3 + 3x_4 = \quad 18$$
$$6x_1 + \quad x_2 + 6x_3 + 8x_4 = -29$$
$$3x_2 + 8x_3 + 7x_4 = \quad 11$$

B.9 Determine the inverse of the matrix shown using the Gauss-Jordan method.

$$\mathbf{A} = \begin{bmatrix} 4 & -3 & -1 \\ -2 & 5 & 1 \\ 6 & -4 & -5 \end{bmatrix}$$

B.10 Determine the inverse of the matrix shown using the Gauss-Jordan method.

$$\mathbf{A} = \begin{bmatrix} 4 & 2 & 0 & -3 \\ 2 & 3 & -4 & 0 \\ 0 & -4 & 2 & -1 \\ -3 & 0 & -1 & 5 \end{bmatrix}$$

C

Computer Software

A CD-ROM containing computer software for analyzing plane framed structures is attached to the back cover of this book. The software, which can be used to analyze plane trusses, continuous beams, and plane frames, is based on the matrix stiffness (displacement) method described in Chapter 18. The software is designed for use on IBM and IBM-compatible personal computers with Microsoft Windows® Operating Systems, and it provides an option for saving input data into files for subsequent modification and/or execution.

INSTALLING THE COMPUTER SOFTWARE

To install the computer software:

1. Close any open programs (except Microsoft Windows®).
2. Insert the software CD-ROM into the appropriate drive. On most computer systems, if the software is being installed for the first time, the setup routine will start automatically. If it does, then skip steps 3–6 and follow the instructions on the screen. If the setup routine does not start automatically, then go to step 3.
3. Click the **Start** button on the taskbar shown on the screen; the start menu will appear.
4. Point to the menu title **Settings** and then click the menu item **Control Panel**; the Control Panel window will appear.

5. Double-click the **Add or Remove Programs** icon, and a dialog box will appear.
6. Click the **Add New Programs** button, and follow the instructions on the screen.

STARTING THE COMPUTER SOFTWARE

To start the computer software:

1. Click the **Start** button on the taskbar.
2. Point to the menu title **Programs** and then click the menu item **Structural Analysis 3.0 by A. Kassimali**; the software's title screen will appear as shown in Fig. C.1.

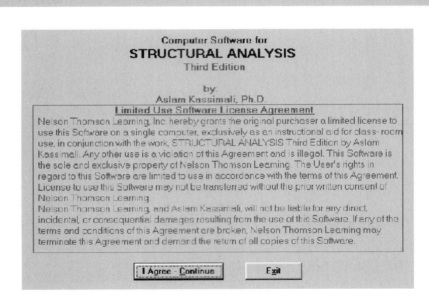

FIG. C.1 Title Screen

INPUTTING DATA

The computer software is interactive in the sense that the user inputs information about the structure by responding to questions and prompts on the screen. The software is designed so that any *consistent* set of units may be used. Thus all the data must be converted into a consistent set of units before being entered into the software. For example, if we wish to

FIG. **C.2** Project Menu

use the units of kips and feet, then the joint coordinates must be defined in feet, the moduli of elasticity in k/ft^2, the areas of cross section in ft^2, the moments of inertia in ft^4, the joint forces and moments in kips and k-ft, respectively, and the uniformly distributed member loads in k/ft.

To start inputting data for a structure, click the menu title **Project**; and then click the menu item **New Project** (Fig. C.2). The input data necessary for the analysis of a structure consist of the following:

1. *General Structural Data*   Input (a) the project title, and (b) the structure type, as shown in Fig. C.3.

2. *Joint Coordinates and Supports*   Input the X and Y coordinates of each joint, and restraints for each support joint, as shown in Fig. C.4. A plot of the joint coordinates and supports will appear on the screen, which can be used to verify that the joint coordinates and restraints have been entered correctly (Fig. C.5).

3. *Material Properties*   Enter the modulus of elasticity (E) for each material (Fig. C.6).

4. *Cross-Sectional Properties*   Enter the cross-sectional area (A) and moment of inertia (I) for each cross-sectional property set (Fig. C.7). For beams, the cross-sectional areas are not needed; whereas for trusses, the moments of inertia are not needed.

5. *Member Data*   For each member, input (a) the beginning joint,

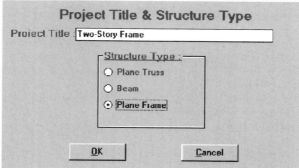

FIG. **C.3** General Structural Data Screen

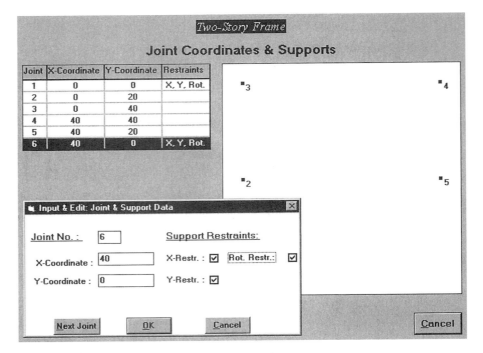

FIG. C.4 Joint Coordinates and Supports Screen

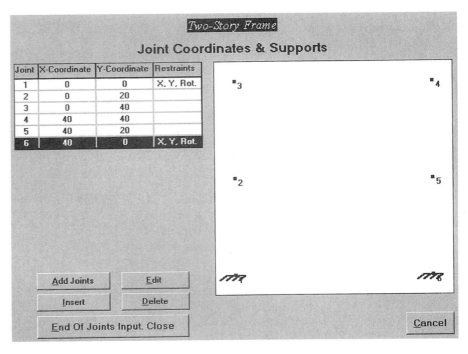

FIG. C.5 A Graphics Display of Joints

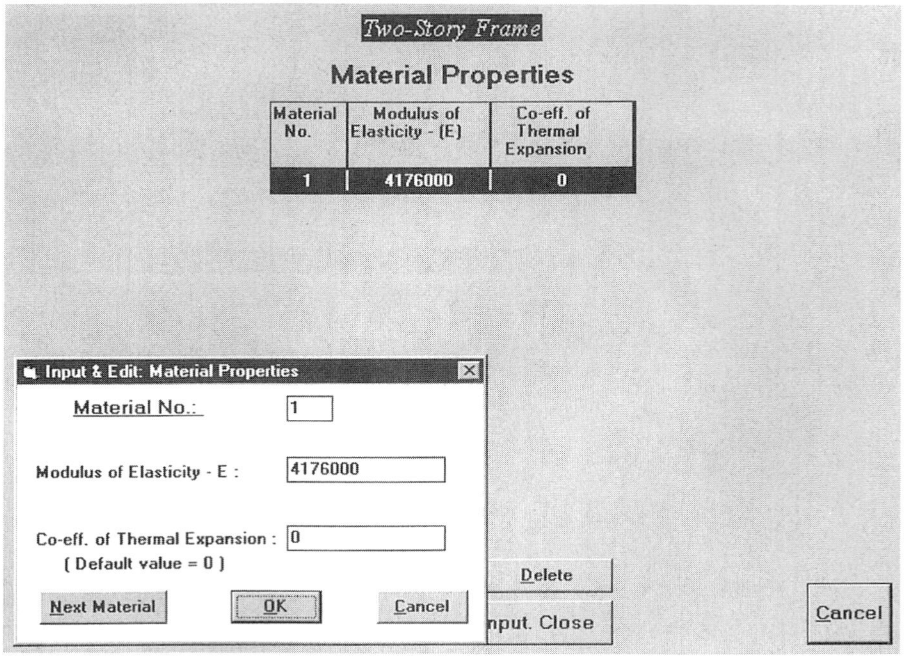

FIG. **C.6** Material Properties Screen

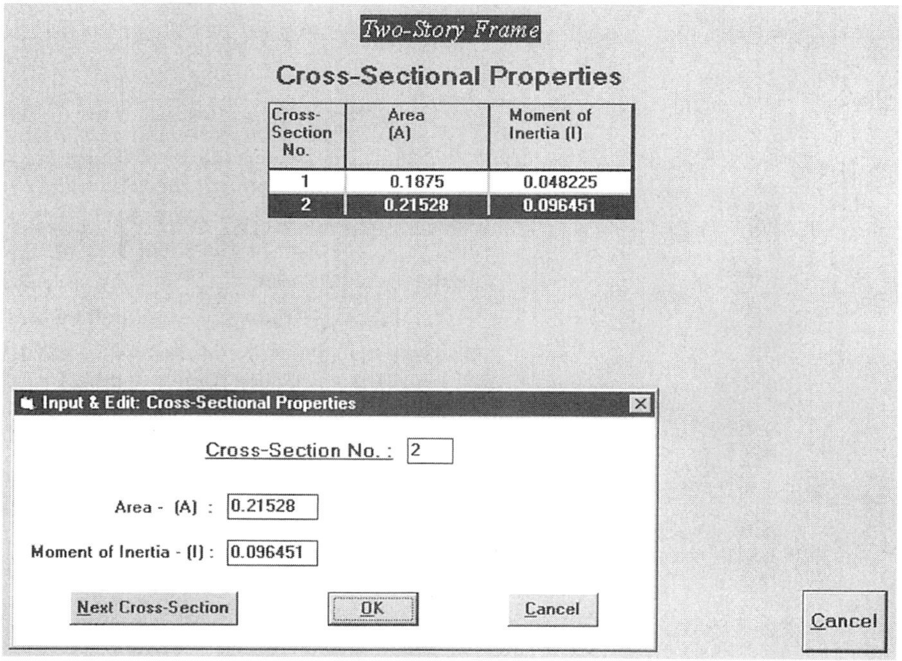

FIG. **C.7** Cross-sectional Properties Screen

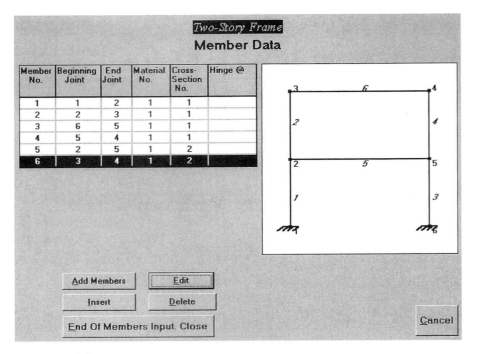

FIG. C.8 Member Data Screen

(b) the end joint, (c) the material number, and (d) the cross-sectional property set number (Fig. C.8). For frames and beams, the member releases option can be used to define any hinges at the member ends. The origin of the local coordinate system for a member is located at the beginning of the member, with the x axis directed from the beginning joint to the end joint. The positive direction of the local y axis is defined by the right-hand rule, with the z axis pointing out of the plane of the page. A plot of the structure appears on the screen, which can be used to verify that the geometry of the structure have been entered correctly.

6. Joint Loads   When analyzing a frame, enter for each joint that is loaded, the joint number, the forces in the global X and Y directions, and the moment (Fig. C.9). In the case of a beam, input only the force in the Y direction and the moment; whereas, for a truss, input only the forces in the X and Y directions. Since the software does not consider member concentrated loads, frame and beam members subjected to such loads must be subdivided into elements (i.e., smaller members) connected together by rigid joints at the locations of the concentrated loads, for the purpose of analysis.

7. Uniformly Distributed Loads on Frame and Beam Members   For each member subjected to uniformly distributed loading, enter the member number and the load intensity (w), as shown in Fig. C.10. Note that

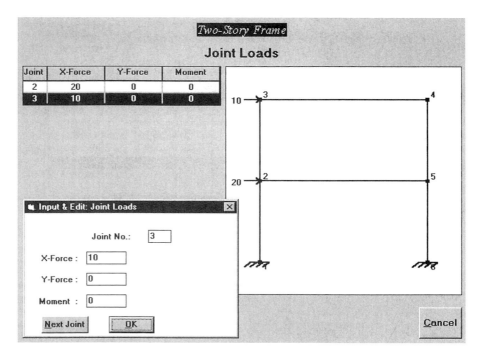

FIG. **C.9** Joint Loads Screen

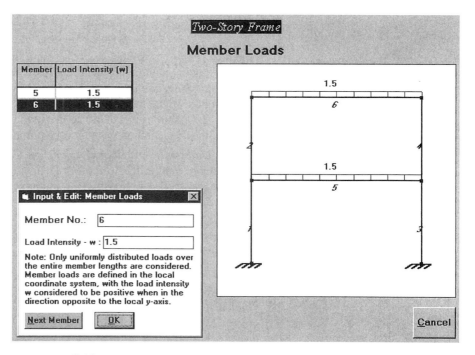

FIG. **C.10** Member Loads Screen

the uniformly distributed load, *w*, is considered to be positive if it acts in the direction *opposite* to the member local *y* axis.

8. *Support Settlements, Temperature Changes and Fabrication Errors* These effects can be input in a manner similar to that for the joint and member loads.

RESULTS OF THE ANALYSIS

Once all the necessary data have been entered, click the menu title **Analysis** of the main screen to analyze the structure (Fig. C.11). The software will automatically compute the joint displacements, member end forces, and support reactions by using the matrix stiffness (displacement) method described in Chapter 18. The results of the analysis are displayed on the screen. The input data as well as the results of the analysis can be printed by clicking on the menu title **Project** and then clicking on the menu item **Print**, of the main screen, as shown in Fig. C.12.

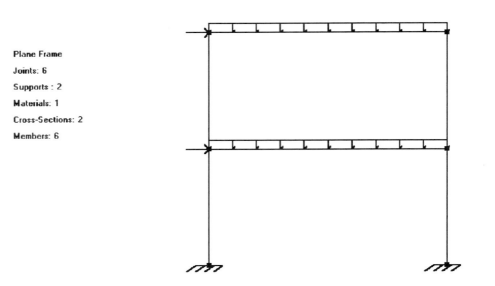

FIG. C.11 Main Screen

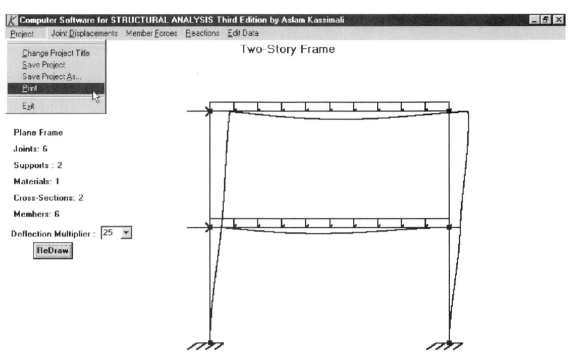

FIG. **C.12** Results of the Analysis

Example C.1

Analyze the two-story frame shown in Fig. C.13(a) using the computer software.

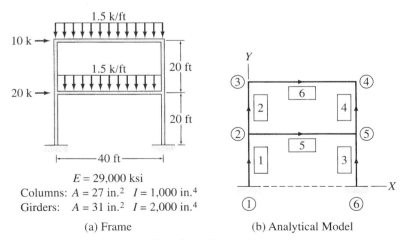

$E = 29,000$ ksi
Columns: $A = 27$ in.^{2} $I = 1,000$ in.^{4}
Girders: $A = 31$ in.^{2} $I = 2,000$ in.^{4}

(a) Frame   (b) Analytical Model

FIG. **C.13**   Two-Story Frame

continued

Solution

This frame was previously analyzed in Example 16.12 by the slope-deflection method, which takes into account only the bending deformations of structures.

The analytical model of the frame is shown in Fig. C.13(b), and the input data are shown on the screen displays given in Figs. C.3 through C.12. The computer printout, which contains the input data and the results of the analysis, is shown in Fig. C.14. Note that the results of the computerized analysis are in agreement with those determined previously by the slope-deflection method.

```
* * * * * * * * * * * * * * * * * * * * * * * * * * *
*        Computer Software          *
*              for                  *
*      STRUCTURAL ANALYSIS          *
*          Third Edition            *
*              by                   *
*       Aslam Kassimali             *
* * * * * * * * * * * * * * * * * * * * * * * * * * *


          =========================
          General Structural Data
          =========================

Project Title : Two-Story Frame
Structure Type : Plane Frame
Number of Joints : 6
Number of Members : 6
Number of Material Property Sets (E) : 1
Number of Cross-Sectional Property Sets : 2


            ==================
            Joint Coordinates
            ==================
```

Joint No.	X Coordinate	Y Coordinate
1	0.0000E+00	0.0000E+00
2	0.0000E+00	2.0000E+01
3	0.0000E+00	4.0000E+01
4	4.0000E+01	4.0000E+01
5	4.0000E+01	2.0000E+01
6	4.0000E+01	0.0000E+00

```
                ========
                Supports
                ========
```

Joint No.	X Restraint	Y Restraint	Rotational Restraint
1	Yes	Yes	Yes
6	Yes	Yes	Yes

FIG. C.14 Computer Printout for Two-Story Frame

```
====================
Material Properties
====================
```

Material No.	Modulus of Elasticity (E)	Co-efficient of Thermal Expansion
1	4.1760E+06	0.0000E+00

```
============================
Cross-Sectional Properties
============================
```

Property No.	Area (A)	Moment of Inertia (I)
1	1.8750E-01	4.8225E-02
2	2.1528E-01	9.6451E-02

```
===========
Member Data
===========
```

Member No.	Beginning Joint	End Joint	Material No.	Cross-Sectional Property No.
1	1	2	1	1
2	2	3	1	1
3	6	5	1	1
4	5	4	1	1
5	2	5	1	2
6	3	4	1	2

```
===========
Joint Loads
===========
```

Joint No.	X Force	Y Force	Moment
2	2.0000E+01	0.0000E+00	0.0000E+00
3	1.0000E+01	0.0000E+00	0.0000E+00

```
============
Member Loads
============
```

Member No.	Load Type	Load Intensity (w)
5	Uniform	1.500E+0
6	Uniform	1.500E+0

************** End of Input Data **************

FIG. C.14 (contd.)

continued

```
* * * * * * * * * * * * * * * * * * * * * * * * * * * * * * * * * * * * * * * * * * * * * * * * *
*                          Results of Analysis                               *
* * * * * * * * * * * * * * * * * * * * * * * * * * * * * * * * * * * * * * * * * * * * * * * * *


                        ====================
                        Joint Displacements
                        ====================

        Joint No.   X Translation    Y Translation    Rotation (Rad)
        ---------   -------------    -------------    --------------
            1        0.0000E+00       0.0000E+00       0.0000E+00
            2        7.6000E-02      -1.2471E-03      -4.0587E-03
            3        1.3024E-01      -1.9299E-03      -3.9537E-03
            4        1.2944E-01      -2.6678E-03       1.7516E-03
            5        7.5934E-02      -1.8180E-03      -1.2035E-03
            6        0.0000E+00       0.0000E+00       0.0000E+00


                ========================================
                Member End Forces in Local Coordinates
                ========================================

        Member   Joint   Axial Force    Shear Force       Moment
        ------   -----   -----------    -----------    -----------
          1        1      4.8825E+01     1.0698E+01      1.4784E+02
                   2     -4.8825E+01    -1.0698E+01      6.6107E+01

          2        2      2.6730E+01    -7.8201E+00     -7.9258E+01
                   3     -2.6730E+01     7.8201E+00     -7.7144E+01

          3        6      7.1175E+01     1.9303E+01      2.0514E+02
                   5     -7.1175E+01    -1.9303E+01      1.8091E+02

          4        5      3.3270E+01     1.7820E+01      1.4844E+02
                   4     -3.3270E+01    -1.7820E+01      2.0796E+02

          5        2      1.4825E+00     2.2095E+01      1.3150E+01
                   5     -1.4825E+00     3.7905E+01     -3.2935E+02

          6        3      1.7820E+01     2.6730E+01      7.7144E+01
                   4     -1.7820E+01     3.3270E+01     -2.0796E+02


                        ==================
                        Support Reactions
                        ==================

        Joint No.    X Force         Y Force          Moment
        ---------   -----------    -----------     -----------
            1       -1.0698E+01     4.8825E+01      1.4784E+02
            6       -1.9303E+01     7.1175E+01      2.0514E+02


        * * * * * * * * * * * * * *  End of Analysis  * * * * * * * * * * * * * *
```

FIG. C.14 (contd.)

PROBLEMS

C.1 and C.2 Using the computer software, determine the smallest cross-sectional area A for the members of the trusses shown in parts (a) through (c) of Figs. PC.1 and PC.2, so that the maximum vertical deflection does not exceed the limit of 1/360 of the span length (i.e., $\Delta_{max} \leq L/360$).

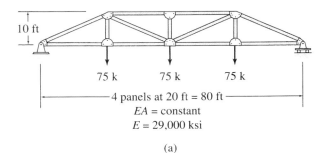

(a)

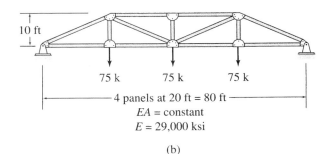

(b)

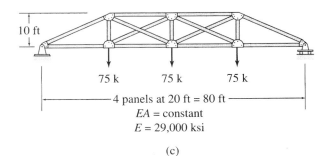

(c)

FIG. **PC.1**

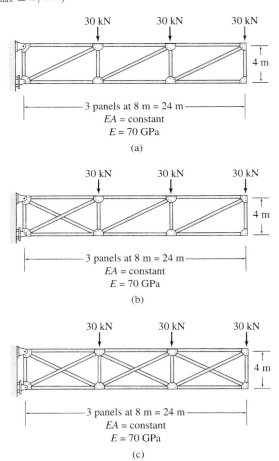

FIG. **PC.2**

C.3 Using the computer software, determine the smallest moment of inertia I required for the frame shown, so that the horizontal deflection of its top right joint does not exceed 1.33 inches.

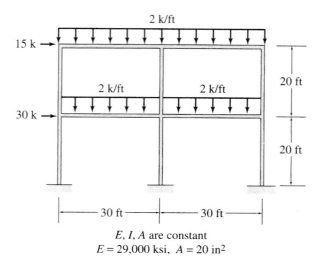

BIBLIOGRAPHY

1. *ASCE Standard Minimum Design Loads for Buildings and Other Structures.* (2003) SEI/ASCE 7-02, American Society of Civil Engineers, Virginia.

2. Arbabi, F. (1991) *Structural Analysis and Behavior.* McGraw-Hill, New York.

3. Bathe, K.J., and Wilson, E.L. (1976) *Numerical Methods in Finite Element Analysis.* Prentice Hall, Englewood Cliffs, N.J.

4. Beer, F.P., and Johnston, E.R., Jr. (1981) *Mechanics of Materials.* McGraw-Hill, New York.

5. Betti, E. (1872) *Il Nuovo Cimento.* Series 2, Vols. 7 and 8.

6. Boggs, R.G. (1984) *Elementary Structural Analysis,* Holt, Rinehart & Winston, New York.

7. Chajes, A. (1990) *Structural Analysis,* 2nd ed. Prentice Hall, Englewood Cliffs, N.J.

8. *Colloquim on History of Structures.* (1982) Proceedings, International Association for Bridge and Structural Engineering, Cambridge, England.

9. Cross, H. (1930) "Analysis of Continuous Frames by Distributing Fixed-End Moments." *Proceedings of the American Society of Civil Engineers* 56, 919–928.

10. Elias, Z.M. (1986) *Theory and Methods of Structural Analysis.* Wiley, New York.

11. Gere, J.M., and Weaver, W., Jr. (1965) *Matrix Algebra for Engineers.* Van Nostrand Reinhold, New York.

12. Glockner, P.G. (1973) "Symmetry in Structural Mechanics." *Journal of the Structural Division, ASCE* 99, 71–89.

13. Hibbler, R.C. (1990) *Structural Analysis,* 2nd ed. Macmillan, New York.

14. Holzer, S.M. (1985) *Computer Analysis of Structures.* Elsevier Science, New York.

15. *International Building Code.* (2003) International Code Council, Falls Church, Virginia.

16. Kassimali, A. (1999) *Matrix Analysis of Structures.* Brooks/Cole, Pacific Grove, California.

17. Kennedy, J.B., and Madugula, M.K.S. (1990) *Elastic Analysis of Structures: Classical and Matrix Methods.* Harper & Row, New York.

18. Laible, J.P. (1985) *Structural Analysis.* Holt, Rinehart & Winston, New York.

19. Langhaar, H.L. (1962) *Energy Methods in Applied Mechanics.* Wiley, New York.

20. Laursen, H.A. (1988) *Structural Analysis,* 3rd ed. McGraw-Hill, New York.

21. Leet, K.M. (1988) *Fundamentals of Structural Analysis.* Macmillan, New York.

22. McCormac, J. (1984) *Structural Analysis,* 4th ed. Harper & Row, New York.

23. McCormac, J., and Elling, R.E. (1988) *Structural Analysis: A Classical and Matrix Approach.* Harper & Row, New York.

24. McGuire, W., and Gallagher, R.H. (1979) *Matrix Structural Analysis.* Wiley, New York.

25. Maney, G.A. (1915) *Studies in Engineering*, Bulletin 1. University of Minnesota, Minneapolis.

26. *Manual for Railway Engineering*. (2003) American Railway Engineering and Maintenance of Way Association, Maryland.

27. Maxwell, J.C. (1864) "On the Calculations of the Equilibrium and Stiffness of Frames." *Philosophical Magazine* 27, 294–299.

28. Noble, B. (1969) *Applied Linear Algebra*. Prentice Hall, Englewood Cliffs, N.J.

29. Norris, C.H., Wilbur, J.B., and Utku, S. (1976) *Elementary Structural Analysis*, 3rd ed. McGraw-Hill, New York.

30. Parcel, J.H., and Moorman, R.B.B. (1955) *Analysis of Statically Indeterminate Structures*. Wiley, New York.

31. Petroski, H. (1985) *To Engineer Is Human—The Role of Failure in Successful Design*. St. Martin's Press, New York.

32. Popov, E.P. (1968) *Introduction to Mechanics of Solids*. Prentice Hall, Englewood Cliffs, N.J.

33. Sack, R.L. (1989) *Matrix Structural Analysis*. PWS-KENT, Boston.

34. Smith, J.C. (1988) *Structural Analysis*. Harper & Row, New York.

35. Spillers, W.R. (1985) *Introduction to Structures*. Ellis Horwood, West Sussex, England.

36. *Standard Specifications for Highway Bridges*, 17th ed. (2002) American Association of State Highway and Transportation Officials, Washington, D.C.

37. Tartaglione, L.C. (1991) *Structural Analysis*. McGraw-Hill, New York.

38. Tezcan, S.S. (1963) Discussion of "Simplified Formulation of Stiffness Matrices." by P.M. Wright. *Journal of the Structural Division*, ASCE 89(6), 445–449.

39. Turner, J.J.; Clough, R.W.; Martin, H.C.; and Topp, L.J. (1956) "Stiffness and Deflection Analysis of Complex Structures." *Journal of Aeronautical Sciences* 23(9), 805–823.

40. Wang, C.K. (1983) *Intermediate Structural Analysis*. McGraw-Hill, New York.

41. West, H.H. (1989) *Analysis of Structures: An Integration of Classical and Modern Methods*, 2nd ed. Wiley, New York.

Answers to Selected Problems

CHAPTER 2

2.1   Beam CD: $w = 662.3$ lb/ft; Girder AE: $w = 111.3$ lb/ft; $P_C = 8,279$ lb; $P_A = P_E = 4,529$ lb

2.3   Beam BF: $w = 16.04$ kN/m; Girder AD: $w = 1.97$ kN/m; $P_B = P_C = 80.2$ kN; $P_A = P_D = 41.85$ kN

2.5   Beam CD: $w = 480$ lb/ft; Girder AE: $P_C = 6,000$ lb; $P_A = P_E = 3,000$ lb

2.7   Beam EF: $w = 180$ lb/ft; Girder AG: $P_C = P_E = 1,800$ lb; $P_A = P_G = 900$ lb; Column A: $P = 2,700$ lb

2.9   Windward side: -76.7 N/m^2 and 191.7 N/m^2; Leeward side: -460.2 N/m^2

2.11   Windward wall: 13.78 psf for $0 \leq z \leq 15$ ft; 15.89 psf for $z = 30$ ft; Leeward wall: -8.3 psf

2.13   Balanced load $= 0.7$ kN/m^2; Unbalanced load $= 1.05$ kN/m^2

CHAPTER 3

3.1   (a) Determinate; (b) Indeterminate, $i_e = 2$; (c) Indeterminate, $i_e = 1$; (d) Unstable

3.3   (a) Unstable; (b) Unstable; (c) Indeterminate, $i_e = 1$; (d) Unstable

3.5 $A_x = 0$; $A_y = 33.75$ k $\uparrow$; $B_y = 56.25$ k $\uparrow$

3.7 $A_x = 22.5$ k $\leftarrow$; $A_y = 0$; $M_A = 225$ k-ft $\circlearrowright$

3.9 $A_x = 0$; $A_y = 220$ kN $\uparrow$; $M_A = 650$ kN $\cdot$ m $\circlearrowright$

3.11 $A_Y = 33.67$ k $\uparrow$; $B_X = 0$; $B_Y = 61.33$ k $\uparrow$

3.13 $R_A = 18.48$ k $\nearrow$; $B_x = 9.24$ k $\leftarrow$; $B_y = 32$ k $\uparrow$

3.15   For $0 \leq x \leq 20$ m: $A_y = 45 - 2x$ kN $\uparrow$; $B_y = 5 + 2x$ kN $\uparrow$

For 20 m $\leq x \leq 25$ m: $A_y = (25 - x)^2/5$ kN $\uparrow$; $B_y = (625 - x^2)/5$ kN $\uparrow$

3.17 $A_Y = 102.75$ k $\uparrow$; $B_X = 0$; $B_Y = 20.25$ k $\uparrow$

3.19 $A_y = 13.33$ k $\uparrow$; $B_x = 25$ k $\leftarrow$; $B_y = 46.67$ k $\uparrow$

3.21 $A_x = 100$ kN $\leftarrow$; $A_y = 216.11$ kN $\uparrow$; $B_y = 183.89$ kN $\uparrow$

3.23 $A_Y = 244.07$ kN $\uparrow$; $B_X = 240$ kN $\leftarrow$; $B_Y = 85.93$ kN $\uparrow$

3.25 $A_Y = 22.5$ k $\uparrow$; $B_X = 0$; $B_Y = 52.5$ k $\uparrow$; $M_B = 1,500$ k-ft $\circlearrowright$

3.27 $A_X = 184.16$ kN $\leftarrow$; $A_Y = 97.48$ kN $\uparrow$; $B_X = 140.82$ kN $\leftarrow$; $B_Y = 130$ kN $\uparrow$

3.29 $A_X = 0$; $A_Y = D_Y = 7.5$ k $\uparrow$; $B_Y = C_Y = 90$ k $\uparrow$

3.31 $A_x = 176.67$ kN $\rightarrow$; $A_y = 356.67$ kN $\uparrow$; $B_x = 23.33$ kN $\rightarrow$; $B_y = 3.33$ kN $\uparrow$

3.33 $A_X = 55$ kN $\leftarrow$; $A_Y = 216.11$ kN $\uparrow$; $B_X = 45$ kN $\leftarrow$; $B_Y = 183.89$ kN $\uparrow$

3.35 $A_X = 8.63$ k $\leftarrow$; $A_Y = 15.46$ k $\uparrow$; $B_X = 11.37$ k $\leftarrow$; $B_Y = 35.45$ k $\uparrow$

3.37 $A_X = 0$; $A_Y = 75.23$ k $\uparrow$; $B_Y = 365.61$ k $\uparrow$; $C_Y = 69.17$ k $\downarrow$; $D_Y = 76.33$ k $\uparrow$

3.39 $A_X = 12.88$ k $\leftarrow$; $A_Y = 18.37$ k $\uparrow$; $B_X = 47.12$ k $\leftarrow$; $B_Y = 106.63$ k $\uparrow$

3.41 $A_X = 52.19$ k $\leftarrow$; $A_Y = 1.25$ k $\downarrow$; $M_A = 421.9$ k-ft $\circlearrowright$; $B_X = 37.81$ k $\leftarrow$; $B_Y = 16.25$ k $\uparrow$; $M_B = 328.1$ k-ft $\circlearrowright$

CHAPTER 4

4.1   (a) Unstable; (b) Determinate; (c) Determinate; (d) Unstable

4.3   (a) Indeterminate, $i = 2$; (b) Indeterminate, $i = 1$; (c) Indeterminate, $i = 1$; (d) Determinate

4.5   (a) Unstable; (b) Unstable; (c) Determinate; (d) Unstable

4.7 $F_{AB} = 0$; $F_{AD} = 100$ kN (T); $F_{BD} = 100$ kN (C); $F_{DE} = 80$ kN (T)

4.9 $F_{AD} = 37.72$ k (C); $F_{AC} = 8.58$ k (C); $F_{CD} = 8.81$ k (T); $F_{DE} = 15$ k (C)

4.11 $F_{AB} = F_{BF} = 20$ kN (T); $F_{AF} = 28.28$ kN (C); $F_{BC} = F_{CD} = F_{CG} = 0$; $F_{BG} = 28.28$ kN (T); $F_{FG} = 20$ kN (C)

4.13 $F_{BC} = 140$ kN (C); $F_{CF} = 108.17$ kN (T); $F_{CG} = 72.11$ kN (C); $F_{FG} = 110$ kN (T)

4.15 $F_{DE} = 195$ k (C); $F_{DJ} = 50$ k (C); $F_{EJ} = 21.21$ k (T); $F_{JK} = 180$ k (T)

4.17 $F_{BC} = 120$ kN (T); $F_{BF} = 60$ kN (C); $F_{BG} = 63.25$ kN (T); $F_{FG} = 189.74$ kN (C)

4.19 $F_{CD} = 17.85$ k (T); $F_{DI} = 4.71$ k (C); $F_{DJ} = 25.69$ k (T); $F_{IJ} = 41.21$ k (C)

4.21 $F_{AC} = F_{BE} = 15.3$ k (C); $F_{AD} = 29.15$ k (C); $F_{CD} = 3$ k (C)

4.23 $F_{AC} = F_{CE} = 4.51$ kN (C); $F_{AD} = 5.59$ kN (C); $F_{BC} = F_{CD} = 9.01$ kN (C)

4.25 $F_{GH} = 27$ kN (C); $F_{GM} = 18$ kN (C); $F_{GN} = 33.33$ kN (T); $F_{HN} = 44.67$ kN (C); $F_{MN} = 7$ kN (T)

4.27 $F_{BC} = 30$ k (T); $F_{BG} = 50$ k (T); $F_{CG} = 50$ k (C); $F_{CH} = 125$ k (T)

4.29 $F_{BC} = 6.1$ k (T); $F_{BE} = 6$ k (C); $F_{BG} = 5$ k (T); $F_{EG} = 2.625$ k (C)

4.31 $F_{BC} = 48$ k-ft/h (T); $F_{GH} = 36$ k-ft/h (C)

4.33 $F_{BC} = 18.75$ k (T); $F_{CF} = 68.94$ k (C); $F_{FG} = 45$ k (T)

4.35 $F_{AD} = 61.85$ kN (C); $F_{CD} = 45.34$ kN (T); $F_{CE} = 6.87$ kN (T)

4.37 $F_{CD} = 113.33$ kN (T); $F_{CH} = 41.67$ kN (C); $F_{GH} = 100$ kN (C)

4.39 $F_{BC} = 72.39$ k (T); $F_{CH} = 16.25$ k (C); $F_{HI} = 9.24$ k (T); $F_{DE} = 27.25$ k (T); $F_{EK} = 0.14$ k (C); $F_{KL} = 2.75$ k (C)

4.41 $F_{CD} = 102.86$ k (C); $F_{DI} = 6.17$ k (C); $F_{DJ} = 35.63$ k (C)

4.43 $F_{CF} = 21.08$ k (T); $F_{CG} = 27.04$ k (T); $F_{EG} = 27.04$ k (C)

4.45 $F_{CH} = F_{EK} = 67.08$ k (C); $F_{HM} = F_{KO} = 67.08$ k (T); $F_{EF} = 20$ k (C); $F_{LM} = 20$ k (T)

4.47 $F_{CD} = 90$ kN (T); $F_{CG} = 26.35$ kN (C); $F_{GJ} = 80$ kN (T); $F_{JK} = 172.45$ kN (C)

4.49 $F_{EF} = 110$ k (T); $F_{EL} = 28.28$ k (T); $F_{LP} = 49.5$ k (T); $F_{OP} = 130$ k (C)

4.51 $F_{AD} = 1.12$ k (T); $F_{BD} = 7.56$ k (C); $F_{CD} = 8.54$ k (C)

4.53 $F_{AB} = 2.18$ k (C); $F_{AC} = 6.24$ k (C); $F_{AD} = 16.63$ k (T); $F_{BC} = 8.61$ k (T)

4.55 $F_{AB} = 29.17$ k (T); $F_{CD} = 15.83$ k (C); $F_{AE} = 4.12$ k (T); $F_{EF} = 28.33$ k (T)

CHAPTER 5

5.1 $Q_A = -40$ kN; $S_A = 32.14$ kN; $M_A = 524.98$ kN · m; $Q_B = 0$; $S_B = -87.14$ kN; $M_B = 261.42$ kN · m

5.3 $Q_A = 86.6$ kN; $S_A = 50$ kN; $M_A = -200$ kN · m; $Q_B = S_B = M_B = 0$

5.5 $Q_A = Q_B = 44.72$ kN; $S_A = 27.6$ kN; $M_A = 855.3$ kN · m; $S_B = -89.44$ kN; $M_B = 268.3$ kN · m

5.7 $Q_A = Q_B = 0$; $S_A = -50$ kN; $M_A = 50$ kN · m; $S_B = -62.5$ kN; $M_B = -150$ kN · m

5.9 $Q_A = Q_B = S_B = 0$; $S_A = 120$ kN; $M_A = -360$ kN · m; $M_B = 120$ kN · m

5.11 $Q_A = 7.5$ kN; $S_A = 12.99$ kN; $M_A = 77.94$ kN · m; $Q_B = S_B = 0$; $M_B = 87.68$ kN · m

5.13   For $0 < x < (L/2)$: $S = P/2$; $M = Px/2$

For $(L/2) < x < L$: $S = -P/2$; $M = P(L - x)/2$

5.15 $S = w(L - 2x)/2$; $M = wx(L - x)/2$

5.17 $S = M/L$

For $0 < x < (L/2)$: Bending Moment $= Mx/L$

For $(L/2) < x < L$: Bending Moment $= M(x - L)/L$

5.19 $S = w(L^2 - 3x^2)/(6L)$; $M = wx(L^2 - x^2)/(6L)$

5.21   For $0 < x < 3$ m: $S = 20$; $M = 20x$

For 3 m $< x < 6$ m: $S = 0$; $M = 60$

For 6 m $< x < 9$ m: $S = -20$; $M = -20x + 180$

5.23   For $0 < x \leq 7$ m: $S = -60$; $M = -60x$

For 7 m $\leq x < 14$ m: $S = 10(1 - x)$; $M = -5x^2 + 10x - 245$

5.25   For $0 < x < 20$ ft: $S = -(x^2/40) - x + 7.5$; $M = -(x^3/120) - (x^2/2) + 7.5x$

For 20 ft $< x < 30$ ft: $S = -(x^2/40) - x + 52.5$; $M = -(x^3/120) - (x^2/2) + 52.5x - 900$

5.27   For $0 < x \leq 5$ m (from A to B): $S = -2x^2 + 83.33$; $M = -(2x^3/3) + 83.33x$

For $0 < x_1 \leq 10$ m (from C to B): $S = x_1^2 - 66.7$; $M = -(x_1^3/3) + 66.67x_1$

5.29 $S_{A,R} = S_{B,L} = 90$ kN; $S_{B,R} = S_{C,L} = -10$ kN; $S_{C,R} = S_{D,L} = -70$ kN; $M_B = 450$ kN · m; $M_C = 350$ kN · m

5.31 $S_{A,R} = S_{B,L} = -10$ k; $S_{B,R} = S_{C,L} = 23.33$ k; $S_{C,R} = S_{D,L} = 3.33$ k; $S_{D,R} = S_{E,L} = -16.67$ k; $M_B = -80$ k-ft; $M_C = 106.67$ k-ft; $M_D = 133.33$ k-ft

5.33 $S_{A,R} = S_{B,L} = -12$ k; $S_{B,R} = S_{C,L} = 24$ k; $S_{C,R} = S_{D,L} = 0$; $S_{D,R} = S_{E,L} = -24$ k; $S_{E,R} = S_{F,L} = 12$ k; $M_B = M_E = -120$ k-ft; $M_C = M_D = 120$ k-ft

5.35 $S_{A,R} = S_{B,L} = 20$ k; $S_{B,R} = S_C = S_D = 0$; $M_A = -125$ k-ft; $M_B = M_C = 75$ k-ft; $M_D = 0$

5.37 $S_{A,R} = S_{B,L} = 225$ kN; $S_{B,R} = S_C = 150$ kN; $S_D = 0$; $M_A = -2,700$ kN · m; $M_B = -1,350$ kN · m; $M_C = -450$ kN · m; $M_D = 0$

5.39 $S_{B,L} = -27$ k; $S_{B,R} = 36$ k; $S_{C,L} = -36$ k; $S_{C,R} = 27$ k; $M_B = M_C = -121.5$ k-ft; $+M_{max} = 94.5$ k-ft, at 21 ft from A

5.41 $S_{A,R} = 109.5$ kN; $S_{B,L} = 37.5$ kN; $S_{B,R} = -112.5$ kN; $S_{C,L} = -184.5$ kN; $S_{C,R} = 120$ kN; $S_{D,L} = 60$ kN; $M_B = 441$ kN · m; $M_C = -450$ kN · m

5.43 $S_{A,R} = S_{B,L} = -8$ k; $S_{B,R} = 23.83$ k; $S_{C,L} = -21.17$ k; $S_{C,R} = S_D = 0$; $M_B = -80$ k-ft; $M_C = M_D = -40$ k-ft; $+M_{max} = 109.3$ k-ft, at 25.89 ft from A

5.45 $S_{A,R} = 30$ k; $S_B = S_{C,L} = -30$ k; $S_{C,R} = S_{D,L} = -50$ k; $M_C = -300$ k-ft; $M_D = -800$ k-ft; $+M_{max} = 225$ k-ft, at 15 ft from A

5.47 $S_{A,R} = S_{B,L} = 33.33$ kN; $S_{B,R} = S_{D,L} = -66.67$ kN; $S_{D,R} = S_{E,L} = -266.67$ kN; $S_{E,R} = 316.67$ kN; $S_{F,L} = 16.67$ kN; $M_B = 333.3$ kN · m; $M_D = -333.3$ kN · m; $M_E = -1,666.7$ kN · m

5.49 $S_{A,R} = 7.5$ k; $S_{C,L} = -22.5$ k; $S_{C,R} = 22.5$ k; $S_{E,L} = -7.5$ k; $S_{E,R} = 4.17$ k; $S_{F,L} = -10.83$ k; $S_{F,R} = 10$ k; $M_C = -225$ k-ft; $M_F = -50$ k-ft; $+M_{max} = 28.13$ k-ft, at 22.5 ft to the left and to the right of C

5.51 $S_{A,R} = 125$ kN; $S_{C,L} = -250$ kN; $S_{C,R} = 187.5$ kN; $S_{D,L} = -187.5$ kN; $S_{D,R} = 250$ kN; $S_{F,L} = -125$ kN; $M_C = M_D = -937.5$ kN · m; $+M_{max} = 312.5$ kN · m, at 5 m from A and F

5.53 (a) $a = 3$ m; (b) $S_{A,R} = S_{B,L} = 50$ kN; $S_{B,R} = S_{C,L} = -100$ kN; $S_{C,R} = S_{D,L} = 150$ kN; $M_B = 450$ kN · m; $M_C = -450$ kN · m

5.55 (a) Determinate; (b) Unstable; (c) Indeterminate, $i = 6$; (d) Indeterminate, $i = 5$

5.57 Member AB: $S_{max} = 16.5$ k; $M_{max} = 247.5$ k-ft; $Q = 0$
Member BC: $S_{max} = -12$ k; $M_{max} = 120$ k-ft; $Q = -8.5$ k

5.59 Member AB: $S_{max} = 48$ kN; $M_{max} = 120$ kN · m; $Q_{max} = -104$ kN
Member BC: $S_{max} = -48$ kN; $M_{max} = 96$ kN · m; $Q = -24$ kN

5.61 Member AB: $S_{max} = -204.97$ kN; $M_{max} = 416.67$ kN · m; $Q = -260.87$ kN
Member BC: $S_{max} = 141.67$ kN; $M_{max} = 416.67$ kN · m; $Q = -300$ kN

5.63 Member AB: $S = 48$ k; $M_{max} = 1,260$ k-ft; $Q = -24$ k
Member BC: $S_{max} = 30$ k; $M_{max} = 300$ k-ft; $Q = 0$

5.65 Member AC: $S_{max} = 108$ kN; $M_{max} = 486$ kN · m; $Q = -7.65$ kN
Member BD: $S = M = 0$; $Q = -217.35$ kN
Member CE: $S_{max} = -142.35$ kN; $M_{max} = 487.95$ kN · m; $Q = 0$

5.67 Member AB: $S = 10$ k; $M_{max} = 200$ k-ft; $Q = -8.83$ k
Member BC: $S_{max} = -30.51$ k; $M_{max} = 225$ k-ft; $Q_{max} = -17.02$ k
Member CD: $S_{max} = 15$ k; $M_{max} = 225$ k-ft; $Q = -27.17$ k

5.69 Member AB: $S_{max} = -24$ k; $M_{max} = 492$ k-ft; $Q = -30$ k
Member BC: $S_{max} = 30$ k; $M_{max} = 492$ k-ft; $Q = -24$ k
Member CD: $S_{max} = 24$ k; $M_{max} = 192$ k-ft; $Q = 0$

5.71 Member AC: $S = 1.25$ k; $M_{max} = 18.75$ k-ft; $Q = -10$ k
Member CE: $S_{max} = -35$ k; $M_{max} = 356.25$ k-ft; $Q = -23.75$ k
Member EG: $S = 23.75$ k; $M_{max} = 356.25$ k-ft; $Q = -35$ k

CHAPTER 6

6.1 $\theta = -\dfrac{M}{6EIL}(3x^2 - 6Lx + 2L^2)$;

$y = -\dfrac{M}{6EIL}(x^3 - 3Lx^2 + 2L^2x)$

6.3   For $0 \leq x \leq a$: $\theta = \dfrac{wx}{2EI}\left[a^2 - L^2 + (L-a)x\right]$;

$y = \dfrac{wx^2}{2EI}\left[\dfrac{a^2 - L^2}{2} + \dfrac{(L-a)x}{3}\right]$

For $a \leq x \leq L$: $\theta = \dfrac{w}{2EI}\left[xL(x-L) - \dfrac{x^3}{3} + \dfrac{a^3}{3}\right]$;

$y = \dfrac{w}{2EI}\left[x^2 L\left(\dfrac{x}{3} - \dfrac{L}{2}\right) - \dfrac{x^4}{12} - \dfrac{a^4}{12} + \dfrac{a^3 x}{3}\right]$

6.5 $\theta = \dfrac{wx}{24EIL}(-x^3 + 6L^2 x - 8L^3)$;

$y = \dfrac{wx^2}{120EIL}(-x^3 + 10L^2 x - 20L^3)$

6.7 $\theta = 0.0174$ rad $\searrow$; $y = 34.8$ mm $\downarrow$

6.9 and 6.35 $\theta_B = 0.00703$ rad $\searrow$; $\Delta_B = 23.4$ mm $\downarrow$

6.11 and 6.37 $\theta_B = Pa^2/EI$ $\searrow$; $\Delta_B = Pa^2(3L-a)/6EI$ $\downarrow$

6.13 and 6.39 $\theta_A = wL^3/8EI$ $\diagdown$; $\Delta_A = 11wL^4/120EI$ $\downarrow$

6.15 and 6.41 $\theta_B = 0.0514$ rad $\searrow$; $\Delta_B = 180$ mm $\downarrow$
$\theta_C = 0.0771$ rad $\searrow$; $\Delta_C = 373$ mm $\downarrow$

6.17 and 6.43 $\theta_B = 0.00304$ rad $\searrow$; $\Delta_B = 67$ mm $\downarrow$
$\theta_C = 0.0122$ rad $\diagdown$; $\Delta_C = 54.8$ mm $\downarrow$

6.19 and 6.45 11,340 (10^6) mm^4

6.21 and 6.47 9,585 (10^6) mm^4

6.23 and 6.49 1.72 in. $\downarrow$

6.25 and 6.51 746 mm $\downarrow$

6.27 and 6.53 1.48 in. $\downarrow$

6.29 and 6.55 0.0139 in. $\downarrow$

6.31 and 6.57 $\theta_D = 0.01134$ rad $\diagup$; $\Delta_D = 2.04$ in. $\uparrow$

6.33 and 6.59 $\theta_B = 0.0099$ rad $\searrow$; $\Delta_B = 0.86$ in. $\downarrow$
$\theta_D = 0.0084$ rad $\diagup$; $\Delta_D = 1.44$ in. $\downarrow$

CHAPTER 7

7.1 and 7.45 $\Delta_{BH} = 0.0225$ in. $\leftarrow$; $\Delta_{BV} = 1.466$ in. $\downarrow$

7.3 and 7.47 $\Delta_{BH} = 9.6$ mm $\rightarrow$; $\Delta_{BV} = 2.13$ mm $\uparrow$

7.5 and 7.49 $\Delta_{BH} = 0.36$ in. $\leftarrow$; $\Delta_{BV} = 1.894$ in. $\downarrow$

7.7   9.1 mm $\downarrow$

7.9   23 mm $\rightarrow$

7.11   3,050 mm^2

7.13   11.91 in^2

7.15   11.07 in^2

7.17   0.175 in. $\leftarrow$

7.19   2.07 in. $\leftarrow$

7.21 and 7.52 $\theta_B = 0.0174$ rad $\searrow$; $\Delta_B = 34.8$ mm $\downarrow$

7.23 and 7.54 373 mm $\downarrow$

7.25 and 7.56 0.0048 in. $\uparrow$

7.27   11,340 (10^6) mm^4

7.29 and 7.57 $\theta_D = 0.01134$ rad $\diagup$; $\Delta_D = 2.04$ in. $\uparrow$

7.31 and 7.59 2.388 in. $\downarrow$

7.33 and 7.60 81.3 mm $\rightarrow$

7.35   2.27 in. $\rightarrow$

7.37   19.2 mm $\downarrow$

7.39 and 7.62 0.182 m $\rightarrow$

7.41   2,225 in^4

7.43   0.00386 rad $\diagdown$

CHAPTER 8

8.1 A_y: 1 at A; 0 at C
C_y: 0 at A; 1 at C
S_B: 0 at A and C; -0.5 at B_L; 0.5 at B_R
M_B: 0 at A and C; 2.5 at B

8.3 A_y: 1 at A; 0 at C
C_y: 0 at A; 1 at C
S_B: 0 at A and C; -0.75 at B_L; 0.25 at B_R
M_B: 0 at A and C; 3.75 at B

8.5 B_y: 1.25 at A; 0 at D
D_y: -0.25 at A; 0 at B; 1 at D
S_C: 0.25 at A; 0 at B and D; -0.5 at C_L; 0.5 at C_R
M_C: -1 at A; 0 at B and D; 2 at C

8.7 A_y: 1 at A; 0 at C
C_y: 0 at A; 1 at C
$S_{A,R}$: 1 at A; 0 at C
M_B: 0 at A and C; 7.5 at B

8.9 A_y: 1 at A and C
M_A $(+\circlearrowleft)$: 0 at A; 20 at C
S_B: 0 at A and B_L; 1 at B_R and C
M_B: 0 at A and B; -8 at C

8.11 S_E: 0 at B, D, and E_L; 1 at E_R and F
M_E: 0 at B, D, and E; -4 at F

8.13 A_y: 1 at A; 0 at C and E
E_y: 0 at A; 1 at C and E
M_E $(+\circlearrowright)$: 0 at A and E; 8 at C

8.15 S_D: 0 at A, D_R, and E; -1 at C and D_L
M_D: 0 at A, D, and E; -4 at C

8.17 A_y: 1 at C; 0 at E; -0.5 at F
B_y: 0 at C; 1.5 at F
S_D: 0 at C and E; -0.5 at D_L and F; 0.5 at D_R
M_D: 0 at C and E; 6 at D; -6 at F

8.19 A_y: 0 at B and E; 2 at D
B_y: 1 at B; 0 at C and E; -1 at D
E_y: 0 at B, C, and D; 1 at E
S_D: 0 at B, C, D_L, and E; 1 at D_R

8.21 S_B: 0 at A, C, D, F, and G; -0.5 at B_L; 0.5 at B_R

M_B: 0 at A, C, D, F, and G; 3 at B
S_C: 0 at A, C_R, D, F, and G; −1 at C_L

8.23 A_y: 1 at A; 0 at B, C, E, F, and G
C_y: 0 at A, E, and G; 1.333 at B; −0.25 at F
E_y: 0 at A, C, and G; −0.333 at B; 1.25 at F
G_y: 0 at A, B, C, E, and F; 1 at G

8.25 S_D: 0 at A, C, E, and G; 0.333 at B; −0.5 at D_L; 0.5 at D_R; −0.25 at F
M_D: 0 at A, C, E, and G; −10 at B; 15 at D; −7.5 at F

8.27 A_y: 1 at A and B; 0 at D and F; −0.667 at E; 0.5 at G
D_y: 0 at A, B, and F; 1.667 at E; −1.25 at G
F_y: 0 at A, B, D, and E; 1.75 at G
$M_A(+\circlearrowleft)$: 0 at A, D, and F; 6 at B; −4 at E; 3 at G

8.29 A_y: 1 at A and B; 0 at D, E, and G
E_y: 0 at A, B, and G; 1.667 at D
G_y: 0 at A, B, and E; −0.667 at D; 1 at G
$M_A(+\circlearrowleft)$: 0 at A, D, E, and G; 20 at B

8.31 A_y: 1 at A and C; 0 at D and F
F_y: 0 at A and C; 1 at D and F
$M_A(+\circlearrowleft)$: 0 at A, D, and F; 10 at C
$M_F(+\circlearrowleft)$: 0 at A, C, and F; −6 at D

8.33 A_y: 1 at A; 0 at B, F, G, and H; −1 at D
B_y: 0 at A, F, G, and H; 2 at D
G_y: 0 at A, B, D, and H; 2 at F
H_y: 0 at A, B, D, and G; −1 at F; 1 at H

8.35 A_X: 0 at C and E; 0.5 at D
A_Y: 1 at C; 0 at E
B_X: 0 at C and E; −0.5 at D
B_Y: 0 at C; 1 at E

8.37 A_y: 1 at B, C, and D; 0 at F
$M_A(+\circlearrowleft)$: −5 at B; 0 at C and F; 5 at D
F_y: 0 at B, C, and D; 1 at F
S_E: 0 at B, C, D, and F; −0.5 at E_L; 0.5 at E_R
M_E: 0 at B, C, D, and F; 2.5 at E

8.39 A_y: 1 at D; 0 at F and H; −0.75 at G
B_y: 0 at D and H; 1 at F; 1.75 at G
C_y: 0 at D, F, and G; 1 at H
S_E: 0 at D, F, and H; −0.5 at E_L; 0.5 at E_R; −0.75 at G
M_E: 0 at D, F, and H; 2 at E; −3 at G

8.41 S_{CD}: 0 at A and E; −0.5 at C; 0.25 at D; −0.5 at G
M_C: 0 at A and E; 15 at C; −15 at G

8.43 S_{BC}: −1 at A and B; 0 at C, D, and E
M_C: −30 at A; 0 at C, D, and E

8.45 F_{AB}: 0 at A and C; 0.5 at B
F_{AD}: 0 at A and C; −0.707 at B
F_{BD}: 0 at A and C; 1 at B

8.47 F_{DH}: 0 at A, B, C, and E; 1 at D
F_{CD}: 0 at A and E; 1 at D
F_{GH}: 0 at A and E; −1.33 at C
F_{CH}: 0 at A and E; 0.833 at C; −0.417 at D

8.49 F_{DE}: 0 at A, B, C, and D; −0.667 at E
F_{CG}: 0 at A and D; −0.401 at B and E; 0.401 at C
F_{GH}: 0 at A and D; −0.889 at C; 0.889 at E
F_{BC}: 0 at A and D; 0.667 at B and C; −0.667 at E

8.51 F_{CD}: −1.6 at A; 0 at C, D, E, F, and G
F_{CI}: −1.8 at A; 0 at C and E; −0.5 at D; 1 at G
F_{DI}: 1.494 at A; 0 at C and E; 0.534 at D; −1.067 at G
F_{DJ}: −0.333 at A and G; 0 at C and E; 0.167 at D

8.53 F_{AB}: 0 at A and G; −1.11 at B
F_{DI}: 0 at A and G; 0.556 at C; −0.833 at D
F_{IJ}: 0 at A and G; 2 at D
F_{CI}: 0 at A and G; −0.333 at C; 0.5 at D

8.55 F_{BC}: 0 at E, F, and G; −4.123 at D
F_{BF}: 0 at E, F, and D; 0.5 at G
F_{BG}: 0 at E, F, and D; −2.236 at G
F_{FG}: 0 at E and F; 2 at G; 4 at D

8.57 F_{AD}: 0 at C and E; −1 at D; 1 at F
F_{BD}: 0 at C, D, and E; −1.67 at F
F_{CD}: 1.33 at C; 0 at D, E, and F

8.59 Δ_B: 0 at A and C; −20.833/(EI) at B
8.61 Δ_D: 0 at A and C; −175/(EI) at D

CHAPTER 9

9.1 −37.5 k-ft
9.3 −8.344 k
9.5 Maximum A_y = 75 k ↑;
Maximum M_A = 1,000 k-ft $\circlearrowright$
9.7 Maximum Positive S_D = 60.417 k;
Maximum Negative S_D = −45.833 k
Maximum Positive M_D = 1,937.5 k-ft;
Maximum Negative M_D = −1,100 k-ft
9.9 Maximum Tensile F_{CH} = 86.1 k (T);
Maximum Compressive F_{CH} = 6.96 k (C)
9.11 Maximum Tensile F_{IJ} = 148.5 k (T);
Maximum Compressive F_{IJ} = 2.76 k (C)
9.13 S_B = 92.5 kN; M_B = 487.5 kN · m
9.15 264 k-ft
9.17 88.56 kN (T)

9.19   42.5 k
9.21   370.1 k-ft
9.23   601.8 k-ft

CHAPTER 10

10.17 $F_{AC} = 26.35$ k (C); $F_{BC} = 36.89$ k (T)
10.19 $F_{AC} = 4.51$ kN (C); $F_{AD} = 5.59$ kN (C); $F_{CD} = 9.01$ kN (C); $F_{CE} = 4.51$ kN (C)
10.21 $A_X^{AD} = 20$ kN →; $A_Y^{AD} = 197.33$ kN ↑; $D_X^{AD} = 20$ kN ←; $D_Y^{AD} = 197.33$ kN ↓; $M_D^{AD} = 240$ kN · m ↻
10.23 $B_X^{BG} = 23.75$ k ←; $B_Y^{BG} = 35$ k ↑; $M_B^{BG} = 356.25$ k-ft ↻; $G_X^{BG} = 23.75$ k →; $G_Y^{BG} = 35$ k ↓; $M_G^{BG} = 356.25$ k-ft ↻

CHAPTER 12

12.1 $S_L = S_R = 90$ kN ↑; $M_L = 48.6$ kN · m ↻; $M_R = 48.6$ kN · m ↺
12.3   Girder DE: $S_L = S_R = 80$ kN; $M_L = 57.6$ kN · m ↻; $M_R = 57.6$ kN · m ↺
Girder EF: $S_L = S_R = 50$ kN; $M_L = 22.5$ kN · m ↻; $M_R = 22.5$ kN · m ↺
12.5   Girder DE: $S_L = S_R = 80$ kN; $M_L = 57.6$ kN · m ↻; $M_R = 57.6$ kN · m ↺
Girder HI: $S_L = S_R = 60$ kN; $M_L = 64.8$ kN · m ↻; $M_R = 64.8$ kN · m ↺
12.7   Member AD: $Q = 12.5$ k (T); $S = 12.5$ k; $M = 125$ k-ft
Member BE: $Q = 0$; $S = 25$ k; $M = 250$ k-ft
Member EF: $Q = 12.5$ k (C); $S = 12.5$ k; $M = 125$ k-ft
12.9   Member AD: $Q = 16.67$ k (C); $S = 15$ k; $M = 120$ k-ft
Member CF: $Q = 16.67$ k (T); $S = 15$ k; $M = 120$ k-ft
Member DE: $Q = 10$ k (C); $S = 13.33$ k; $M = 160$ k-ft
Member HI: $Q = 15$ k (C); $S = 3.33$ k; $M = 40$ k-ft
12.11   Member AD: $Q = 10.5$ k (C); $S = 10$ k; $M = 60$ k-ft
Member CF: $Q = 14$ k (T); $S = 10$ k; $M = 60$ k-ft
Member DE: $Q = 6.25$ k (C); $S = 8.25$ k; $M = 82.5$ k-ft

Member HI: $Q = 11.25$ k (C); $S = 3$ k; $M = 22.5$ k-ft
12.13   Member AE: $Q = 7.33$ k (T); $S = 6.25$ k; $M = 50$ k-ft
Member CG: $Q = 9.67$ k (C); $S = 12.5$ k; $M = 100$ k-ft
Member EF: $Q = 12.5$ k (C); $S = 5.33$ k; $M = 80$ k-ft
Member JK: $Q = 7.5$ k (C); $S = 6$ k; $M = 60$ k-ft
12.15   Member AD: $Q = 12.5$ k (T); $S = 12.5$ k; $M = 125$ k-ft
Member BE: $Q = 0$; $S = 25$ k; $M = 250$ k-ft
Member EF: $Q = 12.5$ k (C); $S = 12.5$ k; $M = 125$ k-ft
12.17   Member AD: $Q = 16.67$ k (C); $S = 15$ k; $M = 120$ k-ft
Member CF: $Q = 16.67$ k (T); $S = 15$ k; $M = 120$ k-ft
Member DE: $Q = 10$ k (C); $S = 13.33$ k; $M = 160$ k-ft
Member HI: $Q = 15$ k (C); $S = 3.33$ k; $M = 40$ k-ft
12.19   Member AD: $Q = 12.66$ k (C); $S = 12.06$ k; $M = 72.4$ k-ft
Member CF: $Q = 11.51$ k (T); $S = 8.24$ k; $M = 49.43$ k-ft
Member DE: $Q = 7.55$ k (C); $S = 9.95$ k; $M = 99.5$ k-ft
Member HI: $Q = 12.1$ k (C); $S = 2.46$ k; $M = 18.45$ k-ft
12.21   Member AE: $Q = 9.18$ k (T); $S = 3.97$ k; $M = 31.75$ k-ft
Member CG: $Q = 2.29$ k (C); $S = 14.78$ k; $M = 118.25$ k-ft
Member EF: $Q = 17.65$ k (C); $S = 5.65$ k; $M = 84.75$ k-ft
Member JK: $Q = 7.49$ k (C); $S = 1.41$ k; $M = 14.1$ k-ft

CHAPTER 13

13.1 and 13.5 $A_y = 99.26$ kN ↑; $M_A = 233.3$ kN · m ↻; $D_y = 60.74$ kN ↑
13.3 and 13.7 $A_y = 28.13$ kN ↑; $C_y = 91.87$ kN ↑; $M_C = 307.4$ kN · m ↺
13.9 and 13.30 $A_y = E_y = 15.625$ k ↑; $C_y = 68.75$ k ↑

13.11 and 13.32 $A_y = E_y = 15.63$ k $\uparrow$; $C_y = 68.75$ k $\uparrow$

13.13 $A_y = 28.36$ k $\uparrow$; $C_y = 85.78$ k $\uparrow$; $D_y = 20.86$ k $\uparrow$

13.15 $A_y = 13.125$ kN $\downarrow$; $M_A = 91.875$ kN $\cdot$ m $\circlearrowright$; $B_y = 223.125$ kN $\uparrow$

13.17 $A_y = (13\ wL)/32$ $\uparrow$; $B_y = (17\ wL)/16$ $\uparrow$; $C_y = (33\ wL)/32$ $\uparrow$

13.19 $A_X = 200$ kN $\leftarrow$; $A_Y = 57.03$ kN $\uparrow$; $M_A = 820.3$ kN $\cdot$ m $\circlearrowright$; $D_Y = 92.97$ kN $\uparrow$

13.21 $A_X = 5.7$ k $\rightarrow$; $A_Y = 50.1$ k $\uparrow$; $C_X = 9.3$ k $\rightarrow$; $C_Y = 24.9$ k $\uparrow$

13.23 $A_X = 0$; $A_Y = 8.23$ kN $\downarrow$; $M_A = 675.8$ kN $\cdot$ m $\circlearrowright$; $B_Y = 98.23$ kN $\uparrow$

13.25 $A_X = 30$ k $\leftarrow$; $A_Y = 0$; $M_A = 160.8$ k-ft $\circlearrowright$; $B_Y = 2.14$ k $\downarrow$; $D_Y = 2.14$ k $\uparrow$

13.27 $A_x = 10$ k $\leftarrow$; $A_y = 11.7$ k $\uparrow$; $C_y = 41.5$ k $\uparrow$; $D_y = 6.8$ k $\uparrow$

13.29 $A_x = 2.7$ kN $\leftarrow$; $A_y = 20$ kN $\downarrow$; $B_x = 57.3$ kN $\leftarrow$; $B_y = 100$ kN $\uparrow$

13.35 $F_{BC} = 119.8$ kN (C); $F_{AD} = 130.2$ kN (T); $F_{AC} = 162.5$ kN (T); $F_{BD} = 170.8$ kN (C)

13.37 $A_y = 92.8$ kN $\uparrow$; $M_A = 114.3$ kN $\cdot$ m $\circlearrowright$; $B_y = 228.6$ kN $\uparrow$; $C_y = 78.6$ kN $\uparrow$

13.39 $A_y = 4.6$ k $\uparrow$; $C_y = 19.3$ k $\uparrow$; $E_y = 44.5$ k $\uparrow$; $G_y = 6.6$ k $\uparrow$

13.41 $A_y = G_y = 23$ k $\uparrow$; $B_y = F_y = 63$ k $\uparrow$; $D_y = 48$ k $\uparrow$

13.43 $A_X = 15$ k $\rightarrow$; $A_Y = 34.13$ k $\uparrow$; $E_X = 10$ k $\rightarrow$; $E_Y = 30.87$ k $\uparrow$; $M_E = 122.4$ k-ft $\circlearrowright$

13.45 $A_X = 4.29$ k $\leftarrow$; $A_Y = 23.25$ k $\uparrow$; $M_A = 107.9$ k-ft $\circlearrowright$; $B_X = 15.71$ k $\leftarrow$; $B_Y = 36.75$ k $\uparrow$; $M_B = 222.1$ k-ft $\circlearrowright$

13.47 $A_x = 10.04$ k $\rightarrow$; $A_y = 13.77$ k $\uparrow$; $C_y = 6.23$ k $\uparrow$; $D_x = 10.04$ k $\leftarrow$; $F_{BD} = 12.16$ k (T)

13.49 $A_y = 179.5$ kN $\uparrow$; $M_A = 955.5$ kN $\cdot$ m $\circlearrowright$; $D_y = 19.5$ kN $\downarrow$

13.51 $A_y = 56.7$ kN $\uparrow$; $B_y = 477.4$ kN $\uparrow$; $C_y = 215.9$ kN $\uparrow$

13.53 $A_y = 165.2$ kN $\uparrow$; $M_A = 449.4$ kN $\cdot$ m $\circlearrowright$; $B_y = 125.8$ kN $\uparrow$; $C_y = 109$ kN $\uparrow$

13.55 $F_{BC} = F_{EF} = 37.34$ kN (C); $F_{BF} = F_{CE} = 46.67$ kN (T)

13.57 $F_{AB} = 3.35$ k (C); $F_{AC} = F_{BC} = 9.46$ k (T); $F_{CD} = 13.38$ k (T)

CHAPTER 14

14.1 $A_y = E_y = 15.63$ k $\uparrow$; $C_y = 68.75$ k $\uparrow$

14.3 $A_y = 28.36$ k $\uparrow$; $C_y = 85.78$ k $\uparrow$; $D_y = 20.86$ k $\uparrow$

14.4 and 14.11 $A_y = 13.125$ kN $\downarrow$; $M_A = 91.875$ kN $\cdot$ m $\circlearrowright$; $B_y = 223.125$ kN $\uparrow$

14.6 and 14.10 $A_y = 4.6$ k $\uparrow$; $C_y = 19.3$ k $\uparrow$; $E_y = 44.5$ k $\uparrow$; $G_y = 6.6$ k $\uparrow$

14.13 $F_{BC} = 119.8$ kN (C); $F_{AD} = 130.2$ kN (T); $F_{AC} = 162.5$ kN (T); $F_{BD} = 170.8$ kN (C)

CHAPTER 15

15.1 and 15.2 A_y: 1 at A; 0.688 at B; 0 at C
M_A: 0 at A and C; 2.25 at B
C_y: 0 at A; 0.313 at B; 1 at C
S_B: 0 at A and C; -0.313 at B_L; 0.687 at B_R
M_B: 0 at A and C; 1.875 at B

15.3 C_y: 0 at A; 0.633 at B; 1 at C; 1.375 at D
S_B: 0 at A and C; -0.633 at B_L; 0.367 at B_R; -0.375 at D
M_B: 0 at A and C; 3.164 at B; -3.125 at D

15.5 A_y: 1 at A; 0 at B and D; -0.047 at C; 0.063 at E
B_y: 0 at A and D; 1 at B; 0.594 at C; -0.625 at E
D_y: 0 at A and B; 0.453 at C; 1 at D; 1.563 at E
S_C: 0 at A, B, and D; -0.453 at C_L; 0.547 at C_R; -0.563 at E
M_C: 0 at A, B, and D; 1.81 at C; -1.75 at E

15.7 F_{BC}: 0 at C; 0.833 at D; 0.938 at E
F_{CD}: 0 at C; 0.667 at D; 1.917 at E

15.9 B_y: 1.643 at A; 1 at B; 0.393 at C; 0 at D and E; -0.054 at $x = 20$ ft
D_y: -0.857 at A; 0 at B and E; 0.767 at C; 1 at D; 0.447 at $x = 20$ ft
S_C: 0.643 at A; 0 at B, D, and E; -0.607 at C_L; 0.393 at C_R; -0.054 at $x = 20$ ft
M_C: -1.79 at A; 0 at B, D, and E; 1.97 at C; -0.27 at $x = 20$ ft

15.11 C_y: 0 at A and D; 0.582 at B; 1 at C
F_{BC}: 0 at A, C, and D; 0.11 at B
F_{CE}: 0 at A, C, and D; -0.252 at B
F_{EF}: 0 at A, C, and D; -0.203 at B

CHAPTERS 16 AND 17

16.1 and 17.1 $M_{AC} = 50.6$ k-ft $\circlearrowright$; $M_{CA} = 58.8$ k-ft $\circlearrowright$; $M_{CE} = 58.8$ k-ft $\circlearrowright$; $M_{EC} = 26.9$ k-ft $\circlearrowright$

16.3 and 17.3 $M_{AB} = 100$ kN $\cdot$ m $\circlearrowright$; $M_{BA} = 200$ kN $\cdot$ m $\circlearrowright$; $M_{BE} = 200$ kN $\cdot$ m $\circlearrowright$; $M_{EB} = 500$ kN $\cdot$ m $\circlearrowright$

16.5 and 17.5 $M_{AB} = M_{CB} = 0$; $M_{BA} = 781$ kN $\cdot$ m $\circlearrowright$; $M_{BC} = 781$ kN $\cdot$ m $\circlearrowright$

16.7 and 17.7 $M_{AB} = 449.4$ kN $\cdot$ m $\circlearrowright$; $M_{BA} = 72.3$ kN $\cdot$ m $\circlearrowright$; $M_{BC} = 72.3$ kN $\cdot$ m $\circlearrowleft$; $M_{CB} = 0$

16.9 and 17.9 $M_{AB} = 103.5$ kN $\cdot$ m $\circlearrowright$; $M_{BA} = 113$ kN $\cdot$ m $\circlearrowleft$; $M_{BC} = 113$ kN $\cdot$ m $\circlearrowright$; $M_{CB} = 85$ kN $\cdot$ m $\circlearrowleft$; $M_{CE} = 85$ kN $\cdot$ m $\circlearrowright$; $M_{EC} = 47.5$ kN $\cdot$ m $\circlearrowleft$

16.11 and 17.11 $M_{BA} = 50$ k-ft $\circlearrowleft$; $M_{BD} = 50$ k-ft $\circlearrowright$; $M_{DB} = 89.3$ k-ft $\circlearrowleft$; $M_{DE} = 89.3$ k-ft $\circlearrowright$; $M_{ED} = 55.4$ k-ft $\circlearrowleft$

16.13 and 17.13 $M_{AB} = M_{ED} = 0$; $M_{BA} = M_{DC} = 57.9$ kN $\cdot$ m$\circlearrowleft$; $M_{BC} = M_{DE} = 57.9$ kN $\cdot$ m $\circlearrowright$; $M_{CB} = 38.6$ kN $\cdot$ m$\circlearrowleft$; $M_{CD} = 38.6$ kN $\cdot$ m $\circlearrowright$

16.15 and 17.15 $M_{AB} = 68.6$ kN $\cdot$ m $\circlearrowright$; $M_{BA} = 183$ kN $\cdot$ m $\circlearrowleft$; $M_{BC} = 183$ kN $\cdot$ m $\circlearrowright$; $M_{CB} = 29$ kN $\cdot$ m $\circlearrowright$; $M_{CE} = 29$ kN $\cdot$ m $\circlearrowleft$; $M_{EC} = 170.2$ kN $\cdot$ m $\circlearrowleft$

16.17 and 17.17 $M_{AC} = 9.4$ kN $\cdot$ m $\circlearrowleft$; $M_{CA} = 187.5$ kN $\cdot$ m $\circlearrowleft$; $M_{CD} = 187.5$ kN $\cdot$ m $\circlearrowright$; $M_{DC} = 0$

16.19 and 17.19 $M_{AD} = M_{CD} = M_{ED} = 0$; $M_{DA} = 65$ k-ft $\circlearrowright$; $M_{DC} = 102.5$ k-ft $\circlearrowleft$; $M_{DE} = 37.5$ k-ft $\circlearrowright$

16.21 and 17.21 $M_{AC} = 58.6$ kN $\cdot$ m $\circlearrowleft$; $M_{CA} = 286$ kN $\cdot$ m $\circlearrowleft$; $M_{CD} = 286$ kN $\cdot$ m $\circlearrowright$; $M_{DC} = 0$

16.23 and 17.23 $M_{AC} = 0$; $M_{DE} = 100$ k-ft $\circlearrowright$; $M_{CA} = 69.6$ k-ft $\circlearrowleft$; $M_{BC} = 301.5$ k-ft $\circlearrowright$; $M_{CB} = 37$ k-ft $\circlearrowright$; $M_{CD} = 32.1$ k-ft $\circlearrowright$; $M_{DC} = 100$ k-ft $\circlearrowleft$

16.25 and 17.25 $M_{AC} = 107.8$ k-ft $\circlearrowright$; $M_{CA} = 20.8$ k-ft $\circlearrowright$; $M_{BD} = 222$ k-ft $\circlearrowright$; $M_{DB} = 249.2$ k-ft $\circlearrowright$; $M_{CD} = 20.8$ k-ft $\circlearrowleft$; $M_{DC} = 249.2$ k-ft $\circlearrowleft$

16.27 and 17.27 $M_{AB} = 127$ k-ft $\circlearrowright$; $M_{BA} = 103.4$ k-ft $\circlearrowright$; $M_{BC} = 103.4$ k-ft $\circlearrowleft$; $M_{CB} = 0$

16.29 and 17.29 $M_{AC} = 11.7$ kN $\cdot$ m $\circlearrowright$; $M_{CA} = 43.9$ kN $\cdot$ m $\circlearrowleft$; $M_{CD} = 43.9$ kN $\cdot$ m $\circlearrowright$; $M_{DC} = 14.7$ kN $\cdot$ m $\circlearrowleft$; $M_{DB} = 14.7$ kN $\cdot$ m $\circlearrowright$; $M_{BD} = 0$

16.31 and 17.31 $M_{AC} = M_{BD} = 119$ k-ft $\circlearrowright$; $M_{CA} = M_{DB} = 83.5$ k-ft $\circlearrowright$; $M_{CE} = M_{DF} = 23.3$ k-ft $\circlearrowright$; $M_{EC} = M_{FD} = 44.2$ k-ft $\circlearrowright$; $M_{CD} = M_{DC} = 106.8$ k-ft $\circlearrowleft$; $M_{EF} = M_{FE} = 44.2$ k-ft $\circlearrowleft$

CHAPTER 18

18.1 $Q_1 = 53.5$ k (T); $Q_2 = 48$ k (C)

18.3 $Q_1 = 102.8$ kN (T); $Q_2 = 28.6$ kN (C); $Q_3 = 145.4$ kN (C)

18.5 $Q_1 = \begin{bmatrix} 104.4 \text{ kN} \\ 394 \text{ kN} \cdot \text{m} \\ -104.4 \text{ kN} \\ 232 \text{ kN} \cdot \text{m} \end{bmatrix}$ $Q_2 = \begin{bmatrix} -45.6 \text{ kN} \\ -232 \text{ kN} \cdot \text{m} \\ 45.6 \text{ kN} \\ -178 \text{ kN} \cdot \text{m} \end{bmatrix}$

18.7 $Q_1 = \begin{bmatrix} 9.36 \text{ k} \\ 14.72 \text{ k} \\ 37.65 \text{ k-ft} \\ -9.36 \text{ k} \\ 15.29 \text{ k} \\ -41.92 \text{ k-ft} \end{bmatrix}$ $Q_2 = \begin{bmatrix} 15.28 \text{ k} \\ 9.36 \text{ k} \\ 41.92 \text{ k-ft} \\ -15.28 \text{ k} \\ 10.64 \text{ k} \\ -54.65 \text{ k-ft} \end{bmatrix}$

18.9 $Q_1 = \begin{bmatrix} 23.26 \text{ k} \\ 4.3 \text{ k} \\ 108 \text{ k-ft} \\ -23.26 \text{ k} \\ -4.3 \text{ k} \\ 21 \text{ k-ft} \end{bmatrix}$ $Q_2 = \begin{bmatrix} 15.7 \text{ k} \\ 23.26 \text{ k} \\ -21 \text{ k-ft} \\ -15.7 \text{ k} \\ 36.74 \text{ k} \\ -249 \text{ k-ft} \end{bmatrix}$

$Q_3 = \begin{bmatrix} 36.74 \text{ k} \\ 15.7 \text{ k} \\ 222 \text{ k-ft} \\ -36.74 \text{ k} \\ -15.7 \text{ k} \\ 249 \text{ k-ft} \end{bmatrix}$

APPENDIX B

B.1 $C = \begin{bmatrix} 18 & -11 & 18 \\ -11 & 19 & 28 \\ 18 & 28 & 4 \end{bmatrix}$

B.3 $C = -26$; $D = \begin{bmatrix} -12 & 8 & -4 \\ 6 & -4 & 2 \\ -30 & 20 & -10 \end{bmatrix}$

B.5 $(AB)^T = B^T A^T = \begin{bmatrix} -6 & -27 & 2 \\ -55 & -14 & -28 \end{bmatrix}$

B.7 $x_1 = -7$; $x_2 = 3$; $x_3 = -5$

B.9 $A^{-1} = \begin{bmatrix} 0.42 & 0.22 & -0.04 \\ 0.08 & 0.28 & 0.04 \\ 0.44 & 0.04 & -0.28 \end{bmatrix}$

APPENDIX C

C.1 (a) 9.12 in^2; (b) 6.33 in^2; (c) 8.66 in^2

C.3 1,257 in^4

Index